APPLIED MATHEMATICS

THIRD EDITION

THE McGRAW-HILL RYERSON MATHEMATICS PROGRAM

LIFE MATH 1
LIFE MATH 2
LIFE MATH 3

INTERMEDIATE MATHEMATICS 1
INTERMEDIATE MATHEMATICS 2
INTERMEDIATE MATHEMATICS 3

TEACHER'S EDITIONS FOR:
INTERMEDIATE MATHEMATICS 1
INTERMEDIATE MATHEMATICS 2
INTERMEDIATE MATHEMATICS 3

BLACKLINE MASTERS FOR:
INTERMEDIATE MATHEMATICS 1
INTERMEDIATE MATHEMATICS 2

APPLIED MATHEMATICS 9
APPLIED MATHEMATICS 10
APPLIED MATHEMATICS 11
APPLIED MATHEMATICS 12

TEACHER'S GUIDES FOR:
AM 9
AM 10
AM 11
AM 12

FOUNDATIONS OF MATHEMATICS 9
FOUNDATIONS OF MATHEMATICS 10
FOUNDATIONS OF MATHEMATICS 11
FOUNDATIONS OF MATHEMATICS 12

TEACHER'S GUIDES FOR:
FM 9
FM 10
FM 11
FM 12

FINITE MATHEMATICS
ALGEBRA AND GEOMETRY
CALCULUS: A FIRST COURSE

APPLIED MATHEMATICS

10

THIRD EDITION

Dino Dottori, B.Sc., M.S.Ed.
George Knill, B.Sc., M.S.Ed.
Robert McVean, B.A.Sc.
John Seymour, B.A., M.Ed.
Darrell McPhail, B.Sc., M.Sc.

McGRAW-HILL RYERSON LIMITED

TORONTO MONTREAL NEW YORK AUCKLAND BOGOTÁ CARACAS HAMBURG
LISBON LONDON MADRID MEXICO MILAN NEW DELHI PARIS SAN JUAN
SÃO PAULO SINGAPORE SYDNEY TOKYO

APPLIED MATHEMATICS 10
THIRD EDITION

ISBN 0-07-548736-5

34567890 BBM 98765

The authors gratefully acknowledge the assistance and contribution of Walter Howard, M.A.Sc., Head of Mathematics, Jarvis Collegiate Institute, Toronto.

Technical illustrations by Frank Zsigo

A complete list of photograph credits and notes appears on page 469.

Printed and bound in Canada

Canadian Cataloguing in Publication Data

Main entry under title:

Applied mathematics 10

(The McGraw-Hill Ryerson mathematics program)
ISBN 0-07-548736-5

1. Mathematics — 1961- . 2. Mathematics — Problems, exercises, etc.
I. Dottori, Dino, date – II. Series.

QA39.2.A77 1990 510 C90-093017-9

Communications Branch, Consumer and Corporate Affairs Canada, has granted permission for the use of the National Symbol for Metric Conversion.

CONTENTS

NUMBER APPLICATIONS

REVIEW AND PREVIEW TO CHAPTER 

GETTING TO KNOW YOUR TEXT

This exercise will introduce you to your textbook. The answer to each question is a number.

1. How many chapters are there in this book?

2. On what page does the Glossary start?

3. How many pages does the Answers section have?

4. How many entries are there in the Glossary that begin with the letter b?

5. What is the page number of the last entry in the Index?

6. Which chapter uses "Geometry, Flight, and Paper Airplanes" as an Extra?

7. Which chapter has "Car Mechanic" as a Career?

8. What was the cost of a concert ticket last year according to Question 14 on page 19?

9. What is the section number teaching "parallel lines"?

10. On what page is the "Test" for the chapter on Statistics?

11. On what pages can you find the "Review Exercise" for the chapter on Borrowing and Saving?

12. On what pages can you find the "Review and Preview" to the chapter on Trigonometry?

13. On what page is the first Mind Bender?

14. How many sections does the chapter on Coordinate Geometry have?

15. What is the largest number on page 55?

POWERS OF TEN

EXERCISE 1

1. Simplify.

(a) 23×10 (b) $23 \div 0.1$
(c) 23×100 (d) $23 \div 0.01$
(e) 23×1000 (f) $23 \div 0.001$
(g) 23×0.1 (h) $23 \div 10$
(i) 23×0.01 (j) $23 \div 100$
(k) 23×0.001 (l) $23 \div 1000$

2. Multiply.

(a) 2.3×100 (b) 0.56×1000
(c) 1.79×10 (d) 0.067×100
(e) 456×0.1 (f) 75.2×0.01
(g) $12\,000 \times 0.001$ (h) 0.078×1000
(i) 0.267×0.1 (j) 3400×100
(k) 100×0.6 (l) 0.1×34.9

3. Divide.

(a) $456 \div 100$ (b) $2000 \div 10$
(c) $16\,200 \div 1000$ (d) $450 \div 0.1$
(e) $200 \div 0.01$ (f) $34 \div 0.001$
(g) $8.9 \div 10$ (h) $0.45 \div 100$
(i) $0.8 \div 0.1$ (j) $1.34 \div 0.001$

EXPONENTS

EXERCISE 2

$$2^4 = 2 \times 2 \times 2 \times 2 = 16$$

1. Evaluate the following.

(a) 3^2 (b) 2^5 (c) 10^3
(d) 5^4 (e) 7^3 (f) 4^4
(g) 1^{10} (h) 6^3 (i) 8^4
(j) 2.5^2 (k) 0.2^3 (l) 3.3^3

2. If you fold a piece of paper in half, you will have 2 thicknesses of paper. If you do it again, you will have 4 thicknesses.
How many thicknesses will there be after folding the paper a total of 10 times?

FRACTIONS AND DECIMALS

George Bell has a batting average of 0.300. To make this decimal number more meaningful we can write it as a fraction.

$$0.300 = \frac{300}{1000} = \frac{3}{10}$$

The average means that for every 10 times at bat George gets an average of 3 hits.

EXERCISE 3

1. Change each decimal to a fraction.

(a) 0.25 (b) 0.7 (c) 0.75
(d) 0.5 (e) 0.56 (f) 0.65
(g) 0.125 (h) 0.875 (i) 0.36

Baseball batting averages are expressed as decimals. If you got 18 hits in 50 times at bat, your average would be

$$\frac{18}{50} = 0.36$$

On a calculator, press

C 1 8 ÷ 5 0 =

2. Change each fraction to a decimal.

(a) $\frac{5}{8}$ (b) $\frac{3}{8}$ (c) $\frac{7}{16}$
(d) $\frac{21}{25}$ (e) $\frac{43}{50}$ (f) $\frac{57}{80}$

3. Change each mixed number to an improper fraction.

(a) $5\frac{1}{4}$ (b) $3\frac{3}{5}$ (c) $1\frac{5}{6}$
(d) $2\frac{3}{8}$ (e) $9\frac{1}{2}$ (f) $6\frac{7}{10}$

1.1 NUMBERS

People are fascinated with numbers and the precise statistics numbers can produce.

For example, the average speed of the winning driver of a recent Le Mans 24 h race was 213.946 km/h.

This number is read

"two hundred thirteen and nine hundred forty-six thousandths"

We can write these numbers on a place value chart.

millions	hundred thousands	ten thousands	thousands	hundreds	tens	ones	.	tenths	hundredths	thousandths	ten thousandths	hundred thousandths
				2	1	3	.	9	4	6		

In the number 213.946, the 2 means 200 because it has a face value of 2 and a place value of 100.

In the number 213.946, the 4 means 0.04 because it has a face value of 4 and a place value of 0.01.

$$\text{Total value} = \text{Face value} \times \text{Place value}$$

$$200 = 2 \times 100$$
$$0.04 = 4 \times 0.01$$

In this section you will review the basic number operations.

EXERCISE 1.1

A

1. Read the following numbers.

(a) 236 (b) 45 670
(c) 456 000 (d) 23.6
(e) 4.12 (f) 2 300 000
(g) 0.234 (h) 0.003
(i) 3001 (j) 3.89

2. State the place value and total value of the digit 7 in each.

(a) 2370 (b) 73 249
(c) 9.76 (d) 0.057
(e) 47 899 (f) 712 000

3. State the face value, place value, and total value of the indicated digit.

(a) 2356 (b) 12 008
(c) 703 (d) 203 9 6
(e) 8.97 (f) 0.145
(g) 3 400 000 (h) 0.98
(i) 0.089 (j) 235 101

B

4. Write each number in words.

(a) 3478
(b) 56 708
(c) 9.58
(d) 0.966
(e) 8 970 000
(f) 0.001
(g) 7.9
(h) 10 101

5. Add.

(a) $\begin{array}{r} 3561 \\ +9002 \\ \hline \end{array}$
(b) $\begin{array}{r} 12\,800 \\ +23\,561 \\ \hline \end{array}$

(c) $\begin{array}{r} 4500 \\ 1230 \\ +7612 \\ \hline \end{array}$
(d) $\begin{array}{r} 560 \\ 879 \\ +104 \\ \hline \end{array}$

(e) $145 + 78 + 3007 + 19$
(f) $2345 + 789 + 1200$
(g) $234 + 100 + 23\,009 + 340$

6. Subtract.

(a) $\begin{array}{r} 6789 \\ -1298 \\ \hline \end{array}$
(b) $\begin{array}{r} 34\,000 \\ -13\,500 \\ \hline \end{array}$

(c) $5600 - 4321$
(d) $45\,602 - 3777$
(e) $4569 - 992$

7. Multiply.

(a) $\begin{array}{r} 340 \\ \times\ 52 \\ \hline \end{array}$
(b) $\begin{array}{r} 7602 \\ \times 37 \\ \hline \end{array}$

(c) 7004×6
(d) 900×52
(e) $56\,000 \times 9$
(f) 87×35

8. Divide.

(a) $1659 \div 7$
(b) $7182 \div 21$
(c) $20\,196 \div 33$
(d) $14\,448 \div 172$

9. It takes 72 weeks to build a large jetliner. An airplane company usually builds ten at one time. The company employs about 9000 people. Each works a 40 h week.
How many working hours are required to build one jetliner?

10. In one year the Toronto police department issued 2 757 612 parking tickets. The average parking fine was $25. How much money was collected during that year?

11. The following skill-testing question was asked in a contest for a new car. "Multiply 45 by 4, divide by 6, add 40, and subtract 25."
Could you win the car?

12. The ancient Egyptians multiplied two numbers using repeated addition. To multiply 21×35, they would set up a table of multiples of 35 as shown.

1	35
2	70
4	140
8	280
16	560

Now we use the numbers in the left column to get 21.

$$16 + 4 + 1 = 21$$

Adding the corresponding numbers in the right column,

$$\begin{array}{r} 16 + 4 + 1 = 21 \\ 560 + 140 + 35 = 735 \end{array}$$

Therefore, $21 \times 35 = 735$

Use this method to perform each multiplication.

(a) 19×27 (b) 34×48 (c) 56×61

MIND BENDER

Write the numbers 1 to 8 in the circles so that no two consecutive numbers are connected by a line segment.

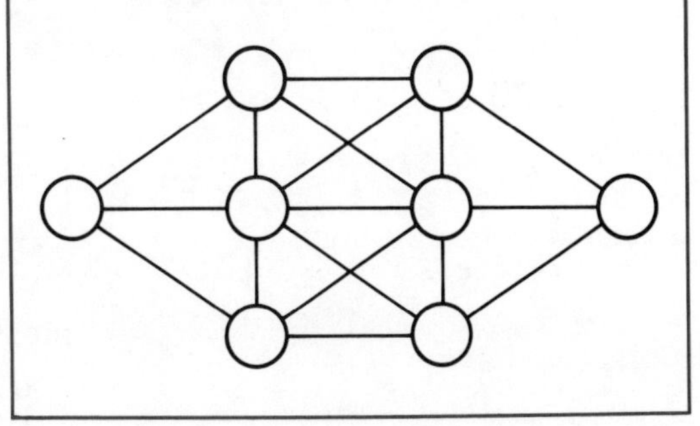

1.2 ORDER OF OPERATIONS

When you perform several operations one after the other, the answer depends on the order of operations you choose.

We can show this by simplifying

three plus two multiplied by four

in two ways.

$$3 + \underbrace{2 \times 4} \qquad\qquad \underbrace{3 + 2} \times 4$$

$$3 + 8 \qquad\qquad 5 \times 4$$

$$= 11 \qquad\qquad = 20$$

Which answer is correct?

To eliminate confusion, mathematicians have agreed to perform operations in the following order. The letters in BEDMAS will help you remember the order.

B	E	D M	A S
Do the computations in brackets first.	Simplify numbers with exponents and "of."	Divide or multiply in the order in which ÷ and × appear from left to right.	Add or subtract in the order in which + and − appear from left to right.

"of" is used with fractions

$\frac{1}{2}$ of 6

$= \frac{1}{2} \times 6$

$= 3$

Therefore, the correct answer to $3 + 2 \times 4$ is

$$3 + 2 \times 4 = 3 + 8$$
$$= 11$$

Enter the following on your calculator.

C 3 + 2 × 4 =

What does the calculator display read?

If your calculator displays 11, then it is programmed to follow the order of operations.

If your calculator displays 20, then it is not programmed to follow the order of operations and you must enter numbers and operations in the correct order.

Example 1.

Simplify. $(7 + 5) + 8 \div 2$

Solution:

$(7 + 5) + 8 \div 2$

$= 12 + 8 \div 2$ B

$= 12 + 4$ D

$= 16$ A

If your calculator does not have brackets,

press C 7 + 5 = M+ C 8 ÷ 2 + MR =

The display is 16.

If your calculator has brackets,

press (7 + 5) + 8 ÷ 2 =

The display is 16.

Example 2.

Simplify. $14^2 + \dfrac{36 \times 4}{5 + 13}$

Solution:

$$14^2 + \frac{36 \times 4}{5 + 13}$$ ← The division bar means brackets.

$$= 14^2 + \frac{(36 \times 4)}{(5 + 13)}$$

$$= 14^2 + \frac{144}{18}$$

$$= 196 + \frac{144}{18}$$

$$= 196 + 8$$

$$= 204$$

Press

C 5 + 1 3 = M+ 3 6 × 4 ÷ MR + 1 4 y^x 2 =

The display is 204.

EXERCISE 1.2

A 1. Simplify.

(a) $4^2 + 5 - 3$
(b) $6 \times 2 - 1^2$
(c) $5 \times 3 + 6$
(d) $16 \div 4 \times 2$
(e) $(7 + 4) \times 3^2$
(f) $20 \div (8 - 4)$
(g) $\dfrac{15 + 6}{7}$
(h) $\dfrac{100}{5 \times 4}$
(i) $\dfrac{9 + 3}{7 - 1} + 2^2$
(j) $\dfrac{18 \div 6}{15 \div 5}$
(k) $\frac{1}{2}$ of 18
(l) $\frac{1}{3}$ of 27

B 2. Simplify.

(a) $15^2 + 9 \times 2 \div 6$
(b) $47 + 72 \div 4$
(c) $108 \div 12 \times 3 + 5^3$
(d) $9^2 + 78 \div 13 - 24$
(e) $34 \times 12 + 41 \times 5$
(f) $216 \div 4 - 11 \times 3$
(g) $65 - 121 \div 11 + 7^3$

3. Simplify.

(a) $34 \times (18 - 8) + 7^2$
(b) $(56 - 49)(123 - 102)$
(c) $(84 + 12) \div 24 - 1$
(d) $65 \div (5 + 24 \div 3)$
(e) $(23 - 3)(15 - 5) \div (6 + 4)$
(f) $63 \div 9 - 5 \times 2 + 3^3$
(g) $(6 \div 3 - 2)(5 + 4 - 7)$
(h) $144 \div 12 \div (5 - 2)$

4. Simplify.

(a) $\dfrac{108 \div 12}{5 - 2} + 7^2$
(b) $\frac{1}{2}$ of $66 + \dfrac{15 \times 5}{11 - 6}$
(c) $\dfrac{14 \times 6}{21 \div 3} + 8^2$
(d) $\dfrac{117 + 33}{75 \div 3} - 4$
(e) $(27 - 2)(12 \div 4) \div (11 + 4)(2 + 7)$
(f) $\frac{1}{3}$ of $63 + \dfrac{136 + 214}{10}$
(g) $\frac{1}{4}$ of $84 - 135 \div 15$
(h) $\dfrac{5 + 3(2 - 7 + 9)}{17}$
(i) $19 - \dfrac{(15 \div 5 + 3)}{(8 - 5)}$
(j) $\dfrac{3(5 + 2) - 3(7 - 5)}{5}$

C 5. Simplify.

(a) $3 + [2 + (7 - 4)]$
(b) $8 - [6 - (10 - 7)]$
(c) $[(28 + 5) \div 3 - 16 \div 4] + 2$

1.3 ROUNDING AND ESTIMATING

There are many times in a day when you use mathematics to solve problems. Before you solve a problem, you should decide what kind of answer you need. Sometimes you will need an exact answer. In other situations an estimate will do.

For example, if someone asks you how long it takes you to get to school, you might say "about 20 min." If you are asked to time the 100 m dash at a track meet, you will have to provide exact times.

Rounded numbers are used when we want to use an estimate as an answer to a problem. Suppose you earn $11.45/h at your part-time job and you worked 28 h last month. An estimate of your earnings is

$\$10 \times 30 = \$300$

We use rounded numbers to find estimates.

The following table gives the rules for rounding numbers.
The key digit is the first digit to the right of the required place value.

Rule	Number	Required Place Value	Key Digit	Rounded Number
If the key digit is less than 5, round down.	4278 3.41	nearest 1000 nearest 0.1	4278 3.41	4000 3.4
If the key digit is 5 or more, round up.	361 0.137	nearest 100 nearest 0.01	361 0.137	400 0.14

Round down if the digit to the right is 4 or less.

To the nearest ten 57.32 is 60

Round up if the digit to the right is 5 or more.

> To find an estimate, first round each number to the highest possible place value.
> Next, calculate mentally.

This is called front-end rounding.

Example.
The airport is 415 km away. To travel to the airport, you can drive at an average speed of 78 km/h.
About how long will it take you to get there?

Solution:
Round, then divide.

$\frac{415}{78} \doteq \frac{400}{80} = 5$

It will take about five hours.

$\doteq$ means "approximately equal to."

When possible we may round to take advantage of compatible numbers. For example, insurance coverage for a house costs $375 per year. To estimate the cost per month, we round 375 to 360 since 360 is easily divided by 12.

$$\frac{375}{12} \doteq \frac{360}{12} = 30$$

The insurance coverage will cost approximately $30 per month.

EXERCISE 1.3

A 1. State whether you need an exact answer or an estimate for each situation.

(a) Temperature: What is the temperature of a hospital patient?
(b) Temperature: Do you need to wear a sweater to go out for a walk?
(c) Money: How much does a new house cost?
(d) Money: How much should you take on a two-week vacation?
(e) Time: When does the plane for Vancouver leave?

B 2. Round to the nearest hundred.

(a) 345 (b) 670
(c) 550 (d) 751

3. Round to the nearest thousand.

(a) 3500 (b) 5689
(c) 6145 (d) 4500

4. Round to the nearest tenth.

(a) 0.67 (b) 0.351
(c) 0.11 (d) 0.35

5. Round to the nearest hundredth.

(a) 0.055 (b) 0.078
(c) 0.024 (d) 0.0452

6. Round to the highest place value.

(a) 23 500 (b) 0.456
(c) 35.67 (d) 7500
(e) 0.56 (f) 0.045
(g) 0.0041 (h) 0.0075
(i) 716 000 (j) 8.47
(k) 65 000 (l) 0.981

7. Estimate.

(a) $345 + 690 + 111$
(b) 45×81
(c) $9.7 \div 2.3$
(d) $48.9 - 32.8$
(e) $0.75 + 0.24 + 0.451 + 0.66$
(f) $24\,000 \div 450$
(g) 11.4×6.1
(h) $101 - 62 - 18$
(i) $6781 + 2103 + 5500$

8. Estimate each of the following.

(a) the cost of eleven T-shirts selling at $19.50 each
(b) the number of lunches sold in the cafeteria in one school year if about 215 people buy their lunch each day
(c) the number of hours you sleep in a year
(d) the time it takes to drive 615 km at 85 km/h
(e) the distance you can travel in 6.5 h at 95 km/h
(f) A long distance call costs $1.95 for the first minute and $1.60 for each additional minute.
How much would a 7 min call cost?
(g) the cost of an overnight trip for 2 people if a hotel room costs $110 per night, dinner costs $35 each, and breakfast costs $9.50 each

9. Henri rents a car for one week. He is charged a total of $290.
How much did he pay on a daily basis?

1.4 DECIMALS

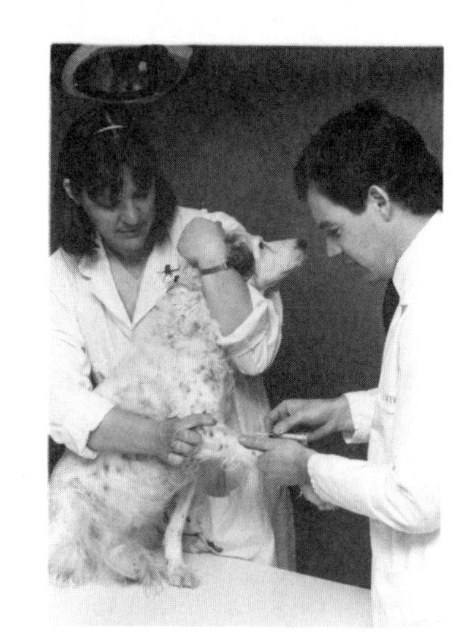

In this section we will review operations with decimals. When multiplying decimals, we determine the number of places in the answer by adding the number of decimal places in the factors

Example 1.
Veterinarians administer specific medicines to an animal according to the body mass of the animal. For allergies, pyrazine is given to a dog or cat in doses of 0.03 mL for each kilogram of mass.
Calculate the amount of pyrazine given to a 4.6 kg Siamese cat, to the nearest hundredth.

Solution:
Multiply the mass of the cat by 0.03

$$\begin{array}{r} 4.6 \\ \times\ 0.03 \\ \hline 0.138 \end{array} \quad \begin{array}{l} \longleftarrow \text{1 decimal place} \\ \longleftarrow \text{2 decimal places} \\ \longleftarrow \text{3 decimal places} \end{array}$$

Estimate.
5×0.03
$= 0.15$

The cat would get approximately 0.14 mL of pyrazine.

When dividing by decimals, multiply the divisor and dividend by the same power of 10 so that the divisor becomes a whole number.

Example 2.
Divide. $110.214 \div 9.42$

Estimate.
$100 \div 10$
$= 10$

Solution:
(i) Multiply the divisor and dividend by 100 so that you divide by a whole number.

$$\frac{110.214 \times 100}{9.42 \times 100} = \frac{11021.4}{942}$$

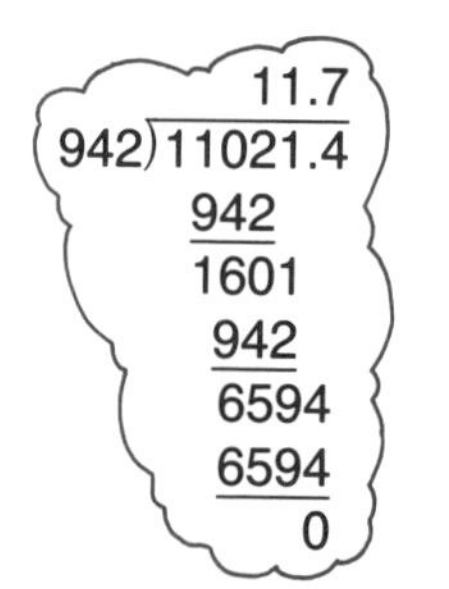

$$\begin{array}{r} 11.7 \\ 942\overline{)11021.4} \\ \underline{942} \\ 1601 \\ \underline{942} \\ 6594 \\ \underline{6594} \\ 0 \end{array}$$

(ii) Move the decimal point in the divisor 2 places to the right.

Move the decimal point in the dividend 2 places to the right.
Divide.

$9.42_{\wedge}\overline{)110.21_{\wedge}4}$

Press
C 1 1 0 · 2 1 4 ÷ 9 · 4 2 =
The display is 11.7

$110.214 \div 9.42 = 11.7$

When division continues, as in $2 \div 3 = 0.666\,666\ldots$
it is necessary to round to an appropriate decimal place.

$2 \div 3 = 0.7$ (nearest tenth)
$2 \div 3 = 0.67$ (nearest hundredth)
$2 \div 3 = 0.667$ (nearest thousandth)

EXERCISE 1.4

A 1. Estimate to locate the decimal point.

(a) $34.65 + 9.7 + 1.212 = 45562$
(b) $5.56 \times 20.3 = 112868$
(c) $169.74 \div 20.7 = 82$
(d) $81.5 - 6.312 = 75188$
(e) $312.6 \times 0.9 = 28134$
(f) $1028.3 \div 182 = 565$

B 2. Estimate, then add.

(a) $\begin{array}{r} 456.7 \\ 33.04 \\ +\ \ 5.6 \\ \hline \end{array}$ (b) $\begin{array}{r} 8.9 \\ 123.22 \\ +\ \ 9.123 \\ \hline \end{array}$

(c) $2.34 + 56.7 + 5.2$
(d) $234.7 + 88.88 + 0.045$
(e) $13.8 + 21 + 0.4 + 1.1$

3. Estimate, then subtract.

(a) $\begin{array}{r} 34.67 \\ -12.56 \\ \hline \end{array}$ (b) $\begin{array}{r} 100.7 \\ -\ \ 23.89 \\ \hline \end{array}$

(c) $5.678 - 1.52$
(d) $678.9 - 43.67$
(e) $81 - 7.96$

4. Estimate, then multiply.

(a) $\begin{array}{r} 234.6 \\ \times\ \ 0.6 \\ \hline \end{array}$ (b) $\begin{array}{r} 2.45 \\ \times 1.7 \\ \hline \end{array}$

(c) 5.6×4.9
(d) 0.671×12
(e) 8.98×7.1

5. Estimate, then divide.

(a) $9.7\overline{)31.04}$ (b) $1.44\overline{)0.432}$
(c) $3.745 \div 0.35$
(d) $483 \div 0.3$
(e) $7 \div 0.035$

6. Calculate to the indicated place value.

(a) 2.3×5.6 (nearest tenth)
(b) $9.91 \div 0.72$ (nearest hundredth)
(c) 0.004×3.6 (nearest thousandth)
(d) $1.04 \div 0.39$ (nearest tenth)
(e) $10 \div 0.013$ (nearest hundredth)

7. Jennifer earns $14.75/h pruning fruit trees. In one week she worked 34.5 h. How much did she earn?

8. The Mach scale is used to express speeds faster than sound. Mach 1 is equal to the speed of sound, Mach 2 is twice the speed of sound, and so on. But the speed of sound is not constant. For land speeds Mach 1 is 1061.78 km/h. For air speeds Mach 1 is 1224.65 km/h.

(a) How much faster is Mach 1 for air speeds than for land speeds?
(b) How fast are you flying if your speed reaches Mach 2?
(c) How fast are you flying if your speed reaches Mach 1.5?
(d) The earth is racing around the sun at speeds that reach Mach 101.1. How fast is this in kilometres per hour?
(e) A Concorde jet can fly at 2320 km/h. Express this speed as a Mach number to the nearest hundredth.

MIND BENDER

Place the numbers from 4 to 9 in the circles so that the sum of the numbers in any side of the triangle is 20.

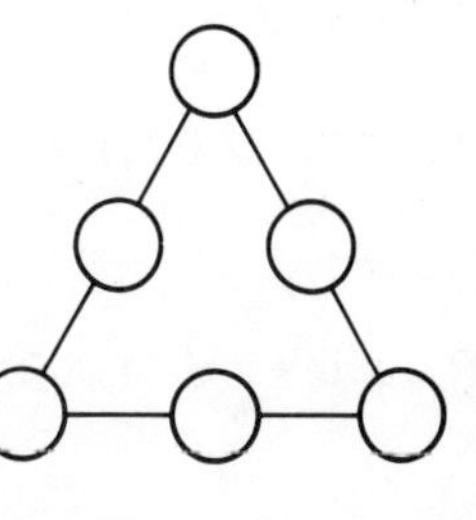

1.5 SUBSTITUTING IN EXPRESSIONS AND FORMULAS

Jack is a physical fitness instructor in the Canadian Armed Forces. One of his duties is to assess the health of recruits. The Body Mass Index (BMI) is recommended by Health and Welfare Canada as the best measure of relative body fatness in adults 20 a of age and over. The following formula shows how BMI is calculated using mass and height measurements.

$$\text{BMI} = \frac{\text{mass in kilograms}}{(\text{height in metres})^2} \quad \text{or} \quad \text{BMI} = \frac{m}{h^2}$$

Justine has a mass of 61 kg and her height is 1.65 m.

$$\text{Justine's BMI} = \frac{m}{h^2}$$
$$= \frac{61}{(1.65)^2}$$
$$\doteq 22.4$$

Estimate.
$$\frac{61}{(1.65)^2} \doteq \frac{60}{4} = 15$$

Press

1 · 6 5 x^2 M+ 6 1 ÷ MR =

The display is 22.405877

The range of BMI values between 20 and 25 indicates good health for most adults.

Expressions, such as $\frac{m}{h^2}$, are called algebraic expressions.

The letters in algebraic expressions are called variables because different numbers can replace the letters.

Example 1.
When a TV station wants to determine the profit for televising a one-hour concert where there will be 12 min of commercial time, this formula is used

$P = 12m - c$ where P is the profit,
m is the charge per commercial minute,
and c is the cost of the production

Find the profit if the production cost is \$250 000 and the commercial minute charge is \$27 000.

Solution:

$P = 12m - c$
$P = 12(\$27\,000) - \$250\,000$
$= \$324\,000 - \$250\,000$
$= \$74\,000$

The profit is \$74 000.

Example 2.
Evaluate $2xy + 4x - y$ for $x = 5$ and $y = 4$

Solution:

$$\begin{aligned} 2xy + 4x - y &= 2(5)(4) + 4(5) - 1(4) \\ &= 40 + 20 - 4 \\ &= 56 \end{aligned}$$

Example 3.
Evaluate $4r + 3t$ for $r = 3.2$ and $t = 5.6$

Solution:

$$\begin{aligned} 4r + 3t &= 4(3.2) + 3(5.6) \\ &= 12.8 + 16.8 \\ &= 29.6 \end{aligned}$$

EXERCISE 1.5

B 1. Evaluate $5x - 4$ when $x = 4, 7, 100$

2. Evaluate $\frac{t + 7}{2}$ when $t = 3, 21, 99$

3. Evaluate $12 - 2y$ for $y = 1, 2, 3$

4. Evaluate $3m - 1$ for $m = 2.3, 3.5, 101$

5. Evaluate $18 - 2s$ for $s = 4.6, 8.2, 0.98$

6. Evaluate $\frac{5m - 7}{2}$ for $m = 3.4, 8.6, 10.2$

7. If $x = 4$ and $y = 3$, evaluate each algebraic expression.
(a) $4x + 5$ (b) $x - y$
(c) $7y - x$ (d) $5xy$
(e) $3(x + y)$ (f) $21 - xy$
(g) $\frac{x + 2y}{5}$ (h) $6x + 11y$

8. Evaluate $9t - 12$ for $t = 21, 22, 25$.

9. Evaluate $5x + 8y$ for the given x- and y-values.
(a) $x = 11$ and $y = 23$
(b) $x = 0$ and $y = 5$
(c) $x = 22$ and $y = 0$
(d) $x = 1.2$ and $y = 3.5$
(e) $x = 100$ and $y = 200$
(f) $x = 0.02$ and $y = 0.03$

10. If $P = 2(\ell + w)$, find P for ℓ and w.
(a) $\ell = 32$ m and $w = 7$ m
(b) $\ell = 5.6$ cm and $w = 2.4$ cm
(c) $\ell = 1.45$ mm and $w = 2.8$ mm
(d) $\ell = 201$ m and $w = 181$ m

11. The formula $E = 300 + 0.05s$ gives the amount Cheryl earns per week selling sound systems.
If s represents her sales per week, calculate Cheryl's earnings for each value of s.
(a) $s = \$9500$ (b) $s = \$14\ 600$
(c) $s = \$26\ 000$ (d) $s = 0$

12. The formula for the cost to rent a car at Big Toy Car Rentals is

$C = 100d + 0.2k$, where

C is the rental cost,
d is the number of days rented,
k is the number of kilometres driven

Calculate the rental cost under the following conditions.
(a) $d = 4$ and $k = 390$
(b) $d = 7$ and $k = 613$
(c) $d = 14$ and $k = 1329$
(d) $d = 9$ and $k = 2323$

MIND BENDER

Place the numbers from 1 to 9 in the circles so that the sum of the numbers in any side of the triangle is 20.

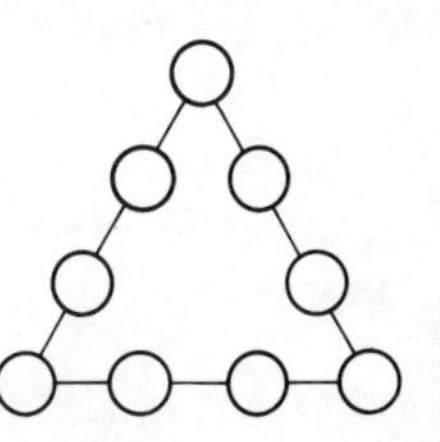

1.6 MULTIPLYING AND DIVIDING FRACTIONS

MULTIPLYING FRACTIONS

A go-cart track is $\frac{7}{10}$ km long. One race consists of $2\frac{1}{2}$ laps of the track.
How long is the race?

To find the length of the race we multiply $2\frac{1}{2}$ by $\frac{7}{10}$.

To multiply fractions we

* write each fraction as an improper fraction $\quad 2\frac{1}{2} \times \frac{7}{10} = \frac{5}{2} \times \frac{7}{10}$
* multiply the numerators and denominators $\quad \frac{5}{2} \times \frac{7}{10} = \frac{5 \times 7}{2 \times 10} = \frac{35}{20} = \frac{7}{4}$
* change the improper fractions back to mixed numbers, if necessary. $\quad \frac{7}{4} = 1\frac{3}{4}$

The go-cart race is $1\frac{3}{4}$ km long.

DIVIDING FRACTIONS

An obstacle course race is $5\frac{1}{4}$ km long. There are three people on each team. Each team member must run the same distance.
How far does each team member run?

To find the distance each member must run we divide $5\frac{1}{4}$ by 3.

To divide fractions we

* write each fraction as an improper fraction $\quad 5\frac{1}{4} \div 3 = \frac{21}{4} \div \frac{3}{1}$
* multiply by the reciprocal of the divisor $\quad \frac{21}{4} \div \frac{3}{1} = \frac{21}{4} \times \frac{1}{3} = \frac{21}{12} = \frac{7}{4}$
* change the improper fraction back to a mixed number, if necessary. $\quad \frac{7}{4} = 1\frac{3}{4}$

Each team member must run $1\frac{3}{4}$ km.

> In general,
> to multiply two fractions, multiply numerators and then the denominators. $\quad \frac{a}{b} \times \frac{c}{d} = \frac{a \times c}{b \times d} = \frac{ac}{bd}$
>
> to divide by a fraction, multiply by the reciprocal of the fraction. $\quad \frac{a}{b} \div \frac{c}{d} = \frac{a}{b} \times \frac{d}{c} = \frac{ad}{bc}$

EXERCISE 1.6

A

1. Express each fraction in lowest terms.

(a) $\frac{16}{20}$ (b) $\frac{25}{30}$ (c) $\frac{14}{21}$
(d) $\frac{13}{39}$ (e) $\frac{8}{40}$ (f) $\frac{9}{54}$

2. State the reciprocal of each of the following.

(a) $\frac{1}{2}$ (b) $\frac{3}{4}$ (c) $\frac{7}{8}$
(d) $\frac{9}{16}$ (e) 2 (f) $2\frac{2}{3}$
(g) $\frac{1}{8}$ (h) $1\frac{1}{2}$ (i) $1\frac{3}{4}$

3. State as mixed numbers.

(a) $\frac{7}{4}$ (b) $\frac{11}{2}$ (c) $\frac{8}{3}$
(d) $\frac{9}{5}$ (e) $\frac{17}{6}$ (f) $\frac{21}{10}$
(g) $\frac{15}{4}$ (h) $\frac{26}{5}$ (i) $\frac{33}{10}$

4. Multiply.

(a) $\frac{1}{2} \times \frac{2}{3}$ (b) $\frac{1}{4} \times \frac{2}{5}$
(c) $\frac{1}{3} \times 2$ (d) $3 \times \frac{3}{4}$
(e) $\frac{1}{8} \times \frac{4}{5}$ (f) $(\frac{1}{2})^2$

5. Divide.

(a) $\frac{1}{2} \div \frac{1}{3}$ (b) $\frac{1}{3} \div \frac{1}{2}$
(c) $\frac{3}{4} \div 2$ (d) $3 \div \frac{2}{5}$
(e) $\frac{3}{5} \div \frac{2}{3}$ (f) $\frac{3}{8} \div \frac{1}{2}$

B

6. Multiply.

(a) $2\frac{3}{4} \times 1\frac{1}{2}$ (b) $3\frac{2}{3} \times 4\frac{1}{5}$
(c) $6 \times 2\frac{1}{8}$ (d) $4\frac{1}{4} \times 8$
(e) $1\frac{1}{5} \times \frac{2}{3}$ (f) $\frac{7}{10} \times 3\frac{1}{2}$
(g) $1\frac{5}{6} \times 2\frac{3}{4}$ (h) $\frac{4}{5} \times 6\frac{1}{4}$

7. Divide.

(a) $1\frac{1}{2} \div \frac{2}{3}$ (b) $3\frac{1}{4} \div 1\frac{1}{2}$
(c) $6 \div 1\frac{2}{3}$ (d) $1\frac{3}{4} \div 2$
(e) $2\frac{3}{5} \div 4\frac{3}{4}$ (f) $1\frac{1}{8} \div \frac{1}{2}$
(g) $1\frac{1}{10} \div \frac{2}{5}$ (h) $3\frac{1}{3} \div 5\frac{1}{3}$

8. Simplify.

(a) $\frac{1}{2}$ of $\frac{3}{4}$ (b) $\frac{1}{2} \div \frac{1}{3} \times \frac{1}{4}$
(c) $\frac{1}{2} \times \frac{3}{5} \div 3$ (d) $5 \times 3\frac{1}{5} \div 2$
(e) $\frac{3}{4}$ of $2\frac{1}{2}$ (f) $\frac{3}{5}$ of $3\frac{1}{3}$

9. If $x = \frac{1}{2}$ and $y = \frac{1}{3}$, evaluate each expression.

(a) $2xy$ (b) $\frac{x}{y}$
(c) $\frac{y}{x}$ (d) $\frac{1}{2}xy$

10. Terry ran $6\frac{1}{2}$ laps every day for seven days.
How many laps did he run?

11. Joanne bought 500 shares of Tridco stock. The cost of one share was $\$2\frac{5}{8}$.
How much did 500 shares cost?

12. The Conservationist magazine sells advertising space by the full page, half page, quarter page, and eighth page. The table gives the amount of advertising sold for the next issue.

Full Page	Half Page	Quarter Page	Eighth Page
8	11	43	55

How many pages of advertising space have been sold?

13. One third of the 60 000 stadium seats are BLUES. One quarter are GOLDS. The rest are GREENS.
How many GREENS are there?

1.7 ADDING AND SUBTRACTING FRACTIONS

ADDING FRACTIONS

Caroline sells advertising for the school yearbook. Advertising is sold in fractions of a page. On Thursday she sold $6\frac{1}{2}$ pages. On Friday she sold $3\frac{3}{4}$ pages and on Saturday she sold $5\frac{7}{8}$ pages.

How many pages did she sell altogether?

To add fractions

* use equivalent fractions to express each fraction with the same denominator $\qquad 6\frac{1}{2} + 3\frac{3}{4} + 5\frac{7}{8} = 6\frac{4}{8} + 3\frac{6}{8} + 5\frac{7}{8}$
* add the numerators and the whole numbers $\qquad 6\frac{4}{8} + 3\frac{6}{8} + 5\frac{7}{8} = 14\frac{17}{8}$
* change the improper fraction back to a mixed number, if necessary $\qquad 14\frac{17}{8} = 16\frac{1}{8}$

Caroline sold $16\frac{1}{8}$ pages of advertising.

SUBTRACTING FRACTIONS

On the stock market the prices of shares are given as fractions of a dollar. On Monday the value of a share of Arco stock was $\$11\frac{1}{8}$. On Tuesday the stock dropped by $\$1\frac{3}{4}$.

ARCO $\$11\frac{1}{8}$

What was the new value of the stock?

To subtract fractions

* use equivalent fractions to express each fraction with the same denominator $\qquad 11\frac{1}{8} - 1\frac{3}{4} = 10\frac{9}{8} - 1\frac{6}{8}$
* subtract the numerators and the whole numbers $\qquad 10\frac{9}{8} - 1\frac{6}{8} = 9\frac{3}{8}$
* change the improper fraction back to a mixed number, if necessary

The new value of the stock was $\$9\frac{3}{8}$.

Example.

Simplify. $\frac{5}{6} + 2\frac{3}{4} - 1\frac{5}{8}$

Solution:

METHOD I

$$\begin{aligned}\frac{5}{6} + 2\frac{3}{4} - 1\frac{5}{8} &= \frac{5}{6} + \frac{11}{4} - \frac{13}{8}\\ &= \frac{20}{24} + \frac{66}{24} - \frac{39}{24}\\ &= \frac{47}{24}\\ &= 1\frac{23}{24}\end{aligned}$$

METHOD II

$$\begin{aligned}\frac{5}{6} + 2\frac{3}{4} - 1\frac{5}{8} &= \frac{20}{24} + 2\frac{18}{24} - 1\frac{15}{24}\\ &= 2\frac{38}{24} - 1\frac{15}{24}\\ &= 1\frac{23}{24}\end{aligned}$$

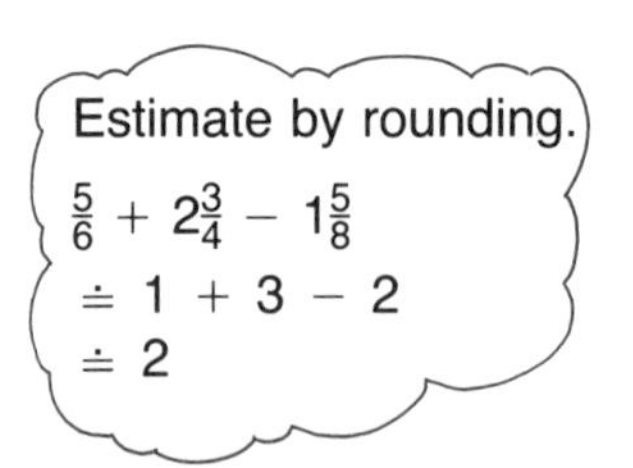

EXERCISE 1.7

A

1. State the lowest common denominator for each.

(a) $\frac{1}{3} + \frac{1}{6}$ (b) $\frac{1}{2} + \frac{3}{4}$ (c) $\frac{2}{3} - \frac{1}{5}$

(d) $\frac{3}{10} - \frac{1}{4}$ (e) $\frac{3}{8} + \frac{5}{6}$ (f) $\frac{1}{9} + \frac{1}{6}$

2. State as improper fractions.

(a) $3\frac{1}{4}$ (b) $2\frac{2}{3}$ (c) $1\frac{3}{5}$

(d) $7\frac{1}{2}$ (e) $1\frac{4}{7}$ (f) $2\frac{3}{10}$

3. Add.

(a) $\frac{3}{8} + \frac{1}{8}$ (b) $\frac{1}{4} + \frac{1}{2}$

(c) $\frac{5}{6} + \frac{1}{3}$ (d) $\frac{7}{10} + \frac{3}{5}$

(e) $6 + 1\frac{1}{2}$ (f) $2\frac{3}{8} + 5$

4. Subtract.

(a) $\frac{5}{6} - \frac{1}{6}$ (b) $\frac{3}{4} - \frac{1}{2}$

(c) $\frac{7}{10} - \frac{2}{5}$ (d) $\frac{7}{12} - \frac{1}{3}$

(e) $5 - \frac{1}{2}$ (f) $3\frac{1}{8} - 1$

B

5. Add.

(a) $3\frac{1}{2} + 4\frac{1}{4}$ (b) $5\frac{2}{3} + \frac{7}{12}$

(c) $4\frac{3}{10} + 2\frac{3}{5}$ (d) $\frac{3}{4} + 1\frac{2}{3} + \frac{1}{2}$

(e) $\frac{5}{6} + 1\frac{1}{3} + \frac{5}{9}$ (f) $2\frac{7}{10} + 3\frac{1}{2} + 4\frac{3}{4}$

(g) $6 + 2\frac{3}{7} + \frac{1}{3}$ (h) $6\frac{1}{6} + 3\frac{4}{9} + 2\frac{1}{2}$

6. Subtract.

(a) $6\frac{1}{2} - 2\frac{1}{4}$ (b) $8\frac{7}{8} - 3\frac{1}{2}$

(c) $5\frac{1}{3} - 3\frac{5}{6}$ (d) $7\frac{1}{4} - \frac{11}{12}$

(e) $7\frac{1}{5} - \frac{3}{4}$ (f) $6 - 3\frac{1}{3}$

(g) $9\frac{1}{8} - 2\frac{5}{6}$ (h) $2\frac{4}{5} - 1$

7. Simplify.

(a) $\frac{3}{4} + 2\frac{1}{2} - \frac{1}{8}$ (b) $3\frac{2}{5} - \frac{1}{4} + 2\frac{7}{10}$

(c) $9 + 4\frac{2}{3} - 3\frac{7}{9}$ (d) $\frac{5}{6} + 3\frac{1}{4} - 1\frac{1}{2}$

(e) $6\frac{3}{4} - 2\frac{5}{6} + \frac{1}{2}$ (f) $\frac{7}{8} - \frac{1}{6} + 2\frac{3}{4}$

(g) $\frac{3}{5} + 4 - 1\frac{3}{4}$ (h) $6\frac{4}{5} - 3\frac{1}{2} - \frac{1}{4}$

8. Simplify.

(a) $\frac{3}{4} \times \frac{1}{2} + \frac{1}{3}$ (b) $\frac{5}{8} + \frac{2}{3} \div \frac{1}{2}$

(c) $3\frac{1}{2} \times 2 - \frac{3}{8}$ (d) $(6\frac{4}{5} - 1\frac{3}{4}) \div \frac{1}{3}$

(e) $\frac{3}{5} \div 3 + 2\frac{1}{3}$ (f) $4\frac{7}{10} - 3\frac{9}{10} + \frac{1}{5}$

(g) $2\frac{3}{4} \times 1\frac{2}{3} - 1$ (h) $20 - (5\frac{1}{4} + 2\frac{3}{5})$

9. If $x = \frac{1}{2}$, $y = \frac{1}{3}$, and $z = \frac{1}{4}$, evaluate each expression.

(a) $x + y + z$ (b) $x - y + z$

(c) $2x + 3y$ (d) xyz

(e) $x - yz$ (f) $(x + y) \div z$

10. On Monday, Roddy bought 300 shares of Devlon stock at $\$3\frac{3}{8}$ per share. He sold it on Friday for $\$5\frac{1}{2}$ per share.
What was his profit on the transaction?

11. It takes $8\frac{1}{4}$ h to drive from Acton to Estherville.
If you have driven for $5\frac{3}{4}$ h, how much longer will it take to get to Estherville?

12. Terry sells advertising for Computer World magazine. He recorded the number of pages he sold in this table.

Day	M	T	W	T	F
Pages	$5\frac{1}{2}$	$6\frac{1}{8}$	$4\frac{1}{4}$	$2\frac{5}{8}$	$7\frac{3}{4}$

Find the total number of pages he sold.

13. Harry spends $\frac{1}{3}$ of his day sleeping and $\frac{1}{4}$ of his day at school.
What fraction of the day does Harry have left for other activities?

1.8 PERCENT

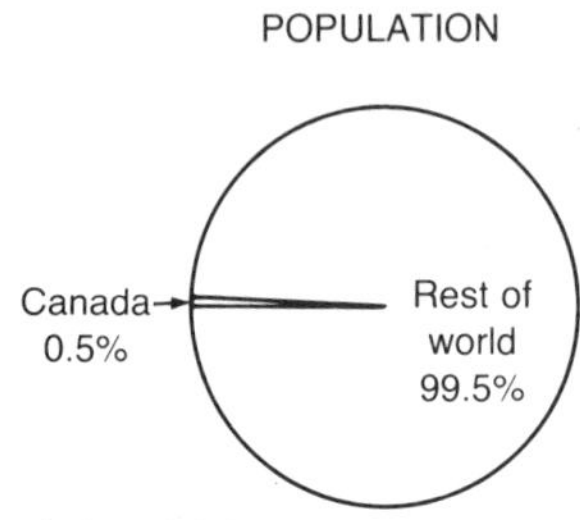

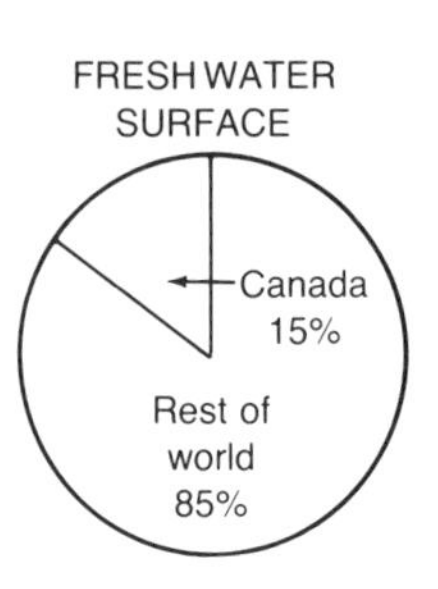

Percents are often used to convey information. The graphs and the percents at the right tell us quickly that Canada has 0.5% of the world's population and 15% of the world's freshwater surface.

Practical problems often involve percent. If 50% of the 800 students in a school will attend the dance, you can expect to sell 400 tickets. If tickets cost \$10 each, you will collect \$4000.

Percent **means** per one hundred

PERCENTS EXPRESSED AS DECIMALS
(Divide by 100)

$34\% = \frac{34}{100}$
$= 0.34$

$7.6\% = \frac{7.6}{100}$
$= 0.076$

DECIMALS EXPRESSED AS PERCENTS
(Multiply by 100)

$0.43 = 0.43 \times 100\%$
$= 43\%$

$0.032 = 0.032 \times 100\%$
$= 3.2\%$

PERCENTS EXPRESSED AS FRACTIONS
(Use a denominator of 100)

$35\% = \frac{35}{100}$
$= \frac{7}{20}$

$7.5\% = \frac{7.5}{100}$
$= \frac{75}{1000}$
$= \frac{3}{40}$

FRACTIONS EXPRESSED AS PERCENTS
(Divide, then multiply by 100)

$\frac{7}{8} = 0.875$
$= 0.875 \times 100\%$
$= 87.5\%$

$6\frac{3}{4} = 6.75$
$= 6.75 \times 100\%$
$= 675\%$

Example.
On a Wednesday night a televised musical received a rating of 14.6. This means that 14.6% of 1 200 000 TV sets were tuned to the show.
How many sets were tuned to the show?

Solution:
Find 14.6% of 1 200 000.

$14.6\% \text{ of } 1\,200\,000 = 0.146 \times 1\,200\,000$
$= 175\,200$

175 200 TV sets were tuned to the show.

Press
C 1 2 0 0 0 0 0 × 1 4 · 6 %

The display is 175 200.

EXERCISE 1.8

B

1. Write each percent as a decimal.

(a) 45% (b) 67% (c) 92%
(d) 6% (e) 1% (f) 100%
(g) 200% (h) 150% (i) 33.6%
(j) 45.9% (k) 7.3% (l) 9.2%
(m) 0.4% (n) 0.1% (o) 0.5%

2. Write each decimal as a percent.

(a) 0.23 (b) 0.56 (c) 0.79
(d) 0.9 (e) 0.7 (f) 0.05
(g) 0.01 (h) 0.137 (i) 0.235
(j) 0.509 (k) 0.075 (l) 0.016
(m) 0.006 (n) 1.2 (o) 2.5

3. Write each percent as a fraction.

(a) 60% (b) 75% (c) 25%
(d) 50% (e) 10% (f) 5%
(g) 80% (h) 100% (i) 200%
(j) 150% (k) 1% (l) $33\frac{1}{3}\%$
(m) $66\frac{2}{3}\%$ (n) 6.5% (o) 1000%

4. Write each fraction as a percent to the nearest tenth.

(a) $\frac{1}{2}$ (b) $\frac{3}{4}$ (c) $\frac{3}{8}$
(d) $\frac{4}{9}$ (e) $\frac{4}{5}$ (f) $\frac{3}{7}$
(g) $1\frac{1}{4}$ (h) $\frac{1}{3}$ (i) $2\frac{2}{3}$
(j) $\frac{5}{6}$ (k) $4\frac{3}{10}$ (l) $\frac{23}{50}$

5. Evaluate.

(a) 20% of 250 (b) 13% of 50 000
(c) 8% of $400 (d) 5% of $7.90

6. Express as a percent.

(a) 6 out of 10 (b) 37 out of 50
(c) 15 out of 60 (d) 8 out of 80

7. A sweater costs $190. The sales tax is 8%.
What is the total cost of the sweater?

8. The regular price of a radio was $230. It was sold at a discount of 10%.
What was the new selling price of the radio?

9. A scientific calculator has a list price of $260. The retailer first gives you a discount of 10% and then adds sales tax of 9%.
What is the cash price of the calculator?

10. Sam has a picture that is 60 cm long. He asked the photo shop to reduce the length by 30%.
What will be the new length of the picture?

11. Coreen got 72 out of the 80 driver's exam questions correct.
What was her mark?

12. Shima deposited $350 in the bank. She earned 9% simple interest in one year.
How much interest did she earn?

13. Fran got 35 hits in 91 times at bat.
What percent of the time did she get a hit, to the nearest tenth?

14. Last year's concert tickets cost $45. The price of this year's ticket will be 150% of last year's.
What is this year's price?

15. Pure gold is said to be 24 karats. A necklace has a mass of 70 g and is 16-karat gold.
How many grams of gold are there in the necklace?

16. A dealer bought a used car for $6000. He marked the price up 50% from what he paid for it. When he couldn't sell the car at this price, he marked it down by 40%. The car was bought 2 weeks later.
Did he make money or lose money?

1.9 ADDING AND SUBTRACTING INTEGERS

If par for a golf course is 72, that means that a good player can complete the course with 72 strokes. If you take 75 strokes, you are 3 over par, or $+3$. If you take 69 strokes, you are 3 under par, or -3. The numbers $+3$ and -3 are examples of integers.

We also use integers to read temperature. A temperature of $+5$ means 5 degrees above zero. A temperature of -6 means 6 degrees below zero.

The set of integers is made up of the positive and negative integers, including zero, which is neither positive nor negative.

$$I = \{\dots -4, -3, -2, -1, 0, +1, +2, +3, +4, \dots\}$$

We can show the set of integers graphically.

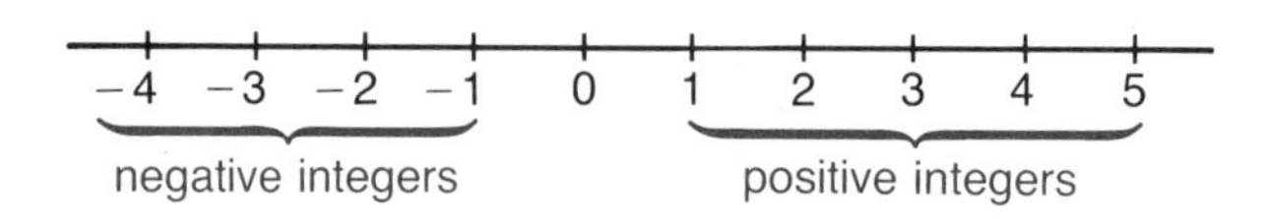

ADDITION

We use a number line to show the addition of integers.

Add. $(+4) + (+3)$

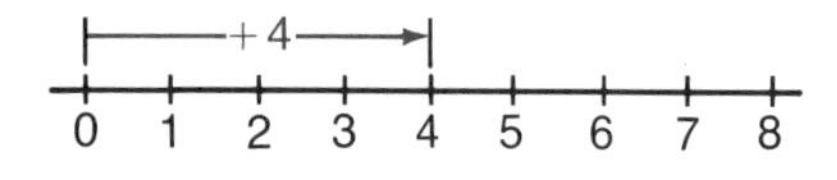

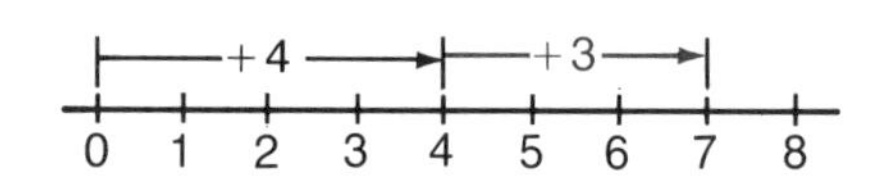

$\therefore 4 + 3 = 7$

Add. $(+3) + (-5)$

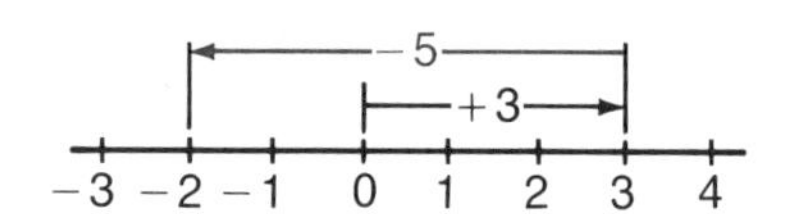

$\therefore 3 + (-5) = -2$

Add. $(-2) + (-4)$

$\therefore -2 + (-4) = -6$

SUBTRACTION

To subtract integers we add the opposite.

$6 - (+4) = 6 + (-4)$
$= 2$

$8 - 5 = 8 + (-5)$
$= 3$

$-7 - (-6) = -7 + (+6)$
$= -1$

EXERCISE 1.9

A 1. State an integer that represents each of the following.

(a) a loss of seven points
(b) five degrees above zero
(c) six under par
(d) a gain of seven dollars
(e) ten points in the hole

2. State the total effect of the following temperature changes.

(a) a rise of 6° followed by a rise of 4°
(b) a rise of 7° followed by a drop of 3°
(c) a drop of 5° followed by a rise of 8°
(d) a drop of 9° followed by a rise of 2°
(e) a rise of 3° followed by a drop of 10°

B 3. Add.

(a) $(+7) + (+6)$ (b) $(+6) + (-4)$
(c) $(-8) + (+5)$ (d) $(-3) + (-5)$
(e) $(+9) + (-7)$ (f) $(-3) + (-11)$
(g) $(-1) + (-12)$ (h) $(-14) + (+16)$

4. Add.

(a) $9 + (-7)$ (b) $-6 + (+11)$
(c) $-3 + 8$ (d) $5 + (-11)$
(e) $16 + 17$ (f) $-15 + 12$
(g) $-17 + (-13)$ (h) $60 + (-40)$

5. Subtract.

(a) $(+8) - (+6)$ (b) $(+4) - (-3)$
(c) $(-4) - (+5)$ (d) $(-7) - (-6)$
(e) $(+9) - (-3)$ (f) $(-1) - (+2)$
(g) $(-20) - (+13)$ (h) $(+6) - (-2)$

6. Subtract.

(a) $9 - (-4)$ (b) $-3 - 6$
(c) $-7 - (-9)$ (d) $6 - (-1)$
(e) $-1 - 8$ (f) $5 - 9$
(g) $-10 - (-12)$ (h) $2 - (-7)$

7. Simplify each of the following.

(a) $8 + 7 + (-2)$ (b) $-5 + (-4) + 6$
(c) $3 + (-6) + (-7)$
(d) $-4 + (-1) + (-8)$
(e) $12 + 6 + (-2) + 7$
(f) $-11 + (-2) + 6 + 5$
(g) $-21 + (-22) + 23 + (-24)$

8. Simplify the following.

(a) $12 - (-3) + (-2)$
(b) $-8 - (-4) - (+3)$
(c) $12 + 6 - 7$
(d) $8 - 9 - 2$
(e) $-3 - 4 - 6$
(f) $10 + 7 - 6 - (-5)$
(g) $-7 + 8 - (+6) - 2$
(h) $-100 + (-200) - 300 - (-500)$

9. Write an integer expression for each of the following, then simplify the expression.

(a) You earn eight points, then lose six points.
(b) The price of one share of Taco stock goes up three dollars in value, then down five dollars.
(c) You deposit forty dollars to your bank account, and then write cheques for fifteen dollars and thirteen dollars.
(d) The elevator goes up seven floors, then down eight floors.
(e) The jet goes up 100 m, then down 200 m, and finally up 400 m.
(f) On the first hole of a golf course you score 2 over par. On the next hole you score 1 under par. You par the next hole.

MIND BENDER

Place 5 dots in the figure so that there is no more than one dot in any row, column, or diagonal.

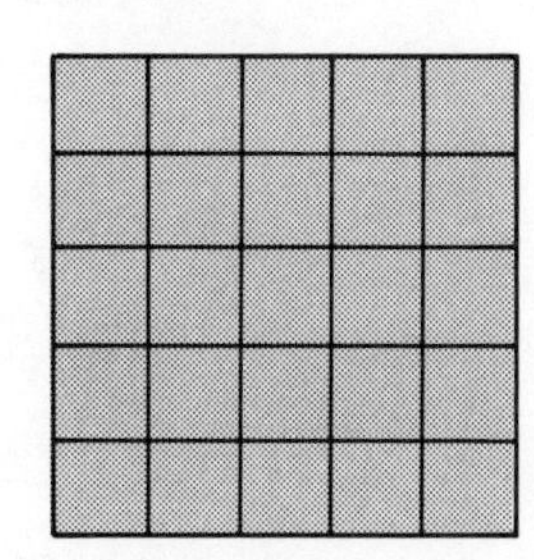

1.10 MULTIPLYING AND DIVIDING INTEGERS

Carla saved \$50 a month for five months.
How much did she save?

$\$50 + \$50 + \$50 + \$50 + \$50 = \$250$ or $5 \times \$50 = \$250$

She saved \$250.

Sam withdrew \$50 a month for five months from his bank account.
What was the effect on his account?

$-\$50 - \$50 - \$50 - \$50 - \$50 = -\$250$ or $5 \times (-\$50) = -\$250$

His account balance went down \$250.

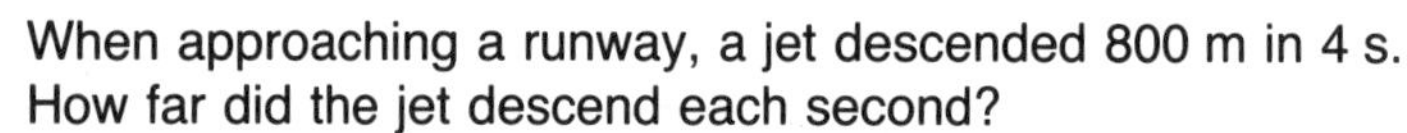

When approaching a runway, a jet descended 800 m in 4 s.
How far did the jet descend each second?

$$\frac{-800}{4} = -200$$

The jet descended 200 m each second.

The following charts summarize the rules for multiplying and dividing integers.

Multiplication

$(+4) \times (+3) = +12$	$(+) \times (+) = (+)$
$(-4) \times (-3) = +12$	$(-) \times (-) = (+)$
$(-4) \times (+3) = -12$	$(-) \times (+) = (-)$
$(+4) \times (-3) = -12$	$(+) \times (-) = (-)$

Division

$(+12) \div (+4) = +3$	$(+) \div (+) = (+)$
$(-12) \div (-4) = +3$	$(-) \div (-) = (+)$
$(-12) \div (+4) = -3$	$(-) \div (+) = (-)$
$(+12) \div (-4) = -3$	$(+) \div (-) = (-)$

Example 1.
Simplify each of the following.
(a) $(-23)(-3)$ (b) -7×8 (c) $56 \div (-8)$ (d) $-144 \div (-4)$

Solution:

(a) $(-23)(-3) = 69$

(b) $-7 \times 8 = -56$

(c) $56 \div (-8) = -7$

(d) $-144 \div (-4) = 36$

Example 2.
Simplify.
(a) $(-4)^2$ (b) -4^2 (c) $(-2)^3 + 7(-3)(-6) - 2^2$

Solution:

(a) $(-4)^2$
$= (-4)(-4)$
$= 16$

(b) -4^2
$= (-1)(4)(4)$
$= -16$

(c) $(-2)^3 + 7(-3)(-6) - 2^2$
$= -8 + 7(-3)(-6) - 4$
$= -8 + 126 - 4$
$= 114$

Example 3. If $x = -2$ and $y = -6$, evaluate the following.
$2xy - 3x - y^2$

Solution:
$$\begin{aligned} 2xy - 3x - y^2 &= 2(-2)(-6) - 3(-2) - 1(-6)^2 \\ &= 24 + 6 - 36 \\ &= -6 \end{aligned}$$

EXERCISE 1.10

A 1. State each product.

(a) $(-3)(+4)$ (b) $(-6)(+8)$
(c) $-2(-5)$ (d) $4(-6)$
(e) $(-12)(-2)$ (f) -3×7
(g) $-2 \times (-6)$ (h) $(-5) \times 0$
(i) $(-1)(-2)(-3)$ (j) $3(-2)(-2)$
(k) $(-11)(-9)$ (l) $1000 \times (-3)$

2. State the quotient.

(a) $\frac{20}{-4}$ (b) $\frac{-15}{-3}$ (c) $\frac{-18}{6}$
(d) $\frac{-36}{-12}$ (e) $\frac{50}{-10}$ (f) $\frac{-72}{-8}$
(g) $(-21) \div 7$ (h) $-42 \div 7$
(i) $60 \div (-5)$ (j) $-80 \div (-8)$

3. Evaluate.

(a) $(-2)^2$ (b) -3^3 (c) $(-4)^3$
(d) $(-10)^4$ (e) -5^2 (f) -1^8

B 4. Simplify.

(a) $(-12)(-4) \div (-2)$
(b) $-36 \div 4 \times (-5)$
(c) $(-18) \div 9 - 11$
(d) $56 - 5(-15) + 3$
(e) $(-100) \div (-4) \div (-5)$
(f) $72 - 3(-6) + 14$
(g) $\frac{-21 - 12}{-6 \div 2}$
(h) $-3 - 4 - 5 \times (-4)$
(i) $\frac{-3(-4)(-5)}{60 \div (-10)}$
(j) $(-27) \div (-3) - 9^2$
(k) $-2(8 - 9) - (-2)(-1)^3$
(l) $-3^2 - 5(5 - 7) + 8(-2)^2$
(m) $\frac{3(-4) - 2(-8)}{(-24) \div (-6)}$

5. If $x = -2$, $y = -1$, and $z = 3$, evaluate each of the following.

(a) $2x + 3y + 4z$
(b) $2xy + 3xz - 2yz$
(c) $x^2 + y^2 + z^2$
(d) $-x^3 - xy + 5z^2$
(e) $-x - y - z$

CALCULATOR MATH

THE [+/−] KEY

Most calculators have a change-sign key: [+/−]. When the key is pressed, the sign of the number in the display is changed. To enter −6 you press [6], then [+/−].

1. To compute $-9 + 5$, press
[C] [9] [+/−] [+] [5] [=]

2. To compute $-5 - 2$, press
[C] [5] [+/−] [−] [2] [=]

3. To compute $(-6) \times (-7)$, press
[C] [6] [+/−] [×] [7] [+/−] [=]

4. To compute $(-8) \div (+2)$, press
[C] [8] [+/−] [÷] [2] [=]

EXERCISE

1. Simplify each of the following.

(a) $-56 + 11 - 35$
(b) $-1518 \div (-66)$
(c) $-74 \times (-29)$
(d) $-56 - 78 + 95$

1.11 RATIONAL NUMBERS

We have worked with integers, both positive and negative fractions, and also decimals. All of these numbers are part of the larger set of rational numbers.

By definition, all rational numbers can be expressed as the ratio of two integers, $\frac{a}{b}$.

The following are examples of rational numbers.

$$\frac{3}{-5}, \quad \frac{-1}{2}, \quad \frac{2}{3}, \quad -\frac{7}{2}, \quad \frac{10}{-5}, \quad \frac{0}{2}$$

The following are also rational numbers.

3 because $3 = \frac{3}{1}$ $\quad -0.5$ because $-0.5 = -\frac{1}{2}$ $\quad -2\frac{1}{3}$ because $-2\frac{1}{3} = \frac{-7}{3}$

Equivalent Forms

$$\frac{-7}{3} = \frac{7}{-3} = -\frac{7}{3}$$

The non-terminating, repeating decimal 0.232 323 ..., or $0.\overline{23}$, is a rational number because it can be expressed in the form $\frac{a}{b}$.

$$\begin{aligned} \text{Let} \quad x &= 0.232\,323\ldots \\ \text{Then} \quad 100x &= 23.232\,323\ldots \\ \text{but} \quad x &= 0.232\,323\ldots \\ \text{Subtract} \quad 99x &= 23 \\ x &= \frac{23}{99}, \text{ which is a rational number.} \end{aligned}$$

Non-terminating, non-repeating decimals such as 0.234 536 912 ..., $\sqrt{2}$, and π are called irrational numbers.

The set of Real numbers, R, is made up of the rational and the irrational numbers.

As with other sets of numbers, the BEDMAS rules for the order of operations apply to rational numbers.

Example.

Simplify. (a) $-1\frac{1}{2} \times \frac{2}{-5} \div \frac{1}{4}$ $\quad$ (b) $1\frac{1}{3} - \frac{1}{2} + \frac{-5}{6}$

Solution:

(a)
$$\begin{aligned} -1\frac{1}{2} \times \frac{2}{-5} \div \frac{1}{4} &= -\frac{3}{2} \times \frac{2}{-5} \times \frac{4}{1} \\ &= \frac{-3}{2} \times \frac{-2}{5} \times \frac{4}{1} \\ &= \frac{24}{10} \\ &= 2\frac{2}{5} \end{aligned}$$

(b)
$$\begin{aligned} 1\frac{1}{3} - \frac{1}{2} + \frac{-5}{6} &= \frac{4}{3} - \frac{1}{2} + \frac{-5}{6} \\ &= \frac{8}{6} - \frac{3}{6} - \frac{5}{6} \\ &= \frac{8 - 3 - 5}{6} \\ &= \frac{0}{6} \\ &= 0 \end{aligned}$$

EXERCISE 1.11

A

1. State two equivalent forms for each of the following rational numbers.

(a) $\frac{2}{3}$ (b) $\frac{-1}{4}$ (c) $1\frac{3}{4}$

(d) 3 (e) -5 (f) $\frac{-2}{3}$

(g) $-\frac{1}{4}$ (h) $\frac{8}{3}$ (i) $\frac{-3}{7}$

B

2. Express each decimal as a fraction in lowest terms.

(a) 0.4 (b) 0.25 (c) 0.1
(d) 0.56 (e) 0.08 (f) 0.125

3. Express each fraction as a decimal.

(a) $\frac{3}{4}$ (b) $\frac{3}{8}$ (c) $-\frac{1}{2}$

(d) $\frac{-7}{10}$ (e) $\frac{5}{-8}$ (f) $\frac{-11}{16}$

4. Express in the form $\frac{a}{b}$.

(a) -0.3 (b) $-2\frac{1}{4}$ (c) $-3\frac{2}{3}$
(d) 0.75 (e) 1.4 (f) -1.6

5. Express each fraction as a decimal.

(a) $\frac{1}{9}$ (b) $\frac{3}{11}$ (c) $\frac{5}{6}$

(d) $\frac{8}{9}$ (e) $\frac{7}{11}$ (f) $\frac{2}{7}$

6. Express each decimal as a fraction.

(a) 0.333 ... (b) $0.\overline{2}$
(c) 0.666 ... (d) $0.\overline{18}$
(e) 0.090 909 ... (f) 0.533 333 ...

7. Simplify.

(a) $1\frac{3}{4} \times \frac{1}{2}$ (b) $2\frac{1}{2} \times \frac{-1}{4}$

(c) $\frac{-2}{5} \times \frac{3}{-7}$ (d) $-1\frac{1}{3} \times \left(-\frac{2}{5}\right)$

(e) $6 \times \frac{-3}{2}$ (f) $-1\frac{3}{4} \times (-4)$

8. Write the reciprocal.

(a) $\frac{2}{3}$ (b) $\frac{-3}{4}$

(c) $\frac{5}{-6}$ (d) 2

(e) -3 (f) $-\frac{1}{3}$

9. Simplify.

(a) $\frac{1}{2} \div \frac{3}{4}$ (b) $\frac{1}{4} \div \frac{-1}{3}$ (c) $1\frac{2}{3} \div \frac{1}{-2}$

(d) $\frac{3}{5} \div \left(-1\frac{2}{5}\right)$ (e) $-5 \div \frac{-3}{10}$ (f) $\frac{-5}{6} \div (-2)$

10. Simplify.

(a) $\frac{2}{3} + \frac{3}{4}$ (b) $\frac{-1}{2} + 1\frac{5}{6}$

(c) $\frac{1}{-3} + \frac{-3}{4}$ (d) $-1\frac{1}{2} + \frac{-1}{8}$

(e) $3 + \frac{-5}{6}$ (f) $-\frac{2}{3} + 2\frac{5}{6}$

11. Simplify.

(a) $\frac{5}{6} - \frac{1}{6}$ (b) $\frac{-3}{4} - \frac{1}{4}$ (c) $\frac{3}{8} - \frac{-5}{4}$

(d) $1\frac{1}{3} - \frac{3}{4}$ (e) $-1\frac{1}{2} - 2$ (f) $-5 - \frac{2}{-3}$

C

12. Simplify.

(a) $\frac{1}{3} + \frac{1}{2} - \frac{1}{4}$ (b) $\frac{2}{3} \times \frac{1}{2} - \frac{1}{4}$

(c) $\frac{2}{3} + \frac{1}{2} \times \frac{-1}{4}$ (d) $1\frac{1}{2} + \frac{-2}{3} - 1\frac{5}{6}$

(e) $\left(\frac{1}{2} + \frac{-1}{3}\right) \div 2$ (f) $-2\frac{1}{4} \times 3 + \frac{3}{-2}$

(g) $\frac{-1}{3} + \frac{1}{-2} \div \frac{-1}{4}$ (h) $\frac{1}{-3} \div \frac{-1}{2} \times \frac{1}{-4}$

(i) $-\frac{2}{3} - \frac{1}{3} - \frac{5}{6}$ (j) $-2\frac{3}{4} + \frac{1}{-5} - 3$

(k) $6 \times \left(\frac{1}{2} - \frac{3}{4}\right)$ (l) $-5 - \frac{3}{5} + 1\frac{4}{10}$

MIND BENDER

Show how to make change for a dollar using exactly fifty standard coins.

1.12 PRINCIPLES OF PROBLEM SOLVING

A problem is a situation that requires a solution. The word itself implies that the solution is not obvious and requires a conscientious effort.

The question "What is 2 + 3?" can hardly be called a problem because in fact it poses no problem. However, the question "How many AM 10 textbooks placed end-to-end would span the distance from Halifax to Vancouver?" would at first be considered a problem, since its solution requires a conscientious approach. If you were to solve many problems similar to this one, their solutions would eventually become so methodical and routine that they would be considered drill exercises rather than problems.

Though there are no hard and fast rules for solving problems, most mathematical problems can be solved using the following four-step approach: READ — PLAN — SOLVE — ANSWER

Whether it be in everyday situations or in a mathematics class, common sense and confidence are the key ingredients to problem solving. The READ — PLAN — SOLVE — ANSWER model presented in this section and used throughout the book will help you to develop both.

READ

Read the problem carefully and devote sufficient time to understanding the problem before trying to solve it. Note key words. Put the problem in your own words. Know what you are asked to find. Be aware of what is given.

PLAN

Think of a plan. Find a connection between the given information and the unknown which will enable you to calculate the unknown.

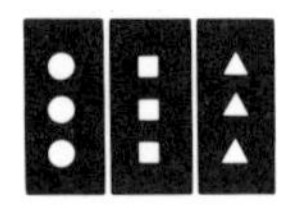

1. Classify information. Study the information carefully to determine what is needed to solve the problem. Identify the relevant and irrelevant information. Some information may be extraneous or redundant.

You may find it helpful to summarize the information or make lists.

2. Search for a pattern. Try to recognize patterns. Some problems are solved by recognizing that some kind of pattern is occurring. The pattern could be geometric, numerical, or algebraic. If you can see there is some sort of regularity or repetition in a problem, then you might be able to guess what the continuing pattern is, and then prove it.

3. Draw a diagram or flow chart. For many problems it is useful to draw a diagram and identify the given and required quantities on the diagram. A flow chart can be used to organize a series of steps that must be performed in a definite order.

4. Estimate, guess, and check. This is a valid method to solve a problem where a direct method is not apparent. You may find it necessary to improve your guess and "zero in" on the correct answer.

1·2·3

5. Sequence operations. To solve some problems, several operations performed in a definite order are needed.

3·2·1

6. Work backwards. Sometimes it is useful to imagine that your problem is solved and work backwards step by step until you arrive at the given data. Then, you may be able to reverse your steps to solve the original problem.

$E=mc^2$

7. Use a formula or an equation. In some problems, after analyzing the data, the problem can be written as an equation, or the data can be substituted into a formula.

8. Solve a simpler problem. A problem can sometimes be broken into smaller problems that can be solved more easily.

9. Account for all possibilities. List all of the cases. Your solution must account for all of these cases. You may sometimes be able to solve your problem using a process of elimination.

10. Make a table. In some problems, it is helpful to organize the data in a table, chart, or grid.

11. Check for hidden assumptions. In some problems, the information concerning what is given is presented in a subtle manner that may not attract your attention. Re-read the problem carefully and look for the implied information.

12. Conclude from assumptions. In some problems, it will be necessary to make assumptions. The conclusions that you draw from these assumptions should be those that you have made in the past, from the same types of information.

13. Introduce something extra. Sometimes it may be necessary to introduce something new, an auxiliary aid, to help make the connection between the given and the unknown. For instance in geometry, the auxiliary could be a new line drawn in the diagram. In algebra, it could be a new unknown which is related to the original unknown.

SOLVE

Before solving the problem, look at the reasons for selecting your strategy. If you have more than one strategy available, you should consider familiarity, efficiency, and ease, in making your choice. In carrying out your strategy, work with care and check each step as you proceed. Remember to present your ideas clearly.

ANSWER

State the answer in a clear and concise manner. Check your answer in the original problem and use estimation to decide if your answer is reasonable. In checking your answer, you may discover an easier way to solve the problem. You may wish to generalize your method of solution so that it can be applied to similar problems.

The following example shows how different strategies can be used to solve a problem.

READ

Example.
There are eight people entered in a round robin handball tournament. Each player must play each of the other players just once.
How many games should be scheduled?

PLAN

Solution:
Start by finding the number of games necessary for 2, 3, and 4 players. Look for a pattern. Diagrams and a table will help.

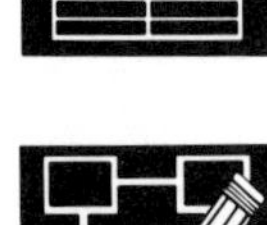

Number of Players	Diagram	Number of Games
2	A B	1
3	A C B	3
4	A D B C	6
5	A E D B C	10

Notice that the number of games needed forms the sequence

1, 3, 6, 10, ...
 2 3 4

We can use the patterns to determine the number of games for more than five players.

SOLVE

Players	2	3	4	5	6	7	8
Games	1	3	6	10	15	21	28

2 3 4 5 6 7

Notice there is another pattern.

Three players: $\frac{3 \times 2}{2} = 3$ Four players: $\frac{4 \times 3}{2} = 6$

Five players: $\frac{5 \times 4}{2} = 10$ Eight players: $\frac{8 \times 7}{2} = 28$

ANSWER

Twenty-eight games should be scheduled.

EXERCISE 1.12

1. When Paul arranged his dollar coins in piles of 6, he had 3 coins left over. If he made piles of 5, there were 4 left over. When he made piles of 8, there were 7 left over. Paul had less than $50 in dollar coins. How many coins would be left over if he made piles of 7?

2. There are 9 boxes. Four contain apples. Five contain oranges. Two contain both apples and oranges.
How many boxes are empty?

3. Put the numbers from 1 to 8 in the rectangles so that no two adjoining rectangles have consecutive numbers.

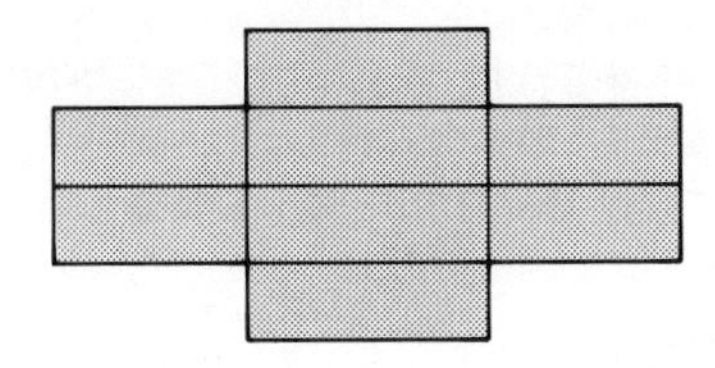

4. Find four consecutive numbers that add up to 274.

5. How many days are there from today to St. Valentine's Day?

6. A rectangular field measures 30 m by 26 m. The field is to be fenced and fence posts are to be 2 m apart. Fence posts cost $21.60 each and fencing costs $18.75/m. It will take three people, each working six hours, to complete the job. They each get paid $11.70/h.
How much will it cost to fence the field?

7. The numbers 2, 3, 5, 7, and 11 are called prime numbers. A prime number can only be divided evenly by 1 and the number itself.
The number 14 can be written as the sum of two prime numbers as follows:

$$11 + 3 = 14$$

Write each of the following numbers as the sum of two other prime numbers.
(a) 24 (b) 38 (c) 54

8. Sixteen more boys than girls went on the art gallery tour. There were 168 on the tour altogether.
How many girls went on the tour?

9. A commuter train left Port Darling for Tardiff. At Tardiff 12 people got off and 18 got on. The next stop was Dundas where 11 people got off and 34 got on. At the next stop, which was Sandhurst, 9 people got off and 21 people got on. The last stop was Carston where 123 people got off, leaving the train empty.
How many people were on the train when it left Port Darling?

10. When the Kitzlers left on their vacation, the odometer on their car read 45 678.3 km. When they returned, the odometer read 49 000.7 km.
How many kilometres had they travelled?

11. Five friends were sitting together in a row at a football game. Bob sat next to Aisha. Natasha sat next to Luc, while Lois sat in the third seat from Aisha.
If Bob sat in the third seat from Natasha, who sat on the other side of Luc?

12. Car motor oil can be recycled and used again. From the used oil of 4 cars you can extract enough good oil for one oil change. How many oil changes can you get from the used oil of 32 cars?

13. You have decided to walk across Canada along the Trans-Canada Highway. Approximately how many steps will you take?

14. Which of the wheels are turning clockwise?

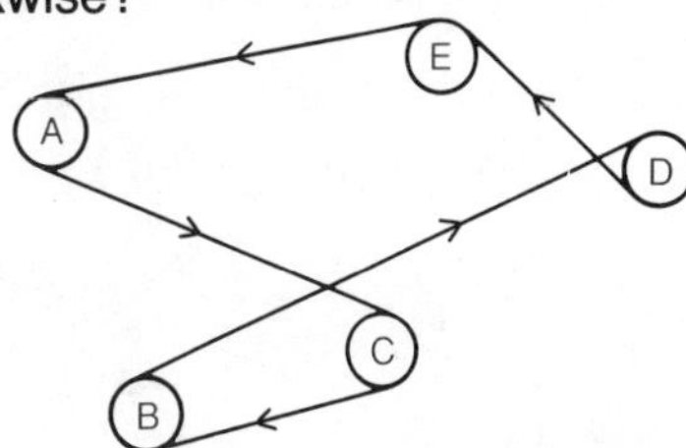

15. Each number in the large squares is found by adding the numbers in the small squares next to it.

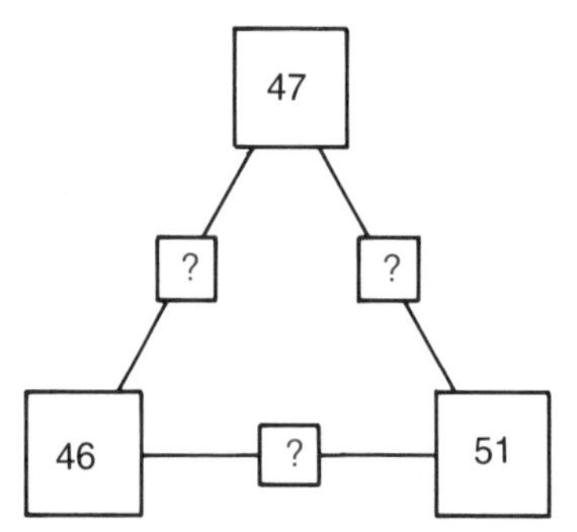

Find the numbers in the small squares.

16. Ari earns \$27.75/h as a mechanic for a racing team. Last week he worked 63 h. His employer deducted \$348.68 from his pay cheque for income tax and other benefits. Ari put his cheque in the bank, where he already had \$5698.78 on deposit. What was Ari's new bank balance?

17. When the time is 02:30 on January 23 in Ottawa, what is the date and time in each of these cities?

(a) Vancouver
(b) Halifax
(c) Honolulu
(d) Perth, Australia
(e) Rome, Italy
(f) Moscow
(g) Cairo, Egypt
(h) St. John's

18. Use the pattern in the following to find the sum of the odd numbers from 1 to 25.

$$1 = 1 \times 1$$
$$1 + 3 = 2 \times 2$$
$$1 + 3 + 5 = 3 \times 3$$
$$1 + 3 + 5 + 7 = 4 \times 4$$

19. The following are the first three figures in a pattern of triangles.
The triangles are made out of toothpicks.

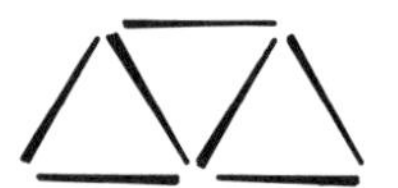

(a) If the pattern is continued, how many toothpicks will be needed for 10 triangles?
(b) How many will be needed for 30 triangles?

20. The sum of two consecutive odd numbers is 244.
What are the numbers?

21. The usual tip to pay on a restaurant meal is 15% of the cost.
Approximately how much of a tip should you leave for a meal that cost
(a) \$23.50?
(b) \$45.70?

22. How many triangles can you draw that have a perimeter of 18 m? (Assume that the lengths of the sides are measured only in whole numbers.)

23. How many days are there in eight consecutive years?

24. There are five towns located as shown in the diagram.

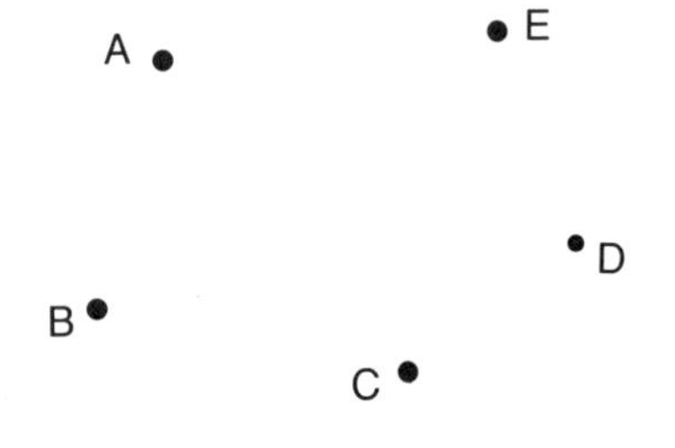

How many roads are needed if each town is to be connected directly to every other town by a road?

25. Each letter stands for a different digit in this cryptogram.
What number does each letter represent?

$$\begin{array}{r} \text{SEND} \\ +\text{MORE} \\ \hline \text{MONEY} \end{array}$$

26. Nicia has 11 bills that total $52. The bills are $2, $5, and $10 denominations. How many $2 bills does she have?

27. A kindergarten teacher bought 24 pencils at $0.26 each and 18 erasers at $0.16 each.
How much change did she receive from a $20 bill?

28. Five men and two boys want to cross a river in a rowboat that can hold two boys or one man.
How many crossings must be made to get everyone to the other side?

29. You are a secret service agent and your instructions are as follows. Your train leaves Istanbul for Paris at exactly 06:15. Your hotel is a 20 min cab ride from the train station. You must arrive at the station 15 min before the train leaves to get your ticket.
It will take you 55 min to get dressed and eat breakfast. You must pick up a package at the embassy on the way to the station. This will take you 20 min. You will need 10 min to lose the agents you spot at the embassy.
For what time should you set your alarm?

30. How many song titles can you name that contain the word "LOVE"?

31. Each pattern starts with the numbers 2, 4, 8.
Predict the next three numbers in each case.

(a) 2, 4, 8, 10, 14, . . .
(b) 2, 4, 8, 16, 32, . . .
(c) 2, 4, 8, 14, 22, . . .

CAREER

ADVERTISING INDUSTRY

You can find advertising in almost every aspect of our life, and the career opportunities are many and varied. Some of them are:

* product label design
* radio and TV commercial production
* sales
* promotion

1. In one year advertisers in Canada spent $9 billion on advertising.

(a) Find the amount spent in each of the following categories:
  (i) about 22% was spent on TV advertising.
  (ii) about 8% was spent on radio advertising.
  (iii) about 26% was spent on newspaper advertising.

(b) Notice that in question 1(a) the percents do not add to 100%. What other types of media are available to advertisers?

2. What kinds of companies would want to advertise nationally?

3. A company spent $200 000 to place a thirty-second TV commercial on a television network.
If 6 million people watch the commercial, how much does the advertiser pay for each person who sees the commercial?

EXTRA

1.13 NETWORKS, MAZES, AND ROOMS

NETWORKS

The study of networks is an important branch of mathematics. Some of the applications resulting from network theory are

* designs for highway systems
* emergency routes for vehicles
* pipelines and phone lines
* electrical circuits

One of the first people to study networks was the Swiss mathematician, Leonhard Euler (1707-1783). One of the network problems he studied involved the bridges of Konigsberg. The townspeople of Konigsberg knew that it was impossible to walk through the town and cross the seven bridges exactly once, without crossing one of the bridges at least *twice*. However, they couldn't explain it.

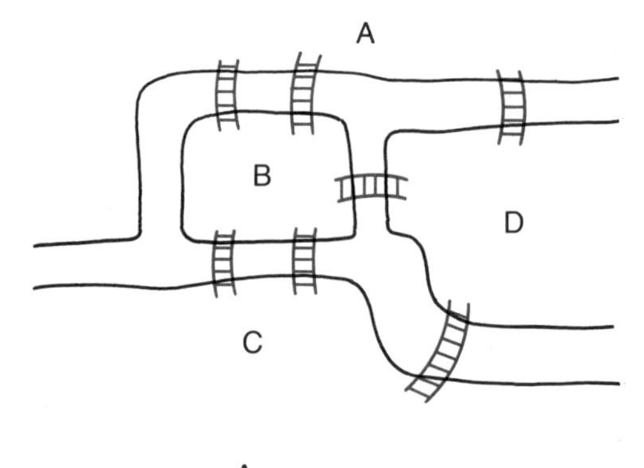

Euler illustrated the problem by drawing a map of the town and replacing the areas A, B, C and D by points. Now the question is to see if it is possible to redraw the figure without lifting a pencil and without retracing any line. If this is possible, we say that the network is traversable.

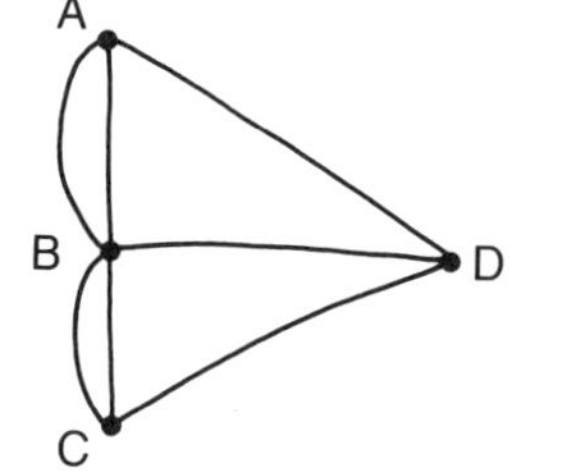

A, B, C and D are vertices. They are called odd as an odd number of lines intersect at each vertex. An even number of lines intersect at an even vertex.

Euler discovered that you can tell if a network is traversable if you count the number of odd and even vertices.

Classify the vertices in these networks as odd or even.
Determine which networks are traversable.
What are Euler's rules for traversability?

(a)

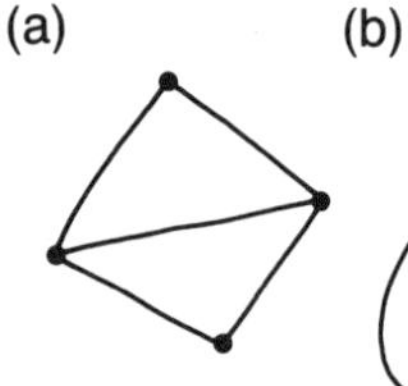

(b)

(c)

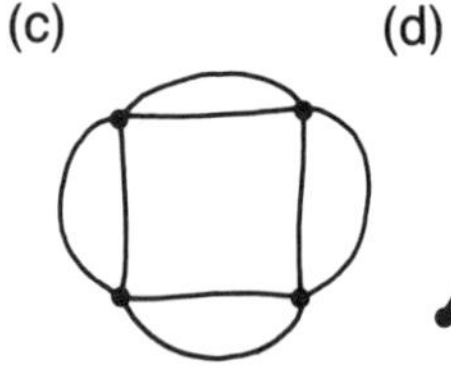

(d)

(e)

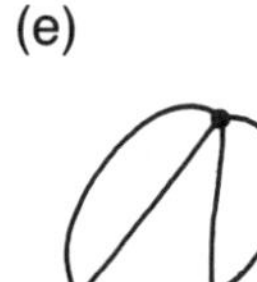

(f)

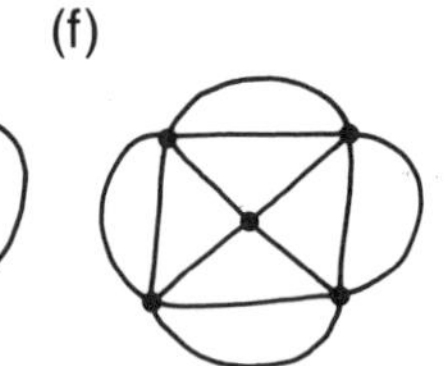

MAZES

According to Greek mythology, the maze at the right is the one supposedly built by Daedalus to house the Minotaur. It was said to be impossible for the terrible Minotaur to escape from the inside. Is this true?

During the seventeenth century, the maze became a fashionable feature in the gardens of many large houses and palaces in Europe. The walls of the mazes were actually hedges. One of the more famous mazes can be visited today in England at Hampton Court palace. This maze was built during the reign of William III.

The object is to get to the middle from the outside.

The maze at the right was designed by a mathematician.
Can you find a way to the red dot?

ROOMS

The diagrams represent the floor plans of houses. The object is to find a path that goes through each door only once.

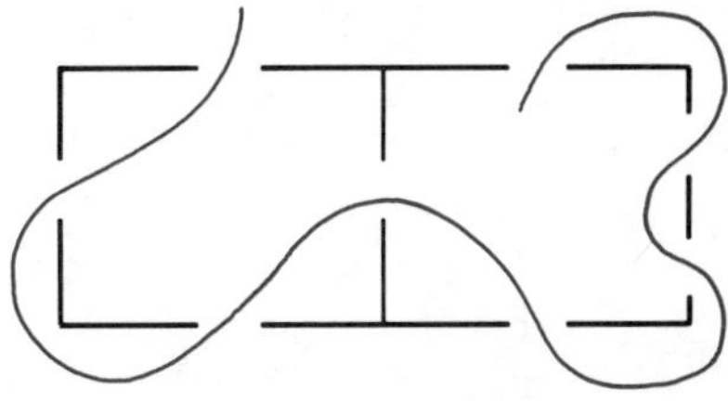

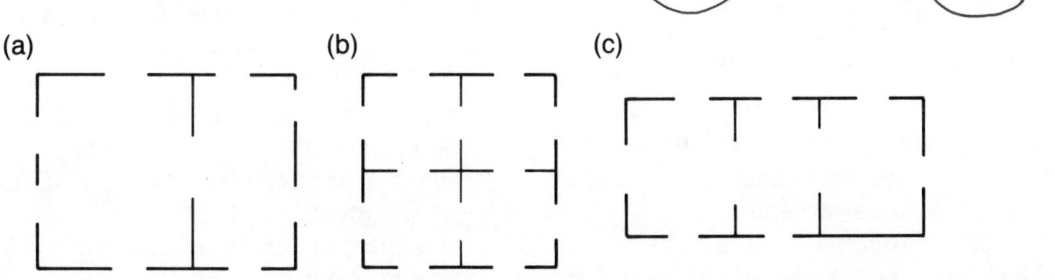

1.14 REVIEW EXERCISE

1. State the face value, place value, and total value of the indicated digit.

(a) 34 708 (b) 0.763
(c) 569 104 (d) 23.04
(e) 56 781 000 (f) 0.062
(g) 3489 (h) 1.496

2. Write each number in words.

(a) 45 780 (b) 5.72
(c) 8 926 000 (d) 0.867
(e) 0.005 (f) 56.801
(g) 3469 (h) 204.2

3. Simplify.

(a) $3450 + 671 + 12\,980 + 101$
(b) $2005 - 1456$
(c) 3403×21
(d) $1302 \div 31$
(e) $45\,678 + 1900 + 7007$
(f) $23\,000 - 19\,456$
(g) $67\,020 \times 9$
(h) $1404 \div 9$

4. Sandra works eight hours a day, five days a week, and fifty weeks a year. How many hours does she work in a year?

5. Simplify.

(a) $45 \times (26 - 16) + 3^2$
(b) $98 \div 14 + 11 \times 13$
(c) $138 - 21 \div 3 - 2^3$
(d) $(34 - 21)(12 + 16)$
(e) $5^2 + 6^2 - 7^2$
(f) $\dfrac{75 + 35}{22 \div 2} + 3^2$
(g) $\dfrac{3}{4}$ of $80 + (11 - 5)^2$
(h) $\dfrac{1}{3}$ of $99 - \dfrac{6 \times 8}{48 \div 4}$
(i) $5^2 + \dfrac{1}{2}$ of $96 - 2^2 + 8 \div 2$

6. Round to the indicated place value.

(a) 6560 (nearest hundred)
(b) 2.34 (nearest tenth)
(c) 57 500 (nearest thousand)
(d) 9.345 (nearest hundredth)

7. Round to the highest place value.

(a) 560 (b) 0.45
(c) 219 000 (d) 0.0055
(e) 4610 (f) 0.0017

8. Estimate.

(a) $231 + 450 + 161 - 101$
(b) 8.7×3.1
(c) $6039 \div 250$
(d) $780 - 350 + 210 - 324$
(e) $45 \times 89 + 23 \times 78$

9. Estimate, then calculate.

(a) $34.67 + 9.28 + 21.5$
(b) $6.002 - 3.456$
(c) 45.7×2.7
(d) $11.34 \div 5.4$
(e) 234.7×0.4
(f) $0.0062 \div 0.31$
(g) $9.76 + 0.781 + 11.17$
(h) $200.9 - 56.2$

10. Estimate the cost of each, then calculate.

(a) seven concert tickets at $45.75 each
(b) nine calculators at $13.67 each
(c) thirteen basketball shirts at $24.60 each
(d) thirty-eight hours of work at $18.20/h

11. Evaluate $24 - 3t$ for $t = 5, 3.2, 6.7$

12. Sarah is a concert promoter. It costs her $100 000 to rent a stadium and set up for a concert. If the band takes 50% of the gate receipts, the formula for Sarah's profit is

$$P = 0.5cn - 100\,000$$

where c is the cost of a ticket and n is the number of people attending.
Find Sarah's profit if

(a) she charges $45 a ticket and 11 000 people attend.
(b) she charges $42.50 a ticket and 15 000 people attend.
(c) she charges $38.75 a ticket and 18 000 people attend.

13. Multiply.

(a) $\frac{3}{4} \times \frac{1}{2}$ (b) $2\frac{1}{2} \times \frac{1}{3}$

(c) $3\frac{2}{5} \times 1\frac{1}{4}$ (d) $5 \times 2\frac{3}{10}$

14. Divide.

(a) $\frac{1}{3} \div \frac{1}{2}$ (b) $2\frac{2}{3} \div \frac{1}{3}$

(c) $3\frac{3}{4} \div 2\frac{1}{2}$ (d) $6 \div 2\frac{2}{3}$

15. Add.

(a) $\frac{5}{12} + \frac{3}{4}$ (b) $2\frac{2}{3} + 3\frac{1}{5}$

(c) $4\frac{5}{6} + 2\frac{1}{9}$ (d) $7 + 3\frac{3}{4}$

16. Subtract.

(a) $\frac{4}{7} - \frac{1}{2}$ (b) $3\frac{1}{8} - 1\frac{1}{4}$

(c) $6 - 3\frac{3}{5}$ (d) $4\frac{1}{6} - 2\frac{3}{4}$

17. Simplify.

(a) $\frac{1}{2} \times \frac{1}{3} + \frac{1}{4}$ (b) $\frac{1}{2} + \frac{1}{3} \times \frac{1}{4}$

(c) $3\frac{1}{2} \div 2 + \frac{3}{4}$ (d) $\frac{4}{5} - 2\frac{1}{4} \div 4$

(e) $2\frac{3}{8} + 1\frac{2}{3} \times \frac{1}{2}$ (f) $6 - 1\frac{3}{4} \times \frac{1}{3}$

18. Evaluate.

(a) 24% of 1200
(b) 45% of $780
(c) 120% of 300 000

19. A large hotel reaches the break-even point, at which costs are equal to revenues, when 38% of the rooms are rented.
If a hotel has 266 rooms, how many must be rented to break even?

20. A survey of 300 people found that 195 read at least one magazine a week.
What percent of the people did not read at least one magazine a week?

21. Simplify.

(a) $9 + (-4) - 6$
(b) $-15 + 8 - 9 - (-2)$
(c) $-7 - 8 - 10$
(d) $14 - (+8) - 11$
(e) $-5 - 6 - 7 + 18$
(f) $-3 - (-3) - (-3) - 3$

22. Simplify.

(a) $-24 \div 6 + 11$
(b) $(-12)(-2) - 3$
(c) $-9 + (-56) \div 7$
(d) $21 \times (-3) + 2$

23. If $s = -3$, $t = -2$, and $r = 2$, evaluate each expression.

(a) $2st - 5r + 6$
(b) $s^2 + t^2 - r^2$
(c) $5t - 3s + 6r - rst$

24. Express each as a decimal.

(a) $\frac{7}{8}$ (b) $\frac{3}{11}$

25. Express each as a fraction.

(a) 0.66 (b) 0.777 777 ...

26. Simplify.

(a) $1\frac{2}{3} \div \frac{1}{-3}$ (b) $-2\frac{1}{2} \times 6$

(c) $-\frac{3}{8} + \frac{-1}{4}$ (d) $\frac{2}{3} \div \frac{-1}{3} + \frac{1}{4}$

(e) $-\frac{1}{3} - \frac{1}{6} - \frac{2}{9}$ (f) $\frac{3}{4} - \frac{1}{2} \times \frac{1}{4}$

(g) $-1\frac{1}{2} \div 2 + \frac{1}{-4}$ (h) $-6 - \frac{3}{5} + 1\frac{3}{10}$

27. Grapefruit are displayed in the shape of a square-based pyramid as shown.

(a) How many grapefruit are in the top layer?
(b) How many are in the second layer from the top?
(c) How many are in the top two layers?
(d) How many are in the top three layers?
(e) How many are in the pyramid?

28. Predict the next three numbers in each case.

(a) 11, 17, 22, 28, 33, . . .
(b) 64, 32, 16, . . .
(c) 30, 29, 27, 24, 20, . . .

1.15 CHAPTER 1 TEST

1. Simplify.

(a) $34\,507 + 4561 + 202 + 3471$
(b) $4001 - 2982$
(c) 347×42
(d) $2890 \div 34$

2. Simplify.

(a) $12 \times 15 - 11 \times 8$
(b) $(76 + 124) \div 40 + 5^2$
(c) $112 \div 14 + 45 \times 2 - 8$
(d) $\frac{1}{4}$ of $36 - \frac{15 \times 3}{30 \div 2}$
(e) $\frac{3(7 + 1)}{2} + 20 \div 5 + 2^2$

3. Round to the highest place value.

(a) 5600	(b) 0.85	(c) 13 456
(d) 750	(e) 0.067	(f) 8.437

4. Calculate.

(a) $34.67 + 9.6 + 234.9$
(b) $30.04 - 19.76$
(c) 4.5×0.68
(d) $2.268 \div 0.42$

5. The formula for the area of a trapezoid is

$A = \frac{1}{2}h\,(a + b)$.

Find A when $h = 5.6$ m, $a = 7.8$ m, and $b = 3.1$ m.

6. Simplify.

(a) $2\frac{1}{2} + 3\frac{2}{3}$ (b) $5\frac{3}{5} - 1\frac{1}{2}$ (c) $4\frac{1}{4} \times 3\frac{3}{5}$ (d) $1\frac{1}{8} \div 2\frac{3}{4}$

7. Sandra received 58% of the 750 votes cast for treasurer of the student council.
How many votes did she get?

8. Simplify the following.

(a) $-36 \div (-9) + 13$
(b) $-7 - 8 + (-4)$
(c) $-6 \div (-2) + 5 \times (-1)$
(d) $13 - (-12) - 9$

9. Simplify.

(a) $2\frac{2}{3} - \frac{1}{2} + \frac{-2}{6}$ (b) $-3\frac{1}{2} \times \frac{-1}{5} \div \frac{1}{2}$

10. The time is 10:00.
What time will it be 219 h from now?

POWERS AND SQUARE ROOTS

REVIEW AND PREVIEW TO CHAPTER 

FRACTIONS AND DECIMALS

ADDITION AND SUBTRACTION

EXERCISE 1

1. Determine the value for ■ to form equivalent fractions.

(a) $\frac{1}{4} = \frac{■}{8}$ (b) $\frac{1}{2} = \frac{■}{10}$ (c) $\frac{2}{3} = \frac{■}{12}$

(d) $\frac{15}{18} = \frac{■}{6}$ (e) $\frac{6}{■} = 3$ (f) $\frac{7}{■} = \frac{14}{18}$

(g) $2\frac{1}{4} = \frac{■}{4}$ (h) $7\frac{3}{5} = \frac{■}{5}$ (i) $12\frac{1}{8} = \frac{■}{8}$

In the problems below, express all answers in lowest terms.

2. Simplify.

(a) $\frac{5}{8} + \frac{1}{8}$ (b) $\frac{4}{5} - \frac{1}{5}$ (c) $\frac{3}{5} + \frac{2}{10}$

(d) $\frac{5}{6} + \frac{1}{6}$ (e) $1\frac{3}{4} + \frac{2}{3}$ (f) $2\frac{1}{2} - 1\frac{1}{3}$

(g) $5\frac{1}{4} + \frac{3}{5}$ (h) $\frac{7}{8} - \frac{5}{6}$ (i) $3\frac{1}{4} - \frac{7}{8}$

(j) $1\frac{1}{6} + \frac{5}{9}$ (k) $4\frac{3}{7} - 2\frac{1}{4}$ (l) $1\frac{1}{8} - \frac{5}{7}$

3. Simplify.

(a) $1\frac{1}{3} + \frac{1}{4} - \frac{5}{6}$ (b) $2\frac{1}{2} - \frac{3}{5} + \frac{5}{6}$

(c) $\frac{2}{3} + \frac{5}{6} - \frac{1}{2}$ (d) $2\frac{3}{4} + \frac{1}{3} + \frac{7}{8}$

(e) $\frac{1}{3} + \frac{1}{5} - \frac{4}{15}$ (f) $5\frac{1}{3} + \frac{3}{10} - 1\frac{1}{6}$

(g) $2\frac{1}{4} + 3\frac{1}{2} - 4\frac{3}{4}$ (h) $5\frac{3}{8} + 2\frac{1}{4} - 3\frac{7}{8}$

4. Simplify.

(a) $5\frac{1}{2} - \left(2\frac{3}{4} + 1\frac{1}{4}\right)$ (b) $\left(2\frac{5}{8} + 3\frac{3}{4}\right) - 1\frac{1}{2}$

(c) $1\frac{1}{4} + \left(2\frac{5}{8} - 1\frac{1}{4}\right)$ (d) $3\frac{5}{8} - \left(2\frac{1}{2} + 1\frac{1}{4}\right)$

(e) $7\frac{1}{2} + 2\frac{1}{4} - 3\frac{5}{8}$ (f) $6\frac{3}{4} - 2\frac{5}{8} + 1\frac{1}{8}$

(g) $5\frac{1}{4} - 2\frac{1}{2} - 1\frac{3}{4}$ (h) $3\frac{2}{3} - 4\frac{5}{6} + 5\frac{1}{2}$

MULTIPLICATION AND DIVISION

EXERCISE 2

1. Simplify.

(a) $\frac{2}{3} \times \frac{5}{7}$ (b) $\frac{1}{4} \times \frac{2}{5}$

(c) $\frac{8}{15} \times \frac{3}{10}$ (d) $1\frac{1}{2} \times \frac{5}{9}$

(e) $3\frac{1}{5} \times \frac{15}{16}$ (f) $2 \times \frac{1}{3}$

(g) $5 \times \frac{5}{8}$ (h) $\frac{3}{4} \times 1\frac{1}{3}$

(i) $3\frac{2}{3} \times 1\frac{1}{8}$ (j) $1\frac{2}{3} \times \frac{1}{5}$

2. Simplify.

(a) $\frac{3}{4} \div \frac{2}{5}$ (b) $\frac{5}{8} \div \frac{2}{3}$ (c) $1\frac{1}{4} \div \frac{3}{5}$

(d) $\frac{5}{12} \div \frac{3}{4}$ (e) $3\frac{1}{5} \div 2\frac{1}{2}$ (f) $2 \div \frac{1}{2}$

(g) $\frac{7}{8} \div 2$ (h) $2\frac{5}{6} \div 1\frac{1}{3}$ (i) $\frac{2}{3} \div 1\frac{1}{2}$

3. Simplify.

(a) $1\frac{1}{2} \times \frac{3}{4} \times 3\frac{1}{2}$ (b) $5\frac{1}{2} \times \frac{5}{22} \times \frac{3}{5}$

(c) $\frac{3}{5}\left(\frac{2}{3} + \frac{1}{4}\right)$ (d) $\left(2\frac{1}{7} - \frac{5}{6}\right) \times \frac{7}{8}$

(e) $3 \times \frac{5}{6} \div 1\frac{1}{3}$

(f) $\left(\frac{2}{5} + \frac{1}{10}\right) \times \left(1\frac{1}{2} - \frac{5}{12}\right)$

(g) $\frac{1}{5} + \frac{2}{3} \times 1\frac{3}{4}$ (h) $\left(\frac{3}{4}\right)^2 - \frac{1}{4}$

(i) $\left(\frac{2}{3} + \frac{1}{5}\right)^2$

4. Simplify.

(a) $\frac{3}{4}\left(\frac{5}{8} + \frac{1}{2}\right)$ (b) $\frac{3}{4} \times \frac{5}{8} + \frac{3}{4} \times \frac{1}{2}$

(c) $3\frac{3}{4}\left(1\frac{1}{2} + 2\frac{5}{8}\right)$ (d) $\left(2\frac{1}{4} + 3\frac{1}{2}\right) 4\frac{1}{2}$

(e) $5\frac{3}{4} \times 2\frac{1}{2} - 2\frac{1}{4} \times 2\frac{1}{2}$

OPERATIONS WITH DECIMALS

EXERCISE 3

Perform the operations.

1. (a) $\begin{array}{r} 4.32 \\ 5.94 \\ +3.05 \\ \hline \end{array}$ (b) $\begin{array}{r} 56.9 \\ 38.2 \\ +97.8 \\ \hline \end{array}$

(c) 3.74 + 2.81 + 3.74
(d) 14.7 + 3.9 + 8.6
(e) 0.750 + 0.086 + 1.920
(f) 0.125 + 2.5 + 3.705
(g) 5.65 + 7.25 + 3.08

2. (a) $\begin{array}{r} 86.32 \\ -31.84 \\ \hline \end{array}$ (b) $\begin{array}{r} 104.7 \\ -\ 85.2 \\ \hline \end{array}$

(c) 35.9 − 26.8
(d) 426.8 − 59.7
(e) 43.725 − 16.54
(f) 625.8 − 37.25
(g) 10.05 − 4.375

3. (a) $\begin{array}{r} 4.73 \\ \times\ 9.5 \\ \hline \end{array}$ (b) $\begin{array}{r} 97.5 \\ \times\ 0.38 \\ \hline \end{array}$

(c) 56.4 × 8.21
(d) 0.157 × 2.54
(e) 2.5 × 3.625
(f) 4.68 × 3.05
(g) 456.3 × 0.004

4. (a) $4.7\overline{)26.461}$
(b) $0.53\overline{)1.325}$
(c) 436.54 ÷ 7.3
(d) 60.918 ÷ 85.2
(e) 38.76 ÷ 5.7
(f) 181.764 ÷ 32.4
(g) 0.701 25 ÷ 0.125

5.

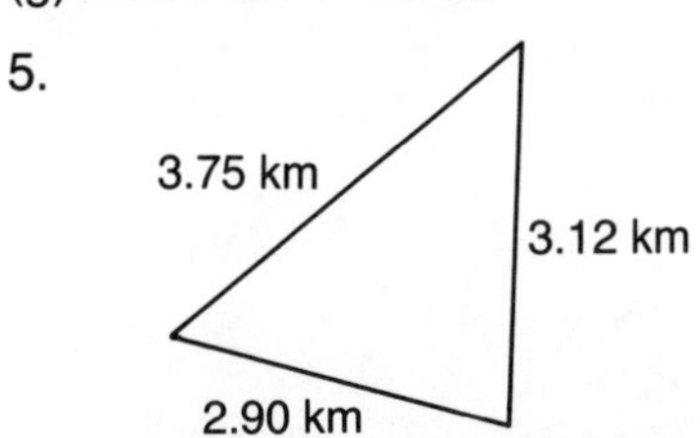

Find the perimeter of the triangle.

ESTIMATION

EXERCISE 4

1. Round each number to the highest place value.

(a) 325 000 (b) 7585
(c) 1 783 000 (d) 52 250
(e) 0.625 (f) 0.0875
(g) 0.149 (h) 2.581

2. Choose the best estimate.

(a) 456 × 12 858 ≐ 586 300, 5 863 000, 58 630 000
(b) 37 588 ÷ 285 ≐ 131.9, 1319, 13.19, 13 190
(c) 25.61 × 82.35 ≐ 2108, 21 080, 218 000
(d) 42.68 ÷ 125 ≐ 0.034, 0.34, 3.4, 34

3. Place the decimal point to make each answer true.

(a) 3.8 × 4.7 ≐ 1786
(b) 1.25 × 83.5 ≐ 1043
(c) 827 500 ÷ 18.20 ≐ 454670
(d) 15.6 ÷ 25.7 ≐ 607

4. Estimate.

(a) 435 × 257
(b) 1683 ÷ 24
(c) 2657 × 8325
(d) 25.65 ÷ 2.14
(e) 3.625 × 6.758
(f) 584.2 × 23.65
(g) 0.589 × 2.504
(h) 0.6534 ÷ 0.022

5. Estimate.

(a) $\frac{26 \times 83}{15 \times 12}$ (b) $\frac{586.7 \times 23.75}{45.32 \times 18.65}$

(c) $\frac{2.565 \times 4.858}{5.95 \div 2.08}$ (d) $\frac{805.2 \div 1.75}{4.75 \times 8.31}$

(e) $\frac{525 \times 838}{4.7 \times 7.8}$ (f) $\frac{648.25}{4.75 \times 9.85}$

(g) $\frac{675\,000 \div 72.57}{9.995}$

2.1 EXPONENTS

Scientists estimate that about 1 000 000 000 a passed after the formation of our planet before the first signs of life on the earth appeared. The number 1 000 000 000 can also be expressed as a power of ten.

$$\underbrace{1\,000\,000\,000}_{\text{9 zeros}} = \underbrace{10 \times 10 \times 10 \times 10 \times 10 \times 10 \times 10 \times 10 \times 10}_{\text{9 factors}}$$

A product of equal factors can be written using exponents.

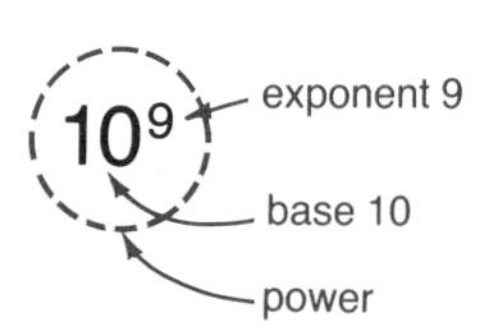

When two or more numbers are multiplied, each of the numbers is called a factor of the product.

32 can also be written as a power of 2.

$$32 = \underbrace{2 \times 2 \times 2 \times 2 \times 2}_{\text{5 factors}} = 2^5$$

3^5 means $3 \times 3 \times 3 \times 3 \times 3 = 243$

5^3 means $5 \times 5 \times 5 = 125$

We can use the constant function of a calculator to multiply by the constant factor 2.

Press 2 × = = = =

Display 2. 2. 4. 8. 16. 32.

We can write a number by listing the prime factors and using exponents.

$$72 = 2 \times 2 \times 2 \times 3 \times 3$$
$$= 2^3 \times 3^2$$

> In general, if $n = 1, 2, 3, \ldots$ then x^n means $x \times x \times x \times x \times \ldots$ to n factors.

Example 1.
If $x = 5$, find the value of each expression.
(a) x^3 (b) $3x^2$ (c) $(3x)^2$ (d) $-x^2$

Solution:

(a) $x^3 = 5^3$
$= 5 \times 5 \times 5$
$= 125$

(b) $3x^2 = 3(5)^2$
$= 3 \times 5 \times 5$
$= 75$

(c) $(3x)^2 = (3 \times 5)^2$
$= 15^2$
$= 225$

(d) $-x^2$
$= -(5)^2$
$= -25$

Example 2.
If $x = -3$ and $y = 4$, find the value of each expression.
(a) x^2y^3 (b) $3x^2 - 2y^3$ (c) $-2x(x + y^2)$

Solution:

(a) x^2y^3
$= (-3)^2(4)^3$
$= 9 \times 64$
$= 576$

(b) $3x^2 - 2y^3$
$= 3(-3)^2 - 2(4)^3$
$= 3 \times 9 - 2 \times 64$
$= 27 - 128$
$= -101$

(c) $-2x(x + y^2)$
$= -2(-3)(-3 + 4^2)$
$= -2(-3)(-3 + 16)$
$= 6 \times 13$
$= 78$

EXERCISE 2.1

A 1. Express each of the following as a power.

(a) $3 \times 3 \times 3 \times 3$
(b) $4 \times 4 \times 4$
(c) $5 \times 5 \times 5 \times 5$
(d) $(-3)(-3)(-3)(-3)(-3)$
(e) $7 \times 7 \times 7 \times 7 \times 7$
(f) $9 \times 9 \times 9$
(g) 5

B 2. Evaluate.

(a) 3^3 (b) 5^2 (c) 7^2
(d) 7^3 (e) 5^3 (f) 2^6
(g) 2^4 (h) 4^2 (i) 8^2
(j) 10^4 (k) 1^5 (l) 0^7

3. List the prime factors of each number using exponents.

(a) 12 (b) 36 (c) 800
(d) 400 (e) 90 (f) 144
(g) 63 (h) 225 (i) 108
(j) 2500 (k) 96 (l) 720

4. Simplify.

(a) $3^2 + 4^2$ (b) $3^3 + 4^3$
(c) $5^2 + 12^2$ (d) $5^3 + 12^3$
(e) $7^2 + 24^2$ (f) $7^3 + 24^3$
(g) $8^2 + 15^2$ (h) $8^3 + 15^3$

5. Write each expression as a product of its factors.

(a) $5x^4$ (b) $3y^2$ (c) $6x^3$
(d) $(xy)^2$ (e) $2(5x)^2$ (f) $7x(2x^2)$
(g) (xy^2) (h) $2x^2y^3$ (i) $-2x^2y^3$

6. Write each product using exponents.

(a) $x \times x \times x \times x$
(b) $y \times y \times y \times y \times y \times y$
(c) $5 \times x \times x \times y \times y \times y$
(d) $5x \times 5x \times 5x$
(e) $-3 \times x \times x \times y \times y$
(f) $(-3xy)(-3xy)(-3xy)$
(g) $(x + y)(x + y)(x + y)$

7. Evaluate each expression for $x = 5$.

(a) x^2 (b) $2x^3$
(c) $x^3 + 2$ (d) $(x + 2)^3$
(e) $x + 2^3$ (f) $(2x)^3$

8. If $y = 1$, state the value of each power.

(a) y^4 (b) $5y^3$
(c) $5y^2$ (d) $-5y^3$
(e) $(-5y)^3$ (f) $(-5y)^2$

9. Evaluate each expression for the given value of the variable.

(a) x^4; 5 (b) x^3; 6
(c) t^4; 4 (d) $5x^2$; 3

10. Evaluate each expression for $x = 2$ and $y = -5$.

(a) $x^2 + y$ (b) $(x + y)^2$
(c) $2(3x - y)^2$ (d) $3x^3 - 2y^2$
(e) $x^3 + y^3$ (f) $x^3 - y^3$

C 11. Given $x = 2$ and $y = -2$, which expression is greater?

(a) $x^2 + y^2$ or $(x + y)^2$
(b) $x^2 - y^2$ or $(x - y)^2$
(c) $x^3 + y^3$ or $(x + y)(x^2 + xy + y^2)$

CALCULATOR MATH

THE EXPONENTIAL KEY y^x

To calculate the value of 1.045^{12}, press

1 · 0 4 5 y^x 1 2 =

The display is 1.6958814

Note that the calculator has given the answer rounded to 7 decimal places.

Calculate the following to six decimal places.

1. 1.05^{12}
2. 1.08^{15}
3. 1.04^{10}
4. 1.015^8
5. 1.012^{10}
6. 1.014^{12}
7. $\frac{1}{1.07^{10}}$
8. $\frac{1}{1.015^{10}}$
9. $\frac{1}{1.045^{15}}$
10. $\frac{1}{1.025^{12}}$

2.2 MULTIPLYING AND DIVIDING POWERS

The North American School of Music awarded 100 scholarships of \$1000 each. To find the total value of the scholarships, we multiply 100×1000.

$$100 \times 1000 = 10 \times 10 \times 10 \times 10 \times 10$$

$$10^2 \times 10^3 = 10^5$$

The product $10^2 \times 10^3$ has 5 factors of 10.

This suggests that

$$10^2 \times 10^3 = 10^{2+3} = 10^5$$

We develop the rule for multiplying powers using exponents.

$$a^2 \times a^3 = \underbrace{\overbrace{a \times a}^{2} \times \overbrace{a \times a \times a}^{3}}_{5}$$

$$= a^5$$

$$a^2 \times a^3 = a^{2+3} = a^5$$

> To find the product of powers with the same base, add the exponents.
> $$x^m \times x^n = x^{m+n}$$

We use a similar process to develop a rule for dividing powers using exponents.

$$a^7 \div a^3 = \frac{\overset{1}{\cancel{a}} \times \overset{1}{\cancel{a}} \times \overset{1}{\cancel{a}} \times a \times a \times a \times a}{\underset{1}{\cancel{a}} \times \underset{1}{\cancel{a}} \times \underset{1}{\cancel{a}}}$$

$$= a^4$$

$$a^7 \div a^3 = a^{7-3} = a^4$$

> To find the quotient of powers with the same base, subtract the exponents.
> $$x^m \div x^n = x^{m-n}$$

Example.

Simplify. (a) $5x^3 \times 3x^4$ (b) $\dfrac{(8a^4b)(3a^3b^2)}{6a^2b}$

Solution:

We arrange the factors so that we deal with the numerical factors first, then with the variables in alphabetical order.

(a) $5x^3 \times 3x^4 = 5 \times 3 \times x^3 \times x^4$
$= 15x^7$

(b) $\dfrac{(8a^4b)(3a^3b^2)}{6a^2b}$

$= \dfrac{8 \times 3 \times a^4 \times a^3 \times b \times b^2}{6a^2b}$

$= \dfrac{24a^7b^3}{6a^2b}$

$= 4a^5b^2$

EXERCISE 2.2

A 1. Evaluate.

(a) 2^3 (b) 3^2 (c) 4^2
(d) 2^4 (e) 5^2 (f) 2^5

2. Find the value of the missing exponent.

(a) $2^5 \times 2^7 = 2^{■}$ (b) $3^6 \times 3 = 3^{■}$
(c) $5^3 \times 5^7 = 5^{■}$ (d) $6^7 \times 6^7 = 6^{■}$
(e) $7^5 \times 7^2 = 7^{■}$ (f) $4^5 \times 4^3 = 4^{■}$
(g) $3 \times 3 = 3^{■}$ (h) $5^4 \times 5 = 5^{■}$

3. Find the value of the missing exponent.

(a) $3^5 \div 3^2 = 3^{■}$ (b) $5^4 \div 5 = 5^{■}$
(c) $4^5 \div 4^3 = 4^{■}$ (d) $7^8 \div 7^3 = 7^{■}$
(e) $6^7 \div 6^4 = 6^{■}$ (f) $4^5 \div 4^5 = 4^{■}$
(g) $5^8 \div 5^3 = 5^{■}$ (h) $2^2 \div 2 = 2^{■}$

4. Find the value of the missing exponent.

(a) $3^2 \times 3^{■} = 3^7$ (b) $2^4 \times 2^{■} = 2^7$
(c) $4^{■} \times 4^5 = 4^8$ (d) $5^{■} \times 5^2 = 5^8$
(e) $6^3 \times 6^{■} = 6^5$ (f) $3^5 \div 3^{■} = 3^2$
(g) $4^7 \div 4^{■} = 4^4$ (h) $5^{■} \div 5^2 = 5^4$

B 5. Simplify.

(a) $x^3 \times x^5$ (b) $a^2 \times a^{10}$
(c) $b \times b \times b$ (d) $m^2 \times m^3 \times m$
(e) $a^2 \times a^3 \times b \times b^4$ (f) $a^5 \times a^2 \times b^3 \times b$

6. Simplify.

(a) $a^7 \div a^3$ (b) $b^3 \div b^2$ (c) $n^{12} \div n^3$
(d) $x^8 \div x^5$ (e) $a^5 \div a$ (f) $x^{10} \div x^9$

7. Simplify.

(a) $(5a^3)(3a)$ (b) $(4x^2)(-2x^3)$
(c) $(3a^2b^3)(2ab^2)$ (d) $(2x)(5x^3)$
(e) $5mn \times 3m$ (f) $(-4x^3)(-3x^2)$
(g) $3y \times 5y \times 2y^2$ (h) $(a)(2a^2)(-3a^5)$
(i) $(-7x^3)(-2x^3)(-x^2)$
(j) $(-2x)(+3x^2)$

8. Simplify.

(a) $12a^5 \div 3a^3$ (b) $21x^2y^5 \div 7xy$
(c) $8a^2 \div 8a$ (d) $48m^2n \div 6mn$
(e) $30m^2 \div (-6m^2)$
(f) $(-18ab^3) \div (-2ab)$
(g) $56pqr \div 8pr$
(h) $(-24a^2b^2) \div (8a^2b^2)$
(i) $(15x^{10}y^2) \div (-5xy)$
(j) $28x^4 \div 7x^2$

9. Simplify.

(a) $6^4 \times 6^3 \times 6^2$ (b) $2^5 \times 2^2 \times 2$
(c) $7^3 \times 7^4 \times 7^3$ (d) $8^2 \times 8 \times 8^3$
(e) $5 \times 5^4 \times 5^4$ (f) $2^1 \times 2^2 \times 2^3$
(g) $3^4 \times 3^3 \div 3^2$ (h) $2^5 \div 2^2 \times 2^3$
(i) $4^6 \div 4^2 \times 4^2$ (j) $5^4 \times 5^2 \div 5^3$

10. Simplify.

(a) $x^4 \times x^3 \times x^2$ (b) $y^2 \times y \times y^3$
(c) $m^5 \div m^2 \times m^2$ (d) $t^3 \div t \times t^2$
(e) $x^6 \div x^2 \times x^3$ (f) $y^5 \div y^3 \div y$
(g) $t^4 \times t^2 \times t^3$ (h) $x^6 \div x^2 \times x^3$

11. Evaluate.

(a) $2^2 \times 5^2$ (b) $3^3 \times 2$
(c) 3×2^3 (d) 4×3^2
(e) $28 \div 2^2$ (f) $2^6 \div 8$
(g) $175 \div 5^2$ (h) $40 \div 2^3$

INVESTIGATING SQUARES AND CUBES

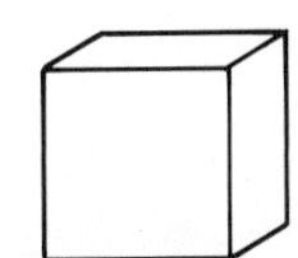

1. Take the numbers 1, 2, and 3 and find the sum of their cubes.
2. Now find the sum of the numbers 1, 2, and 3 and square the sum.
3. What do you notice about the results of steps 1 and 2?
4. Repeat steps 1 to 3 for the numbers 1, 2, 3, 4.
5. Repeat steps 1 to 3 for the numbers 1, 2, 3, 4, 5, 6, 7.
6. Describe the pattern in step 5. Does this pattern apply to any set of consecutive whole numbers?
7. Make a general statement about your observations.

2.3 POWER LAWS FOR EXPONENTS

When we write the power $5x^3$, we mean $5 \times x \times x \times x$

In the expression $5x^3$, the exponent 3 applies only to the factor x. If the exponent 3 is to apply to both factors, the expression must be written $(5x)^3$.

Hence, $5x^3$ means $(5)(x)(x)(x)$

and $(5x)^3$ means $(5x)(5x)(5x)$

If we simplify the power $(5x)^3$, the result is $(5x)(5x)(5x) = 125x^3$

Setting $x = 2$, $5x^3 = 5 \times 2 \times 2 \times 2 = 40$

but $(5x)^3 = (5 \times 2)(5 \times 2)(5 \times 2) = 1000$

The following table illustrates the power laws for exponents.

$(x^2)^3 = x^2 \times x^2 \times x^2$ $= x^6$ $(x^2)^3 = x^{2\times 3} = x^6$ To find the power of a power, we multiply the exponents. $(x^a)^b = x^{a\times b} = x^{ab}$	$(ab)^2 = (ab)(ab)$ $= a \times a \times b \times b$ $= a^2b^2$ $(ab)^2 = a^{1\times 2} \times b^{1\times 2}$ $= a^2b^2$ To find the power of a product, we multiply the exponents. $(xy)^a = x^ay^a$ $(x^ay^b)^c = x^{ac}y^{bc}$	$\left(\frac{a}{b}\right)^3 = \frac{a}{b}\frac{a}{b}\frac{a}{b}$ $= \frac{a^3}{b^3}$ $\left(\frac{a}{b}\right)^3 = \frac{a^{1\times 3}}{b^{1\times 3}} = \frac{a^3}{b^3}$ To find the power of a quotient, we multiply the exponents. $\left(\frac{x}{y}\right)^a = \frac{x^a}{y^a}$ $\left(\frac{x^a}{y^b}\right)^c = \frac{x^{ac}}{y^{bc}}$

Example 1.

Simplify.

(a) $(2x^4)^3$

(b) $\frac{(4x^3)^2}{-8x}$

(c) $\left(\frac{5x^3}{y}\right)^2$

Solution:

(a) $(2x^4)^3 = 2^3 \times x^{4\times 3}$
$= 8x^{12}$

(b) $\frac{(4x^3)^2}{-8x} = \frac{4^2 \times (x^3)^2}{-8x}$
$= \frac{16x^6}{-8x}$
$= -2x^5$

(c) $\left(\frac{5x^3}{y}\right)^2 = \frac{5^2x^{3\times 2}}{y^2}$
$= \frac{25x^6}{y^2}$

Example 2.

Simplify.

(a) $(3x^3)^2(2x^5)$

(b) $(x^2y^3)^5$

Solution:

(a) $(3x^3)^2 \times (2x^5) = 3^2(x^3)^2 \times (2x^5)$
$= 9x^6 \times 2x^5$
$= 18x^{11}$

(b) $(x^2y^3)^5 = x^{2x5}y^{3\times 5}$
$= x^{10}y^{15}$

EXERCISE 2.3

A 1. Find the value of the missing exponent.

(a) $(2^2)^3 = 2^{■}$
(b) $(3^4)^3 = 3^{■}$
(c) $(5^3)^5 = 5^{■}$
(d) $(7^4)^4 = 7^{■}$
(e) $(4^5)^2 = 4^{■}$
(f) $(10^4)^5 = 10^{■}$
(g) $(3x^3)^2 = 9x^{■}$
(h) $(2x^2)^3 = 8x^{■}$

2. Find the value of the missing exponent.

(a) $(x^2y^3)^2 = x^4y^{■}$
(b) $(x^4y^5)^3 = x^{■}y^{15}$
(c) $(x^4y^3)^3 = x^{■}y^9$
(d) $(x^{■}y^3)^2 = x^6y^6$
(e) $(x^4y^{■})^2 = x^8y^6$
(f) $(x^7y^4)^{■} = x^{14}y^8$
(g) $(x^5y^{■})^3 = x^{15}y^3$
(h) $(x^2y^4)^2 = x^{■}y^8$

3. Find the value of the missing exponent.

(a) $\left(\frac{x^4}{y^3}\right)^{■} = \frac{x^8}{y^6}$
(b) $\left(\frac{x^5}{y^4}\right)^3 = \frac{x^{■}}{y^{12}}$
(c) $\left(\frac{x^4}{y^3}\right)^3 = \frac{x^{12}}{y^{■}}$
(d) $\left(\frac{x}{y^3}\right)^3 = \frac{x^{■}}{y^9}$
(e) $\left(\frac{x^5}{y^2}\right)^3 = \frac{x^{■}}{y^6}$
(f) $\left(\frac{x^4}{y^4}\right)^{■} = \frac{x^4}{y^4}$
(g) $\left(\frac{x^7}{y^2}\right)^{■} = \frac{x^{21}}{y^6}$
(h) $\left(\frac{x^{■}}{y^{■}}\right)^3 = \frac{x^9}{y^9}$

B 4. Simplify.

(a) $(x^4)^2$
(b) $(a^3)^3$
(c) $(a^2b)^3$
(d) $(xy^3)^5$
(e) $(abc)^5$
(f) $(b^8)^3$
(g) $(2x^3)^3$
(h) $(a^5b^2)^3$
(i) $(3a^5)^2$
(j) $(3xy^2)^3$
(k) $(5a^8)^3$
(l) $(4x^2yz)^3$

5. Simplify.

(a) $\left(\frac{x}{y}\right)^5$
(b) $\left(\frac{a^2}{3}\right)^2$
(c) $\left(\frac{x^2}{y}\right)^3$
(d) $\left(\frac{a}{b^5}\right)^4$
(e) $\left(\frac{3x}{y}\right)^2$
(f) $\left(\frac{2x^2}{w}\right)^3$
(g) $\left(\frac{5a^2}{2b^3}\right)^2$
(h) $\left(\frac{3a}{b^3}\right)^2$

6. Simplify.

(a) $(-2a^3)^3$
(b) $(3x \times x^2)^2$
(c) $\left(\frac{12x^5}{4x^3}\right)^2$
(d) $(2a^3b \times 3a^4)$
(e) $\left(\frac{a^5b}{c^4}\right)^3$
(f) $\left(\frac{-32a^5b}{-16ab}\right)^2$
(g) $\left(\frac{48x^3y^7}{-12xy}\right)^3$
(h) $\left(\frac{(5a^2)(-12a^5)}{15a^2}\right)^3$
(i) $\left(\frac{-51a^4b^2c}{(17a^2b)(b^2c)}\right)^3$

C 7. Simplify.

(a) $(2x^2)^3 \times (2x^5)$
(b) $\frac{(8ab^2)(3a^2b)}{12a^3}$
(c) $\frac{(3x^2y^5)^3}{9xy^2}$
(d) $\frac{(12m^2n^5)(-5mn^3)}{15m^3n^2}$
(e) $\frac{(-12x^2)(-4x^2y)}{(-6x^3)}$
(f) $\frac{-32m^{10}n^3}{(8m^5)(mn)}$

INVESTIGATING SQUARES, CUBES, AND FACTORS

1. List all the factors of 6.

1, 2, 3, 6

2. List the number of factors for each factor of 6.

1	2	3	6
↓	↓	↓	↓
1	2	2	4

3. Find the sum of the cubes of these numbers.

$$1^3 + 2^3 + 2^3 + 4^3$$
$$= 1 + 8 + 8 + 64$$
$$= 81$$

4. Find the square of the sum of these numbers.

$$1 + 2 + 2 + 4 = 9 \text{ and } 9^2 = 81$$

5. Repeat steps 1 to 4 for the number 12.

6. Repeat the steps for the number 20.

7. Summarize the results of this investigation for any whole number.

2.4 INTEGRAL EXPONENTS

Heather and Mario each write a computer program to divide 32 by 2 eight times and each work on the problem in a different way.
Heather starts with 32 and then divides. Mario expresses 32 as 2^5 and then divides. The following shows the results that each obtains.

Heather's Method	Mario's Method
$32 \div 2 = 16$	$2^5 \div 2 = 2^4$
$16 \div 2 = 8$	$2^4 \div 2 = 2^3$
$8 \div 2 = 4$	$2^3 \div 2 = 2^2$
$4 \div 2 = 2$	$2^2 \div 2 = 2^1$
$2 \div 2 = 1$	$2^1 \div 2 = 2^0$
$1 \div 2 = \frac{1}{2}$	$2^0 \div 2 = 2^{-1}$
$\frac{1}{2} \div 2 = \frac{1}{4}$	$2^{-1} \div 2 = 2^{-2}$
$\frac{1}{4} \div 2 = \frac{1}{8}$	$2^{-2} \div 2 = 2^{-3}$

By calculating the same answers in different ways, Heather and Mario find values for the zero and negative exponents.

THE **ZERO** EXPONENT

To evaluate 2^0, we simplify $2^3 \div 2^3$ in two ways.

$$\frac{2^3}{2^3} = 2^{3-3} = 2^0 \quad \text{and} \quad \frac{2^3}{2^3} = \frac{8}{8} = 1$$

THE **NEGATIVE** EXPONENT

To evaluate 2^{-3}, we simplify $2^4 \div 2^7$ in two ways.

$$\frac{2^4}{2^7} = 2^{4-7} = 2^{-3} \quad \text{and} \quad \frac{2^4}{2^7} = \frac{\not{2}^1 \times \not{2}^1 \times \not{2}^1 \times \not{2}^1}{2 \times 2 \times 2 \times \not{2}_1 \times \not{2}_1 \times \not{2}_1 \times \not{2}_1} = \frac{1}{2^3}$$

We can generalize these results for a zero and a negative exponent.

$$\boxed{x^0 = 1 \qquad \text{and} \qquad x^{-n} = \frac{1}{x^n}}$$

Example 1.

Simplify. $a^5 \div a^{-3}$

Solution:

$$\begin{aligned} a^5 \div a^{-3} &= a^{5-(-3)} \\ &= a^{5+3} \\ &= a^8 \end{aligned}$$

Example 2.

Simplify. $(x^{-3}y^2) \times (x^3y^{-4})$

Solution:

$$\begin{aligned} (x^{-3}y^2) \times (x^3y^{-4}) &= x^{-3} \times x^3 \times y^2 \times y^{-4} \\ &= x^0 \times y^{-2} \\ &= 1 \times y^{-2} \\ &= y^{-2} \end{aligned}$$

EXERCISE 2.4

A

1. Simplify.

(a) $x^3 \times x^{-5}$
(b) $a^{-1} \times a^8$
(c) $y^0 \times y^4$
(d) $b^{-1} \times b^{-3}$
(e) $a^{10} \times a^3 \times a^{-5}$
(f) $x \times x^0$
(g) $b^{-5} \times b^5$
(h) $m \times m^{-1} \times m^0$

2. Simplify.

(a) $x^{10} \div x^5$
(b) $b^9 \div b^{12}$
(c) $a^0 \div a^3$
(d) $x^{-3} \div x^2$
(e) $m^{-5} \div m^0$
(f) $n^4 \div n^{-3}$
(g) $0 \div y^5$
(h) $m^{-8} \div m^{-4}$

3. Evaluate.

(a) 4^0
(b) 2^{-1}
(c) $(-3)^0$
(d) 10^{-1}
(e) 3^{-2}
(f) 10^0
(g) 4^{-2}
(h) 2^{-3}
(i) 2^{-4}
(j) 2^{-5}

4. Simplify.

(a) $3x^5 \times 5x^{-2}$
(b) $(a^2b^5)(a^3b^{-8})$
(c) $(2y^5)(3y^{-5})$
(d) $x^{10} \times x^{-3} \times x^{-5}$
(e) $(3m^{-1})^2$
(f) $(3a^2)(5a^{-8})$
(g) $(m^2n)(m^5n^{-1})$
(h) $(2a^5b^{-3})^3$
(i) $(x^{-5}y^{-2})^{-1}$
(j) $(5xy^{-1})(7x^3y^{-1})$
(k) $(7m^3)(m^{-5}n^{-2})$
(l) $(5a^{-3})(3a^3)$

5. Simplify.

(a) $a^{-3} \div a^{-5}$
(b) $12x^5 \div 4x^{10}$
(c) $24b^{-5} \div 6b^5$
(d) $m^2n \div m^5n^0$
(e) $a^{12} \div a^{15} \times a^3$
(f) $15a^4b^5 \div 5a^2b^7$
(g) $b^0 \div b^{-4}$
(h) $(y^4 \times y^2) \div y^{10}$
(i) $(4x^{-2})^2$
(j) $(3x^{-5})^{-1} \times 3x^2$
(k) $\dfrac{12b^2 \times 8b^{-4}}{6b^{-10}}$
(l) $(2a^3)^{-3} \times 4a^{-5}$

6. Evaluate.

(a) $5^0 + 5^{-1}$
(b) $3^{-1} + 4^{-1}$
(c) $(5^{-1})^2$
(d) $(4^{-3} \times 4^2)^2$
(e) $(2^{-1})^{-1}$
(f) $(\frac{1}{4})^{-1}$
(g) $(\frac{2}{3})^0$
(h) $5^{-1} + 2^{-2}$
(i) $(\frac{1}{10})^{-1}$
(j) 10^{-2}
(k) $(\frac{1}{3})^{-2}$
(l) $10^3 \times 10^{-5}$

MICRO MATH

The following program prints a table of powers for the exponents
$-5, -4, -3, ..., 4, 5$
when you enter the value of the base.

```
NEW
10 PRINT"EXPONENTS"
20 INPUT"ENTER THE BASE:";B
30 PRINT"BASE","EXPONENT","POWER"
40 FOR X = -5 TO 5
50 Y=B^X
60 PRINT B,X,Y
70 NEXT X
80 END
RUN
```

The printout which follows is the result when the program is RUN and we INPUT the value 2 for the base.

```
EXPONENTS
ENTER THE BASE:? 2
BASE      EXPONENT    POWER
2         -5          .03125
2         -4          .0625
2         -3          .125
2         -2          .25
2         -1          .5
2          0          1
2          1          2
2          2          4
2          3          8
2          4          16
2          5          32
```

EXERCISE

Run the program to compute tables for each base.

1. 3 2. 4 3. 5 4. 6
5. 7 6. 8 7. 9 8. 10

To make this program reiterative, add these lines.

```
74 PRINT"ANOTHER QUESTION?"
75 INPUT"Y OR N";Z$
76 IF Z$="Y" THEN 20
```

2.5 POWERS OF TEN

Dr. Edward Kasner of Columbia University is credited with giving a special name to the number 10^{100}. He called it a "googol," which can be written in expanded form as follows:

10 000 000 000 000 000 000 000 000 000 000 000
000 000 000 000 000 000 000 000 000 000 000
000 000 000 000 000 000 000 000 000 000 000

What is a googolplex?

Using exponents helps us to write large or small numbers without having to write all the zeros.

The computer program and printout at the right show the expanded form of powers of 10, 10^n, for $n = -5, -4, -3, ..., 3, 4, 5$

```
10 PRINT"POWERS OF 10"
20 LPRINT"BASE","EXPONENT",
   "POWER"
30 FOR X = -5 TO 5
40 Y = 10^X
50 LPRINT 10,X,Y
60 NEXT X
70 END
```

```
POWERS OF 10
BASE      EXPONENT       POWER
10        -5             0.000 01
10        -4             0.000 1
10        -3             0.00 1
10        -2             0.01
10        -1             0.1
10        0              1
10        1              10
10        2              100
10        3              1000
10        4              10 000
10        5              100 000
```

Example 1.
Simplify by first writing the powers using exponents.
(a) 10 000 × 1000 (b) 10 000 000 ÷ 10 000 (c) 1000^4

Solution:
(a) 10 000 × 1000
$= 10^4 \times 10^3$
$= 10^7$

(b) 10 000 000 ÷ 10 000
$= 10^7 \div 10^4$
$= 10^3$

(c) 1000^4
$= 10^3 \times 10^3 \times 10^3 \times 10^3$
$= 10^{12}$

The metre, gram, and litre are units of measurement in the metric system as shown in the table at the right.

Length	metre	m
Mass	gram	g
Capacity	litre	L

In the metric system powers of 10 are indicated by using prefixes, as shown in the table.

mega- (M)	kilo- (k)	hecto- (h)	deca- (da)	Unit	deci- (d)	centi- (c)	milli- (m)	micro- (μ)
1 000 000 10^6	1000 10^3	100 10^2	10 10^1	1 10^0	0.1 10^{-1}	0.01 10^{-2}	0.001 10^{-3}	0.000 001 10^{-6}

Example 2.
Express each distance in metres.
(a) 10 000 km (b) 100 000 mm

Solution:
(a) 10 000 km = 10 000(1000) m
= 10 000 000 m
= 10^7 m

(b) 100 000 mm = 100 000(0.001) m
= 100 m
= 10^2 m

EXERCISE 2.5

A

1. Read in expanded form.

(a) 10^3 (b) 10^8 (c) 10^5
(d) 10^2 (e) 10^0 (f) 10^7
(g) 10^1 (h) 10^{11} (i) 10^6

2. Read each number as a power of 10.

(a) 10 000 (b) 100 000
(c) 100 (d) 1 000 000
(e) 10 (f) 1
(g) 100 000 000 (h) 1000

3. Read each power in expanded form.

(a) 10^{-2} (b) 10^{-1} (c) 10^{-6}
(d) 10^{-8} (e) 10^{-4} (f) 10^0
(g) 10^{-5} (h) 10^{-3} (i) 10^{-9}

4. Read each decimal as a power of 10.

(a) 0.001 (b) 0.0001
(c) 0.1 (d) 0.000 001
(e) 0.01 (f) 0.000 01

B

5. Simplify, and express in expanded form.

(a) $10^3 \times 10^4$ (b) $10^7 \div 10^2$
(c) $(10^3)^4$ (d) $10^{12} \div 10^8$
(e) $(10^4)^2$ (f) $10^4 \times 10$
(g) $\dfrac{10^5 \times 10^7}{10^9}$ (h) $\dfrac{10^{10}}{10^3 \times 10^5}$

6. Simplify, and express in expanded form.

(a) $10^{-2} \times 10^{-4}$ (b) $10^{-5} \times 10^5$
(c) $10^{-3} \times 10^{-2}$ (d) $10^{-7} \times 10^{-1}$
(e) $10^{-4} \div 10^{-3}$ (f) $10^{-5} \div 10^{-7}$
(g) $(10^{-2})^3$ (h) $(10^{-3})^{-2}$
(i) $\dfrac{10^{-3} \times 10^{-4}}{10^{-10}}$ (j) $\dfrac{10^{-5}}{10^{-3} \times 10^{-2}}$

7. Simplify, and express as a power of 10.

(a) $\dfrac{10^{-4} \times 10^{-5}}{10^2}$ (b) $\dfrac{10^{-3} \times 10^{-2}}{10}$
(c) $\dfrac{10^3 \div 10^4}{10^{-2}}$ (d) $\dfrac{10^5 \times 10^{-3}}{10^2}$
(e) $\dfrac{(10^2)^3}{10^5}$ (f) $\dfrac{(10^{-3})^0}{10^{-3}}$
(g) $\dfrac{(10^3)^3}{10^6}$ (h) $\dfrac{(10^0)^3}{(10^{-3})^{-1}}$

8. Complete the following statements using powers of 10.

(a) 1000 km = ■ m
(b) 0.1 km = ■ m
(c) 100 m = ■ km
(d) 10 000 m = ■ km
(e) 0.001 km = ■ m
(f) 10 km = ■ m

9. Express in grams.

(a) 10 kg (b) 0.01 kg
(c) 100 mg (d) 1000 kg
(e) 0.001 kg (f) 10 000 mg
(g) 1000 mg (h) 0.1 kg

10. Express in litres.

(a) 1000 mL (b) 100 mL
(c) 10 kL (d) 1000 kL
(e) 1 mL (f) 1 kL
(g) 100 000 mL (h) 1 000 000 kL

C

11. Complete the following statements using powers of 10.

(a) 10 000 g = ■ kg
(b) 10 000 mL = ■ L
(c) 100 000 m = ■ km
(d) 1 000 000 mg = ■ kg
(e) 1 000 000 L = ■ mL
(f) 100 kg = ■ g
(g) 1 000 000 mL = ■ kL
(h) 0.001 kg = ■ g
(i) 0.01 kL = ■ L
(j) 0.001 kg = ■ mg
(k) 10 000 mm = ■ km

MIND BENDER

John is a golfer. He has 3 shirts, 4 pairs of slacks, and 2 pairs of shoes for golfing.
How many different outfits can he wear?

2.6 OPERATIONS WITH POWERS OF TEN

Jason has 1.5 kg of potassium to share with three other students in his biology class. How many grams of potassium will each student receive?

Jason can solve this problem using the strategy of sequencing operations:

1·2·3

(i) change kilograms to grams (multiply by 1000)
(ii) divide by four

METHOD I

Divide the 1.5 kg of potassium:

$1.5 \text{ kg} \div 4 = 0.375 \text{ kg}$

Change 0.375 kg to grams:

$0.375 \text{ kg} = 0.375 \times 1000 \text{ g}$
$= 375$

METHOD II

Change 1.5 kg to grams:

$1.5 \text{ kg} = 1.5 \times 1000 \text{ g}$
$= 1500 \text{ g}$

Divide the 1500 g of potassium:

$1500 \text{ g} \div 4 = 375 \text{ g}$

Each student would receive 375 g of potassium.

Multiplying and dividing by powers of ten can be simplified by moving the decimal point in the number. This chart compares place value with powers of 10.

millions	hundred thousands	ten thousands	thousands	hundreds	tens	ones		tenths	hundredths	thousandths	ten thousandths
1 000 000	100 000	10 000	1000	100	10	1	.	0.1	0.01	0.001	0.0001
10^6	10^5	10^4	10^3	10^2	10^1	10^0		10^{-1}	10^{-2}	10^{-3}	10^{-4}

To multiply by a power of 10 greater than 1, move the decimal point to the right as many places as there are zeros in the power.

$1000 = 10^3$

$$\begin{array}{r} 25.68 \\ \times 1000 \\ \hline 2568000 \end{array}$$

$25.68 \times 1000 = 25680.00$

3 zeros — 3 places to the right

To divide by a power of 10 greater than 1, move the decimal point to the left as many places as there are zeros in the power.

$$1000\overline{)25.68000} = 0.02568$$

$25.68 \div 1000 = 0.025\,68$

3 zeros — 3 places to the left

To multiply by a power of 10 less than 1, move the decimal point to the left as many places as there are decimal places in the power.

$0.01 = 10^{-2}$

$$\begin{array}{r} 256.8 \\ \times\ 0.01 \\ \hline 2.568 \end{array}$$

$256.8 \times 0.01 = 2.568$

2 places — 2 places to the left

To divide by a power of 10 less than 1, move the decimal point to the right as many places as there are decimal places in the power.

$$0.01\overline{)256.8} \rightarrow 1\overline{)25\,680} = 25\,680$$

$256.8 \div 0.01 = 25\,680.$

2 places — 2 places to the right

EXERCISE 2.6

A 1. Multiply each number by 1000.

(a) 425 (b) 5.64 (c) 156.75
(d) 0.025 (e) 0.75 (f) 0.0012
(g) 3650 (h) 45.75 (i) 0.0003

2. Divide each number by 100.

(a) 6741 (b) 56.2 (c) 389.75
(d) 365 (e) 0.0025 (f) 5.25
(g) 39 275 (h) 125.375 (i) 275 000

3. Multiply each number by 0.01

(a) 64.25 (b) 157.3 (c) 345.836
(d) 25 000 (e) 0.0035 (f) 0.0005
(g) 0.004 (h) 0.000 24 (i) 23 645

4. Divide each number by 0.1

(a) 45.6 (b) 256.8 (c) 657 290
(d) 2.758 (e) 35.007 (f) 2.7045
(g) 0.025 (h) 0.0004 (i) 0.1243

B 5. Simplify.

(a) 1000×275 (b) $10\,000 \times 152$
(c) $57 \times 10\,000$ (d) $325 \times 10\,000$
(e) 36×100 (f) 65×1000
(g) $539 \div 100$ (h) $51 \div 10\,000$
(i) $625 \div 10\,000$ (j) $125 \div 100$
(k) $\dfrac{450 \times 10\,000}{100}$
(l) $\dfrac{27\,000 \div 10\,000}{100\,000}$

6. Simplify.

(a) 100×2.75 (b) 5.65×1000
(c) $10\,000 \times 56.3$ (d) $5.2 \times 100\,000$
(e) 0.25×1000 (f) 0.036×100
(g) $68.5 \div 1000$ (h) $5.38 \div 100$
(i) $645.2 \div 100$ (j) $0.256 \div 10$
(k) $\dfrac{52.85 \div 100\,000}{10\,000}$
(l) $\dfrac{0.527 \times 100\,000}{1000}$

7. Simplify.

(a) 5.25×0.01 (b) 53.27×0.001
(c) 658.2×0.01 (d) 4.9×0.0001
(e) 63.9×0.001 (f) 0.0001×3.6
(g) $45.7 \div 0.1$ (h) $26.5 \div 0.001$
(i) $0.25 \div 0.001$ (j) $5.7 \div 0.000\,1$
(k) $\dfrac{5.75 \times 0.000\,01}{0.001}$
(l) $\dfrac{3.75 \div 0.000\,001}{10\,000}$
(m) $\dfrac{0.002\,75 \div 0.000\,01}{1000 \times 0.000\,000\,1}$

8. Multiply 0.005 26 by each of the following.

(a) 1000 (b) 10 000 (c) 100 000
(d) 0.001 (e) 0.0001 (f) 0.1

9. Divide 1.675 by each of the following.

(a) 100 (b) 10 000 (c) 1 000 000
(d) 0.1 (e) 0.001 (f) 0.000 001

10. Multiply 356.8 by each power.

(a) 10^4 (b) 10^1 (c) 10^5
(d) 10^{-2} (e) 10^{-1} (f) 10^{-5}

11. Divide 0.2358 by each power.

(a) 10^2 (b) 10^4 (c) 10^0
(d) 10^{-1} (e) 10^{-5} (f) 10^{-2}
(g) 10^{-3} (h) 10^3 (i) 10

12. Express each measure in metres.

(a) 2450 cm (b) 0.25 km
(c) 125 mm (d) 0.0125 km
(e) 35 550 cm (f) 1.5 km
(g) 3.75 cm (h) 250 000 mm

13. Express each measure in litres.

(a) 2500 mL (b) 6.5 kL
(c) 375 mL (d) 0.0025 kL
(e) 0.875 mL (f) 475 kL

14. Express each mass in kilograms.

(a) 3500 g (b) 3.5 g
(c) 2 500 000 mg (d) 45 000 g
(e) 0.5 g (f) 500 000 g

15. Scientists estimate that there are about 5 500 000 blood corpuscles in one drop of blood.

(a) Estimate the number of drops of blood in 1 L.
(b) How many corpuscles would you expect to find in 4.7 L of blood?

2.7 SCIENTIFIC NOTATION

Scientists have calculated that the distance from Earth to the planet Neptune is about 4 350 000 000 km.
How long would it take a radio signal to travel from Neptune to Earth at the speed of light, which is 300 000 km/s?

Calculations with such numbers are made easier using scientific notation.

> A number is expressed in scientific notation when it is in the form $a \times 10^n$ where a is a number between 1 and 10 ($1 \leqslant a \leqslant 10$), and n is an integer (..., −3, −2, −1, 0, 1, 2, 3, ...).

We first express the numbers in the problem in scientific notation.

Distance = $4\underbrace{350\ 000\ 000}_{9 \text{ places}} = 4.35 \times 10^9$

Speed = $3\underbrace{00\ 000}_{5 \text{ places}} \doteq 3.0 \times 10^5$

We now solve the problem by dividing.

$$\text{time} = \frac{\text{distance}}{\text{speed}}$$

$$\text{time} = \frac{4.35 \times 10^9}{3.0 \times 10^5}$$

$$= 1.45 \times 10^4$$

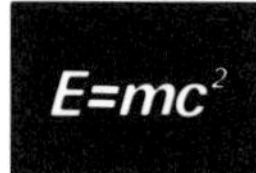

The time taken for the signal to travel from Neptune to Earth is 1.45×10^4 s.

$1.45 \times 10^4 \text{ s} = 14\ 500 \text{ s}$
$\doteq 4 \text{ h}$

Example 1.
Express the following numbers in scientific notation.
(a) 15 750 000 000 (b) 0.000 000 72

Solution:
(a) 15 750 000 000
$= 1.575 \times 10\ 000\ 000\ 000$
$= 1.575 \times 10^{10}$

(b) $0.000\ 000\ 72 = 7.2 \times 0.000\ 000\ 1$
$= 7.2 \times 10^{-7}$

Numbers in scientific notation can be multiplied or divided.

Example 2.
Multiply 4.2×10^{-4} by 1.5×10^7 and give your answer in scientific notation.

Solution:
$(4.2 \times 10^{-4}) \times (1.5 \times 10^7) = (4.2 \times 1.5) \times (10^{-4} \times 10^7)$
$= 6.30 \times 10^3$
$= 6.3 \times 10^3$

EXERCISE 2.7

B 1. Express each number in scientific notation.

(a) 576
(b) 24.35
(c) 0.0248
(d) 4 265 000 000
(e) 0.000 25
(f) 0.789
(g) 0.081
(h) 528 000
(i) 0.0360
(j) 507.5
(k) 2500
(l) 0.000 000 275
(m) 5.75
(n) 403 000
(o) 0.000 205
(p) 52 050

2. Express the following in decimal form.

(a) 5.3×10^3
(b) 5.1×10^{-1}
(c) 9.5×10^{-4}
(d) 5.265×10^5
(e) 4.1×10^{-2}
(f) 7.5×10^0
(g) 7.5×10^{-1}
(h) 1.08×10^{-8}
(i) 1.005×10^4
(j) 2.08×10^{-4}
(k) 3.025×10^{-2}
(l) 5.125×10^{-3}
(m) 5.25×10^4
(n) 4.14×10^{-4}
(o) 2.005×10^7
(p) 5.008×10^{-4}

3. Complete the following table.

	Value	Scientific Notation
(a)	2 500 000	
(b)	95.3	
(c)		1.25×10^3
(d)	0.0025	
(e)		5.625×10^{-1}
(f)	5.1	
(g)	0.025 70	
(h)		5.8×10^{-5}

C 4. A light year is the distance light travels in one year. This distance is about 9.5×10^{12} km.

(a) How many kilometres will light travel in 50 a?
(b) How far is a galaxy if its light takes 2.0×10^6 a to reach the earth?

CALCULATOR MATH

SCIENTIFIC NOTATION

To enter very large or very small numbers on a scientific calculator, you should use scientific notation. In some calculations, the numbers become so large or so small that they automatically appear in scientific notation in the display.

If you enter

2 5 × = = = = = = =

the display is 1.5258 11

(The number of digits varies with the calculator.)

This means that

$25^8 \doteq 1.5258 \times 10^{11}$

To enter the number 2.65×10^4 in scientific notation, you have to use the exponential key exp or EE and press

2 . 6 5 exp 4

The display is 2.65 04

EXERCISE

1. Enter the following numbers in scientific notation.

(a) 1.625×10^5
(b) 5.75×10^{-3}
(c) 68 500
(d) 0.003 25

2. Write the following displays in scientific notation.

(a) 8.205 03
(b) 5.265 02
(c) 2.325 04
(d) 7.035 03

3. Simplify using a calculator.

(a) $3.125 \times 10^4 \times 4.65 \times 10^{-2}$
(b) $4.52 \times 10^4 \times 5.65 \times 10^{-6}$
(c) $3.25 \times 10^{-3} \div 5.75 \times 10^3$

2.8 SQUARE ROOTS

Roman covers the square foyer of the new school with 400 carpet tiles. There are no carpet tiles left over. Each carpet tile is a 1 m by 1 m square. What are the dimensions of the foyer?

? $A = 400\ m^2$?

To find the area, Roman uses the formula $A = s^2$

In this case, he substitutes to get $400 = s^2$

To find the length of a side of the square, we take the square root of each side. $s = \sqrt{400}$

The expression $\sqrt{400}$ is called "the square root of 400."

$\sqrt{400} = 20$ because $20 \times 20 = 400$

It should also be noted that $(-20) \times (-20) = 400$

However, when we use the symbol $\sqrt{400}$, we mean "the positive square root of 400" or 20.

The symbol $\sqrt{}$ is called the radical sign or the square root sign, and is used to represent the positive square root of any positive number.

Example.

Evaluate.

(a) $\sqrt{49}$ (b) $\sqrt{36}$

Solution:

(a) $\sqrt{49} = \sqrt{7 \times 7}$
$= 7$

(b) $\sqrt{36} = \sqrt{2 \times 2 \times 3 \times 3}$
$= \sqrt{(2 \times 3) \times (2 \times 3)}$
$= 2 \times 3$
$= 6$

If the factors of a number can be grouped into two identical sets, then the number is a perfect square and has a rational square root.

$$\begin{aligned}\sqrt{144} &= \sqrt{2 \times 2 \times 2 \times 2 \times 3 \times 3}\\ &= \sqrt{(2 \times 2 \times 3) \times (2 \times 2 \times 3)}\\ &= 2 \times 2 \times 3\\ &= 12\end{aligned}$$

EXERCISE 2.8

B

1. Find the value of the following by grouping the factors.

(a) $\sqrt{81}$ (b) $\sqrt{25}$ (c) $\sqrt{9}$
(d) $\sqrt{121}$ (e) $\sqrt{100}$ (f) $\sqrt{4}$
(g) $\sqrt{64}$ (h) $\sqrt{49}$ (i) $\sqrt{1}$
(j) $\sqrt{36}$ (k) $\sqrt{16}$ (l) $\sqrt{144}$

2. Determine which of the following are perfect squares by grouping factors.

(a) 72 (b) 75 (c) 400
(d) 196 (e) 324 (f) 500
(g) 256 (h) 225 (i) 125

The following program prints the positive square roots of the perfect squares from 1 to 144.

```
NEW
10 PRINT"PERFECT SQUARES"
20 PRINT"NUMBER","SQUARE ROOT"
30 FOR X = 1 TO 12
40 PRINTX*X,X
50 NEXT X
60 END
RUN
```

2.9 FINDING SQUARE ROOTS WITH A CALCULATOR

What would be the dimensions of a square room which has an area of 40 m^2?

Rewriting the formula $A = s^2$, we have

$s = \sqrt{A}$

Substituting $A = 40$ gives

$s = \sqrt{40}$

A = 40 m^2

?

?

The number 40 is not a perfect square, but we can find an approximate square root of 40 by comparing the square roots of perfect squares.

$\sqrt{36} < \sqrt{40} < \sqrt{49}$

$\therefore \quad 6 < \sqrt{40} < 7$

This means that $\sqrt{40}$ lies between 6 and 7.

Using the square root key on the calculator,

press 4 0 √ and the display is 6.3245553

Note that you do not press the = key after pressing the √ key.

Example.

Find the approximate square root without using the square root key.

(a) $\sqrt{7}$ (b) $\sqrt{30}$ (c) $\sqrt{78}$

Solution:

(a) Estimate.

$\sqrt{4} < \sqrt{7} < \sqrt{9}$

$\therefore 2 < \sqrt{7} < 3$

Since 7 is closer to 9 than 4, estimate 2.7

(b) Estimate.

$\sqrt{25} < \sqrt{30} < \sqrt{36}$

$\therefore 5 < \sqrt{30} < 6$

Since 30 is about midway between 25 and 36, estimate 5.5

(c) Estimate.

$\sqrt{64} < \sqrt{78} < \sqrt{81}$

$\therefore 8 < \sqrt{78} < 9$

Since 78 is closer to 81 than 64, estimate 8.8

These results can be checked using the square root key or by squaring the roots.

Using √ $\sqrt{7} \doteq 2.646$ | Using √ $\sqrt{30} \doteq 5.477$ | Using √ $\sqrt{78} \doteq 8.832$

EXERCISE 2.9

A 1. Choose the best estimate of the square root from the given list.

(a) $\sqrt{247}$ = 13.6, 15.7, or 17.2
(b) $\sqrt{9.53}$ = 3.1, 3.4, or 3.6
(c) $\sqrt{0.384}$ = 0.58, 0.6, or 0.62
(d) $\sqrt{74.2}$ = 8.6, 8.8, or 9.0
(e) $\sqrt{5790}$ = 76.1, 78.4, or 80.5
(f) $\sqrt{0.0784}$ = 0.24, 0.26, or 0.28
(g) $\sqrt{372}$ = 17.8, 19.3, or 20.9
(h) $\sqrt{3.6}$ = 1.7, 1.8, or 1.9

B 2. Find the approximate square root of each number.

(a) 32 (b) 60 (c) 21
(d) 38 (e) 593 (f) 750
(g) 1480 (h) 5920 (i) 0.42

3. Estimate the square root of each number.

(a) 87.4 (b) 12.7 (c) 58.4
(d) 6.25 (e) 574 (f) 1382

2.10 USING SQUARE ROOTS: FORMULAS

A hot-air balloon is 600 m above the earth. The pilot can calculate the oblique distance to the horizon using the formula

$d = 3.6\sqrt{h}$

where d is the distance to the horizon in kilometres,
h is the height of the balloon in metres.

To find the distance that the pilot can see to the horizon, we substitute in the formula.

$d = 3.6\sqrt{h}$
$d = 3.6\sqrt{600}$
$\doteq 3.6 \times 24.495$
$\doteq 88.182$

Using a calculator, press

The display is 88.181631

The pilot can see about 88 km to the horizon.

EXERCISE 2.10

B 1. Using the formula $d = 3.6\sqrt{h}$, find the distance you could see from the top of the following towers or buildings.

(a) Eiffel Tower, Paris	300 m
(b) World Trade Center, New York	411.5 m
(c) Sears Tower, Chicago	443.2 m
(d) CN Tower, Toronto	553.2 m

2. The time for an object to fall a distance, h, is given by the formula

$t = \sqrt{\frac{h}{4.9}}$ t = time in seconds, h = height in metres.

Calculate to the nearest tenth of a second the time for an object to fall through each distance.

(a) 95 m (b) 115.3 m (c) 150 m
(d) from the top of the Eiffel Tower which is 300 m high
(e) from the top of the World Trade Center which is 411.5 m high
(f) from the top of the Great Pyramid in Giza, Egypt, which is about 140 m high

3. The period of a simple pendulum is the time required for the pendulum to make one complete swing and is given by the formula

$T = 6.28\sqrt{\frac{\ell}{9.8}}$

where T is the period in seconds, and ℓ is the length of the pendulum in metres.

(a) A clockmaker is making a grandfather clock.
If the pendulum has a length of 1 m, how long will the pendulum take to make one complete swing?
(b) What is the period of a wall clock with a pendulum which is 0.25 m in length?

C 4. The radius of the base of a cone is given by the formula

$r = \sqrt{\frac{3V}{3.14h}}$

where
r is the radius,
V is the volume, and
h is the height.

A conical storage bin can be up to 3.0 m high to fit in a shed, and it must hold 5.0 m^3 of material.
Calculate the radius of the bin.

2.11 APPLICATIONS OF SQUARE ROOT: THE PYTHAGOREAN THEOREM

The school gymnasium is a rectangle with dimensions 30 m by 40 m. In order to decorate the gymnasium with streamers for a pep rally, the spirit committee needs to stretch a wire diagonally from corner to corner.
How can they calculate the length of the wire?

By making a diagram, we can see that the unknown length is the side of a right triangle. Carpenters, designers, and surveyors often have to calculate unknown dimensions of right triangles.

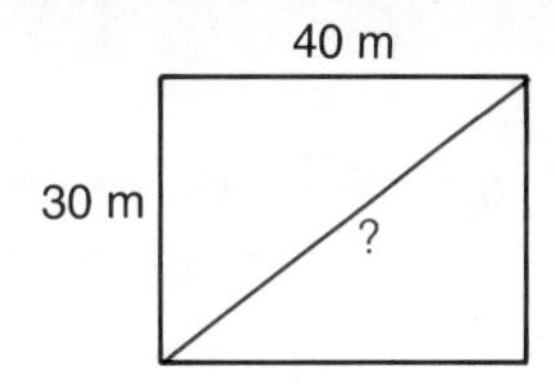

A useful rule, or theorem, named for the Greek philosopher and mathematician, Pythagoras (c. 580 B.C. – c. 500 B.C.), is still used today.

> Pythagorean Theorem
>
> In any right triangle, with sides a, b, c and where c is the hypotenuse, $c^2 = a^2 + b^2$
>
> 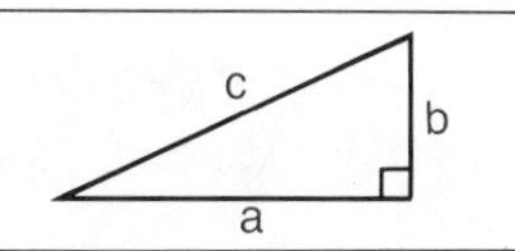
>

We can now apply the theorem to finding the length of wire in the gymnasium.

$$\begin{aligned} c^2 &= a^2 + b^2 \\ c^2 &= 30^2 + 40^2 \\ &= 900 + 1600 \\ &= 2500 \\ \sqrt{c^2} &= \sqrt{2500} \\ c &= 50 \end{aligned}$$

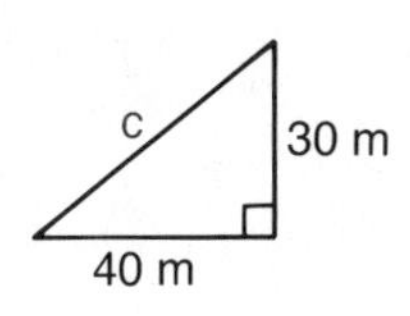

The length of the wire is 50 m.

From the given formula, we see that

$$\begin{aligned} c^2 &= a^2 + b^2 \quad \text{and} \quad c = \sqrt{a^2 + b^2} \\ a^2 &= c^2 - b^2 \quad \text{and} \quad a = \sqrt{c^2 - b^2} \\ b^2 &= c^2 - a^2 \quad \text{and} \quad b = \sqrt{c^2 - a^2} \end{aligned}$$

Example 1.

(a) Find c if a = 7 cm and b = 4 cm.

(b) Find b if c = 2.3 cm and a = 1.9 cm.

Solution:

(a)
$$\begin{aligned} c^2 &= a^2 + b^2 \\ c &= \sqrt{a^2 + b^2} \\ &= \sqrt{7^2 + 4^2} \\ &= \sqrt{49 + 16} \\ &= \sqrt{65} \\ c &\doteq 8.062 \text{ cm} \\ &\doteq 8 \text{ cm} \end{aligned}$$

c, a = 7 cm, b = 4 cm

(b)
$$\begin{aligned} b^2 &= c^2 - a^2 \\ b &= \sqrt{c^2 - a^2} \\ &= \sqrt{2.3^2 - 1.9^2} \\ &= \sqrt{5.29 - 3.61} \\ &= \sqrt{1.68} \\ b &\doteq 1.3 \text{ cm} \end{aligned}$$

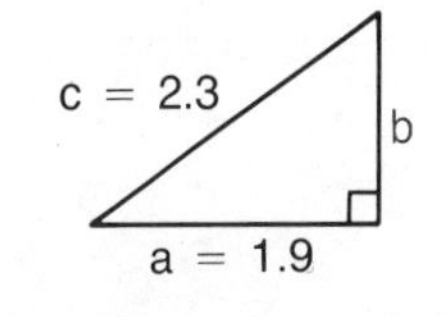

Example 2.
What is the length of the height, or rise, of the roof if the run is 4.25 m and the rafter length is 5.40 m?

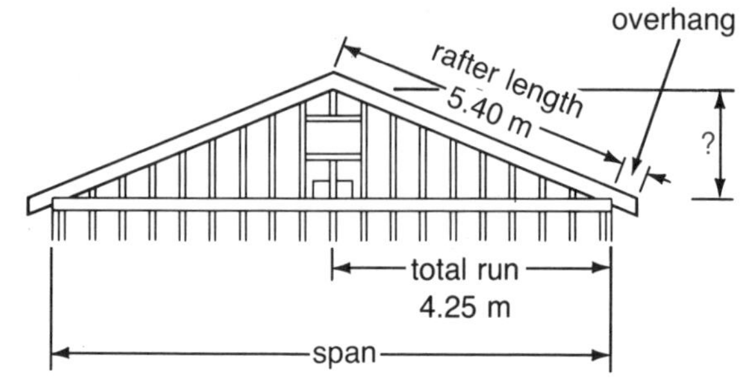

Solution:
In applying the Pythagorean Theorem to construction problems, we often assume that there are right angles.

For this problem, we assume that the wall studs are at right angles to the top plate of the wall frame. We can label the diagram to show a = 4.25 m, b = ? m, and c = 5.40

METHOD I

$b^2 = 5.40^2 - 4.25^2$
$= 29.16 - 18.0625$
$= 11.0975$
$b \doteq 3.33$

METHOD II

$b = \sqrt{5.40^2 - 4.25^2}$
$= \sqrt{29.16 - 18.0625}$
$= \sqrt{11.0975}$
$\doteq 3.33$

Using a calculator, press

5 · 4 x^2 − 4 · 2 5 x^2 = $\sqrt{\ }$

The display is 3.3312910

The rise of the roof is 3.33 m, to the nearest hundredth.

EXERCISE 2.11

B 1. Use a calculator to find the unknown dimensions in each right triangle. Give answers to two decimal places.

	Length of side a	Length of side b	Length of side c
(a)	4 cm	3 cm	■
(b)	5.2 m	4.7 m	■
(c)	4.5 cm	■	5.7 cm
(d)	2.0 m	5.0 m	■
(e)	2.25 cm	■	4.75 cm
(f)	4.5 km	3.8 km	■
(g)	■	5.75 m	8.65 m
(h)	■	12.4 cm	18.8 cm
(i)	4.5 cm	■	6.25 cm
(j)	2.5 m	2.5 m	■

2. Find the length of the diagonal in each of the following to the nearest tenth.

(a) a 50 cm by 125 cm ceiling tile

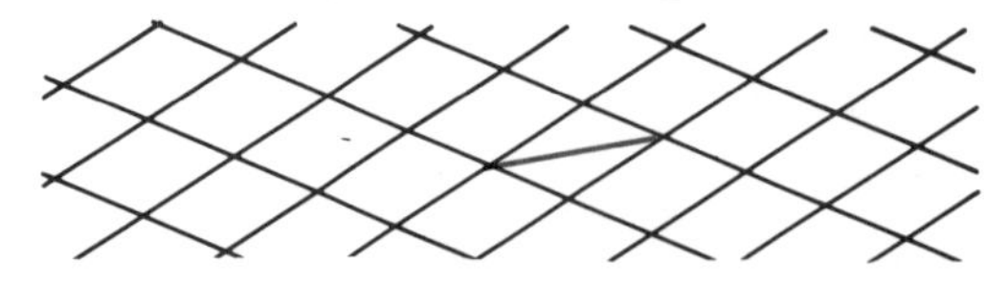

(b) a 1.3 m by 1.3 m rug

(c) the longest magician's cane which could lie on the bottom of a trunk which measures 50 cm by 100 cm

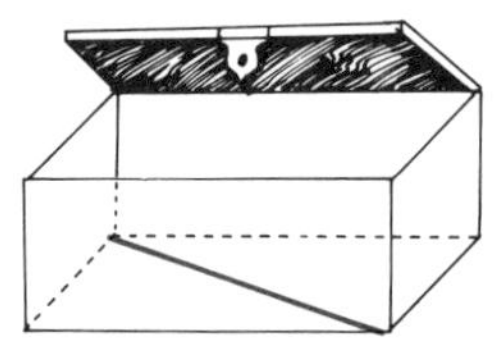

(d) A square timber 10 cm by 10 cm is to be cut from a single log.
What is the smallest diameter of the log which could be used to cut the timber?

3. What is the length of the rafter?

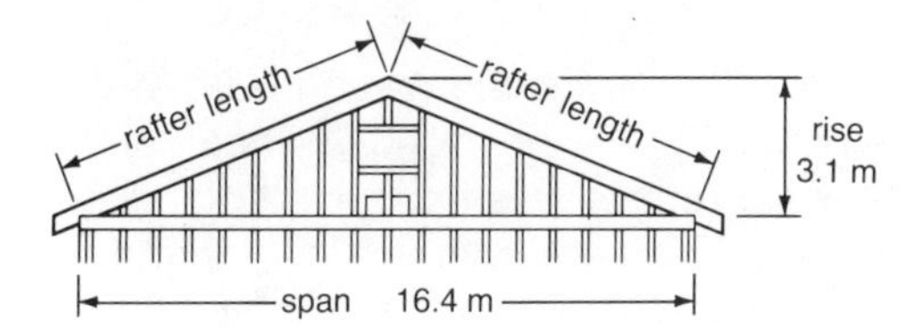

4. Find the missing dimensions to the nearest tenth, given the information in the table.

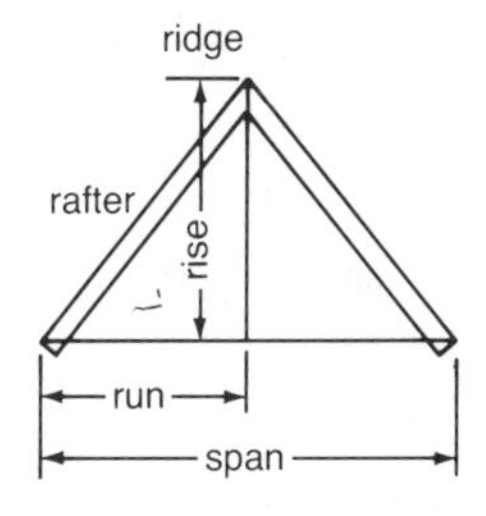

	Rise (m)	Run (m)	Rafter (m)
(a)	2	4.5	
(b)	2.4	7.2	
(c)	1.6	2.4	
(d)	2		12
(e)	1.2		9.6

5. A golfer has estimated the distance from himself to two places as shown in the diagram below.

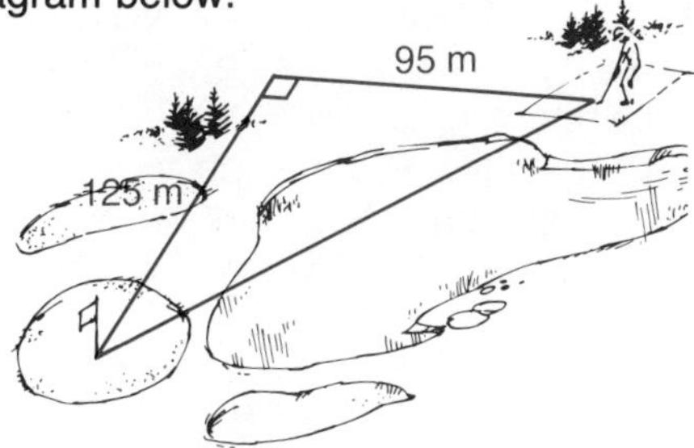

What is the distance over the pond?

6. A ramp for wheelchairs must span a horizontal distance of 6.2 m and a vertical distance of 1.2 m.
How long is the ramp?

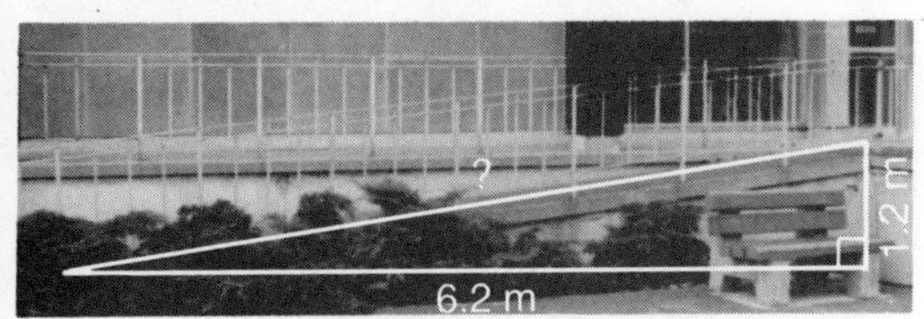

7. When a foundation is planned, it is important that the corners be square. The diagram illustrates the layout for a foundation.

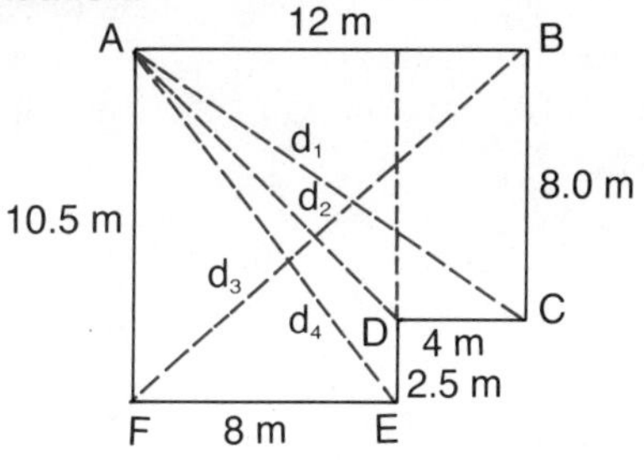

(a) Calculate the lengths of the coloured diagonals to check the dimensions specified in the layout.

(b) Make a scale drawing using a scale of 1 cm = 1 m (or use paper and compasses).
   (i) Measure and lay out AB.
   (ii) Calculate d_1 and place C the correct distance from A and B.
   (iii) Calculate d_2 and place D the correct distance from A and C.
   (iv) Calculate d_3 and place F the correct distance from A and B.
   (v) Calculate d_4 and place E the correct distance from A and F. Check your angles using a protractor.

MICRO MATH

The following program will compute the length of any side of a right triangle, given the other two sides.

```
NEW
 10 PRINT"RIGHT TRIANGLES"
 20 PRINT"IS HYPOTENUSE KNOWN? Y OR N"
 21 INPUT Z$
 22 IF Z$ = "Y" THEN 80
 30 INPUT"LENGTH OF ONE SIDE";A
 40 INPUT"LENGTH OF OTHER SIDE";B
 50 C = SQR(A*A+B*B)
 60 PRINT"THE HYPOTENUSE IS";C
 70 GOTO 120
 80 INPUT"ENTER THE HYPOTENUSE";H
 90 INPUT"ENTER THE OTHER SIDE";S
100 X = SQR(H*H - S*S)
110 PRINT"THE THIRD SIDE IS";X
120 END
RUN
```

2.12 PROBLEM SOLVING

1. A clock gains 2 min/d for 7 d. The regulator is adjusted, but the clock loses 1 min/d for the next 3 d.
How much time is gained or lost over the 10 d period?

2. The daily number of hours of sunshine at Ellie Lake recorded over a 7 d period are as follows:

7.3, 4.8, 1.9, 6.4, 5.9, 7.6, 6.9

What was the average daily number of hours of sunshine during this period?

3. The average number of red blood cells in 1 mm^3 of human blood is about 4 500 000. In order to perform blood tests, a laboratory technician dilutes 1 part blood with 300 parts saline solution.
How many red blood cells would you expect to find in 0.25 mm^3 of the solution?

4. Find three consecutive numbers whose sum is 111.

5. Find two numbers whose sum is 27 and whose product is 180.

6. In 18 home games, the school basketball team won 4 more games than they lost. There were no tie games.
How many games did the team win?

7. The economics club at your school went to Montreal on a field trip by bus. The bus travelled at 90 km/h on the way to Montreal and at 80 km/h returning to the school.
(a) Make an assumption about the distance from your school to Montreal.
(b) Calculate the time required to drive to the city at the indicated speed.
(c) Calculate the time required to return from this city at the indicated speed.

8. A train that is 1 km long takes 2 min to pass through a tunnel that is 1 km long. Calculate the speed of the train in kilometres per hour.

9. In a badminton tournament with four players, Rance and Ingrid finished behind Flavia. Frances finished behind both Rance and Ingrid. Rance was ahead of Ingrid.
Which player finished in third place?

10. A train is 1 km long and takes 5 min to pass a signal next to the track.
What is the speed of the train?

11. What is the units digit in the following power?

25^{125}

12. How many people can ride an elevator that can carry a maximum load of 800 kg if you can assume that the average person weighs 77 kg?

13. The human heart pumps about 80 mL of blood per second.
How much blood would such a heart pump in one week?
Express your answer in scientific notation.

14. Jean is a travelling sales person who is centrally located in Drouin. On Monday morning, Jean leaves her office, visits all of the towns shown in the map and finishes on Friday morning at Bisonville.
How can Jean leave Drouin on Monday, visit all of the other towns, and arrive in Bisonville with a minimum amount of travel?

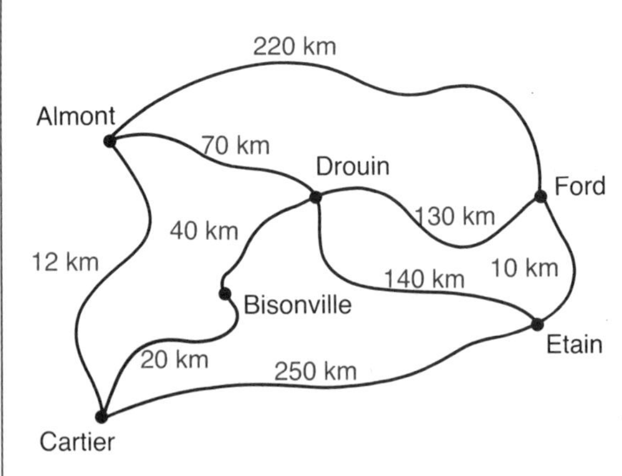

If you require more information to solve some of the following problems, you will be able to find the missing information in general reference books, which are available in your classroom, school library or public library.

15. How many days are there in the next five consecutive years?

16. How long would it take for you to drive from your provincial capital to New York City at an average speed of 80 km/h?

17. What is the time in Moscow when it is 09:00 in your school?

18. How many times must a tennis ball turn to travel the distance that a basketball travels in one turn?

19. What is the difference between the highest and lowest temperatures ever recorded in Canada?

20. How many times must a tennis ball turn to roll 1 m?

21. How many times will a long-playing record turn during the playing of "Sweet Georgia Brown," which is 3 min 34 s long?

22. How many textbooks placed end to end would reach from Calgary, Alberta to Moncton, New Brunswick by the Trans-Canada Highway?

(a) Take a guess.
(b) Make an estimate.
(c) Make a calculation.

CAREER

MACHINIST

Machinists are skilled workers who use lathes, milling machines, cutters, grinders, and drill presses to manufacture precision parts for industrial uses.

Many machines are now controlled by computers so that some machinists must also have knowledge of computers. These machinists have specialized and unique skills in the computer aided technology and are referred to as CNC (computer numerically controlled) machinists.

Machinists must be able to read blueprints and to understand written job specifications. In addition, they must be completely familiar with all the measuring tools of the trade in order to check for precision.

1. Three holes are drilled in a steel plate. Find the length of the slot which must be milled between the indicated holes.

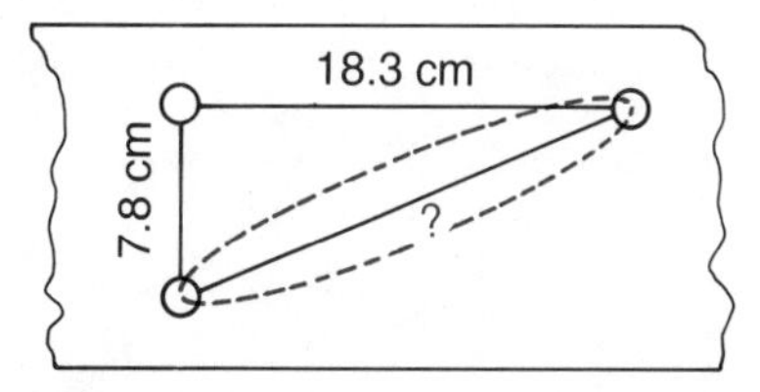

2. A steel plate in the shape of a washer has an inner diameter of 10 cm and an outer diameter of 30 cm. What is the area of the steel plate?

EXTRA

2.13 INVESTIGATING THE PATH OF A ROLLING BALL

For a project, the computer science class designed a simulation of a game table to investigate the path of a rolling ball. The dimensions of the table can be changed to suit the experiment. When a ball is rolled on a flat surface and it strikes another surface, the angle at which the ball strikes the surface will be equal to the angle at which it rebounds.

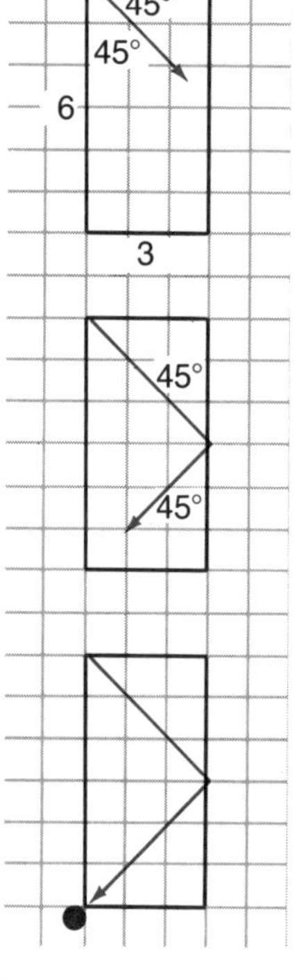

The diagram at the right shows the direction that a ball takes when it is rolled from the upper left corner of the game table. The ball travels at an angle of 45° to the sides of the table. The dimensions of the table are 6 units by 3 units (or simply 6 by 3).

The second diagram shows that the ball has hit the side of the table at an angle of 45°. It then rebounds at the same angle.

This diagram shows that the ball goes into the lower left corner. The ball stops when it comes to a corner.

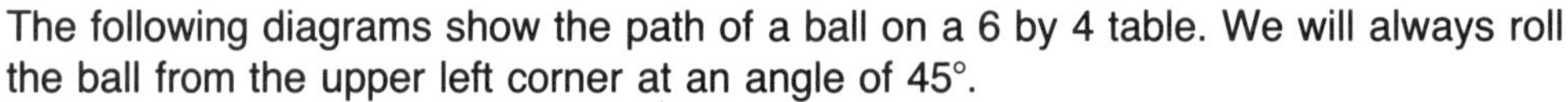

The following diagrams show the path of a ball on a 6 by 4 table. We will always roll the ball from the upper left corner at an angle of 45°.

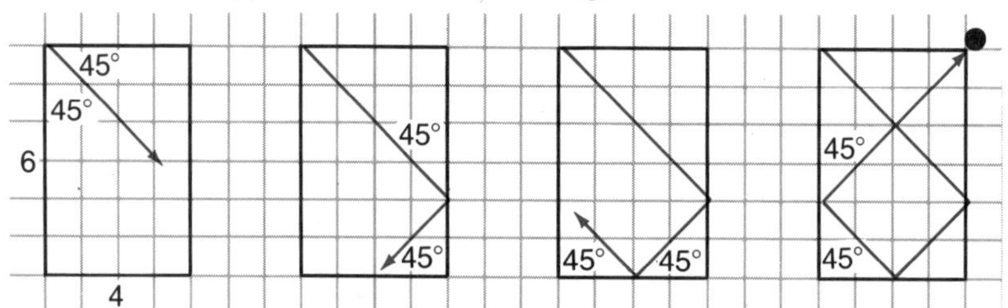

In the exercise you will study how the ball would travel on differently shaped tables.

EXERCISE 2.13

1. Draw the following tables on graph paper and draw the path of a ball, which is rolled from the upper left corner.

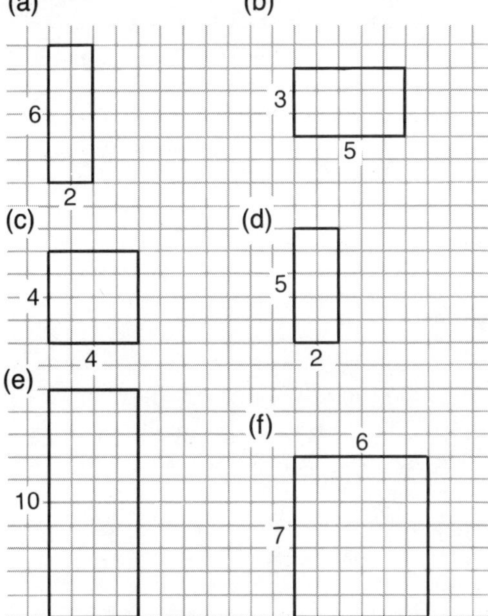

2. The tables below measure 7 by 1, 7 by 2, 7 by 3, and 7 by 4. Notice that the path of the ball crosses every square.

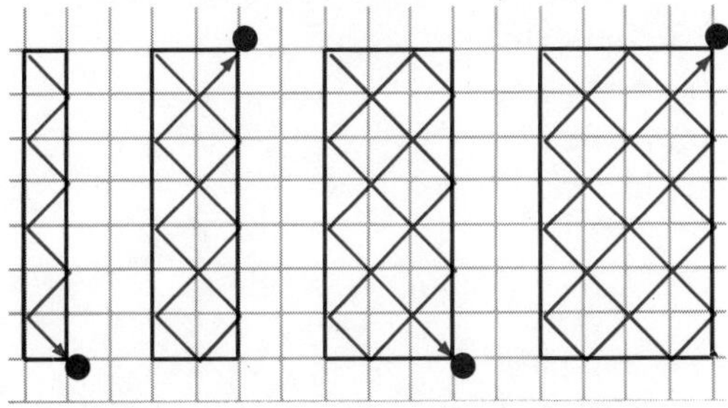

(a) Draw tables with dimensions 7 by 5 and 7 by 6. Draw the path of the ball.
Does the path cross each square for both tables?

(b) Will the ball cross each square on a 7 by 7 table?

3. What do the paths of balls on all square tables have in common?

4. For the following tables the length is twice as long as the width. Sketch the path of a ball on each table.
What do the paths of the balls on all tables shaped like this have in common?

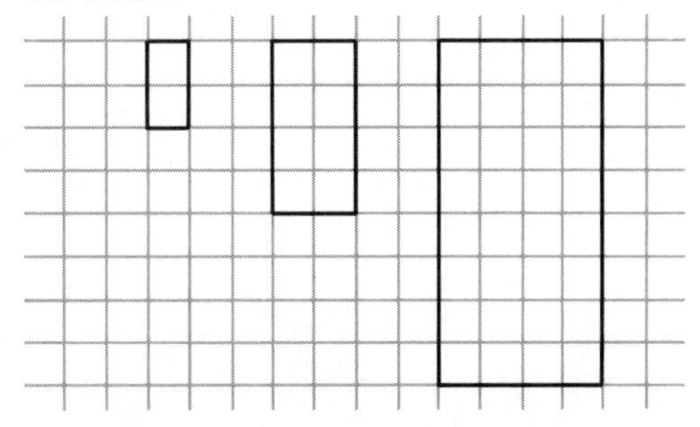

5. (a) What do the shapes of the following tables have in common?

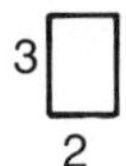

9

6

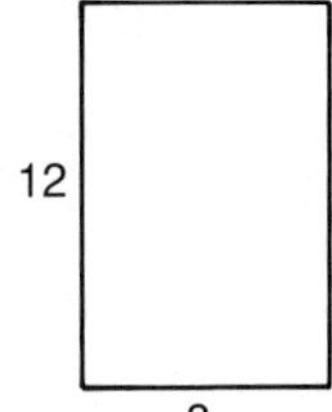

(b) Sketch the path of a ball on each table.
Are the paths for the balls the same?

6. (a) What do the shapes of the following tables have in common?

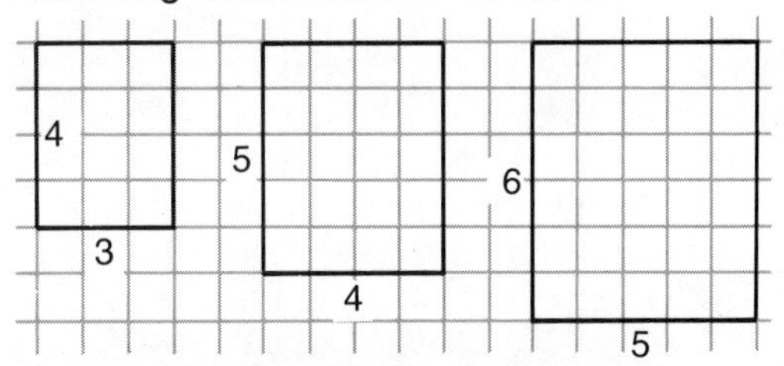

(b) Sketch the path of a ball on each table.
What is similar about the paths?

2.14 REVIEW EXERCISE

1. Evaluate.

(a) 3^3 (b) 2^3 (c) 4^3
(d) 5^2 (e) 4^2 (f) 2^4
(g) 6^2 (h) 2^5 (i) 3^2

2. Evaluate.

(a) 2^{-1} (b) 2^{-3} (c) 2^{-5}
(d) 3^{-1} (e) 3^{-2} (f) 3^{-3}
(g) 4^{-2} (h) 2^{-4} (i) 5^{-2}
(j) 5^0 (k) -3^0 (l) $(-2)^0$
(m) $2^{-1} + 3^{-1}$
(n) $3^{-2} + 2^{-3}$

3. Simplify and express in expanded form.

(a) $10^3 \times 10^2$ (b) $10^0 \times 10^3$
(c) $\frac{10^3 \times 10^4}{10^5}$ (d) $\frac{10^5}{10^2 \times 10}$
(e) $\frac{10^{-2} \times 10^5}{10^{-4}}$ (f) $\frac{10^{-3} \times 10^{-2}}{10^{-6}}$

4. Simplify.

(a) 3.265×1000 (b) $4.624 \div 100$
(c) $10^3 \times 54.6$ (d) $368 \div 10^{-2}$
(e) $10^{-3} \times 682.5$ (f) 45.3×10^3
(g) $456.2 \div 10^{-3}$ (h) 325×10^{-2}

5. Express in scientific notation.

(a) 345.6 (b) 34.78
(c) 256 000 (d) 42 000 000
(e) 12 000 (f) 63
(g) 0.002 56 (h) 0.000 000 235
(i) 0.0246 (j) 0.575
(k) 0.000 456 (l) 0.000 025 1

6. Express in decimal notation.

(a) 2.25×10^4 (b) 3.75×10^5
(c) 9.3×10^6 (d) 5.25×10^4
(e) 5.25×10^{-3} (f) 1.25×10^{-4}
(g) 6.25×10^{-5} (h) 7.75×10^{-1}
(i) 1.25×10^0 (j) 4.125×10^{-2}

7. Evaluate to the nearest tenth.

(a) $\sqrt{65}$ (b) $\sqrt{72}$ (c) $\sqrt{85}$
(d) $\sqrt{125}$ (e) $\sqrt{150}$ (f) $\sqrt{50}$
(g) $\sqrt{6.8}$ (h) $\sqrt{7.2}$ (i) $\sqrt{19}$
(j) $\sqrt{0.3}$ (k) $\sqrt{0.1}$ (l) $\sqrt{0.75}$

8. Evaluate for $x = 3$.

(a) x^4 (b) x^5 (c) $2x^2$
(d) x^3 (e) x^2 (f) x^1
(g) x^{-2} (h) x^{-1} (i) x^{-3}
(j) x^0 (k) $3x^0$ (l) $(3x)^0$

9. Evaluate for $x = 3$ and $y = -3$.

(a) $x^2 + y^3$ (b) $2x^3 + 3y^2$
(c) $-x^2 - y^3$ (d) $3x^2 - 2y^3$
(e) $x^{-2}y^{-2}$ (f) $(3x^{-2})(2y^{-2})$
(g) $(3x^0)(2x)$ (h) $(-3x^0) \div x^{-2}$

10. Evaluate for $x = 5$.

(a) $(3x^2) \div 3x^2$
(b) $2x^3 + (2x^2)^3$
(c) $3x^{-2} + 5x^{-1}$
(d) $(3x^{-2})^0 \div (-2x)^0$

11. Find the value of each exponent.

(a) $3^3 \times 3 = 3^{■}$ (b) $2^3 \div 2^2 = 2^{■}$
(c) $5^8 \div 5^4 = 5^{■}$ (d) $4^5 \times 4^3 = 4^{■}$
(e) $2^3 \times 2^2 = 2^{■}$ (f) $3^8 \div 3^4 = 3^{■}$
(g) $2^{-1} \times 2^3 = 2^{■}$ (h) $3^0 \times 3^5 = 3^{■}$
(i) $3^{-4} \times 3^4 = 3^{■}$ (j) $2^{-3} \div 2^{-2} = 2^{■}$

12. Simplify.

(a) $x^3 \times x^3$ (b) $y^4 \times y^3$
(c) $x^8 \div x^4$ (d) $x^5 \div x^2$
(e) $3x^3 \times 5x^3$ (f) $2y^7 \div y^3$
(g) $20x^4 \div 5x^3$ (h) $3y^3 \times 2y^2$

13. Simplify.

(a) $(2x^3)(3x^4)$ (b) $(2x^3)(-2x^2)$
(c) $(-3x^3)(-2x^5)$ (d) $(-5x^4)(3y^4)$
(e) $(5x^3)(2x^4)$ (f) $(-2x^4)(-3x^3)$
(g) $(2x^2)^4$ (h) $(5x^2)^3$
(i) $(x^3y)^3$ (j) $2(x^4)^2$

14. Simplify.

(a) $(3x^3)(2x^2)(x^4)$
(b) $(-2x^3)(3x^3)(-2y^4)$
(c) $(-4x^3)(3x^2) \div (6x^3)$
(d) $(18x^4)(2x^2) \div (6x)$
(e) $(-5x^3)(4x^5) \div (3x^6)$
(f) $(12x^2y^4) \div (2xy^2)(3x^2y^3)$
(g) $(4x^3y^2) \div (2x^2y^3)(5x^3y^3)$
(h) $(24x^6y^9) \div (-2xy) \div (-3x^2y^3)$

15. Simplify.

(a) $\left(\frac{x}{y}\right)^3$ (b) $\left(\frac{2x}{3y}\right)^3$

(c) $\left(\frac{x^3}{y^2}\right)^2$ (d) $\left(\frac{2x^4}{3y^2}\right)^3$

(e) $\left(\frac{-5x}{3y}\right)^2$ (f) $\left(\frac{-3x^2}{4y^2}\right)^3$

(g) $-\left(\frac{2x^3}{5y}\right)^2$ (h) $-\left(\frac{-3x^3}{2y^2}\right)^2$

(i) $-\left(\frac{-2x}{-3y}\right)^2$ (j) $-\left(\frac{-2x^2}{-5y}\right)^3$

16. Simplify.

(a) $(6xy^{-2})(3xy^{-2})$ (b) $(3x^{-2})(2x^{-3})$

(c) $(2xy)^0(-3x^{-1}2y^{-1})$ (d) $(15x^{-3})(3x^3)$

(e) $\frac{32x^3y^4}{8x^2y^2}$ (f) $\frac{12x^{-2}y^4}{4x^3y^3}$

(g) $\frac{7x^{-3}}{14x^{-3}}$ (h) $\frac{25x^0}{5x^4}$

(i) $\frac{12x^2y^{-3}}{4x^3y^3}$ (j) $\frac{24x^4y^5}{8x^3y^3}$

17. The length of the diagonal of a square is given by the formula

$$d = \sqrt{2s^2}$$

Find the length of the diagonals in each square to the nearest tenth.

(a)

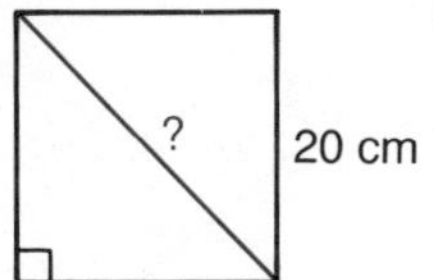

(b)

Area is 25 cm²

18. Find the length of the missing dimension in each of the following triangles to the nearest tenth.

(a)

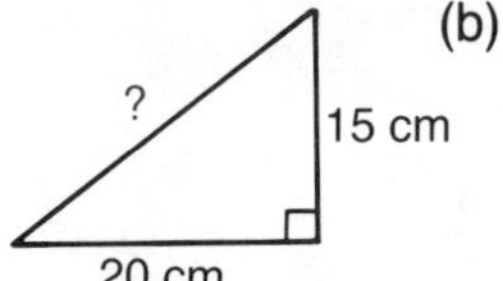

(b)

10 cm
?
10 cm

(c) (d)

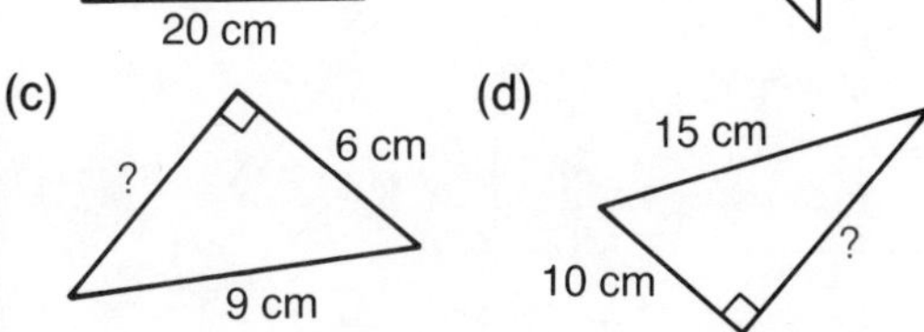

19. Find the length of the rafter in the following drawing.

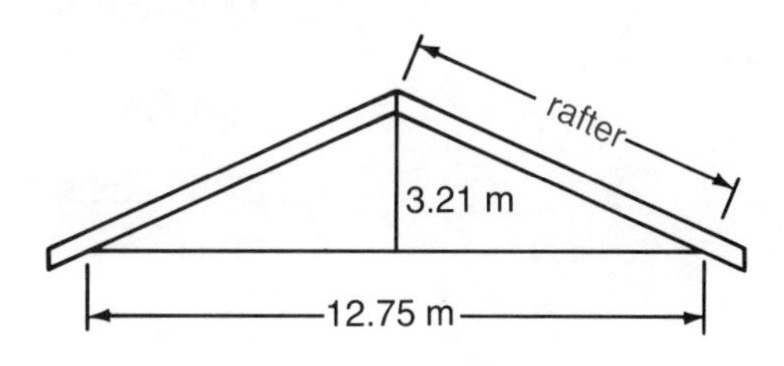

20. A householder builds a fence from the corner of his house to the street corner to stop pedestrians from cutting across his lawn.

If the house is 3.5 m from the side street and 10 m from the front street, how long is the fence?

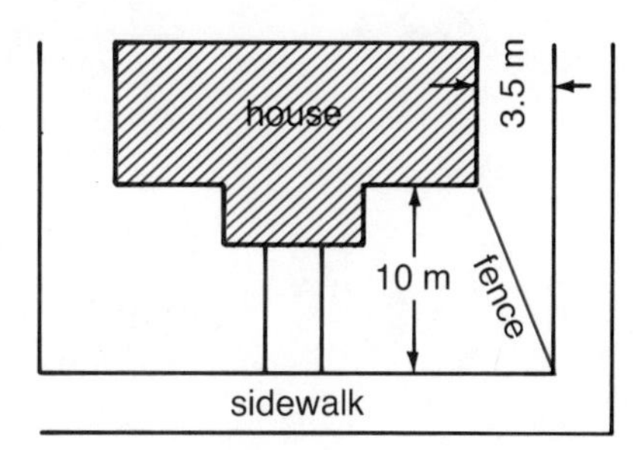

21. The Sulphur Mountain Gondola lift in the Canadian Rocky Mountains rises 700 m in a horizontal distance of 1370 m. The actual length of the cable from the bottom to the top is 1560 m.

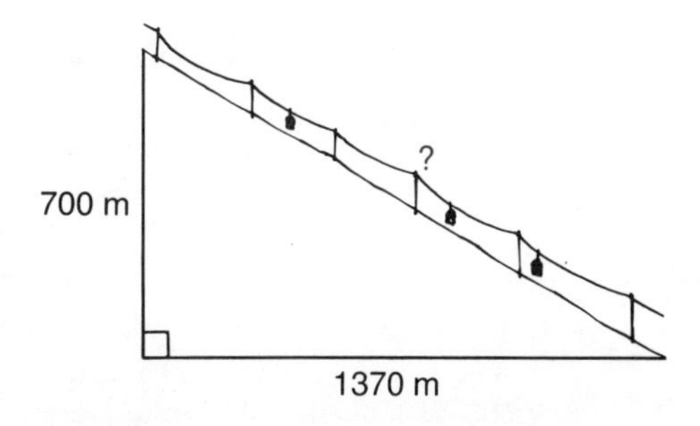

(a) What is the actual length of the slope?

(b) How much slack is there in the cable?

2.15 CHAPTER 2 TEST

1. Evaluate.

(a) 5^3 (b) 2^{-3} (c) 12^0
(d) $\sqrt{25}$ (e) $\sqrt{81}$ (f) $\sqrt{121}$

2. Evaluate to the nearest tenth.

(a) $\sqrt{75}$ (b) $\sqrt{125}$ (c) $\sqrt{300}$

3. Express in scientific notation.

(a) 325 000 (b) 56.75 (c) 0.002 45

4. Express in decimal notation.

(a) 5.15×10^4 (b) 6.39×10^{-3}

5. Simplify.

(a) $(3x^2)(-5x^3)$ (b) $(-5x^2y^3)^2$
(c) $(-2x^3)^0$ (d) $(-3x^2y)^{-1}$
(e) $\dfrac{32x^4y^3}{(-2x^2y^{-1})^2}$ (f) $\dfrac{(3x^0)(-12x^3y^3)}{4x^2y^2}$

6. A cylindrical storage tank is 5 m high and has a volume of 62.8 m^3. The formula for the radius of the tank is

$$r = \sqrt{\frac{V}{3.14\,h}}$$

where V is the volume and h is the height.
Find the radius of the tank.

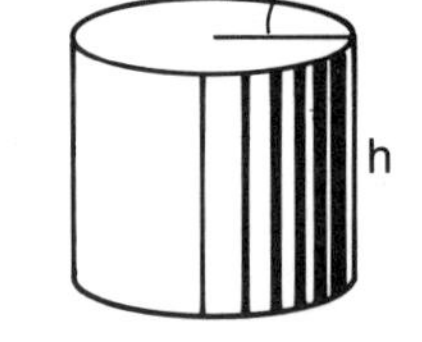

7. Find the length of the missing side to the nearest tenth.

(a)

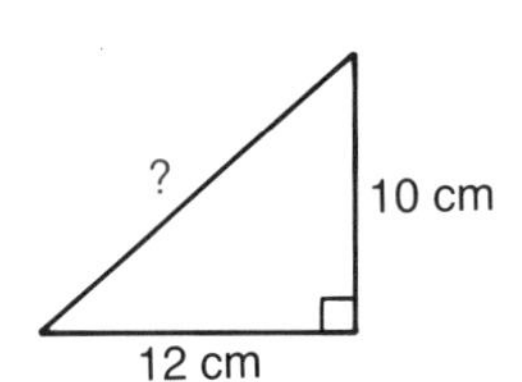

(b)

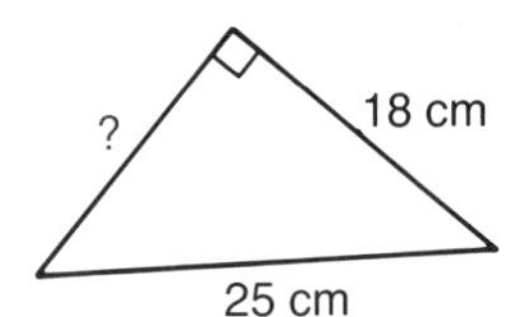

8. A triangular scaffolding support has sides 1.2 m and 0.8 m.

(a) How long must the bracing pieces be?
(b) The scaffold is 2.5 m above the ground and the scantling holding it against the wall is 3 m long.
How far should the foot of the scantling be from the wall?

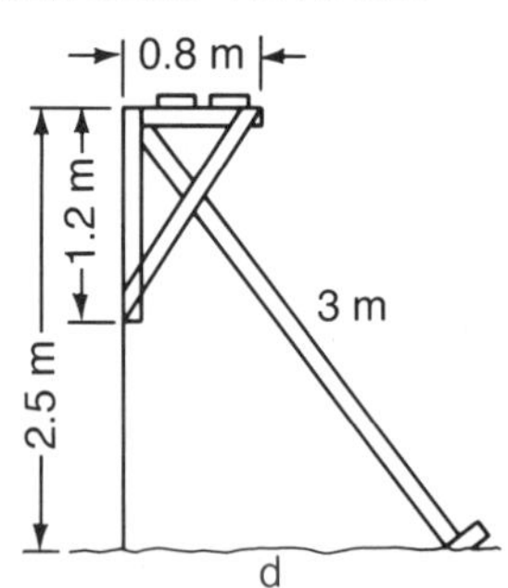

ALGEBRA

REVIEW AND PREVIEW TO CHAPTER 3

INTEGERS

EXERCISE 1

1. Simplify.

(a) $8 + (-2)$
(b) $-7 + (-4)$
(c) $-5 - 3$
(d) $4 - 2 + 5$
(e) $9 + 7 - 12$
(f) $(-3)(-4)$
(g) $-5(-6)$
(h) $20 \div (-2)$
(i) $-18 \div (-3)$
(j) $-6 \times (-5)$
(k) $-3 - 4 - 2$
(l) $-4 - 6 + 8 - 1$
(m) $9 + 6 - 13$
(n) $(-5) \times (-3)$
(o) $6 \div (-3)$
(p) -3×8

2. If $x = -1$ and $y = -2$, evaluate each of the following.

(a) $x + y$
(b) $x - y$
(c) $y - x$
(d) $2xy$
(e) $2x + 3y$
(f) $x + 2y - 4$
(g) $-2xy$
(h) $4x - 3y + 7$

3. If $a = 3$, $b = -3$, and $c = -4$, evaluate each of the following.

(a) $a + b + c$
(b) $a - b - c$
(c) $3a + 4b - 6c$
(d) $2ab - 3bc$
(e) $c - 2a - 3b$
(f) $9 - a - b - c$
(g) $ab + bc + ac$

THE DISTRIBUTIVE PROPERTY

EXERCISE 2

1. Expand.

(a) $2(x + 6)$
(b) $3(m + 7)$
(c) $4(t - 1)$
(d) $5(1 - m)$
(e) $2(3x - 5)$
(f) $-2(s + 4)$
(g) $-6(4t - 7)$
(h) $-(3m - 2)$
(i) $6(4w + 3t)$
(j) $-5(2m - 3n)$
(k) $-5(a - b - c)$

EXPONENTS

EXERCISE 3

1. Simplify.

(a) $x^{12} \div x^3$
(b) $a^5 \div a^4$
(c) $b^6 \div b^0$
(d) $x^8 \div x^{13}$
(e) $z \div z^4$
(f) $a^{-7} \div a^{10}$
(g) $b^8 \div b$
(h) $w^3 \div w^5$

2. If $a = 2$, $b = 3$, and $c = 0$, evaluate.

(a) $2ab$
(b) $a^2 + b^2$
(c) $2bc + b$
(d) $-3ab$
(e) $4b^2$
(f) $b^2 - a^2$

3. Simplify.

(a) $3a^2 \times 5a^5$
(b) $4y^2 \times 7y$
(c) $(6x^3)(-2x^5)$
(d) $(-7a^3)(-5a^5)$
(e) $(a^7)(7ab)$
(f) $(3a^2b)(-5ab^4)$
(g) $(4xy)(-5xy^2)$
(h) $(5a^{-7})(3a^{-2})$

4. Simplify.

(a) $12b^3 \div 4b$
(b) $45x^7 \div 9x$
(c) $24x^3y^3 \div 8x^3y$
(d) $12x^5 \div 3x^{-2}$
(e) $54ab^3 \div 6ab$
(f) $12x^3 \div 2x^{-2}$
(g) $51xy^5 \div 3xy$
(h) $(56w^3) \div (-7w^{-2})$

5. Simplify.

(a) $\dfrac{15x^3 \times 3x^5}{9x^4}$

(b) $(x^3y^5)^2$

(c) $(8x^{-3})(2x^7) \div 4x$

(d) $\left(\dfrac{15a^2b}{3ab}\right)^2$

(e) $\dfrac{12a^2 \times 5a^3}{4a}$

6. Evaluate.

(a) 3^{-2}
(b) $(\frac{1}{4})^{-1}$
(c) $2^{-1} + 3^{-1}$
(d) $(\frac{1}{2})^{-2}$
(e) $(\frac{3}{4})^{-1}$
(f) $4^0 + 2^{-1}$

SUBSTITUTION

EXERCISE 4

1. If $a = 2$, $b = 1$, and $c = 0$, evaluate the following.

(a) $2a + 3b$ (b) $6a - 3b$
(c) $5ab$ (d) $6ab - 2bc$
(e) $7abc$ (f) $a^2 + b^2$

2. If $I = Prt$, find the value of I under each set of conditions.

(a) $P = 1000$, $r = 1$, $t = 3$
(b) $P = 500$, $r = 0.5$, $t = 2$
(c) $P = 200$, $r = 0.04$, $t = 6$
(d) $P = 2000$, $r = 0.06$, $t = 10$

3. The formula for the perimeter of a rectangle is $P = 2(\ell \times w)$. Calculate the perimeter of each rectangle.

(a) $\ell = 35$ m, $w = 17$ m
(b) $\ell = 107$ m, $w = 82$ m
(c) $\ell = 256$ m, $w = 142$ m

EQUATIONS

EXERCISE 5

1. Solve the following equations.

(a) $x + 3 = 12$ (b) $y + 7 = 10$
(c) $t + 4 = 8$ (d) $2 + m = 16$
(e) $b - 3 = 4$ (f) $x - 7 = 12$

2. Solve.

(a) $3x = 15$ (b) $4a = 20$
(c) $7t = 28$ (d) $-2t = 8$
(e) $-3x = -21$ (f) $-4m = -20$

3. Solve.

(a) $\frac{x}{2} = 4$ (b) $\frac{m}{3} = 5$
(c) $\frac{t}{4} = 6$ (d) $\frac{r}{6} = 1$
(e) $\frac{e}{6} = 5$ (f) $\frac{b}{2} = 7$

MENTAL MATH

EXERCISE 6

Perform the following calculations mentally and record your answers.

1. (a) 23×1000 (b) 43×100
(c) 100×1.2 (d) 0.45×1000
(e) 0.06×10 (f) 100×7.65
(g) 0.343×100 (h) $12 \times 10\,000$

2. (a) 45×0.1 (b) 450×0.1
(c) 0.01×1200 (d) 2000×0.1
(e) 7×0.01 (f) 0.001×300
(g) 0.01×0.4 (h) 5.6×0.1
(i) $70\,000 \times 0.001$ (j) 0.001×66

3. (a) $78 \div 100$ (b) $98 \div 10$
(c) $48 \div 1000$ (d) $1.2 \div 100$
(e) $0.4 \div 10$ (f) $0.09 \div 10$
(g) $12\,000 \div 1000$ (h) $200 \div 1000$

4. (a) $43 \div 0.1$ (b) $200 \div 0.1$
(c) $700 \div 0.1$ (d) $650 \div 0.01$
(e) $3000 \div 0.001$ (f) $0.4 \div 0.1$
(g) $8.9 \div 0.001$ (h) $0.1 \div 0.1$
(i) $20\,000 \div 0.01$ (j) $0.07 \div 0.01$

5. (a) 3×200 (b) 22×300
(c) 12×500 (d) 700×2
(e) 40×50 (f) 400×30
(g) 23×200 (h) 450×20

6. (a) $1200 \div 20$ (b) $360 \div 12$
(c) $1800 \div 9$ (d) $555 \div 5$
(e) $888 \div 4$ (f) $44\,000 \div 11$
(g) $65 \div 5$ (h) $242 \div 2$
(i) $969 \div 3$ (j) $4004 \div 2$

7. (a) 10% of 450 (b) 10% of 200
(c) 50% of 88 (d) 25% of 800
(e) 75% of 40 (f) 50% of 660
(g) 25% of 4 (h) 75% of 1000

8. (a) What percent of 50 is 25?
(b) What percent of 80 is 20?
(c) What percent of 200 is 150?
(d) What percent of 80 is 8?

3.1 DEVELOPING FORMULAS

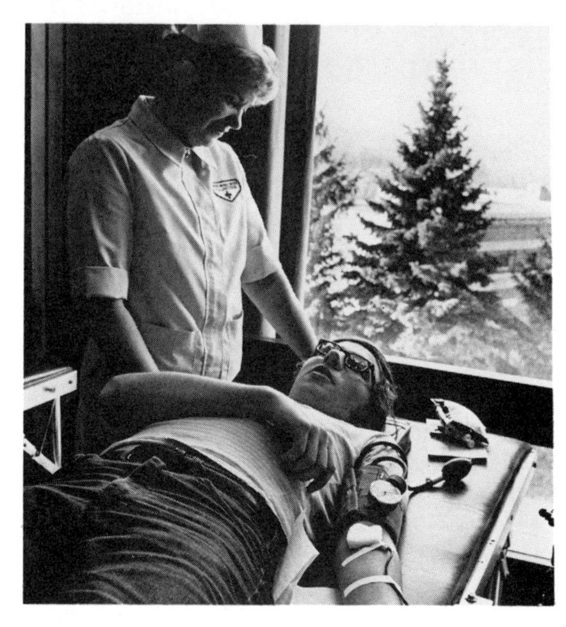

When you donate blood, a nurse takes your blood pressure. For young adults, a normal blood pressure reading is between 110/70 and 120/80. The first number in each reading is a measure of systolic pressure. The heart muscle works like a strong fist, which opens and closes to pump blood through the body. When the fist closes, or the heart contracts, it forces blood to flow through the arteries. The pressure on the walls of the arteries, caused by the contraction of the heart muscle, is systolic pressure. A person's systolic pressure, P, can be found by adding 100 to half the person's age, a.

$$P = 100 + \frac{a}{2}$$

The above equation, which describes the relationship between systolic pressure and age, is called a formula.

> A formula shows the relationship between two or more variables using the operations of arithmetic.

Example.

The collision impact of a car is a measure of the degree of impact a car can produce when it hits an object. The collision impact depends on the speed of the car. The table gives the collision impact for a specific car at several speeds.

Speed (km/h)	10	20	40	60	80
Collision Impact	100	400	1600	3600	6400

Write a formula that shows the relationship between the speed of this car and the collision impact.

Solution:

Look for a pattern in the table.

The collision impact of this car is the square of the speed of the car.

The formula for this relationship is

$$I = s^2$$

where I is the collision impact and s is the speed of the car.

EXERCISE 3.1

B 1. Write a formula for each situation.

(a) The cost of n doughnuts if doughnuts cost 75 cents each.

(b) The distance travelled in t hours at 80 km/h.

(c) The batting average is the number of hits divided by the number of times at bat.

(d) The cost for a banquet is $400 for the room plus $35 for each meal.

2. A car rental agency charges $60 a day, and 150 free km are allowed for each day. A distance of over 150 km is charged at a rate of $0.22/km.

(a) Write a formula to determine the cost per day.

(b) Use the formula to calculate what you would pay if you rented a car for one day and drove 350 km.

3. In hospitals, physicians always prescribe the medication for patients, but the nurses must be aware of the usual dose for frequently used medications. For this reason nurses often use the following rules to check the prescription.

Clark's Rule (newborn to 12 a)

Estimate a child's dose by multiplying the adult dose by a fraction. The numerator of the fraction is the child's mass in kilograms; the denominator is 70 kg.

Young's Rule (1 or 2 to 12 a)

Estimate a child's dose by multiplying the adult dose by a fraction. The numerator of the fraction is the child's age in years; the denominator is the child's age in years plus 12.

(a) Write a formula for Clark's rule.

(b) Write a formula for Young's rule.

(c) If the adult dose is 10 grains, use both rules to calculate the dose for a two-year-old child who has a mass of 10 kg.

(d) The adult dose is 80 mg. Use both rules to calculate the dose for a five year old who has a mass of 22 kg.

4. The table gives Colleen's wages, according to hours worked, that she earns as a horse trainer's assistant.

Hours	3	7	9	14
Wages	28.50	66.50	85.50	133.00

(a) Write a formula that relates Colleen's salary to the number of hours worked.

(b) How much does she earn in 21 h?

5. Write a formula that represents the perimeter of each figure.

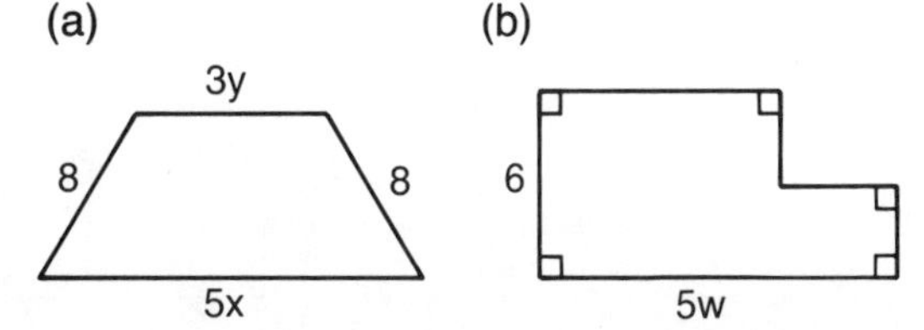

6. Write a formula that represents the area of each figure.

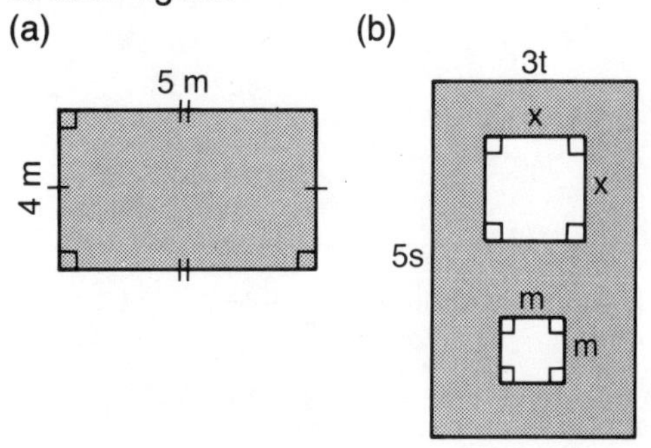

MIND BENDER

Arrange 14 chairs against the walls of a square room so that the same number of chairs are along each wall.

3.2 SIMPLIFYING POLYNOMIALS

Algebra is the language of mathematics. By learning this simple language you will be able to solve many problems.

The height, h, of a rocket with initial velocity, v, at t seconds after blast-off is given by the formula,

$$h = vt - 5t^2$$

$vt - 5t^2$ is called an algebraic expression
This expression has two terms, namely, vt and $5t^2$.

The simplest algebraic expression is a term. A term is either
(i) a number such as 9,
(ii) a variable such as x, or
(iii) the product of numbers and variables, such as 12xy, where 12 is called the coefficient of the term 12xy.

Terms, such as $3x^3$ and $5x^3$, that share the same variable raised to the same exponent are called like terms. Terms, such as $2x^2$ and $4x^3$, are called unlike terms.

We name expressions by the number of unlike terms they have.

Number of Unlike Terms	Example	Name	Example
1	$5x$	monomial	..... mono is also found in the word monorail
2	$4s - 5t$	binomial	..... bi is also found in the word bicycle
3	$3x^2 + 5x - 7$	trinomial	..... tri is also found in the word tricycle

Monomials, binomials, trinomials, and expressions with more than three terms are all called polynomials.

Some polynomials can be simplified by combining like terms.

Example 1.
Simplify. $8xy + 9 - 3xy - 2$

Solution:
Combine like terms.

$$\begin{aligned} 8xy + 9 - 3xy - 2 &= 8xy - 3xy + 9 - 2 \\ &= 5xy + 7 \end{aligned}$$

Example 2.
Simplify. $3x^2 - 4x + 5 - x^2 - 6x$

Solution:
Combine like terms.

$$\begin{aligned} 3x^2 - 4x + 5 - x^2 - 6x &= 3x^2 - x^2 - 4x - 6x + 5 \\ &= 2x^2 - 10x + 5 \end{aligned}$$

EXERCISE 3.2

A

1. State the coefficient of each term.

(a) $7x$ (b) $8xy$ (c) $-4ab$
(d) $-5t^2$ (e) mn (f) $-c^2$

2. State the variable in each term.

(a) $6a$ (b) $2x^2$ (c) $-9t$
(d) $-11m$ (e) 8 (f) $-n^3$

3. State whether the terms are like or unlike.

(a) $3x, 5x$ (b) $-3t, 4t$
(c) $6y^2, 7y$ (d) $3m, 2n$
(e) $4xy, -xy$ (f) $-9a^3, 6a^4$

4. State which terms can be combined.

(a) $3x, 4y, 8x, -3y$
(b) $4x^2, -3x, 6x^2, 7, 5x$
(c) $7a, 5b, -2ab, 4a, 6ab, 3b$
(d) $-2m^2, 3n^2, 3mn, -mn, -n^2, 7m^2$
(e) $12 + x - 3xy + 5x - 4 + xy$

5. Simplify.

(a) $5x + 6x$ (b) $7y - 4y$
(c) $-3t + 8t$ (d) $4b - 8b$
(e) $3x + 6x$ (f) $-5d - 3d$
(g) $-x - 3x$ (h) $4a - a$

B

6. Simplify.

(a) $3x + 6x + 5x$ (b) $5y - y - 3y$
(c) $-3a + a - 6a$ (d) $-3y - 2y - 4y$
(e) $6t + 10t - t$ (f) $4r - 5r - r$

7. Simplify.

(a) $4x + 5y + 3x + 6y$
(b) $7a + 8b - 2b + 3a$
(c) $5s - 3t - 2s + 8t$
(d) $11m - 9m - 3n - 2n$
(e) $5x - 3y - x - y$
(f) $-3a - 4a + 5b - 6b$
(g) $3a + 7b - 3 + 7a - 6a$

8. Simplify.

(a) $4x - 3y + 5z - x + y - 5z$
(b) $6r + 4t - 3s - 6s + r - t$
(c) $7a + 3b - c - 2b - a + 5c$
(d) $5x - 9 + 4y - 7 - x + 5x$
(e) $a - 2b - c - 3b + 6b + 5c$

9. Simplify.

(a) $3xy - 4x + 5y - 2xy$
(b) $-3x^2 + 4x - 5x^2 - 7x$
(c) $5t^3 - 3t - 4t^3$
(d) $2x - x^2 + 7 - 5x^2 + 2x^2$
(e) $4a + a^2 - 3a^2 + 7a - 11$

10. Simplify where possible.

(a) $4x - 3xy + 5y + 2$
(b) $7a + 3b - 4 - 6a + 11$
(c) $2x^2 + 5x + 6$
(d) $5x + 3$

11. Write an expression for the perimeter of each figure.

(a)

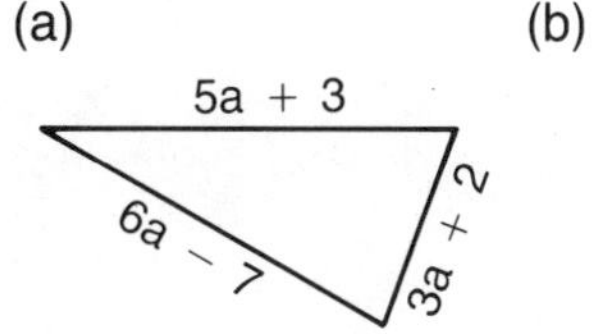

(b)

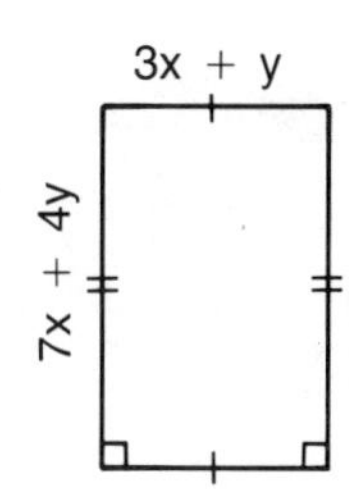

(c)

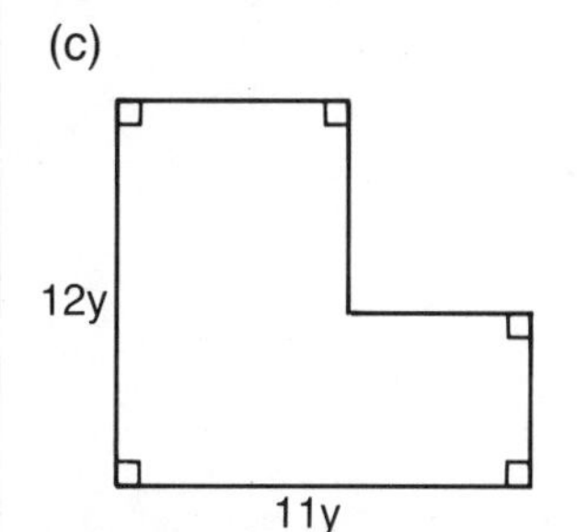

MIND BENDER

Draw 2 straight lines on the face of the clock to divide the clock into 3 parts. The numbers in each part must add to give the same number.

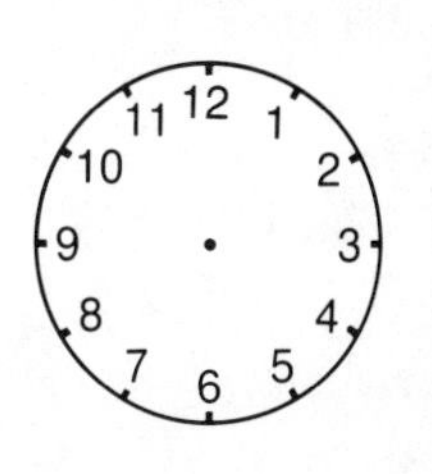

3.3 EVALUATING POLYNOMIALS

Fred Charles makes Tamarack geese to sell.
A consultant has determined that Fred's profit formula is

$$P = 40n - 0.2n^2$$

where P is the profit in dollars and n is the number of geese sold.

What is the profit if Fred sells 20 geese?

To determine the profit we evaluate the polynomial by substituting.

$$\begin{aligned} P &= 40n - 0.2n^2 \\ P &= 40(20) - 0.2(20)^2 \\ &= 800 - 80 \\ &= 720 \end{aligned}$$

The profit for selling 20 geese is \$720.

Example.
If $x = 2$, $y = -1$, and $m = -2$, evaluate $2xy^2 - 3xym$

Solution:

$$\begin{aligned} 2xy^2 - 3xym &= 2(2)(-1)^2 - 3(2)(-1)(-2) \\ &= 2(2)(1) - 3(2)(2) \\ &= 4 - 12 \\ &= -8 \end{aligned}$$

EXERCISE 3.3

A 1. If $a = 2$, $b = 3$, and $c = 1$, evaluate.

(a) $2a + 3b$ (b) $2bc$
(c) $6a - 6c$ (d) $bc - a$

B 2. If $a = 1$, $b = 2$, and $c = 3$, evaluate.

(a) $2a^2 + b^2$ (b) $3abc + 2c^2$
(c) $3ab^2 - 2$ (d) $6bc - a^2$

3. If $x = 2$, $y = -2$, and $m = -1$, evaluate.

(a) $2x + 3y - 2m$ (b) $3xy + 6m + 5$
(c) $x - y - m$ (d) $-2xm - 3ym + 2$
(e) $2xym - m + 7$ (f) $-2y - 3xm - m$

4. If $a = -1$, $b = -2$, and $d = 3$, evaluate.

(a) $2a^2 + b^2$ (b) $3b^2 - 2ad + 3$
(c) $7 - 3d^2 + 2a^2b^2$ (d) $2a^2b^2 - 3 + bd^2$

5. If $x = -2$, $m = -3$, $c = 0$, and $d = -1$, evaluate.

(a) $\dfrac{2xm}{d}$ (b) $\dfrac{3m - 2x}{2m}$

(c) $\dfrac{2x^2 + 2d^2 + m^2}{2x}$ (d) $\dfrac{3xmd - 2mc}{3}$

6. Justine makes and sells souvenir scarves for her college. The formula for her profit is

$$P = 30n + 0.1n^2 - 0.01n^3$$

where P is the profit in dollars and n is the number of scarves sold.
What is her profit for each number of scarves sold?

(a) 20 (b) 30 (c) 40

3.4 ADDING AND SUBTRACTING POLYNOMIALS

The Sea World Company makes waterproof watches.
The cost of making the watches is found using the formula

$$C = 80n + 3000$$

where C represents the cost in dollars, and
n represents the number of watches.

The money Sea World Company receives from selling the watches is called income. The income is found using the formula

$$I = 120n$$

where I represents the income in dollars, and
n represents the number of watches.

The cost of making 100 watches is

$$\begin{aligned} C &= 80n + 3000 \\ C &= 80(100) + 3000 \\ &= 8000 + 3000 \\ &= 11\,000 \end{aligned}$$

The income from selling these 100 watches is

$$\begin{aligned} I &= 120n \\ I &= 120(100) \\ &= 12\,000 \end{aligned}$$

The profit is \$12 000 − \$11 000 or \$1000.

To find the profit from selling 200, 3000, or any number of watches, it is easier to find a formula for the profit first, then substitute. To find the profit formula we will subtract polynomials. We have already added polynomials to simplify.

Example 1.
Find the sum of $(3x^2 + 5x - 2)$ and $(x^2 - 8x - 7)$

Solution:

$$\begin{aligned} (3x^2 + 5x - 2) + (x^2 - 8x - 7) &= 3x^2 + 5x - 2 + x^2 - 8x - 7 \\ &= 3x^2 + x^2 + 5x - 8x - 2 - 7 \\ &= 4x^2 - 3x - 9 \end{aligned}$$

To subtract an integer, we add its opposite. $\quad -4 - (+5) = -4 + (-5) = -9$

We use the same method to subtract polynomials.
You can find the opposite of a number or polynomial by multiplying by −1.

Polynomial	Multiply by −1	Opposite
$3y$	$-1(3y)$	$-3y$
$5y - 4$	$-1(5y - 4)$	$-5y + 4$
$3xy + 7t$	$-1(3xy + 7t)$	$-3xy - 7t$
$2x^2 - 3x - 1$	$-1(2x^2 - 3x - 1)$	$-2x^2 + 3x + 1$

Example 2.
Find the difference.

$$(4x - 5y) - (7x - 3y)$$

Solution:
Add the opposite.

$$\begin{aligned}(4x - 5y) - (7x - 3y) &= (4x - 5y) + (-1)(7x - 3y)\\ &= 4x - 5y - 7x + 3y\\ &= -3x - 2y\end{aligned}$$

Example 3.
Find the difference.

$$(x^2 - 2x - 3) - (2x^2 + 5x - 4)$$

Solution:
Add the opposite.

$$\begin{aligned}(x^2 - 2x - 3) - (2x^2 + 5x - 4) &= (x^2 - 2x - 3) + (-2x^2 - 5x + 4)\\ &= x^2 - 2x - 3 - 2x^2 - 5x + 4\\ &= -x^2 - 7x + 1\end{aligned}$$

Example 4.
The cost, C, for the Sea World Co. to make waterproof watches, n, is given by the formula

$$C = 80n + 3000$$

The income, I, from the sale of the watches is given by the formula

$$I = 120n$$

(a) Find a formula for the profit if the company makes and sells n watches.
(b) Find the profit if 2000 watches are sold.
(c) What is the profit if 50 watches are sold?

Solution:
(a) To find the profit, P, subtract cost from income.

$$\begin{aligned}P &= I - C\\ P &= 120n - (80n + 3000)\\ &= 120n + (-80n - 3000)\\ &= 120n - 80n - 3000\\ &= 40n - 3000\end{aligned}$$

(b)
$$\begin{aligned}P &= 40n - 3000\\ P &= 40(2000) - 3000\\ &= 80\,000 - 3000\\ &= 77\,000\end{aligned}$$

The profit is \$77 000.

(c)
$$\begin{aligned}P &= 40n - 3000\\ P &= 40(50) - 3000\\ &= 2000 - 3000\\ &= -1000\end{aligned}$$

The negative profit means the company lost \$1000.

EXERCISE 3.4

A 1. Simplify.

(a) $(2x + 3) + (5x + 6)$
(b) $(4s + 6) + (2s + 5)$
(c) $(3w + 9) + (2w - 7)$
(d) $(4m + 3n) + (5m - 2n)$
(e) $(8x - 5y) + (3x - 4y)$
(f) $(2y - 3w) + (7y - 5w)$
(g) $(-3x - 9) + (-5x + 2)$
(h) $(-6t - 1) + (-t - 3)$
(i) $(-4s + 5) + (-3s - 5)$

2. State the opposite of each polynomial.

(a) $2x + 5$
(b) $7 - 3m$
(c) $6t - 8$
(d) $x^2 + 3x + 4$
(e) $2t^2 - 4t - 5$
(f) $-5 - 3m - m^2$

3. Add.

(a) $\begin{array}{r} 5x + 8 \\ \underline{2x + 1} \end{array}$
(b) $\begin{array}{r} 4t + 3 \\ \underline{5t - 7} \end{array}$
(c) $\begin{array}{r} 8s + 4t \\ \underline{5s - 6t} \end{array}$
(d) $\begin{array}{r} 3m - 2n \\ \underline{6m - 3n} \end{array}$

4. Subtract.

(a) $\begin{array}{r} 6x + 12 \\ \underline{2x + 3} \end{array}$
(b) $\begin{array}{r} 8m + 6 \\ \underline{5m + 1} \end{array}$
(c) $\begin{array}{r} 6x + 8y \\ \underline{8x - 2y} \end{array}$
(d) $\begin{array}{r} 9m - 3n \\ \underline{7m - 5n} \end{array}$

B 5. Simplify.

(a) $(3x + 5) - (2x + 6)$
(b) $(4m + 8) - (3m + 9)$
(c) $(3s + 5t) - (2s + 6t)$
(d) $(4w + 3) - (2w - 5)$
(e) $(6a - 7) - (2a - 8)$
(f) $(4x - 3y) - (5x - y)$
(g) $(6m - 2n) - (7m - 3n)$

6. Simplify.

(a) $(2x^2 + 3x + 4) + (5x^2 + 4x + 9)$
(b) $(3m^2 - 2m - 1) + (4m^2 - 6m - 2)$
(c) $(4t^2 + 7t + 6) - (3t^2 + 4t - 2)$
(d) $(3y^2 - 2y - 1) - (5y^2 - 4y + 1)$
(e) $(m^2 + 7m - 9) + (3m - 9)$
(f) $(5x^2 - 2x - 6) - (2x^2 - 6)$
(g) $(7t - 4) + (3t^2 - 2t - 1)$
(h) $(m^2 + 2) - (5m^2 + 2m - 6)$

7. Simplify.

(a) $(3x + 5) + (2x + 4) + (5x + 1)$
(b) $(6t + 1) + (4t - 3) - (t + 8)$
(c) $(4a - 3) - (7a - 6) - (6a - 9)$
(d) $(4m + 3n) - (5m - n) + (3m + 4n)$
(e) $(7x + 3y) + (x - 7y) - (8x - 7y)$
(f) $(4s - 5t) - (s + 3t) + (2s - 8t)$
(g) $(5x + 6y) - (3y + 6x) - (y - 3x)$

8. If $x = 3$, find the value of each.

(a) $(3x + 8) + (x + 7)$
(b) $(2x - 4) + (x - 9)$
(c) $(5x + 1) - (2x + 3)$
(d) $(4x - 3) - (5x - 7)$

9. If $x = 2$ and $y = 4$, find the value of the following.

(a) $(3x + 4y) + (9x + 2y)$
(b) $(x - 6y) + (6x + y)$
(c) $(5x - y) - (7x + 2y)$
(d) $(8x - 2y) - (x - y)$

10. The cost, C, of blow-up skeletons for Hallowe'en is found using the formula

$$C = 2n + 700$$

where n is the number of skeletons.
The income, I, from sales of the skeletons is found using the formula

$$I = 4n$$

(a) Find a formula for the profit if n skeletons are sold.
(b) Find the profit for selling 1200 skeletons.
(c) Find the profit for selling 100 skeletons.
(d) Is there a profit for selling 400 skeletons? 250 skeletons?

11. The cost, C, to make T-shirts, n, is given by the formula $C = 3n + 5000$. The income, I, from T-shirt sales is given by the formula $I = 20n$.

(a) Find a formula for the profit, P.
(b) Find the profit for selling 400 T-shirts.
(c) Find the profit for selling 6000 T-shirts.
(d) Is there a profit for selling 200 T-shirts?
(e) For what number of shirts is cost equal to income?

3.5 MULTIPLYING A POLYNOMIAL BY A MONOMIAL

The area of the rectangle at the right is found by multiplying the length, $2x + 7$, by the width, 3.

Separate the original rectangle into two smaller rectangles. The area of the large rectangle is equal to the sum of the areas of the two smaller rectangles.

$$\begin{aligned} 3(2x + 7) &= (3)(2x) + (3)(7) \\ &= 6x + 21 \end{aligned}$$

We get the same result using the distributive property

$$\begin{aligned} 3(2x + 7) &= 3(2x + 7) \\ &= 6x + 21 \end{aligned}$$

The area of the rectangle at the right is found by multiplying $3x + 2y$ by $4x$.

$$\begin{aligned} 4x(3x + 2y) &= 4x(3x + 2y) \\ &= 12x^2 + 8xy \end{aligned}$$

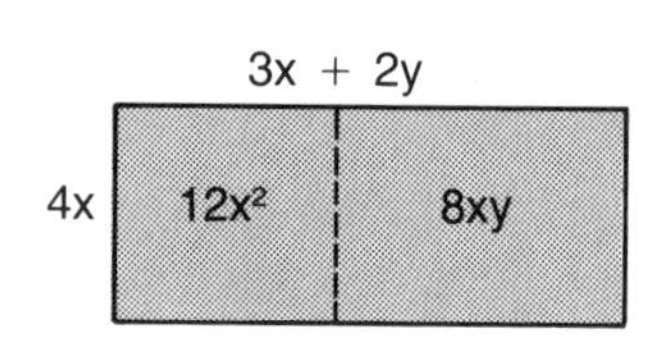

Example 1.
Expand and simplify. $2(x + 3y) + 4(2x - y)$

Solution:

$$\begin{aligned} 2(x + 3y) + 4(2x - y) &= 2(x + 3y) + 4(2x - y) \\ &= 2x + 6y + 8x - 4y \\ &= 10x + 2y \end{aligned}$$

Example 2.
Expand. (a) $-2(3x + 2)$ (b) $-(3a - 2)$

Solution:

(a) $-2(3x + 2)$

$$\begin{aligned} &= -2(3x + 2) \\ &= -6x - 4 \end{aligned}$$

(b) $-(3a - 2)$

$$\begin{aligned} &= -1(3a - 2) \\ &= -1(3a - 2) \\ &= -3a + 2 \end{aligned}$$

Example 3.
Expand and simplify. $2a(a - 3b) - 2a(4a - 5b)$

Solution:

$$\begin{aligned} 2a(a - 3b) - 2a(4a - 5b) &= 2a(a - 3b) - 2a(4a - 5b) \\ &= 2a^2 - 6ab - 8a^2 + 10ab \\ &= -6a^2 + 4ab \end{aligned}$$

EXERCISE 3.5

A 1. State each product.

(a) $2x \times 3y$
(b) $(4a^4)(5a^3)$
(c) $(-4x^2)(-2x^3)$
(d) $-3a(7a^5)$
(e) $(-x)(-5x^3)$
(f) $7m(-3n)$
(g) $-6s(-4t)$
(h) $3x \times (-5x^4)$
(i) $(-y)(-4x^2)$
(j) $-7m(-4n)$
(k) $-6a^3(5a^3)$
(l) $(3x)^2$

2. State each product.

(a) $3 \times 5xy$
(b) $-8ab(-3a^2b^2)$
(c) $(-3xy)(2xy)$
(d) $-7xy(4x^2y)$
(e) $(5ab)(-3a^3b^4)$
(f) $(-2x^3y^2)(8xy^3)$
(g) $(6s^4t)(-2s^2t^2)$
(h) $7abc(-2a^2b^2c^3)$
(i) $-5pq(-4p^2q)$
(j) $(-7ab^3)(-5a^4b^2)$
(k) $3abc(-8abc)$
(l) $-6x^2y^2z^2(-xyz)$

3. Expand.

(a) $2(x + 3)$
(b) $3(2a + 1)$
(c) $4(1 + 2b)$
(d) $5(3x^2 - 2)(-1)$
(e) $-2(x - 5)$
(f) $-6(1 - 3a)$
(g) $-(x^2 + 6x + 7)$
(h) $-3(x - y)$

B 4. Expand.

(a) $x(3x + 1)$
(b) $2a(1 + 3a)$
(c) $-2b(b^2 + 3b - 1)$
(d) $-3m(1 - 2m)$
(e) $-a(2a - 4)$
(f) $ab(1 - c)$
(g) $2x(x^2 - 2x + 1)$

5. Expand.

(a) $2(x^2 - x - 3)$
(b) $-3(a^2 - 2a - 7)$
(c) $-3x(2x^2 - 2x - 1)$
(d) $4b(b^2 + b + 2)$
(e) $-(3m^2 - m - 4)$
(f) $2x(1 - x - x^2)$

6. Expand and simplify.

(a) $2(x + 3) + 3(x + 5)$
(b) $4(2a - 3) + 3(1 + 5a)$
(c) $2(m + 2) + 3(m + 5) + 4(m - 6)$
(d) $3(d - 4) + 2(5 - 3d)$
(e) $6(1 - 2x + x^2) + (2x^2 - 6x + 5)$
(f) $5(a - 3b) + 6(2b - a) + 2(a - b)$
(g) $4(a - 2b + c) + 3(2a + 5b - 6c)$

7. Expand and simplify.

(a) $3(x - 5) - 2(x + 7)$
(b) $4(2x - 3y) - 2(x - y) + 3(x - 2y)$
(c) $5(3a - 1) - 4(2 - 3a) - (5a + 1)$
(d) $-2(x^2 - 3x - 1) - 5(2x^2 + x - 3)$
(e) $-2(m - 3) + 7m - 3(1 - 2m) + 6$
(f) $3(a - 2b + c) - 3(2b - 3a - c) + a$

8. Expand and simplify.

(a) $2x(x - 7) - 3x(2x + 1)$
(b) $a(2a - 3) - 3a(1 - 2a) - a(3a - 2)$
(c) $2x(x^2 - 2x + 1) - 3(x - 2) - x(1 - 3x)$
(d) $3b(2b - 7) - b(1 + 3b) - 2b(b + 3)$
(e) $2a(a - 3) - 4(a^2 - 2a - 2) - a(1 - 3a)$
(f) $3b(a - b) - 2a(2a - 3b) - b(a - 3b)$

9. Find an expression for the perimeter of each figure.

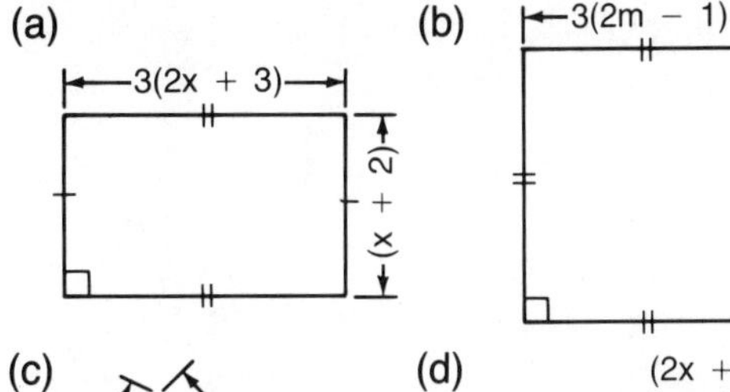

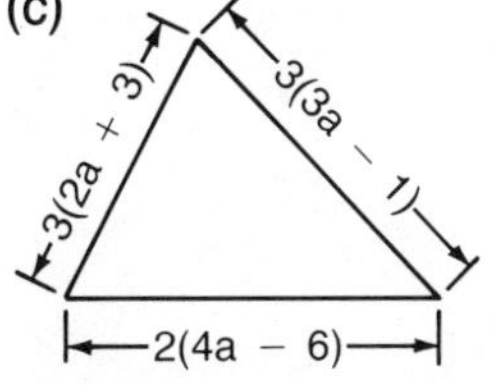

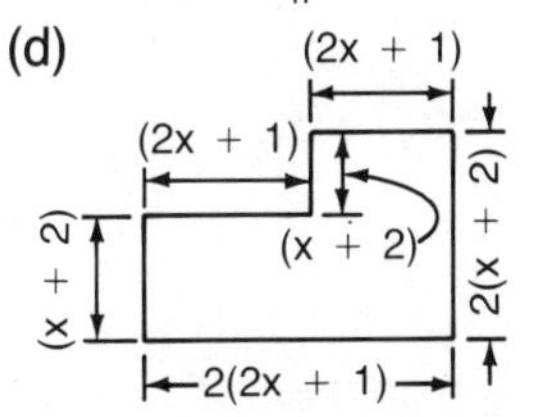

10. Find an expression for the area of each figure.

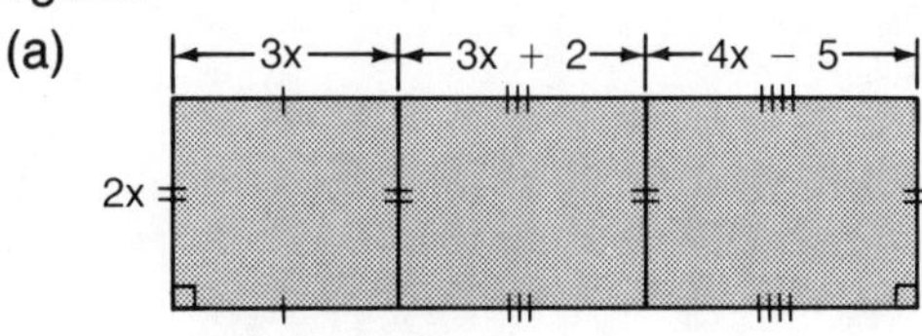

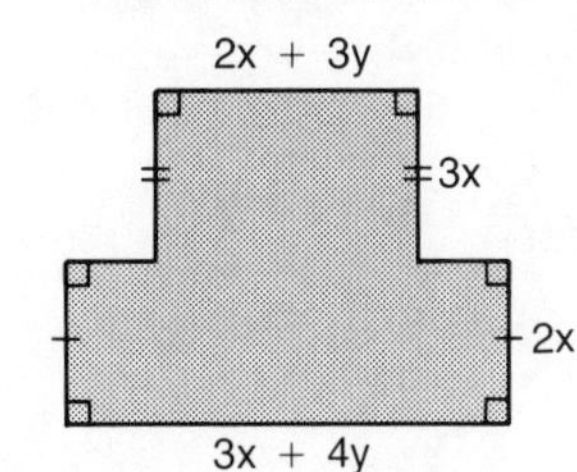

3.6 SOLVING EQUATIONS

Problems often arise that require you first to write an equation or formula to represent the situation.
Then use your algebra skills to solve the equation.

For example, Carl went shopping and spent \$135.60. He had \$76.25 left.
How much money did he have before shopping?

If Carl started with \$m and spent \$135.60, then the expression for what he has left is $m - 135.60$

We are told he has \$76.25 left.
Therefore, the equation to solve is $m - 135.6 = 76.25$

Adding 135.60 to each side gives

$$\begin{aligned} m - 135.60 + 135.60 &= 76.25 + 135.60 \\ m &= 211.85 \end{aligned}$$

Carl started with \$211.85.

Example 1.
Solve and check. $7x - 8 = 4x + 10$

Solution:

$$\begin{aligned} 7x - 8 &= 4x + 10 \\ 7x - 8 + 8 &= 4x + 10 + 8 \quad \longleftarrow \text{Add 8 to each side.} \\ 7x &= 4x + 18 \\ 7x - 4x &= 4x + 18 - 4x \quad \longleftarrow \text{Subtract 4x from each side.} \\ 3x &= 18 \\ \frac{3x}{3} &= \frac{18}{3} \quad \longleftarrow \text{Divide each side by 3.} \\ x &= 6 \end{aligned}$$

Check: $x = 6$

$$\begin{aligned} \text{L.S.} &= 7x - 8 \\ &= 7(6) - 8 \\ &= 42 - 8 \\ &= 34 \end{aligned}$$

$$\begin{aligned} \text{R.S.} &= 4x + 10 \\ &= 4(6) + 10 \\ &= 24 + 10 \\ &= 34 \end{aligned}$$

The solution is $x = 6$.
The number 6 is also called a root of the equation.

Example 2.
Solve and check. $2(x + 1) + 3 = 19$

Solution:

$$\begin{aligned} 2(x + 1) + 3 &= 19 \\ 2x + 2 + 3 &= 19 \quad \longleftarrow \text{Remove brackets.} \\ 2x + 5 &= 19 \\ 2x + 5 - 5 &= 19 - 5 \quad \longleftarrow \text{Subtract 5 from each side.} \\ 2x &= 14 \\ \frac{2x}{2} &= \frac{14}{2} \quad \longleftarrow \text{Divide each side by 2.} \\ x &= 7 \end{aligned}$$

Check: $x = 7$

$$\begin{aligned} \text{L.S.} &= 2(x + 1) + 3 \\ &= 2(7 + 1) + 3 \\ &= 2(8) + 3 \\ &= 16 + 3 \\ &= 19 \end{aligned}$$

$$\text{R.S.} = 19$$

The solution is $x = 7$.

Example 3.
Solve and check. $2(2x - 1) - (x + 3) = x + 1$

Solution:

$$\begin{aligned}
2(2x - 1) - (x + 3) &= x + 1 \\
4x - 2 - x - 3 &= x + 1 \\
3x - 5 &= x + 1 \\
3x - 5 + 5 &= x + 1 + 5 \\
3x &= x + 6 \\
3x - x &= x - x + 6 \\
2x &= 6 \\
\frac{2x}{2} &= \frac{6}{2} \\
x &= 3
\end{aligned}$$

$\therefore$ The root of the equation is 3.

Check: $x = 3$

$$\begin{aligned}
\text{L.S.} &= 2(2x - 1) - (x + 3) \\
&= 2(6 - 1) - (3 + 3) \\
&= 10 - 6 \\
&= 4
\end{aligned}$$

$$\begin{aligned}
\text{R.S.} &= x + 1 \\
&= 3 + 1 \\
&= 4
\end{aligned}$$

EXERCISE 3.6

A

1. Solve.

(a) $x + 3 = 7$
(b) $a + 2 = 5$
(c) $m - 3 = 6$
(d) $t - 5 = 12$
(e) $3 + r = 11$
(f) $x - 4 = -11$
(g) $b + 3 = -4$
(h) $-2 + m = 0$

2. Solve.

(a) $2x = 6$
(b) $3m = 27$
(c) $4a = -32$
(d) $-2a = 6$
(e) $-3a = -18$
(f) $5d = 0$
(g) $-10t = -70$
(h) $-x = 3$

B

3. Solve and check.

(a) $2x + 3 = 7$
(b) $3a + 7 = 2a - 4$
(c) $5b = 7b - 4$
(d) $4t - 7t + 3 - 8 + 2t = 0$
(e) $5m - 7m + 3 = 4m + 6$
(f) $5x + 1 = 3x + 7$
(g) $3t + 2t - 7 = 8$
(h) $4m - 5m = 6 - 3$
(i) $-2s + s - 4 = 0$

4. Solve and check.

(a) $2(x - 1) = 6$
(b) $3(t + 4) = 0$
(c) $5 - (m - 1) = 4$
(d) $5(x - 3) - 2 = 7$
(e) $8 = 5(y - 4) - 2$

5. Solve and check.

(a) $2(x - 1) = x + 6$
(b) $3(a + 1) + 4 = 2(a - 1)$
(c) $2(m + 3) + 3(m - 1) = 3(m - 2) + 1$
(d) $3(2b - 1) - 4 + 2(b + 3) = 7$

6. Solve.

(a) $3 + 2(2x - 1) + 6 = 5(x - 3)$
(b) $7 - 2(1 - 3a) + 16 = 8a + 11$
(c) $10(2m - 7) - 3(m + 2) = 20m - 70$
(d) $(2t + 3) - (1 - 3t) - 4t = 3t$

7. Solve.

(a) $3x - 2(x - 3) + 6 = 4x - (x - 1)$
(b) $4a - 3 = 2(a - 3) - 4(2a - 1)$
(c) $2m - 6 = -(1 - 3m) + m$
(d) $7t - 3(2t - 4) = -3t + 6(1 - 4t) - 4$

8. Tanya had $158.60 left in her chequing account after she wrote a cheque for $46.73.
Write an equation to determine how much she had in her account before she wrote the cheque. Solve the equation.

9. After making a deposit of $95.50, Anthony had $238.65 in his savings account.
Write an equation to determine how much he had in the account before the deposit. Solve the equation.

3.7 SOLVING EQUATIONS USING MULTIPLICATION

When purchasing a car, it is important to know how much gasoline it uses, or the fuel consumption. The consumption is expressed as a ratio which tells you how many litres of gasoline the car will consume to travel 100 km. A fuel consumption of 14 L/100 km means a car will consume 14 L of gasoline for every 100 km.

A test car burned 12.4 L of fuel in travelling 80 km. To find the number of litres, n, it would use to travel 100 km, we must solve the equation

$$\frac{12.4}{80} = \frac{n}{100}$$

To solve equations such as this one, we first eliminate the fractions.

Example 1.

Solve and check. $\frac{x}{2} - \frac{x}{3} = 4$

Solution:

$$\frac{x}{2} - \frac{x}{3} = 4$$

6 is a common denominator of 2 and 3. Multiply each side of the equation by 6.

$$6 \times \left[\frac{x}{2} - \frac{x}{3}\right] = 6 \times 4$$

$$6 \times \frac{x}{2} - 6 \times \frac{x}{3} = 6 \times 4$$

$$3x - 2x = 24$$

$$x = 24$$

Check:

$$\text{L.S.} = \frac{x}{2} - \frac{x}{3} = \frac{24}{2} - \frac{24}{3} = 12 - 8 = 4$$

R.S. = 4

$\therefore x = 24$ is the solution.

Example 2.

Solve. $\frac{4x - 2}{5} - \frac{3x - 1}{2} = 5$

Solution:

$$\frac{4x - 2}{5} - \frac{3x - 1}{2} = 5$$

← 10 is a common denominator of 5 and 2.

$$10 \times \left[\frac{(4x - 2)}{5} - \frac{(3x - 1)}{2}\right] = 10 \times 5$$

← Multiply each side of the equation by 10.

$$2(4x - 2) - 5(3x - 1) = 50$$

$$8x - 4 - 15x + 5 = 50$$

$$-7x + 1 = 50$$

$$-7x + 1 - 1 = 50 - 1$$

$$-7x = 49$$

$$\frac{-7x}{-7} = \frac{49}{-7}$$

← Divide by -7.

$$x = -7$$

To find the amount of fuel a test car consumes to travel 100 km, we solve the equation

$$\frac{12.4}{80} = \frac{n}{100}$$

400 is a common denominator of 80 and 100.

$$400 \times \left(\frac{12.4}{80}\right) = 400 \times \left(\frac{n}{100}\right)$$

$$5 \times 12.4 = 4 \times n$$

$$62 = 4n$$

$$15.5 = n$$

5 × 12 · 4 ÷ 4 =

The display is 15.5

The car's fuel consumption is 15.5 L/100 km.

EXERCISE 3.7

A 1. State a common denominator for each pair of fractions.

(a) $\frac{1}{3}, \frac{1}{4}$ (b) $\frac{1}{2}, \frac{1}{6}$ (c) $\frac{1}{5}, \frac{3}{4}$

(d) $\frac{5}{6}, \frac{2}{5}$ (e) $\frac{1}{2}, \frac{3}{7}$ (f) $\frac{1}{4}, \frac{7}{10}$

(g) $\frac{1}{8}, \frac{2}{3}$ (h) $\frac{4}{5}, \frac{2}{3}$ (i) $\frac{17}{20}, \frac{1}{4}$

2. By what number would you multiply each side to clear the fractions?

(a) $\frac{x}{3} - \frac{x}{5} = 2$ (b) $\frac{m}{4} = \frac{m}{2} + 5$

(c) $\frac{s}{3} + \frac{s}{4} = 1$ (d) $\frac{3}{7} + \frac{r}{5} = 1$

(e) $\frac{x + 1}{2} = \frac{2x - 1}{6}$ (f) $\frac{x}{15} = \frac{1}{3} + \frac{x}{5}$

B 3. Solve.

(a) $\frac{x}{2} = 6$ (b) $\frac{m}{3} = 5$

(c) $\frac{t}{4} = 1$ (d) $\frac{s}{5} = 3$

(e) $\frac{y}{2} = 4$ (f) $\frac{x}{7} = 2$

4. Solve and check.

(a) $\frac{x}{2} = \frac{5}{3}$ (b) $\frac{2a}{3} = \frac{3}{2}$

(c) $\frac{2m}{3} + \frac{1}{2} = \frac{m}{2}$ (d) $\frac{b}{2} + 1 = \frac{b}{3}$

(e) $\frac{x + 1}{2} = 4$ (f) $\frac{x - 2}{3} = 5$

5. Solve and check.

(a) $\frac{2m - 1}{3} = 5$

(b) $\frac{x - 2}{3} - 5 = 4$

(c) $\frac{x + 3}{2} - 1 = 0$

(d) $\frac{a + 2}{2} = \frac{a - 1}{3}$

(e) $\frac{t + 7}{4} = \frac{2t - 1}{3}$

(f) $\frac{b}{3} = \frac{2b - 1}{2}$

6. Solve and check.

(a) $\frac{x + 1}{2} + \frac{x + 1}{3} = 5$

(b) $\frac{2m + 1}{3} - \frac{m + 1}{4} = 3$

(c) $\frac{1 - 3b}{4} - \frac{2 + b}{6} = \frac{1}{3}$

(d) $\frac{4t - 3}{2} - \frac{1}{4} = \frac{t}{8}$

(e) $\frac{1 - 3a}{4} - \frac{a - 1}{3} = -a$

7. A car consumed 11.2 L of gasoline to travel 80 km.
Find the car's fuel consumption in L/100 km.

8. A car consumed 16.2 L of gasoline to travel 120 km.
Find the car's fuel consumption in L/100 km.

3.8 SOLVING PROBLEMS USING EQUATIONS

In this section we will use equations to simplify and solve problems. When solving problems, we translate language into mathematics.

Jennifer deposited \$127.00 in the bank. Her balance after the deposit is \$445.70.
How much did she have in her account before the deposit?

Jennifer started with \$x and deposited \$127.00. The expression is $x + 127.00$
Since the balance is \$445.70, the equation is

$$x + 127.00 = 445.70$$

and $$x + 127.00 - 127.00 = 445.70 - 127.00$$

$$x = 318.70$$

Jennifer originally had \$318.70 in her account.

READ

Example 1.
There are 8 more dolphins than sharks and 66 dolphins and sharks altogether at the aquarium.
How many sharks and dolphins are there?

PLAN

Solution:
Let the number of sharks be n. Then the number of dolphins is $n + 8$. Since there are 66 altogether, then

SOLVE

$$n + n + 8 = 66$$
$$2n + 8 = 66$$
$$2n + 8 - 8 = 66 - 8$$
$$2n = 58$$
$$\frac{2n}{2} = \frac{58}{2}$$
$$n = 29$$

Check: Since $37 - 29 = 8$, there are 8 more dolphins than sharks. Also, $29 + 37 = 66$

ANSWER

There are 29 sharks and 37 dolphins.

Example 2.
Simone needs \$1350 so she can attend a summer basketball camp. She has \$200 in the bank, and ten months to save the rest.
How much should she save each month?

Solution:
Let x be the amount Simone should save each month. Then she will save 10x in ten months.
The amount of \$200 and her savings must equal \$1350.

$$10x + 200 = 1350$$
$$10x + 200 - 200 = 1350 - 200$$
$$10x = 1150$$
$$\frac{10x}{10} = \frac{1150}{10}$$
$$x = 115$$

Simone should save \$115 each month.

Check:

$115 \times 10 =$	1150
Plus	200
Total	1350

EXERCISE 3.8

Write an equation to represent each statement.

B

1. A number increased by eight is twenty.
2. A number decreased by four is thirty.
3. Twelve times a number is seventy-two.
4. A number divided by four is eleven.
5. Alan is two years older than Tom. The sum of their ages is thirty-two.
6. Two consecutive integers have a sum of sixty-five.
7. The sum of eight and three times a number is forty-one.
8. Melinda ran twice as far as Zee. Together they ran a distance of seventy-eight kilometres.

Solve the following problems.

9. A number increased by 11 is 52.
Find the number.
10. A number decreased by 14 is 45.
Find the number.
11. Six times a number is 108.
Find the number.
12. A number divided by 7 is 21.
Find the number.
13. There are 15 more cows than horses on the Lee farm. Altogether there are 57 cows and horses.
How many of each are there?
14. Nizar paid off a bill of $585 in three instalments. The second payment was $50 more than the first. The third payment was $85 more than the first.
How much was each payment?
15. The sum of three consecutive whole numbers is 246.
Find the numbers.
16. Frank wrote a cheque for $34.75. This left a balance of $234.64 in his account.
How much was in his account before he wrote the cheque?
17. The perimeter of a square field is 232 km.
What is the length of one side?
18. Paul earns $500 a week for selling swimming pools. He earns a bonus of $75 for every pool he sells.
How many pools must he sell in a week to earn $1100?

19. The Blue Jays won 14 more games than they lost. They played a total of 162 games.
How many games did they win?
20. Marco had 4 new tires installed on his car. The tires cost $92 each. The total bill was $401.
What was the installation charge?
21. The length of a rectangular patio is 5 m longer than the width. The perimeter of the patio is 94 m.
Find the length and width.
22. The Aiellos plan to have their son's seventh birthday party at a hotel so the children can use the swimming pool. The hotel charges $80 for the room plus $10 for each child.
How many children can attend if the Aiellos have $250 to spend?
23. A board is 4.5 m long. It is cut into 2 pieces. One piece is 4 times as long as the other.
How long is each piece?

3.9 MULTIPLYING BINOMIALS

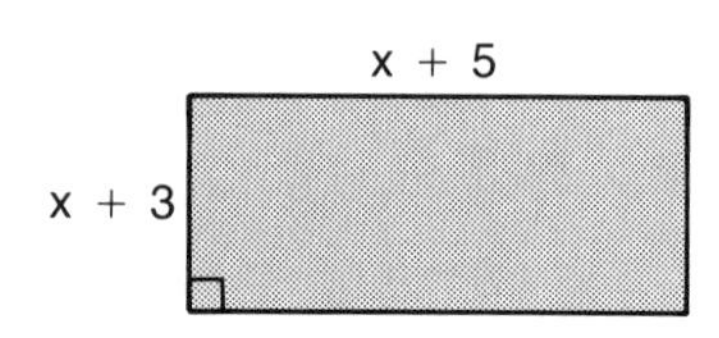

A binomial is a polynomial with two terms.
The length of the rectangle at the right is $(x + 5)$
The width is $(x + 3)$
We can find the area of the rectangle by multiplying the length by the width: $(x + 5) \times (x + 3)$

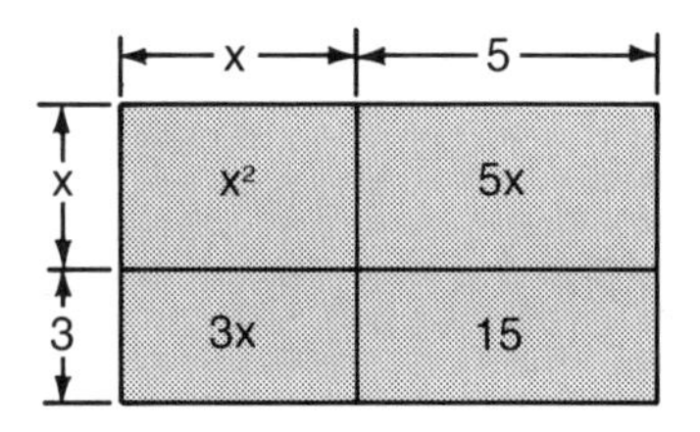

We can divide the original rectangle into four smaller rectangles. The area of the original rectangle is equal to the sum of the areas of the four smaller rectangles.

$$\begin{aligned}(x + 5)(x + 3) &= x^2 + 5x + 3x + 15 \\ &= x^2 + 8x + 15\end{aligned}$$

We can get the same result using the distributive property.

$$\begin{aligned}(x + 5)(x + 3) &= (x + 5)(x + 3) \\ &= (x + 5)x + (x + 5)3 \\ &= x^2 + 5x + 3x + 15 \\ &= x^2 + 8x + 15\end{aligned}$$

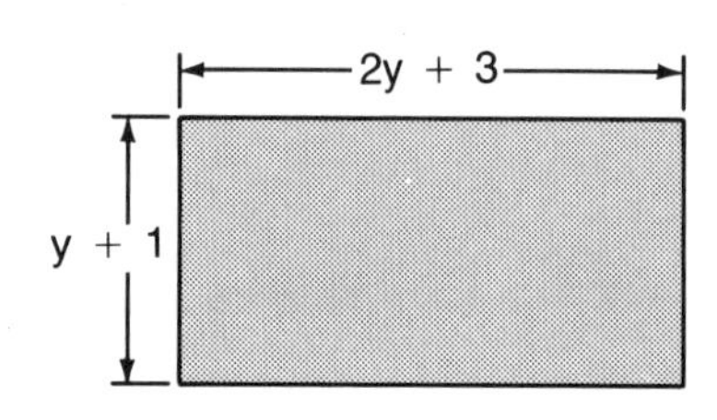

Another way to multiply two binomials is to multiply each term in the first binomial by each term in the second binomial.
For example,

$$\begin{aligned}(2y + 3)(y + 1) &= (2y + 3)(y + 1) \\ &= 2y^2 + 2y + 3y + 3 \\ &= 2y^2 + 5y + 3\end{aligned}$$

2y 3
y 2y² 3y
1 2y 3

Using the word FOIL as follows can help you remember the process.

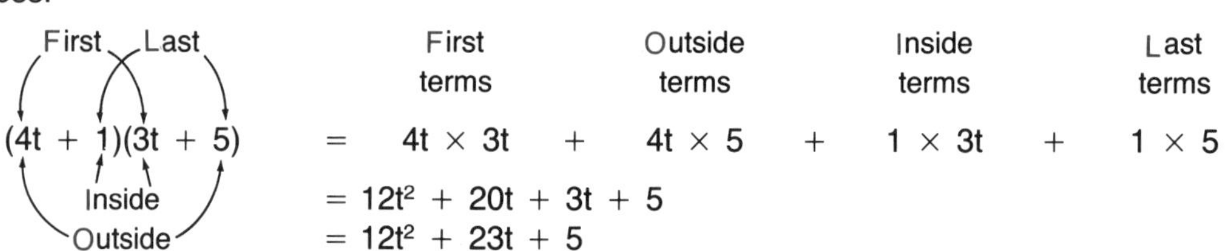

$$\begin{aligned}&= 12t^2 + 20t + 3t + 5 \\ &= 12t^2 + 23t + 5\end{aligned}$$

Example 1.
Find the product. $(4x - 3y)(2x - 7y)$

Solution:

$$\begin{aligned}(4x - 3y)(2x - 7y) &= (4x - 3y)(2x - 7y) \\ &= 8x^2 - 28xy - 6xy + 21y^2 \\ &= 8x^2 - 34xy + 21y^2\end{aligned}$$

Example 2.

Find the products. (a) $(4a - 3)^2$ (b) $(3s - 1)(3s + 1)$

Solution:

(a) $(4a - 3)^2 = (4a - 3)(4a - 3)$
$= 16a^2 - 12a - 12a + 9$
$= 16a^2 - 24a + 9$

(b) $(3s - 1)(3s + 1) = (3s - 1)(3s + 1)$
$= 9s^2 + 3s - 3s - 1$
$= 9s^2 - 1$

EXERCISE 3.9

A 1. Find each product.

(a) $(x + 3)(x + 1)$ (b) $(a + 5)(a + 7)$
(c) $(m + 3)(m + 6)$ (d) $(t + 11)(t + 3)$
(e) $(d + 6)(d + 7)$ (f) $(4 + m)(2 + m)$
(g) $(r + 6)(r + 10)$ (h) $(1 + s)(s + 12)$

2. Find each product.

(a) $(m - 3)(m + 2)$ (b) $(x - 5)(x - 6)$
(c) $(a - 5)(a + 6)$ (d) $(b - 3)(b + 6)$
(e) $(y - 3)(y - 3)$ (f) $(m - 10)(m + 6)$
(g) $(d - 3)(d + 3)$ (h) $(s - 7)(s - 8)$

3. Find each product.

(a) $(x + 3)^2$ (b) $(y - 2)^2$
(c) $(t + 4)^2$ (d) $(a - 5)^2$
(e) $(s + 5)^2$ (f) $(x + 7)^2$
(g) $(m - 1)^2$ (h) $(n + 1)^2$

B 4. Find each product.

(a) $(2x + 1)(x + 6)$ (b) $(3a - 4)(2a - 1)$
(c) $(2m - 7)(m + 2)$ (d) $(3x - 1)(3x + 1)$
(e) $(2b - 3)(2b - 3)$ (f) $(d + 6)(2d - 5)$
(g) $(1 - 3r)(5 + 2r)$ (h) $(4r + 5)(r - 3)$
(i) $(3t - 5)(2t + 3)$ (j) $(1 - 3x)(5x - 7)$

5. Find each product.

(a) $(7x - 5)(x - 2)$ (b) $(3m + 7)(m - 2)$
(c) $(1 - 5b)(1 + 5b)$ (d) $(3x - y)(3x + y)$
(e) $(2a + b)(3a + b)$ (f) $(a + b)(a + b)$
(g) $(3x - 2y)(x - y)$
(h) $(2b + c)(3b - 2c)$

6. Find each product.

(a) $(3m + 2)^2$ (b) $(5a + 1)^2$
(c) $(3a - 2)^2$ (d) $(4b + 2)^2$
(e) $(2x - 7)^2$ (f) $(6x - 1)^2$
(g) $(2y + 3)^2$ (h) $(5t - 3)^2$
(i) $(6m + 1)^2$ (j) $(1 - 2r)^2$
(k) $(1 + 6a)^2$ (l) $(2s - 1)^2$

7. Find an expression for each area and simplify.

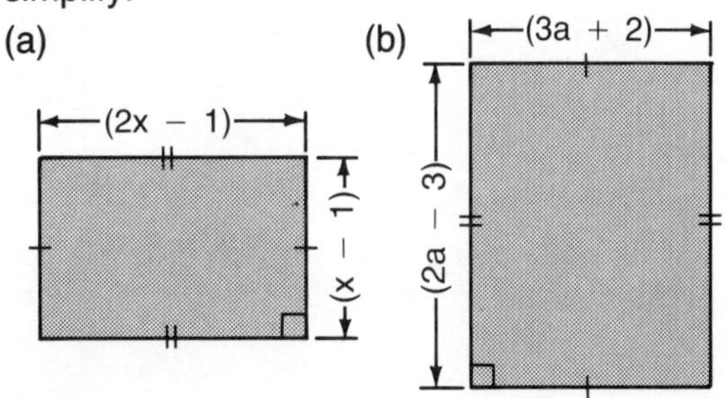

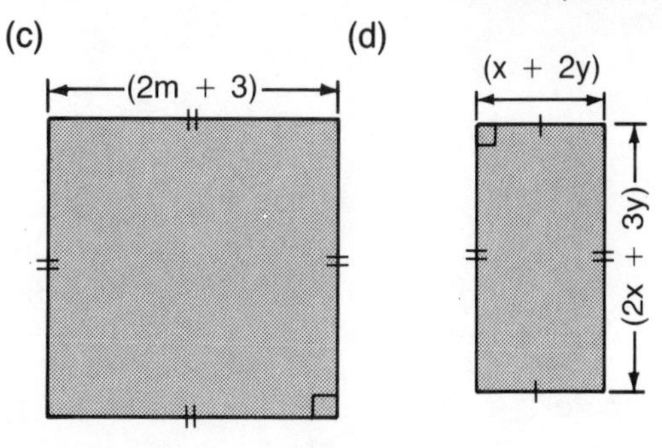

C 8. Expand and simplify.

(a) $(x + 3)(x^2 - 2x + 2)$

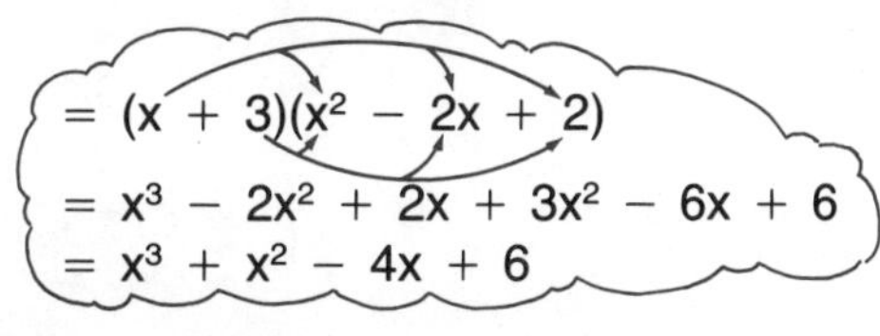

(b) $(x + 2)(x^2 + 3x + 3)$
(c) $(m - 2)(2m^2 - m - 1)$
(d) $(s - 3)(2s^2 + 3s - 3)$
(e) $(2x - 1)(x^2 - 3x - 4)$
(f) $(2b + 3)(2b^2 - 3b + 6)$
(g) $(2a^2 - 3a + 2)(3a + 5)$
(h) $(x^2 - 2x - 1)(x^2 + 3x + 1)$
(i) $(2m^2 - 3m + 1)(m^2 + m - 3)$

3.10 SIMPLIFYING EXPRESSIONS

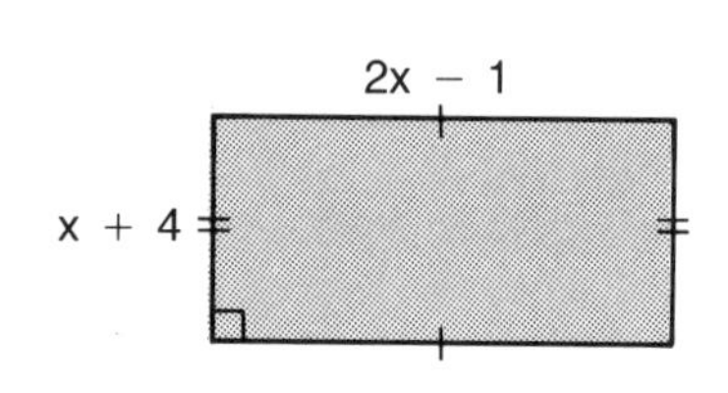

The area of the rectangle can be expressed as

$(2x - 1)(x + 4)$

Expanding, we have

$$\begin{aligned}(2x - 1)(x + 4) &= (2x - 1)(x + 4)\\ &= 2x^2 + 8x - x - 4\\ &= 2x^2 + 7x - 4\end{aligned}$$

To expand an expression, such as $3(t + 1)(2t - 5)$, you multiply the binomials first.

$$\begin{aligned}3(t + 1)(2t - 5) &= 3(t + 1)(2t - 5)\\ &= 3(2t^2 - 5t + 2t - 5)\\ &= 3(2t^2 - 3t - 5)\\ &= 6t^2 - 9t - 15\end{aligned}$$

We use the distributive property and the rules for addition and subtraction to simplify polynomial expressions.

Example 1.

Simplify. $(x + 2)(x - 3) - 2(x - 7)$

Solution:

$$\begin{aligned}(x + 2)(x - 3) - 2(x - 7) &= (x + 2)(x - 3) - 2(x - 7)\\ &= (x^2 - 3x + 2x - 6) - 2x + 14\\ &= (x^2 - x - 6) - 2x + 14\\ &= x^2 - x - 6 - 2x + 14\\ &= x^2 - 3x + 8\end{aligned}$$

Example 2.

Simplify. $2(3a - 1)(a + 2) - (2a + 3)^2$

Solution:

$$\begin{aligned}2(3a - 1)(a + 2) - (2a + 3)^2 &= 2(3a - 1)(a + 2) - (2a + 3)(2a + 3)\\ &= 2(3a^2 + 6a - a - 2) - (4a^2 + 6a + 6a + 9)\\ &= 2(3a^2 + 5a - 2) - (4a^2 + 12a + 9)\\ &= 6a^2 + 10a - 4 - 4a^2 - 12a - 9\\ &= 2a^2 - 2a - 13\end{aligned}$$

EXERCISE 3.10

A 1. Simplify.

(a) $2(x + 2)$ (b) $3(2x - 1)$
(c) $(a - 7)$ (d) $2(1 - 3c)$
(e) $-4(2x - 1)$ (f) $-(x - 7)$
(g) $-3(2x + 4y)$ (h) $2x(x - 4)$
(i) $4(a - 5)$ (j) $-y(2y + 3)$

B 2. Expand and simplify.

(a) $3(x + 2) + 4(x + 5)$
(b) $4(a - 3) - 5(a + 7)$
(c) $8(t - 7) - (2t + 4)$
(d) $5(2x - 1) - 3(2x + 5)$
(e) $3(4m - 1) - (2m + 3) + 2(m + 4)$

3. Expand.

(a) $(x + 3)(x + 4)$ (b) $(x - 5)(x - 6)$
(c) $(t + 8)(t - 6)$ (d) $(a + 7)(a - 2)$
(e) $(m - 3)(m - 5)$ (f) $(y - 1)(y - 6)$
(g) $(w + 2)^2$ (h) $(a - 5)^2$

4. Expand.

(a) $(3x + 2)(x + 1)$ (b) $(4y - 5)(y - 5)$
(c) $(4y + 7)(2y + 1)$ (d) $(5m + 3)(m + 4)$
(e) $(3a - 1)(3a + 1)$ (f) $(6s + 5)(2s + 3)$
(g) $(6x - 7)(2x - 1)$ (h) $(2m + 3)(m + 8)$

5. Expand.

(a) $3(x + 4)(x + 5)$
(b) $2(a - 1)(a - 2)$
(c) $4(3b - 1)(b + 2)$
(d) $5(m + 2)(3m + 1)$
(e) $-2(x - 6)(x - 2)$
(f) $-3(x + 1)(x + 5)$

6. Simplify.

(a) $(a + 7)(a - 2) + (a - 1)(a + 2)$
(b) $(x - 3)(x - 6) + (x + 2)^2$
(c) $(2m - 1)(m + 3) - (m - 4)(m + 2)$
(d) $3(2d - 7) + (2d + 3)(d - 7)$
(e) $(b - 4)^2 - (b + 3)^2$
(f) $2(m - 6) - (3m + 1)^2 + 6$

7. Simplify.

(a) $2(x - 3)(x + 2) + 3(2x - 1)$
(b) $3(2a + 1)(a - 2) - 4(3a + 6)$
(c) $2(t + 5)^2 - 3(2t + 1)(t + 5)$
(d) $3(2b - 1)(3b + 2) - (3b - 1)^2$
(e) $2(x - 3)(x + 3) - 3(x - 7)^2$

8. Simplify.

(a) $2(x + 3)(2x - 1) - 3(2x - 4) + 7$
(b) $(2m - 1)^2 + 3(m - 2)(2m + 1) + 6$
(c) $(1 - 2x)(1 + 3x) - (1 - 2x)^2 + 3x$
(d) $4(3x - 1) - (2x + 1)^2 + (2x - 1)(2x + 1)$
(e) $2(3m^2 - 2m + 1) - (1 - 3m) + (m + 1)^2$
(f) $3(1 - 2b) + 2(b - 3)^2 - (2b - 1)(b + 3)$

9. Expand and simplify.

(a) $(x + 3)^2 + (x - 2)^2$
(b) $(a - 3)^2 + (a + 5)^2$
(c) $(2x + 1)^2 + (3x - 1)^2$
(d) $(m + 7)^2 + (2m - 3)^2$
(e) $(2b + 5)^2 + (3b - 1)^2$
(f) $(x + 3)^2 + (x + 1)^2 + (x + 2)^2$

10. The area of a ring is found using the formula

$$A = \pi(R + r)(R - r)$$

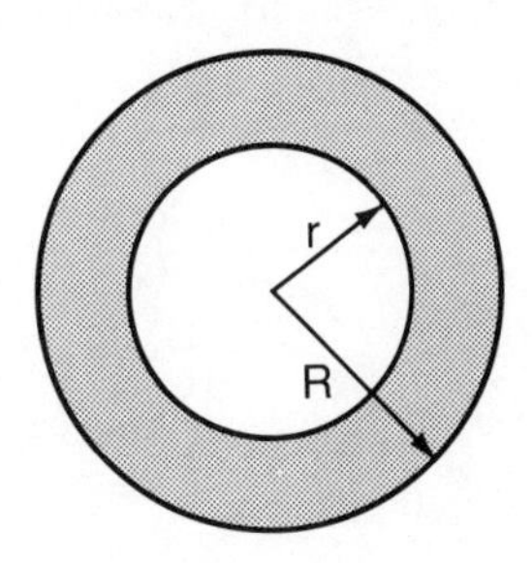

Find the area of a ring with inside radius 3 cm and outside radius 7 cm.
(Use $\pi = 3.14$)

MIND BENDER

What is the name of the first person in each of these famous pairs?

▬▬ and Hardy
▬▬ and Costello
▬▬ and Garfunkel
▬▬ and Wagnall
▬▬ and Bailey

3.11 DIVIDING POLYNOMIALS BY MONOMIALS

We use the exponent rule for division to divide a monomial by a monomial.

Example 1.

Divide. (a) $\frac{6x^3}{3x^2}$ (b) $\frac{-24s^4t^2}{-8s^2t}$

Solution:

(a) $\frac{6x^3}{3x^2} = 2x^{3-2}$

$= 2x$

(b) $\frac{-24s^4t^2}{-8s^2t} = 3s^{4-2}t^{2-1}$

$= 3s^2t$

To divide a polynomial by a monomial we use the distributive property. The expression $2x(x + 3)$ means that each term in the brackets is multiplied by $2x$.
For example,

$$2x(x + 3) = 2x(x + 3)$$
$$= 2x^2 + 6x$$

In a division, such as $\frac{6x^2 + 8x}{2x}$, the division line indicates that the terms above and below the line can be grouped using brackets. So,

$$\frac{6x^2 + 8x}{2x} = \frac{(6x^2 + 8x)}{2x}$$

The expression $\frac{(6x^2 + 8x)}{2x}$ means that each term in the brackets is divided by $2x$.

$$\frac{(6x^2 + 8x)}{2x} = \frac{6x^2}{2x} + \frac{8x}{2x}$$
$$= 3x + 4$$

Example 2.

Simplify. (a) $\frac{9x^3 + 12}{3}$ (b) $\frac{4ab - 8b}{4b}$

Solution:

(a) $\frac{9x^3 + 12}{3} = \frac{9x^3}{3} + \frac{12}{3}$

$= 3x^3 + 4$

(b) $\frac{4ab - 8b}{4b} = \frac{4ab}{4b} - \frac{8b}{4b}$

$= a - 2$

Example 3.

Simplify. $\frac{15t^4 - 10t^3 + 25t^2}{5t}$

Solution:

$$\frac{15t^4 - 10t^3 + 25t^2}{5t} = \frac{15t^4}{5t} - \frac{10t^3}{5t} + \frac{25t^2}{5t}$$
$$= 3t^3 - 2t^2 + 5t$$

Example 4.
The area of the rectangle is $14x^3y^2 + 21x^2y^3$.
The width is $7xy$.
Find the length.

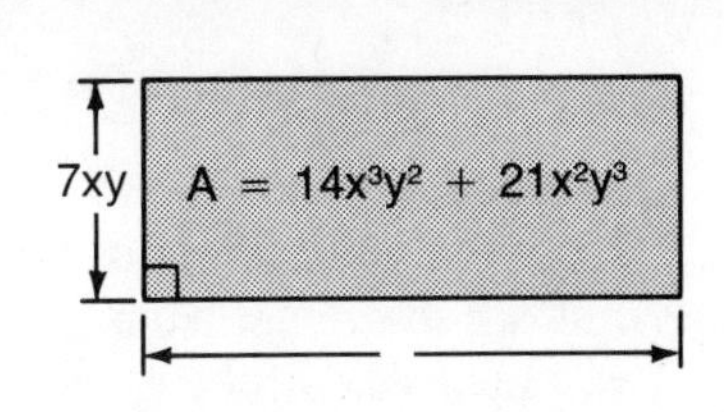

Solution:
For a rectangle, Area = length × width

$$A = \ell \times w$$

$$\text{and } \ell = \frac{A}{w}$$

$$\ell = \frac{14x^3y^2 + 21x^2y^3}{7xy}$$

$$= \frac{14x^3y^2}{7xy} + \frac{21x^2y^3}{7xy}$$

$$= 2x^2y + 3xy^2$$

EXERCISE 3.11

1. Simplify.

(a) $\frac{4ab}{2}$ (b) $\frac{6xyt}{6}$ (c) $\frac{10ab}{a}$

(d) $\frac{12abc}{6a}$ (e) $\frac{7mnt}{7m}$ (f) $\frac{6dem}{-3}$

2. Divide.

(a) $\frac{6x^2y}{6x}$ (b) $\frac{3a^2b}{a^2}$ (c) $\frac{6x^2y}{3xy}$

(d) $\frac{-12abc^2}{3abc}$ (e) $\frac{3m^2n^2}{3mn}$ (f) $\frac{x^6}{x^2}$

(g) $\frac{18x^2m^2n}{-3x^2m}$ (h) $\frac{12x^6y^4}{3x^2y^3}$ (i) $\frac{21a^4b^2c}{-7a^2bc}$

3. Simplify.

(a) $\frac{8x + 6}{2}$ (b) $\frac{7a - 14}{7}$

(c) $\frac{3m + 3}{3}$ (d) $\frac{2ab - 6b}{2}$

(e) $\frac{10a - 5b}{5}$ (f) $\frac{12x + 8}{-4}$

4. Simplify.

(a) $\frac{6x^2 - 4x}{2x}$ (b) $\frac{4mn + 6mt}{2m}$

(c) $\frac{10m^3 - 5m}{5m}$ (d) $\frac{9y^2 - 6y}{3y}$

(e) $\frac{8r^4 + 16r^2}{4r}$ (f) $\frac{3t^2 + 5t}{t}$

5. Simplify.

(a) $\frac{4x^3 - 6x^2 + 8x}{2x}$

(b) $\frac{16 + 4a - 12a^2}{4}$

(c) $\frac{3ab - 6ab^2 + 9a^2b}{3ab}$

(d) $\frac{-7t + 14t^2 - 21t^3}{7t}$

6. Find the unknown dimension in each rectangle.

(a)

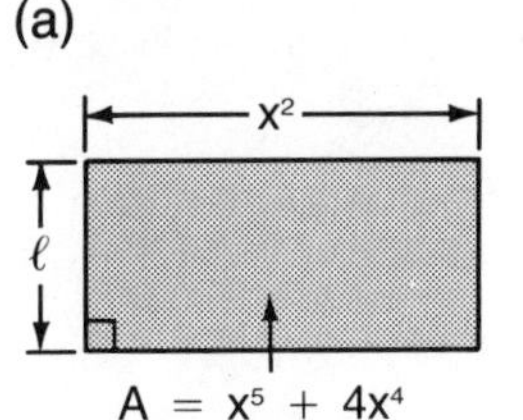

(b)

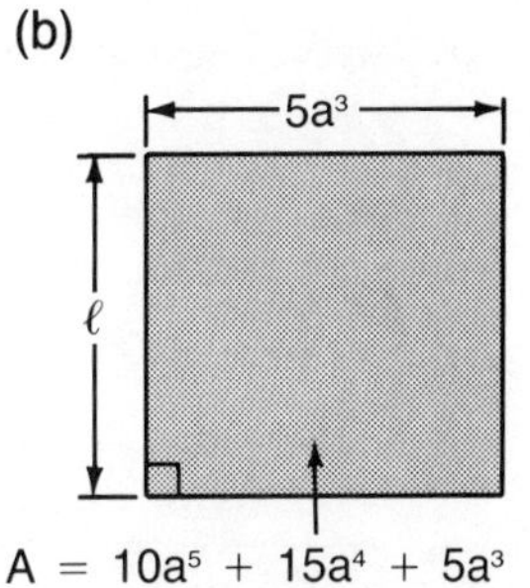

(c)

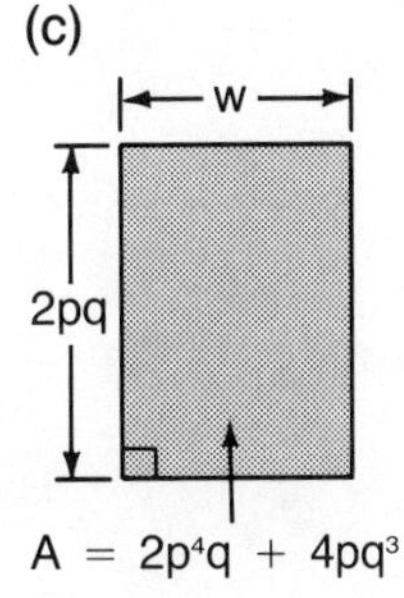

(d)

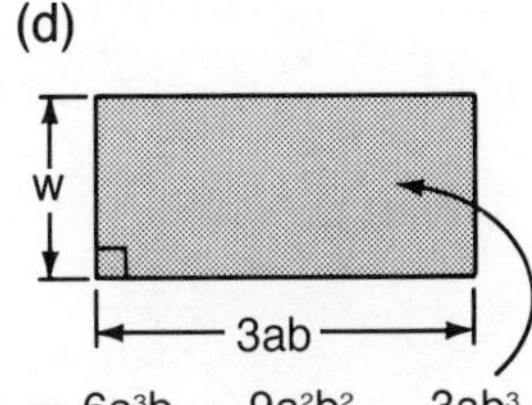

3.12 COMMON FACTORING AND POLYNOMIALS

Suppose a sunroom floor requires 24 m^2 of carpet.
What are the dimensions of the floor?
Two possibilities are:

8 m by 3 m → 24 m^2
6 m by 4 m → 24 m^2

8 and 3 are factors of 24 because they both divide into 24 evenly. Similarly, 6 and 4 are also factors of 24.
24 can be factored completely.

$24 = 8 \times 3$	$24 = 6 \times 4$
$= 4 \times 2 \times 3$	$= 3 \times 2 \times 2 \times 2$
$= 2 \times 2 \times 2 \times 3$	

The factors 2 and 3 are called prime numbers. A prime number is an integer, greater than 1, whose only factors are itself and 1.

If two or more numbers have the same factor, it is called a common factor.
To find the common factors of the numbers 18, 30, and 42, we first write each number as a product of prime factors.

$18 = 2 \times 3 \times 3$
$30 = 2 \times 3 \times 5$
$42 = 2 \times 3 \times 7$

2, 3, and 2×3 or 6 are the common factors of 18, 30, and 42. The largest or greatest common factor (GCF) of the three numbers is 6.

We can find the greatest common factor of monomials in the same way. To find the GCF of $18y^4$ and $30y^2$ we first express each term as a product of prime factors.

$18y^4 = 2 \times 3 \times 3 \times y \times y \times y \times y$
$30y^2 = 2 \times 3 \times 5 \times y \times y$

The GCF is $2 \times 3 \times y \times y$ or $6y^2$.

We have used the distributive property to multiply a binomial by a monomial.

$2x(x + 3y) = 2x^2 + 6xy$

To factor $2x^2 + 6xy$ we first find the GCF of $2x^2$ and $6xy$.

$2x^2 = 2 \times x \times x$
$6xy = 2 \times 3 \times x \times y$

The GCF of $2x^2$ and $6xy$ is $2x$. Now divide by $2x$ to find the other factor.

$$\frac{2x^2 + 6xy}{2x} = \frac{2x^2}{2x} + \frac{6xy}{2x}$$
$$= x + 3y$$

$\therefore 2x^2 + 6xy = 2x(x + 3y)$

——Factoring——→
$2x^2 + 6xy = 2x(x + 3y)$
←——Expanding——

Example.

Factor. $4a^5 + 10a^3 + 6a^2$

Solution:

Find the GCF.

$$4a^5 = 2 \times 2 \times a \times a \times a \times a \times a$$
$$10a^3 = 2 \times 5 \times a \times a \times a$$
$$6a^2 = 2 \times 3 \times a \times a$$

The GCF is $2a^2$ and dividing

$$\frac{4a^5 + 10a^3 + 6a^2}{2a^2} = \frac{4a^5}{2a^2} + \frac{10a^3}{2a^2} + \frac{6a^2}{2a^2}$$
$$= 2a^3 + 5a + 3$$

$\therefore 4a^5 + 10a^3 + 6a^2 = 2a^2(2a^3 + 5a + 3)$

Check by expanding: $2a^2(2a^3 + 5a + 3) = 2a^2(2a^3 + 5a + 3)$
$= 4a^5 + 10a^3 + 6a^2$

EXERCISE 3.12

A

1. Find the greatest common factor.

(a) 6, 10, 12 (b) 12, 18, 30
(c) $6x$, $8x$, $10x$ (d) $18a^3$, $36a^4$, $24a^2$
(e) $15b^3$, $25b^2$, $20b$ (f) $36a^2b^3$, $24a^3b$
(g) $20x^2y^3$, $30x^3y^2$ (h) $21s^4$, $14s^3$, $28s^2$

2. State the missing factor.

(a) $3a + 9 = ■(a + 3)$
(b) $4x - 20 = ■(x - 5)$
(c) $6t - 14 = ■(3t - 7)$
(d) $9ab + 6ac = ■(3b + 2c)$
(e) $-8x^2 - 10x = ■(4x + 5)$

B

3. State the quotient.

(a) $\frac{20x + 35}{5}$ (b) $\frac{18a - 9b}{9}$
(c) $\frac{10t^2 - 4t}{2t}$ (d) $\frac{3m^2 + 4m}{m}$
(e) $\frac{4x^4 + 5x^3 - 6x^2}{x^2}$ (f) $\frac{8xy - 4xz + 2xy}{2x}$
(g) $\frac{5x^4 - 10x^3}{5x^2}$ (h) $\frac{24x^5 - 8x^4 - 16x^3}{4x^3}$

4. Factor.

(a) $2x + 6$ (b) $3a + 9$
(c) $5y - 10$ (d) $4s - 12$
(e) $12m + 6n$ (f) $7y - 35w$
(g) $8a + 6b + 2c$ (h) $3x - 12y + 15z$
(i) $20d - 30e + 60f$ (j) $ab + ac$

5. Factor. Check by expanding.

(a) $4x^2 + 6x$ (b) $5ab - 6ac$
(c) $6xy + 3xt + 9ax$ (d) $2a^2 - 4a$
(e) $12a^2 - 6ab$ (f) $10x - 5x^2$
(g) $8b^3 - 4b^2 + 6b$ (h) $3m - 6m^2 - 9m^3$

6. Factor. Check by expanding.

(a) $3xy + 7xyt$
(b) $7mn + 6m^2n$
(c) $12a^2b - 6ab^2 + ab$
(d) $2rt^2 + 6rt^3 - 4r^2t$

7. Evaluate by mentally factoring first before calculating.

(a) $9 \times 6 + 9 \times 4$
(b) $5 \times 16 + 5 \times 4$
(c) $16 \times 3 + 16 \times 7$
(d) $24 \times 14 - 24 \times 4$
(e) $63 \times 35 - 63 \times 25$

MIND BENDER

Do you need some Rest and Relaxation? Here's some R and R. Name four famous people whose initials are R.R.

3.13 FACTORING THE DIFFERENCE OF SQUARES

Suppose you want to find the product of a sum, $(x + 3)$, and a difference, $(x - 3)$.

$$\begin{aligned}(x + 3)(x - 3) &= (x + 3)(x - 3) \\ &= x^2 - 3x + 3x - 9 \\ &= x^2 - 9\end{aligned}$$

Notice that x^2 is the square of x, and 9 is the square of 3. For this reason, $x^2 - 9$ is called a difference of squares.

The pattern in the product of this difference is

$$(\blacksquare + \bullet)(\blacksquare - \bullet) = \blacksquare^2 - \bullet^2$$

We use this pattern to quickly factor a difference of squares.

Example.

Factor. (a) $t^2 - 25$ (b) $16r^2 - 49s^2$

Solution:

(a) $t^2 - 25 = t^2 - 5^2$
$= (t + 5)(t - 5)$

(b) $16r^2 - 49s^2 = (4r)^2 - (7s)^2$
$= (4r + 7s)(4r - 7s)$

EXERCISE 3.13

A

1. State the positive square root of each monomial.

(a) 16 (b) 36 (c) t^2
(d) $9m^2$ (e) $25x^2$ (f) $49k^2$
(g) $4m^6$ (h) $64x^8$ (i) $100y^4$

2. State the product.

(a) $(x + 2)(x - 2)$ (b) $(t + 5)(t - 5)$
(c) $(a + 4)(a - 4)$ (d) $(m + 6)(m - 6)$
(e) $(r + 7)(r - 7)$ (f) $(y + 8)(y - 8)$

3. State the remaining factors.

(a) $x^2 - 4 = (x + 2)(\blacksquare)$
(b) $a^2 - 9 = (a - 3)(\blacksquare)$
(c) $m^2 - 36 = (m + 6)(\blacksquare)$
(d) $m^2 - 25 = (\blacksquare)(m - 5)$
(e) $4x^2 - 1 = (\blacksquare)(2x + 1)$
(f) $16 - n^2 = (4 - n)(\blacksquare)$

4. State the factors.

(a) $x^2 - 49$ (b) $m^2 - 16$ (c) $a^2 - 144$
(d) $b^2 - 121$ (e) $c^2 - 100$ (f) $d^2 - 64$
(g) $m^2 - 1$ (h) $x^2 - 4$ (i) $t^2 - 49$
(j) $r^2 - 9$ (k) $36 - x^2$ (l) $100 - x^2$
(m) $121 - t^2$ (n) $1 - b^2$ (o) $9 - a^2$

B

5. Factor.

(a) $4x^2 - 16$ (b) $1 - y^2$ (c) $16m^2 - 1$
(d) $9a^2 - 25$ (e) $100 - a^2$ (f) $9t^2 - 36$
(g) $4m^2 - 49$ (h) $1 - 36x^2$ (i) $49 - 9a^2$

6. Factor.

(a) $4a^2 - 9b^2$ (b) $16x^2 - 25y^2$
(c) $a^2 - 16b^2$ (d) $144m^2 - 16b^2$
(e) $25b^2 - 49x^2$ (f) $1 - 400t^2$
(g) $25a^2 - 16b^2$ (h) $144b^2 - 36t^2$

7. Factor.

(a) $16a^4 - 9$ (b) $25x^6 - 1$
(c) $36y^8 - 49x^2$ (d) $9x^2y^2 - 25$
(e) $4x^2 - 9x^6$ (f) $100t^8 - 1$

8. Factor and find the value of each.

(a) $9^2 - 4^2$ (b) $6^2 - 3^2$ (c) $2^2 - 1$
(d) $7^2 - 2^2$ (e) $10^2 - 8^2$ (f) $12^2 - 2^2$
(g) $99^2 - 98^2$ (h) $75^2 - 73^2$ (i) $27^2 - 20^2$

C

9. Calculate using difference of squares.

(a) 32×28 (b) 44×36
(c) 105×95 (d) 310×290

3.14 FACTORING TRINOMIALS

A polynomial with three terms, such as $x^2 + 5x + 6$, is called a trinomial. The product of two binomials is very often a trinomial. For example,

$$
\begin{aligned}
(x + 2)(x + 3) &= (x + 2)(x + 3) \\
&= x^2 + 3x + 2x + 6 \\
&= x^2 + 5x + 6
\end{aligned}
$$

Factoring is the reverse process of finding a product or expanding. As with the difference of squares, there is a pattern in factoring trinomials.

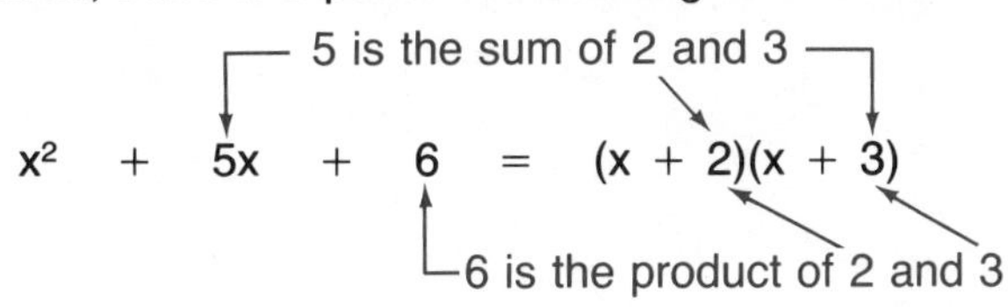

We use this pattern to factor trinomials.

Example 1.
Factor. $x^2 + 8x + 15$

Solution:
To factor $x^2 + 8x + 15$, we need to find two numbers whose sum is 8 and product is 15. We can list pairs of numbers in a table.

From the table at the right, the numbers are 5 and 3.

$$x^2 + 8x + 15 = (x + 5)(x + 3)$$

$$
\begin{aligned}
\text{Check: } (x + 5)(x + 3) &= (x + 5)(x + 3) \\
&= x^2 + 3x + 5x + 15 \\
&= x^2 + 8x + 15
\end{aligned}
$$

Factors of 15	Sum of Factors
15, 1	16
−15, −1	−16
5, 3	8

Example 2.
Factor and check. $t^2 - 7t + 12$

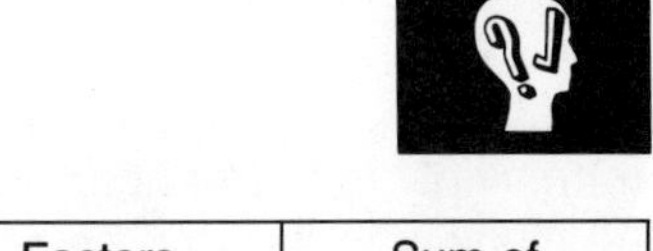

Solution:
To factor $t^2 - 7t + 12$ we need to find two numbers whose sum is -7 and product is 12.

From the table, the numbers are -3 and -4.

$$t^2 - 7t + 12 = (t - 3)(t - 4)$$

$$
\begin{aligned}
\text{Check: } (t - 3)(t - 4) &= (t - 3)(t - 4) \\
&= t^2 - 4t - 3t + 12 \\
&= t^2 - 7t + 12
\end{aligned}
$$

Factors of 12	Sum of Factors
12, 1	13
−12, −1	−13
6, 2	8
−6, −2	−8
4, 3	7
−4, −3	−7

Example 3.
Factor and check. $m^2 - 3m - 28$

Solution:
For $m^2 - 3m - 28$ we need two numbers whose sum is -3 and product is -28.
List the pairs of numbers that multiply to give -28. Watch the signs.

Factors of -28	Sum of Factors
$-28, 1$	-27
$28, -1$	27
$-14, 2$	-12
$14, -2$	12
$-7, 4$	-3

From the table, the numbers are -7 and 4.

$$m^2 - 3m - 28 = (m - 7)(m + 4)$$

Check: $(m - 7)(m + 4) = (m - 7)(m + 4)$
$= m^2 + 4m - 7m - 28$
$= m^2 - 3m - 28$

Example 4.
Factor and check. $b^2 - 10b + 25$

Solution:
For $b^2 - 10b + 25$ we need two numbers whose sum is -10 and product is 25.

Factors of 25	Sum of Factors
$25, 1$	26
$-25, -1$	-26
$5, 5$	10
$-5, -5$	-10

From the table, the numbers are -5 and -5.

$$b^2 - 10b + 25 = (b - 5)(b - 5)$$
$$= (b - 5)^2$$

Check: $(b - 5)^2 = (b - 5)(b - 5)$
$= b^2 - 5b - 5b + 25$
$= b^2 - 10b + 25$

EXERCISE 3.14

A 1. Find the pair of numbers that satisfy each set of conditions.
(a) multiply to 10 and add to 7
(b) multiply to 3 and add to 4
(c) multiply to 6 and add to 5
(d) multiply to -8 and add to 2
(e) multiply to -8 and add to -2
(f) multiply to -12 and add to -1
(g) multiply to -12 and add to 1

2. Which pair of binomials are the factors of $x^2 + 8x + 12$?
(a) $(x + 4)(x + 3)$ (b) $(x + 12)(x + 1)$
(c) $(x + 6)(x + 2)$ (d) $(x - 4)(x - 3)$

3. Which pair of binomials are factors of $t^2 + 7t + 10$?
(a) $(t + 10)(t + 1)$ (b) $(t + 5)(t + 2)$
(c) $(t - 5)(t - 2)$ (d) $(t - 10)(t - 2)$

4. Which pair of binomials are factors of $x^2 - x - 12$?
(a) $(x - 6)(x + 2)$ (b) $(x + 4)(x - 3)$
(c) $(x - 2)(x + 6)$ (d) $(x - 4)(x + 3)$

5. Which pair of binomials are factors of $m^2 - 5m - 6$?
(a) $(m - 6)(m + 1)$ (b) $(m - 3)(m + 2)$
(c) $(m - 1)(m + 6)$ (d) $(m - 2)(m - 3)$

B 6. Complete the following.

(a) $x^2 + 7x + 12 = (x + 3)(x + \blacksquare)$
(b) $x^2 + 8x + 15 = (x + \blacksquare)(x + 3)$
(c) $x^2 - x - 6 = (x - \blacksquare)(x + 2)$
(d) $x^2 - 7x + 12 = (x - 3)(x - \blacksquare)$
(e) $a^2 + 2a - 15 = (a + \blacksquare)(a - 3)$
(f) $m^2 - 6m + 9 = (\blacksquare - 3)(m - 3)$
(g) $b^2 + 2b + 1 = (b + 1)(\blacksquare + 1)$
(h) $c^2 - 10c + 25 = (c - 5)(c - \blacksquare)$
(i) $t^2 + 3t - 28 = (t + 7)(\blacksquare - 4)$

7. Complete the following.

(a) $x^2 + 5x + 6 = (\blacksquare + 3)(x + \blacksquare)$
(b) $x^2 + 9x + 14 = (x + \blacksquare)(\blacksquare + 2)$
(c) $a^2 - 2a - 15 = (a - \blacksquare)(\blacksquare + 3)$
(d) $m^2 - 8m + 16 = (\blacksquare - 4)(m - \blacksquare)$
(e) $b^2 - 10b + 21 = (b - \blacksquare)(\blacksquare - 7)$
(f) $t^2 + 4t + 4 = (t + \blacksquare)(\blacksquare + 2)$
(g) $d^2 - 3d - 18 = (\blacksquare - 6)(d + \blacksquare)$
(h) $x^2 + 13x + 22 = (x + \blacksquare)(\blacksquare + 2)$
(i) $n^2 + 8n - 20 = (n + \blacksquare)(\blacksquare - 2)$

8. State the remaining factors.

(a) $x^2 + 7x + 10 = (x + 2)(\rule{2em}{1ex})$
(b) $a^2 - a - 12 = (a - 4)(\rule{2em}{1ex})$
(c) $b^2 + 8b + 7 = (\rule{2em}{1ex})(b + 7)$
(d) $m^2 - 6m + 5 = (m - 1)(\rule{2em}{1ex})$
(e) $x^2 + x - 20 = (\rule{2em}{1ex})(x - 4)$
(f) $r^2 + 4r + 4 = (\rule{2em}{1ex})(r + 2)$
(g) $t^2 + 2t - 8 = (t - 2)(\rule{2em}{1ex})$
(h) $b^2 - 3b - 18 = (b - 6)(\rule{2em}{1ex})$

9. Complete the following table.

	Trinomial	a + b	ab	a, b	Factors
(a)	$x^2 + 5x + 6$	5	6	3, 2	
(b)	$a^2 - a - 12$	−1	−12	−4, 3	
(c)	$m^2 - 6m + 9$	−6	9	−3, −3	
(d)	$r^2 - 3r - 18$	−3	−18		
(e)	$n^2 - 10n + 25$	−10			
(f)	$b^2 + 2b - 15$				
(g)	$t^2 + 5t + 4$				
(h)	$x^2 + 9x + 14$				
(i)	$c^2 - 4t - 21$				
(j)	$y^2 - 8y + 16$				

10. Factor.

(a) $x^2 + 3x + 2$
(b) $a^2 + 8a + 15$
(c) $x^2 + 10x + 25$
(d) $m^2 + 7m + 6$
(e) $r^2 + 3r - 18$
(f) $d^2 - 7d + 12$
(g) $a^2 - 2a - 15$
(h) $r^2 - 4r + 4$
(i) $t^2 + 6t + 8$
(j) $n^2 - 6n + 5$

11. Factor.

(a) $x^2 - 9x + 20$
(b) $a^2 + 14a + 24$
(c) $t^2 + 13t + 36$
(d) $d^2 + 20d + 100$
(e) $x^2 + 2x - 3$
(f) $a^2 - 3a - 88$
(g) $t^2 - t - 20$
(h) $h^2 - 11h + 30$
(i) $m^2 - 9m - 22$
(j) $m^2 + m - 20$

12. Factor.

(a) $x^2 - 3x - 70$
(b) $a^2 - 14a + 49$
(c) $m^2 - 11m + 24$
(d) $b^2 + 8b + 16$
(e) $d^2 + 2d - 35$
(f) $m^2 - 8m - 33$
(g) $h^2 - 15h + 50$
(h) $r^2 + r - 56$
(i) $r^2 + 24r + 144$
(j) $x^2 + 6x - 40$

C 13. Factor completely. (Find the common factor first.)

(a) $2x^2 + 10x + 12$
(b) $5a^2 - 10a - 40$
(c) $10x^2 + 20x - 150$
(d) $7x^2 - 14x - 21$
(e) $4r^2 - 16r + 16$
(f) $3m^2 + 21m + 36$
(g) $2t^2 - 12t + 18$
(h) $3b^2 + 30b + 75$

MIND BENDER

Francine threw six darts at the target and each one hit the target.
Which of the numbers could be her score?

3 11 60 26 41 27

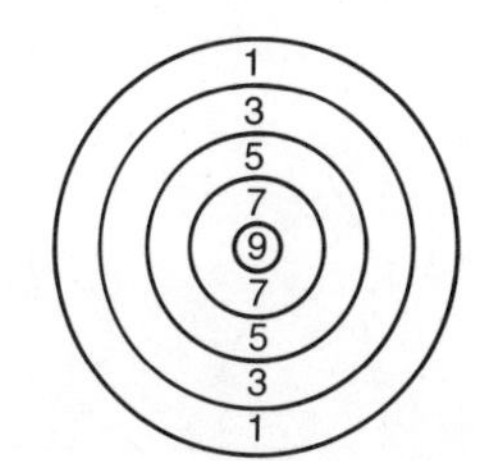

3.15 PROBLEM SOLVING

1. What is the next number in the following sequence?

   19, 18, 16, 15, 13, ■

2. Four students must enter the haunted house at the county fair in single file.
In how many different ways can they do this if Leroy is one of the four and he must go in last?

3. The figure is made up of congruent squares. The area of the figure is 144 cm^2.
What is the perimeter of the figure?

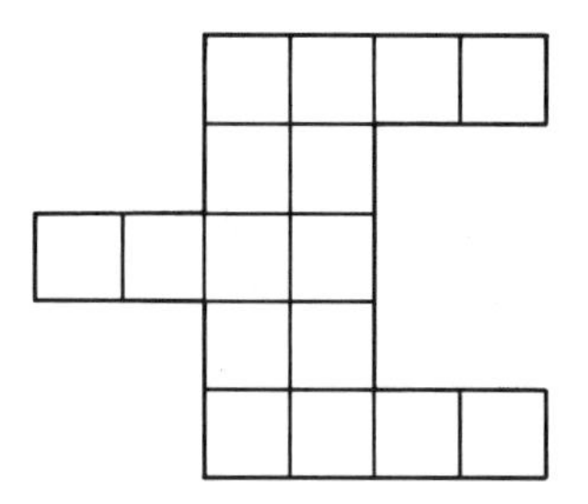

4. In a magic square all the rows, columns, and diagonals have the same sum.
Use the numbers 3, 6, 9, and 10 to complete the magic square.

		8
5	7	
	11	4

5. Find the missing number.

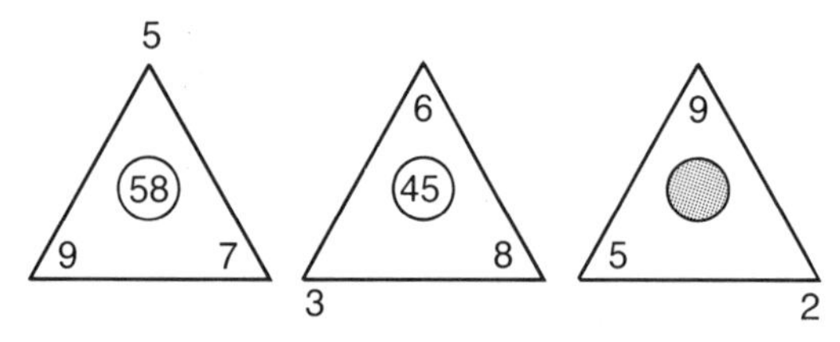

6. Each year Susan's new car depreciates by 20% of what it was worth the year before. The car originally cost Susan $30 000.
What was the value of her car after four years?

7. This package was wrapped and tied with cord as shown.

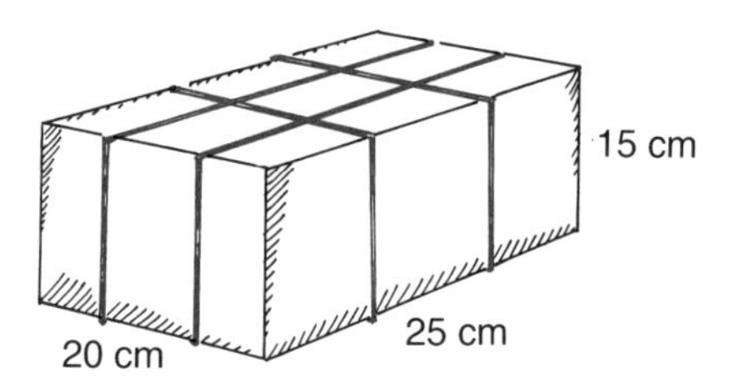

How much cord was needed if you allow 20 cm extra for knots?

8. For the first three games that she bowled, Karen got scores of 210, 207, and 235.
What will she have to bowl in her fourth game to have an average of 217 for all four games?

9. A plane is flying over an airport. The altitude of the plane is 7000 m. The temperature at the airport is -3°C. For every 1000 m increase in altitude, the temperature drops 6°C.
What is the temperature outside the plane?

10. It takes Carl 2.5 min to jog one lap of the track. It takes Tony 3 min. They both start jogging at the same place at the same time.
How many times will Carl pass Tony if they jog for 16 min?

11. A train is 2 km long. It takes 5 min for the train to pass through a level crossing.
What is the speed of the train?

12. Find three consecutive numbers whose product is 148 824.

13. What is the largest area of a rectangular playground that can be enclosed with 24 m of fence if one wall of the school is used?

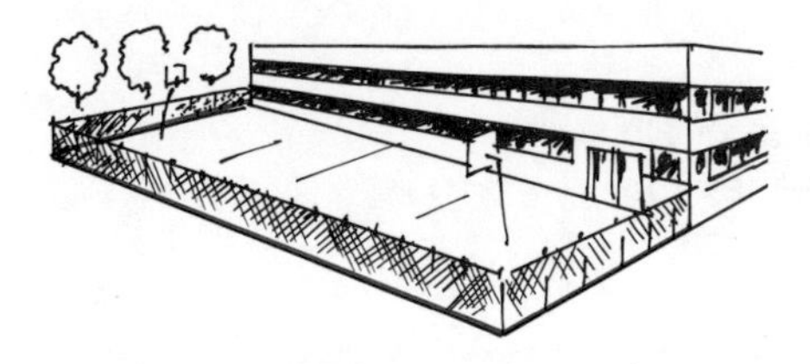

14. Janet's digital watch shows hours and minutes. The watch shows 8 p.m. as 20:00. How many times during one day will all the digits on her watch be the same?

15. (a) Divide the number 234 234 by 13. Divide the result by 11, then divide the result by 7.
What do you notice about the answer?
(b) Repeat part (a) for the number 125 125.
(c) Find the product of 13, 11, and 7.
Use this result to explain your answers to parts (a) and (b).

16. A bag contains 41 gold coins. One coin is counterfeit and lighter than the other coins. To find the counterfeit coin you can use a two-pan balance.
What is the least number of times that you would have to weigh the coins to find this coin?

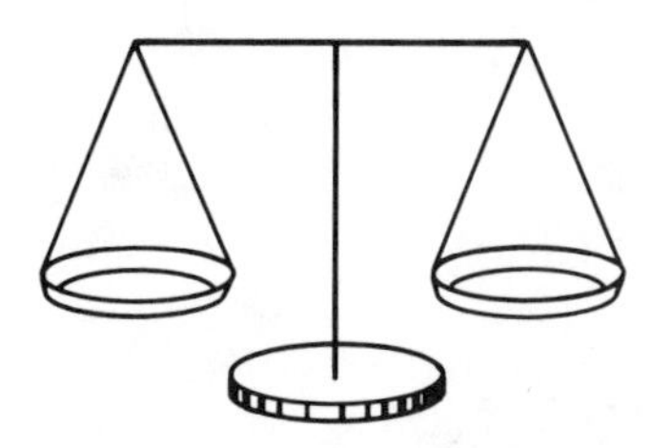

17. The picture measures 30 cm by 40 cm. The frame is 5 cm wide. What is the area of the frame?

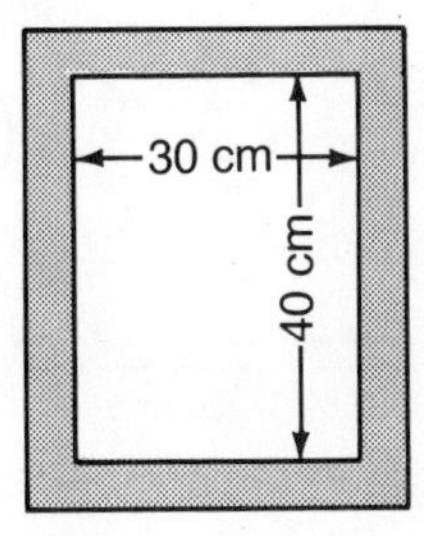

CAREER

CAR RENTAL AGENCY MANAGER

Because of the geography of Canada and the United States, many air and bus travellers, both business and recreational, rent cars when they arrive at their destination.
Most of the rental agencies are franchise operations.

1. The table gives the rental rates for an economy car and a luxury car.

Car	Weekly Rate ($)	Daily Rate ($)	Additional km Charge (¢)
Economy	202	36	17
Luxury	358	59	24

Rates are based on 1000 free km per week or 150 free km per day. Any additional kilometres will be calculated at the rate shown.
(a) What would you charge to rent an economy car for one week if the driver drove 1235 km?
(b) What would you charge to rent a luxury car for five days if the driver drove 2138 km?

2. If you choose to buy a franchise, what benefits can you expect to receive from the parent company?

EXTRA

3.16 GEOMETRY, FLIGHT, AND PAPER AIRPLANES

Geometry has played an important part in the design of all aircraft. If you were to take a close look at an airplane, you should be able to identify triangles, curves, straight lines, angles, and symmetry.

Many flight characteristics of aircraft can be simulated with paper airplanes. In fact, some believe that the design ideas for many modern aircraft were first tested with paper airplanes. The First International Paper Airplane Contest, held by *Scientific American* in 1967, attracted 12 000 entries. Airplanes were judged according to four categories:

1. time in the air
2. distance flown
3. aerobatics
4. origami (beauty)

Try making one or all of the following paper airplanes and judge it on the basis of the above categories. You will need scissors, tape, and fairly stiff 21.5 cm × 28 cm paper.

AN AEROBATIC PLANE

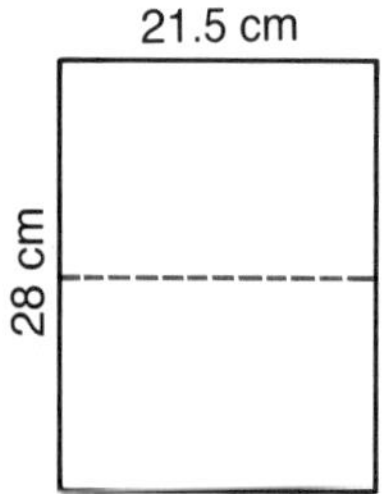

1. Crease across centre.

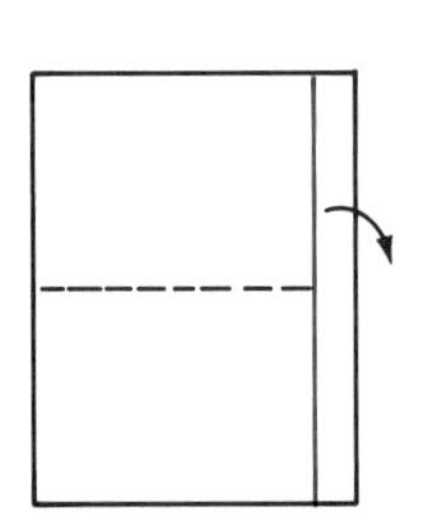

2. With top of crease facing up, fold about 1 cm down.

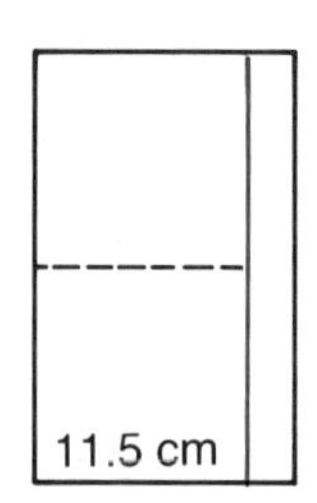

3. Continue folding in this way until 10 to 11 cm remains.

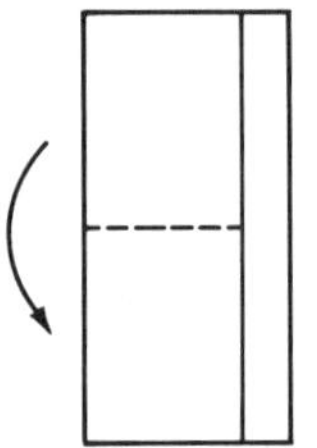

4. Fold in half along crease.

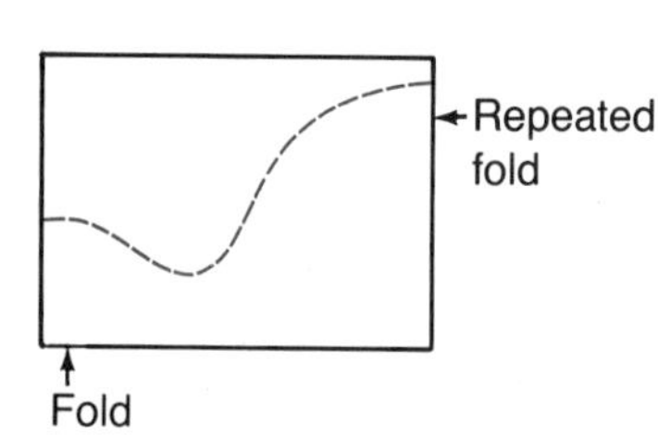

5. Cut along indicated line.

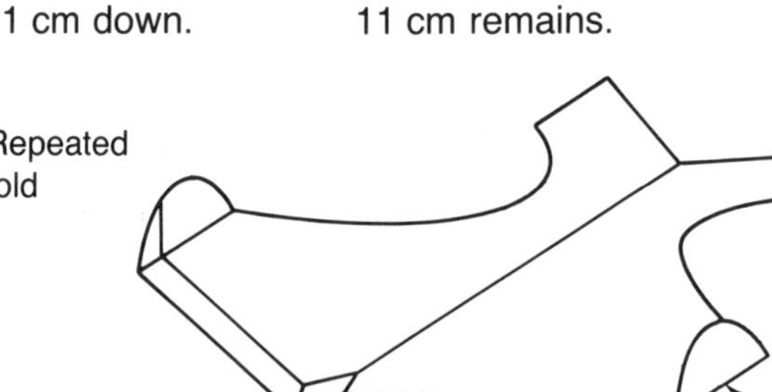

6.

A DISTANCE PLANE

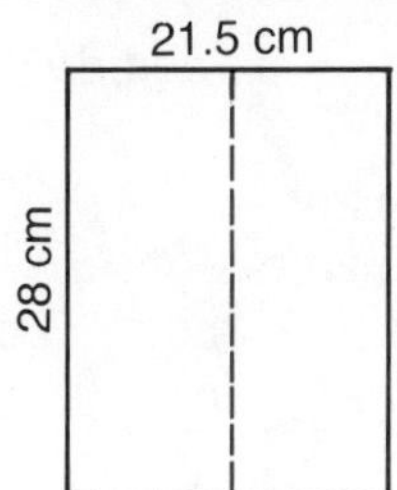

1. Crease on centre line.

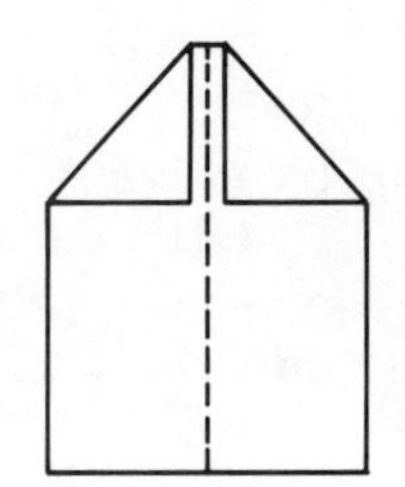

2. Fold corners in.

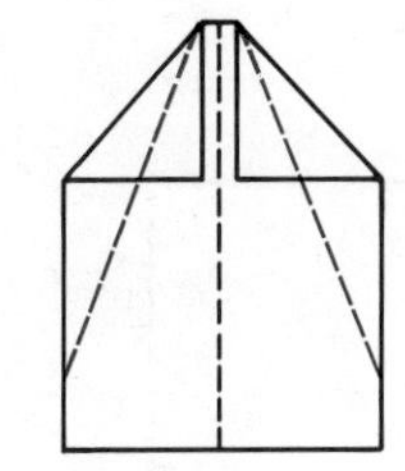

3. Fold in on dotted lines.

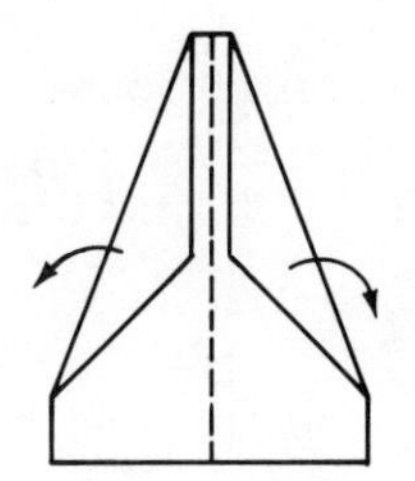

4. Fold away from you on centre fold.

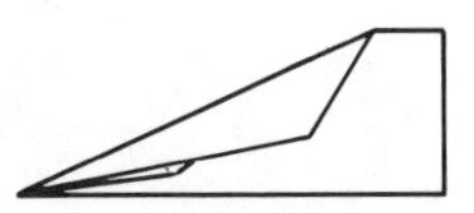

5. Turn sideways.

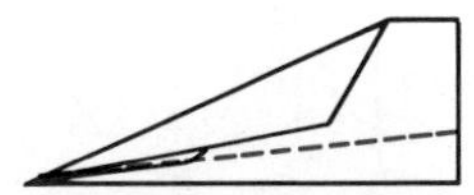

6. Fold each side down on dotted line.

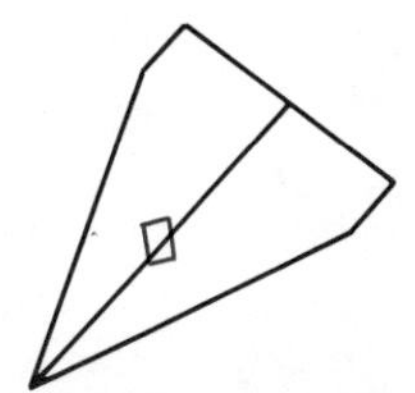

7. Tape top as shown to hold the wings together and fly by pushing forward.

A HELICOPTER

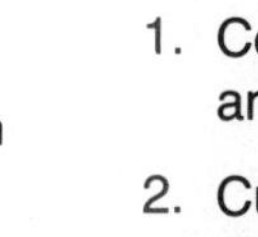

5 cm
A
B
22 cm
C
D
11 cm
E
2.5 cm
1.5 cm
1.5 cm

1. Copy the design onto a sheet of paper and cut it out.
2. Cut along all solid lines.
3. Fold A forward.
4. Fold B backwards.
5. Fold C forward.
6. Fold C in.
7. Fold D in.
8. Fly by dropping from your raised hand.

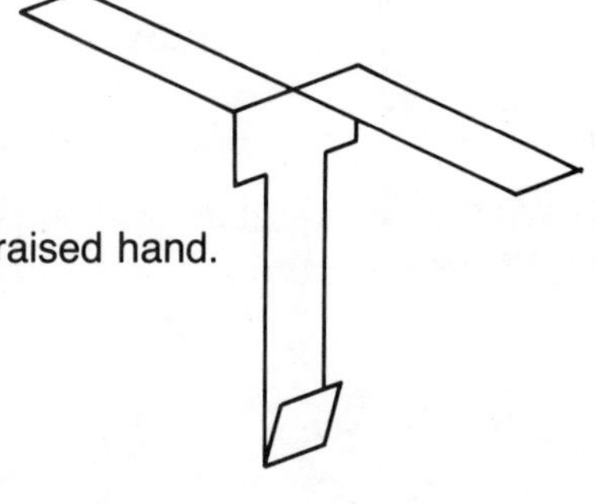

3.17 REVIEW EXERCISE

1. Write a formula that represents the perimeter of each figure.

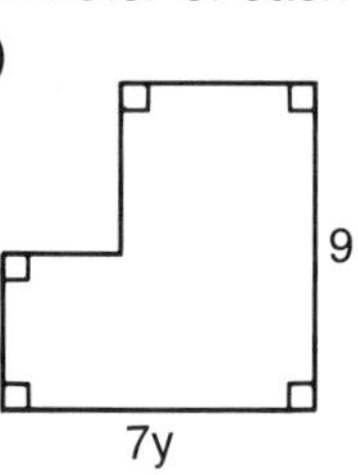

(b)
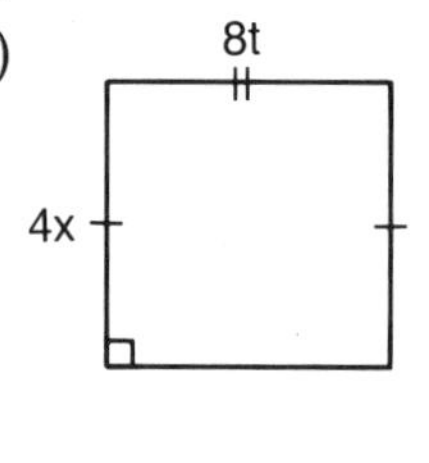

2. Write a formula that represents the area of each figure.

(a)
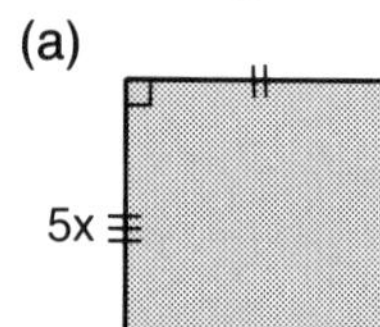

(b)
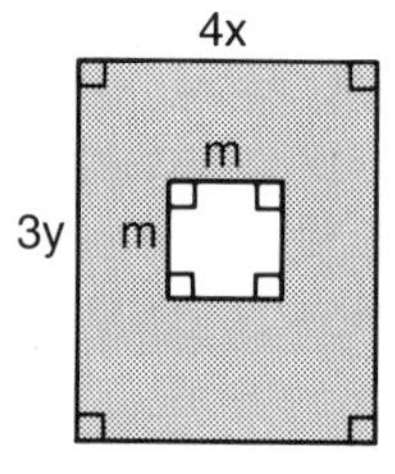

3. A long distance telephone call from Stoney Creek to Freelton costs \$0.80 for the first minute and \$0.56 for each additional minute.

(a) Write a formula for the cost of a call according to the length of the call.

(b) Use the formula to determine the cost of a 9 min call.

4. Simplify.

(a) $3a + 4b - 2a + 7b$

(b) $4x - 2y - 5x - y$

(c) $5x^2 + 3x - 1 - 7x^2 - 5x + 1$

(d) $-4a - 6b + 7b - 2a$

(e) $2xy + 4y + 3x - 4x - y - 5xy$

(f) $4m - n - 6 - 5m + 7 - 4n$

5. Write an expression for the perimeter of each figure.

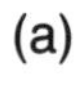
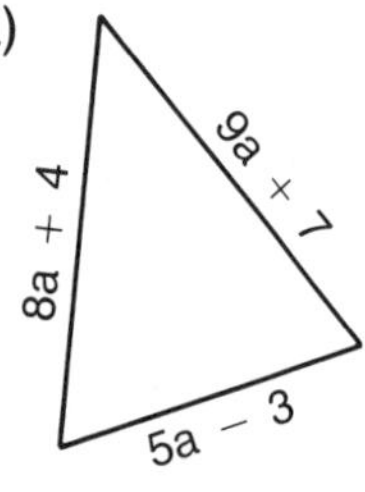

(b)
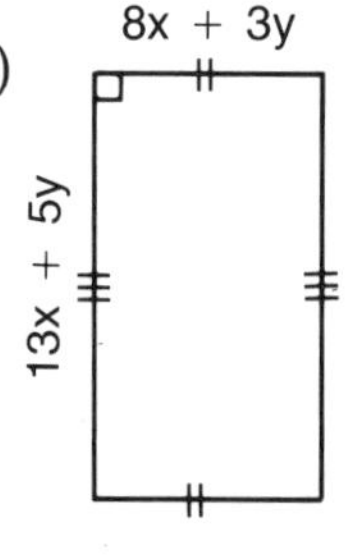

6. If $a = -1$, $b = 3$, $c = -2$, and $d = 0$, evaluate.

(a) $2b^2 + 3ac$ (b) $2a^2 - c$

(c) $-6abd$ (d) $2a^2b - 3b^2c$

(e) $4a^2 - 3 + 6abd$ (f) $1 - 3abc + 2c^2$

(g) $\dfrac{6ab - 2}{-5}$ (h) $\dfrac{a^2 + b^2 + c^2}{-2}$

7. Simplify.

(a) $(3x + 7) + (4x + 2)$

(b) $(5a - 3) + (7a - 11)$

(c) $(3x^2 + 2x + 4) + (5x^2 - x - 7)$

(d) $(4m - 7) - (3m - 2)$

(e) $(6x + 5) - (x - 3)$

(f) $(3a - 4b) + (5a + 6b) - (a - 2b)$

(g) $(7x - y) - (4x - 3y) + (2x + 6y)$

(h) $(3t^2 - t - 4) - (5t^2 + 6t + 1)$

8. Simplify.

(a) $2(x - 3) - 3(1 - 2x) - (x - 4)$

(b) $2a(1 - a) - 3(2a^2 - 2) + 3(2a - 1)$

(c) $2(2a - 3b + c) - (3a - b - c) - 2(a - 2b)$

(d) $m(3m - 1) - 2m(1 - m) + 6$

(e) $2(3b^2 - 2b - 1) - b(1 + 3b) - 5$

(f) $2x(1 - 2x) - x(3x - 2) + 7x$

(g) $5 - 2c(3c - 4) + 6(c^2 - 3c + 2)$

9. Find expressions for the perimeters of the following.

(a)
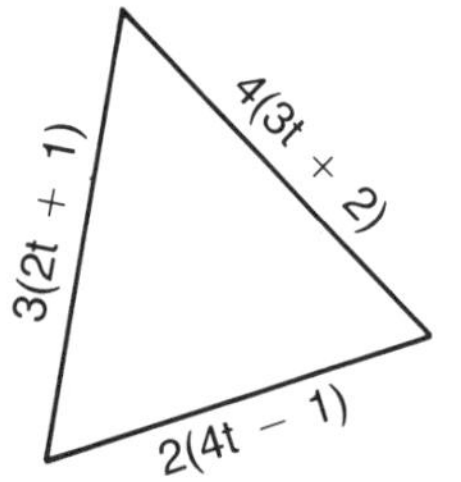

(b)

2(m + 4)

3(2m + 5)

10. Solve and check.

(a) $2(a - 1) = 8$

(b) $3(x + 1) = 2x - 5$

(c) $2(x - 2) + 4x = 3(x + 1) + 5$

(d) $5(t + 3) - (t - 2) = 1$

(e) $4m - (m - 7) = 2(m - 4)$

11. Solve and check.

(a) $\frac{x-1}{2} = 7$ (b) $\frac{m+1}{2} - \frac{m+1}{3} = 2$

(c) $\frac{t+3}{4} = \frac{t-2}{2}$ (d) $\frac{2y-3}{3} = 5$

12. A number increased by 17 is 54.
Find the number.

13. The sum of three consecutive whole numbers is 165.
Find the numbers.

14. Expand.

(a) $(x + 2)(x + 3)$ (b) $(m + 3)^2$
(c) $(b - 2)(b - 4)$ (d) $(t - 3)(t + 3)$
(e) $(r - 2)^2$ (f) $(1 + m)^2$
(g) $(b - 4)(b + 3)$ (h) $(d - 4)(d - 5)$
(i) $(x - 1)^2$ (j) $(a - 5)(a + 5)$

15. Expand and simplify.

(a) $(2a - 3)(a + 6)$
(b) $3(2x - 1)(x + 3)$
(c) $2(m + 7)^2$
(d) $-2(1 - 3x)(1 + 5x)$
(e) $4(2b - 1)(b - 3) + (b + 5)^2$
(f) $2(x - 3)(x + 3) - (x - 1)^2 + 5x$
(g) $2(a - 3) - (a + 3) + 2(a - 1)(a - 5) - 3(a + 3)^2$
(h) $3(2x^2 - x - 3) - (x + 1)^2 - (2x - 1)(3 - 2x)$

16. Find the area of each figure.

(a)

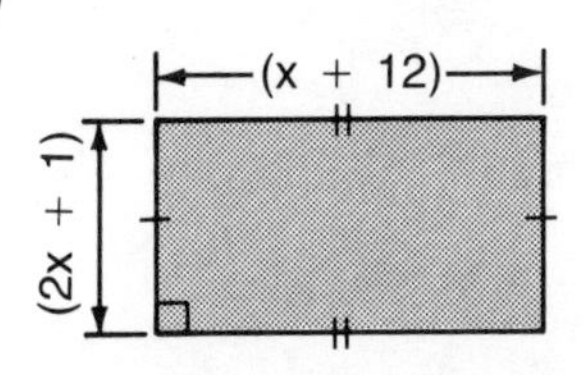

(b)

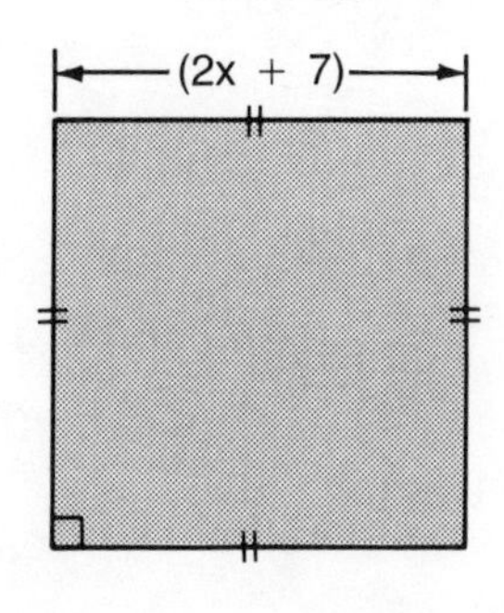

(c)

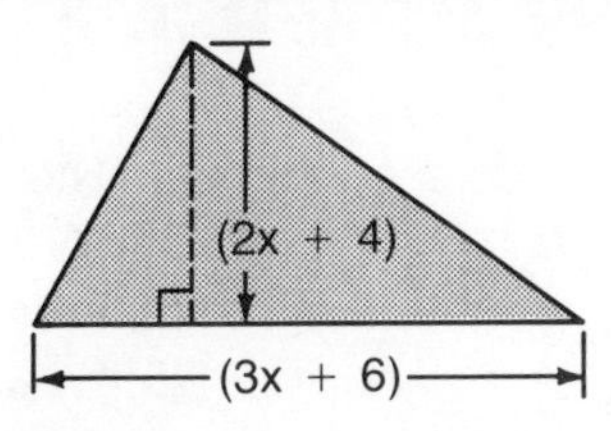

17. Divide.

(a) $\frac{14x^2 + 7x}{7x}$

(b) $\frac{9ab - 3ac + 6ad}{3a}$

(c) $\frac{12x^4 - 8x^3 + 4x^2}{4x}$

(d) $\frac{25a^2b^2 - 10ab + 15ab^2}{5a}$

18. Factor.

(a) $4x - 12$
(b) $3a^2 + 9a$
(c) $20m^2 - 10a$
(d) $8xy - 7xyz$
(e) $2b^4 - 6b^3 + 2b^2$
(f) $12a^2bc - 6ab^2c + 3abc$
(g) $3xy - 6xy^2 + 12x^2y$

19. Factor.

(a) $m^2 - 121$ (b) $x^2 + 9x + 18$
(c) $b^2 - 12b + 36$ (d) $4r^2 - 49$
(e) $c^2 - 6c - 27$ (f) $a^2 + 4a - 12$
(g) $1 - 49a^2b^2$ (h) $1 - 8d + 12d^2$
(i) $b^2 + 10b - 56$ (j) $y^2 + 21y + 80$
(k) $36a^2 - 121b^2$ (l) $16x^4 - y^2$
(m) $x^2 - x - 90$ (n) $a^2 + 19a - 92$

20. Each expression in the diagrams represents the area of the individual rectangle or square.
Give the dimensions and expression for the total area of each square.

(a)

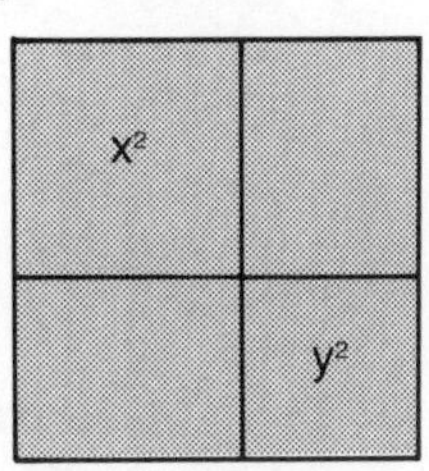

(b)

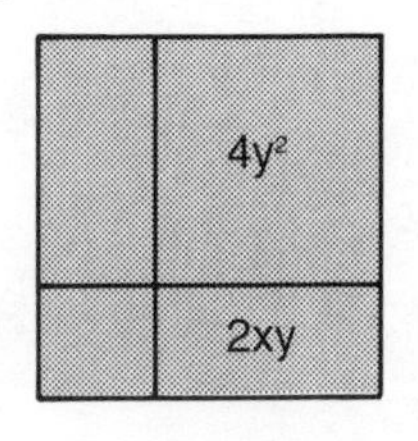

3.18 CHAPTER 3 TEST

1. The cost for a banquet is \$500 for the room plus \$40 per person.

(a) Write an equation for the cost of a banquet.
(b) What would a banquet for 340 people cost?

2. Simplify.

(a) $3x + 5y - 2x - 4y$
(b) $5t - 3s - 4s + 6t$
(c) $8a - 2b + 3a - 4c - c - 6b$

3. If $a = -2$ and $b = -1$, evaluate each of the following.

(a) $a^2 + 2ab + b^2$
(b) $-2ab + 3a^2b - 7$

4. Simplify.

(a) $(4x + 7) + (2x - 1) - (5x - 4)$
(b) $(3a - 2b) - (a - 3b) - (2a - b)$
(c) $(4x + 3y) - (x - 6y) - (3x - 5y)$

5. Expand and simplify.

(a) $2(2x - 1) + 5(3x + 4) - 2(x - 2)$
(b) $3(x^2 - 2x - 1) - 4(x^2 + 3x + 5)$
(c) $2t(3t - 4) - t(t - 1)$

6. Find each product.

(a) $(t - 4)(t + 5)$
(b) $(2x + 3)(x - 5)$
(c) $(3a + 2b)(4a - 3b)$
(d) $(3x - 2)(3x + 2)$

7. Expand and simplify.

(a) $2(x - 4)(x + 5) - 3(x - 5)$
(b) $3(t - 2)(t + 1) + 4(t + 2)(t - 2)$

8. Simplify.

(a) $\dfrac{4t^3 - 8t^2}{4t}$
(b) $\dfrac{6x^4 - 9x^3 + 12x}{3x}$

9. Factor.

(a) $3xy - 6yz$
(b) $8x^3 - 4x^2 + 16x$

10. Factor.

(a) $x^2 - 9$
(b) $x^2 + 7x + 12$
(c) $t^2 + t - 20$
(d) $s^2 - 12s + 36$

RELATIONS

REVIEW AND PREVIEW TO CHAPTER 

ESTIMATING

EXERCISE 1

1. Estimate the totals for the following grocery purchases.

(a) $3.70
3.50
1.75
2.02
▬

(b) $5.63
4.12
2.28
0.56
▬

(c) $1.56
1.09
0.87
3.78
1.78
▬

(d) $0.78
0.23
2.67
0.34
2.67
▬

(e) $4.56
0.78
0.98
2.68
▬

(f) $1.89
2.22
3.08
0.81
▬

2. Estimate the cost of each of the following.

(a) three tickets @ $35.50 each
(b) eight hot dogs @ $3.75 each
(c) five litres of cleaner @ $1.23/L
(d) ten books @ $6.50 each
(e) seven pens @ $0.87 each

3. Estimate the change you should receive from a $20 bill if you made the following purchases.

(a) newspaper for $1.50
magazine for $3.00
(b) milk for $2.67
bread for $1.56
cheese for $4.09
(c) cereal for $2.87
bacon for $6.78
eggs for $3.20
grapefruit for $2.05

TABLES OF VALUES

EXERCISE 2

1. Frank earns $12.50/h as a ski instructor.
Complete the table for the pay earned.

Hours Worked	2	3	5.5	9.5	13
Pay ($)					

2. A freight train averages 45 km/h while travelling across the prairies. Complete the table for the distance travelled.

Hours	2	4	8.5	14	21.5
Distance Travelled (km)					

3. Oktoberfest fudge sells for $4.80/kg. Complete the table for the cost of the fudge.

Amount Purchased (kg)	3.0	4.5	9.2	12.7	34.0
Cost ($)					

4. Paula earns $16.50/h training dolphins. She works 35 h each week and is paid time and a half for each hour over 35 h.
Calculate her pay.

Week	1	2	3	4
Hours Worked	33	29.5	39	42.5
Pay ($)				

INVESTIGATING TWIN PRIMES

Two factors of 6 are 3 and 2 because both 3 and 2 divide 6 evenly. The number 6 has a total of 4 whole number factors, namely 1, 2, 3, and 6.

A prime number, like 7, has only two factors, namely itself and 1.

Some other prime numbers are 2, 3, 5, 11, and 13.

Mathematicians have noticed that some prime numbers, like 5 and 7, differ by 2. These are called twin primes.

EXERCISE 3

1. Find all the pairs of twin primes that are less than 50.

2. There are two pairs of twin primes that are between 50 and 100. What are they?

3. One pair of twin primes is 5 and 7. The number between these two primes is 6. 5, 6, 7

The next largest pair of twin primes is 11 and 13. The number between these two primes is 12. 11, 12, 13

(a) Continue to find similar pairs of twin primes and list them in the same way as above.

(b) What do these twin primes have in common? Describe the pattern in your own words.

INVESTIGATING DIAGONALS

A diagonal is a straight line that joins two vertices of a polygon. For example, a quadrilateral has four sides and two diagonals.

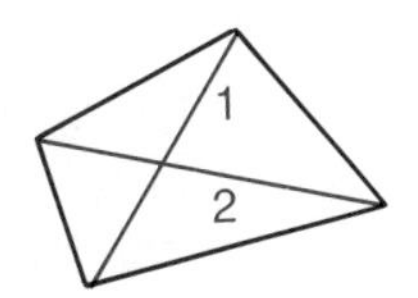

EXERCISE 4

1. A pentagon has five sides. Draw a pentagon and determine the number of diagonals.

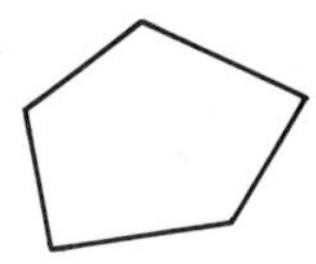

2. A hexagon has six sides. Draw a hexagon and determine the number of diagonals.

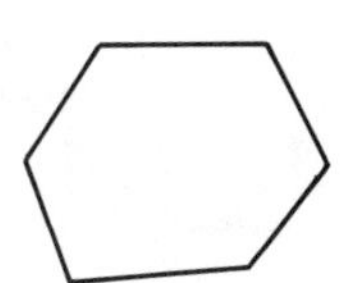

3. Complete the table.

Name	Number of Sides	Number of Diagonals
quadrilateral	4	2
pentagon	5	
hexagon	6	
heptagon	7	
octagon	8	
nonagon	9	
decagon	10	

4. Write a formula that relates the number of diagonals to the number of sides in any polygon.

4.1 GENERAL RELATIONS

There are many examples of how two things relate to or depend on each other. Some examples are:

- the number of people at the beach depends on the outside temperature
- the value of a painting depends on the popularity of the artist
- the distance a plane travels depends on how long it flies

Mathematics can be described as the study of how numbers relate to each other. Each of the above examples is a relation and can be shown in many ways.

The table shows the relation between the distance travelled by a plane flying at 500 km/h and time. The relation can also be shown on a graph.

Time (h)	Distance (km)
0	0
1	500
2	1000
3	1500
4	2000

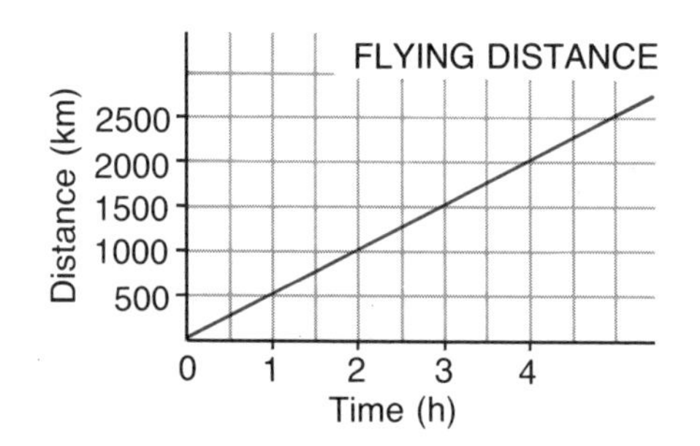

A graph provides a better picture of a relation. In this section we will look at applications of relations and sketch a graph on a pair of axes.

Example.
Misha filled an electric kettle. The kettle has a safety switch that turns the power off a few seconds after the water starts to boil.
Sketch a graph of the water temperature versus time.

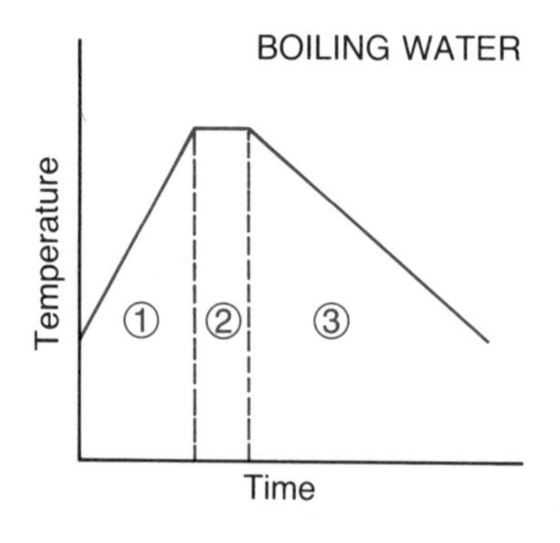

Solution:
Draw and label the axes.

1. The starting temperature of the water is close to room temperature since the water has been sitting in the pipes.

2. The temperature of the water rises to boiling and stays at this temperature until the power shuts off.

3. Once the power shuts off, the water starts to cool gradually to room temperature.

EXERCISE 4.1

1. The amount Anton earns each week as a short order cook depends on the number of hours he works.
Which graph best describes Anton's earnings?

(a) (b)

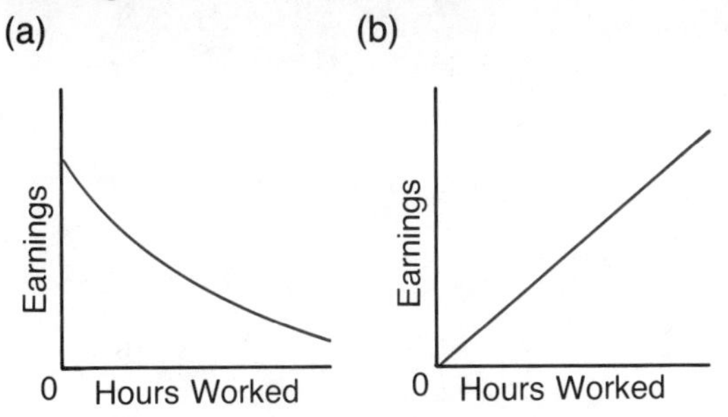

2. The number of litres of gasoline consumed by a car depends on the distance driven.
Which graph best describes gasoline consumption?

(a) (b)

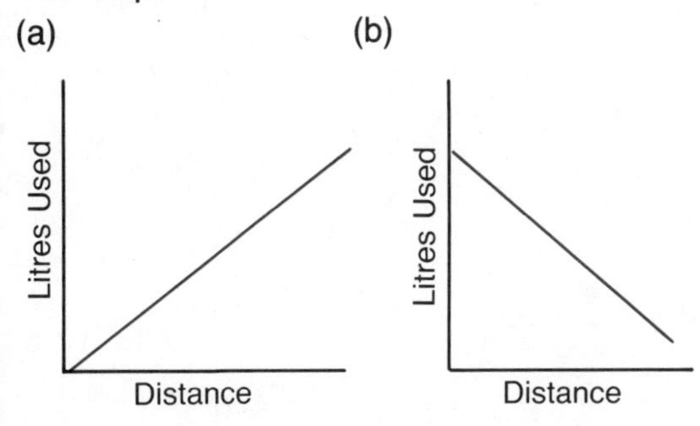

3. When a football is punted, the path of the ball follows a curve and the height of the ball changes over time.
Which graph best shows this situation?

(a) (b)

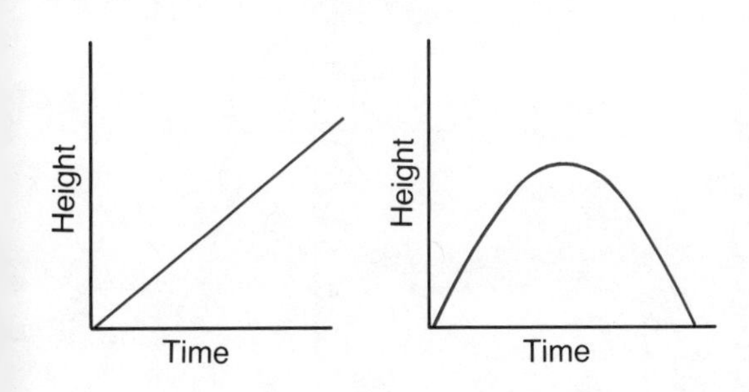

4. The graph shows the temperature of water running from a hot water faucet over a period of time.
Explain the graph.

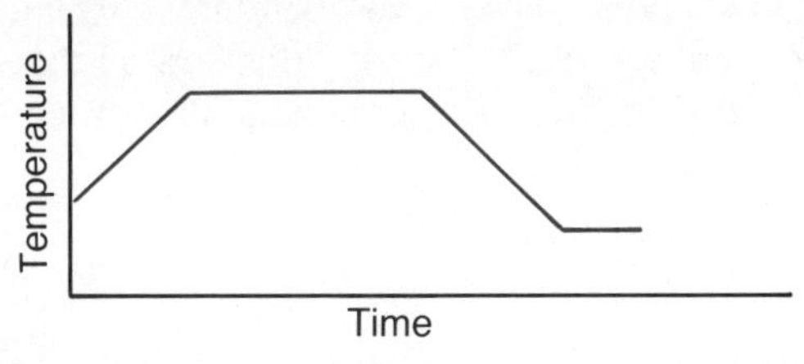

5. The number of minutes of daylight per day depends on the time of year.
Draw a graph of this relationship.

6. You are the first person to get on a Ferris wheel. All the other seats must be filled, one after the other, before the ride starts.
Sketch a graph of your height above the ground versus the time it takes for the ride to start.

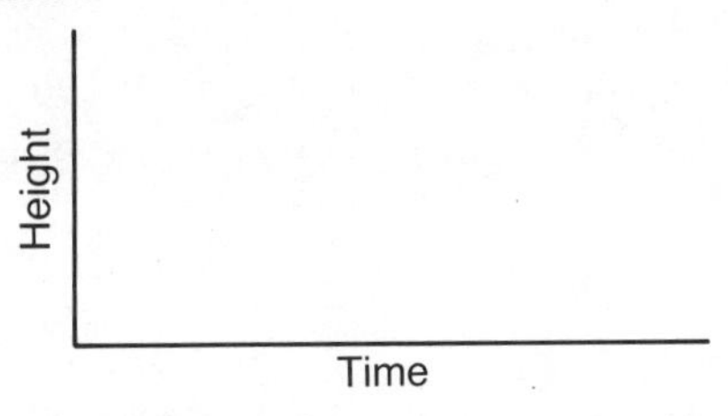

7. Sketch a graph that best shows each of the following situations.

(a) The price you pay for oranges depends on the number you buy.

(b) The value of a pick-up truck depends on its age.

(c) A water bomb is dropped from a plane over a forest fire.
Sketch a graph of the height of the water versus time.

4.2 GRAPHING ORDERED PAIRS

In order to direct aircraft, air traffic controllers must know the positions of the planes in their area. The location of each plane is given using coordinates.

Latitude and longitude are coordinates. Using latitude and longitude, we can locate any position on the earth's surface. For example, the approximate location of Ottawa is

Longitude: 76° West
Latitude: 45° North

In mathematics we use coordinates called ordered pairs to locate points on a plane.
Intersecting and perpendicular number lines divide the plane into 4 quadrants: I, II, III, and IV.
The horizontal number line is called the x-axis.
The vertical number line is called the y-axis.
They intersect at the origin.

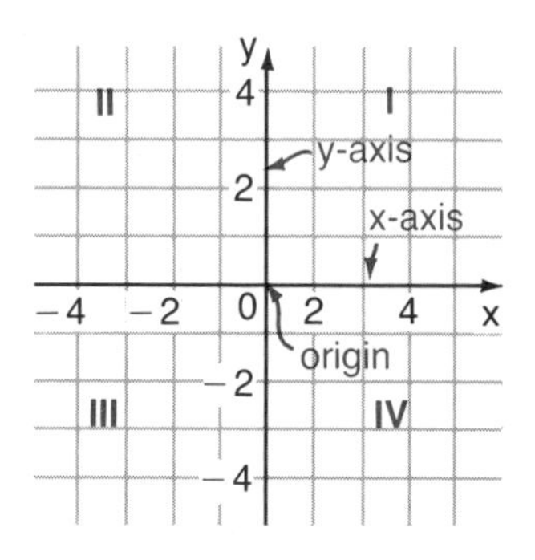

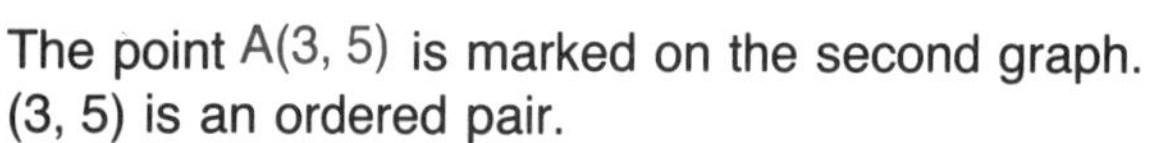

The point A(3, 5) is marked on the second graph.
(3, 5) is an ordered pair.
We use a letter, in this case A, to name a point on the graph.

The first number 3 is called the x-coordinate.
It indicates the number of units to the left or the right of the origin.

The second number 5 is called the y-coordinate.
It indicates the number of units above or below the origin.

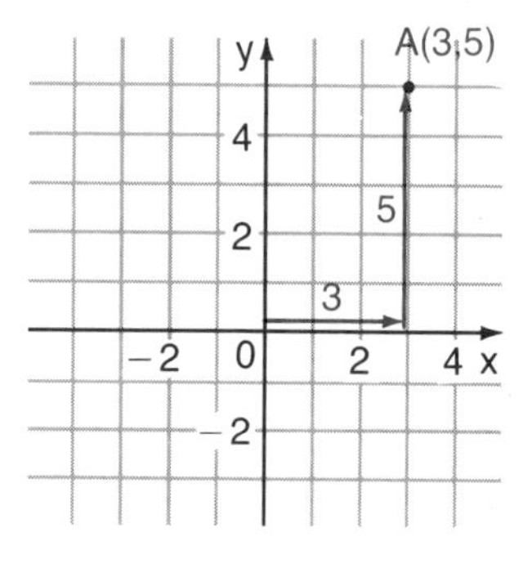

Example.

Graph the points.
(a) P(−3, 2) (b) Q(−4, −2) (c) R(5, −4)

Solution:
(a) For P(−3, 2) start at the origin. Move 3 units to the left. Then move 2 units up.

(b) For Q(−4, −2) start at the origin. Move 4 units to the left. Then move 2 units down.

(c) For R(5, −4) start at the origin. Move 5 units to the right. Then move 4 units down.

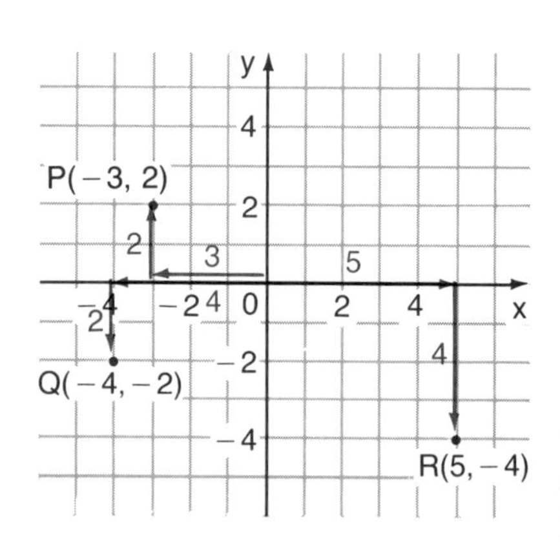

EXERCISE 4.2

A 1. State the coordinates of the points in the graph.

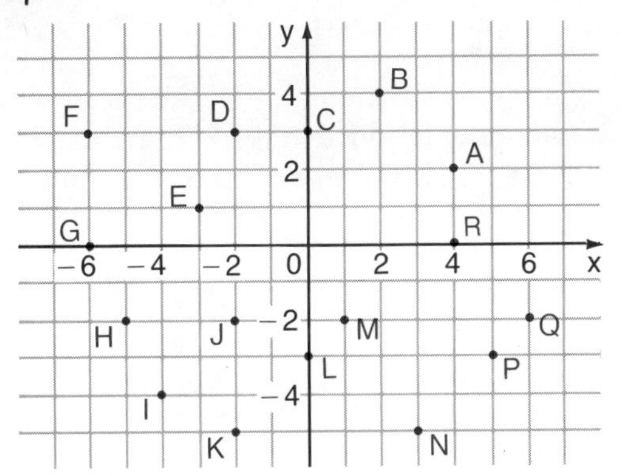

2. Find a point to match each pair of coordinates.

(a) $(-5, -2)$ (b) $(-3, 0)$ (c) $(2, -2)$
(d) $(-1, 2)$ (e) $(3, 1)$ (f) $(2, 5)$
(g) $(0, 3)$ (h) $(0, -4)$ (i) $(5, 0)$
(j) $(-4, -4)$ (k) $(5, -2)$ (l) $(-5, 4)$

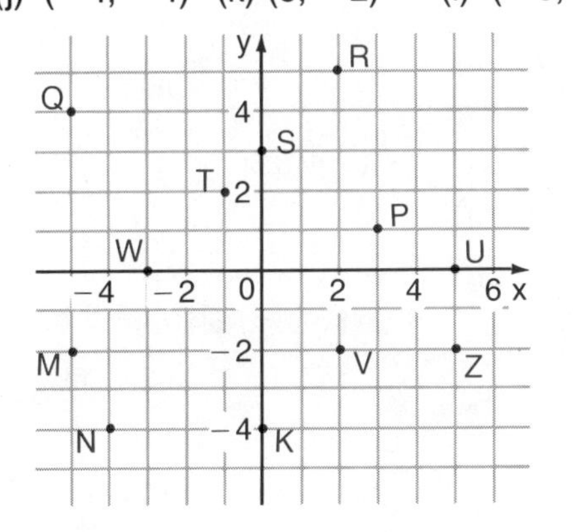

3. State the quadrant, or axis, where the point is located.

	Coordinate		Quadrant or Axis
	x	y	
(a)	positive	negative	
(b)	negative	negative	
(c)	zero	negative	
(d)	negative	zero	
(e)	positive	positive	
(f)	negative	positive	

B 4. Plot the following points on a graph and join them in the order A-B-C-D-A. Identify the resulting geometric figure.

(a) $A(-4, 4)$, $B(-6, -2)$, $C(3, -2)$, $D(5, 4)$
(b) $A(0, 6)$, $B(-3, -4)$, $C(5, -3)$
(c) $A(-2, 3)$, $B(-4, 3)$, $C(-4, 1)$, $D(-2, 1)$
(d) $A(0, 0)$, $B(0, -6)$, $C(4, -6)$, $D(4, 0)$
(e) $A(3, -5)$, $B(2, 3)$, $C(-3, 3)$, $D(-8, -5)$

5. (a) Plot the points $A(2, 2)$, $B(-4, 2)$, and $C(-4, -3)$.
(b) Find the coordinates of D so that ABCD is a rectangle.

6. (a) Plot the points $P(7, 2)$, $Q(1, 2)$, and $R(-3, -3)$.
(b) Find the coordinates of S so that PQRS is a parallelogram.

7. Given (2, 4), (5, 7), and (8, 10), what relation exists between these ordered pairs?

C 8. What are the approximate latitude and longitude coordinates for each city?

(a) Winnipeg (b) Toronto
(c) Miami (d) Moscow
(e) Honolulu (f) Hong Kong
(g) Sydney (Australia)

INVESTIGATIONS

For each of the following
(a) list 4 ordered pairs with this characteristic;
(b) graph the ordered pairs and notice the pattern.

9. The first coordinate equals the second.

10. The first coordinate is 4 less than the second.

11. The sum of the coordinates is 6.

12. The first coordinate minus the second is 2.

13. The first coordinate is 1 more than twice the second.

4.3 GRAPHING RELATIONS

The time you take to drive a certain distance depends on the speed at which you drive.
The slower the speed, the longer the time.
The faster the speed, the shorter the time.

The distance around Cold Creek Park is 60 km. The park rangers drive in jeeps to patrol the border of the park. The time taken to complete one patrol at different speeds is given in the table.

Average Speed (km/h)	10	20	30	40	50	60
Time (h)	6	3	2	1.5	1.2	1

Each pair of values for speed and time can be thought of as an ordered pair and then can be plotted on a grid to give a picture of the relation. Follow these steps to graph a relation.

1. Draw and label the axes.
2. Plot the ordered pairs.
3. Join the ordered pairs with a smooth curve.
4. Give the graph a title.

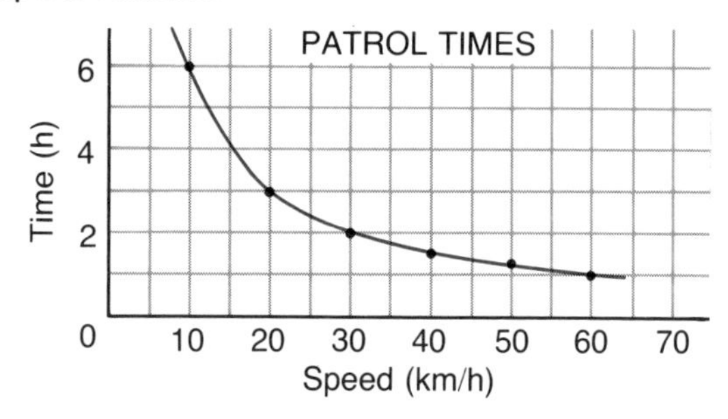

Example.
Some rescue ships have Safety Buoy launchers. A launcher propels a buoy, which is attached to a rope, to a person who has gone overboard. The table gives the height of the buoy versus the distance it travels, when the buoy is projected at an angle of 45°.
Draw a graph of this relation.

Distance (m)	0	5	10	15	20	25	30	35
Height (m)	1	3	7	8	7	4	2	0

Solution:
1. Draw and label the axes.
2. Plot the ordered pairs.
3. Join the points with a smooth curve.
4. Give the graph a title.

EXERCISE 4.3

B

1. The graph of a relation is shown below.

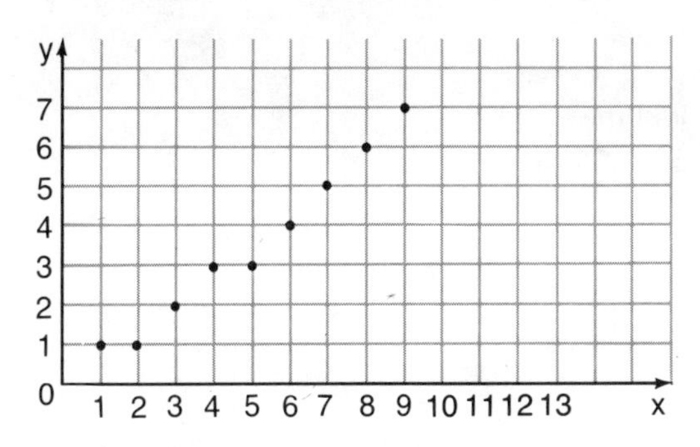

Copy and complete the table for this relation.

x	1	2		5			8	
y			2		4	5		7

2. The table gives the number of litres of gasoline used by a truck and the number of kilometres driven.

Litres of Gasoline	2	5	8	10	15
Kilometres	10	25	40	50	75

Draw the graph of this relation.

3. Veronica parachuted over an open field. The table gives her height above the ground at specific times.

Height (m)	Time (min)
500	0
400	3
300	5
200	6
100	8
0	10

Draw the graph of the relation between height and time.

4. The deeper you dive into a lake, the less sunlight penetrates the water. The table gives the percent of sunlight at different depths.

Depth (m)	Percent of Sunlight
0	100
10	80
30	50
50	30
70	30
100	10

Draw a graph of the relation.

5. The dimensions and areas of different squares are shown.

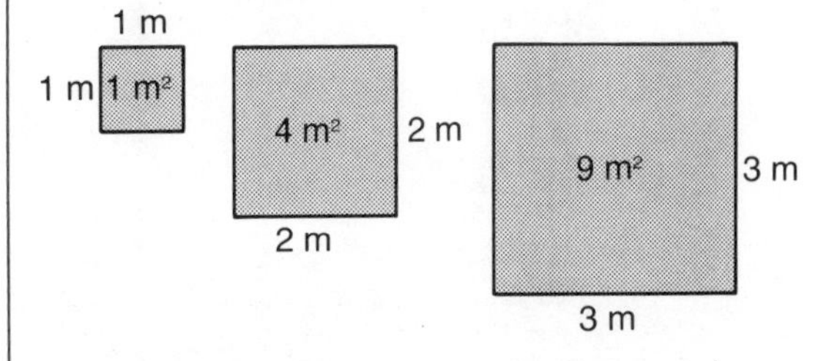

(a) Set up a table with Length of Side and Area as the headings.
(b) Find six ordered pairs for the relation.
(c) Draw a graph of the relation.

4.4 READING INFORMATION FROM GRAPHS

A light plane can carry a load of 1000 kg. Of this, a maximum of 560 kg of fuel can be carried. This leaves 440 kg for people or cargo. If the people and cargo load is over 440 kg, then less fuel can be carried. Less fuel means less flying time. The table gives the amount of fuel needed to fly a specific number of hours.

Flying Time (h)	0	1	2	3	4
Fuel (kg)	0	80	160	240	320

The ordered pairs in the table are graphed. From this graph we can determine either the amount of fuel or the number of flying hours. For example, to find the amount of fuel required to fly for 2.5 h, locate 2.5 on the x-axis. Draw a vertical line that intersects the graph. The corresponding value on the y-axis is 200. Therefore, 200 kg of fuel are required to fly 2.5 h. This is called interpolation because the values are between given ordered pairs.

To find values that exceed the given values we can extend the line of the graph. For example, a flight of 4 h and 15 min takes 340 kg of fuel. This is called extrapolation.

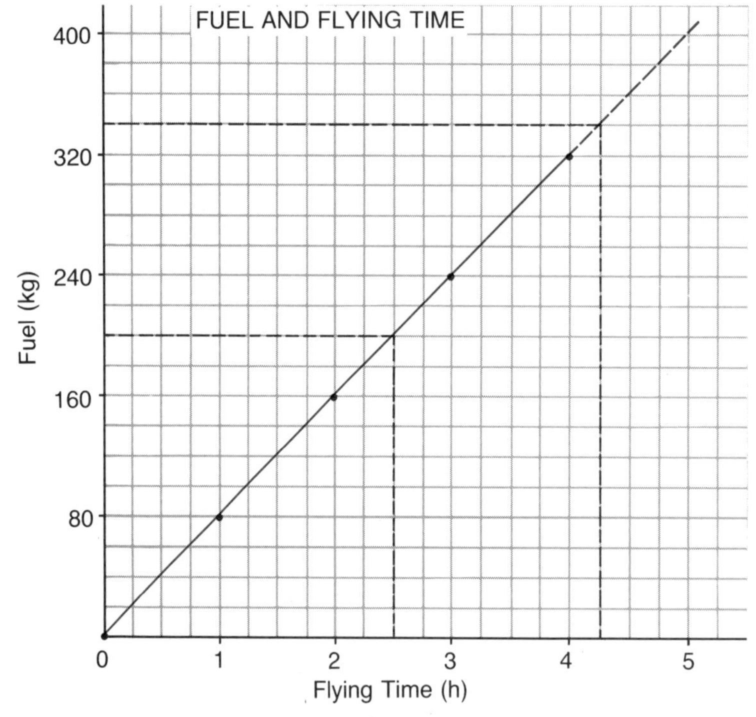

How much fuel is needed for a flight lasting 3 h and 45 min?
How long can you fly on 100 kg of fuel?
How long can you fly on 380 kg of fuel?

EXERCISE 4.4

B 1. The table gives the distance travelled by a speedboat moving at 200 km/h.

Time (h)	1	2	3	4
Distance (km)	200	400	600	800

(a) Display this information on a graph.
(b) Interpolate to find the distance travelled in 2.25 h.
(c) Extrapolate to find the distance travelled in 4.5 h.

2. By measuring the length of the tire skid marks at a traffic accident, police officers can determine the approximate speed of a vehicle before braking. The table gives the length of a skid mark for a vehicle travelling at a given speed on dry concrete.

Length (m)	Speed (km/h)
9	45
12	50
15	55
18	60
21	65
24	70

(a) Display this information on a graph.
(b) How fast was a car travelling if the skid mark measured 14 m?
(c) How fast was a car travelling if the skid mark measured 6 m?
(d) How fast was a car travelling if the skid mark measured 28 m?

3. The table gives the cost of peanuts according to mass.

Amount (kg)	2	4	6
Cost ($)	9	17	24

(a) Display this information on a graph.
(b) Interpolate to find the cost of 2.5 kg of peanuts.
(c) How many kilograms of nuts can you buy for $22?

C 4. A wristwatch was dropped from various heights to test shock resistance. The table gives the total distance travelled by the falling watch after each of the first three seconds.

Time (s)	0	1	2	3
Distance (m)	0	4.9	19.6	44.1

(a) Draw the graph of this relation.
(b) Find the total distance travelled after 2.5 s.
(c) Find the total distance travelled after 3.5 s.

MIND BENDER

Place the numbers 1, 2, 3, 4, 5, 6, 9, and 10 on the edges of the pyramid so that the sum of the edges meeting at each vertex is 16.

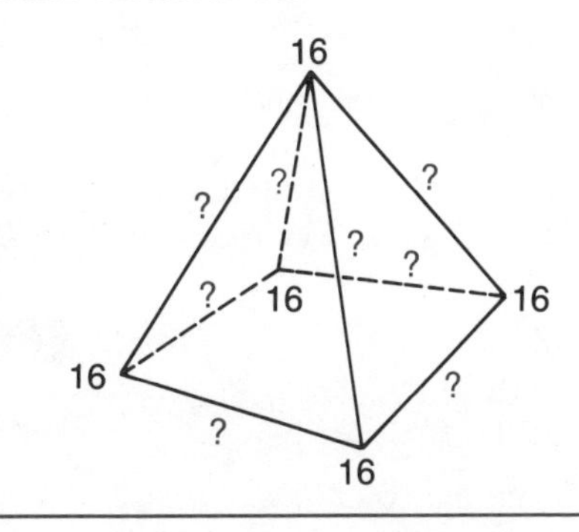

4.5 DIRECT VARIATION

During the summer, Marie-Hélène worked as a lifeguard at a wave pool. She was paid by the hour. The table gives her earnings for four weeks.

	Week 1	Week 2	Week 3	Week 4
Hours	10	15	20	25
Pay ($)	90	135	180	225

The relation between hours worked and pay is graphed at the right.
The graph starts at the origin (0, 0).
As the number of hours increases, Marie-Hélène's pay also increases.
When the hours are doubled, her pay is doubled.
When the hours are cut in half, her pay is likewise cut in half.
Notice that when the pay is divided by the hours worked, the result is the same number, or a constant.

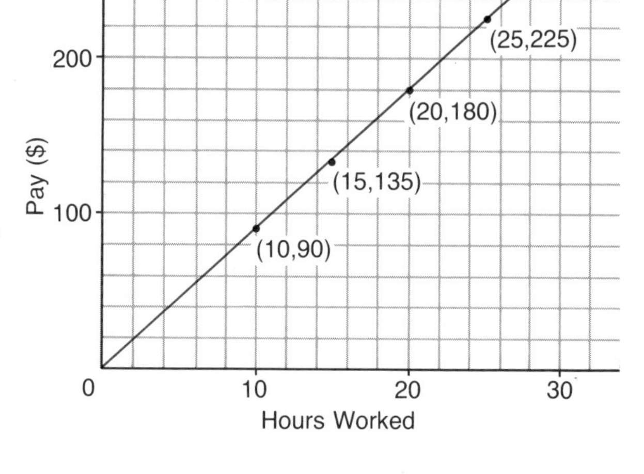

$$\frac{90}{10} = 9, \quad \frac{135}{15} = 9$$
$$\frac{180}{20} = 9, \quad \frac{225}{25} = 9$$

Here the constant is 9.

We can say that $\dfrac{\text{pay}}{\text{hours worked}}$ = a constant

or $\dfrac{p}{h} = k$, or $p = kh$

Mathematically we say that the pay varies directly as the number of hours worked. This is an example of direct variation. The constant, k, is called a constant of variation.

In the example above, since $k = 9$,

we can write $\dfrac{p}{h} = 9$ and $p = 9h$

We can now use this formula to solve other problems. For example, to calculate how much Marie-Hélène would earn in 18.5 h,

$p = 9h$
$p = 9(18.5)$
$= 166.50$

Press C 9 × 1 8 · 5 =

The display is 166.5

She would earn $166.50.

Example 1.
The cost of a hotel room varies directly with the number of days the room is rented. The Lees rented a room at $285 for three days.
How much will the room cost for eight days?

Solution:
First find the constant, k.

$\frac{c}{d} = k$ where c is the cost and d is the number of days

$\frac{285}{3} = k$

$k = 95$

Therefore, $\frac{c}{d} = 95$ and $c = 95d$.

When $d = 8$,
$c = 95(8)$
$= 760$

The room will cost $760 for eight days.

Example 2.

READ

When a powerboat is travelling at a certain speed and the motor is cut, the boat continues to move before coming to a complete stop. The distance the boat moves varies directly as the speed the boat is travelling just before the motor is turned off. The faster the speed, the longer the distance. The testing of a new powerboat showed that it travelled 12 m before stopping when the motor was cut at 15 km/h.
If the boat is moving at 40 km/h and the motor is cut, what distance will the boat travel before coming to a stop?

Solution:

PLAN

Find the constant, k.

E=mc²

$\frac{d}{v} = k$, where d is the distance before stopping in metres and v is the speed in kilometres per hour

$\frac{12}{15} = k$

SOLVE

$k = 0.8$

Therefore, $\frac{d}{v} = 0.8$ and $d = 0.8v$

When $v = 40$ km/h,
$d = 0.8(40)$
$= 32$

ANSWER

The boat will move 32 m before coming to a stop.

EXERCISE 4.5

B 1. Does each set of data represent a direct variation? If your answer is yes, state the constant of variation.

(a)

Distance (km)	70	140	210	350
Time (h)	1	2	3	5

(b)

Number of Cars	2	3	5	7	9
Number of Spark Plugs	12	24	30	42	72

(c)

Number of Shirts	2	3	5	8
Cost of Shirts ($)	70	135	175	280

2. Write an equation with a constant of variation to describe each relation.

(a)

Pay ($)	8	16	24	56	72
Hours Worked	1	2	3	7	9

(b)

Number of Students	30	90	120	150	270
Number of Buses	1	3	4	5	9

(c)

Distance Travelled (km)	160	400	480	640
Hours Driving	2	5	6	8

3. In the following, y varies directly as x.

(a)

x	y
2	4
3	
	10
8	

(b)

x	y
3	9
4	
	21
9	

(c)

x	y
8	4
10	
	3
14	

(i) Find the missing numbers.
(ii) Find the constant, k.
(iii) Write an equation relating y to x.

4. x varies directly as y.
When x = 45, y = 15.

(a) Find the value of k and write an equation relating x to y.
(b) Find x when y = 19.

5. The cost of building a bridge varies directly as the length of the bridge.
If it costs $4 000 000 to build a 50 m bridge, how much would it cost to build a 125 m bridge?

6. At a department store the number of Christmas packages wrapped varies directly as the number of people wrapping.
If three people can wrap 225 presents in a day, how many presents could seven people wrap in a day?

7. The cost for parking varies directly as the number of hours parked.
If it costs $11.25 to park for five hours, how much would it cost to park for eight hours?

8. The cost of a whale watch cruise depends on the length of the cruise. If a three-hour cruise costs $10.50, how much would a five-hour cruise cost?

C

9. If the area of a cone's base remains constant, the volume of a cone varies directly as the height.
If the volume of a cone is 132 cm^3 and the height is 14 cm, calculate the volume if the height increases to 21 cm.

4.6 PARTIAL VARIATION

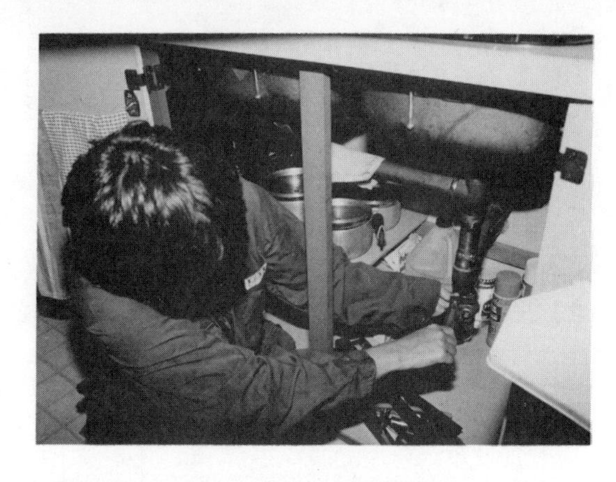

There are many familiar examples of partial variation.

A car rental agency charges \$50/d plus \$0.20/km. The \$50 is called the fixed part. Whether you drive 500 km or 50 km, the fixed cost would not change. The \$0.20/km is the variable part. This cost varies depending on how many kilometres you drive. How much would you pay if you rented a car for one day and drove 300 km?

A plumber charges \$40 for a service call to your home plus \$55/h. The \$40 charge for the service call is called the fixed part. Despite any number of hours the plumber may work, the fixed cost remains the same, or constant. The \$55/h is called the variable part. How much would you pay if the plumber worked 4 h?

The cost of publishing a novel has two parts. Examples of the fixed part are the cost of printing, staff salaries, and so on. The variable part depends on how many copies of the book are printed. Variable costs include paper, ink, publishing, and printing charges.

The fixed costs for publishing a book are \$50 000 and the variable costs are \$20 000 for every 1000 books printed. The table and the graph show how the total cost varies according to the number of copies printed.

Number of Copies	Total Cost (\$)
1000	70 000
2000	90 000
3000	110 000
4000	130 000
5000	150 000

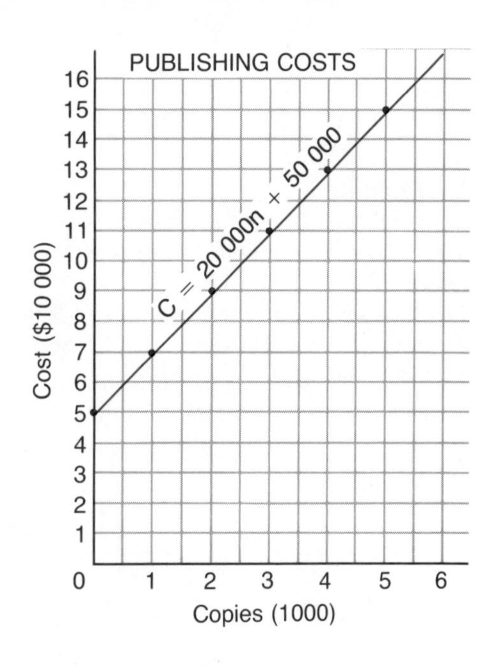

This is an example of partial variation. The graph is a straight line similar to the graph for a direct variation. It is not a direct variation because the total cost divided by the number of copies is not a constant.

A formula for this relation is

$$C = 20\,000n + 50\,000$$

where n is the number of copies in thousands and C is the total cost

C — Total cost; 20 000n — Variable cost; 50 000 — Fixed cost

Example 1.

The Fraser Company runs deep-sea fishing trips. The fixed cost to rent a boat and captain for a day is \$300. The charge for each passenger is \$50.

(a) Draw a graph of the relation.
(b) Write an equation relating the cost and the number of passengers.
(c) If there are 8 passengers, how much will each pay?

Solution:

(a) Set up a table of values and then draw the graph.

Number of Passengers	0	1	2	3	4
Cost (\$)	300	350	400	450	500

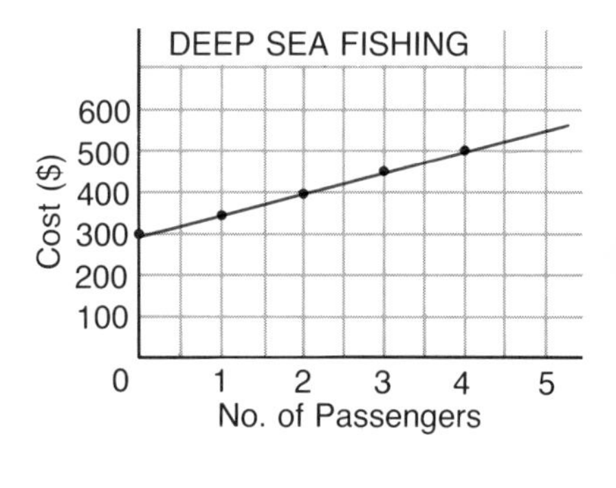

(b) The equation relating the cost, C, to the number of passengers, n, is

$$C = 50n + 300$$

(c) For 8 passengers the total cost is

$$\begin{aligned} C &= 50n + 300 \\ C &= 50(8) + 300 \\ &= 400 + 300 \\ &= 700 \end{aligned}$$

Each will pay $\frac{700}{8}$ or \$87.50.

Example 2.

The fuel tank of a small car holds 50 L of gasoline. The engine burns gasoline at a rate of 10 L/100 km. Suppose the tank is full and the car is driven around a test track until the tank is empty.

(a) Draw a graph of the number of litres remaining, L, in the tank versus the distance travelled in hundreds of kilometres, d.
(b) Write an equation relating L and d.
(c) How much fuel remained in the tank after 350 km?

Solution:

(a)

Distance (km)	0	100	200	300	400	500
Litres Remaining	50	40	30	20	10	0

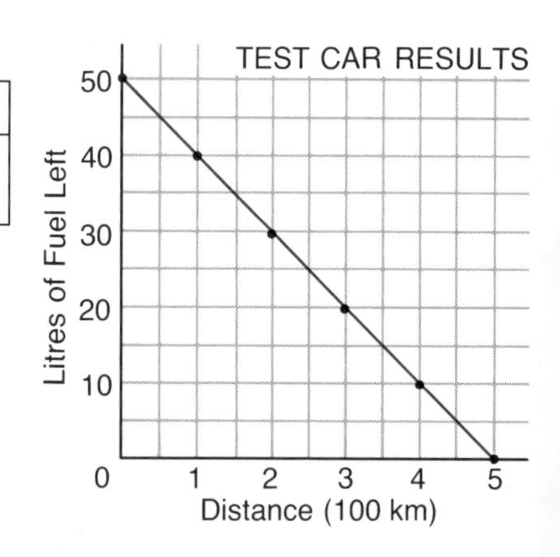

(b) $L = 50 - 10d$

(c)
$$\begin{aligned} L &= 50 - 10d \\ L &= 50 - 10(3.5) \\ &= 50 - 35 \\ &= 15 \end{aligned}$$

Fifteen litres remained in the tank.

EXERCISE 4.6

B 1. The table shows that the cost of a southern vacation varies with the length of the vacation in days.

Number of Days	Cost ($)
1	700
2	900
3	1100
4	1300
5	1500

(a) Draw a graph of this partial variation.
(b) What is the fixed cost? What might this cost represent?
(c) Write an equation relating cost to number of days.
(d) How much would a 14 d vacation cost?

2. The Big Toy Car Limousine will rent you a limousine and driver for an initial fee of $80 plus $20/h.
(a) Draw a graph of this relation.
(b) Write an equation relating cost and the number of hours you rent the car.
(c) How much would it cost to rent the limousine for 6.5 h?

3. The River Rangers Band can be hired with a minimum of three musicians for $600 an evening. Each additional musician costs $125.
(a) Write an equation relating the cost to the number of musicians.
(b) How much would you pay for a 14 member band?

4. The Harbour View Inn offers Murder Mystery Weekends. The cost for two nights accommodation, participating in the mystery, and meals is $600 for the group plus $225 per person in the group.
(a) Write an equation relating the total cost and the number of people attending.
(b) What is the total cost if 37 people attend?

5. An electrician charges $56 for a house call and $41/h for each hour of work.
(a) Write an equation relating cost to the number of hours.
(b) What would a 4.5 h house call cost?

6. The Silverado Ring Company specializes in rings for teams and schools. The members of the Eastwick High School Marching Band won a provincial championship and decide to purchase rings. It costs the Silverado Company $800 to cast the die for the ring. The material and time to make each ring will cost $150.
(a) Write an equation relating total cost to the number of rings made.
(b) How much would it cost to make 10 rings?
How much would each person pay to buy a ring?
(c) How much would it cost to make 80 rings?
How much would each person pay to buy a ring?

MIND BENDER

The number 371 has the property as shown

$$371 = 3^3 + 7^3 + 1^3$$
$$= 27 + 343 + 1$$
$$= 371$$

(a) A three-digit number that begins with one has the same property. What is this number?
(b) A three-digit number that begins with four has the same property. What is this number?

4.7 GRAPHING LINEAR RELATIONS

In the table at the right the second number is 2 more than the first number for each and every number.

We can write the numbers in the table as ordered pairs

(0, 2), (1, 3), (2, 4), (3, 5)

First Number	Second Number
0	2
1	3
2	4
3	5

We can graph the ordered pairs on a grid as shown. If we let x represent the first number and y the second number, we can write an equation that relates x and y.

$$y = x + 2$$

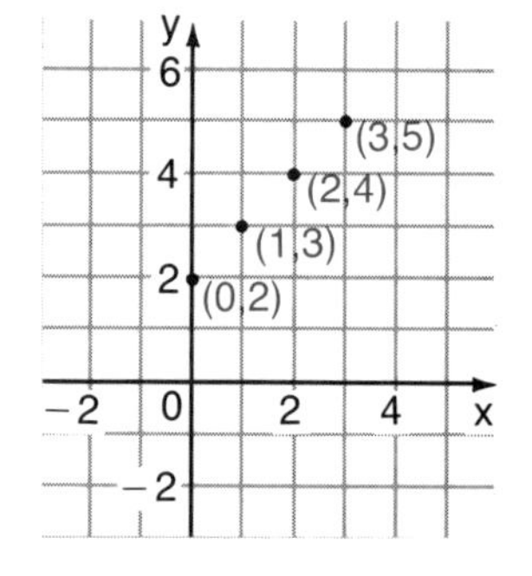

Equations written in this form provide us with a simple way of calculating more ordered pairs that satisfy the equation.

Equation: $y = x + 2$

Ordered Pairs

First Number	Second Number	Ordered Pair
x	$y = x + 2$	(x, y)
4	$y = 4 + 2$	(4, 6)
5	$y = 5 + 2$	(5, 7)
8	$y = 8 + 2$	(8, 10)
9	$y = 9 + 2$	(9, 11)

Example 1.
Draw a graph of the equation $y = 2x + 3$.

Solution:

1. Construct a table of values for at least 4 ordered pairs.

Equation: $y = 2x + 3$

Ordered Pairs

x	$y = 2x + 3$	(x, y)
1	$y = 2(1) + 3$	(1, 5)
2	$y = 2(2) + 3$	(2, 7)
3	$y = 2(3) + 3$	(3, 9)
4	$y = 2(4) + 3$	(4, 11)

2. Plot the ordered pairs.

Since x can be any number, including fractions and decimals, we can draw a line through the points.

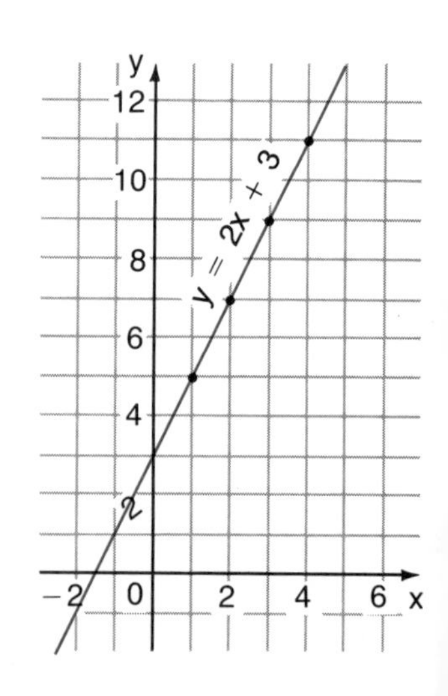

Example 2.
Draw the graph of the equation $x + y = 5$.

Solution:
Construct a table of values for 5 ordered pairs. Plot and then join the ordered pairs.

Ordered Pairs

x	x + y = 5	(x, y)
1	1 + (4) = 5	(1, 4)
2	2 + (3) = 5	(2, 3)
3	3 + (2) = 5	(3, 2)
4	4 + (1) = 5	(4, 1)
5	5 + (0) = 5	(5, 0)

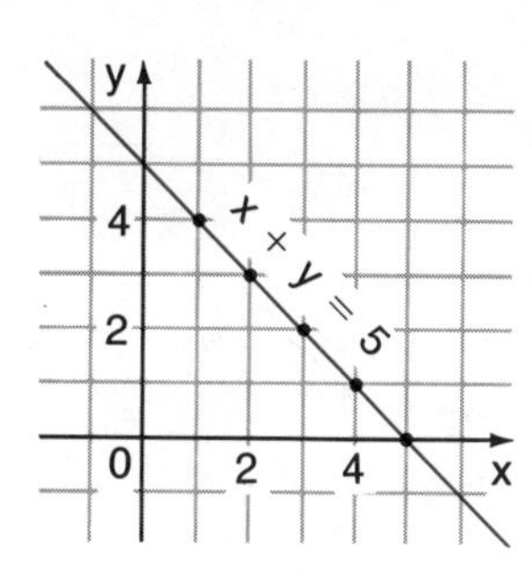

EXERCISE 4.7

B

1. For each table of values, determine the pattern and find the missing numbers.

(a)

x	y
1	5
2	4
3	
6	
	1

(b)

x	y
4	3
6	5
2	
7	
	4

2. Copy and complete each table.

(a)

x	y = 3x + 2
2	
4	
5	
8	

(b)

x	y = 2x − 1
1	
3	
6	
7	

(c)

x	y = 4x − 3
3	
0	
2	
5	
7	

(d)

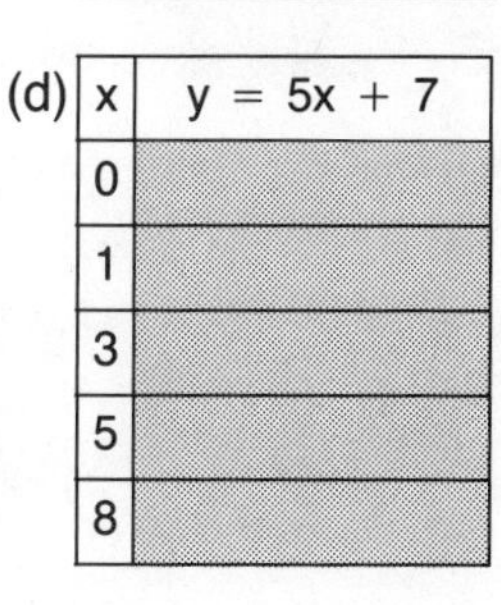

x	y = 5x + 7
0	
1	
3	
5	
8	

3. Which of the following points are on the line $x + y = 7$?

(a) (3, 4) (b) (5, 3) (c) (0, 7)
(d) (1, 6) (e) (4, 4) (f) (−1, 8)

4. Which of the following points are on the line $y = 3x - 1$?

(a) (1, 2) (b) (2, 6) (c) (3, 8)
(d) (0, −1) (e) (−1, −5) (f) (−2, −7)

5. Draw the graph of each equation.

(a) $y = x + 4$ (b) $y = x + 1$
(c) $y = x - 3$ (d) $y = x - 2$
(e) $y = 2x + 1$ (f) $y = 2x - 3$
(g) $y = 3x + 2$ (h) $y = 4x + 3$
(i) $y = 5x + 1$ (j) $y = 2x + 8$

6. Graph the following equations.

(a) $x + y = 7$ (b) $x + y = 3$
(c) $x + y = 4$ (d) $x + y = 6$
(e) $x - y = 1$ (f) $x - y = 2$
(g) $x - y = 3$ (h) $x - y = 5$

C

7. Graph the following equations.

(a) $y = -2x + 3$ (b) $y = -3x - 2$
(c) $y = -x + 4$ (d) $y = -x - 1$

4.8 GRAPHING USING $y = mx + b$

When linear relations are written in this way, $x + y = 4$ or $y = 2x + 1$, finding ordered pairs that satisfy the relation is not difficult. For relations, such as $4x + 2y = 7$, finding ordered pairs is simplified if we rewrite the relation as follows.

$$4x + 2y = 7$$
$$2y = -4x + 7$$
$$\frac{2y}{2} = \frac{-4x}{2} + \frac{7}{2}$$
$$y = -2x + \frac{7}{2}$$

The relation is now expressed in the general form

$y = mx + b$ where m, the coefficient of x, is -2 and b, the constant term, is $\frac{7}{2}$

We will investigate what the values of m and b mean in the Exercise.

We now find ordered pairs by substituting values for x.

x	$y = -2x + \frac{7}{2}$	(x, y)
0	$y = -2(0) + \frac{7}{2}$ $= 0 + \frac{7}{2}$ $= 3\frac{1}{2}$	$(0, 3\frac{1}{2})$
2	$y = -2(2) + \frac{7}{2}$ $= -4 + \frac{7}{2}$ $= -\frac{1}{2}$	$(2, -\frac{1}{2})$
−2	$y = -2(-2) + \frac{7}{2}$ $= 4 + \frac{7}{2}$ $= 7\frac{1}{2}$	$(-2, 7\frac{1}{2})$

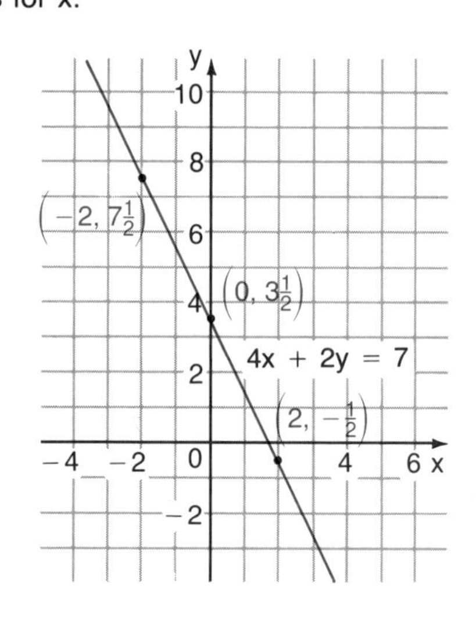

The graph of $4x + 2y = 7$ is shown at the right.

Example.
Write the relation $2x - 6y = 7$ in the form $y = mx + b$, and state the values of m and b.

Solution:

$$2x - 6y = 7$$
$$-6y = -2x + 7$$
$$\frac{-6y}{-6} = \frac{-2x}{-6} + \frac{7}{-6} \quad \longleftarrow \text{Divide by } -6$$
$$y = \frac{1}{3}x - \frac{7}{6} \qquad m = \frac{1}{3} \text{ and } b = -\frac{7}{6}$$

EXERCISE 4.8

A

1. State the value of m and b in each of the following.

(a) $y = 2x + 3$ (b) $y = -3x - 2$
(c) $y = 4x + 6$ (d) $y = -\frac{1}{2}x$
(e) $y + 3x = 8$ (f) $y - 2x + 7 = 0$

B

2. Write in the form $y = mx + b$

(a) $y + 3x - 7 = 0$ (b) $y - 4x = 6$
(c) $3x + y = 9$ (d) $4x + 6 = y$
(e) $5x = 3 + y$ (f) $2 = -3x + y$

3. Write in the form $y = mx + b$

(a) $2y = 4x + 6$ (b) $3y = -15x + 12$
(c) $4y + 6x = 8$ (d) $2y + 7x = 12$
(e) $5y - 10x = 15$ (f) $6x + 2y = 8$

4. Write in the form $y = mx + b$

(a) $3x - 2y = 8$ (b) $4x + 5y = 7$
(c) $-3y + 6 = 8x$ (d) $5y - 3x = 6$
(e) $4x + 6 = -5y$ (f) $3x + 2y = 17$

5. Write in the form $y = mx + b$

(a) $-y = 3x + 2$ (b) $-y = -4x - 7$
(c) $-y - 2x = 6$ (d) $3 - y = 8x$
(e) $3x - y = 12$ (f) $5x = 7 - y$

6. Write in the form $y = mx + b$

(a) $4x + 2y - 8 = 0$
(b) $3x - 2y - 6 = 0$
(c) $7x - 3y - 6 = 0$
(d) $-3x - 4y + 6 = 0$
(e) $-2 + 3y = 8x$
(f) $-4y + 3x = 12$

7. Express the following in the form $y = mx + b$, and then sketch the graph. Find at least 3 ordered pairs for each.

(a) $3x + y = 6$ (b) $2y + x = 6$
(c) $2y - x = 3$ (d) $2x - y = 4$
(e) $x - 3 = 2y$ (f) $3x + 2y = 1$
(g) $x - 3y = 2$ (h) $x + 3y = 0$
(i) $4y + 3 = x$ (j) $x - y + 2 = 0$
(k) $x + y + 3 = 0$ (l) $x - y - 1 = 0$
(m) $x + y = 0$ (n) $x - y = 0$
(o) $4x + 2y = 6$ (p) $y - x = 0$

INVESTIGATIONS

8. On the same set of axes sketch the graphs of the following. Use at least 3 ordered pairs to sketch each relation.

(a) $y = 2x + 1$

x	$y = 2x + 1$	y
0	$y = 2(0) + 1$	1
1	$y = 2(1) + 1$	3
2	$y = 2(2) + 1$	5

(b) $y = 2x + 5$
(c) $y = 2x - 3$
(d) $y = 2x$
(e) $y = 2x + 4$

These relations are in the form $y = mx + b$, where $m = 2$ and $b = -3, 0, 1, 4, 5$.

9. On the same set of axes sketch the graphs of the following. Use at least 3 ordered pairs to sketch each relation.

(a) $y = -2x + 3$

x	$y = -2x + 3$	y
0	$y = -2(0) + 3$	3
1	$y = -2(1) + 3$	1
2	$y = -2(2) + 3$	-1

These relations are in the form $y = mx + b$, where $m = -2$ and $b = 6, 3, 0, -1, -3$.

(b) $y = -2x + 6$
(c) $y = -2x - 3$
(d) $y = -2x - 1$
(e) $y = -2x$

10. Using the graphs in questions 8 and 9, determine the role of b in $y = mx + b$.

11. On the same set of axes sketch the graphs of the following. Use at least 3 ordered pairs from each relation.

(a) $y = 2x + 3$

x	y = 2x + 3	y
0	$y = 2(0) + 3$	3
1	$y = 2(1) + 3$	5
2	$y = 2(2) + 3$	7

(b) $y = x + 3$
(c) $y = -2x + 3$
(d) $y = -\frac{1}{2}x + 3$
(e) $y = 3x + 3$

These relations are in the form $y = mx + b$, where $b = 3$ and $m = -2, -\frac{1}{2}, 1, 2, 3$.

12. On the same set of axes sketch the graphs of the following. Use at least 3 ordered pairs from each relation.

(a) $y = 2x - 2$

x	y = 2x − 2	y
0	$y = 2(0) - 2$	−2
1	$y = 2(1) - 2$	0
2	$y = 2(2) - 2$	2

(b) $y = -2x - 2$
(c) $y = x - 2$
(d) $y = 3x - 2$
(e) $y = -\frac{1}{2}x - 2$

These relations are in the form $y = mx + b$, where $b = -2$ and $m = 3, 2, 1, -\frac{1}{2}, -2$.

13. (a) Using the graphs in questions 11 and 12, discuss the role of m in $y = mx + b$.
(b) For what values does the line rise
  (i) upwards to the right?
  (ii) upwards to the left?
(c) How does the value of m affect the steepness of the graph?

14. Match the equations with the appropriate graphs.

(a) $y = 2x - 2$ (b) $y = -2x$
(c) $y = 2x + 2$ (d) $y = -\frac{1}{2}x + 3$
(e) $y = -3x - 4$

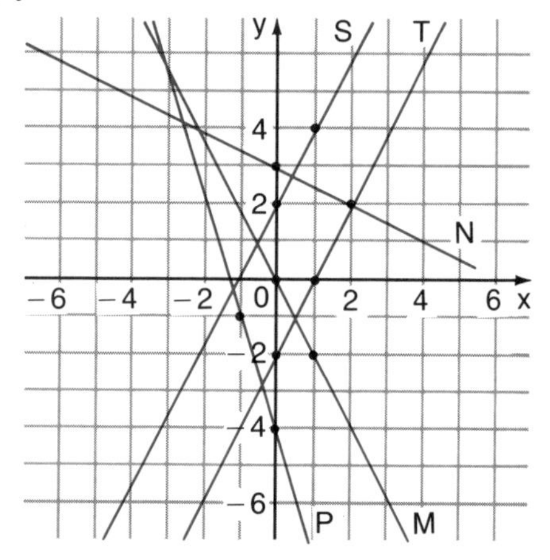

15. Match the equations with the appropriate graphs.

(a) $y = -3x + 1$ (b) $y = x$
(c) $y = 2x + 1$ (d) $y = \frac{1}{2}x - 2$

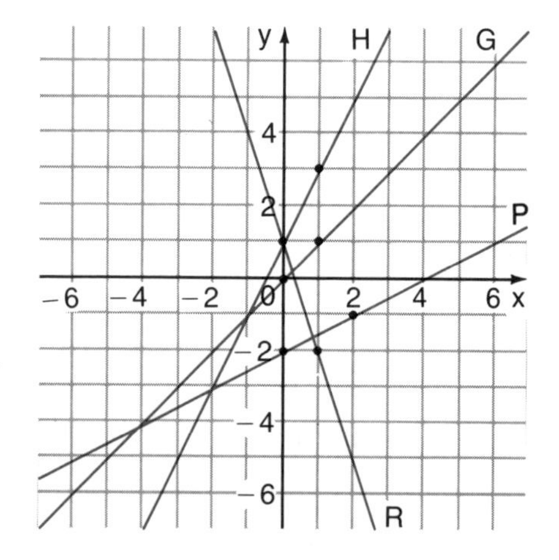

4.9 GRAPHING RELATIONS USING INTERCEPTS

Another way to graph a relation is to use intercepts. For any relation, such as $3x + 2y = 12$, the y-intercept is the distance from the origin to where the line of the graph crosses the y-axis. At the y-intercept, $x = 0$ since $x = 0$ for all points on the y-axis.

If $x = 0$,

$$\begin{aligned} 3x + 2y &= 12 \\ 3(0) + 2y &= 12 \\ 2y &= 12 \\ y &= 6 \end{aligned}$$

6 is called the y-intercept.
One ordered pair for the line is then (0, 6).

The x-intercept is the distance from the origin to where the line of the graph crosses the x-axis.
At the x-intercept, $y = 0$ since $y = 0$ for all points on the x-axis.

If $y = 0$,

$$\begin{aligned} 3x + 2y &= 12 \\ 3x + 2(0) &= 12 \\ 3x &= 12 \\ x &= 4 \end{aligned}$$

4 is called the x-intercept.
Another ordered pair for the line is (4, 0).

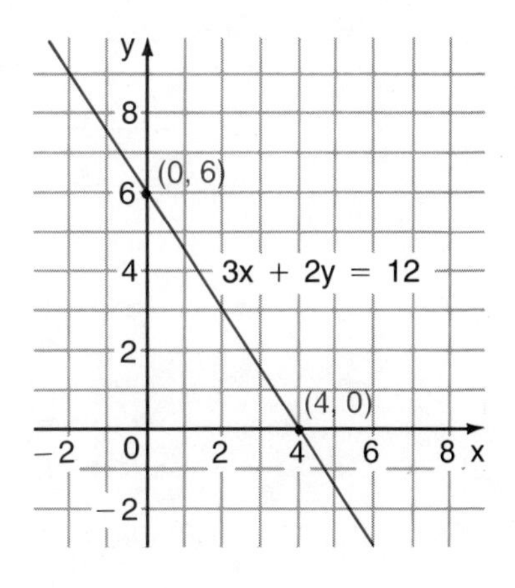

EXERCISE 4.9

A 1. State the x- and y-intercepts of the following.

(a) $3x + 4y = 12$
(b) $2x + 3y = 6$
(c) $5x + 2y = 10$
(d) $3x + 5y = 15$
(e) $x + 2y = 4$
(f) $3x + y = 6$
(g) $2x - y = 6$
(h) $3x - 2y = 6$
(i) $4x - 3y = 12$
(j) $x - 3y = 9$

B 2. Graph the lines which have these intercepts.

(a) x-intercept 2, y-intercept 3
(b) x-intercept 3, y-intercept −1
(c) x-intercept −4, y-intercept 5
(d) x-intercept −2, y-intercept −3

3. Graph the following relations, using the x- and y-intercepts.

(a) $x + y = 6$
(b) $x - y = 2$
(c) $2x + y = 6$
(d) $x - 2y = 4$
(e) $2x - y = 4$
(f) $2x + 3y = 6$
(g) $2x - 3y = 12$
(h) $3x - 5y = 15$
(i) $-3x + 2y = 6$
(j) $-4x + 3y = -12$
(k) $3x + y = -9$
(l) $-x - y = -1$

MIND BENDER

Put these units of length in order from shortest to longest.

1. centimetre
2. furlong
3. fathom
4. hand
5. nautical mile
6. rod
7. league
8. millimetre
9. inch
10. mile

4.10 GRAPHING NON-LINEAR RELATIONS

In the previous section we graphed linear relations, or relations that represent straight lines. In this section we will graph relations that appear as curved lines. These are called non-linear relations.

Example 1.
Graph the equation. $y = x^2$

Solution:
First make a table of values. Start with $x = 0$.

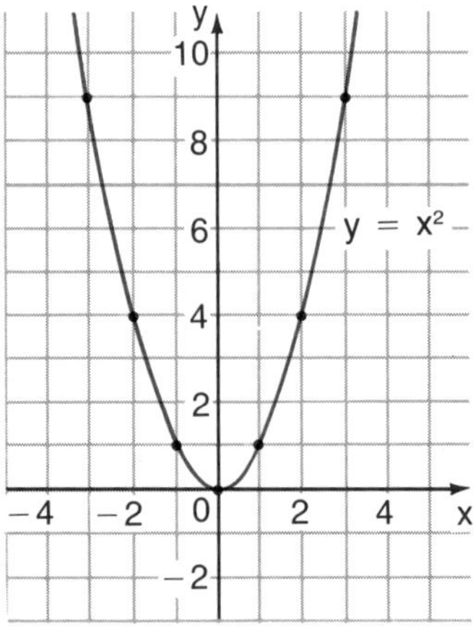

Equation

$y = x^2$

Ordered Pairs

x	$y = x^2$	(x, y)
0	$y = (0)^2$	(0, 0)
1	$y = (1)^2$	(1, 1)
−1	$y = (-1)^2$	(−1, 1)
2	$y = (2)^2$	(2, 4)
−2	$y = (-2)^2$	(−2, 4)
3	$y = (3)^2$	(3, 9)
−3	$y = (-3)^2$	(−3, 9)

Join the points with a smooth curve.
This curve is called a parabola.

Example 2.
Graph the equation. $x^2 + y^2 = 16$

Solution:
Make a table of values, starting with $x = 0$.

Equation

$x^2 + y^2 = 16$

Ordered Pairs

x	Calculating y	(x, y)
0	$0^2 + y^2 = 16$ $y^2 = 16$ $y = \pm 4$	(0, 4) (0, −4)
1	$1^2 + y^2 = 16$ $y^2 = 15$ $y = \pm\sqrt{15}$ $y \doteq 3.9$	(1, 3.9) (1, −3.9)
−1	$(-1)^2 + y^2 = 16$ $1 + y^2 = 16$ $y \doteq \pm 3.9$	(−1, 3.9) (−1, −3.9)

x	Calculating y	(x, y)
2	$2^2 + y^2 = 16$ $4 + y^2 = 16$ $y^2 = 12$ $y \doteq \pm 3.5$	(2, 3.5) (2, −3.5)
−2	$(-2)^2 + y^2 = 16$ $y = \pm 3.5$	(−2, 3.5) (−2, −3.5)
3	$3^2 + y^2 = 16$ $y^2 = 7$ $y \doteq \pm 2.6$	(3, 2.6) (3, −2.6)
−3	$(-3)^2 + y^2 = 16$ $y \doteq \pm 2.6$	(−3, 2.6) (−3, −2.6)

x	Calculating y	(x, y)
4	$4^2 + y^2 = 16$ $y^2 = 0$ $y = 0$	(4, 0)
−4	$(-4)^2 + y^2 = 16$ $y^2 = 0$ $y = 0$	(−4, 0)

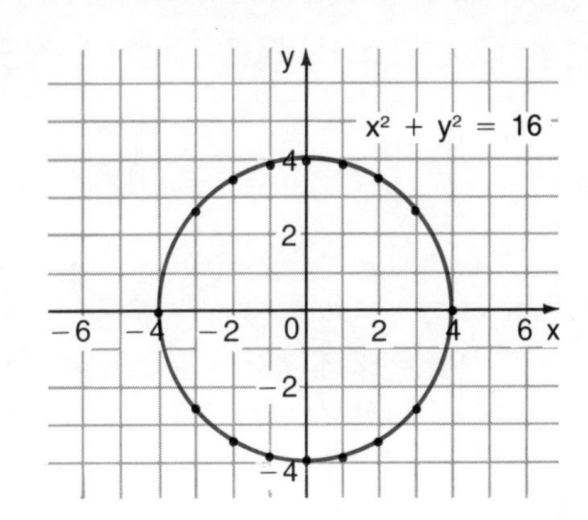

The graph of the relation is a circle.

EXERCISE 4.10

B 1. Sketch the graphs of the following parabolas.

(a) Equation $y = x^2 + 2$

Ordered Pairs

x	$y = x^2 + 2$	(x, y)
0		
1		
−1		
2		
−2		
3		
−3		

(b) Equation $y = x^2 - 3$

Ordered Pairs

x	$y = x^2 - 3$	(x, y)
0		
1		
−1		
2		
−2		
3		
−3		

(c) $y = x^2 + 4$
(d) $y = x^2 - 1$
(e) $y = x^2 + 5$
(f) $y = x^2 - 4$
(g) $y = x^2 + 6$
(h) $y = x^2 - 5$

2. Sketch the graphs of the following circles.

(a) Equation $x^2 + y^2 = 25$

Ordered Pairs

x	Calculating y	(x, y)
0		
3		
−3		
4		
−4		
5		
−5		

(b) Equation $x^2 + y^2 = 36$

Ordered Pairs

x	Calculating y	(x, y)
0		
2		
−2		
4		
−4		
6		
−6		

(c) $x^2 + y^2 = 9$
(d) $x^2 + y^2 = 100$
(e) $x^2 + y^2 = 49$
(f) $x^2 + y^2 = 64$

EXTRA

4.11 GALLUP HISTORY TEST

The Gallup Corporation, a large polling company, gave the following multiple-choice history test to a sample of 1016 Canadian adults.

1. What is the name of Canada's first Prime Minister?

(a) W. L. Mackenzie King (b) Sir John A. Macdonald (c) Samuel de Champlain
(d) Sir Wilfrid Laurier (e) Louis-Joseph Papineau (f) Sir Robert Borden

2. Who was Canada's longest serving Prime Minister?

(a) Sir John A. Macdonald (b) Sir Wilfrid Laurier (c) W. L. Mackenzie King
(d) John Diefenbaker (e) Pierre Trudeau (f) Louis St. Laurent

3. Who was the leader of the North-West Rebellion of 1885?

(a) Sir John A. Macdonald (b) Gabriel Dumont (c) Louis Riel
(d) William Lyon Mackenzie (e) Louis-Joseph Papineau (f) Sir Charles Tupper

4. In what year did Confederation take place?

(a) 1776 (b) 1812 (c) 1837
(d) 1865 (e) 1867 (f) 1885

5. Who was the leader of the English army at the Battle of the Plains of Abraham?

(a) Joseph Howe (b) Edward Cornwallis (c) James Wolfe
(d) John Durham (e) Louis-Joseph de Montcalm (f) Sir Isaac Brock

6. Who was the leader of the French Army at the Battle of the Plains of Abraham?

(a) Louis-Joseph de Montcalm (b) Louis-Joseph Papineau (c) James Wolfe
(d) Samuel de Champlain (e) Jacques Cartier (f) Louis de Buade Frontenac

Sir Charles Tupper

Sir John A. Macdonald

The results of the test are displayed in the table.

CANADIAN REPORT CARD

		Correct Answer	Other Answer	Don't Know
1.	What is the name of Canada's first Prime Minister?	40%	16%	44%
2.	Who was Canada's longest serving Prime Minister?	18%	56%	26%
3.	Who was the leader of the North-West Rebellion of 1885?	37%	2%	61%
4.	What year was Confederation?	45%	15%	40%
5.	Who was the leader of the English army at the Battle of the Plains of Abraham?	31%	5%	64%
6.	Who was the leader of the French army at the Battle of the Plains of Abraham?	25%	8%	67%

This table shows how many questions were answered correctly.

6 Correct	5 Correct	4 Correct	3 Correct	2 Correct	1 Correct
5%	9%	10%	13%	15%	17%

Representation of the Battle of the Plains of Abraham, 1759
by G.B. Campion
(water colour)

EXERCISE 4.11

1. (a) If getting 3 questions correct is considered a pass, what percent of the sample passed?
(b) How many of the sample did not pass?

2. How many of the sample answered question 1 correctly?

3. How many did not know the answer to question 3?

4. How many people in the sample did not answer any questions correctly?

5. Is this quiz a good test of Canadians' knowledge of history?

6. How do you compare to the Canadians that took the test?

4.12 PROBLEM SOLVING

1. The Eagle Theatre has 725 seats. There are 510 orchestra seats and 215 balcony seats. Tickets for Shakespeare's *Julius Caesar* cost \$70 each for orchestra seats and \$45 each for balcony seats. On opening night the total receipts were \$42 240. There were 480 orchestra seats sold.
How many balcony seats were sold?

2. If there was a sidewalk built along the equator, how long would you take to walk around the world?

3. The sum of the first two even numbers is $2 + 4 = 6$, or 2×3.
The sum of the first three even numbers is $2 + 4 + 6 = 12$, or 3×4.
The sum of the first four even numbers is $2 + 4 + 6 + 8 = 20$, or 4×5.
Determine the pattern in the above and then use it to find the sum of the first 20 even numbers.

4. The number 144 is a perfect square because $12^2 = 144$.
What is the largest three-digit perfect square?

5. The prize money for a golf tournament was divided among four golfers as follows: Smith received one-half of the money; Lee, one-quarter of the money; Roche, one-fifth of the money; and Bevan, \$7000.
What was the total prize money?

6. Place the numbers from 11 to 19 in the circles so that the sum of each side in the triangle is 60.

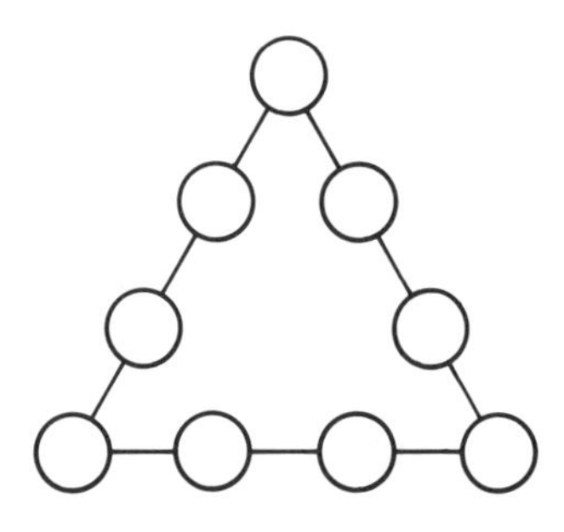

7. Carol got her pay cheque on Friday. She bought a book for \$30. She then used half of what she had left for a car payment. She bought groceries for \$75 and put \$200 in the bank. This left her with \$40.
How much was her pay cheque?

8. Determine the pattern of the products and then use the pattern to find the last two products, without multiplying.

$15 \times 15 = 225$
$25 \times 25 = 625$
$35 \times 35 = 1225$
$45 \times 45 = 2025$
$55 \times 55 = 3025$
$65 \times 65 = ■$
$75 \times 75 = ■$

9. How many times will you see the digit 7 in the page numbers of the first 100 pages of this book?

10. At the beginning of a game three coins are laid on a table as shown. Turning over two coins next to each other is a move. How can you get three heads in two moves?

11. What is the next year that February 14 will fall on a Monday?

12. What is the maximum number of times four straight lines can intersect at once?

13. In how many ways can you make change for a quarter?

14. Twenty fence posts are evenly spaced around a rectangular field. There are eight posts along the length of the field.
How many posts are along the width?

15. The volume of a cube is 729 cm^3.
What is the length of each edge?

16. Which wheels are turning clockwise?

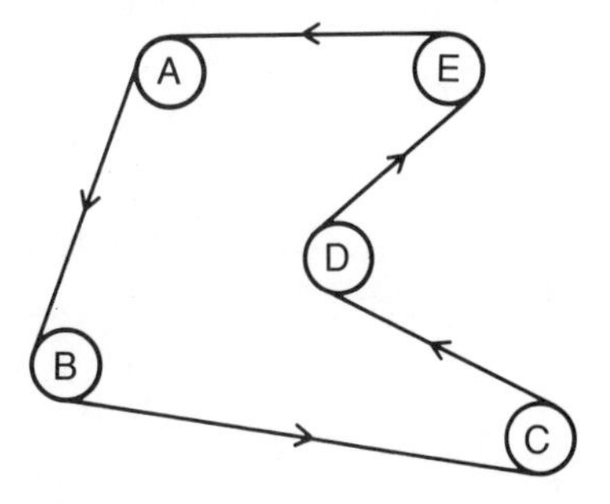

17. How old will you be in minutes on your next birthday?

18. The numbers in the large squares are found by adding the numbers in the small squares.
What are the numbers in the small squares?

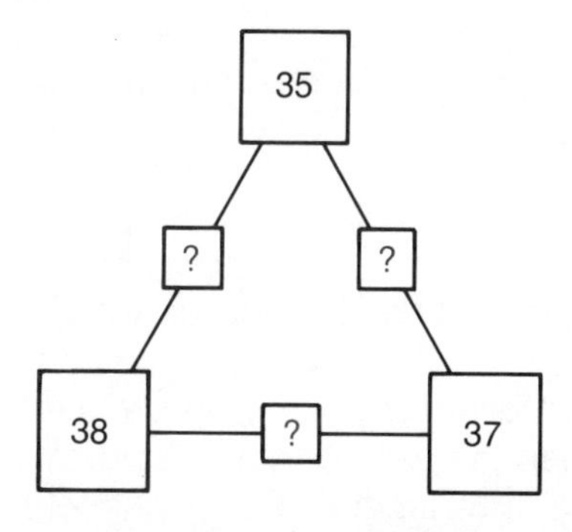

19. The gasoline consumption on Lou's car is 16 L/100 km. He lives 45 km from work.
(a) How much gasoline will he use driving to and from work in a five-day week?
(b) What will the gasoline cost?

CAREER

DENTAL HYGIENIST

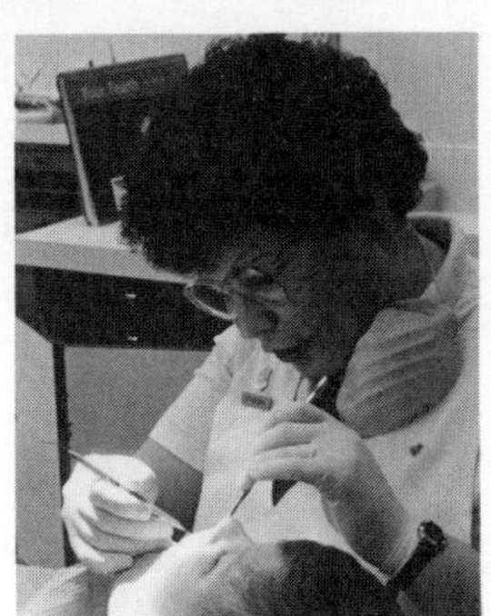

A dental hygienist works with other members of the dental health team to assist patients in achieving and maintaining good oral health.

The diagram shows the upper and lower teeth and is divided into four quadrants.

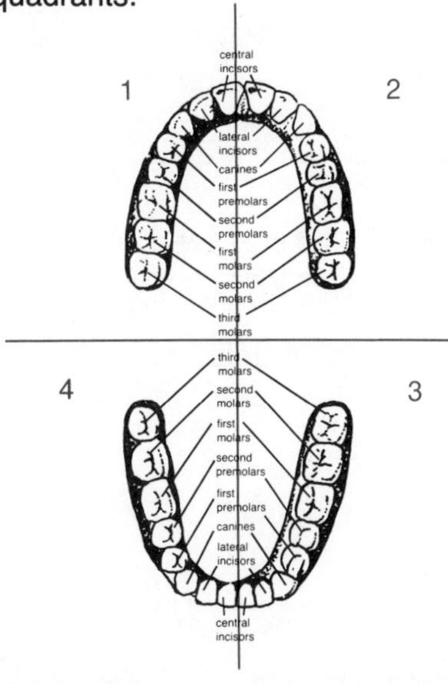

The teeth are numbered in a quadrant starting with 1 as the central incisor and 8 as the third molar. Teeth are named as ordered pairs starting with the quadrant number. For example, the lateral incisor in the lower left of the mouth is named four two.

1. Name all the teeth in the mouth using ordered pairs.

2. What colleges or universities offer courses in dental hygiene?

4.13 REVIEW EXERCISE

1. As most cars get older, their value decreases.
Sketch a graph of this relation.

2. The cost of riding in a taxi cab depends on the distance you travel.
Sketch a graph of this relation.

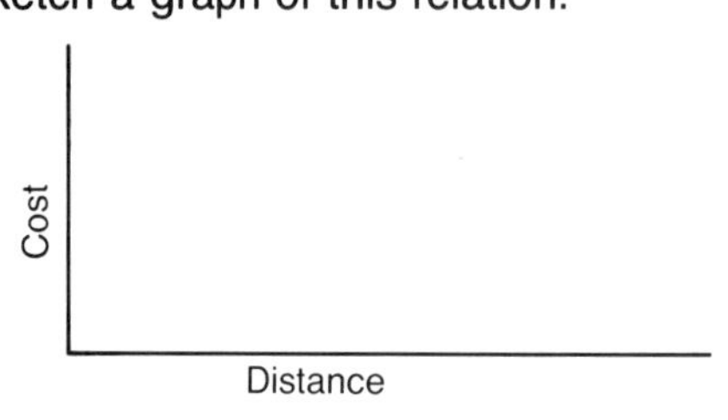

3. The length of the shadow a telephone pole casts on a sunny day depends on the time of day.
Sketch a graph of this relation.

4. Write the coordinates of each point.

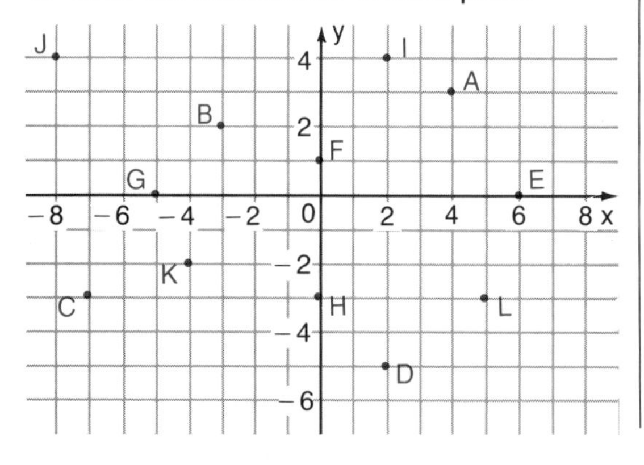

5. Plot each set of points on a pair of axes. Join the points in the order A-B-C-D-A and identify the resulting figure.

(a) A(5, 0), B(5, 8), C(−2, 8), D(−2, 0)
(b) A(3, −1), B(1, 7), C(−1, −1)
(c) A(4, 4), B(−4, 4), C(−4, −4), D(4, −4)

6. What are the coordinates of the vertices of each figure?

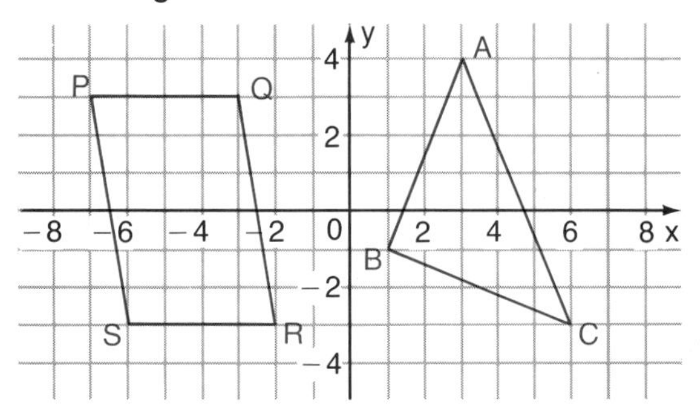

7. The graph of a relation is shown below.

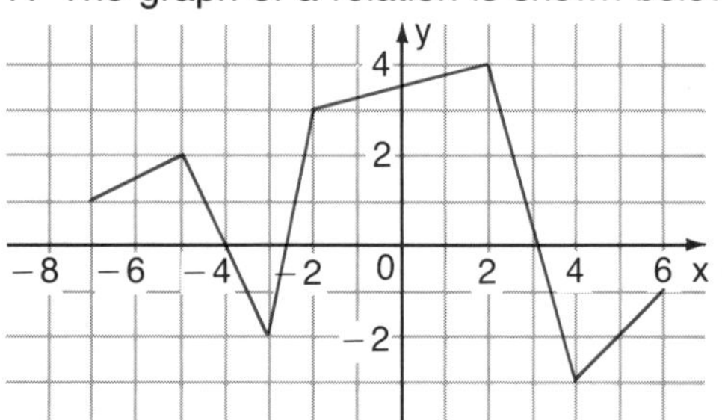

Copy and complete the table for this relation.

x	−7	−5		2	4	
y			−2			0

8. The following graph shows the relation between the distance travelled by a car driving at 50 km/h and the length of time spent travelling.

(a) How far has the car travelled after 3 h?
(b) How far has the car travelled after 4.25 h?
(c) How long will it take to travel 275 km?

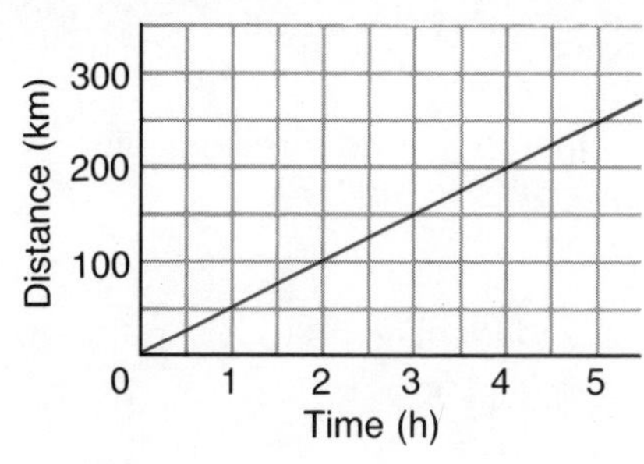

9. The table gives the total distance travelled by a falling star after each of the first three seconds.

Time (s)	Distance (m)
0	0
1	5
2	20
3	44

(a) Display this information on a graph.
(b) Use the graph to determine how far the star has fallen after 3.5 s.

10. The table gives the distance travelled by a bicycle travelling at 8 km/h.

Time (h)	Distance (km)
0	0
1	8
2	16
3	24
4	32

(a) Draw a graph of this relation.
(b) Find the distance travelled after 3.5 h.
(c) Find the distance travelled after 6.5 h.

11. Write an equation with a constant of variation that describes each relation.

(a)

Hours Worked	1	2	3	7
Pay ($)	13	26	39	91

(b)

Time Cycling (h)	2	4	6	9
Distance Cycled (km)	36	72	108	162

12. x varies directly as y.
When $x = 56$, $y = 8$.
Find y when $x = 98$.

13. The amount of sales tax you pay varies directly with the cost of the item.
If there is a tax of $4 on a $50 item, what will be the tax on a $125 item?

14. Barbara sells books to computer stores. She receives a salary of $125 per day plus a commission of 5% of all sales.
(a) Write an equation relating her daily earnings and the value of the books sold.
(b) How much would she make if she sold $2000 worth of books in one day?
(c) How much would she earn if she sold $3500 worth of books in one day?

15. The cost of owning a car has two parts. There is a fixed part that you must pay whether or not you drive the car. The variable part depends on how far you drive the car. For her car, Alicia spends $1200 each year in fixed charges plus $85 for every 1000 km she drives.
(a) Draw a graph of this relation.
(b) Write an equation relating the cost per year and the number of kilometres driven.
(c) How much would it cost her to drive 11 500 km in a year?
(d) What charges would make up the fixed amount of owning a car?

16. Graph the following equations.
(a) $y = x + 5$ (b) $y = x - 1$
(c) $x + y = 8$ (d) $x - y = 4$

17. Graph the following equations.
(a) $y = 2x + 3$ (b) $4x + 2y = 1$

18. Graph the following using intercepts.
(a) $x + y = 5$ (b) $3x - 4y = 12$

19. Sketch the graph of the following.
(a) $y = x^2 + 2$ (b) $x^2 + y^2 = 4$
(c) $x^2 + y^2 = 25$ (d) $y = x^2 - 3$

4.14 CHAPTER 4 TEST

1. The number of listeners that tune in to a radio station depends on the time of day.
Sketch a graph of this relation.

2. Plot the points on a grid and identify the resulting figure.
A(−3, 2), B(4, 2), C(2, −1), D(−5, −1)

3. The table gives the distance travelled by an ocean ship in rough seas from 09:00, when it left port, to 13:00.

Time (h)	09:00	10:00	11:00	12:00	13:00
Distance from Port (km)	0	10	20	25	35

(a) Plot the points on a grid.
(b) Join the points.
(c) How far was the ship from port at 10:30?

4. x varies directly as y. When $x = 7$, $y = 91$.
Find y when $x = 12$.

5. The amount of interest earned on a savings account deposit varies directly with the amount on deposit.
Caroline earned \$91 on a \$700 deposit.
How much interest would she earn on a \$1100 deposit?

6. A software management system package costs \$500. Training to use the package costs \$50/h.
(a) Write an equation relating the total cost to the price of the software and the number of hours needed for training.
(b) What would the total cost be if 8 h of training are needed?

7. Graph the following.
(a) $y = x + 6$ (b) $x + y = 7$

8. Graph the following.
(a) $2x + y = 1$ (b) $6x - 3y = 9$

9. Graph the following using intercepts.
(a) $x - y = 4$ (b) $5x + 2y = 10$

10. Sketch the graphs.
(a) $y = x^2 + 3$ (b) $x^2 + y^2 = 100$

4.15 CUMULATIVE REVIEW FOR CHAPTERS 1 TO 4

1. State the place value, face value, and total value of the indicated digit.

(a) 65 789 (b) 3 560 000
(c) 7892 (d) 43.56
(e) 0.789 (f) 6.78

2. Simplify.

(a) $2004 + 4500 + 222 + 133$
(b) $4003 - 3456$
(c) 1004×16
(d) $1125 \div 25$
(e) $12\,455 + 563 + 6666$
(f) $56\,000 - 3509$

3. Simplify.

(a) $56 + 21 \times 2$
(b) $23 \times 3 - 11 \times 5$
(c) $4^2 + 5^2 - 3^2$
(d) $3(9 - 4) + 15$

4. Round to the highest place value.

(a) 720 (b) 762
(c) 45 000 (d) 35 000
(e) 0.053 (f) 0.065

5. Estimate the answer.

(a) $45.7 + 61.6 + 18.1$
(b) 45.89×9.2
(c) $1200 \div 6.7$
(d) $345.7 - 49.1$

6. Simplify.

(a) $\frac{2}{3} \times \frac{1}{4}$ (b) $\frac{1}{3} + \frac{1}{12}$
(c) $\frac{1}{2} \div 1\frac{1}{2}$ (d) $1\frac{1}{4} - \frac{7}{8}$
(e) $3\frac{5}{6} + 4\frac{1}{3}$ (f) $6 \div \frac{3}{4}$
(g) $2\frac{1}{4} \times 3$ (h) $5 - 1\frac{3}{5}$

7. Evaluate.

(a) 23% of 4600 (b) 67% of 700
(c) 130% of 50 (d) 2% of 68

8. Simplify.

(a) $8 + (-3) - 4$ (b) $-36(-4) + 2$
(c) $-9(-3) - 9$ (d) $-6 - 9 - 12$

9. Express each as a decimal.

(a) $\frac{3}{8}$ (b) $\frac{1}{7}$

10. Evaluate.

(a) 2^4 (b) 2^{-1} (c) 4^{-2}
(d) 3^0 (e) 4^3 (f) 2^{-3}
(g) 6^3 (h) 5^{-1} (i) $(-5)^0$

11. Express in scientific notation.

(a) 34 000 (b) 0.000 098
(c) 120 000 (d) 0.000 76

12. Evaluate to the nearest tenth.

(a) $\sqrt{60}$ (b) $\sqrt{145}$ (c) $\sqrt{8.6}$

13. Simplify.

(a) $y^3 \times y^2$ (b) $t^5 \div t^3$
(c) $4m^4 \times 3m^5$ (d) $12r^6 \div 2r^2$
(e) $(x^3y^2)^2$ (f) $(2m^2n^3)^3$

14. Simplify.

(a) $(4x^2)(2x^5)(5x^2)$
(b) $(-3t^2)(-2t^4)(t^5)$
(c) $(5m^4)(-4m^5) \div (2m^2)$
(d) $(10x^3y^2)(-2x^4y^3)$
(e) $(24m^5n^4) \div (-12m^4n^4)$
(f) $(-6s^6t^5)(-3s^2t^2) \div (9s^4t^4)$
(g) $\frac{45m^5n^3}{9m^2n^3}$ (h) $\frac{-54x^8y^6}{6x^6y^5}$
(i) $\left(\frac{3m^2}{2n}\right)^2$ (j) $\left(\frac{-3x^2}{5y}\right)^3$

15. Simplify.

(a) $(-3m^{-3})(6m^5)$ (b) $(2x^{-2}y^3)(4x^4y^{-3})$
(c) $\frac{10x^5y^{-2}}{5x^2y^3}$ (d) $\frac{-20m^6n^{-1}}{-5m^4n^{-3}}$

16. Find the length of the missing dimension in each figure to the nearest tenth.

(a)

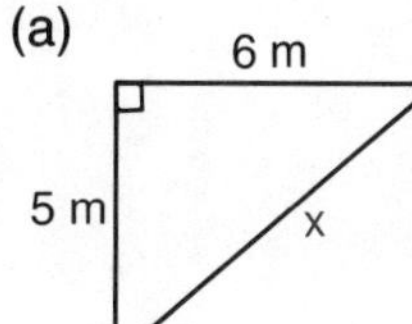

(b)

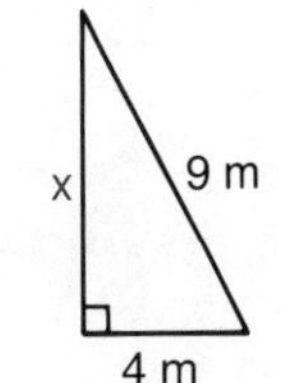

17. If $x = 3$, $y = -2$, and $z = -4$, evaluate each of the following.

(a) $2xy + 3xz$ (b) $2x^2y - 3z$
(c) $4xyz - 6yz$ (d) $x^2 + y^2 - z^2$

18. Simplify.

(a) $3(x - 4) - 2(x + 5) - 6x$
(b) $2a(a + 3) - a(2a - 1) - a(a + 2)$
(c) $4t(3t - 2) - (4t - 1) - t(5 - 4t)$

19. Find the perimeter of the following figures.

(a)

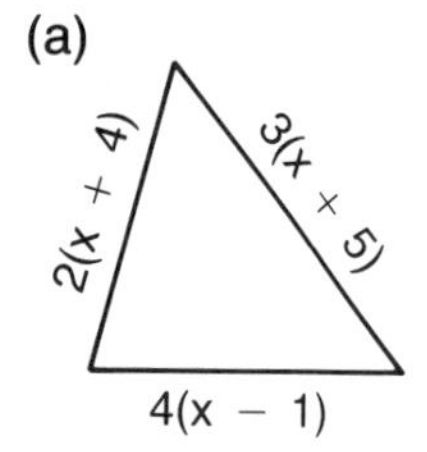

(b)

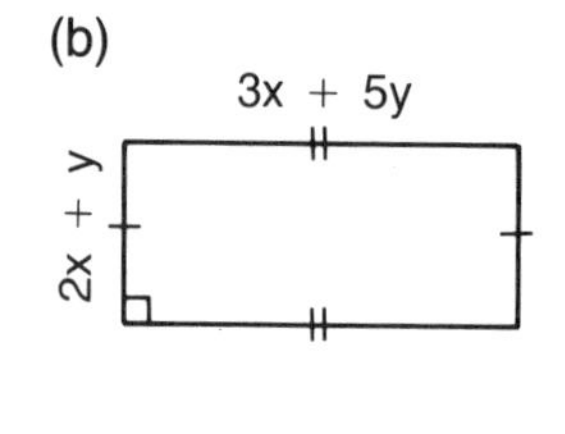

20. Solve and check.

(a) $3(m - 1) = 15$
(b) $4(x - 3) = 3x - 5$
(c) $5(t - 2) - 2(t + 1) = 15$

21. Expand and simplify.

(a) $(x + 3)(x + 4) + 2(x - 3)$
(b) $(t + 1)(t - 3) - (t + 4)$
(c) $2(m - 3)(m + 2) - 3(m + 4)(m - 1)$

22. Find the area of each figure.

(a)

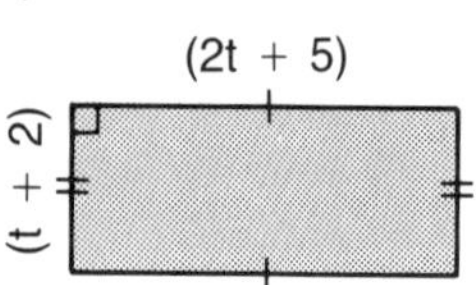

(b)

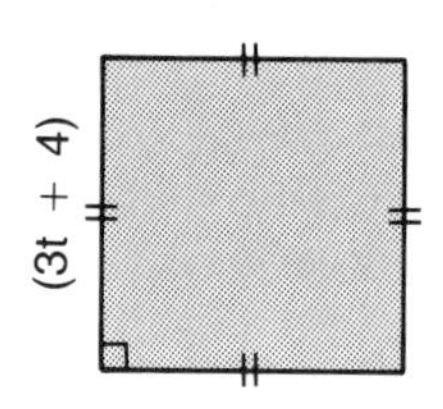

23. Divide.

(a) $\dfrac{8x^2 + 16x^3}{8x}$

(b) $\dfrac{3ab - 9a^2b + 12ab^3}{3ab}$

24. Factor.

(a) $m^2 + 2m - 3$
(b) $t^2 - 8t + 16$
(c) $y^2 - 16$
(d) $x^2 - 8x + 15$

25. The number of daily minutes from sunset to sunrise depends on the time of year.
Sketch a graph of this relationship.

26. The height, h, of a rocket in metres after blast-off is given in the following table, where t is the time in seconds after blast-off.

h	0	6	14	24	36	50
t	0	1	2	3	4	5

Sketch a graph of this relation.

27. The table gives the cost of sea shells.

Amount (kg)	Cost ($)
2	9
4	16
6	21
8	24

(a) Display this information on a graph.
(b) Interpolate to find the cost of 5 kg.
(c) Extrapolate to find the cost of 9 kg.
(d) How many kilograms can you buy for $14?

28. The cost of renting a sailboat varies directly with the number of hours the boat is rented. It costs $72 to rent a boat for 4 h. How much will it cost to rent the boat for 9 h?

29. The Sunrise Hotel will put on a banquet for an initial fee of $100 plus $20 per person.

(a) Draw a graph of this relation.
(b) Write an equation relating the cost and the number of people attending.
(c) How much would a banquet cost for 73 people?

30. Graph each of the following.

(a) $3x + y = 7$ (b) $2x + 3y = 18$
(c) $y = x^2 - 2$ (d) $x^2 + y^2 = 4$
(e) $x + y = 0$ (f) $y = 3x - 4$
(g) $y = x^2 + 3$ (h) $x^2 + y^2 = 25$

SOLVING PAIRS OF EQUATIONS

REVIEW AND PREVIEW TO CHAPTER 

ALGEBRA: ADDITION AND SUBTRACTION

EXERCISE 1

1. Add.

(a) $\begin{array}{r} 3x + 2y \\ \underline{4x + 7y} \end{array}$

(b) $\begin{array}{r} 2a + 3b + 5c \\ \underline{6a + 7b + 2c} \end{array}$

(c) $\begin{array}{r} 5x - 3y \\ \underline{8x + 2y} \end{array}$

(d) $\begin{array}{r} 6p - 2q + 5r \\ \underline{7p - 5q - 8r} \end{array}$

(e) $\begin{array}{r} -3x - 5y \\ \underline{4x + 3y} \end{array}$

(f) $\begin{array}{r} 5a - 4b - c \\ \underline{2a - 3b - 5c} \end{array}$

(g) $\begin{array}{r} -3s - 2t - 5 \\ 5s + 3t + 6 \\ \underline{3s + 2t - 4} \end{array}$

(h) $\begin{array}{r} 7a + 4b + 6c \\ 2a - 5b - 7c \\ \underline{5a - 3b - 2c} \end{array}$

2. Simplify.

(a) $3x + 2y + 4z + 3x + 5y + 2z$
(b) $4a - 2b + 4c - 3a - 2b + 9c$
(c) $4x + 5z - 3y + 6z + 2x + 3x$
(d) $5a + 3b + 4a - 2c + 6c - 2a$

3. Subtract.

(a) $\begin{array}{r} 6x + 9y \\ \underline{3x + 7y} \end{array}$

(b) $\begin{array}{r} 5a + 4b + 6c \\ \underline{2a + 3b + 3c} \end{array}$

(c) $\begin{array}{r} 8x + 4y \\ \underline{9x - 3y} \end{array}$

(d) $\begin{array}{r} 3p + 5q - 6r \\ \underline{2p - 3q - 7r} \end{array}$

(e) $\begin{array}{r} 5a - 6b \\ \underline{-3a - 4b} \end{array}$

(f) $\begin{array}{r} 4x - 2y - 5z \\ \underline{4x - y - 6z} \end{array}$

(g) $\begin{array}{r} -4x - 3y \\ \underline{-2x - 5y} \end{array}$

(h) $\begin{array}{r} 9a - 7b - 2c \\ \underline{-7a + 6b - 5c} \end{array}$

4. Subtract $(3x + 7y)$ from $(2x - 5y)$.

5. Subtract $(2a - b)$ from $(a - 2b)$.

SOLVING EQUATIONS

EXERCISE 2

1. Solve.

(a) $x + 5 = 12$
(b) $2a = 14$
(c) $t - 2 = 8$
(d) $-3m = 12$
(e) $6 + b = 6$
(f) $3d = -3$
(g) $-t = 4$
(h) $3b + 3 = 6$
(i) $3 + x = -6$
(j) $\frac{1}{2}m = 8$
(k) $\frac{y}{3} = 9$
(l) $\frac{m}{-5} = 4$
(m) $a - 2 = -9$
(n) $-3x = 0$
(o) $2t - 2 = 2$

2. Solve and check.

(a) $5x - 10 = 15$
(b) $6a = 4a + 8$
(c) $4b = 5b + 3$
(d) $3(x - 1) = 9$
(e) $2(m - 3) - (m + 1) = 8$
(f) $4(2t - 3) - (1 + 3t) = -3$
(g) $2(1 - 3x) - 3(x - 7) = 2(x + 1) - 1$
(h) $3m - (m - 3) + 2(1 - 2m) = 3(1 + m) - 18$

3. Solve.

(a) $\frac{x - 1}{3} = 6$
(b) $\frac{m}{3} - \frac{1}{2} = \frac{1}{6}$
(c) $\frac{b}{4} - \frac{b}{6} = 3$
(d) $\frac{2b + 1}{3} = 3$
(e) $\frac{2a - 1}{4} = \frac{3a + 2}{6} - \frac{7}{12}$
(f) $\frac{x - 7}{6} + 3 = \frac{x + 1}{2}$
(g) $\frac{m + 7}{6} + \frac{1}{2} = \frac{m - 2}{4}$
(h) $\frac{3m + 7}{4} - \frac{m}{3} = -\frac{1}{3}$

MENTAL MATH

EXERCISE 3

Perform the following calculations mentally and record your answers.

1. Your hockey team boards a bus to travel to Ottawa where the team will play in a tournament. Ottawa is 250 km away and the bus will travel at an average of 100 km/h.
How long will the trip be if the bus stops one-half hour for lunch?

2. Movie tickets cost $8 each. The bus fare one way is $1 and food at the show will cost $3 per person.
What is the total cost for you and a friend to travel to the theatre and see a movie?

3. An advertisement on TV announces that you may receive a 52 week subscription of a magazine at 50% off the cover price.
If the cover price is $4 per week, what is the cost of the TV special?

4. A tow ticket at the ski hill costs $15 per day. Transportation costs $7 each way and lunch, at least $8.
How much would you spend to go skiing for a day?

5. A clothing store offers you a job that will pay $8/h for 15 h of work a week. A shoe store offers you $7/h and 20 h of work a week.
Which job will let you earn the most money in a week?

6. The air show will start at noon. The air field, where the show will take place, is 250 km from your house. The roads are all two-lane and the speed limit is 50 km/h. The line-up for tickets is always one hour long.
At what time should you leave for the airport to arrive at the air show on time?

EXTRA

YOU BE THE DETECTIVE

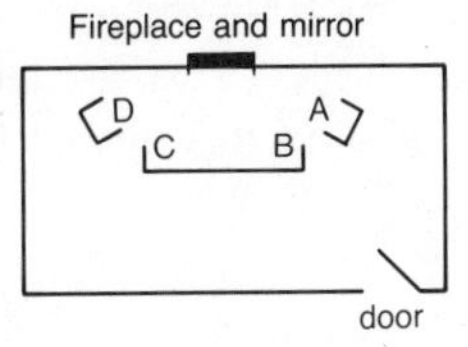

Adams the doctor has been found unconscious in the kitchen. Someone put knock-out drops in his milk.

There are four people seated on the sofa and armchairs, as shown in the diagram, talking about Adams. Their names are Beeton, Jarvis, Steele, and Roberts. They are, but not necessarily in this order, a lawyer, a dentist, an accountant, and a writer.

1. Beeton and Jarvis are drinking milk.
2. The writer is allergic to milk.
3. Beeton does not have a sister.
4. Steele does not have a sister.
5. Steele is sitting in one of the armchairs.
6. The writer is sitting next to Steele, on Steele's left.
7. Steele is the accountant's brother-in-law.
8. The lawyer looks in the mirror over the fireplace and sees a waiter walk past the open door.
9. Roberts is sitting next to the lawyer.
10. Then a hand is seen putting knock-out drops in Jarvis' milk. It is the same person who drugged Adams.
11. No one has left their seat.
12. There is no one else in the room.

Who is the culprit?
What is the profession of each person and where are they seated?

Hint: Draw a chart, listing the people on one side and the professions along the top.

5.1 ORDERED PAIRS AND SOLUTION SETS

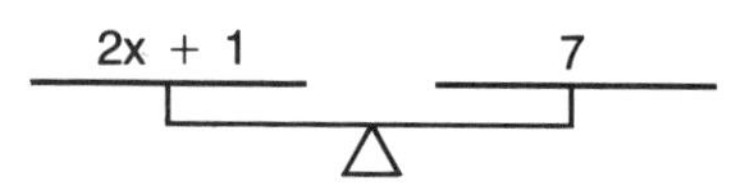

For an equation, such as $2x + 1 = 7$, there is one number for x that makes the left side equal to the right side. We find this number by solving the equation.

$$\begin{aligned} 2x + 1 &= 7 \\ 2x + 1 - 1 &= 7 - 1 \\ 2x &= 6 \\ x &= 3 \end{aligned}$$

We verify that 3 is the solution, or root, by substituting.

$$2x + 1 = 7$$

$$\begin{aligned} \text{L.S.} &= 2x + 1 \qquad \text{R.S.} = 7 \\ &= 2(3) + 1 \\ &= 6 + 1 \\ &= 7 \end{aligned}$$

Since L.S. = R.S., the solution is 3.

The school dance committee must have 9 people on it. There are several combinations of boys and girls that can make up the committee.

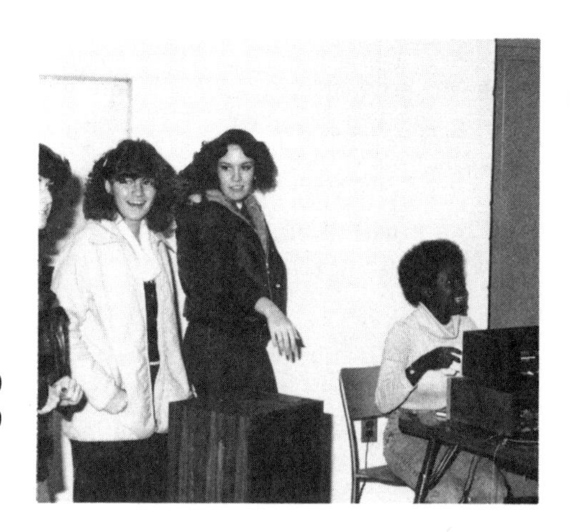

3 boys and 6 girls = 9 committee members
5 boys and 4 girls = 9 committee members

The equation $x + y = 9$ has two variables, x and y. There are many values for x and y that will make the left side equal to the right side.

When $x = 2$ and $y = 7$, $x + y = 2 + 7 = 9$
When $x = 4$ and $y = 5$, $x + y = 4 + 5 = 9$

The solutions for $x + y = 9$ are expressed as ordered pairs, or (x, y).
There are an infinite number of solutions for $x + y = 9$ and some are:

(1, 8), (2, 7), (3, 6), (4, 5), (5, 4), (6, 3), (7, 2)

These ordered pairs, or points, can be graphed. Notice that all points on the line satisfy the equation

$$x + y = 9$$

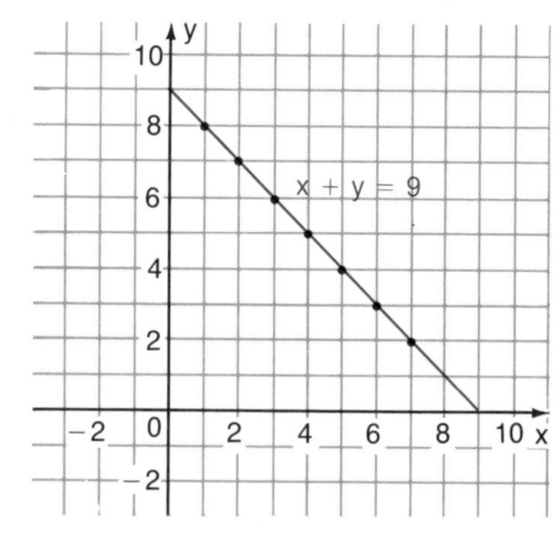

Example 1.
Are (1, 5) and (2, 8) solutions of $y = 2x + 3$?

Solution:

For (1, 5), $x = 1$ and $y = 5$

$$\begin{aligned} \text{L.S.} &= y & \text{R.S.} &= 2x + 3 \\ &= 5 & &= 2(1) + 3 \\ & & &= 2 + 3 \\ & & &= 5 \end{aligned}$$

(1, 5) is a solution of $y = 2x + 3$ and this point is on the graph of $y = 2x + 3$.

For (2, 8), $x = 2$ and $y = 8$

$$\begin{aligned} \text{L.S.} &= y & \text{R.S.} &= 2x + 3 \\ &= 8 & &= 2(2) + 3 \\ & & &= 4 + 3 \\ & & &= 7 \end{aligned}$$

(2, 8) is not a solution of $y = 2x + 3$ and this point is not on the graph of $y = 2x + 3$.

Example 2.
Does the ordered pair (3, 2) satisfy both $x + y = 5$ and $x - y = 1$?

Solution:
Substitute $x = 3$ and $y = 2$ in both equations.

For $x + y = 5$

$$\begin{aligned} \text{L.S.} &= x + y & \text{R.S.} &= 5 \\ &= 3 + 2 \\ &= 5 \end{aligned}$$

(3, 2) satisfies $x + y = 5$.

For $x - y = 1$

$$\begin{aligned} \text{L.S.} &= x - y & \text{R.S.} &= 1 \\ &= 3 - 2 \\ &= 1 \end{aligned}$$

(3, 2) satisfies $x - y = 1$.

(3, 2) satisfies both equations.

The results of Example 2 are shown on the graph at the right. The point (3, 2) is on both graphs and is also the point of intersection of the graphs.

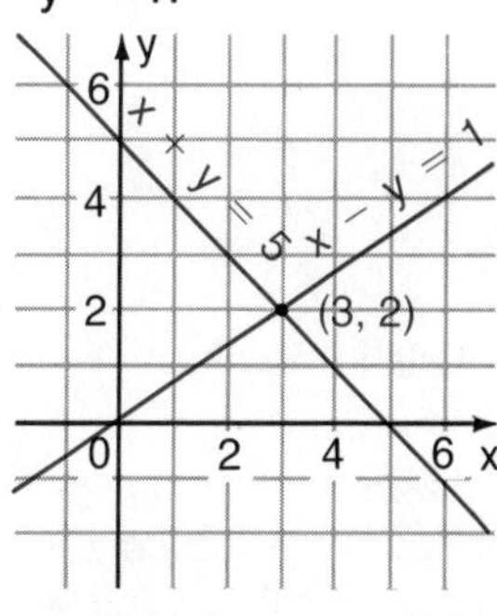

EXERCISE 5.1

A 1. Determine the ordered pairs that satisfy each equation.

(a) $x + y = 8$; (3, 5), (1, 8), (6, 2), (7, 1)
(b) $x - y = 3$; (5, 2), (7, 4), (6, 2), (4, 0)
(c) $y = 2x - 1$; (1, 0), (2, 3), (4, 6), (5, 9)

2. State the missing number in each ordered pair so that the equation is satisfied.

(a) $x + y = 6$
(3, ■), (■, 6), (■, 1), (■, −2), (−3, ■)
(b) $x - y = 3$
(4, ■), (■, 2), (■, 5), (9, ■), (2, ■)

B 3. Find the ordered pair that satisfies both equations.

(a) $x + y = 4$
$x - y = 2$
(b) $2x + y = 7$
$x + y = 6$
(c) $y = 3x - 1$
$y = 2x + 1$
(d) $x + y = 6$
$x - y = 2$

4. Find the ordered pair that satisfies both equations.

(a) $x + y = 7$
$x - y = 1$
(b) $x + y = 6$
$x - y = 4$
(c) $x - y = 3$
$x + y = 5$
(d) $x + y = 3$
$x - y = 3$

5.2 SOLVING PAIRS OF LINEAR EQUATIONS USING GRAPHS

Frank Edwards rents a booth at a market every Saturday and sells college sweatshirts. The shirts cost him \$20 each and the booth rent is \$400.

The equation for his total cost, C, is

$C = 20n + 400$

where n is the number of sweaters he sells.

Frank sells the shirts for \$30 each. The equation for his sales income, S, is

$I = 30n$

If Frank sells 25 shirts in one day, his total cost is

$$\begin{aligned} C &= 20(25) + 400 \\ &= 500 + 400 \\ &= 900 \end{aligned}$$

His income for 25 shirts is

$$\begin{aligned} I &= 30(25) \\ &= 750 \end{aligned}$$

For selling 25 shirts, Frank loses \$900 − \$750 or \$150.

If Frank sells 50 shirts in one day, his total cost is

$$\begin{aligned} C &= 20(50) + 400 \\ &= 1000 + 400 \\ &= 1400 \end{aligned}$$

His income is

$$\begin{aligned} I &= 30(50) \\ &= 1500 \end{aligned}$$

For selling 50 shirts, Frank makes a profit of \$1500 − \$1400 or \$100.

If Frank sells more than a certain number of shirts, he will make a profit. If he sells less than this number, he will take a loss. For this specific number of shirts, cost equals income and Frank will break even.

How many shirts must Frank sell in order to break even?

To find this number of shirts, we draw the graph of each equation on the same pair of axes.

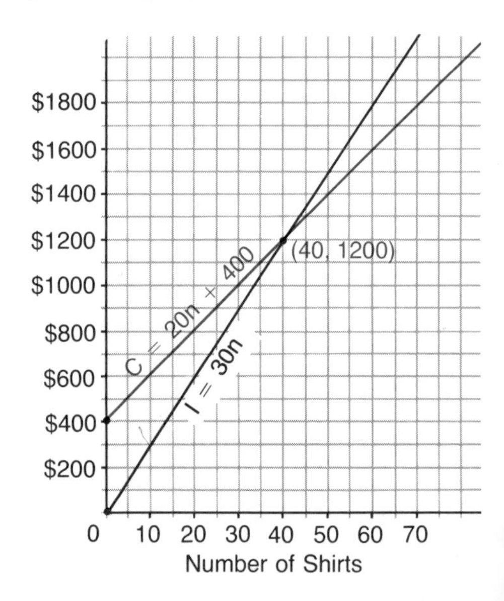

From the graph we see that the point of intersection of the two lines is (40, 1200). This is the only point that satisfies both equations. We can also express this in terms of total cost and sales income: if 40 shirts are sold, then the total cost is \$1200 and the sales income is also \$1200.

We solve a pair of equations graphically by finding the point of intersection. One x- and one y-value will be common to both equations.

Example 1.
Solve the following pair of equations graphically.

$y = 2x - 1$
$y = 4 - 3x$

Solution:
Set up a table for the x- and y-values in each equation.
Use the points in the table to draw an accurate graph.

Equation
$y = 2x - 1$
Ordered Pairs

x	y = 2x − 1	(x, y)
0	$y = 2(0) - 1$ $= -1$	(0, −1)
1	$y = 2(1) - 1$ $= 1$	(1, 1)
−1	$y = 2(-1) - 1$ $= -3$	(−1, −3)
2	$y = 2(2) - 1$ $= 3$	(2, 3)

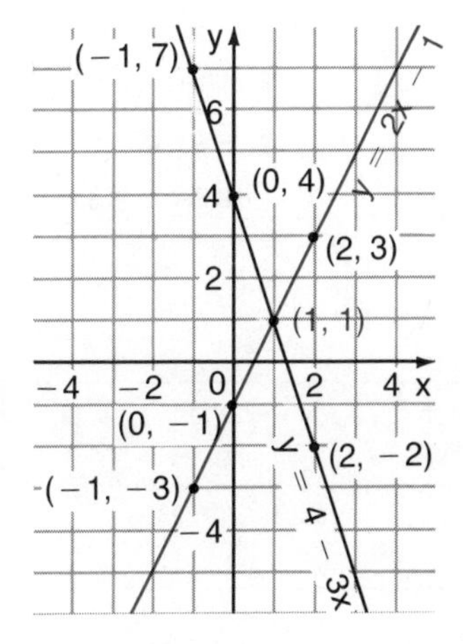

Equation
$y = 4 - 3x$
Ordered Pairs

x	y = 4 − 3x	(x, y)
0	$y = 4 - 3(0)$ $= 4$	(0, 4)
1	$y = 4 - 3(1)$ $= 1$	(1, 1)
−1	$y = 4 - 3(-1)$ $= 7$	(−1, 7)
2	$y = 4 - 3(2)$ $= -2$	(2, −2)

When graphing:
1. Label and scale the axes.
2. Plot the necessary points.
3. Draw and label the lines.

Check: $y = 2x - 1$

L.S. $= y$
$= 1$

R.S. $= 2x - 1$
$= 2(1) - 1$
$= 2 - 1$
$= 1$

Check: $y = 4 - 3x$

L.S. $= y$
$= 1$

R.S. $= 4 - 3x$
$= 4 - 3(1)$
$= 4 - 3$
$= 1$

The point of intersection is (1, 1) and the solution to the system is $x = 1$ and $y = 1$.

Example 2.

Solve the following graphically.

$$y + 2x = -7$$
$$y - 3x = 3$$

Solution:

Write each equation in the form $y = mx + b$ and set up a table of values for each equation.

$y + 2x = -7$
$y = -2x - 7$

$y - 3x = 3$
$y = 3x + 3$

Equation

$y = -2x - 7$

Ordered Pairs

x	$y = -2x - 7$	(x, y)
0	$y = -2(0) - 7$ $= -7$	(0, −7)
1	$y = -2(1) - 7$ $= -9$	(1, −9)
−1	$y = -2(-1) - 7$ $= -5$	(−1, −5)

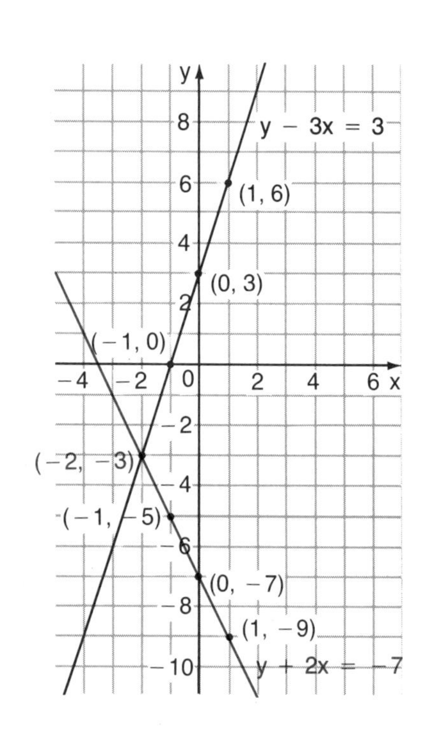

Equation

$y = 3x + 3$

Ordered Pairs

x	$y = 3x + 3$	(x, y)
0	$y = 3(0) + 3$ $= 3$	(0, 3)
1	$y = 3(1) + 3$ $= 6$	(1, 6)
−1	$y = 3(-1) + 3$ $= 0$	(−1, 0)

Check: Always check a solution using the original equation.

The lines intersect at (−2, −3).

$y + 2x = -7$

L.S. $= y + 2x$ R.S. $= -7$
$= -3 + 2(-2)$
$= -3 - 4$
$= -7$

$y - 3x = 3$

L.S. $= y - 3x$ R.S. $= 3$
$= -3 - 3(-2)$
$= -3 + 6$
$= 3$

Since L.S. = R.S. for both equations, the solution is $x = -2$ and $y = -3$.

Example 3.

Solve the following graphically.

$$3x - 2y = 12$$
$$x - 2y = 8$$

Solution:

Use the intercept method to graph the equations.

$3x - 2y = 12$

When $x = 0$, $3(0) - 2y = 12$
$-2y = 12$
$y = -6$

When $y = 0$, $3x - 2(0) = 12$
$3x = 12$
$x = 4$

Using these values, two points are $(0, -6)$ and $(4, 0)$.

$x - 2y = 8$

When $x = 0$, $(0) - 2y = 8$
$-2y = 8$
$y = -4$

When $y = 0$, $x - 2(0) = 8$
$x = 8$

Using these values, two points are $(0, -4)$ and $(8, 0)$.

In the graph, the point of intersection is $(2, -3)$.

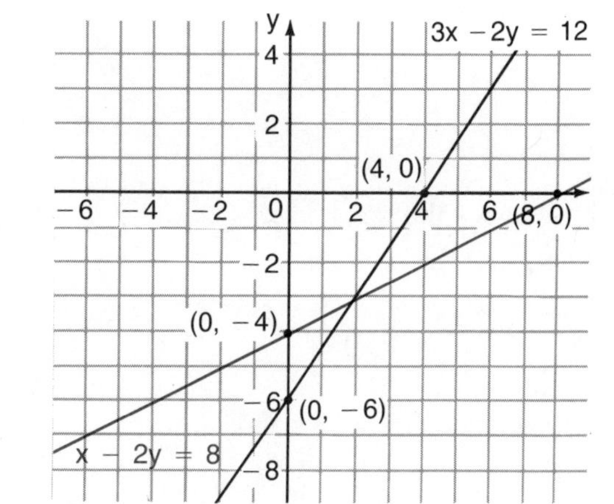

Check: $3x - 2y = 12$

L.S. $= 3x - 2y$ R.S. $= 12$
$= 3(2) - 2(-3)$
$= 6 + 6$
$= 12$

Check: $x - 2y = 8$

L.S. $= x - 2y$ R.S. $= 8$
$= (2) - 2(-3)$
$= 2 + 6$
$= 8$

The solution is $x = 2$ and $y = -3$.

EXERCISE 5.2

B 1. Solve each pair of equations graphically.

(a) $y = x + 3$, $y = 5 - x$
(b) $y = 2x + 1$, $y = 4 - x$
(c) $y = 2x - 3$, $y = 5 - 2x$
(d) $y = 5 + 3x$, $y = 1 + x$
(e) $y = x + 8$, $y = -3x$
(f) $y = 2x$, $y = 3 - x$

2. Solve the following graphically.

(a) $x + 2y = 4$, $x - y = 1$
(b) $x + y = 6$, $3x - y = 6$
(c) $3x - 2y = 18$, $2x + y = 12$
(d) $5x - 2y = 10$, $x + y = 9$
(e) $2x + 3y = -12$, $2x - y = -4$
(f) $x + 2y = 2$, $x - y = -4$

C 3. Solve the following graphically.

(a) $y = 2x + 3$, $y = 5 - 2x$
(b) $y = 1 - x$, $y = \frac{4x - 1}{2}$
(c) $y = 1 - 2x$, $y = 4x + 4$
(d) $y = 2 - 2x$, $y = 4x - 13$

4. Solve the following graphically and answer the question below.

(a) $y = x + 5$, $y = x - 1$
(b) $y = 3x + 2$, $y = 3x - 1$
(c) $y = -2x + 2$, $y = -2x - 3$
(d) $2x + 3y = 6$, $4x + 6y = 12$

How many points of intersection are there for each pair of equations?

5.3 SOLVING PAIRS OF EQUATIONS: PART 1 — BY SUBTRACTION

Pairs of linear equations can also be solved using algebra. In this section we solve pairs of equations by subtracting one equation from the other.

Example.

Solve. $5x + 2y = 23$
$x + 2y = 11$

Solution:

Because the y terms are identical we can subtract to eliminate the y terms. Remember to subtract, we add the opposite.

$$\begin{aligned} 5x + 2y &= 23 \quad ① \\ x + 2y &= 11 \quad ② \\ \hline \text{Subtracting.} \quad 4x \quad &= 12 \\ x \quad &= 3 \end{aligned}$$

Substitute $x = 3$ in ①.

$$\begin{aligned} 5x + 2y &= 23 \\ 5(3) + 2y &= 23 \\ 15 + 2y &= 23 \\ 2y &= 8 \\ y &= 4 \end{aligned}$$

Now check the solution $x = 3$ and $y = 4$ for both equations.

Check in ①.

L.S. $= 5x + 2y$ R.S. $= 23$
$= 5(3) + 2(4)$
$= 23$

Check in ②.

L.S. $= x + 2y$ R.S. $= 11$
$= 3 + 2(4)$
$= 11$

The solution is $x = 3$ and $y = 4$.

EXERCISE 5.3

A

1. Subtract.

(a) $\begin{array}{r} 3x + 4y \\ \underline{3x + 2y} \end{array}$ (b) $\begin{array}{r} 2s + 5t \\ \underline{s + 5t} \end{array}$

(c) $\begin{array}{r} 6a - 7b \\ \underline{4a - 7b} \end{array}$ (d) $\begin{array}{r} 3p - 2q \\ \underline{p - 2q} \end{array}$

(e) $\begin{array}{r} 4d - 3e \\ \underline{6d - 3e} \end{array}$ (f) $\begin{array}{r} 6x - y \\ \underline{2x - y} \end{array}$

(g) $\begin{array}{r} 3m - 2n \\ \underline{-5m - 2n} \end{array}$ (h) $\begin{array}{r} 5x + 8y \\ \underline{6x + 8y} \end{array}$

B

2. Solve and check.

(a) $3x + 2y = 5$
$x + 2y = 3$

(b) $7m + 3n = 23$
$2m + 3n = 13$

(c) $2x + 5y = 26$
$2x + y = 10$

(d) $5s - 3t = 14$
$2s - 3t = 2$

(e) $4m - n = 22$
$2m - n = 10$

(f) $4x + y = 19$
$4x + 3y = 25$

(g) $2r + 5s = -12$
$2r + 3s = -8$

(h) $2a - 3b = -13$
$4a - 3b = -17$

(i) $x - y = 1$
$x + y = 3$

(j) $2x + 3y = 8$
$x + 3y = 7$

3. Solve and check.

(a) $x + y = 10$
$x - y = 8$

(b) $4a + 3b = 7$
$4a - b = -5$

(c) $r + s = -7$
$3r + s = -9$

(d) $x + 3y = 19$
$x - y = -1$

5.4 SOLVING PAIRS OF EQUATIONS: PART 2 — BY ADDITION

Subtracting will not always eliminate one of the variables.

For $5x + 2y = 19$
$3x - 2y = 5$

Subtracting.
$$\begin{array}{rl} 5x + 2y = 19 & ① \\ 3x - 2y = 5 & ② \\ \hline 2x + 4y = 14 & \end{array}$$

In this section we solve pairs of equations by addition.
Adding the opposite of $-2y$ to $2y$ gives $4y$. However, since $-2y$ and $2y$ are opposites, we can add to eliminate the y terms.

$$\begin{array}{rl} 5x + 2y = 19 & ① \\ 3x - 2y = 5 & ② \\ \hline 8x \quad = 24 & \\ x \quad = 3 & \end{array}$$

Adding.

Substitute $x = 3$ in ①.
$$\begin{aligned} 5x + 2y &= 19 \\ 5(3) + 2y &= 19 \\ 15 + 2y &= 19 \\ 2y &= 4 \\ y &= 2 \end{aligned}$$

Check the solution $x = 3$ and $y = 2$ in both equations.

Check in ①.

L.S. $= 5x + 2y$
$= 5(3) + 2(2)$
$= 15 + 4$
$= 19$

R.S. $= 19$

Check in ②.

L.S. $= 3x - 2y$
$= 3(3) - 2(2)$
$= 9 - 4$
$= 5$

R.S. $= 5$

The solution is $x = 3$ and $y = 2$.

EXERCISE 5.4

A

1. Add.

(a) $\begin{array}{r} 3x + 4y \\ \underline{4x - 4y} \end{array}$ (b) $\begin{array}{r} 7s - 6t \\ \underline{9s + 6t} \end{array}$

(c) $\begin{array}{r} 2a + 9b \\ \underline{-2a - 7b} \end{array}$ (d) $\begin{array}{r} -8m - 9n \\ \underline{8m - 3n} \end{array}$

(e) $\begin{array}{r} 4x - 10y \\ \underline{-7x + 10y} \end{array}$ (f) $\begin{array}{r} 5c + d \\ \underline{-3c - d} \end{array}$

B

2. Solve and check.

(a) $4x + y = 14$
$2x - y = 4$

(b) $a + 3b = 5$
$4a - 3b = 5$

(c) $3m - 2n = 7$
$5m + 2n = 33$

(d) $-4x - 6y = 4$
$4x - 3y = 14$

(e) $5s - t = 4$
$-5s + 3t = -12$

(f) $5x - 6y = 9$
$2x + 6y = -30$

(g) $-3m - n = 13$
$-6m + n = 23$

(h) $2x + y = -5$
$-2x + 3y = 1$

(i) $3r + s = -1$
$-3r + 2s = 7$

(j) $3m + 2n = 5$
$-3m - n = -4$

(k) $7c - 4d = -5$
$-2c + 4d = -10$

5.5 SOLVING PAIRS OF EQUATIONS BY ELIMINATION

For some pairs of equations such as

$$2x + 3y = 16$$
$$5x - 2y = 2$$

neither addition nor subtraction will eliminate one of the variables as the equations are written. Elimination is possible when the x terms or the y terms are identical or opposites. We can multiply both sides of an equation to make terms that can be eliminated. The following shows how to eliminate either variable.

METHOD I — Eliminating y first

$$2x + 3y = 16 \quad ①$$
$$5x - 2y = 2 \quad ②$$

Multiply ① by 2 and ② by 3.

① × 2 $\quad 4x + 6y = 32$

② × 3 $\quad 15x - 6y = 6$

Adding. $\quad 19x = 38$

$$x = 2$$

Substitute $x = 2$ in ①.

$$2x + 3y = 16$$
$$2(2) + 3y = 16$$
$$4 + 3y = 16$$
$$3y = 12$$
$$y = 4$$

METHOD II — Eliminating x first

$$2x + 3y = 16 \quad ①$$
$$5x - 2y = 2 \quad ②$$

Multiply ① by 5 and ② by 2

① × 5 $\quad 10x + 15y = 80$

② × 2 $\quad 10x - 4y = 4$

Subtracting. $\quad 19y = 76$

$$y = 4$$

Substitute $y = 4$ in ①.

$$2x + 3y = 16$$
$$2x + 3(4) = 16$$
$$2x + 12 = 16$$
$$2x = 4$$
$$x = 2$$

Check the solution $x = 2$ and $y = 4$ in both equations.

Check in ①.

L.S. $= 2x + 3y$ $\qquad$ R.S. $= 16$

$= 2(2) + 3(4)$

$= 4 + 12$

$= 16$

Check in ②.

L.S. $= 5x - 2y$ $\qquad$ R.S. $= 2$

$= 5(2) - 2(4)$

$= 10 - 8$

$= 2$

The solution is $x = 2$ and $y = 4$ or (2, 4).

The solution can be shown on a graph.

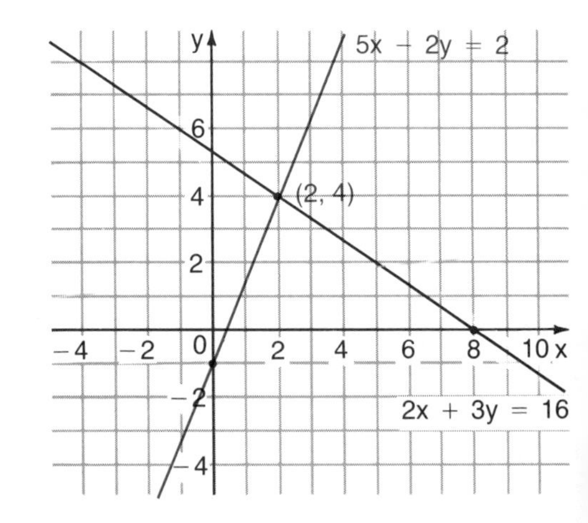

EXERCISE 5.5

A

1. Solve for x and y.

(a) $x - y = 2$
$x + y = 4$

(b) $x + y = 10$
$x - y = 2$

(c) $x - y = 0$
$x + y = 4$

(d) $2x - y = 3$
$x + y = 6$

(e) $2x + y = 4$
$3x - y = 6$

(f) $-x + y = 0$
$x + y = 2$

(g) $-x + 2y = 1$
$x + y = 2$

(h) $2x + y = 5$
$x + y = 4$

B

2. Solve and check.

(a) $2x + y = 10$
$3x - 2y = 8$

(b) $a + 2b = 7$
$a - b = 1$

(c) $3b + c = 12$
$2b + 5c = 21$

(d) $m + 3n = 10$
$3m + 2n = 16$

(e) $r - 3s = 1$
$3r - 2s = 17$

(f) $2x + 5y = 19$
$3x - y = 3$

(g) $5x - 3y = 4$
$4x - y = 6$

(h) $4a - b = 7$
$6a + 5b = 17$

3. Solve and check.

(a) $3a - 2b = 5$
$2a + 3b = 12$

(b) $2x + 3y = 8$
$5x + 2y = 9$

(c) $4m + 3n = 17$
$5m + 2n = 16$

(d) $5x - 2y = 1$
$3x + 4y = 11$

(e) $4r - 3s = 2$
$3r + 5s = 16$

(f) $7t - 3m = 11$
$3t + 4m = 10$

(g) $3b - 4c = 11$
$2b - 3c = 7$

(h) $2c - 3d = 2$
$5c - 2d = 16$

4. Solve.

(a) $x - 2y = 3$
$2x - 3y = 4$

(b) $3a - 2b = -8$
$4a + 3b = -5$

(c) $2x + 3y = 11$
$3x - 2y = -16$

(d) $5x + 2y = 5$
$2x + 3y = 13$

(e) $5m + 3n = -19$
$2m - 5n = 11$

(f) $3b - 4c = 5$
$5b + 3c = -11$

(g) $3r - 4t = 10$
$5r - 12t = 6$

(h) $4c - 3d = -3$
$5c - 2d = -9$

(i) $x - y = -3$
$2x + y = 6$

(j) $x - y = 8$
$3x + 4y = 10$

C

5. Solve. (Clear fractions first.)

(a) $\frac{x}{2} + \frac{y}{3} = 2$
$\frac{x}{2} - \frac{y}{3} = 0$

(b) $\frac{x}{2} + \frac{y}{5} = 2$
$\frac{3x}{2} - \frac{y}{5} = 2$

(c) $x - y = 6$
$\frac{2x}{3} + \frac{y}{3} = 1$

(d) $\frac{1}{2}x + y = -4$
$\frac{x}{2} - \frac{3y}{2} = 1$

(e) $\frac{x}{3} + \frac{y}{2} = 2$
$\frac{3x + 1}{2} + \frac{y + 1}{3} = 6$

(f) $\frac{1}{2}x - \frac{1}{3}y = 1$
$\frac{x - 1}{3} - \frac{y - 1}{4} = \frac{1}{2}$

6. Solve.

(a) $x + y = 7$
$3(x - y) = 9$

(b) $2(x - y) = 3 + x$
$x = 3y + 4$

(c) $5(x + y) = 55$
$19 - 5x = y$

MIND BENDER

Fill in the blanks to make the following true.

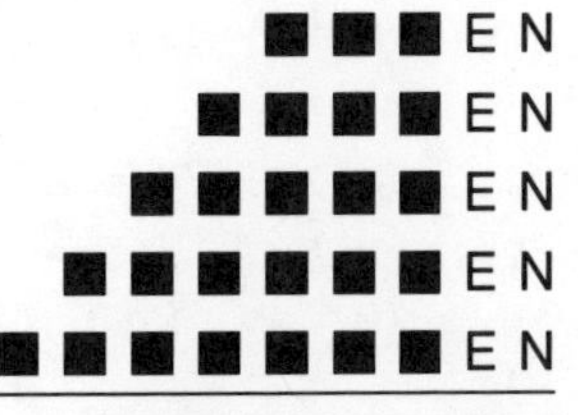

5.6 SOLVING PAIRS OF EQUATIONS BY SUBSTITUTION

Another way to solve a system is by substitution

Example.
Solve by substitution.
$$x - y = 2 \quad ①$$
$$2x + 3y = 19 \quad ②$$

Solution:
Solve equation ① for x.

$x - y = 2$
$x = 2 + y$

Replace x in equation ② by 2 + y.

$$\begin{aligned} 2x + 3y &= 19 \\ 2(2 + y) + 3y &= 19 \\ 4 + 2y + 3y &= 19 \\ 4 + 5y &= 19 \\ 5y &= 15 \\ y &= 3 \end{aligned}$$

Substitute y = 3 in ①.

$$\begin{aligned} x - y &= 2 \\ x - (\ ) &= 2 \\ x - 3 &= 2 \\ x &= 5 \end{aligned}$$

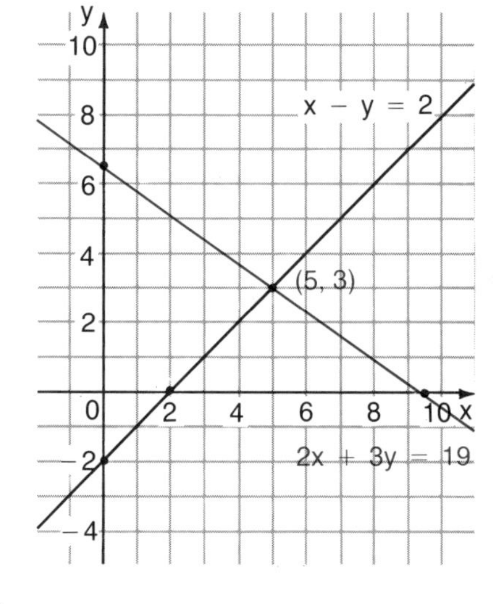

Check the solution x = 5 and y = 3 in both equations.

Check in ①.

L.S. $= x - y$ R.S. $= 2$
$= 5 - 3$
$= 5 - 3$
$= 2$

Check in ②.

L.S. $= 2x + 3y$ R.S. $= 19$
$= 2(5) + 3(3)$
$= 10 + 9$
$= 19$

The solution is x = 5 and y = 3 or (5, 3).

EXERCISE 5.6

B

1. Solve each of the following by substitution.

(a) $x - y = 2$, $x + y = -4$

(b) $x + y = 10$, $x - y = -8$

(c) $a + 2b = 10$, $3a + 4b = 8$

(d) $x + 2y = 20$, $x - y = 2$

(e) $3s - 4t = 5$, $s + 7t = 10$

(f) $a - 2b = 6$, $3a + 2b = 4$

(g) $3m - n = 15$, $m + n = 1$

(h) $2a + 5b = -22$, $a + 2b = -9$

(i) $x + 3y = -5$, $3x - 2y = 7$

(j) $y = 2x - 3$, $y = x + 4$

C

2. Solve by substitution.

(a) $2x - 3y = -4$, $-x + 2y = 3$

(b) $\frac{m}{2} + \frac{3n}{2} = 2$, $\frac{m}{5} - \frac{n}{2} = 3$

5.7 SOLVING PROBLEMS USING PAIRS OF EQUATIONS

There are some problems that can be solved using one variable and one equation. Other problems are solved using two variables and two equations. The following problem is solved in both ways.

Baseball Hall of Fame

Example 1.
The Red Sox baseball team retired two numbers. The difference between the two numbers is 4. The sum of two times the larger number and three times the smaller number is 43.
Find the two numbers.

Solution 1 — Using One Variable
Let the larger number be x.
Let the smaller number be $(x - 4)$.

$$\underbrace{\text{two times the larger}}_{2x}\ \underbrace{\text{plus}}_{+}\ \underbrace{\text{three times the smaller}}_{3(x-4)}\ \underbrace{\text{equals}}_{=}\ \underbrace{43}_{43}$$

$$\begin{aligned} 2x + 3(x - 4) &= 43 \\ 2x + 3x - 12 &= 43 \\ 5x - 12 &= 43 \\ 5x &= 55 \\ x &= 11 \end{aligned}$$

If the larger number, x, is 11, then the smaller number is $11 - 4$, or 7.
The numbers are 7 and 11.

Solution 2 — Using Two Variables
Let the larger number be x. Let the smaller number be y.

The difference between the two numbers is 4. $\quad x - y = 4 \quad ①$
Two times the larger plus three times the smaller is 43. $\quad 2x + 3y = 43 \quad ②$

$$\begin{aligned} 3 \times ① \quad 3x - 3y &= 12 \\ 2x + 3y &= 43 \\ \hline \text{Adding.} \quad 5x \quad\quad &= 55 \\ x \quad\quad &= 11 \end{aligned}$$

Substitute $x = 11$ in ①.

$$\begin{aligned} x - y &= 4 \\ 11 - y &= 4 \\ -y &= -7 \\ y &= 7 \end{aligned}$$

The two numbers are 11 and 7.

If you have a choice between using one or two variables, it is often easier to solve problems using two variables.

Example 2.

The Oceans Alive Co. rents helicopters for \$300 a day plus \$70/h of flying time. Sea View Equipment rents helicopters for \$600 a day plus \$40/h.

A university whale research team wants to rent a helicopter for a day.

For what number of hours are the rental costs equal?

Solution:

Let c represent the total rental cost per day and t the number of hours rented.

The cost to rent from Oceans Alive Co. is:

$$C = 300 + 70t \quad ①$$

The cost to rent from Sea View Equipment is:

$$C = 600 + 40t \quad ②$$

The costs will be the same at the point where the graphs of the two equations intersect.

We can find this point by substitution.

From equation ①

$$C = 300 + 70t$$

Substituting for C in ②

$$\begin{aligned} C &= 600 + 40t \quad ② \\ 300 + 70t &= 600 + 40t \\ 70t - 40t &= 600 - 300 \\ 30t &= 300 \\ t &= 10 \end{aligned}$$

The cost will be the same for a 10 h rental and we find the cost by substituting.

Oceans Alive Co.: $\begin{aligned} C &= 300 + 70t \\ &= 300 + 70(10) \\ &= 300 + 700 \\ &= 1000 \end{aligned}$

Sea View Equipment: $\begin{aligned} C &= 600 + 40t \\ &= 600 + 40(10) \\ &= 600 + 400 \\ &= 1000 \end{aligned}$

For 10 h, the cost is \$1000.

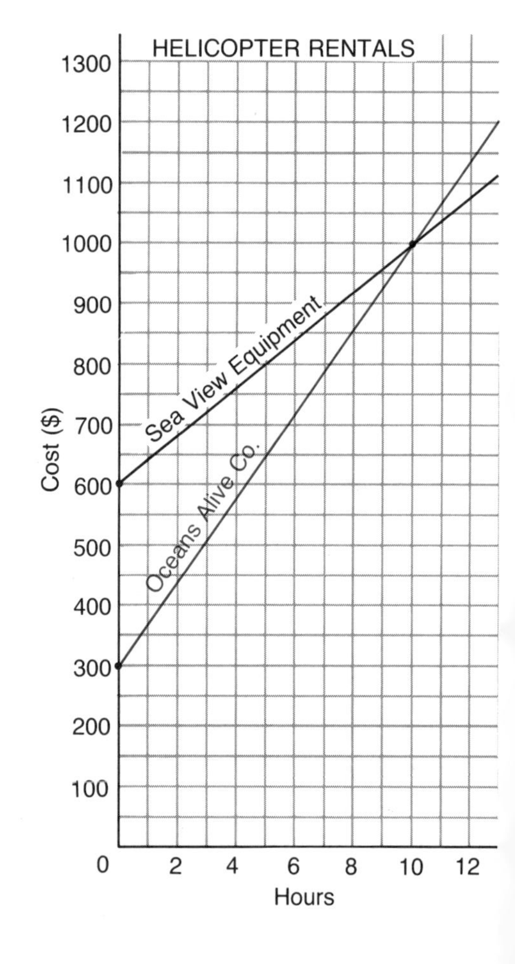

From the graph, which company is cheaper if the team rents a helicopter for 4 h?

From the graph, which company is cheaper if the team rents a helicopter for 12 h?

EXERCISE 5.7

A 1. Express each of the following as an algebraic expression.
(a) three times a number
(b) one-half a number
(c) 7 more than x
(d) 12 more than y
(e) b increased by 12
(f) twice m
(g) m diminished by 4
(h) 8 less than x
(i) 4 times a number, increased by 6
(j) twice a number diminished by 7
(k) 7 more than b
(l) 3 times a number less 6

2. Express each of the following as an equation in two variables.
(a) The sum of two numbers is 12.
(b) The sum of two numbers is 26.
(c) The difference between two numbers is 6.
(d) There are 28 boys and girls in the class.
(e) The store sold a total of 156 sweaters and shirts.
(f) There were 4 more cars than vans bought.
(g) Twice a number plus 6 times another number is 88.
(h) 5 times a number less 3 times another number is 18.
(i) The difference between two numbers is 56.
(j) The store sold a total of 77 basketballs and footballs.
(k) Enzo had a total of 96 goals and assists.

B 3. The sum of two numbers is 41. The difference between the numbers is 7.
Find the numbers.

4. The sum of two numbers is 52, and their difference is 6.
Find the numbers.

5. The sum of two numbers is 112. Their difference is 14.
Find the numbers.

6. On a busy Saturday morning the market bakery sold a total of 172 pies and strudels. There were 76 more pies sold than strudels. How many of each were sold?

7. The sum of two numbers is 24. Twice the smaller number plus three times the larger is 62.
Find the numbers.

8. The difference between two numbers is 6. Three times the larger number minus twice the smaller is 21.
Find the numbers.

9. The Lobster Landing Restaurant employs 120 people. There are 66 more full-time staff than part-time staff.
How many people work part time?

10. The sum of two numbers is 131 and their difference is 25.
Find the numbers.

11. A sporting goods store sold 3 more footballs than basketballs. Three times the number of basketballs sold plus four times the number of footballs sold is 96.
How many footballs were sold?

12. There are 30 students in Shelagh's math class. There are 2 more girls than boys.
How many girls are in the class?

13. I am thinking of two numbers. Five times the smaller plus 3 times the larger is 19. Four times the smaller plus 7 times the larger is 29.
What are the numbers?

14. The difference between two numbers is 24. The sum of 3 times the larger and 8 times the smaller is 160.
Find the numbers.

15. The cost of 5 oranges and 6 grapefruits is $2.17. The cost of 3 oranges and 4 grapefruits is $1.39.
Find the cost of one orange.

16. The difference between the length and width of a rectangle is 8 m. The perimeter is 44 m.
Find the length and width.

17. The Brown farm is 24 ha smaller than the Jorgenson farm. The 2 farms together contain 312 ha.
How large is each farm?

18. At the parks commission meeting there were 6 more men than women. There were a total of 70 people at the meeting.
How many women were present?

19. The perimeter of a soccer field is 346 m. The length is 27 m longer than the width.
What are the dimensions of the field?

20. A computer repair service charges a flat fee for a service call plus an hourly rate. A 4 h call costs $98. A call that takes 8 h costs $166.
What are the flat fee and the hourly rate?

21. There were a total of 120 student and adult tickets sold for the lunch hour school play. Five times the number of student tickets sold decreased by 3 times the number of adult tickets sold equals 160.
How many student tickets were sold?

22. Pauline earned a total of 170 marks in chemistry and history. Twice her history mark, decreased by 40, equals her chemistry mark, increased by 33.
What was her chemistry mark?

23. On Saturday a beverage company sold 40 more cases of cola than ginger ale. If the company had sold 30 more cases of cola and 70 more cases of ginger ale, a total of 700 cases would have been sold.
How many cases of cola were actually sold?

24. The sum of two angles is 90°. Three times the smaller angle is 10 more degrees than two times the larger angle.
What is the size of each angle?

25. The length of a rectangular flower bed at a botanical gardens is 50% longer than the width. The perimeter of the flower bed is 60 m.
What are the dimensions of the flower bed?

26. A member of a search-and-rescue team is plotting the courses of a rescue ship and a disabled ship on a grid. The course of the rescue ship is described by the equation

$$y = 3x + 40$$

The course of the disabled ship is described by the equation

$$y = 5x + 10$$

What are the coordinates of the point where the two ships are expected to meet?

27. The Rockton Company makes souvenir painter caps. The cost of making the caps is described by the equation

$y = 2x + 1400$

where y is the total cost and x is the number of caps.
The income from selling the caps is described by the formula

$y = 6x$

where y is the income and x is the number of caps.
How many painter caps must the Rockton Company make and sell to break even?

28. A company makes fountain pens. The cost of making the pens is described by the equation

$y = 8x + 400$

where y is the total cost and x is the number of pens.
The income from selling the pens is described by the equation

$y = 10x$

where y is the income and x is the number of pens.
If the company sells all the pens that it makes, how many pens must be made for the company to break even?

29. The cost of making team jackets is given by

$y = 80x + 300$

where y is the total cost and x is the number of jackets.
The manufacturer's income from selling the jackets is given by

$y = 100x$

where y is the income and x is the number of jackets.
The jackets' manufacturer demands full payment of the order before the jackets are made.
How many jackets must be sold in one order for the manufacturer to break even?

30. The Star Computer Company makes computer discs. The discs cost $2.50 each to make and the daily overhead costs for the company are $40 000. The company sells the discs for $5.00 each.

(a) Write an equation for the company's cost to make the discs each day.
(b) Write an equation for the amount the company receives if it sells all the discs it makes each day.
(c) Determine how many discs the company has to sell each day to break even.

MICRO MATH

Computers are often used to solve systems of equations.

$ax + by = c$
$dx + ey = f$

The following BASIC program can be used to solve a linear system.

```
10 INPUT A,B,C,D,E,F
20 IF A * E - B * D = 0 THEN 70
30 LET X = (C * E - B * F) /
   (A * E - B * D)
40 LET Y = (A * F - C * D) /
   (A * E - B * D)
50 PRINT "(";X;",";Y;") IS A
   SOLUTION"
60 GOTO 80
70 PRINT "NO SOLUTION"
80 END
```

Use the program to solve each system.

1. $3x + 4y = 23$
   $5x + 6y = 35$
2. $5x - 3y = 11$
   $7x + 2y = 34$
3. $2x + 5y = 12$
   $x - 6y = -11$
4. $4x - 7y = 17$
   $6x + 5y = -21$
5. $3.7x + 2.8y = 7.4$
   $6.3x - 2.1y = 5.1$
6. $0.25x - 3.75y = 4.50$
   $1.75x + 2.50y = -0.95$

EXTRA

5.8 ILLUSIONS

We know that the sun does not actually rise. It is the rotation of the earth that gives the sun the illusion of rising and setting. Yet every day we talk about the sun setting and rising. Can we take for granted that what we see is always reality?

EXERCISE

1. Study the pattern below.

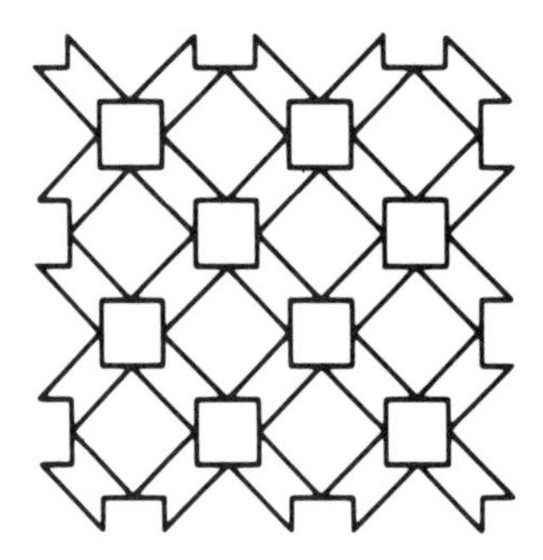

At first glance it appears that this pattern is made up of small squares and large squares. Or, taking a second look, we can see that the pattern is made up of the following shape.

Sketch the shape that makes up the following pattern.

2. Unless you squint, you cannot see what is directly in front of your nose. We can prove this with the fence drawing below.
Move the space in the fence closer and closer to your nose and the space will disappear.

3. You can fool your sense of touch with Aristotle's illusion. Cross your fingers as shown and touch your nose.

You should get the impression that you are touching two noses.

4. Both squares are the same size and made with straight lines, but they look very different.

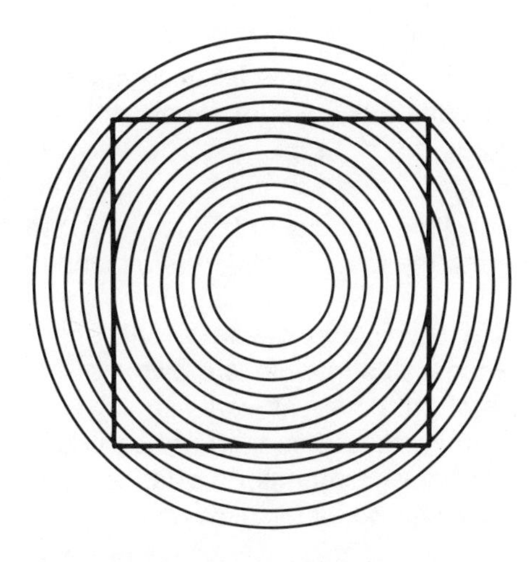

Using your knowledge of shapes, angles, and lines, try creating your own illusion using a square the same size as the one above.

5. The following riddle is based on illusion.
What does it mean?

Two legs sat upon three legs
With one leg in his lap;
In comes four legs
And runs away with one leg;
Up jumps two legs,
Catches up with three legs,
Throws it after four legs,
And makes him bring back one leg.

6. Are the two horizontal bars the same length?

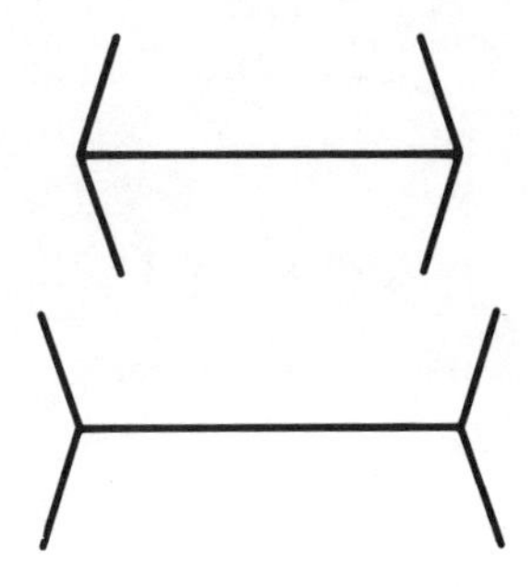

7. The figure below is called the Hermann grid illusion.
Can you explain why you see grey spots at the intersections?

8. The areas of the three figures are the same.
Which one looks taller?
Which one looks wider? Why?

5.9 PROBLEM SOLVING

1. A circle has seven points on it.

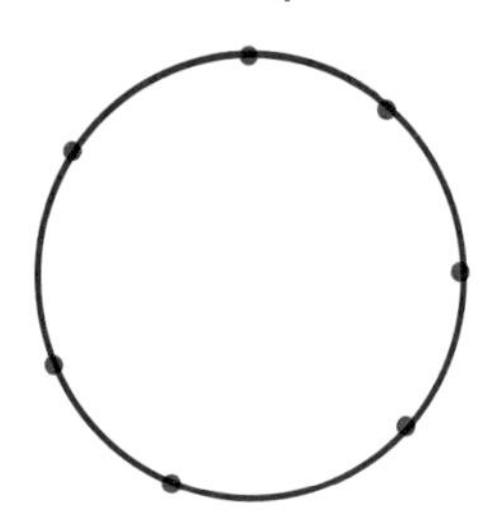

How many different chords can be drawn joining any two points?

2. Determine the pattern and complete the table.

6	11	5	7		8
54	77		63	36	
9	7	8		4	
60	88				104

3. During the first three laps of a race a boat averaged 20 km/h.
How fast does it have to travel on the fourth and last lap to average 30 km/h during all four laps of the race?

4. What fraction of the rectangle is shaded?

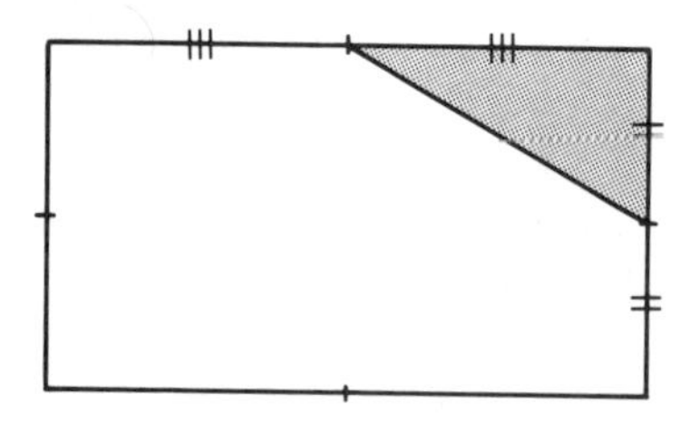

5. The power 4^3 means $4 \times 4 \times 4$. The value of 4^3 is 64.

To find the last digit in the value of 4^{152} you cannot use a calculator because the number is too large for the calculator to display every digit.

To find the last digit, look for a pattern in the following.

$4^1 = 4$
$4^2 = 16$
$4^3 = 64$
$4^4 = 256$
$4^5 = 1024$

When the exponent is an even number, the last digit is 6. When the exponent is an odd number, the last digit is 4. Therefore, the last digit of 4^{152} is 6.

(a) Find the last digit if 2^{437} is expanded.
(b) Find the last digit if 3^{209} is expanded.

6. The 9 points below are evenly spaced. One isosceles triangle is drawn, having three of the points as vertices.
How many isosceles triangles can be drawn altogether? (Remember that an equilateral triangle can be also isosceles.)

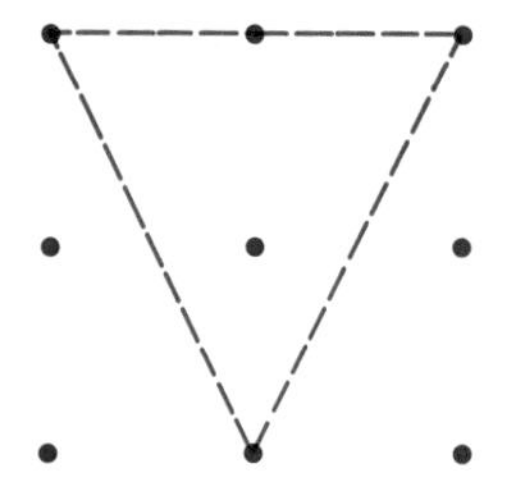

7. What is the seventh letter in this sequence?

F, S, T, F, F, S, ■

8. A field measures 130 m by 50 m. During a storm, 1 cm of rain fell.
How many litres of water theoretically fell on the field?

9. Alicia bought several items at the hardware store. The costs are listed below.

Nails	$13.50
Paint	$22.00
Fertilizer	$14.45
Screwdriver	$11.80

(a) If the sales tax is 8%, what is the total cost of her purchases?
(b) How much change should she receive from four $20 bills?

10. The number 96 can be written as the difference of two squares; one example is

$$25^2 - 23^2 = 625 - 529$$
$$= 96$$

Find three other examples of how to express the number 96 using the difference of two squares. The squares are less than 200.

11. You work as an air traffic controller. You must report for work by 15:30. It takes you five minutes to get from the airport parking lot to the tower and thirty-five minutes to get from home to the parking lot. On this day you have to stop on the way to work to arrange a loan for a new car. You estimate that this will take forty-five minutes. You also have to stop for some gas which will take another ten minutes.

At what time should you leave home?

12. Bob bought a pencil for fifteen cents. He gave the clerk a $1 bill, who then gave him 9 coins in change.
What are the possible combinations of coins that he received?

CAREER

METEOROLOGIST

Meteorologists at Environment Canada use information from satellites, 2500 weather observing stations, aircraft, ships, charts, and graphs to predict the weather. The table gives the winter and summer highs and lows for several Canadian cities.

City	Temperature (°C)			
	Winter		Summer	
	High	Low	High	Low
Vancouver	5	0	22	13
Edmonton	−11	−19	23	12
Regina	−13	−23	26	12
Winnipeg	−14	−24	26	13
Toronto	−1	−8	27	17
Ottawa	−6	−15	26	15
Montreal	−6	−15	26	16
Fredericton	−4	−15	26	13
Halifax	−2	−10	23	13
Charlottetown	−3	−11	23	14
St. John's	−1	−7	20	11

1. Use the 'Summer High' data and rank the cities from warmest to coldest.

2. On average Environment Canada's forecasts are 80% accurate.
How many days are they wrong each year?

3. It costs 2 cents per person per day for the Canadian weather service.
How much is that each year?

5.10 REVIEW EXERCISE

1. Find the missing number in each ordered pair so that the ordered pair satisfies the equation.

(a) $x + y = 8$; (3, ■), (■, 7), (−1, ■)
(b) $x - y = 3$; (5, ■), (■, 1), (■, −1)
(c) $y = 2x + 5$; (4, ■), (−3, ■), (0, ■)
(d) $y = 3x - 7$; (3, ■), (−4, ■), (−6, ■)

2. Solve the following pairs of equations graphically.

(a) $y = 2x + 1$, $y = x + 3$ (b) $y = 3x - 4$, $y = x + 6$
(c) $y = -2x + 4$, $y = 3x - 6$ (d) $y = -3x - 6$, $y = 2x + 4$
(e) $x + y = 5$, $x - y = 3$ (f) $x + y = 12$, $x - y = 4$
(g) $2x + y = 6$, $3x + y = -3$ (h) $x + y = 14$, $x - y = 8$

3. Solve the pairs of equations by subtraction.

(a) $3x + y = 9$, $x + y = 5$ (b) $a + 3b = 10$, $a - 2b = 5$
(c) $3x + 4y = 10$, $-x + 4y = 2$ (d) $8s + 3t = 25$, $2s + 3t = 13$
(e) $9x + 4y = 43$, $3x + 4y = 25$ (f) $2x - 3y = 4$, $-x - 3y = -11$

4. Solve the pairs of equations by addition.

(a) $x + y = 10$, $x - y = 8$ (b) $2a + b = 6$, $3a - b = 9$
(c) $-x + y = 4$, $x + y = 2$ (d) $s - 3t = 6$, $4s + 3t = 9$
(e) $2x - 4y = -16$, $-2x + 2y = 8$

5. Solve and check the pairs of equations.

(a) $4x + y = 10$, $3x + 2y = 5$ (b) $3m - 2n = 3$, $4m + 3n = 4$
(c) $5s - 2t = 11$, $2s - 3t = 0$ (d) $4x + 5y = 18$, $3x + 7y = 20$
(e) $3x - 2y = -4$, $4x + 5y = 10$ (f) $5m - 3n = 3$, $6m - 4n = 2$
(g) $2x + 5y = 21$, $4x - y = 9$ (h) $6a - 5b = -21$, $7a + 2b = -1$

6. Solve the pairs of equations by substitution.

(a) $4x + y = 6$, $3x + 2y = 7$ (b) $a + b = 4$, $2a - b = -1$
(c) $2m + n = 4$, $m - n = -1$ (d) $r - s = -3$, $2r + 3s = -6$
(e) $x + y = -2$, $x - y = 6$ (f) $2x + y = 10$, $4x - y = 5$

7. The sum of two numbers is 39. Their difference is 11.
Find the numbers.

8. The sum of two numbers is 52 and their difference is 8.
Find the numbers.

9. The sum of two numbers is 23. Three times the first added to four times the second is 83.
Find the numbers.

10. One number is three more than another. Two times the larger added to three times the smaller is 81.
Find the numbers.

11. There are a total of 56 students and teachers on the graduation committee. There are 28 more students than teachers. How many teachers are on the committee?

12. The perimeter of a rectangular field is 220 m. The length is 10 m longer than the width.
Find the length and width.

13. The perimeter of a basketball court is 80 m. The length is 12 m longer than the width.
What are the dimensions of the court?

14. The Sunburst Company makes and sells sunglasses. The cost of making the glasses is described by the equation

$$y = 5x + 20\,000$$

where y is the total cost and x is the number of pairs of sunglasses.
The income from selling the sunglasses is described by the formula

$$y = 15x$$

where y is the income.
How many pairs of sunglasses must the Sunburst Company make and sell to break even?

15. The Alpha Epsilon Fraternity makes and sells college scarves. The cost of making the scarves is given by the formula

$$y = 10x + 4500$$

where y is the total cost and x is the number of scarves.
The income from selling the scarves is given by the formula

$$y = 25x$$

where y is the income.
If the fraternity members can sell all the scarves that they make, how many scarves must they sell to break even?

EXTRA

LET THE STICKS FALL WHERE THEY MAY

Some sticks are placed on a flat table. All the sticks have the same width. The thickness of the sticks is the same as the width. As you can see, the sticks have different lengths.

Some of the sticks are on the table. Other sticks rest partly on the table and partly on other sticks. Some do not even touch the table. Others have one end higher than the other.

1. Write the number of each level stick.
2. Write the number of each sloping stick.
State which end of each stick is lower, the numbered end or the other end.
3. Which level stick is the highest above the table?
4. Which stick has the greatest slope?
5. Does stick 1 touch stick 12?
6. Does stick 4 touch stick 9?
7. Does stick 10 touch stick 1?

5.11 CHAPTER 5 TEST

1. Find the missing numbers so that each ordered pair satisfies the indicated equation.

(a) $x + y = 7$
   (i) (3, ■)
   (ii) (■, 6)
   (iii) (−1, ■)

(b) $y = x - 3$
   (i) (6, ■)
   (ii) (■, 5)
   (iii) (−3, ■)

2. Solve the following system graphically.

$$y = 2x - 1$$
$$y = x + 2$$

3. Solve each pair of equations by elimination.

(a) $3x - 2y = 2$
$2x + 3y = 23$

(b) $3x + 4y = -15$
$5x - 2y = 1$

4. Solve the pair of equations by substitution.

$$4x + y = 14$$
$$3x - 2y = 5$$

5. The sum of two numbers is 21. Their difference is 5. Find the numbers.

6. The perimeter of a rectangular field is 60 m. The length of the field is 6 m longer than the width.
Find the dimensions of the field.

7. A company makes bracelets. The cost of making the bracelets is described by the formula

$$C = 6n + 5000$$

where C is the total cost and n is the number of bracelets. The income from selling the bracelets is described by the formula

$$I = 10n$$

where I is the income and n is the number of bracelets.
If the company sells all the bracelets it makes, how many bracelets must be sold for the company to break even?

STATISTICS

REVIEW AND PREVIEW TO CHAPTER 6

PERCENT

EXERCISE

1. Express each of the following percents as decimals.

(a) 36% (b) 25%
(c) 10% (d) 8%
(e) 123% (f) 345%
(g) 100% (h) 200%
(i) 0.5% (j) 0.1%

2. Express the following decimals as percents.

(a) 0.25 (b) 0.75
(c) 0.5 (d) 0.1
(e) 0.56 (f) 0.78
(g) 0.07 (h) 1.3
(i) 0.003 (j) 0.051

3. Express each fraction as a percent.

(a) $\frac{1}{2}$ (b) $\frac{1}{4}$
(c) $\frac{3}{10}$ (d) $\frac{3}{20}$
(e) $\frac{4}{25}$ (f) $\frac{17}{25}$
(g) $\frac{156}{200}$ (h) $\frac{3}{8}$

4. Express each percent as a fraction in lowest terms.

(a) 25% (b) 50%
(c) 75% (d) 100%
(e) 10% (f) 23%
(g) 9% (h) 3%

5. Evaluate the following.

(a) 20% of 300 (b) 10% of 456
(c) 50% of 270 (d) 80% of 16
(e) 5% of 140 (f) 1% of 7000
(g) 100% of 67 (h) 200% of 88
(i) 0.5% of 500 (j) 0.1% of 1300
(k) 1.5% of 60 (l) 4.7% of 300

6. Calculate.

(a) What percent of 40 is 20?
(b) What percent of 60 is 15?
(c) What percent of 90 is 9?

7. Calculate the total cost of each of the following purchases given the sales tax.

(a) sweater: $78, sales tax: 7%
(b) painting: $640, sales tax: 6%
(c) candles: $15.70, sales tax: 8%

8. Joseph scores 68 out of 85 on a math test.
What is his mark, as a percent?

9. The swim team sold T-shirts to 244 out of the 360 students in the school. What percent of the students bought T-shirts, to the nearest tenth?

10. Marie found from her survey that a candidate would need 42% of the votes to win an election to the university's Students Council.
If there were 9200 voters, how many voters were needed to win?

11. Clayton spends between 25% and 30% of his earnings on clothes.
If he earns $260 each week, how much does he spend on clothes?

12. The following diagram shows the results of a survey of students who watch baseball and hockey games.

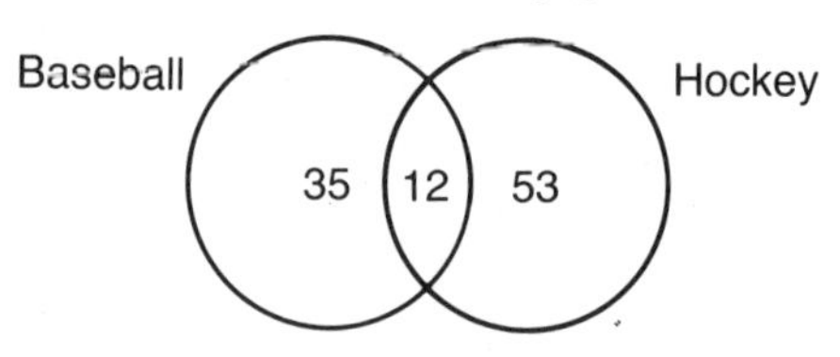

(a) What percent of students watch only baseball?
(b) What percent watch only hockey?
(c) What percent watch both baseball and hockey?

13.

(a) If a cruise ship is going to serve 5000 dinners, how many orders of milk can be expected?
(b) Why do the percents not add to 100?
(c) How may a company gather the information shown in the graph?

14.

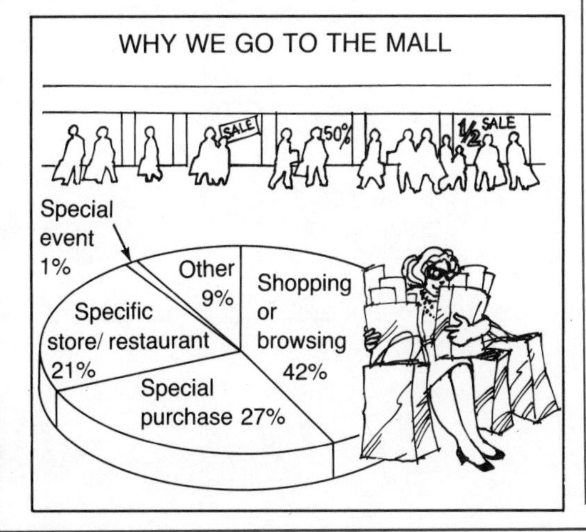

(a) If 300 000 people go to the High Street Mall every week,
  (i) how many go to shop or browse?
  (ii) how many go to special events?
  (iii) how many go to make special purchases?
(b) The Other sector represents reasons for going to the mall besides those which are given. Suggest what these other reasons may be.

15.

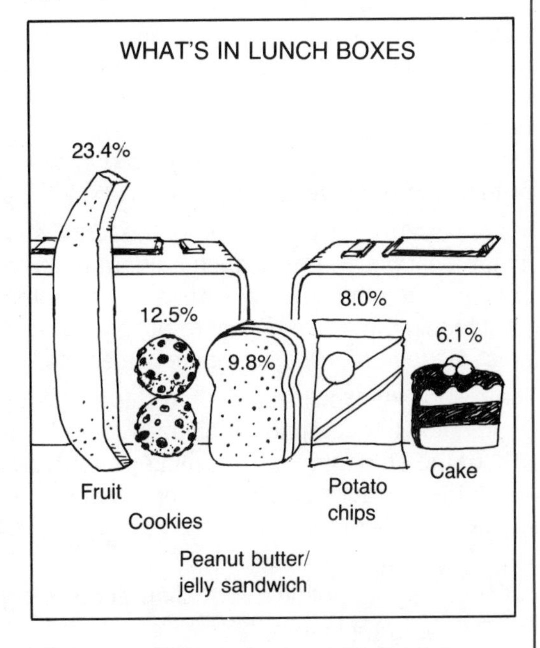

There are 500 students at Clarksdale school who bring their lunches to school. Assume that you can apply the information above to this school.

(a) How many sandwiches can you expect to find in these lunches?
(b) How many pieces of cake can you expect to find?
(c) Why don't the percents add to 100?
(d) What are some items that are missing?

6.1 STATISTICS AND SAMPLING

People from all walks of life use statistics to obtain information. For example, manufacturers make use of statistics in a number of ways: quality control, production planning, and marketing information.

A manufacturer makes winter jackets for high school and college students. The manufacturer wants to know which colours of jackets students prefer. To find out, the manufacturer hires a polling company to ask or survey the students.

There are thousands of high school and college students in the manufacturer's province. The polling company can afford to interview only a portion or sample of the entire group of students, or student population. One thousand is a good number for this kind of survey.

It is important that the sample of students is chosen randomly so that each student in the population has an equal chance of being chosen. Once the sample is chosen, the company conducts interviews. This is one way of collecting information, but there are many others: by questionnaire, test, telephone survey, direct measurement, or computer database.

The responses which are gathered from the survey are called data. To make the data meaningful, the company organizes it, and discovers certain patterns. The survey shows that 80% of the students in the sample prefer blue jackets. You can conclude or infer that 80% of the student population also prefer blue jackets. How may this inference affect the jacket manufacturer's production plans?

In summary, to obtain information about a population:

1. Decide on a sample.
2. Choose a sampling method.
3. Collect data from the sample.
4. Organize and interpret the data.
5. From the data, make decisions (inferences) about the entire population.

Example.
The students in grades nine and ten who buy their lunch at school are divided into groups as follows.

	Grade 9	Grade 10
Boys	90	81
Girls	120	69

If you want a random sample of 50 for a food survey, how many from each group should you ask?

Solution:
There are 360 students in all. There are 90 grade nine boys.

These boys should be $\frac{90}{360}$ of the sample.

Number of grade nine boys: $\frac{90}{360} \times 100 = 25$

Number of grade nine girls: $\frac{120}{360} \times 100 \doteq 33$

Number of grade ten boys: $\frac{82}{360} \times 100 \doteq 23$

Number of grade ten girls: $\frac{68}{360} \times 100 \doteq 19$

A sample of 100 students should include 25 grade nine boys, 33 grade nine girls, 23 grade ten boys, and 19 grade ten girls, selected at random.

EXERCISE 6.1

A 1. A survey of Canadians' opinions on a controversial issue is to be taken.
What problems might researchers have if they use the following survey methods?

(a) stopping people on their way to work at a busy subway station in Toronto or Montreal
(b) calling people at random during a weekday afternoon in Victoria
(c) giving questionnaires to all grade five students to take home to their parents
(d) placing ads in national newspapers and asking people to write letters expressing their opinions

2. State the methods you would use to gather the following data.

(a) the need for a new Canadian flag
(b) the most popular car in your province
(c) the names of the colleges in Canada that offer computer technology courses
(d) the number of pets per family in your town

3. The Devron Company makes waterproof matches for campers. They test one out of every 10 000 matches made.

(a) Why can they not test all of the matches?
(b) What other types of manufacturer could not test 100% of their products?
(c) What types of manufacturers should test 100% of their products?

B 4. There are 156 boys and 165 girls in the graduating class who will attend the dance. You have been asked to determine where to have the dance. You plan to survey 50 graduating students.
How many from each group should you survey?

5. How would you choose a random sample in each of the following situations?

(a) A 200 person sample in a town of 50 000 people; the issue is the cost of mail service.
(b) A 50 person sample in a school of 900 students; the issue is school bus routes.
(c) A 2000 person sample across Canada; the issue is federal gasoline tax.
(d) A 1000 person sample in a city of 300 000; the issue is the popularity of a new TV program.

6.2 MAKING PREDICTIONS FROM A RANDOM SAMPLE

There are companies whose sole business is to conduct surveys or polls for a variety of clients: governments, corporations, and the media. These companies or pollsters use information from samples to make inferences or predictions. For example, there are 100 000 homes with TV sets in Maple Hills. Twelve TV stations broadcast to Maple Hills. A survey of 200 homes conducted between 21:05 and 21:25 found that 104 homes had their sets tuned to TV station CHFN.

Since $\frac{104}{200}$ is 52% of the homes surveyed, the owners of CHFN can infer that 52% of the 100 000 homes with TV sets were tuned to CHFN.

$$\begin{aligned} 52\% \text{ of } 100\,000 &= 0.52 \times 100\,000 \\ &= 52\,000 \end{aligned}$$

CHFN had 52 000 viewers in Maple Hills between 21:05 and 21:25.
How could the station owners use this information?

Example.
The table gives the results of a survey, paid for by local business people, which determines how many people buy the three daily newspapers in a city. The tally column gives the number of people who buy each paper.

READ

If there are 300 000 of these three newspapers sold each day, how many of each are bought?

Newspaper	Tally
The *Record*	𝍸 𝍸 𝍸 𝍸 𝍸 𝍸 𝍸 𝍸 𝍸 II
The *Standard*	𝍸 𝍸 𝍸 𝍸 𝍸 𝍸 𝍸 𝍸 𝍸 𝍸 𝍸 III
The *Times*	𝍸 𝍸 𝍸 𝍸 𝍸 𝍸 𝍸
Total	140

Solution.

PLAN

Convert each number of newspaper buyers to a percent.
Apply these percents to the total population.

SOLVE

Newspaper	Frequency	Percent	Calculations	Buyers
The *Record*	47	$\frac{47}{140} \doteq 34\%$	$0.34 \times 300\,000 = 102\,000$	102 000
The *Standard*	58	$\frac{58}{140} \doteq 41\%$	$0.41 \times 300\,000 = 123\,000$	123 000
The *Times*	35	$\frac{35}{140} \doteq 25\%$	$0.25 \times 300\,000 = 75\,000$	75 000

ANSWER

There are about 102 000 people who buy the *Record*, 123 000 who buy the *Standard*, and 75 000 who buy the *Times*.

EXERCISE 6.2

1. The traffic at the Main and King Streets intersection is always backed up during the evening rush hours. The number of vehicles travelling from the east at the intersection was recorded for 15 min during these rush hours. The data show which way the vehicles travelled through the intersection.
R = right, S = straight, L = left

R, S, S, L, S, L, R, S, L, L, R, L, S,
L, S, R, L, S, S, S, S, S, L, R, L, R,
L, S, S, L, L, R, L, L, L, S, S, L, R,
L, L, S, S, S, S, R, L, S, L, R

(a) Calculate the percent of vehicles for each way.
(b) If 2000 vehicles travel from the east during the rush hours, how many travel each way?
(c) How could the Roads department use this information?

2. Quality control inspectors sample a shipment of 1000 computer disks each day. The table gives the number of defective disks found on each day during one week.

Day	Tally
Mon	𝍸 III
Tues	𝍸 𝍸 II
Wed	III
Thurs	𝍸
Fri	𝍸 I

If the company produces 100 000 computer disks each day, predict how many of these disks were defective on each day.

3. A 1% sample of 150 000 readers was taken by a magazine to determine the popularity of a new rock group.
(a) How many readers were sampled?
(b) If 833 liked the group, what percent of the readers could be inferred to like the group?
(c) Similarly, how many readers did not like the group?

EXTRA

THE RATINGS GAME

Television shows are graded with two numbers: a rating and a share. A rating of 11.5 means that 11.5% of the nation's households with televisions were tuned to the show. The share of 14 means that 14% of the sets in use were tuned to the show.

A rating is always lower than a share because the rating is a percent of all TV sets in homes, including sets that are not turned on. The share is a percent of the TV sets in homes that are turned on.
If there are 8 million households with televisions, then a rating of 11.5 means that an estimated 920 000 televisions in homes were tuned to the show.

11.5% of 8 000 000 = 920 000

To calculate the share you must know how many sets were in use at the time of the poll. Suppose the polling company determined that 70% of the TV sets were in use.
So, 70% of 8 000 000 = 5 600 000
Then, 0.14 × 5 600 000 = 784 000

1. A rating of 15.2 and a share of 17 were recorded one evening for a new comedy show. Of the 8 million household TV sets tuned to the show, 75% were in use at the time.
Change the rating and share to numbers of TV sets.

2. The previous afternoon, a football game had a rating of 21.4 and a share of 26. Of the 88 million household TV sets tuned to the show, 69% were in use at the time.
Change the rating and share to numbers of TV sets.

3. Why do you need both numbers to rate a show?

6.3 READING GRAPHS

Graphs are used to display data in an eye-appealing and informative way. Different graphs display different types of data.

THE BAR GRAPH

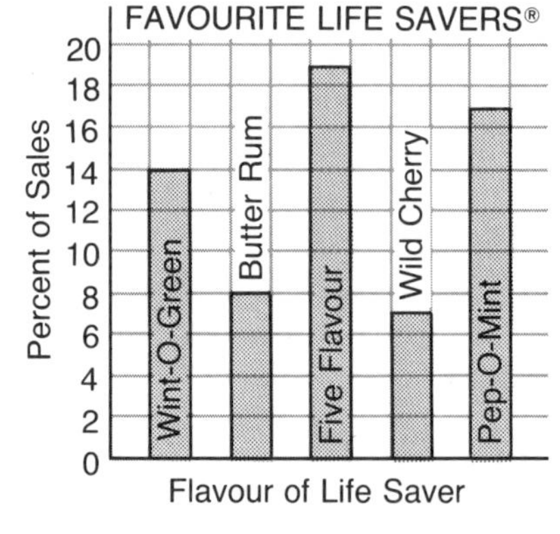

The bar graph is used to compare similar things. This graph shows the favourite flavours of Life Savers®, according to sales.

(a) What percent of consumers prefer each of these flavours?
 (i) Pep-O-Mint (ii) Wild Cherry (iii) Butter Rum
 (iv) Five Flavour (v) Wint-O-Green

(b) Rank the Life Saver® flavours in popularity from highest to lowest.

(c) Why is the total of the percents only 65?

THE BROKEN-LINE GRAPH

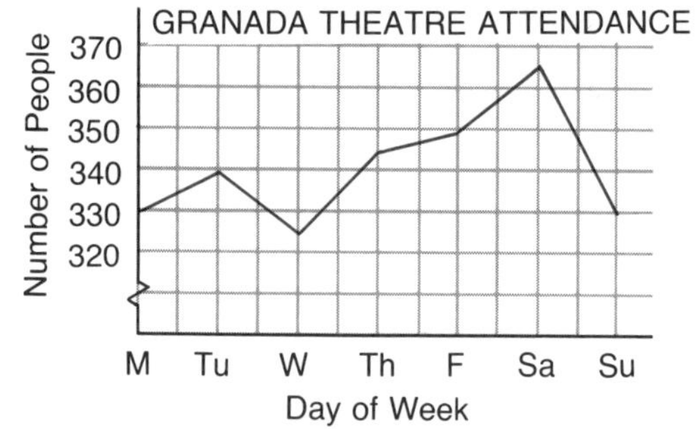

A broken-line graph is used to show how one thing changes. This line graph shows the attendance at the Granada Theatre for a one-week period.

(a) What was the attendance on Friday?

(b) What day had the lowest attendance?

(c) What day had the highest attendance?

(d) What was the total attendance for the week?

THE CIRCLE GRAPH

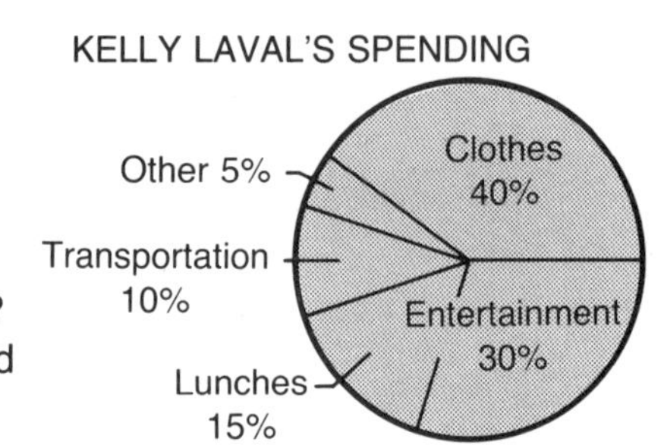

A circle graph is used to show how one thing is divided. This circle graph shows how Kelly Laval spends what she earns from her part-time job.

(a) What is the greatest expense?

(b) What is the least expense?

(c) What expenses could be included in the Other section?

(d) If Kelly earns \$200 a month, how much does she spend on each item?

THE PICTOGRAPH

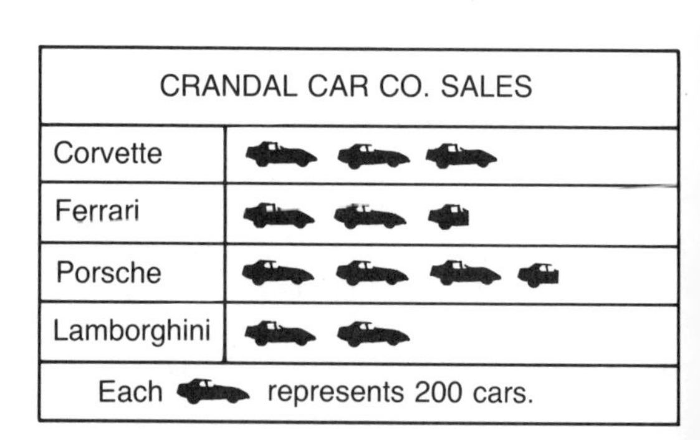

Pictographs use pictures or drawings to represent numbers. This pictograph shows the number of custom cars sold during one year by the Crandal Car Company.

(a) Approximately how many Corvettes were sold?

(b) Approximately how many Ferraris?

(c) Approximately how many Lamborghinis?

(d) What were the total sales for the year?

EXERCISE 6.3

1. The bar graph shows the loudness, in decibels, of different sounds.

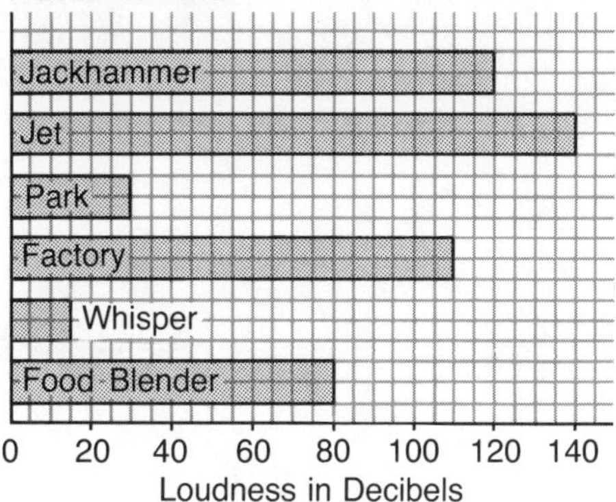

(a) How loud is a whisper?
(b) How many times louder than a park is a jackhammer?
(c) List the noises in order from least noisy to most noisy.

2. The broken-line graph shows the pulse rate of a patient in a hospital. The pulse is taken once each morning and each evening during every day.

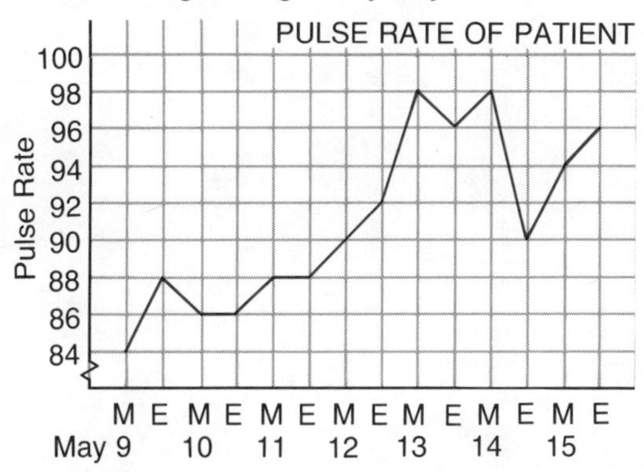

(a) What was the pulse rate on the morning of May 12?
(b) What was the pulse rate on the evening of May 14?
(c) On what days did the pulse rate stay the same?
(d) When did the greatest increase in pulse rate occur?
(e) When did the greatest decrease occur?

3. The circle graph shows how Dunnville taxes were spent last year. If the total expenditure was $20 000 000, how much was spent for

(a) education?
(b) public works?
(c) sanitation?
(d) protection?
(e) recreation?
(f) government?

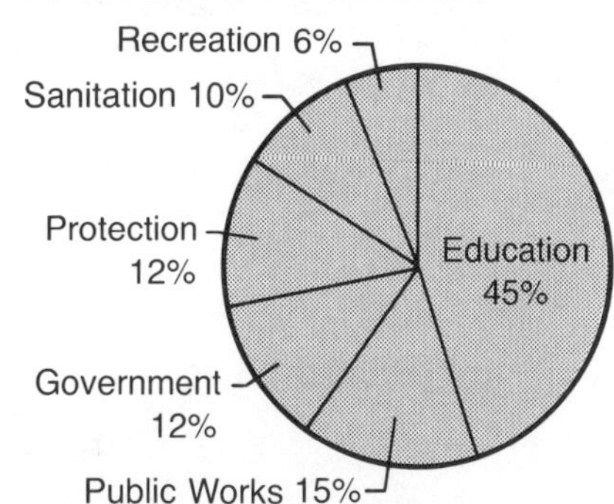

4. The graph shows the type of vehicles owned by families.

FAMILY VEHICLES

Mid-size car 35%
Full-size car 29%
Compact car 21%
Pick-up truck 8%
Sub-compact car 4%
Van 3%

If there are 5000 families in Estevan, how many of each type of vehicle would you expect them to own?

6.4 DRAWING GRAPHS

The temperatures of a patient in a hospital are taken every morning and evening. Body temperature is one indication of a patient's well-being. The table gives the temperature of a patient for five days.

Date	Temperature (°C)	
	Morning	Evening
July 1	37.7	38.5
July 2	37.4	38.3
July 3	37.2	38.0
July 4	36.8	37.9
July 5	36.8	37.4

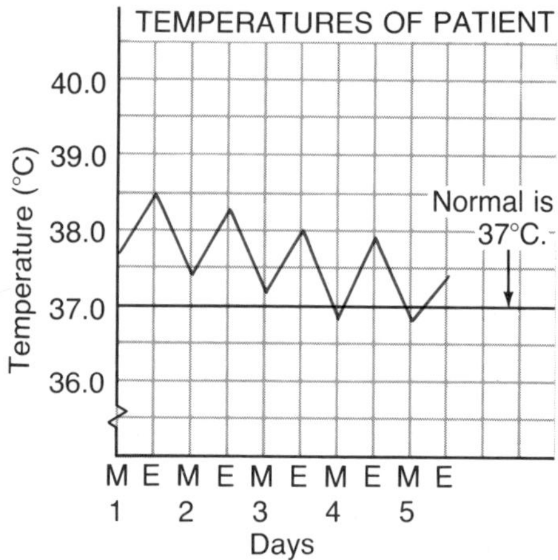

This information becomes much more meaningful when it is displayed on a broken-line graph.

Example.
A baseball team spent the sponsor's money as follows:

Uniforms	$ 960
Travel	1440
Park rental	1200
Equipment	480
Miscellaneous	720
Total	$4800

Construct a circle graph to show this information.

Solution:
First express each expense as a percent of the total.

Uniforms: $\frac{960}{4800} \times 100 = 20\%$

Travel: $\frac{1440}{4800} \times 100 = 30\%$

Park rental: $\frac{1200}{4800} \times 100 = 25\%$

Equipment: $\frac{480}{4800} \times 100 = 10\%$

Miscellaneous: $\frac{720}{4800} \times 100 = 15\%$

Check: 100%

To convert each percent to a central angle:
(i) write the percent as a decimal
(ii) multiply by 360° (the number of degrees in a circle)

Uniforms	$0.2 \times 360° = 72°$	Equipment	$0.1 \times 360° = 36°$
Travel	$0.3 \times 360° = 108°$	Miscellaneous	$0.15 \times 360° = 54°$
Park rental	$0.25 \times 360° = 90°$		

To make the circle graph, use a compass to draw a circle and mark off the required central angles with a protractor. Label each section and give the graph a title.

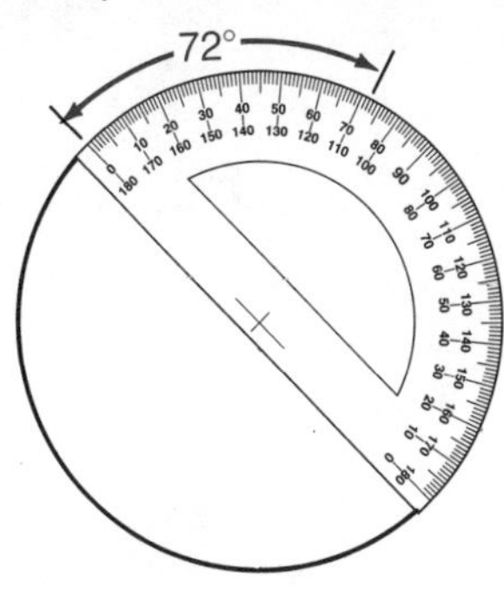

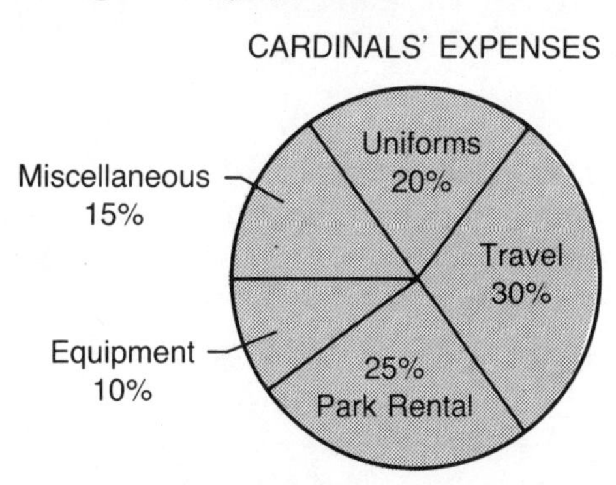

EXERCISE 6.4

To illustrate the data in this exercise, consider all four graphs: bar, line, and circle graph and pictograph.

A 1. What type of graph would you use to illustrate each of the following?

(a) the temperature change during a day
(b) the favourite type of TV program for the members of your class
(c) how the O'Meara family spends its weekly income
(d) the world's population growth since 1900

B 2. Construct graphs to illustrate the following data.

(a) The number of points scored during one game by the starting five members of a basketball team.

Player	Points
Adams	26
Smith	18
Chang	21
Kolchak	17
Brown	23

(b) The cost of producing an LP record is divided as follows.

Manufacturing	15%
Distribution and Selling	40%
Songwriter, Singer, Musicians	20%
Record Company Profit	25%

(c) Eric has a total body mass of 75 kg.
Muscle 37.5 kg   Fat 15 kg
Bone 13.5 kg   Other 9 kg

(d) the lengths of the world's five longest rivers

(e) Hot dogs contain 85% meat, 10% water, 2.5% salt, and 2.5% spices by mass.

(f) The average precipitation for Dry Gulch, South Dakota, was recorded for a period of one year.

Month	J	F	M	A	M	J	J	A	S	O	N	D
Precip. (cm)	1	1	2	4	5.5	7.5	5	4	3.5	2.5	2	1.5

6.5 THE MEAN, MEDIAN, AND MODE

A forest manager measures the heights, in metres, of a sample of trees after two years of growth for a reforestation project.

2.1, 2.3, 2.7, 2.6, 2.9, 2.8, 2.6, 2.6,
2.3, 2.2, 2.2, 2.3, 2.4, 2.4, 2.3, 2.8,
2.9, 2.4, 2.2

To interpret data like this, it is often useful to use the middle of the data as the representative value of the data. Something that tells us about the middle is called a measure of central tendency. There are three measures of central tendency: mean, median, and mode We will calculate the mean, median, and mode for the tree heights.

1. **Mean**

   The mean is the arithmetical average. It is the sum of all the values divided by the number of values in the set.

   $$\text{mean} = \frac{\text{sum of the heights of the trees}}{\text{number of trees}} = \frac{47}{19} \doteq 2.47$$

   The mean height is approximately 2.5 m to the nearest tenth.

2. **Median**

   The median is the middle value when the values are arranged in order. If there is an even number of values, the median is the average of the two middle values.
   There are 19 values and the middle one is the tenth value.

   2.1, 2.2, 2.2, 2.2, 2.3, 2.3, 2.3, 2.3, 2.4, (2.4), 2.4, 2.6, 2.6, 2.6, 2.7, 2.8, 2.8, 2.9, 2.9

   The median height is 2.4 m.

3. **Mode**

   The mode is the value that occurs most often. There may be more than one mode; or no mode at all.

   2.1, 2.2, 2.2, 2.2, (2.3), (2.3), (2.3), (2.3), 2.4, 2.4, 2.4, 2.6, 2.6, 2.6, 2.7, 2.8, 2.8, 2.9, 2.9

   The mode is 2.3 m.

EXERCISE 6.5

A 1. Determine the modes of the following sets of data.

(a) 6, 9, 13, 4, 6, 8, 9, 6, 9, 9
(b) 9, 11, 10, 9, 13, 10, 8, 16
(c) 6, 12, 11, 4, 9, 3, 8, 7
(d) 21, 24, 21, 30, 29, 24, 21, 26, 20
(e) 15, 25, 20, 8, 14, 17, 15, 16, 15

B 2. Determine the mean, median, and mode of each data set.

(a) 10, 9, 13, 14, 9
(b) 36, 43, 55, 41, 31, 40
(c) 31, 34, 36, 35, 41, 31, 30
(d) 33, 9, 40, 68, 59, 9, 68, 38
(e) 91, 90, 84, 86, 91

3. The chart gives numbers that have been retired by some National Hockey League teams.

BOSTON BRUINS
2 Eddie Shore
3 Lionel Hitchman
4 Bobby Orr
5 Dit Clapper
7 Phil Esposito
9 Johnny Bucyk
15 Milt Schmidt

CHICAGO BLACKHAWKS
9 Bobby Hull
21 Stan Mikita

DETROIT RED WINGS
6 Larry Aurie
9 Gordie Howe

EDMONTON OILERS
3 Al Hamilton

HARTFORD WHALERS
2 Rick Ley
9 Gordie Howe
19 John McKenzie

LOS ANGELES KINGS
30 Rogie Vachon

MINN. NORTH STARS
19 Bill Masterton

MONTREAL CANADIENS
2 Doug Harvey
4 Jean Beliveau
7 Howie Morenz
9 Maurice Richard
10 Guy Lafleur
16 Henri Richard

NEW YORK RANGERS
7 Rod Gilbert

PHILADELPHIA FLYERS
1 Bernie Parent
4 Barry Ashbee
16 Bobby Clarke

PITTSBURGH PENGUINS
21 Michel Briere

QUEBEC NORDIQUES
3 J.C. Tremblay
8 Marc Tardif

ST. LOUIS BLUES
3 Bob Gassoff
8 Barclay Plager

TORONTO MAPLE LEAFS
5 Bill Barilko
6 Ace Bailey

VANCOUVER CANUCKS
11 Wayne Maki

WASHINGTON CAPITALS
7 Yvon Labre

(a) Find the mode of these numbers.
(b) Why do the mean and median not have any significance?
(c) Which player has had his number retired by two teams?

4. Calculate Sam's bowling average if his scores for 6 games are:

198, 187, 192, 199, 189, and 192

5. A survey of the speeds of motor boats in a narrow canal was recorded.

Speed (km/h)	5	6	7	8	9
Number of Boats	7	10	12	9	3

Calculate the mean, median, and mode of this data.

6. The following are the noon temperatures, in °C, for Port Carling, Ontario, during the month of February.

$-4, -5, -3, 0, 2, 3, -1, 4, 2, 3,$
$-4, -6, 2, 0, 1, -2, -4, 6, 6, 3,$
$-1, 0, 3, 4, 5, 7, 3, 4$

Find the mean and median temperatures.

7. Roger is a hockey goalie. In 63 games 145 goals were scored against him. What was his goals-against average?

Some calculators have statistical keys that let you calculate the mean of a set of numbers.
The data entry key on many calculators is the $\Sigma+$ key.

To find the mean of 7, 8, and 12,

press 7 $\Sigma+$ 8 $\Sigma+$ 1 2 $\Sigma+$

The display is 3. or 12.

The displays vary depending on the calculator. After pressing $\Sigma+$, some calculators display the last number entered while others display the number of entries.

To find the mean of 7, 8, and 12,

press $\bar{x}$.

The display is 9. The mean is 9.

EXERCISE

1. Calculate the mean of the following test marks for five students.

Mary: 78, 89, 56, 78, 66, 88, 91, 44
Jason: 56, 78, 77, 99, 98, 42, 65, 81
Paul: 56, 57, 98, 76, 74, 70, 60, 84
Carol: 88, 82, 80, 90, 96, 47, 68, 55
Susan: 99, 59, 92, 89, 79, 83, 78, 55

6.6 CHOOSING THE MEAN, MEDIAN, OR MODE

The mean, median, and mode are all measures of central tendency of data. Each one can be described as an average. The one that you use should give an idea of a typical number in the set.

Example.
The Buried Sea Treasures Company paid its employees the following bonuses after a successful treasure hunt.

1 President	\$600 000
1 Vice President	\$160 000
1 Treasurer	\$140 000
1 Researcher	\$80 000
3 Divers	\$60 000 each
4 Ship's Crew	\$40 000 each

Calculate the mean, median, and mode of these bonuses.

Solution:
(a) To calculate the mean, first find the sum of the bonuses:

1 President	\$600 000
1 Vice President	\$160 000
1 Treasurer	\$140 000
1 Researcher	\$80 000
3 Divers (3 × \$60 000)	\$180 000
4 Ship's Crew (4 × \$40 000)	\$160 000
Total	\$1 320 000

$$\frac{\$1\ 320\ 000}{11} = \$120\ 000$$

There are 11 people and the mean is \$120 000.

(b) The median is the middle bonus, which is \$60 000.
(c) The mode is \$40 000.

\$600 000
\$160 000
\$140 000
\$80 000
\$60 000
\$60 000
\$60 000
\$40 000
\$40 000
\$40 000
\$40 000

If we use the mean, \$120 000, as the average, we get a distorted picture of the bonuses because 8 of the 11 employees got less than \$120 000. The mean is affected by extreme values.

A better way to describe the bonuses is to use the median, \$60 000, as the middle number. There are just as many people with bonuses above \$60 000 as there are below it. We say the median is more meaningful because it is not affected by extreme values.

The mean is used for data such as:

(a) sports averages: bowling, batting, pitching, scoring
(b) the average yearly rainfall for an area
(c) the fuel consumption for a car

The median is used for:

(a) test marks
(b) length of stay in a hospital
(c) salaries

The mode is useful for data like:

(a) shoe sizes, hat sizes, dress sizes, shirt sizes
(b) number of paperclips in a box
(c) the number of cylinders in an engine

EXERCISE 6.6

B 1. Dennis and Tyson played nine holes of golf on a par 3 course. The scorecard gives their scores.

Name	Hole									
	1	2	3	4	5	6	7	8	9	Total
Dennis	3	8	5	4	2	3	5	1	5	36
Tyson	2	15	3	3	2	2	5	1	4	37

(a) Who had the best score at the end of the 9 holes?
(b) Calculate the mean, median, and mode for each player's scores.
(c) If you use the mean score to decide the best golfer, who is the best player?
(d) If you use the mode, who is the best?
(e) If you use the median, who is the best?
(f) Which measure, the mean, median, or mode, gives the best indication of the abilities of Dennis and Tyson?

2. The table gives the number of cars damaged in accidents at a highway interchange for a 12-month period.

J	F	M	A	M	J	J	A	S	O	N	D
18	6	8	9	5	6	7	9	8	6	7	34

(a) Calculate the mean, median, and mode for the data.
(b) Which measure best represents the average of the data? Why?
(c) Which measure is least representative of the data? Why?
(d) If you were asked to estimate the number of accidents for next year, what answer would you give?

3. Caroline is a hockey goaltender. In the first eleven games of the season she had the following numbers of goals scored on her.

2, 0, 4, 7, 9, 0, 1, 5, 4, 0, 1

(a) Calculate the mean, median, and mode of the data.
(b) Which measure is the best indication of Caroline's ability as a goaltender? Why?
(c) Which measure is the worst indicator of her ability? Why?
(d) How many goals would you predict Caroline will have scored against her in a sixty-game season?

4. The heights of students, in centimetres, of a grade ten class are as follows:

166, 167, 160, 160, 161, 162, 163,
164, 167, 161, 199, 195, 167, 168,
169, 170, 161, 162, 164, 172, 174,
180, 177, 162, 168, 162, 168

(a) Calculate the mean, median, and mode.
(b) Which measure best represents the data? Explain.

5. A set of data contains 15 numbers. The largest number is increased by 10. The smallest is decreased by 10.
What changes may occur to the three measures of central tendency?

6. Terry has written 5 tests. Her mean or average mark is 82.
What will she have to get on the next test to raise her average to 84?

C 7. Research the meaning of a set of data which is bi-modal.
What does the word bi-modal mean?

6.7 STEM-AND-LEAF AND BOX-AND-WHISKER PLOTS

In order to preserve the habitat of whales, conservationists must study their feeding habits. The following is the number of whales sighted each day, for a number of days, at a whale feeding area.

20, 21, 34, 38, 40, 16, 41, 17, 17, 24, 42,
20, 16, 31, 30, 27, 28, 22, 39, 17, 23,
18, 19, 25, 27, 40, 32, 16, 21, 36, 41

This data can be quickly organized on a stem-and-leaf plot. Think of each tens digit as a stem, and each ones digit as a leaf.

WHALE SIGHTINGS

Stem	Leaf
1	6 7 7 6 7 8 9 6
2	0 1 4 0 7 8 2 3 5 7 1
3	4 8 1 0 9 2 6
4	0 1 2 0 1

The data in each leaf are now ordered from smallest to largest. Since there are 31 pieces of data, the median number of whale sightings is 25.

Stem	Leaf
1	6 6 6 7 7 7 8 9
②	0 0 1 1 2 3 4 ⑤ 7 7 8
3	0 1 2 4 6 8 9
4	0 0 1 1 2

To display this data on a box plot, also known as a box-and-whisker plot, first draw a number line that includes each piece of data. Then mark the median on the number line.

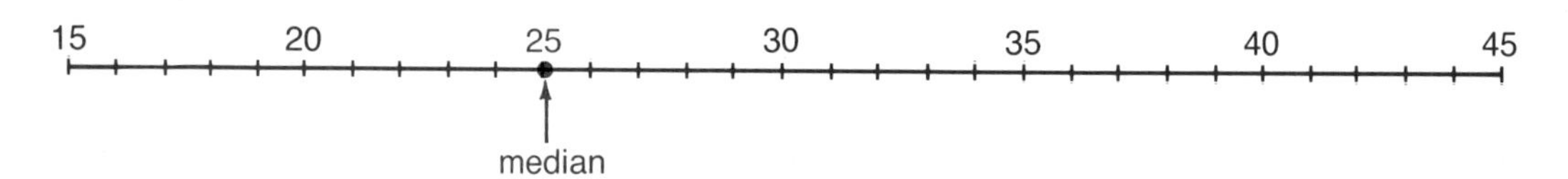

There are fifteen numbers above and below the median. Treat each set of numbers as a set of data. Locate the medians of each set, namely 19 and 36, and mark them on the number line. These points are called the hinges.
Also mark the highest and lowest values, namely 16 and 42, on the number line.

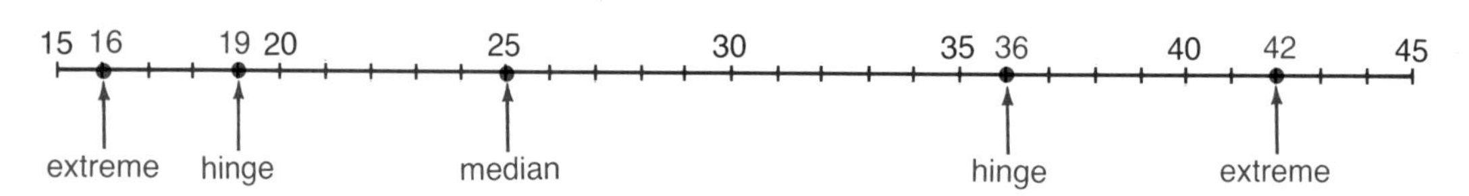

Draw a rectangle between the hinges. Draw whiskers from the hinges to the highest and lowest values.

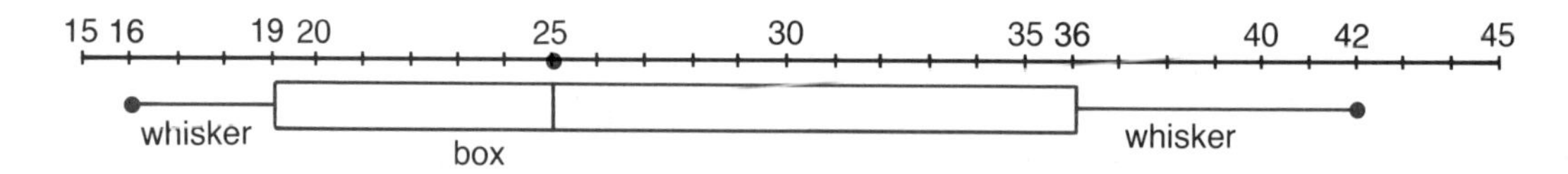

The median and the two hinges divide the data into four equal parts. Half the data is represented by the rectangle; the other half, by the whiskers.

EXERCISE 6.7

B 1. Display the data on a box-and-whisker plot.

SHIP ARRIVALS IN AUGUST

4	7 7 8 9 9
5	1 3 4 4 5 6 6 8
6	0 0 1 1 2 3 4 6 6 9 9
7	0 0 0 1 2 2 3

2. The following plot gives the number of forest fires being fought each day during a summer dry spell.

FOREST FIRES IN JULY

12	9 8 3 3 6 7 3
13	0 6 7 2 8 8 9 2 1 0
14	5 7 9 1 0 2 3 4 1
15	2 7 8 2 1

(a) Display this data on a box-and-whisker plot.
(b) Between what two numbers is half the data?

3. A survey was conducted to determine how much money students under sixteen years spend on rides, food, and souvenirs during one day at Wonderworld. Thirty students were surveyed, with the following results.

$21, $19, $25, $17, $32, $34, $16, $29, $18, $24, $26, $28, $31, $18, $20, $23, $15, $33, $26, $30, $26, $29, $39, $42, $30, $28, $20, $10, $30, $25

Display this information on a box-and-whisker plot.

4. During an ideal time for seeing shooting stars, the local astronomy society conducted a shooting star count on thirty-five nights with the following star count.

11, 45, 32, 62, 51, 24, 12, 43, 41, 25, 27, 28, 14, 29, 42, 21, 48, 54, 20, 15, 40, 41, 55, 53, 61, 62, 21, 21, 13, 25, 33, 53, 60, 33, 48

(a) Display this information on a box-and-whisker plot.
(b) Half the number of sightings are between what two values?

C 5. Conduct a survey in your class to determine the number of minutes it takes each student to get to school on a normal day.
Display the information on a box-and-whisker plot.

MIND BENDER

Here's a different way to multiply two numbers such as 38×57.

Write the numbers beside each other. Call one the halves column; the other, the doubles column.
Halve the 38, ignoring any remainder, until you get to 1.
Double the 57 the same number of times.
Cross out every even number in the halves column and its partner in the doubles column.
Add the numbers that remain in the doubles column.

$$\begin{aligned} & 114 + 228 + 1824 \\ &= 2166 \\ &= 38 \times 57 \end{aligned}$$

Halves	Doubles
~~38~~	~~57~~
19	114
9	228
~~4~~	~~456~~
~~2~~	~~912~~
1	1824

6.8 THE USE AND MISUSE OF STATISTICS

We use graphs to represent data because they give a simple picture of what the data represent. However, graphs can be drawn to give a false impression of what the data really represent. When someone uses statistics to support an argument or sell you something, you must study the data and the way the data are presented very carefully.

For example, the CFAN TV station receives money to operate, like most TV stations, from advertising. Advertisers want their ads to be seen by as many people as possible.

The owners of CFAN TV had a survey taken to show how the number of viewers had increased over the last six months. A graph of the results of the survey is shown at the right. We see from the scale that the viewers increased by 10 000.

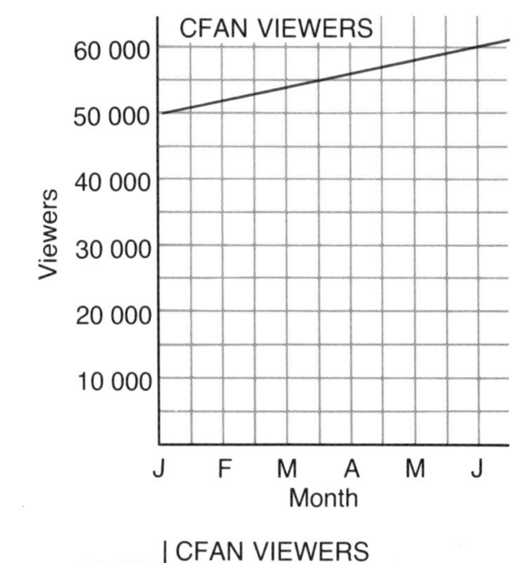

If we shorten the horizontal scale and cut off the bottom of the vertical scale, we get a different graph. Someone looking at this graph without checking the scale very closely might conclude that CFAN TV was gaining many viewers very quickly.

CFAN VIEWERS
60 000
50 000
40 000
Viewers
J F M A M J
Month

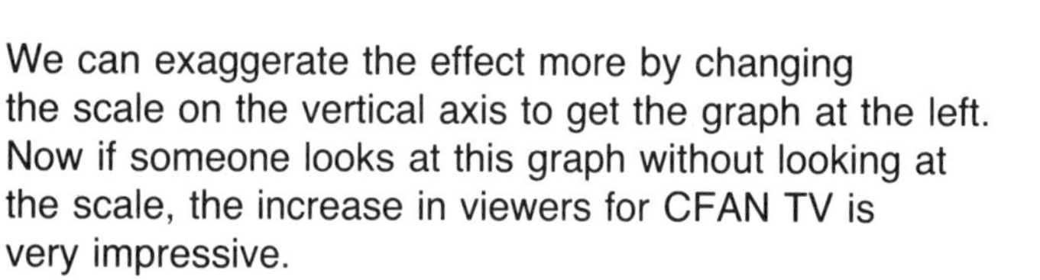
We can exaggerate the effect more by changing the scale on the vertical axis to get the graph at the left. Now if someone looks at this graph without looking at the scale, the increase in viewers for CFAN TV is very impressive.

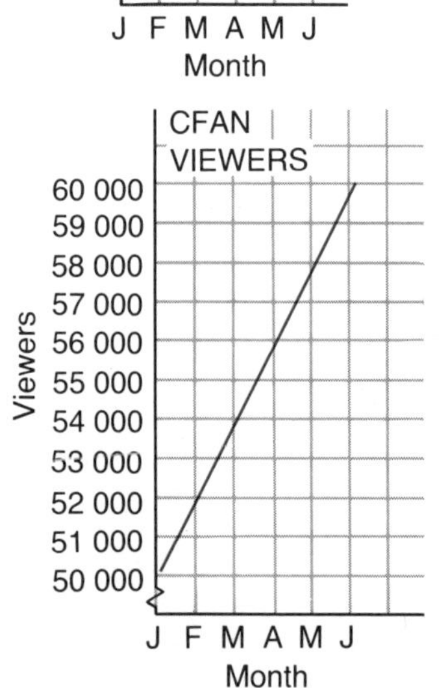

EXERCISE 6.8

B 1. The graph shows the number of student absences at Westmount High School on four consecutive Fridays. The last Friday was the one before Thanksgiving Monday.

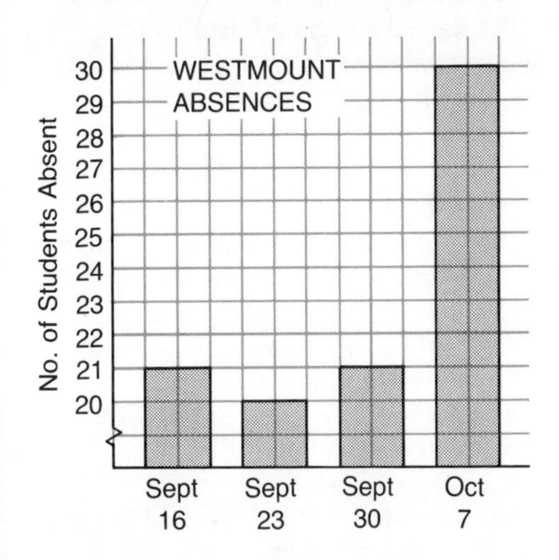

(a) What are the number of absences on each Friday?
(b) What impression is the graph trying to give to the reader?
(c) How was the graph drawn in order to try and distort the information?

2. The manager of Clown Carnivals is trying to decrease monthly maintenance expenses. The table gives the expenses for the past five months.

Month	Expenses ($)
May	44 000
June	42 000
July	41 000
August	40 000
Sept	38 000

(a) Draw a graph that exaggerates the decrease in expenses.
(b) Explain how the graph distorts the information.

3. A luggage company doubled its sales in one year. The following graph was displayed at a sales meeting.

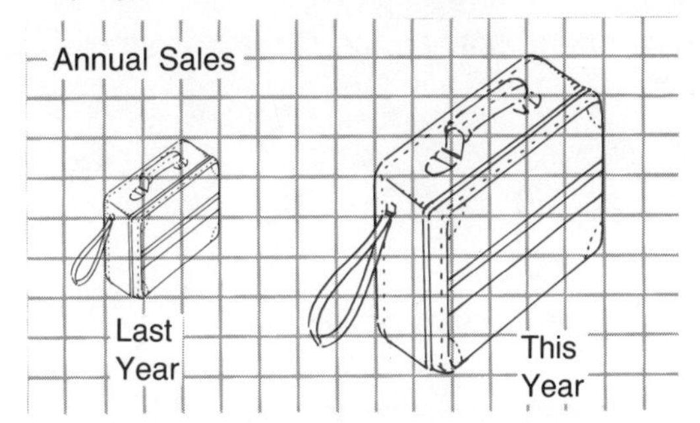

(a) What impression does the diagram give the reader about this year's and last year's sales?
(b) The person who drew the diagram can say the diagram shows that this year's sales are double last year's. Why?
(c) Draw another diagram to show that the sales doubled from last year to this year.

4. There are 300 grade ten students and 150 of them attended the play. There are 100 grade twelve students and 75 of them attended the play.
(a) Display this data on a graph so that the grade ten students appear to have supported the play better than the grade twelves.
(b) Display the information so that the grade twelves appear to be the best supporters of the play.

C 5. Find three examples in newspapers and magazines where a graph misrepresents data.

MIND BENDER

What was the last year when the Friday before Thanksgiving Monday was October 7?

6.9 PROBLEM SOLVING

1. The drawing shows three views of the same die. The die is *not* numbered as other dice usually are.
How many spots are on the face on the bottom of the die in the third view?

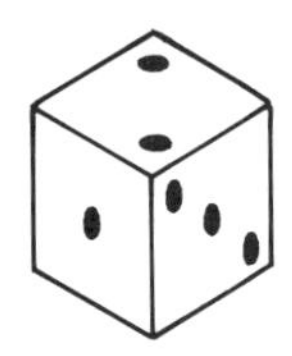

2. The diagram below is a map of a railway system. There are five stations: A, B, C, D, and E.

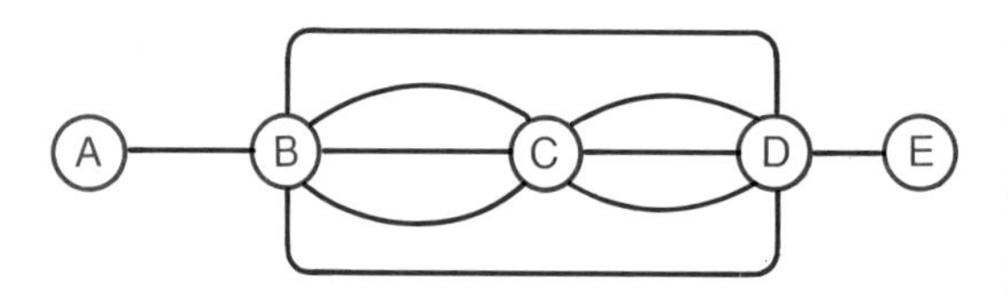

How many different ways can you travel from A to E, during one trip, if you can never travel more than once along any section of track and you can never enter a station more than once?

3. The coins in each of the three rows and three columns add to 41¢. It does not appear so because there may be a second coin under any or all of the coins (except for the dimes). No stack has more than two coins and no stack contains two coins of the same denomination.

Where are the hidden coins?

4. You have a nine o'clock call for rehearsal. You will need to spend 50 min in make-up at the theatre. Dressing will take 15 min. The bus trip from your place to the theatre takes 25 min. Breakfast at the Theatre Coffee House takes 15 min. You take 45 min to get ready to leave your place once you get up.
For what time should you set your alarm?

5. There are four houses in a row along a dead-end street. The Wongs live next to the Jeffreys but not next to the Roberts. The Beetons do not live next to the Roberts.
Who are the Roberts' next door neighbours?

6. The numbers in the square at the left have been placed according to some rule. The numbers in the square at the right were placed according to the same rule.
What is the missing number?

3	1	3
2	2	1
6	2	3

6	2	3
4	2	2
■	3	5

7. The following equation is true if you match the correct units to the numbers. For example,

3 + 5 = 1

if you write

3 quarters + 5 nickels = 1 dollar

Add the units to make these equations true.
(a) 5 + 48 = 1
(b) 1 − 60 = 23

8. Find two integers with a sum of 36 and a difference of 6.

9. You are offered discounts of 25% and 20% on a pair of jeans. You can take the discounts in any order.
Which order of discounts will give the lowest price?

10. A hotel has three buses to pick up guests when they arrive at the airport. The buses hold 10, 12, and 15 passengers. Sixty people arrive at the airport.
In how many different ways can these people be transported to the hotel if each bus used must be full?

11. When you travel to a different time zone it has been said that for every hour's difference in time your body needs one day to adjust.
How long will your body take to adjust if you took a trip to Hawaii?

12. A train leaves Montreal for Vancouver every day at 06:00. A train leaves Vancouver for Montreal every day at 07:00. The trip between Montreal and Vancouver takes exactly 4 d. Suppose you got on the train for Vancouver in Montreal.
How many trains would you pass going from Vancouver to Montreal on your four-day trip?

13. What is the area of the shaded figure?

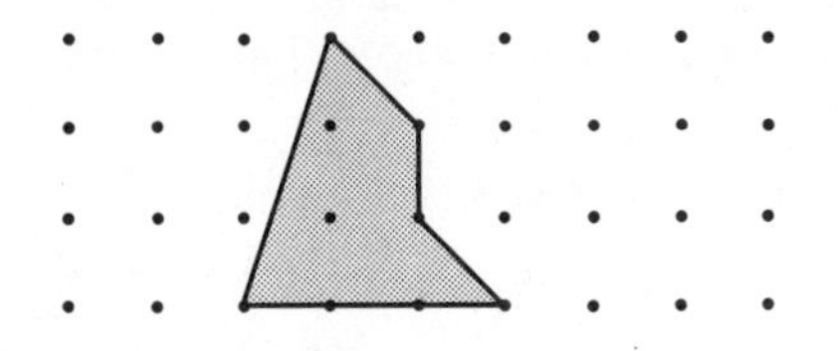

14. There are 5 faces that can be counted in the cube at the right.

There are 8 faces here.

There are 11 faces here.

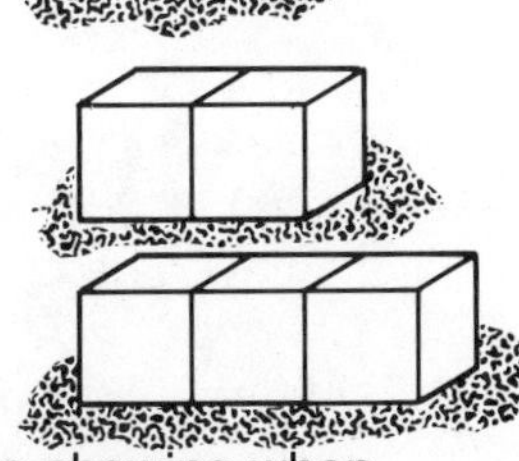

How many faces will be showing when there are 10 cubes in a row?

CAREER

BROADCASTER

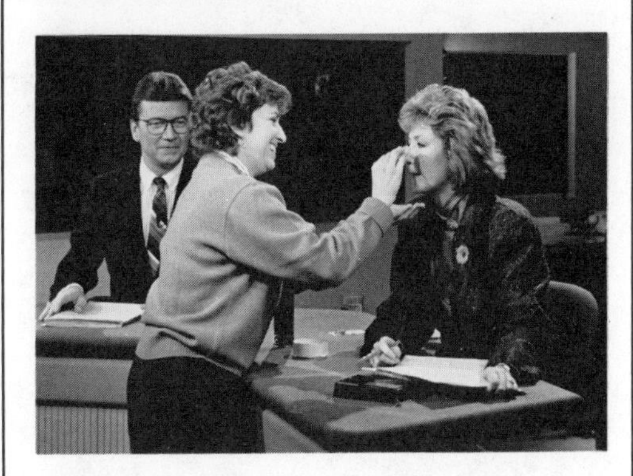

The world of radio and television broadcasting offers many exciting careers. One of them is a television announcer.

EXERCISE

1. The television crew in the photograph is preparing to tape a spot for the evening news. In the crew are two announcers and a make-up artist.
What other career people would get involved before the spot is seen on television?

2. Many television announcers have agents. Carol Jones has an agent who charges her 15% of the money she earns for speaking engagements and commercial work. Last year she earned $26 500 doing this type of work.
How much did her agent receive?

3. Television announcers don't start at the top of their profession. They usually follow a career path of increasing experience before working with major networks.
Describe a career path.

4. What colleges or universities offer courses in journalism or media arts?

EXTRA

6.10 RADIO STATIONS AND RANDOM SAMPLING

There are basically three groups of people that are involved in the production of a record:

(a) the song writer or writers;
(b) the artists and technicians who record the music; and
(c) the publisher who provides the finances and expertise to hire and pay the people necessary to get the record recorded, packaged, promoted, and into the record stores.

Writing and publishing a song is similar to writing and publishing a book. If you write a book and a publisher agrees to publish it, you receive money for each book that the publisher sells. This money is called royalties. The process works a little differently in the music business.

If you write a song and it is published or recorded, then each time a radio station plays the song you, the writer, and the publisher are paid a royalty. In general, vocalists and instrumentalists are not paid royalties unless they also write the music.

Radio stations pay the royalties to the Performing Rights Organization of Canada which in turn distributes them to the song writers and publishers. Royalties are based on the number of times a song is played in Canada that year and this number can be very large.

Across Canada there are about 400 radio stations. We can estimate that each broadcasts 365 d a year, 24 h a day, so that there are $400 \times 365 \times 24$ or 3 504 000 h of broadcasting time to be filled. If each station plays an average of 10 songs an hour, 8 to 9 songs each hour during news hours, or 12 to 13 songs each hour during non-news hours, then there are approximately $10 \times 3\ 504\ 000$ or 35 040 000 songs played each year.

As you can see from the size of this number, the task of logging each and every song would be virtually impossible. To simplify the task, the organization makes use of a random sampling technique. Each radio station is sampled once a year for seven days. During this time the station is asked to record all the songs it plays. This number represents $\frac{1}{52}$ of the total number of songs the station plays in a year.

The organization divides the year into quarters: January to March, April to June, July to September, and October to December. One-quarter of the radio stations are sampled in each quarter. Stations are divided into groups or strata, according to region, type, and AM or FM:

Region	Type	AM/FM
Atlantic Provinces Quebec Ontario Manitoba/Saskatchewan Alberta British Columbia	Country Rock MOR (Middle of the Road)	AM FM

For example, one stratum would be Alberta – MOR – AM, and if there are 28 stations in this group, then seven of them would be sampled in each quarter of the year. This sampling ensures that all types of music are represented.

Now suppose at the end of the year the sample shows that the song "Rio" published by Tritec music was played 930 times.

Since this number represents $\frac{1}{52}$ of the total number of times the song was played across Canada during the year, we can infer the song was played 52×930 or 48 360 times altogether.

The song writer and the publisher each receive a fixed amount, for example 36¢, for each time the song is played. The writer receives $48\,360 \times \$0.36$ or \$17 409.60 and the publisher receives the same amount.

EXERCISE 6.10

1. The survey shows that the song "Blue Knight" recorded by Range Publishing was played 814 times. How much did the writer and publisher receive in royalties?

2. How many groups can radio stations be placed into?

3. Why is it necessary to sample every quarter instead of taking the sample for all stations in one quarter, say January to March?

4. What types of radio stations would not fall into any of the groups?

6.11 REVIEW EXERCISE

1. The following table gives the numbers of boys and girls by grade in a school.

Grade	Boys	Girls
9	210	200
10	175	180
11	155	164
12	135	130

You have been asked to conduct a survey of 100 students to determine whether the two-month summer vacation should be moved to December and January to save on heating costs.
Approximately how many students should you include from each group in the survey?

2. One hundred students were surveyed to determine who would buy school T-shirts and sixty-three said they would.
If there are 657 students in the school, approximately how many would be expected to buy T-shirts?

3. The Silver Fox Company makes 100 000 whistles each day. A sample of 200 whistles was tested and 3 were found to be defective.
How much of the day's production can you predict will be defective?

4. The circle graph shows the causes of automobile accidents.

HOW ACCIDENTS HAPPEN

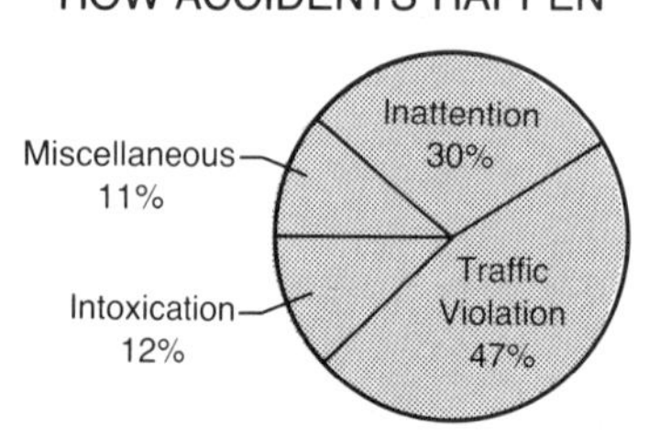

Out of 500 accidents, how many are caused by:
(a) inattention?
(b) traffic violation?
(c) intoxication?

5. The graph shows the number of yachts sold by the Big Boat Company during one year.

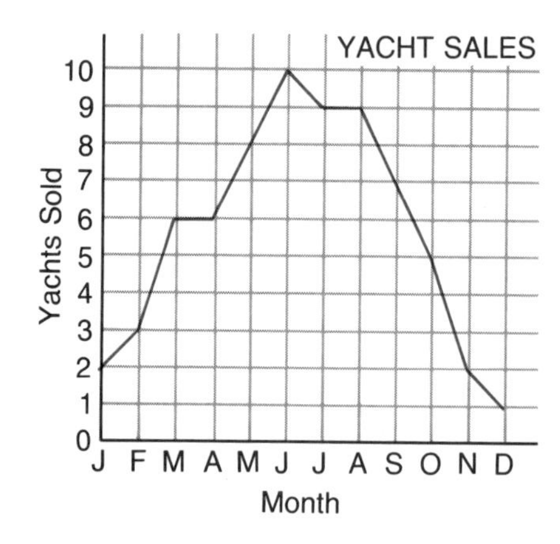

(a) During which month were sales the highest?
(b) During which months were the sales the same?
(c) By how much did the sales in September exceed those in February?
(d) What were the total sales for the year?

6. The maximum speeds of several animals are given in the table.

Animal	Speed (km/h)
Cheetah	110
Rabbit	55
Greyhound	65
Human	40
Lion	80
Giraffe	50

Display this information on a graph.

7. It cost Robert $10 000 to attend college last year. His expenses were as follows.

Tuition:	$2000	Food:	$2000
Travel:	$500	Books:	$460
Room:	$3600	Personal:	$1560

Display this information on a graph.

8. The temperature in Brandon, Manitoba, on a day in April was recorded each hour as follows.

Time	Temp. (°C)	Time	Temp. (°C)
06:00	5	13:00	17
07:00	7	14:00	16
08:00	8	15:00	15
09:00	12	16:00	13
10:00	13	17:00	12
11:00	16	18:00	11
12:00	17		

Display this information on a graph.

9. A group of employers was asked what subjects, other than languages, are essential for their employees to understand. The following are the results of the survey.

Mathematics	83%
Computer skills	64%
Science	38%
Geography	37%
History	36%

(a) Display this information on a graph.
(b) Why do the percents add to more than 100?

10. Determine the mean, median, and mode of the following sets of data.
(a) 15, 19, 15, 12, 20, 25, 20, 17, 16, 13, 15
(b) 149, 127, 113, 137, 134, 131, 105
(c) 36, 50, 46, 44, 42, 50, 41, 51

11. The table gives Mark's bowling scores for a five-week period.
Determine his average after each of these weeks.
(a) Week 1 (b) Week 2 (c) Week 3
(d) Week 4 (e) Week 5

Week	Game 1	Game 2	Game 3
1	200	201	205
2	195	215	220
3	219	194	187
4	204	212	196
5	191	201	235

12. (a) The mean of a set of data is 43. Does the number 43 have to be one of the data? Use an example to explain your answer.
(b) The median of a set of data is 43. Does the number 43 have to be one of the data? Use an example to explain your answer.
(c) The mode of a set of data is 43. Does the number 43 have to be one of the data? Why?

13. The following are the marks earned by 30 students in a geography test.

39 42 48 41 52 55
59 62 62 63 63 68
69 69 70 73 74 75
76 77 78 79 81 83
84 85 86 87 92 95

(a) Display the data on a stem-and-leaf plot.
(b) Display the data on a box-and-whisker plot.
(c) What is the median of the data?

14. Tom had the following marks on his first five mathematics tests.

Test	Mark
1	73
2	75
3	76
4	80
5	81

(a) Draw a graph that exaggerates Tom's increase in marks.
(b) Draw an honest graph of the data.

MIND BENDER

Write any word on a piece of paper. Fold the paper in half and stand on it. The Answers at the back of the book will tell you what is on the paper.

6.12 CHAPTER 6 TEST

1. A survey was conducted to determine the favourite radio station in a Canadian city. Three hundred people were interviewed. The following are the results:

90 preferred CFCF
60 preferred CEED
150 preferred CJAM

If there are an estimated 400 000 radio listeners, predict how many listen to each station.

2. The table gives the number of TV sets sold during one week.
Display this information on a graph.

Day	M	T	W	T	F	S
Sets sold	7	8	11	10	16	21

3. Determine the mean, median, and mode of the data.

34, 42, 47, 50, 66, 47, 40, 41, 39,
38, 48, 41, 37, 47, 45, 51, 52

4. The following are the marks earned by 35 students on a sailing test.

73 81 42 55 64 91 63
71 87 54 65 79 80 89
35 50 66 78 67 52 54
86 87 78 77 51 64 70
74 75 88 81 91 95 65

(a) Display this information on a stem-and-leaf plot.
(b) Display this information on a box-and-whisker plot.

5. The broken-line graph gives the hourly temperature during part of a day in Bracebridge.

(a) What was the highest temperature reached?
(b) During what time was the temperature increase the highest?
(c) What was the temperature at 12:00?

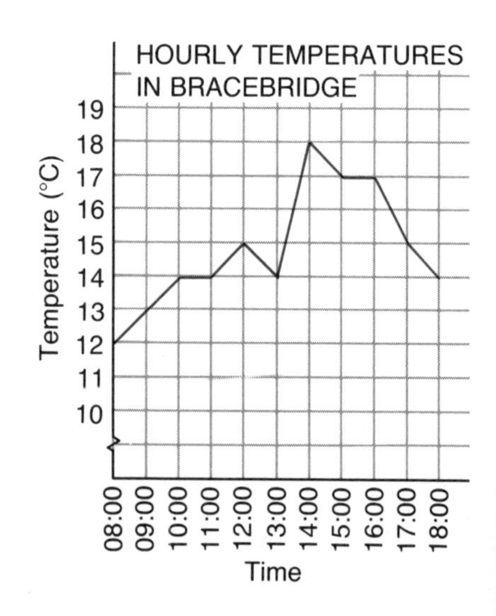

PERSONAL FINANCE AND BANKING

REVIEW AND PREVIEW TO CHAPTER 

PERCENT

EXERCISE 1

1. When working with a calculator, enter percents as decimals.
Convert the following percents to decimals.

Example. $6\frac{3}{4}\% = 6.75\%$
$= \frac{6.75}{100}$
$= 0.0675$

(a) 7% (b) 5%
(c) 12.5% (d) 100%
(e) 8.5% (f) $3\frac{1}{4}\%$
(g) $\frac{3}{4}\%$ (h) $2\frac{1}{2}\%$
(i) $1\frac{1}{2}\%$ (j) 4.75%

2. Convert each number to a percent.

Example. $0.045 = 0.045 \times 100\%$
$= 4.5\%$
$= 4\frac{1}{2}\%$

(a) $\frac{1}{4}$ (b) $\frac{3}{4}$
(c) 0.075 (d) 0.14
(e) 0.0475 (f) 0.061
(g) 0.0125 (h) 0.037

3. Find each amount to the nearest cent.

Example. $5\frac{1}{2}\%$ of \$200
$= 0.055 \times \$200$
$= \$11.00$

(a) 6% of \$470 (b) 3% of \$1500
(c) $4\frac{1}{2}\%$ of \$1200 (d) $6\frac{3}{4}\%$ of \$500
(e) 12% of \$74.50 (f) $4\frac{1}{2}\%$ of \$15.50
(g) 5.25% × \$25 (h) 7.5% × \$142
(i) $106\frac{1}{2}\%$ × \$250 (j) 103% × \$650

CIRCLE GRAPHS

EXERCISE 2

1. The following circle graph shows how Mark used his time in a 24 h period.

MARK'S TIME

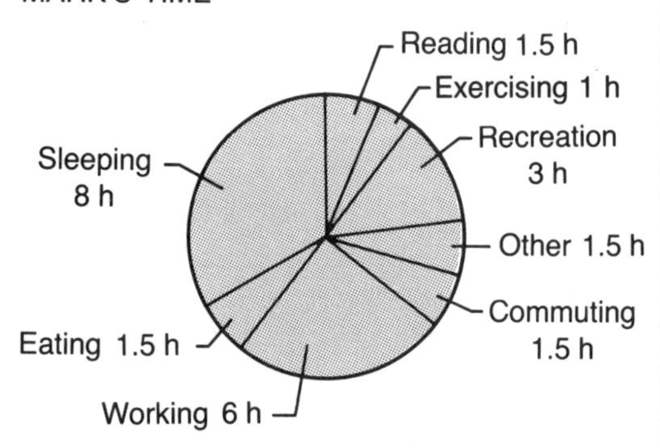

(a) What fraction of Mark's time was spent sleeping?
(b) What fraction was spent working?
(c) What fraction was spent exercising?
(d) Express each period of time in the graph as a percent.

2. In a class of 30 students, 12 walk to school, 6 take the school bus, and 6 take the city bus. The others are given rides to school by their parents.
Show this information in a circle graph.

3. What type of car do drivers prefer? A researcher surveyed 500 drivers and compiled these findings:

luxury cars	50
standard cars	200
compact cars	150
import cars	100

(a) Express these preferences as percents.
(b) Draw a circle graph to display this information.

SUBSTITUTING INTO FORMULAS

EXERCISE 3

1. The formula for the area of a rectangle is $A = \ell \times w$.
The formula for the perimeter of a rectangle is $P = 2(\ell + w)$.
Find the area and perimeter of each rectangle.

(a)

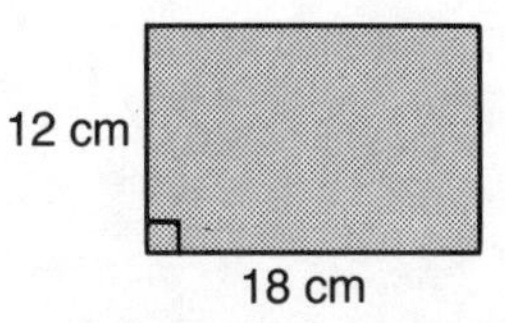

(b) $\ell = 25$ cm, $w = 15$ cm
(c) $\ell = 30$ cm, $w = 25$ cm

2. The formula for the area of a trapezoid is

$$A = \frac{(a + b)}{2} \times h$$

Find the area if $a = 2.5$ cm, $b = 3.5$ cm, and $h = 3.0$ cm.

3. Given the formula

$$c^2 = a^2 + b^2$$

find c if $a = 9$ and $b = 12$.

4. The formulas for the circumference and area of a circle are

$$C = \pi d \quad \text{and} \quad A = \pi r^2$$

Find the circumference and the area of a circle given each diameter. (Use $\pi = 3.14$)

(a) 12 cm (b) 20 cm
(c) 100 cm (d) 50 cm

5. The formula for distance, d, is

$$d = r \times t \qquad \begin{array}{l} r = \text{speed} \\ t = \text{time} \end{array}$$

(a) Find the distance a car travels in 7 h at 90 km/h.
(b) Find the distance a spacecraft travels in 36 h at 30 000 km/h.
(c) Find the distance a runner travels in 10 s at 9.6 m/s.

6. (a) If $a*b = a(a + b)$, find the value of $2*3$.
(b) If $a* = a(a + 2)$, find the value of $3*$.
(c) If $a*b*c = a(b + c)$, evaluate $2*3*4$.

MENTAL MATH

EXERCISE 4

Find the value of the unknown variable.

1. $6x = 24$
2. $8 + x = 30$
3. $7x = 28$
4. $9 + x = 25$
5. $3x + 1 = 13$
6. $x + 9 = 12$
7. $5x = 35$
8. $9x + 5 = 41$
9. $x - 7 = 10$
10. $3x = -15$
11. $3x - 2 = 16$
12. $5x - 1 = 29$
13. $6 - x = 7$
14. $8 - x = 3$
15. $7x + 0 = 42$
16. $5x - 5 = 0$
17. $5y + 2 = 22$
18. $3t + 2 = 14$
19. $z + z = 18$
20. $3r + 4r = 28$
21. $\frac{x}{5} = \frac{18}{15}$
22. $\frac{y}{3} = \frac{14}{12}$
23. $\frac{a}{6} = \frac{12}{18}$
24. $\frac{a}{15} = \frac{11}{3}$
25. $\frac{5}{b} = \frac{15}{20}$
26. $\frac{18}{b} = \frac{3}{2}$
27. $\frac{6}{10} = \frac{c}{20}$
28. $\frac{10}{24} = \frac{c}{48}$
29. $\frac{7}{5} = \frac{14}{d}$
30. $\frac{36}{15} = \frac{12}{d}$

7.1 GROSS INCOME

The money a person is paid for doing a job, or performing a service, is called income. Most income will fall into one of the categories below. The income before deductions, such as income tax and unemployment insurance, is called gross income, or gross earnings.

Wages and Salary. A wage is a rate paid according to the hour, day, week, or month. A salary is similar to a wage and is paid on an annual, monthly, or bi-weekly basis. Read the classified or employment section in a newspaper.
What kinds of jobs offer a wage? a salary?
Overtime, which is work in addition to the specified time period, is often paid at a higher rate.

Piecework. In manufacturing industries, many employers pay according to the number of units, or pieces, produced. Producing more units results in greater pay. People who work for piece rates may be self-employed or may work for someone else.

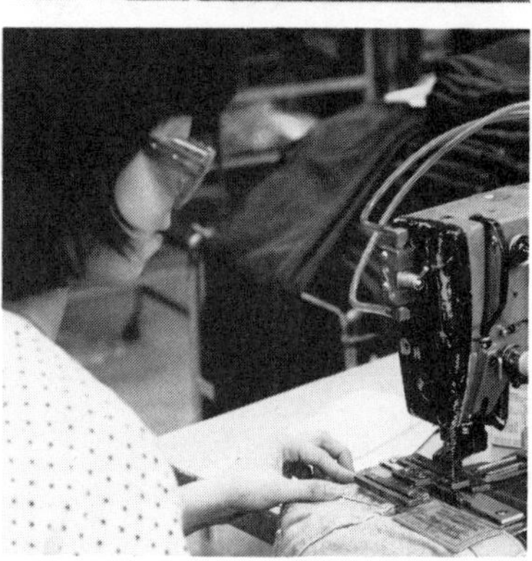

Tips and Commission. Tips are a large part of the income for many service employees, such as taxi drivers or restaurant servers. Commission is a percent of the total value of goods or services sold. Both tips and commission are incentives for the employee to do a good job.

Fees. Medical, legal, and many technical services are based on a flat rate, which is called a fee. The fee is usually charged according to the service which has been provided. The fee does not include time to perform the service.

Example 1.
Henri Berbier is a security guard who works a 40 h week and earns a wage of \$14.50/h, plus time and a half for overtime.
What is Henri's gross income for a 45 h week?

Solution:

$$\begin{aligned} \text{Hourly rate} &= \$14.50 \\ \text{Time-and-a-half rate} &= \$14.50 + \tfrac{1}{2} \times \$14.50 \\ &= \$14.50 + \$7.25 \\ &= \$21.75 \end{aligned}$$

Press

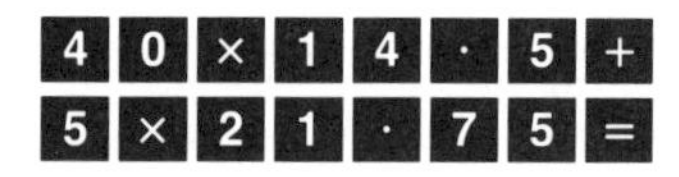

For a 45 h week, Henri's gross income is:

$$\begin{aligned} 40 \text{ h} @ \$14.50 &= 40 \times \$14.50 = \$580.00 \\ 5 \text{ h} @ \$21.75 &= 5 \times \$21.75 = \$108.75 \\ \text{Gross income} &= \$688.75 \end{aligned}$$

∴ Henri Berbier's gross income is \$688.75.

The display is

688.75

Example 2.
Claudette Smith is a drapery sales consultant for an interior decorator. She earns a straight $100/week salary plus 12% sales commission.
In a two-week period, Claudette had sales of $7465.80.
What are Claudette's gross earnings for the two-week period?

Solution:

Gross income = salary + commission
Salary = 2 weeks @ $100 = $200
Commission = 12% of $7465.80
= 0.12 × $7465.80
≐ $895.90
Gross income = $200 + $895.90
= $1095.90

∴ Claudette's gross earnings for the two-week period are $1095.90.

[·] [1] [2] [×] [7] [4] [6] [5] [·] [8] [=]

The display is 895.896

EXERCISE 7.1

A 1. Find the gross income for each of the following part-time jobs.
(a) $6.65/h for 10 h
(b) $12.50/h for 20 h
(c) $9.75/h for 10 h
(d) $7.50/h for 20 h

2. What is the overtime rate of pay at time and a half for each hourly rate of pay?
(a) $6.00/h (b) $10.00/h
(c) $14.00/h (d) $12.00/h
(e) $6.50/h (f) $9.00/h
(g) $9.50/h (h) $15.50/h

B 3. Find the gross earnings for each of the following based on a standard 40 h week, with overtime at time and a half.
(a) 40 h @ $9.10/h
(b) 48 h @ $12.80/h
(c) 52 h @ $18.50
(d) 26 h @ $8.40/h

4. During the month of March, Lucia earned $400 in salary plus 12% commission on sales of $18 697.80.
What were Lucia's gross earnings for this month?

5. Some companies use step commission as an incentive for even greater sales performance. A step commission increases as sales increase. The Magic Looks Cosmetic Company pays

10% on the first $2500 of sales
12.5% on the next $2500
15% on sales over $5000

Calculate the commission, to the nearest cent, for each total sales.

(a) $1575.50 (b) $6200.00
(c) $4725.00 (d) $3520.00
(e) $975.00 (f) $7585.00
(g) $1068.00 (h) $2590.00
(i) $1742.00 (j) $7480.00
(k) $45 653.00 (l) $95 720.00

6. Jessie sells light construction equipment. He is paid a salary of $950/month plus a commission of 2% on the first $25 000 of sales, 3% on the next $25 000, and 4% on total sales over $50 000.
Calculate Jessie's gross income for sales of $92 000 in one month.

7. Company A pays $25 200/a to a computer operator. Company B pays $13.05/h for the same position. Both companies require the computer operator to work a 36-h week.
Compare the weekly salaries offered by the two companies.

8. Shirley Sayer works as a tour bus driver. She earns a $40 bonus for every 100 km over an average of 7500 km that she drives each month. In June, she drove 10 625 km. Calculate Shirley's bonus for June.

9. Gerald works as a server in the restaurant of a large bank building. He earns $4.75/h plus tips. The following table shows Gerald's hours and tips for one week.
Calculate Gerald's gross income for the week.

Day	Hours	Tips
Monday	6	$72.15
Tuesday	6.5	$75.90
Wednesday	7.5	$88.40
Thursday	9	$107.25
Friday	4	$52.60

10. Freida works as a tour guide for the Banff Bus Tours. She is paid $5/h and receives tips from the tourists. Freida's earnings for a 9 d period are given in the table.

Day	Hours	Tips
Sunday	6	$42
Monday	8	$55
Tuesday	8	$48
Thursday	8	$53
Friday	10	$64
Saturday	12	$88
Sunday	4	$24
Monday	8	$52
Tuesday	8	$36

(a) Calculate Freida's gross earnings for the 9 d period.
(b) What were Freida's average daily earnings for the 9 d period?

11. Andy Travich has been offered two different jobs as an accountant. He must consider costs of transportation in both cases. The job at Downtown Motors pays $3200/month, and the train ticket to work will cost $3.60/d return. The job at Country Leasing pays $740/week and driving his car will cost Andy $4.40/d for gas.
Compare the salaries at the two jobs on an annual basis.

12. Bill and Michelle Bourke are self-employed: they provide word processing and photocopying services. Their rates are given.

Word processing	$2.25/page (double-spaced)	$3.00/page (single-spaced)
Copying	$0.12 per page	
Mailing labels	$0.05 per label	
Stuffing envelopes	$0.03 per envelope	

(a) Calculate the charge for preparing a 6-page, double-spaced report and making 12 copies.
(b) Calculate the charge for preparing a 3-page report, if 75 copies are to be mailed.
(i) double-spaced (ii) copying
(iii) preparing labels (iv) stuffing envelopes

C 13. Mariette works for a stockbroker and earns a straight commission of 1.1% of stock sold. Her record for one day of selling stock is given in the table.

Stock	Number of Shares Sold	Price per Share
Abitibi	100	19\frac{1}{2}$
Asbestos	200	$12
Bombardier	100	27\frac{5}{8}$
Campeau	200	16\frac{3}{4}$
Dofasco	100	27\frac{3}{8}$
Inco	50	32\frac{5}{8}$
Lac Minerals	200	$16
Maple Leaf G	10	$42
Pamour	500	6\frac{1}{4}$
Seagram	10	72\frac{1}{2}$

7.2 NET INCOME

Murray DeJong earns $310/week in a restaurant. His take-home pay, also called net pay or net income, is $250.58.

PAY SLIP FOR:	MURRAY DEJONG
TIME PERIOD:	AUG. 4–10
GROSS PAY:	310.00
INCOME TAX:	47.95
CANADA PENSION PLAN:	5.42
UNEMPLOYMENT INSURANCE:	6.05
TOTAL DEDUCTIONS:	59.42
NET PAY:	250.58

There are several deductions from Murray's gross pay, or gross income. Deductions for income tax, Canada Pension Plan (C.P.P.), and Unemployment Insurance (U.I.), as well as a provincial medical plan, are required by law and are called statutory deductions. Other deductions, described as employee benefits, include extended medical insurance, long-term disability, life insurance, dental insurance, and private pension plan.

Net income = Gross income − Deductions

From Murray's gross pay, his employer deducts $47.95 for income tax, $5.42 for C.P.P., and $6.05 for U.I. Murray's employer calculates the income tax for each employee on the basis of a net claim code. Because Murray is single and has no dependants, his net claim code is 1. The employer uses tables to make the appropriate deductions. Portions of these tables are shown.

WEEKLY TAX DEDUCTIONS **TABLE 1**
Basis — 52 Pay Periods per Year

WEEKLY PAY Use appropriate bracket — PAIE PAR SEMAINE Utilisez le palier approprié		IF THE EMPLC SI LE CODE DE DEMANDE			
		0	1	2	3
From - De	Less than Moins que	DEDUCT FROM E.			
296.-	300.	78.50	44.90	41.10	33.60
300.-	304.	79.60	45.90	42.10	34.60
304.-	308.	80.65	46.90	43.15	35.60
308.-	312.	81.70	47.95	44.15	36.65
312.-	316.	82.75	48.95	45.15	37.65
316.-	320.	83.80	49.95	46.15	38.65
320.-	324.	84.85	50.95	47.20	39.65
324.-	328.	85.90	52.00	48.20	40.70
328.-	332.	86.95	53.00	49.20	41.70
332.-	336.	88.00	54.00	50.20	42.70
336.-	340.	89.05	55.00	51.25	43.70

CANADA PENSION PLAN CONTRIBUTIONS

Remuneration *Rémunération* From-*de*	To-*à*	C.P.P. *R.P.C.*
306.45 -	306.91	5.35
306.92 -	307.39	5.36
307.40 -	307.87	5.37
307.88 -	308.34	5.38
308.35 -	308.82	5.39
308.83 -	309.30	5.40
309.31 -	309.77	5.41
309.78 -	310.25	5.42
310.26 -	310.72	5.43
310.73 -	311.20	5.44
311.21 -	311.68	5.45
311.69 -	312.15	5.46
312.16 -	312.63	5.47
312.64 -	313.11	5.48

UNEMPLOYMENT INSURANCE PREMIUMS

Remuneration *Rémunération* From-*de*	To-*à*	U.I. Premium *Prime d'a.-c.*
304.88 -	305.38	5.95
305.39 -	305.89	5.96
305.90 -	306.41	5.97
306.42 -	306.92	5.98
306.93 -	307.43	5.99
307.44 -	307.94	6.00
307.95 -	308.46	6.01
308.47 -	308.97	6.02
308.98 -	309.48	6.03
309.49 -	309.99	6.04
310.00 -	310.51	6.05
310.52 -	311.02	6.06
311.03 -	311.53	6.07
311.54 -	312.05	6.08
312.06 -	312.56	6.09
312.57 -	313.07	6.10
313.08 -	313.58	6.11
313.59 -	314.10	6.12

For calculating income tax, C.P.P. and U.I. in this chapter, use the tables on pages 400 to 404. (Current tables can be obtained from your local District Taxation Office.)

Example 1.

Cheryl Manitou works as a cosmetician and is paid a salary of $350/week. For income tax purposes, her net claim code is 1.

Complete a pay slip showing the weekly deductions from her cheque for income tax, C.P.P., and U.I., and calculate her take-home pay.

Solution:

STATEMENT OF EARNINGS FOR: CHERYL MANITOU	
GROSS PAY:	350.00
DEDUCTIONS INCOME TAX: 58.05 ← from the table on page 400 C.P.P.: 6.26 ← from the table on page 402 U.I.: 6.83 ← from the table on page 404 TOTAL DEDUCATIONS:	 71.14
NET PAY:	278.86

∴ Cheryl's take-home pay is $278.86.

Example 2.

Barry Cormier is a computer technician. His pay for one week is $300 in regular earnings, plus $44.80 in overtime pay. The deductions are income tax, C.P.P., U.I., a provincial medical premium of $13.80, and group life insurance of $6.75. His net claim code is 1. Complete a pay slip for Barry, and calculate his take-home or net pay.

Solution:

THE COMPUTER CLINIC					
STATEMENT OF EARNINGS FOR: BARRY CORMIER					
WEEK END JUNE 27	EMPLOYEE #AX4253	CHEQUE # 123456	REGULAR EARNINGS 300.00	OVERTIME EARNINGS 44.80	GROSS PAY 344.80
EMPLOYEE DEDUCTIONS					
C.P.P.			6.15		
U.I.			6.72		
INCOME TAX			57.05		
PROVINCIAL MEDICAL			13.80		
GROUP LIFE INSURANCE			6.75		
TOTAL DEDUCTIONS			90.47	NET PAY:	254.33

EXERCISE 7.2

Use the tables on pages 400 to 404 to answer the questions in this exercise. Assume that each employee in this exercise has net claim code 1.

B 1. Find the C.P.P. deduction for each weekly gross earnings.

(a) \$250.00 (b) \$276.80
(c) \$345.00 (d) \$325.50

2. Find the U.I. deduction for each weekly gross income.

(a) \$275.00 (b) \$308.75
(c) \$278.90 (d) \$335.00

3. Find the income tax deduction for each weekly gross earnings.

(a) \$255.00 (b) \$325.00
(c) \$298.00 (d) \$345.00

4. Copy and complete the pay slip below for Edgar Malone.

STATEMENT OF EARNINGS FOR:		
GROSS EARNINGS:		315.00
DEDUCTIONS		
C.P.P.:	(a)	
U.I.:	(b)	
INC.TAX:	(c)	
PROV.MED.:	13.50	
TOTAL:	(d)	
NET PAY:		(e)

5. Lorraine Maxwell has weekly gross earnings of \$296.

(a) Find her C.P.P. deduction.
(b) Find her U.I. deduction.
(c) Find her income tax deduction.
(d) What are Lorraine's net earnings?

6. Albert Swain works in a book store and earns \$312/week. His other deductions include:

Provincial medical	\$13.50
Dental plan	\$2.85
Extended health	\$3.50
Company pension	\$7.50

(a) Find the amount deducted for C.P.P., U.I., and income tax.
(b) Find Albert's total deductions.
(c) Prepare a pay slip and calculate Albert's net income.

7. Copy and complete the pay slip below for Alda McIntyre.

STATEMENT OF EARNINGS FOR:		
REGULAR EARNINGS:		275.50
OVERTIME EARNINGS:		44.80
GROSS PAY:		(a)
DEDUCTIONS		
C.P.P.:	(b)	
U.I.:	(c)	
INC.TAX:	(d)	
PROV.MED.:	13.50	
GRP.LIFE:	5.25	
SVGS.BND.:	9.25	
TOTAL:	(e)	
NET PAY:		(f)

8. Janice Jonas earns \$7.40/h plus time and a half after 36 h. Janice has the following deductions: C.P.P., U.I., income tax, provincial medical plan of \$13.50, life insurance of \$2.20, union dues of \$1.50, and dental plan of \$2.10. Last week, Janice worked 42 h.
Copy and complete the following pay slip.

HARVEY SPORTS COMPANY STATEMENT OF EARNINGS FOR: ______		
HOURLY RATE OF PAY:	(a)	
REGULAR HOURS:	(b)	
OVERTIME HOURS:	(c)	
GROSS PAY	(d)	
DEDUCTIONS		
TOTAL:	(e)	
NET PAY:		(f)

7.3 PERSONAL BUDGETING

Rafaella works part-time on Fridays after school and on Saturdays as a cashier in a convenience store and her take-home pay is $132. Her plan for managing the money she earns is called a budget. She uses this budget as the key to spending her money wisely. The circle graph at the right shows Rafaella's budget for one month.

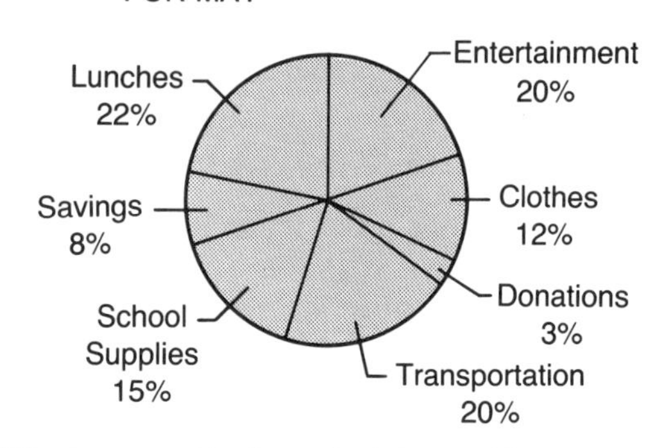

To what part is the least amount allocated?
To what item is allocated the greatest amount?
What percent is for transportation?
What percent is for entertainment?
What percent is for savings?

Expense	Percent	Calculation	Rounded Amount
Clothes	12%	0.12 × 132 = 15.84	$16
Charitable donations	3%	0.03 × 132 = 3.96	$4
Entertainment	20%	0.20 × 132 = 26.40	$26
Lunches	22%	0.22 × 132 = 29.04	$29
Savings	8%	0.08 × 132 = 10.56	$11
School supplies	15%	0.15 × 132 = 19.80	$20
Transportation	20%	0.20 × 132 = 26.40	$26
Check by adding	100%		$132

Example.
Net income is the amount of money remaining after taxes and other deductions are made.
Variance is the difference between the budget amount and the actual amount.
Calculate the variance of each item in the Willford family's budget and prepare the budget in terms of percent.

Solution:

Item	Budget	Actual	Variance	Percent
Rent	$450	$450.00	0.00	450 ÷ 1610 × 100% ≐ 28%
Food	$280	$287.50	− 7.50	280 ÷ 1610 × 100% ≐ 17%
Clothing	$170	$155.60	14.40	170 ÷ 1610 × 100% ≐ 11%
Entertainment	$110	$152.50	−42.50	110 ÷ 1610 × 100% ≐ 7%
Car expenses	$225	$218.75	6.25	225 ÷ 1610 × 100% ≐ 14%
Car loan	$325	$325.00	0.00	325 ÷ 1610 × 100% ≐ 20%
Savings	$50	$20.65	29.35	50 ÷ 1610 × 100% ≐ 3%
Total	$1610	$1610.00	0.00	100%

Note that the budget and actual columns add to the total amount of the budget.
The percent column adds to 100%.

EXERCISE 7.3

B 1. The following circle graph shows the budget of a high school student who plays hockey and buys his own sticks. His net income is $224.50/month.

SAL MAGUIRE'S BUDGET

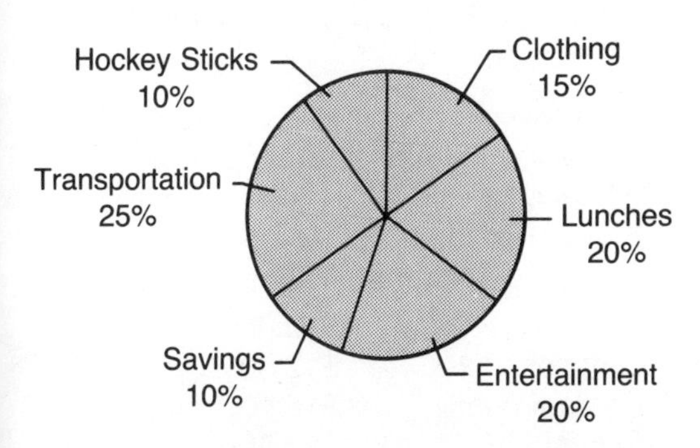

(a) Estimate the amount of money budgeted for each item.
(b) Calculate the amount budgeted by Sal Maguire for each of the following items.
(i) clothes (ii) entertainment
(iii) lunches (iv) transportation
(v) savings (vi) hockey sticks

2. Sherri Acton studies drama and works as an actress in a dinner theatre. She has a net income of $1485 during the month of May.
(a) Express each budget item as a percent and find the variance for each item.
(b) Draw a circle graph to show Sherri's budget for this month.
(c) By how much did Sherri overspend or underspend on her budget?

Item	Budget	Actual
Rent/utilities	$615	$605.83
Food	$200	$188.60
Clothes	$140	$152.80
Personal	$75	$84.20
Acting lessons	$250	$250.00
Furniture loan	$188	$188.00
Contributions	$12	$15.50
Savings	$5	$0.00

3. The following circle graph shows the budget for the Lima family. The total combined income of the two wage earners in the family is $89 000.

LIMA FAMILY BUDGET

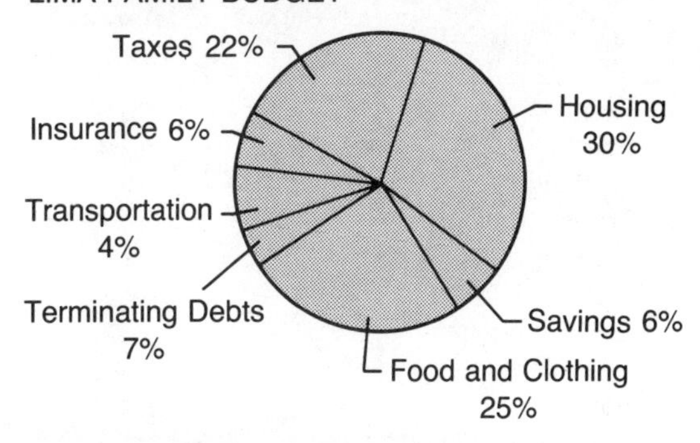

(a) Estimate the amount budgeted by the Lima family for each item.
(b) Calculate the amount budgeted for each of the following items.
(i) food and clothing (ii) savings
(iii) insurance (iv) taxes
(v) terminating debts (vi) housing

4. Reba works part-time and earns $60/week. Her expenses are as follows:

Item	Week 1	Week 2	Week 3	Week 4
Bus	$10.00	$10.00	$10.00	$10.00
Lunches	$5.25	$4.75	$5.50	$5.65
Clothes	$15.20	$27.75	$15.95	$18.50
Entertainment	$18.00	$10.70	$21.50	$9.20
Miscellaneous	$9.20	$5.05	$3.80	$0.00
Savings	$2.35	$1.75	$3.25	$16.65

(a) Find the total expenses for each week and also for each item for all four weeks.
(b) Find the average amount spent each week for each item, to the nearest cent.
(c) Prepare a budget, giving each item to the nearest percent.
(d) Show the budget in a circle graph.

5. Bill and Amy graduated as nurses from the same college. Bill works in a senior citizens' home and his net annual income is $32 000. Amy works in a hospital and her net annual income is $36 000.

The following table shows how they budget their incomes.

Item	Bill's Budget	Amy's Budget
Housing	24%	20%
Food	28%	23%
Clothing	8%	12%
Transportation	11%	10%
Insurance	8%	7%
Loans	12%	14%
Entertainment	5%	6%
Miscellaneous	2%	4%
Savings	2%	4%

(a) Calculate how much Bill spends on each item in one year.
(b) Calculate how much Amy spends on each item in one year.
(c) Make circle graphs to show each of the above budgets.

6. The following table shows Terry Hammada's income and expenses for one month.

Income	September	
Take-home pay	$1478.90	
Bank interest	$212.00	
Other income	$0.00	
Total Income		
Expenses	September	%
Rent	$534.20	
Utilities	$72.55	
Food	$301.22	
Transportation	$42.00	
Insurance	$33.75	
Clothing	$121.50	
Loan payments	$238.00	
Savings	$120.00	
Miscellaneous	$58.65	
Total Expenses		
Balance +/−		

(a) Complete the table, calculating all percentages.
(b) By how much are Terry's total expenses over or under her total income?
(c) When Terry's expenses vary, she balances the budget for the month by adding to her savings or taking from savings.
What are Terry's savings for the month of September?

7. Allison works part-time in a fast-food restaurant and has net earnings of $88/week. She lives at home and pays $15/week for room and board. Her average expenditures for one week are:

Clothing	$18.00
Lunches	$12.50
Entertainment	$15.00
School	$5.00

Allison saves the rest of her earnings in a bank account.

(a) What percent of her net earnings does Allison save?
(b) What percent of Allison's net earnings does she contribute for room and board?

8. Josie works in the family business and receives $46.25/week in net pay. The following table shows Josie's budget and actual expenses for one week. Unspent money is saved for clothes.

Expenses	Budget	Actual
Lunches	$15.00	$12.75
Entertainment	$12.00	$5.00
Transportation	$5.00	$5.00
Magazines	$3.25	$4.00
Save for clothes	$11.00	$12.50

(a) What is the variance of each item?
(b) How much was Josie able to save for clothes this week?

7.4 BANKING SERVICES

If Charlie wants to save money for a purchase, such as a VCR or a motorbike, or simply to save for the future, he would likely open a savings account. If, on the other hand, Charlie has insufficient money, he may borrow the funds he needs to make this purchase. The depositing of money into an account, or borrowing to make a purchase, require the services of a financial institution.

You may already have a savings account at a bank, and, if you do, you know that the bank pays interest for the use of your money. On the other hand, if you borrow money from a bank, you will pay the bank interest. The interest rate a bank charges to lend money is higher than the interest rate paid on a savings account and other types of savings. The money the bank earns from the difference between these two sets of interest rates is used to pay the bank's operating costs. A small percentage represents profits. You can also rent safety deposit boxes, pay utility bills, cash cheques, buy money orders and bank drafts at a bank, which can also act as a broker to buy and sell savings bonds and stocks. To extend the hours of service, most banks have automated banking machines.

Trust companies are licensed by the federal and provincial governments. They offer a wide range of financial services, including trust, real estate and property investment, in addition to savings and loans. Importantly, trust companies are the only financial institutions that can act as a trustee — a party who is responsible for the affairs of a person or company.

Credit unions, or caisse populaires — people's banks — in Quebec, also offer deposit, chequing, and lending services, in addition to other financial services, but only to members. Members of individual credit unions are part of a common interest group, for example, they may come from the same neighbourhood or from the same business, or social group. Members are also shareholders. Whether a member has as little as one share or hundreds of shares, she is entitled to only one vote. Members are in effect the owners of a credit union and can participate in its general operation by electing other members to a Board of Directors.

While banking services are provided by a variety of financial institutions, the procedures for withdrawing, depositing, and borrowing money are very similar.

Most financial institutions offer a variety of ways to save money: the most common type of savings is a savings account. Before opening an account, you should investigate the kind of account you require. Most financial institutions offer several kinds of accounts.

SAVINGS ACCOUNTS

In a true savings account, cheques are not allowed, so that withdrawals may be made only by the depositor on the forms provided. Interest is paid at a rate established by the institution, and you usually have a choice of terms such as daily or monthly interest. This account tends to offer a slightly higher rate of interest.

CHEQUING/SAVINGS ACCOUNTS

Chequing/savings accounts are similar to the true savings accounts with the added feature that you can write a small number of cheques without a service charge. The number of cheques that can be written depends on the amount of money on deposit. There is a service charge for extra cheques which are written and the interest rates for these accounts are lower than those for true savings accounts.

CHEQUING ACCOUNTS

Chequing accounts allow you to deposit money, make withdrawals, and write cheques. These accounts usually pay no interest. There is a service charge based on a flat rate, or on a given number of cheques that can be written. Some free cheques may be allowed, depending on the minimum balance. Cancelled cheques are returned to the customer each month so that they serve as receipts.
The depositor is provided with a passbook or monthly account statement, depending on the institution. These items are records of all transactions a depositor makes, including withdrawals and cheques (debits), and also deposits (credits). The financial institution keeps perpetual records via computers. The use of computers enables branch to branch transactions to take place very quickly.
Most financial institutions use similar forms as cheques, deposit slips, or withdrawal slips.

PROVINCIAL BANK		X 1			Cheques and Coupons (List on reverse if necessary)		
		X 2					
		X 5					
Date	Account Number	X 10					
		X 20					
Credit Account of		X					
		Coin					
		Total Cash			▶		
Signature for cash received (Please sign in front of Teller)					Sub Total		
					Less Cash Received		
Deposit all Personal Accounts at this Branch			Depositor's Initials		Net Deposit ▶		

PROVINCIAL BANK	Date
	Account Number
Branch	
Received from Provincial Bank	
Dollars	
Withdrawal All Savings Accounts	Signature (Please sign in front of Teller)

NAME:
ADDRESS:
CITY/TOWN:

19

PAY TO THE
ORDER OF $

DOLLARS
100

PROVINCIAL BANK

⑆11422⑈519⑆ 508094⑈4⑈

Most of the services described here require the payment of a service charge. The following list is an example of the charges to the customer for some of these services.

Safety deposit box rental.................................$27.50 per year and up
Payment of each utility bill $1.00
Selling money orders.....................................$2.50
Selling bank drafts...$5.50
Buying and selling stocks ....commission negotiated
Selling Canada Savings Bonds ....no charge to the customer
Buying and selling foreign currency ....$2.00 per cheque
Traveller's cheques ....1% commission, $3.00 minimum

Most banks will combine these services in a package and charge a flat rate.

Kumar has a chequing account. A page in Kumar's cheque book is illustrated. We can use a calculator to check the balance.

DATE	NO.	PARTICULARS	CHEQUES	√	DEPOSITS	BALANCE
						1204.32
Feb 3	065	To Big Bear Sports For new skates	256.37			Cheque or Dep. 256.37 Bal. 947.95
Feb 3	066	To Adams Mens Wear For shirt and tie	72.83			Cheque or Dep. 72.83 Bal. 875.12
Feb 8		To Deposit Pay Cheque For			356.20	Cheque or Dep. 356.20 Bal. 1231.32
Feb 10	067	To Monast Gift Shop For Valentine present	17.82			Cheque or Dep. 17.82 Bal. 1213.50
Feb 10	068	To Ernesto's Rest. For Dinner with Jim	37.25			Cheque or Dep. 37.25 Bal. 1176.25
Feb 11		To Sam's Retail For Refound			26.58	Cheque or Dep. 26.58 Bal. 1202.83

Press	Display
1 2 0 4 . 3 2	1204.32
− 2 5 6 . 3 7 =	947.95
− 7 2 . 8 3 =	875.12
+ 3 5 6 . 2 0 =	1231.32
− 1 7 . 8 2 =	1213.50
− 3 7 . 2 5 =	1176.25
+ 2 6 . 5 8 =	1202.83

Example.
Jennifer has a chequing/savings account at the bank. She is allowed one free cheque for every $100 minimum monthly balance in her account and pays a service charge of $0.30 for each additional cheque. Interest of 0.3%/month is paid on the minimum monthly balance. Interest and service charges are applied on the first day of the month. Jennifer started the month of November with a balance of $687.52 and made the following transactions.

Date	Cheques	Withdrawals	Deposits
Nov 5	$65.80, $54.25	Nov 5 $100.00	Nov 15 $875.40
Nov 8	$16.24, $74.60	Nov 20 $125.00	Nov 30 $875.40
Nov 9	$75.00		
Nov 15	$88.25		
Nov 20	$475.00		
Nov 25	$38.50		

Prepare a sample bank statement and find the balance on December 1.

Solution:

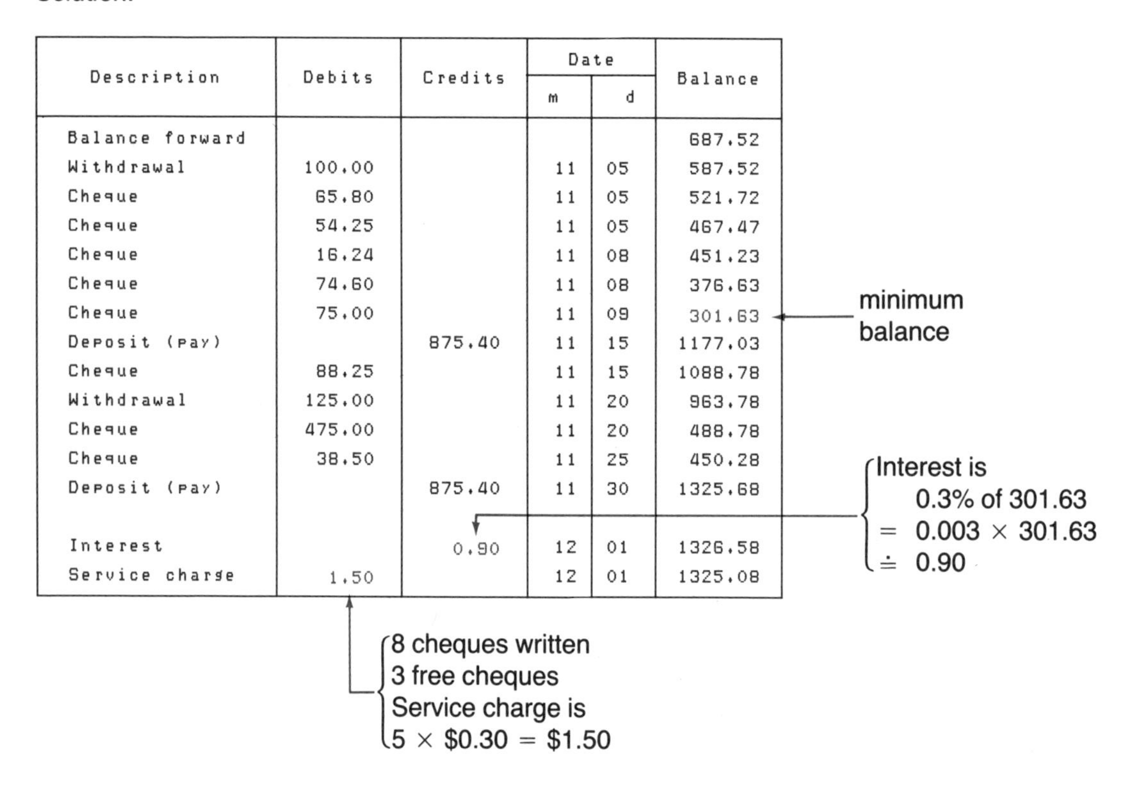

Description	Debits	Credits	Date m	Date d	Balance
Balance forward					687.52
Withdrawal	100.00		11	05	587.52
Cheque	65.80		11	05	521.72
Cheque	54.25		11	05	467.47
Cheque	16.24		11	08	451.23
Cheque	74.60		11	08	376.63
Cheque	75.00		11	09	301.63
Deposit (pay)		875.40	11	15	1177.03
Cheque	88.25		11	15	1088.78
Withdrawal	125.00		11	20	963.78
Cheque	475.00		11	20	488.78
Cheque	38.50		11	25	450.28
Deposit (pay)		875.40	11	30	1325.68
Interest		0.90	12	01	1326.58
Service charge	1.50		12	01	1325.08

The balance on December 1 was $1325.08.

EXERCISE 7.4

B

1. An account has a balance of $643.50. What is the new balance after each of the following?
(a) a withdrawal of $50
(b) writing a cheque for $100

2. An account has a balance of $633.50.
(a) What withdrawal will reduce the balance to $500?
(b) What deposit will raise the balance to $933.50?
(c) How much should be deposited so that the balance is $1000?

3. Write each amount in the words that would appear on a cheque.
(a) $245.50 (b) $1635.75
(c) $25 000.00 (d) $16 535.00

4. Write the balance in each account after the following transactions.
(a) Cheques of $16.50 and $13.50 are drawn on a balance of $87.50.
(b) An account has a balance of $125.75; a deposit of $25.50 and a withdrawal of $40.00 are made.

5. Here is a page from Jules Leverne's savings account passbook.
Complete the balance column to determine the final balance in the account.

Date	Withdrawals	Deposits	Balance
Opening balance			318.24
10 01	52.50		
10 03		125.00	
10 07	75.80		
10 10	15.00		
10 12		47.50	
10 13		24.00	
10 15	65.00		
10 16	56.00		
10 18		15.00	
10 20	110.00		
10 24		24.00	
10 28	55.00		
10 31	25.00		

6. On the first of the month, Henry has a balance of $346.80 in his chequing account. During the month, he makes deposits of $746.80 and $849.75. He also makes a withdrawal of $400.00 and writes cheques for $250.00, $200.00, $145.50, $165.00, and $425.00. The total service charge for the month is $7.50.
Calculate Henry's balance at the end of the month.

7. On June 30, the balance in Jeremy's savings account is $1257.80. Interest is paid monthly at a rate of 0.9%/month on the minimum monthly balance. Jeremy made no deposits or withdrawals until October 12.
What is the balance in the account on October 1?

8. The following is a page from Sylvia Morello's chequing/savings passbook. Interest is paid at a rate of 0.25%/month on the minimum monthly balance and Sylvia is allowed one free cheque for every $100 of the minimum balance in July. Additional cheques cost $0.30 each.
Complete the balance column to determine the balance in the account on August 1.

Date	Cheques	With-drawals	Deposits	Balance
Opening balance				456.20
07 02	35.47		125.00	
07 08		50.00		
07 09			125.00	
07 10	57.80			
07 12		25.00		
07 12	15.75			
07 14		40.00		
07 16			125.00	
07 20	240.00			
07 22		35.00		
07 23			75.00	
07 23	42.50			
07 26	25.00			
07 28	45.00			
07 30			125.00	

9. A chequing/savings account allows one free cheque for every $100 of a minimum balance and charges of $0.30 for all other cheques written. Interest at a rate of 0.3%/month is paid on the minimum monthly balance. Interest and service charges are applied on the first day of the month.

(a) Miles started the month of March with a balance of $896.47 and made the following transactions.

Date	Cheques	Withdrawals	Deposits
Opening balance			896.47
Mar 3	25.50		85.00
Mar 5		244.50	
Mar 9	8.75	100.00	
Mar 14	63.75		
	75.00		
Mar 15		355.00	1285.00
Mar 16	32.00		
Mar 17	53.75		
Mar 20	25.00	120.00	
Mar 22	145.00		85.00

Prepare a sample bank statement for Miles and find the balance on April 1.

(b) Alwyn started the month of May with a balance of $324.63 and made the following transactions.

Date	Cheques	Withdrawals	Deposits
Opening balance			324.63
May 4		200.00	
May 5	75.00		1445.00
May 10		200.00	
May 12	275.80		
	58.00		
May 16	165.00	200.00	
May 17	232.00		175.00
May 19	115.50		
May 20			208.65
May 23	208.00	200.00	
May 28	94.65		
May 31	750.00	200.00	1285.00

Prepare a sample bank statement and find the balance on June 1.

10. Morgan Clement starts the month of July with $1421.76 in his chequing/savings account. He is allowed one free cheque for every $100 of a minimum monthly balance; additional cheques cost $0.33 each. Interest is paid at a rate of 0.3%/month on the minimum monthly balance. Both interest and service charges are applied on the first day of the month.
Complete the balance column and find the final balance on August 1.

Date	Cheques	With-drawals	Deposits	Balance
Opening balance				1421.76
07 02	73.25			
07 03		200.00		
07 06	35.00			
07 06	58.50			
07 09	12.75			
07 10		200.00		
07 12	42.27			
07 12	216.34			
07 15	75.00		1237.68	
07 17		125.00		
07 18	25.00			
07 18	35.65			
07 19	18.80			
07 20	690.00			
07 22	85.90			
07 23	235.00			
07 24	25.63	225.00		
07 28		100.00		
07 30	118.82			
07 31			1237.68	

C

11. A frequent user of banking services can save money by using the flat fee system.
Investigate the meaning of the following terms which are available in a flat fee package.

(a) unlimited no-charge chequing
(b) personalized cheques
(c) overdraft protection
(d) utility bill payment
(e) traveller's cheques, money orders, and bank drafts

7.5 SAVINGS ACCOUNTS

Marnie Batoski wants to save money to pay for a vacation. She deposits her money in a savings account in a bank and her money earns interest. To deposit money into a savings account, she fills out a deposit slip. Marnie receives her pay cheque of $135.80 and wishes to keep $50 in cash and deposit the rest. Marnie presents this deposit slip and passbook to the teller to make her deposit. Note that there is an area to list cheques and an area to record cash.

PROVINCIAL BANK		X 1			Cheques and Coupons (List on reverse if necessary)	135	80
		X 2					
		X 5					
Date Jan 5	Account Number 81636	X 10					
		X 20					
Credit Account of		X					
M. BATOSKI		Coin					
		Total Cash			→		
Signature for cash received (Please sign in front of Teller) Marnie Batoski					Sub Total	135	80
					Less Cash Received	50	00
Deposit all Personal Accounts at this Branch.			Depositor's Initials M.B.		Net Deposit	85	80

When Marnie hands the cheque, deposit slip, and passbook to the bank teller, the teller enters the information on the computer and places the passbook in the printer to be updated.

On January 12, Marnie gives the teller her passbook and this withdrawal slip for $42.00.

PROVINCIAL BANK	Date Jan 12
	Account Number 81636
Branch 6522	
Received from Provincial Bank	
forty-two ——————————————— xx	Dollars
Withdrawal All Savings Accounts	Marnie Batoski Signature (Please sign in front of Teller)

Again the teller enters the information into the computer and places the passbook in the printer. The computer uses this information to update Marnie's passbook. Part of a page in Marnie's passbook is shown below. Marnie can check her balance using a calculator.

Date	Item	With-drawals	Deposits	Balance
Previous	Balance			628.93
Dec 27	WD	18.00		610.93
Dec 30	WD	24.00		586.93
Dec 31	INT		1.20	588.13
Jan 05	DEP		85.80	673.93
Jan 12	WD	42.00		631.93

Press	Display
6 2 8 . 9 3	628.93
− 1 8	610.93
− 2 4	586.93
+ 1 . 2	588.13
+ 8 5 . 8 0	673.93
− 4 2 =	631.93

EXERCISE 7.5

B 1. Complete the following deposit slips.

(a)

PROVINCIAL BANK		X 1			Cheques and Coupons (List on reverse if necessary)	685	30
		X 2				21	16
		X 5				18	24
Date: Jan 7	Account Number: 68412	X 10					
		X 20					
Credit Account of: M. BISSETT		X					
		Coin					
		Total Cash			→		
Signature for cash received (Please sign in front of Teller): Michele Bissett					Sub Total		
					Less Cash Received	100	00
Deposit all Personal Accounts at this Branch			Depositor's Initials: M.B.		Net Deposit ▶		

(b)

PROVINCIAL BANK		X 1			Cheques and Coupons (List on reverse if necessary)	253	24
		8 X 2				86	50
		12 X 5					
Date: Jan 7	Account Number: 37284	2 X 10					
		X 20					
Credit Account of: SALVATORE SMITH		X					
		Coin					
		Total Cash			→		
Signature for cash received (Please sign in front of Teller): Salvatore Smith					Sub Total		
					Less Cash Received		
Deposit all Personal Accounts at this Branch			Depositor's Initials: S.S.		Net Deposit ▶		

(c)

PROVINCIAL BANK		2 X 1			Cheques and Coupons (List on reverse if necessary)	756	85
		6 X 2				167	42
		4 X 5				150	00
Date Jan 7	Account Number 30785	8 X 10					
		5 X 20					
Credit Account of PIERRE ARCHAMBEAULT		3 X 50					
		Coin					
		Total Cash			→		
Signature for cash received (Please sign in front of Teller) Pierre Archambeault					Sub Total		
					Less Cash Received		
Deposit all Personal Accounts at this Branch		Depositor's Initials PA			Net Deposit		

2. Sketch or obtain blank deposit slips and complete a slip for each of the following deposits.

(a) On April 12, Joe McLean deposits two cheques for $586.50 and $725.00, and receives $400.00 in cash.

(b) On September 5, Sally Galipeau deposits a cheque for $875.00 along with the following cash:

4	$50 bills	92	quarters
7	$20 bills	124	dimes
12	$10 bills	32	nickels
35	$5 bills	78	pennies
7	$2 bills		
27	$1 bills		

3. Complete the balance column.

Date	Item	With-drawals	Depos-its	Balance
Opening	Balance			1256.83
Jul 02	DEP		250.00	
Jul 04	WD	100.00		
Jul 10	DEP		125.00	
Jul 13	WD	575.00		
Jul 15	WD	20.50		
Jul 18	DEP		50.00	
Jul 19	DEP		125.00	
Jul 21	WD	85.00		
Jul 24	WD	45.00		
Jul 27	DEP		250.00	
Jul 31	DEP		175.00	

4. Find the errors in the balance column and make the corrections.

Date	Item	With-drawals	Depos-its	Balance
Opening	Balance			3276.48
Aug 01	INT		6.00	3282.48
Aug 06	WD	50.00		3332.48
Aug 07	DEP		250.00	3082.48
Aug 10	WD	125.00		2957.48
Aug 12	WD	200.00		2757.48
Aug 15	DEP		225.00	3182.48
Aug 16	DEP		55.00	3237.48
Aug 18	WD	75.00		3162.48
Aug 20	WD	50.00		3112.48
Aug 22	WD	250.00		2912.48
Aug 27	DEP		125.00	2957.48
Aug 31	SC	0.99		2956.49
Aug 31	DEP		200.00	3156.46

MIND BENDER

Determine a pattern in these numbers and complete the chart.

7	■	10	■	8	■
3	2	4	3	2	■
4	5	6	6	■	5
12	10	■	■	■	20

7.6 CHEQUING ACCOUNTS

Jacob McGill has a personal chequing account to write cheques for purchases instead of paying cash. For the privilege of writing cheques, paying tax and utility bills, cashing cheques, and purchasing money orders, Jacob pays a total monthly service charge of $7.50. Jacob writes the $86.35 cheque below for a basketball. He uses the deposit slip to the right to put $150.00 into the account.

The cheque states the amount in both words and numerals, as with a withdrawal slip, in addition to the name of the party who will cash the cheque.

Note that the same deposit slip is used for both chequing and savings accounts.

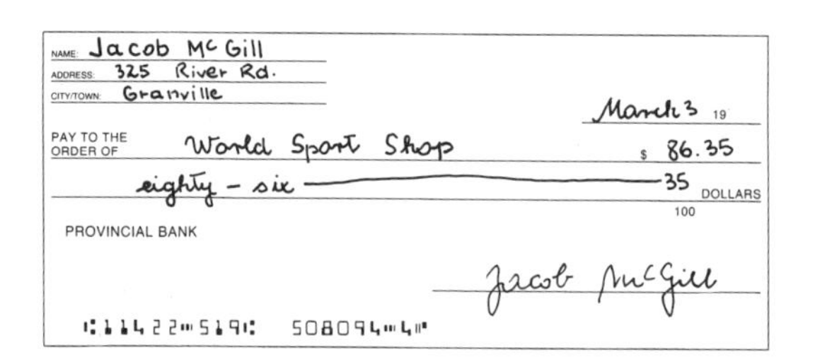
NAME: Jacob McGill
ADDRESS: 325 River Rd.
CITY/TOWN: Granville
March 3 19
PAY TO THE ORDER OF World Sport Shop $ 86.35
eighty-six ——— 35/100 DOLLARS
PROVINCIAL BANK
Jacob McGill
⑆11422⑈519⑆ 508094⑈4⑈

PROVINCIAL BANK		X 1		Cheques and Coupons (List on reverse if necessary)	150	00
		X 2				
		X 5				
Date: March 4	Account Number: 508094	X 10				
		X 20				
Credit Account of: Jacob McGill		X				
		Coin				
		Total Cash				
Signature for cash received (Please sign in front of Teller): Jacob McGill				Sub Total	150	00
				Less Cash Received	0	00
Deposit all Personal Accounts at this Branch		Depositor's Initials: JM		Net Deposit	150	00

Jacob keeps a personal record that gives a running balance of the transactions in his cheque book so that he can check the account statement from the bank. Here is a page from Jacob's personal chequing account record.

DATE	NO.	PARTICULARS	CHEQUES	✓	DEPOSITS		BALANCE
							275 80
Mar 3	048	To World Sport Shop / For Basketball	86 35	✓		Cheque or Dep / Bal	86 35 / 189 45
Mar 15		To Deposit / For from Pay		✓	150 00	Cheque or Dep / Bal	150 00 / 339 45
Mar 18	049	To Joe's Garage / For Auto repair	42 00	✓		Cheque or Dep / Bal	42 00 / 297 45
Mar 31		To Deposit / For from Pay		✓	135 00	Cheque or Dep / Bal	135 00 / 432 45
Apr 4	050	To Smith Clothes / For jacket	149 00	✓		Cheque or Dep / Bal	149 00 / 283 45
Mar 31		To Service Charge / For	7 50	✓		Cheque or Dep / Bal	7 50 / 275 95

Notice that the bank has assessed a service charge of $7.50 on the last day of the month. For this charge, Jacob receives banking services including a monthly statement showing all transactions, along with cancelled cheques. Jacob receives this statement from the bank.

Jacob McGill FROM: Feb 28 TO: Mar 31
CHEQUING ACCOUNT

Description	Debits	Credits	Date	Balance
Balance Forward				275.80
Cheque 048	86.35		03 03	189.45
Deposit		150.00	03 15	339.45
Cheque 049	42.00		03 18	297.45
Deposit		135.00	03 31	432.45
Service Charge	7.50		03 31	424.95

He must account for cheques not cashed and deposits that are made after the statement date. Jacob can reconcile his current balance with the account statement. Most financial institutions have a reconciliation form on the back of the account statement that can be used to check the personal record with the bank record.

STATEMENT RECONCILIATION

You can use the following steps to reconcile your own records with the account statement.

1. Check off all items in your cheque book that are shown in the statement.
2. Subtract the service charges from the statement in your cheque book register to get the cheque book balance.
3. Complete the reconciliation form on the back of the statement.
4. Check that the two balances agree.

The following is the reconciliation for Jacob's statement.

CLOSING BALANCE ON THIS STATEMENT	424.95	A	Write in the closing balance of $424.95.
PLUS DEPOSITS MADE AFTER STATEMENT CLOSING DATE	∅	B	No deposits were made after the closing date.
SUB TOTAL	424.95	T	A + B
LESS OUTSTANDING CHEQUES	149.00	C	Write in cheques not cashed, $149.00.
EQUALS	275.95	D	Subtract T − C.
CHEQUE BOOK BALANCE	275.95	E	Write in the cheque book balance of $275.95.
DIFFERENCE (IF ANY)	∅		If D − E = 0, then the balances are equal and the reconciliation is complete.

EXERCISE 7.6

B 1. Complete the balance column in each cheque book below.

(a)

DATE	NO.	PARTICULARS	CHEQUES	√	DEPOSITS	BALANCE 562.83
Jun 5	124	To Delux Dry Cleaning For	21.40			Cheque or Dep. Bal.
Jun 9		To Withdrawal For Pocket money	50.00			Cheque or Dep. Bal.
Jun 10		To Deposit For From Paycheque			425.00	Cheque or Dep. Bal.
Jun 16		To Visa For Credit Card	257.28			Cheque or Dep. Bal.
Jun 25		To Deposit For From Paycheque			400.00	Cheque or Dep. Bal.
Jun 30		To Service Charge For	7.50			Cheque or Dep. Bal.

(b)

DATE	NO.	PARTICULARS	CHEQUES	√	DEPOSITS	BALANCE 412.56
Aug 4		To George's Cycle For bicycle wheel	135.00			Cheque or Dep. Bal.
Aug 12		To Holiday Inn For Room	109.00			Cheque or Dep. Bal.
Aug 15		To Deposit For Paycheque			972.40	Cheque or Dep. Bal.
Aug 24		To Joe's Garage For Car tuneup	242.70			Cheque or Dep. Bal.
Sep 1		To Floatwell Canoe For Kayak	216.00			Cheque or Dep. Bal.

2. Helen Germain's personal cheque book balance is $429.37. The closing balance on her monthly account statement is $754.83. Since the statement date, Helen made a deposit of $250.00 and she wrote cheques totalling $575.46 after the statement date. These cheques are not listed on the bank statement.

(a) Prepare a reconciliation to show that Helen's balance agrees with the bank statement.

(b) Summarize the reconciliation with the equation (closing balance on statement) + (deposits made after statement closing date) − (outstanding cheques) = (cheque book balance).

3. (a) Below is a page from Muriel's personal chequing book. Complete the balance column.

DATE	NO.	PARTICULARS	CHEQUES	√	DEPOSITS	BALANCE
						406.15
Oct 4	122	To Finlee Fish	178 60			Cheque or Dep.
		For freezer order				Bal.
Oct 6		To Deposit			1563 20	Cheque or Dep.
		For from sale of stock				Bal.
Oct 10		To Deposit			1324 00	Cheque or Dep.
		For Pay cheque				Bal.
Oct 15	123	To Acme T.V.	432 00			Cheque or Dep.
		For new V.C.R				Bal.
Oct 24	124	To Hodgkins Investment	1250 00			Cheque or Dep.
		For buy stock				Bal.
Oct 31	125	To Cable T.V.	32 00			Cheque or Dep.
		For service				Bal.
Nov 3	126	To A & B Records	22 15			Cheque or Dep.
		For two L.P.s				Bal.
Nov 8	127	To Carl's Books	28 34			Cheque or Dep.
		For 4 novels				Bal.
Oct 31		To Service charge	7 50			Cheque or Dep.
		For				Bal.
		To				Cheque or Dep.
		For				Bal.

(b) Muriel's bank statement is given below.
Prepare a reconciliation form as on page 213, and check your figures with the bank statement.

Chequing Account — From: Sept 30 — To: Oct 31

Description	Debits	Credits	Date	Balance
Balance Forward				406.15
Cheque 122	178.60		10 04	227.55
Deposit		1563.20	10 06	1790.75
Deposit		1324.00	10 10	3114.75
Cheque 124	1250.00		10 24	1864.75
Cheque 123	432.00		10 25	1432.75
Cheque 125	32.00		10 31	1400.75
Service Charge	7.50		10 31	1393.25

4. Here is a page from Karl Kauffman's personal chequing book and his monthly bank statement.

(a) Prepare a reconciliation form and check Karl's figures with the bank statement.

(b) Correct Karl's calculations if necessary, and complete a new reconciliation.

Karl's personal record:

DATE	NO.	PARTICULARS	CHEQUES	√	DEPOSITS		BALANCE
							617.24
July 8	068	To Fairway Sports	270 00			Cheque or Dep.	270 00
		For Golf clubs				Bal.	347 24
July 12	069	To Northern Lights	68 25			Cheque or Dep.	68 25
		For clothes				Bal.	415 49
July 15		To Deposit			508 12	Cheque or Dep.	508 12
		For Pay cheque				Bal.	923 61
July 20	070	To Mrs. Bertha Klink	250 00			Cheque or Dep.	250 00
		For cottage rental				Bal.	673 61
July 27	071	To Round Lake Marine	75 82			Cheque or Dep.	75 82
		For Boat rental & gas				Bal.	597 79
July 31		To Deposit			472 16	Cheque or Dep.	427 16
		For Pay cheque				Bal.	1024 95
Aug 3	072	To Deposit			300 00	Cheque or Dep.	300 00
		For Interest on T.D.				Bal.	1324 95
Aug 5	073	To Greenwood College	240 00			Cheque or Dep.	240 00
		For Tuition deposit				Bal.	1084 95
July 31		To Service Charge	7 50			Cheque or Dep.	7 50
		For				Bal.	1077 45

Bank Statement:

Chequing Account	From: Jun 30	To: Jul 31		
Description	**Debits**	**Credits**	**Date**	**Balance**
Balance Forward				617.24
Cheque 068	270.00		07 08	347.24
Cheque 069	68.25		07 12	278.99
Deposit		508.12	07 15	787.11
Cheque 070	250.00		07 20	537.11
Cheque 071	75.82		07 27	461.29
Deposit		472.16	07 31	933.45
Service Charge	7.50		07 31	925.95

5. Investigate the following bank services and charges.

(a) What is the bank charge when a chequing account is overdrawn?

(b) What is a certified cheque?
When would you use a certified cheque?

(c) What are traveller's cheques?
For what purpose would you use traveller's cheques?

(d) What is inter-branch banking?
Are there any charges for this service?

(e) What is overdraft protection in a chequing account?

7.7 AUTOMATED BANKING MACHINES

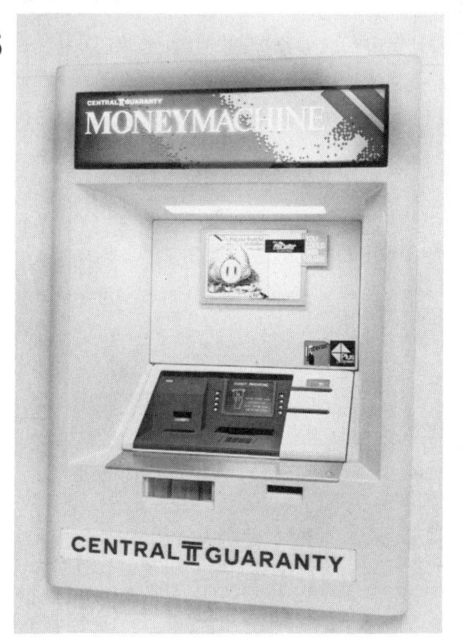

Like most businesses, such as restaurants and stores, banks and other financial institutions have peak times. In order to reduce long line-ups at the tellers' wickets and also to make banking services available 24 h a day, 7 d a week, most financial institutions have automated banking machines (ABMs or ATMs) across Canada.

Most personal banking can be done at an ABM. You activate an ABM by inserting a card in the machine and by also entering a personal identification number (PIN). ABM cards can also be credit cards.

An ABM is electronically sophisticated and expensive. The average cost is $250 000. An ABM is not entirely electronic. When you deposit cash or a cheque into an account using an ABM, the bank or financial company must verify manually the amount of the deposit. For this reason, among others, it is important to keep the paper vouchers that the machine issues for each transaction so that you can check the transactions against an account passbook or credit card statement.

EXERCISE 7.7

A 1. What is a PIN?

2. If possible, describe the steps to perform each transaction at an automated banking machine.

(a) withdrawing cash
(b) making deposits
(c) transferring funds between accounts
(d) making payments on loans
(e) paying utility bills
(f) cashing cheques
(g) obtaining account information

B 3. Find the final balance.

(a) The balance is $824.63 and the sum of $220.00 is withdrawn.
(b) The balance is $628.71 and the sum of $488.25 is deposited.
(c) The balance is $284.99. A cheque for $581.30 is deposited and the sum of $350.00 is withdrawn.
(d) The balance is $615.21. The sum of $52.80 is deposited.

4. Find the final balance in each of the following accounts after the given transactions.

(a) The balance is $657.50. The sum of $300.00 is transferred to another account at the same bank.
(b) The balance is $408.25. The sum of $525.80 is transferred into the account from another account.
(c) The balance is $592.84. The bank withdraws $200.00 for repayment of a loan.
(d) The balance is $928.65. A cheque for $774.50 is deposited and the sum of $500.00 is withdrawn.
(e) The balance is $576.82. A cheque for $586.45 is deposited, $300.00 is withdrawn in cash, and $650.00 is transferred to another account.
(f) The balance is $5280.26. A pay cheque is deposited for $721.40 and $200 is withdrawn.

7.8 BUYING WITH CREDIT CARDS

Anne and Bill Zablonski went on a vacation. They paid for their expenses using cash and credit cards. To avoid the risk in carrying large amounts of cash, they paid for hotels, meals, gasoline, and other purchases with credit cards.

Most businesses will accept payment by credit card and cash rather than by personal cheque.

There is a credit limit which is the maximum amount that can be charged. When a credit card account is not paid in full by the statement due date, a minimum interest charge is added to the account bill.

After Anne and Bill return from their trip, they receive the following statement for purchases made with one of their credit cards.

Account no. 4285 040 835 734

Name: Anne Zablonski
Bill Zablonski

Date	Reference Number	Debits Credits (-)	Particulars	
Jul 02	001	135.00	Chateau Laconte	Edmonton
Jul 03	002	42.00	Willy's Dining Lounge	Edmonton
Jul 03	003	38.25	Woodbine Stores	Edmonton
Jul 04	004	108.00	Western Tack Shop	Calgary
Jul 04	005	62.35	Dobrowski's Steak House	Calgary
Jul 05	006	88.65	Cotton Ben's Clothes	Calgary
Jul 06	007	221.00	Stampede Motel	Calgary
Jul 07	008	38.58	Mellissa's Restaurant	Banff
Jul 08	009	235.00	The Jewel Box	Banff
Jul 08	010	74.00	Banff Springs Dining	Banff
Jul 09	011	40.00	Holiday Tours	Banff
Jul 09	012	18.65	Banff Camera & Film	Banff
Jul 10	013	56.00	Nancy Ski Rental	Banff
Jul 10	014	384.00	Mountainview Hotel	Banff
Jul 10	015	34.00	Willy's Dining Lounge	Edmonton
Jul 11	016	135.00	Chateau Laconte	Edmonton
Jul 25	017	255.00	Hollywood Men's Wear	Red Deer
Jul 31	018	1358.63-	Payment	

Date of Last Statement	Balance on Last Statement	Interest	Total Debits This Month	Total Credits This Month	New Balance
Jul 10	358.63	+ 7.17	+ 1965.48	- 1358.63	= 972.65

Statement Date: AUG 10 Due Date: AUG 25	Minimum Payment	49.00
	Amount of Payment	

EXERCISE 7.8

A

1. Answer the following from the statement on page 217.

(a) How many purchases were made in Calgary?
(b) How much was charged at Nancy Ski Rental?
(c) On what day was the charge made at the Western Tack Shop?
(d) What was the total amount charged in July?
(e) What was the balance on the last statement?
(f) What is the minimum payment that must be made?
(g) When is the due date for the payment?

INVESTIGATION

B 2. The minimum monthly payment on a credit card depends on the balance at the end of the month.

MINIMUM MONTHLY PAYMENT CHART

Balance	Minimum Monthly Payment
Up to $200.00	$10.00
$200.01 to $220.00	$11.00
$220.01 to $240.00	$12.00
$240.01 to $260.00	$13.00
$260.01 to $280.00	$14.00
$280.01 to $300.00	$15.00
$300.01 to $320.00	$16.00
$320.01 to $340.00	$17.00
$340.01 to $360.00	$18.00
$360.01 to $380.00	$19.00
$380.01 to $400.00	$20.00
$400.01 to $420.00	$21.00
$420.01 to $440.00	$22.00
$440.01 to $460.00	$23.00
$460.01 to $480.00	$24.00
$480.01 to $500.00	$25.00

Use the table at the bottom left to determine the minimum monthly payment for each of the following new balances.

(a) $356.40 (b) $58.65
(c) $425.25 (d) $165.80
(e) $495.95 (f) $308.26
(g) $175.65 (h) $215.25
(i) $400.00 (j) $395.00

3. A credit card company calculates the minimum monthly payment by taking the greater of 5% of the new balance (taken to the next highest dollar) or $10.
Calculate the minimum monthly payment for each of the following new balances.

(a) $972.65 (b) $1288.94
(c) $674.46 (d) $739.45
(e) $1080.65 (f) $1856.75
(g) $1534.54 (h) $1177.55

4. The interest rate on the unpaid balance is set by the credit card companies. Calculate the monthly interest charge for each of the following unpaid balances if the rate is 2%/month.

(a) $375.65 (b) $625.98
(c) $659.34 (d) $48.65
(e) $165.24 (f) $325.00

5. Joe Leung has a credit card balance of $163.45. This month, he charged purchases totalling $85.25 to his card and did not make any payments. Interest is 2%/month.

(a) Calculate the new balance.
(b) Use the table in question 2 to find the minimum monthly payment.

6. Elaine Coniglio had a balance of $208.16 on her last statement. During February, she made a payment of $24.00 and charged purchases of $65.24 and $56.85, and was charged $4.16 interest.

(a) Calculate the new balance.
(b) Use the table in question 2 to find the minimum monthly payment.

7. The balance on Mike Hynes' last statement was $608.75. Mike paid $200.00 after the due date, and charged purchases of $165.00, $26.58, and $78.24.

(a) Calculate the interest at 2%/month on the balance on the last statement.

(b) Calculate the new balance.

8. The balance on Bob Hoskins' June statement was $88.64. During the month of July, Bob paid $88.64 before the due date, and charged the following purchases to his card.

Jul 06	$85.60	Hillcrest Shoes
Jul 10	$45.00	House of Beef
Jul 15	$16.20	Merryman Flowers
Jul 22	$21.55	Hardware Corners

(a) Calculate the amount of interest on the unpaid previous balance. The rate is 2%/month.

(b) Calculate the new balance.

(c) Use the table in question 2 to find the minimum monthly payment.

9. Complete the summary at the bottom of the statement.

(a) Calculate the interest. The interest rate is 2%/month.

(b) Calculate the total debits and total credits.

(c) Calculate the new balance.

(d) What is the minimum monthly balance?

Account no. 5185 030 738 279 Name: Bert Armistan

Date	Reference Number	Debits Credits(-)	Particulars	
Oct 04	001	345.00	Albert's Men's Wear	Truro
Oct 06	002	38.90	Tru-Gas	Truro
Oct 09	003	17.45	Dean's Diner	Truro
Oct 11	004	31.08	Quick Clean	Truro
Oct 16	005	35.00	Tru-Gas	Truro
Oct 17	006	628.50-	Payment	
Oct 31	007	147.50	Truro Auditorium	Truro
Oct 31	008	136.45	Halifax Motel	Halifax

Date of Last Statement	Balance on Last Statement	Interest	Total Debits This Month	Total Credits This Month	New Balance
OCT 10	$628.50 +	+	-	=	

Statement Date: Nov 10	MINIMUM PAYMENT	
Due Date: Nov 25	AMOUNT OF PAYMENT	

10. Complete the summary at the bottom of the statement.

(a) Calculate the interest. The rate is 2%/month on the unpaid balance.

(b) Calculate the total debits and total credits.

(c) Calculate the new balance.

(d) Use the table in question 2 to find the minimum monthly payment.

(i)

Account no. 4645 070 836 927 Name: H. Applestein

Date	Reference Number	Debits Credits(-)	Particulars	
Aug 04	001	65.00	Beaver Clothing	Overbrook
Aug 04	002	215.95	Tip Topper	Westfield
Aug 04	003	72.75	Towne Shoes	Westfield
Aug 06	004	26.75	Corner Drugs	Westfield
Aug 06	005	75.87	Junior Clothes	Westfield
Aug 08	006	125.16	Uniform Shop	Westfield
Aug 10	007	35.72	Charles Men's Wear	Westfield
Aug 10	008	63.00	Mario's	Overbrook
Aug 12	009	235.00-	Payment	
Aug 12	010	93.00	Smith's Tire & Bat	Westfield
Aug 16	011	43.20	Tower Sports	Westfield
Aug 16	012	215.95	Fred Brown TV	Westfield
Aug 17	013	45.00	Westfield Flowers	Westfield
Aug 18	014	24.65	Bert's Clothes	Westfield
Aug 19	015	16.20	KC Enterprises	Westfield
Aug 19	016	19.40	Triple A Records	Westfield
Aug 25	017	135.00	Fine Leathers	Westfield
Aug 31	018	700.00-	Payment	

Date of Last Statement	Balance on Last Statement	Interest	Total Debits This Month	Total Credits This Month	New Balance
JUL 31	$863.25 +	+	-	=	

Statement Date: SEP 10	MINIMUM PAYMENT	
Due Date: SEP 25	AMOUNT OF PAYMENT	

(ii)

Account no. 5704 050 285 835 Name: Bea Regis			
Date	**Reference Number**	**Debits Credits(-)**	**Particulars**
Sep 04	001	45.00	Weed Controller King
Sep 04	002	48.64	Buyworth King
Sep 04	003	85.82	Official Sports King
Sep 06	004	36.50	House of Golf King
Sep 06	005	88.47	Wallpaper Place King
Sep 08	006	300.00-	Payment
Sep 10	007	35.72	Charles Men's Wear Westfield
Sep 10	008	261.08	Mark's Clothes King
Sep 12	009	237.24-	Payment
Sep 12	010	35.00	Encore King
Sep 16	011	51.27	Encore King
Sep 16	012	261.08-	Mark's Clothes King
Sep 17	013	52.00	O'Hares of King King
Sep 18	014	75.00	King's Court Din King
Sep 19	015	21.60	King Video King
Sep 19	016	32.40	Wallpaper Place King
Sep 31	017	208.24	Queens Clothes Queens
Sep 31	018	126.64	Marie's Readywear Queens
Sep 31	019	86.36	Marty's Men's Wear Queens

Date of Last Statement	Balance on Last Statement	Interest	Total Debits This Month	Total Credits This Month	New Balance
AUG 31	$537.24 +	+	-	=	

Statement Date: OCT 10	MINIMUM PAYMENT	
Due Date: OCT 25	AMOUNT OF PAYMENT	

MIND BENDER

You place four cards as shown below on a table. You ask someone to pick a number from 1 to 15 and tell you *only* which cards contain this number. You can quickly add the top left hand numbers on these cards, and the sum will be the same as the chosen number.
Why does this work?

A

8 9 10
11 12 13
14 15

B

4 5 6
7 12 13
14 15

C

2 3
6 7
10 11
14 15

D

1 3 5
7 9 11
13 15

7.9 PROBLEM SOLVING

1. Adam bought a 15¢ stamp and gave the postal clerk a one-dollar coin. The clerk gave Adam the stamp and eight coins in change.
What are the possible combinations of coins that the clerk gave Adam?

2. Find the values of the variables in the following subtraction.

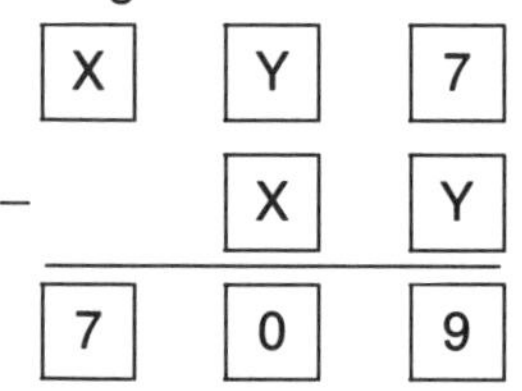

3. Anne has to drive to the airport at an average speed of 80 km/h to catch a plane in 1 h. Half-way to the airport, the car has a flat tire and Anne takes 0.5 h to change the tire.
Can Anne drive quickly, and also safely, and still meet the plane?

4. The number 54 can be written as the sum of 3 consecutive numbers.

$17 + 18 + 19 = 54$

Write 54 as the sum of 4 consecutive numbers.

5. Place the numbers from 1 to 9 in the circles so that each side of the triangle adds to 23.

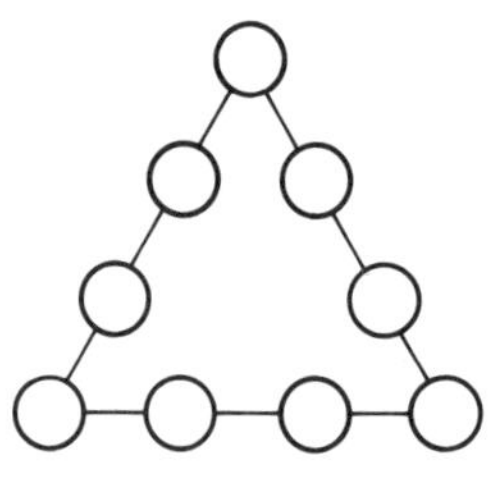

6. Investigate the following pattern.

2
4 + 6
8 + 10 + 12
14 + 16 + 18 + 20
* * * * *

What is the sum of the numbers in the tenth row?

7. Are there four consecutive integers

$a, \quad b, \quad c, \quad d$

such that

$a^2 + b^2 + c^2 = d^2$?

8. What is the maximum number of times that 5 straight lines can intersect?

9. The digits in the number 79 add to 16 and the number 16 is a perfect square.
For what numbers, from 20 to 99, do the digits add to a perfect square?

10. ABCDE is a five-digit number. By adding a 1 to make a six-digit number, we have

$ABCDE1 = 3 \times 1ABCDE$

What is the original number?

11. On the way to the fall fair, Jim drove two thirds of the way at 90 km/h. Jim's father drove one third of the way through a construction zone at 60 km/h.
What was the average speed during the trip?

12. Tommy's Sports pays $75 each wholesale for team jackets and increases the price by one-third to sell to customers. One jacket is left over and is reduced to sell at one-third off.
What is the selling price of this last jacket?

13. Hockey sticks cost $24.95 each. When you buy 12, you get one free.
What is the cost per stick if you buy all 12 and get a free one?

14. The disc jockey for a dance charges $450. Tickets to the dance are sold at $5 each.
How many tickets must be sold to make a profit of $2000?

15. Doug has $180 in $2 bills and $5 bills from the sale of concert tickets.
How many $2 bills does he have if he has 60 bills in total?

16. A cake pan 20 cm by 40 cm and 5 cm deep holds two boxes of prepared cake mix.
How many boxes of cake mix are required to bake a cake in a pan 30 cm by 60 cm and 5 cm deep?

17. The front seat of a van has room for 3 people.
In how many ways can Tom, Hal, and Hubert travel in the front seat if only Tom and Hal can drive?

18. Paula and Monique buy a raffle ticket and each pays 35¢ and 65¢, respectively. Their ticket was drawn for a prize of $50. How should they divide the winnings?

19. The Montreal Expos play their home games in the Olympic Stadium.
About 40 000 fans attended one particular game.

(a) Every fan was asked to complete a questionnaire on food and drinks they bought at the stadium. The results of the survey are as follows:

Food and drinks	Number
Pop (only)	8500
Ice cream (only)	4700
Hot dog (only)	7200
Pop and ice cream	4400
Pop and hot dog	4100
Hot dog and ice cream	4200
Pop, hot dog, and ice cream	3500

How many fans did not have any food or drink?

(b) There were four National Baseball League games that day. Three sportswriters picked the teams that would win:
Jim Dorici's picks:
Expos, Mets, Astros, Braves
Sharon Washington's picks:
Expos, Astros, Reds, Giants
Dan Perkins's picks:
Expos, Reds, Pirates, Mets
No one picked the Chicago Cubs! Which teams played in these four games?

CAREER

BANKER

After completing a college course in accounting, Janice was hired by a bank as a management trainee. Part of Janice's training includes working at a variety of jobs in the customer service department.

EXERCISE

1. Nunzio Macco writes 5 cheques per month and pays three utility bills every month. The bank charges are $1 per utility bill and $0.50 per cheque, or a flat service charge of $7.50 per month. Which is the least expensive charge?

2. Mac Spooner wants to invest $2000 in Bankfund mutual shares. The price for a mutual fund share on the day of sale is $12.50.
How many shares can Mac buy if there is no commission charge for these shares?

7.10 MATCHING PAIRS

There are 39 pairs of figures that match and two figures that do not match.
Find these two figures.

	1	2	3	4	5	6	7	8
A								
B								
C								
D								
E								
F								
G								
H								
I								
J								

7.11 MAGIC IN MATHEMATICS

The trick of tying two people's hands together with loops of string and then separating the hands without cutting the string is done with the help of mathematics.

A and B represent two people. Attach A and B together with loops of string as shown. The trick is to separate the strings without cutting them.

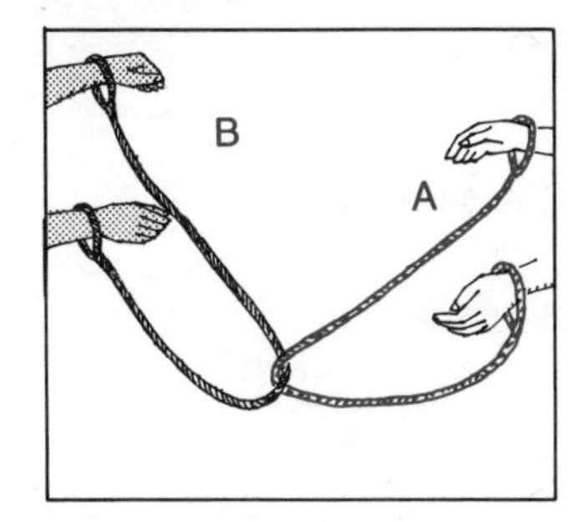

A loop is made in A's string and passed through the loop at B's wrist from B's upper arm toward the hand.

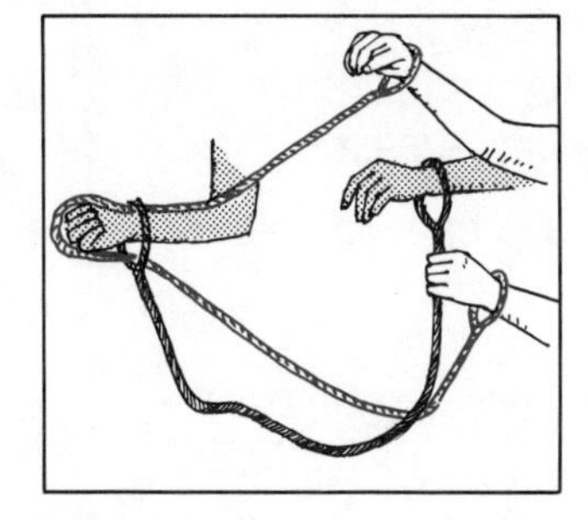

B's hand is passed through the loop, then the loop is pulled back under the string around B's wrist.
The strings are separated!

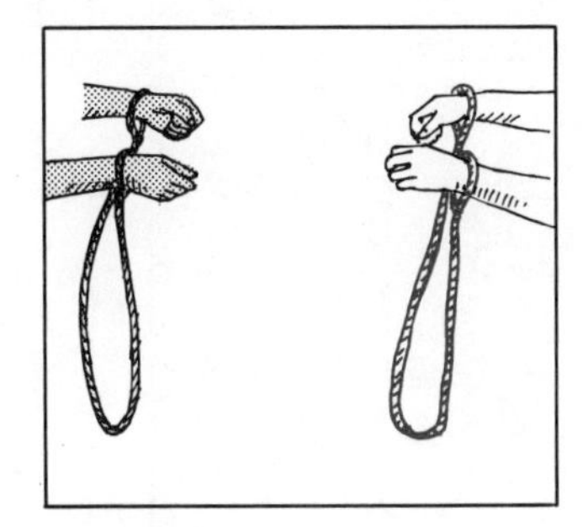

Now, separate the button from the scissors without untying or cutting the string.

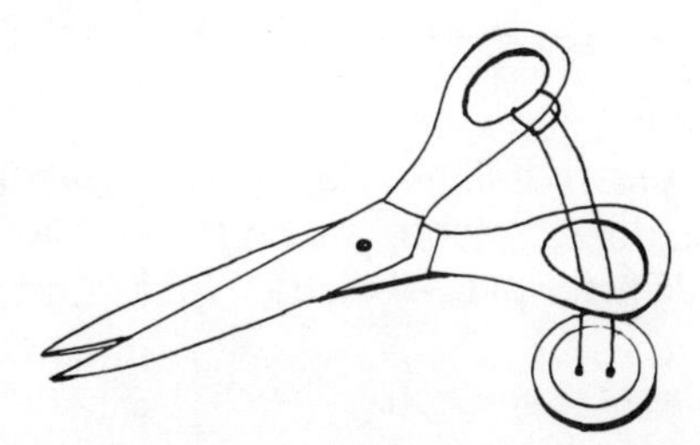

7.12 REVIEW EXERCISE

1. Find the gross earnings for each of the following.
(a) $12.55/h for 23 h
(b) 36 h at $14.50/h plus 5 h at time and a half
(c) 20 h at $7.50/h plus 1.5% commission on $2565 of sales
(d) $8/h for 36 h plus $12 per unit for 60 units

2. Using the tables and assuming a net claim code of 1, find:
(a) Canada Pension Plan deduction
(b) Unemployment Insurance deduction
(c) income tax deduction

Then, calculate the net earnings given each of the following weekly gross earnings.

(i)	$288.00	(ii)	$275.50
(iii)	$304.25	(iv)	$358.02
(v)	$442.68	(vi)	$389.20

3. Copy and complete the pay slip below for Terry Singer.

STATEMENT OF EARNINGS FOR: T. SINGER		
REGULAR EARNINGS:		305.00
OVERTIME EARNINGS:		120.00
GROSS EARNINGS:		(a)
DEDUCTIONS		
C.P.P.:	(b)	
U.I.:	(c)	
INC.TAX:	(d)	
PROV.MED.:	13.50	
DENTAL:	7.60	
SVGS.BND.:	10.00	
TOTAL:	(e)	
NET PAY:		(f)

4. Lincoln Jones earns $800/month part-time as an accountant trainee. He spends 35% on rent, 35% on food, 5% on transportation, and 25% for miscellaneous items.
How much does he spend on each category?

5. The following circle graph shows the budget for Jeremy Kovitius, a college student. Jeremy has budgeted $875/month.

JEREMY'S MONTHLY BUDGET

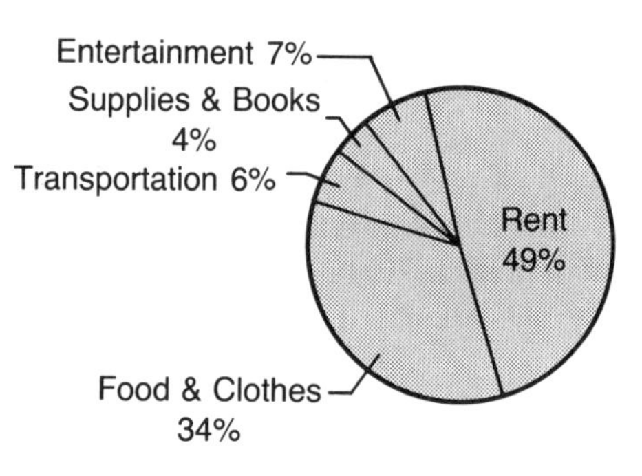

(a) Estimate the amount budgeted for each item.
(b) Calculate the amount budgeted for each item.

6. The following chart shows Morani's expenses for a four-week period.

Item	Week 1	Week 2	Week 3	Week 4
Lunches	$6.25	$5.75	$4.50	$5.80
Clothes	$14.00	$18.65	$10.20	$9.72
Supplies	$8.25	$7.40	$16.00	$11.25
Leisure	$7.00	$11.25	$8.40	$14.50
Bus	$4.50	$5.00	$4.00	$6.00
Savings	$10.00	$1.95	$6.90	$2.73

(a) Find the average amount he spent every week for each item.
(b) Prepare a budget, rounding figures to the nearest percent.
(c) Show the budget in a circle graph.

7. Leslie's personal records for her chequing/savings account in the month of February are as follows.

Opening Balance: $1245.63
Cheques: Feb 5: $56.80, $26.85, $135.90
Feb 12: $65.86, $44.80, $88.63
Feb 19: $235.67, $128.45

Withdrawals: Feb 26: $300.00
Deposits: Feb 10: $625.00
Feb 25: $618.00

There is one free cheque for every $100.00 of the minimum monthly balance, and $0.30 is charged for each additional cheque. Interest of 0.3%/month is calculated on the minimum monthly balance.

Prepare a sample bank statement and find the balance on March 1.

8. Below is a page from José Rivero's personal chequing book.

(a) Complete the balance column in the personal chequing book.
(b) Prepare a reconciliation form and check your figures against the bank statement.

José's Personal Record:

DATE	NO.	PARTICULARS	CHEQUES	√	DEPOSITS	BALANCE 575.74
Jan 6	304	To Herman's Flowers For Gift	12 00			Cheque or Dep. Bal.
Jan 12	305	To Andy's Tire For Snow Tires	165 24			Cheque or Dep. Bal.
Jan 15		To Deposit For From pay			312 00	Cheque or Dep. Bal.
Jan 20	306	To Georges Appliances For repair TV	72 00			Cheque or Dep. Bal.
Jan 22	307	To A & B records For Buy CD	26 46			Cheque or Dep. Bal.
Jan 31		To Deposit For Pay Cheque			325 00	Cheque or Dep. Bal.
Feb 1	308	To G. Gordini For Rent	260 00			Cheque or Dep. Bal.
Feb 4	309	To House Hardware For Craft Supplies	18 00			Cheque or Dep. Bal.
Jan 31		To Service Charge For	7 50			Cheque or Dep. Bal.

Bank Statement:

CHEQUING ACCOUNT FROM: Jan 01 91
TO: Jan 31 91

Description	Debits	Credits	Date	Balance
Balance Forward				575.74
Cheque 304	12.00		01 06	563.74
Cheque 305	165.24		01 12	398.50
Deposit		312.00	01 15	710.50
Cheque 306	72.00		01 20	638.50
Cheque 307	26.46		01 22	612.04
Deposit		325.00	01 31	937.04
Service Charge	7.50		01 31	929.54

Reconciliation:

CLOSING BALANCE THIS STATEMENT	(i)
DEPOSITS MADE AFTER STATEMENT CLOSING DATE	(ii)
SUB TOTAL	(iii)
LESS OUTSTANDING CHEQUES	(iv)
EQUALS	(v)
CHEQUE BOOK BALANCE	(vi)
DIFFERENCE (IF ANY)	(vii)

9. The balance on Hermes Boutet's last credit card statement was $807.26. During the month of December, Hermes made two payments: the first of $600.00 on December 12 and the second of $1000.00 on December 31. Payment is due by the 25th day of the month. He charges the following purchases to his card. Payment is due on the 15th.

Dec 04	001	$35.64	Chippie's Toys
Dec 05	002	$63.80	Allen's Gifts
Dec 05	003	$97.71	Import Shoes
Dec 06	004	$35.26	Merrill Cosmetics
Dec 06	005	$85.00	Empress Dining
Dec 08	006	$110.00	Travellers Lodge
Dec 08	007	$18.65	Muriel's Gifts
Dec 08	008	$185.00	BC Optical
Dec 08	009	$253.80	Kim Fong Clothes
Dec 08	010	$68.04	Jo Chow Jewellers
Dec 09	011	$58.00	Ciro's Dining
Dec 09	012	$86.00	Pacific Sounds
Dec 15	013	$35.10	Games Store
Dec 15	014	$21.55	Evoy Card Shop
Dec 17	015	$43.15	Fur Traders
Dec 18	016	$64.80	Silk Imports
Dec 20	017	$18.31	Tie Shop
Dec 22	018	$65.00	Classy Formal

(a) Calculate the interest at the rate of 2%/month on the unpaid balance.
(b) Calculate the total debits and total credits.
(c) Calculate the new balance.
(d) What is the minimum monthly payment?

7.13 CHAPTER 7 TEST

1. Complete the pay slip.

STATEMENT OF EARNINGS FOR:		
REGULAR EARNINGS:		346.00
OVERTIME EARNINGS:		77.85
GROSS EARNINGS:		(a)
DEDUCTIONS		
C.P.P.:	(b)	
U.I.:	(c)	
INC.TAX:	(d)	
PROV.MED.:	13.50	
DENTAL:	8.10	
SVGS.BND.:	10.00	
TOTAL:	(e)	
NET PAY:		(f)

2. The circle graph shows Mary's budget for a month. Mary is a student and has $800 for expenses each month.
What percent of her budget does Mary spend on each item? Round to the nearest percent.

MARY'S MONTHLY BUDGET

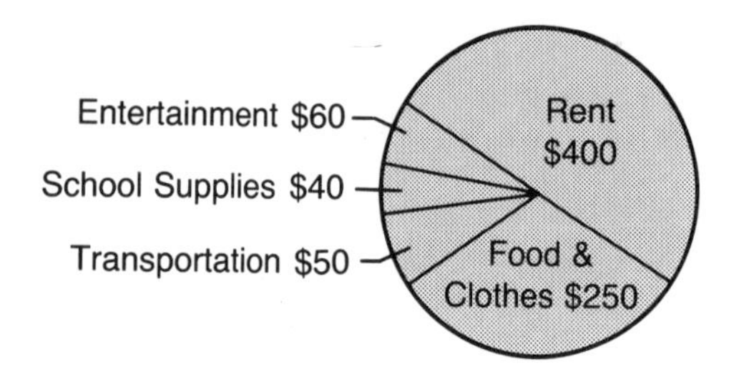

3. At the beginning of April, Bill Mack starts with $647.22 in his chequing account. During the month, he writes $526.37 in cheques and incurs a service charge of $7.50. He deposits $600 after April 30. The closing balance on the bank statement is $113.72. There are no outstanding cheques.
Complete the reconciliation form.

CLOSING BALANCE THIS STATEMENT	(a)
DEPOSITS MADE AFTER STATEMENT CLOSING DATE	(b)
SUB TOTAL	(c)
LESS OUTSTANDING CHEQUES	(d)
EQUALS	(e)
CHEQUE BOOK BALANCE	(f)
DIFFERENCE (IF ANY)	(g)

4. The balance on Ardeth Bozzer's last credit card statement is $185.26. During March, Ardeth puts a total of $204.65 additional charges on her card. On March 13, she makes a payment of $185.26. Interest is charged at a rate of 2%/month on the unpaid balance.
Complete the summary at the bottom of the page and calculate the new balance.

Date of Last Statement	Balance on Last Statement	Interest	Total Debits This Month	Total Credits This Month	New Balance
Feb 28	+	+	−	=	
Statement Date: Mar 10			MINIMUM PAYMENT		
Due Date: Mar 25			AMOUNT OF PAYMENT		

BORROWING AND SAVING

REVIEW AND PREVIEW TO CHAPTER 8

PERCENT

EXERCISE 1

When working with a calculator, remember to enter percents as decimals.

1. Convert the percents to decimals.

(a) 45% (b) $6\frac{1}{2}$% (c) $7\frac{3}{4}$% (d) 75% (e) $9\frac{1}{8}$% (f) $5\frac{1}{3}$%

2. Convert the following to percents.

(a) $\frac{5}{8}$ (b) 0.06 (c) 0.055 (d) 0.55 (e) 0.455 (f) 0.003 75

3. Find the following amounts.

(a) 2% of $750 (b) $10\frac{3}{4}$% of $1000 (c) 8.5% of $400 (d) 110% × $700 (e) 112.75% × $1000

4. Solve for x to the nearest cent.

(a) $3\frac{1}{2}$% of x = $1.25

(b) $2\frac{3}{4}$% of x = $9.30

(c) 15% of x = $35.50

5. Find the total amount and then calculate 8% sales tax.

(a) $6.49, $12.84, $7.35, $47.92
(b) $137.50, $256.14, $583.12, $142.19
(c) $2324.18, $7215.93, $1597.64, $1482.95

6. Calculate the average amount and find 10% of this average.

(a) $475.50, $84.20
(b) $532.50, $49.60
(c) $5842.00, $394.70
(d) $84.96, $7.50

PROFIT AND LOSS

EXERCISE 2

Profit (or Loss) = Selling price − Cost price

1. Complete the following table to determine the profit or loss.

	Selling Price	Cost Price	Profit or Loss
(a)	$25.95	$15.32	
(b)	$159.95	$77.07	
(c)	$575.00	$311.52	
(d)	$69.50	$35.00	
(e)	$15.95	$7.49	

2. A flashlight is sold for $15.95. The cost price of the flashlight is $7.50.

(a) What is the profit on the sale of the flashlight?
(b) What percent of the selling price is profit?

3. A leather jacket is sold for $295. The profit is 40% of the selling price.

(a) Calculate the profit on the jacket.
(b) What was the cost price of the jacket?

4. A car is sold by the dealer at 3% over cost. The cost price of the car is $22 500.

(a) What is the selling price of the car?
(b) What profit is made on the sale of the car?

5. A pair of skis are sold for $79.50. The cost price of the skis is $110. Calculate the loss on the sale of the skis.

SUBSTITUTION

EXERCISE 3

1. Evaluate $2x - 3y$ for the given x- and y-values.
(a) $x = 5, y = 3$
(b) $x = 6, y = 4$
(c) $x = -3, y = 4$

2. Evaluate $x^2 + 2x - 3$ for each value of x.
(a) $x = 5$ (b) $x = 7$
(c) $x = 0$ (d) $x = -1$
(e) $x = -3$ (f) $x = -5$

3. Find the value of x that makes each expression equal 0.
(a) $x - 5$ (b) $x + 3$
(c) $3 - x$ (d) $2x - 6$
(e) $x - 1$ (f) $x - 4$

4. Evaluate $x^2 + y^2$ for the given values.
(a) $x = 5, y = 2$
(b) $x = 3, y = 4$
(c) $x = -3, y = 4$
(d) $x = -3, y = -4$

5. If x can be only $-2, -1, 0, 1$, or 2, choose the values of x that make each expression equal 0.
(a) $x^2 - 3x + 2$ (b) $x^2 - x$
(c) $x^2 - x - 2$ (d) $x^2 + x - 2$

6. Evaluate each expression for the given value.
(a) $(x + 2)^2$ for $x = 3$
(b) $(x - 1)^2$ for $x = -1$
(c) $(x + 1)^2$ for $x = -1$
(d) $(x - 2)^2$ for $x = 2$
(e) $(x + 3)^2$ for $x = 0$

7. Evaluate using a calculator.
(a) 1.25^x for $x = 3$ (b) 1.15^x for $x = 2$
(c) 1.75^x for $x = 4$ (d) 1.12^x for $x = 5$
(e) 1.1^x for $x = 12$ (f) 1.2^x fo $x = 12$

MENTAL MATH

EXERCISE 4

1. Find the value of each unknown.
(a) $x + 5 = 10$ (b) $x - 5 = 10$
(c) $2x = 18$ (d) $5x = 20$
(e) $y + 2 = 11$ (f) $t - 3 = 12$
(g) $s + 6 = 6$ (h) $4w = 28$
(i) $\frac{x - 1}{2} = 5$ (j) $\frac{y + 4}{4} = 7$
(k) $\frac{y + 3}{2} = -3$ (l) $\frac{x - 5}{7} = 0$
(m) $a + 3.5 = 5.5$ (n) $b - 1.6 = 4.6$
(o) $1.2 + c = 6.2$ (p) $0.7 + d = 1.4$
(q) $x + 2x = 0.9$ (r) $5y - 2y = 1.2$
(s) $m + 4.6 = 10$ (t) $n - 2.5 = 8$

2. Perform the following calculations mentally and record your answers.
(a) Find 10% of each of the following.
(i) \$25 (ii) \$125 (iii) \$5250
(b) Find 25% of each.
(i) \$20 (ii) \$100 (iii) \$4400
(c) Find 50% of each.
(i) \$60 (ii) \$15 (iii) \$500
(d) Find 110% of each.
(i) \$100 (ii) \$500 (iii) \$800

3. What is the result if each of the following is reduced by 25%?
(a) \$20 (b) \$80 (c) \$200
(d) \$240 (e) \$480 (f) \$600

4. Each number is 10% of what other number?
(a) 5 (b) 12 (c) 241
(d) 3.5 (e) 4.7 (f) 6.8
(g) 4.95 (h) 62.25 (i) 10.95

5. Which amount is greater?
(a) 10% of \$42.75 or \$5
(b) 10% of \$6.95 or \$0.75
(c) 20% of \$49.95 or \$5
(d) 20% of \$99.95 or \$20
(e) 25% of \$95 or \$25

8.1 BUYING ON CREDIT

Anna went to buy furniture for her new apartment at a local store. The furniture cost $2160, including sales tax. She could pay for the furniture with cash, or by monthly payments. Each payment would include an interest charge. She decides to make a down payment of $300 and agrees to pay the balance in 24 equal monthly payments of $89.13 each.

Her choice is called buying on credit.

> Credit is the exchange of goods or services for a promise to pay at a later date.

We can use the example of Anna buying furniture to illustrate the terms used in credit buying.

Instalment Plan: a number of payments over a specific time period.	Anna pays $300 in cash and the balance will be paid according to an instalment plan of 24 equal monthly payments.
Down Payment: a partial payment made when the goods or service is purchased.	Down payment = $300
Principal: the difference between the selling price, including tax, and the down payment.	Principal = $2160 − $300 = $1860
Instalment Price: the sum of the down payment and the monthly payments. It represents the total cost of the goods or service.	Down payment = $300 Total instalments 24 × $89.13 = $2139.12 Instalment price = $2439.12
Finance Charge: (also called the carrying charge) the additional cost the consumer pays for buying on credit. The carrying charge can be thought of as interest.	Instalment price = $2439.12 Cash price = $2160.00 Finance charge = $279.12

The following formulas are used in the above calculations for instalment plans.

> Principal to be financed = Cash price (including tax) − Down payment
> Total instalments = Number (of months) × Payment (monthly)
> Instalment price = Down payment + Total instalments
> Finance charge = Instalment price − Cash price (including tax)

EXERCISE 8.1

A

1. Calculate the principal for each of the following.

(a) A stove is purchased for $750.60, including tax, and the down payment is $300.

(b) A refrigerator is purchased for $1074.60, including tax, and the down payment is $500.

(c) A television and stereo home entertainment centre is purchased for $4314.60, including tax, and the down payment is $1200.

2. Calculate the total instalments for each credit purchase.

(a) 12 payments of $200 each
(b) 24 payments of $50 each
(c) 18 payments of $200 each
(d) 50 payments of $400 each
(e) 36 payments of $50 each

3. Calculate the instalment price for each credit purchase.

(a) total instalments of $1800 and a down payment of $400
(b) a down payment of $250 with total instalments of $1200
(c) a down payment of $500 with total instalments of $2750
(d) total instalments of $1850 with a down payment of $450
(e) a down payment of $500 with 12 instalments of $100 each

4. Calculate the finance charge on each of the following.

(a) The instalment price is $3500 and the cash price, including tax, is $3000.
(b) The cash price, including tax, is $4500, and the instalment price is $5300.
(c) The instalment price is $1300, and the cash price, including tax, is $1080.
(d) The instalment price is $2500, and the cash price, including tax, is $1775.
(e) The instalment price is $2500, and the cash price is $1250.

B

5. A washer and dryer set sells for $1495, plus 8% sales tax. Mark buys the appliances for $500 down and 12 instalments of $103.60 each.
Calculate the following.

(a) the cash price including tax
(b) the principal
(c) the total instalments
(d) the instalment price
(e) the finance charge

6. Gerry buys a new Roadhandler for $19 500 plus 8% sales tax. Gerry is allowed $5000 on the trade-in as a down payment. The balance is to be paid in 36 instalments of $539.80 each.
Calculate the following.

(a) the principal
(b) the total instalments
(c) the instalment price
(d) the finance charge

7. A microwave oven costs $495, plus 8% sales tax. It can be purchased for $100 down and 12 instalments of $39.84.
Calculate the finance charge.

8. The cash price of a compact disk player is $534.60, including tax. It can be purchased for $100 down and 24 equal payments with a carrying charge of $86.68.
Calculate the monthly payment.

9. A portable stereo costs $395, plus 9% sales tax. Sam buys it for $75 down and 12 payments of $33.20 each.
Calculate the finance charge.

10. The selling price of a house is $375 000. Ted and Florence pay $200 000 down from the sale of their first house, and make monthly payments of $1555 each for the next 25 years.
Calculate the following.

(a) the principal
(b) the total instalments
(c) the instalment price
(d) the finance charge

8.2 INSTALMENT PAYMENTS

Kyler purchases a piano keyboard and the total cost is $1795, plus 8% sales tax. Kyler pays $500 as a down payment, and the balance on an instalment plan.
What is the monthly payment?
Kyler first calculates the amount to be financed.

Selling price = $1795.00
Sales tax (8%) = 8% of $1795.00
= 0.08 × $1795.00
= $143.60
Total cost = $1795.00 + $143.60
= $1938.60

Total payment = $1938.60
Down payment = $500.00
Amount to be financed = $1938.60 − $500.00
= $1438.60

Kyler's next step is to calculate the monthly payment. The size of this payment is determined by the interest rate and time required to pay off the loan. The table gives the monthly payment for every $100 borrowed, according to the number of payments and the interest rate.

Monthly Payment per $100 Borrowed						
Annual Rate	Number of Equal Monthly Payments					
	6	12	18	24	30	36
12%	17.255	8.885	6.098	4.707	3.875	3.321
13%	17.304	8.932	6.145	4.754	3.922	3.369
14%	17.354	8.979	6.192	4.801	3.970	3.418
15%	17.403	9.026	6.238	4.849	4.019	3.467
16%	17.453	9.073	6.286	4.896	4.066	3.516
17%	17.503	9.120	6.333	4.944	4.115	3.565
18%	17.553	9.168	6.381	4.992	4.164	3.615
19%	17.602	9.216	6.428	5.041	4.213	3.666
20%	17.652	9.263	6.476	5.090	4.263	3.716
21%	17.702	9.311	6.524	5.139	4.313	3.768
22%	17.752	9.359	6.573	5.188	4.363	3.819
23%	17.802	9.408	6.621	5.237	4.414	3.871
24%	17.853	9.456	6.670	5.287	4.465	3.923

To calculate the actual monthly payment, find the monthly payment, per $100 borrowed, in the table and use the following formula.

$$\text{Monthly payment} = \text{Monthly payment per \$100 borrowed} \times \frac{\text{Amount borrowed}}{100}$$

The interest rate that Kyler agrees to pay is 18%, and he plans to repay the loan in 12 payments. Using the formula to calculate the monthly payment,

$$\text{Monthly payment} = 9.168 \times \frac{\$1438.60}{100}$$
$$\doteq \$131.89$$

∴ Kyler's monthly payments are $131.89 each.

Press:
9 · 1 6 8
× 1 4 3 8 · 6 0 ÷ 1 0 0 =
The display is 131.89085

EXERCISE 8.2

A

1. Use the table on the opposite page to find the monthly payment for each of the following equal payment loans.
(a) $100 at 15% for 12 months
(b) $100 at 20% for 6 months
(c) $100 at 16% for 36 months
(d) $100 at 12% for 18 months
(e) $100 at 17% for 30 months
(f) $100 at 20% for 24 months

2. Compare the monthly payments for the following case.
Which monthly payment is greater?

$100 at 12% for 24 months
or
$100 at 24% for 12 months

(a) Make a guess.
(b) Calculate using the values in the table.

3. Use the table on the opposite page to calculate each of the following monthly payments.
(a) $1000 at 12% for 12 months
(b) $1000 at 24% for 30 months
(c) $1000 at 18% for 24 months
(d) $1000 at 21% for 12 months
(e) $10 000 at 14% for 24 months
(f) $10 000 at 19% for 36 months
(g) $100 000 at 15% for 36 months

4. The current interest rate is 18%.
Eric wants to pay off a loan of $100 in the shortest possible time.
How long will Eric take to pay off the loan if he can only pay up to $5/month?

5. The current interest rate is 14%.
Cheryl wants to pay off her loan of $1000 in the shortest possible time.
How long will Cheryl take to pay off the loan if she can only pay up to $90/month?

6. What is the monthly payment on a principal of $1 at 18% for each number of payments?
(a) 6 (b) 12 (c) 18
(d) 24 (e) 30 (f) 36

B

7. Roberto Benlolo buys a four-wheel, all-terrain vehicle (ATV) for $2592, including tax. He makes a down payment of $800. What are his monthly payments over an 18 month period with 13% interest?

8. Agnes McIntosh buys a new television set and VCR. With tax, the package costs $1453. Agnes pays $400 down and finances the balance over 18 months. Calculate the monthly payment if the current interest rate is 12%.

9. The selling price of a cello is $2495 and sales tax on the instrument is 8%. André pays $694.60 as a down payment and finances the balance at 15% for 30 months.
(a) What is the principal?
(b) What is the monthly payment?

10. Shirley buys a camera on sale for $595. The sales tax is 8%. She makes a down payment of $142.60, and pays for the balance in 12 monthly payments at a rate of 14%.
(a) What is the principal?
(b) Calculate the monthly payment.
(c) What is the total instalment price?
(d) What is the finance charge?

11. Mac buys the following new computer equipment:

computer	$995.00
colour monitor	$495.00
printer	$379.00
modem	$289.00
diskettes	$49.50
software	$558.50

The sales tax is 8%. Mac pays $987.28 as a down payment and finances the balance at 12% for 18 months.
(a) What is the total cost of the equipment, including tax?
(b) What is the amount financed?
(c) What is the monthly payment?
(d) What is the total instalment price?
(e) What is the finance charge?

8.3 THE EFFECTIVE INTEREST RATE

Mark purchases luggage for \$648, including tax. He pays \$148 as a down payment, and pays for the balance of \$500 with 12 monthly payments of \$45.37 each. While the finance charge may appear to be reasonable, the actual effect of the finance charge can only be determined by finding the effective interest rate.

> Effective Interest Rate:
>
> $$r = \frac{2NI}{P(n + 1)}$$
>
> where r = effective interest rate per annum (expressed as a decimal)
> N = number of payments per year
> I = finance or carrying charge
> P = principal
> n = total number of payments

Using this formula for Mark's case,

$$r = \frac{2 \times 12 \times \$44.44}{\$500(12 + 1)}$$
$$\doteq 0.164$$

Finance charge:
$12 \times \$45.37 - \$500$
$= \$44.44$

∴ The effective interest rate is 16.4%.

Example.
A TV is advertised at \$862.92 including tax, with no down payment, and payable at an annual rate of 12% over 18 months with equal monthly payments of \$52.62 each. What is the effective rate of interest?

Solution:

Instalment price $= 18 \times \$52.62$
$= \$947.16$
Finance charge $= \$947.16 - \$862.92$
$= \$84.24$

Using the effective interest rate formula,

$$r = \frac{2NI}{P(n + 1)}$$

where N = 12, I = \$84.24, P = \$862.92, and n = 18

$$r = \frac{2 \times 12 \times \$84.24}{\$862.92(18 + 1)}$$
$$= 0.123\ 312$$

∴ The effective interest rate is 12.3%.

Press

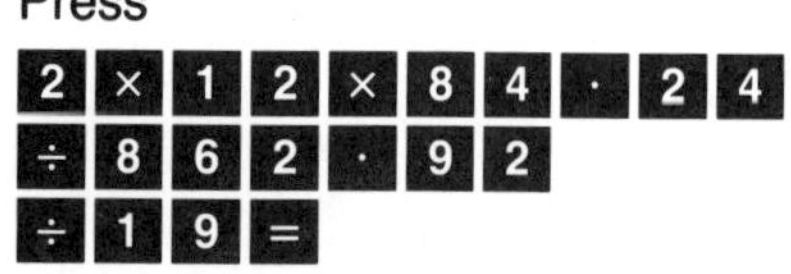

The display is 0.123312

EXERCISE 8.3

A 1. State the total instalment price for each of the following.

(a) $200 down, $100/month for 12 months
(b) $1000 down, $200/month for 6 months
(c) $500 down, $100/month for 24 months

B 2. State the finance charge for each of the following. Payments are monthly.

(a) A principal of $1800 is paid in 18 payments of $114.86 each.
(b) A principal of $1000 is paid in 12 payments of $89.79 each.
(c) A principal of $3000 is paid in 24 payments of $152.70 each.
(d) A principal of $500 is paid in 12 payments of $47.28 each.

3. Find the effective interest rate of a loan of $1000, which is repaid in 12 monthly instalments of $91.68 each.

4. Marino borrows $8000 to purchase a used car. The loan is to be paid in 30 monthly payments at an effective interest rate of 18%.
Use the effective interest rate formula to calculate the finance charge.

5. Find the effective interest rate to the nearest tenth of a percent for each of the following.

(a) A large-screen television set costs $1800, which is financed over 18 months with a $140 finance charge.
(b) A balance of $2000, owing on the purchase of a home stereo, is paid in 18 monthly instalments of $127.62 each.
(c) After trade-in, the amount owing on a new car is $15 800, which is to be paid in 36 monthly payments of $555.53 each.
(d) A loan of $5000 is repaid in 24 monthly payments of $240 each.

6. Adam borrows $7200 to buy a used car. He pays off the loan in 30 monthly payments of $303.34.
Calculate the effective interest rate.

7. A portable stereo sells for $399.95, plus 8% tax. Francine paid $100 down and financed the balance at $19.42 per month for 18 months.
Calculate each of the following.

(a) instalment price
(b) finance charge
(c) effective interest rate

8. Don borrows $8000 to buy a boat. He pays $9645.60 in 30 equal, monthly payments.
Calculate each of the following.

(a) monthly instalment
(b) finance charge
(c) effective interest rate

9. A new car costs $19 995 plus $400 for Pre-delivery Inspection (PDI) and 8% sales tax. Ed and Janice Toscanni pay $10 000 down and plan to finance the balance. They have a choice of borrowing money from a chartered bank or from the car dealership. Compare the following plans to determine which one offers the lower effective interest rate.
Chartered bank:
24 monthly payments of $528.44
Car dealership:
36 monthly payments of $383.72

MIND BENDER

A & B Sports buys skates from a manufacturer at a cost of $112.00 a pair. The price of a pair of skates is increased by 50%, and the skates are displayed for sale. At the end of the season, the price is reduced by 50%. The reduced price of a pair of skates is not $112.00.
Explain.

8.4 SIMPLE INTEREST

When Steve borrows money, he has to pay interest. Interest is the service charge paid for the use of someone else's money. The amount of interest is determined by the principal, the rate, and time. Steve borrowed \$500 for 60 d at 14%/a.

Interest = Principal × Annual rate × Time in years

$$\text{Interest} = \$500.00 \times 14\% \times \frac{60}{365}$$
$$= \$500 \times 0.14 \times \frac{60}{365}$$
$$\doteq \$11.51$$

The interest on \$500 borrowed for 60 d is \$11.51.

Press

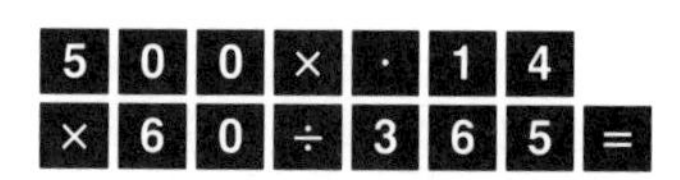

The display is 11.506849

$I = Prt$
$A = P + I$

Interest = Service charge
A = Total amount repaid
I = Rent paid for the use of someone else's money
P = The amount of money borrowed
r = The rate at which the interest is charged, expressed as a percent of the principal per time unit
t = Number of time units

Example 1.
The Go Gas Co. charges 1.5%/month interest on overdue accounts. Ms. Lafleur allows a bill of \$94 to go one month overdue.
How much interest will be charged?
What amount must she repay?

Solution:

$P = \$94, r = 1.5\%, t = 1$ month
$= 0.015$

$I = Prt$
$I = \$94 \times 0.015 \times 1$
$= \$1.41$

$A = P + I$
$A = \$94 + \$1.41$
$= \$95.41$

Ms. Lafleur is charged \$1.41 interest and repays \$95.41.

Example 2.
\$1200 is borrowed for three months. The interest charged is \$63.
(a) What monthly interest rate is charged?
(b) What is the annual interest rate charged?

Solution:
(a) $I = \$63, P = \$1200, t = 3$ months

$I = Prt$
$$\frac{I}{Pt} = \frac{Prt}{Pt}$$
$$r = \frac{I}{Pt}$$

$$r = \frac{\$63}{\$1200 \times 3}$$
$= 0.0175$
$= 1.75\%$

The interest rate is $1\frac{3}{4}$%/month.
What tells us the time unit is months?

(b) ($1\frac{3}{4}$%/month) × (12 months/a) = 21%/a
The interest rate is 21%/a.

Example 3.
$6.75 interest is paid on a loan of $450 borrowed at 6%/a. Find the time period of the loan.

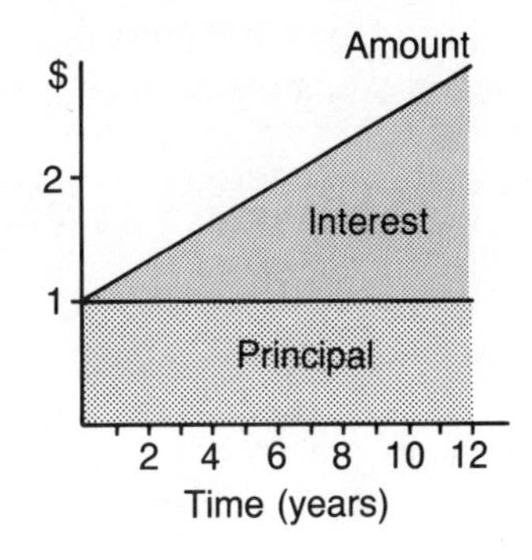

Solution:
I = $6.75, P = $450, r = 6%
= 0.06

$$I = Prt$$
$$\frac{I}{Pr} = \frac{Prt}{Pr}$$
$$t = \frac{I}{Pr}$$

$$t = \frac{\$6.75}{\$450 \times 0.06} = 0.25$$

The money was borrowed for 0.25 a.

What tells us that the unit of time is a year?

EXERCISE 8.4

B 1. Find the simple interest on each of the following loans.
(a) $550 for 1 a at 10%
(b) $750 for 2 a at 15%
(c) $7500 for 3 a at 12%
(d) $2500 for 1 a at 18%
(e) $4000 for 2 a at 14%
(f) $1500 for 1.5 a at 15%
(g) $3500 for 3 a at 9%
(h) $8000 for 1.5 a at 10%
(i) $925 for 1.25 a at 11%

2. Heidi has $1274.30 in her savings account. Simple interest is paid monthly at a rate of 7%/a.
What is the interest for 1 month?

3. How long will it take for $100 to double at simple interest of 10%/a?

C 4. The interest paid on a loan of $500 at 8%/a simple interest is $30.
What is the time period of the loan?

5. A small boat and motor costs $12 500, including tax. Wilson pays $5000 in cash and the balance in 4 months with simple interest charged at 18%/a.
(a) What amount must Wilson pay after 4 months?
(b) What is the total cost of the boat?

6. What sum of money must be invested today at 12% simple interest in order to have $1000 at the end of one year?

MICRO MATH

The following program calculates the simple interest on a principal, P, at a rate of interest, R, and for a period of time, T, in years.

```
NEW
100 PRINT"SIMPLE INTEREST"
110 INPUT"ENTER PRINCIPAL P = "; P
120 INPUT"ENTER THE RATE R = "; R
130 INPUT"ENTER THE TIME T = "; T
140 PRINT
150 I=P*.01*R*T
160 PRINT"INTEREST = $"; I
170 A = P + I
180 PRINT"AMOUNT A = $"; A
190 END
RUN
```

Enter the program and RUN it for the following values.
1. $52 550 for 218 d at 12%
2. $1259.75 for 178 d at 10.5%
3. $7232.50 for 4 months at 12.7%
4. $525 for 6 months at 8%
5. $1845 for 475 d at 18%

8.5 COMPOUND INTEREST

Morry deposited \$1000.00 in a savings account which earns 9%/a compounded annually.	$I = Prt$ $A = P + I$
After one year, the \$1000.00 increases to \$1090.00, with 9% interest.	$I = \$1000.00 \times 0.09 \times 1$ $= \$90.00$ $A = \$1000.00 + \$90.00$ $= \$1090.00$
After the second year, the \$1090.00 increases to \$1188.10, with 9% interest.	$I = \$1090.00 \times 0.09 \times 1$ $= \$98.10$ $A = \$1090.00 + \$98.10$ $= \$1188.10$
After the third year, the \$1188.10 increases to \$1295.03, with 9% interest.	$I = \$1188.10 \times 0.09 \times 1$ $\doteq \$106.93$ $A = \$1188.10 + \$106.93$ $= \$1295.03$

After three years, the amount of money in Morry's account is \$1295.03. The interest which Morry's money earned was added to the account and it also earned interest. This is called compound interest. In this case we say that the interest is compounded annually since the interest is calculated once per year.

Interest can be compounded twice per year, or semi-annually, and also quarterly, monthly, or even daily.

Example.
A sum of \$2000 is invested at 8%/a compounded semi-annually.
Calculate the compounded amount after 3 a.

Solution:
Since the interest is compounded semi-annually, there are 6 interest periods in 3 a.

Interest Period	Principal P	Rate r	Time t	Interest $I = Prt$	Amount $A = P + I$
1	\$2000.00	8%	0.5	\$80.00	\$2080.00
2	\$2080.00	8%	0.5	\$83.20	\$2163.20
3	\$2163.20	8%	0.5	\$86.53	\$2249.73
4	\$2249.73	8%	0.5	\$89.99	\$2339.72
5	\$2339.72	8%	0.5	\$93.59	\$2433.31
6	\$2433.31	8%	0.5	\$97.33	\$2530.64

Using a calculator,

2 0 0 0 × · 0 8 × · 5 + 2 0 0 0 =

The display is 2080.

After 3 a the amount is \$2530.64.

We can calculate the same investment under simple interest.

$I = Prt$, where $P = \$2000.00$, $r = 8\% = 0.08$, and $t = 3$ a

$I = \$2000.00 \times 0.08 \times 3$
$= \$480.00$

$A = P + I$
$= \$2000 + \$480$
$= \$2480$

The amount after 3 a under simple interest is $2480.

Amount with compound interest	$2530.64
Amount with simple interest	2480.00
Difference	$50.64

The amount under compound interest is $50.64 more than that under simple interest.

EXERCISE 8.5

B 1. Find the simple interest on each of the following deposits.
(a) $300 at 8%/a for 6 months
(b) $500 at 7%/a for 8 months
(c) $250 at 6%/a for 15 months

2. Find the amount for each of the following at simple interest.
(a) $500 at 8%/a for 4 months
(b) $750 at 9%/a for 18 months
(c) $1000 at 6%/a for 5 a

3. Marion deposits $1000 in an account that pays 10%/a interest compounded annually.
How much is in the account after 3 a?

4. Barbara deposits $2000 into an account that pays interest at a rate of 10%/a compounded quarterly.
Calculate the amount in the account after one year.

5. Monty invests $2000 at 8%/a.
Calculate the amount in the account after 2 a if the interest is compounded:
(a) annually
(b) semi-annually
(c) quarterly

6. Calculate the amount of each of the following investments.
(a) $1000 is invested for 2 a at 8%/a compounded annually.
(b) $2000 is invested for 3 a at 9%/a compounded semi-annually.
(c) $3000 is invested for 2 a at 10%/a compounded quarterly.

C 7. (a) Use a table to show the amount at the end of each period for $100 invested at 10%/a compounded semi-annually.
(b) How long will it take for the $100 to double?

8. Find the difference between the investments.
(a) $1000 invested for 2 a at 10%/a compounded quarterly
or
$1000 invested for 3 a at 8%/a compounded semi-annually
(b) $2000 invested for 3 a at 12% simple interest
or
$2000 invested for 3 a at 10%/a compounded semi-annually

MICRO MATH

The following program calculates and prints the amount of a loan or an investment under compound interest.

```
100 PRINT"COMPOUND INTEREST"
110 INPUT"ENTER PRINCIPAL P ="; P
120 INPUT"ENTER THE ANNUAL RATE R% ="; R
130 INPUT"TIME IN YEARS "; T
140 PRINT"CHOOSE THE COMPOUNDING NUMBER"
150 PRINT"MONTHLY      (12) QUARTERLY (4)"
160 PRINT"SEMI-ANNUALLY (2)  ANNUALLY (1)"
170 INPUT"ENTER THE COMPOUNDING NUMBER": N
180 A=P*(1+.01*R/N)^(N*T)
190 PRINT"THE AMOUNT IS "; A
200 END
```

8.6 BORROWING MONEY

Joe McLean decides to buy a stereo and, to pay for it, he must borrow money from a source of credit.

Two sources of credit are the retailer who sells the goods and financial institutions.

THE RETAILER

One of the more popular types of credit arrangements for customers of large department stores and oil companies is a charge account. The customer is billed usually once a month. The bill may be paid in full or by instalment plan.

FINANCIAL INSTITUTIONS

Financial institutions are very similar with regard to lending money:

* an interest rate is charged
* you will be required to make an application
* your credit history will be checked before a loan is approved
* a loan is repaid according to a predetermined schedule

When applying for a loan, you are required to tell or disclose information about yourself: where you live, your employment history, other loans you may have, and so on.
The application will give the financial institution an indication of your credit history and will answer these questions:

* Can you afford to borrow money?
* Will you be able to pay the loan?
* Have you paid debts on time in the past?

Joe McLean needs to borrow $2500 to buy the stereo. He investigates the types of loans available at the financial institutions in his community and finds the most suitable loan at the Provincial Bank.

When Joe applies for a loan, the bank investigates his credit rating. The rating is a record of Joe's previous loans, credit card payments, and so on. If Joe has made payments on time, he will have a good credit rating.

Joe's application is approved and the interest rate is 12%. The loan is due in ten months. He is required to sign a written promise to pay a specific sum of money on a certain date. This is called a promissory note, part of which is shown below. Joe increased the security of his loan by having Peter Webster co-sign the loan. The co-signer is responsible for the loan if Joe should fail to pay, or default. Another way to lower the interest rate is with a secured loan. To secure a loan, the borrower must put up something valuable as collateral, such as a house, that can be used to repay the loan if needed.

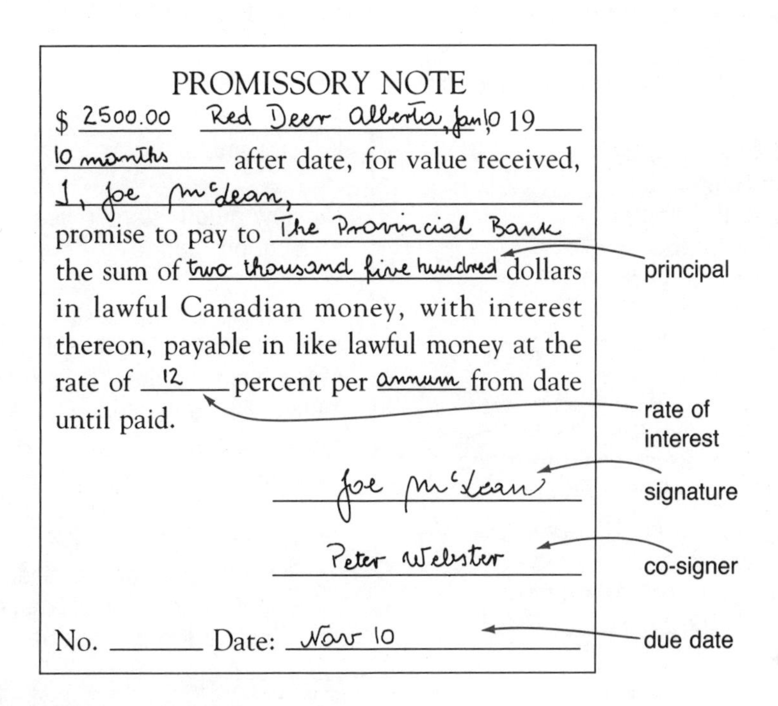

PROMISSORY NOTE

$ 2500.00 Red Deer Alberta, Jan 10 19___

10 months after date, for value received,

I, Joe McLean,

promise to pay to The Provincial Bank

the sum of two thousand five hundred dollars in lawful Canadian money, with interest thereon, payable in like lawful money at the rate of 12 percent per annum from date until paid.

Joe McLean

Peter Webster

No. ______ Date: Nov 10

How much will Joe McLean have to repay for the loan?

Finding the interest,

$I = Prt$

$P = \$2500$, $r = 12\%$, $t = 10$ months

$r = 0.12$

$I = \$2500 \times 0.12 \times \frac{10}{12}$

$= \$250$

Finding the total amount,

$A = P + I$

$P = \$2500$, $I = \$250$

$A = \$2500 + \$250$

$= \$2750$

∴ Joe McLean will pay $2750 for a loan of $2500 for 10 months.

EXERCISE 8.6

A 1. How do financial lending institutions make money to operate at a profit?

2. What are the factors that contribute to a good credit rating?

3. What is a promissory note?

4. How does a co-signer help to lower the interest rate on a loan?

5. (a) What is collateral?
(b) What are some examples of collateral?

6. What is a revolving charge account?

B 7. How many days are there from the first date to the second date? For example, from May 15 to May 16 is one day.
(a) March 15 to August 24
(b) May 14 to September 7
(c) March 17 to December 24
(d) October 1 to February 25
(e) March 1 to September 30
(f) July 15 to January 17
(g) November 1 to February 1

8. Compare the interest charges for each of the following. Assume simple interest is charged and the loan is repaid with one payment at the end of the period.
(a) \$11 600 at 12%/a for 12 months
or
\$11 000 at 11%/a for 12 months
(b) \$615 at 9%/a for 2 a
or
\$510 at 12%/a for 2 a
(c) \$5000 at 12%/a for 30 months
or
\$4700 at 14%/a for 30 months
(d) \$10 000 at 12.5%/a for 2 a
or
\$12 000 at 11%/a for 2 a

9. Jerome Burnier borrows \$3000 from the Provincial Bank to pay for a complete overhaul of his boat. He agrees to pay the interest every month at a rate of 1%/month, and to pay the principal in 6 months.

(a) Copy and complete the following note for the above loan.

Date: ____________
On demand after date,
____
promise to pay to Provincial Bank
____________________/100 dollars
with interest payable monthly at the rate of ______ percent per ________ from date until paid.

____

____

(b) How much interest was paid for the 6 month period?

10. On April 7, Suzie Okawi borrows \$1200 from the Provincial Bank to pay for automobile repairs. The interest rate is 12%/a. Suzie will repay the loan on July 15.
(a) Prepare a promissory note to record the loan if the co-signer is Bert Okawi.
(b) How much interest does Suzie pay on the loan?
(c) Prepare another promissory note for this loan if the rate of interest is 14%/a and calculate the interest paid.

11. Below is a demand note signed by Mary Lima. According to this note, Mary can make payments at any time and pays interest on the principal, including the day of the payment.

> Date: March 24
>
> On demand after date, I, Mary Lima, promise to pay to Credit Union four thousand two hundred /100 dollars with interest payable at the rate of 14 percent per annum from date until paid.
>
> Mary Lima

(a) Complete the following schedule of payments for this loan.

Loan Payment Record				
Date	Pay't	Interest Period	Interest Owing	Balance Owing
		from to		
03 24	-	-	-	$4200.00
03 31	$500	03 24 03 31	11.38	$3711.28
04 15	$200	03 31 04 15	21.35	$3532.63
04 30	$500			
05 15	$200			
05 31	$500			
06 15	$200			
06 30	$500			
07 15	$900			
07 31				$0.00

(b) What is the balance owing after the July 15 payment?

(c) What is the last payment on July 31?

(d) What is the finance charge?

(e) What is the instalment price of the furniture?

12. On March 12, Jean Marc Ether trades in his car and borrows $10 000 from Consumers Trust at 15%/a to pay off the balance to buy a new car. Jean Marc pays the loan promptly with $1000/month payments on the last day of each month. Here is the demand note which Jean Marc signed.

> Date: March 12
>
> On demand after date, I, Jean Marc Ether, promise to pay to Consumers Trust ten thousand /100 dollars with interest payable monthly at the rate of 15 percent per annum from date until paid.
>
> Jean Marc Ether

(a) Complete the following schedule of payments for this loan.

Loan Payment Record				
Date	Pay't	Interest Period	Interest Owing	Balance Owing
		from to		
03 12	-	-	-	$10 000.00
03 31	$1000	03 12 03 31		
04 30	$1000			
05 31	$1000			

(b) What is the date of the last payment?

(c) How much is the last payment?

(d) What is the finance charge?

(e) What is the instalment price of the new car if the value of the trade-in was $12 775?

8.7 SAVING MONEY

Yuri Yakuchev works part-time at a gas bar and saves $5000. He would like to save the money and earn interest. One choice is to deposit the money in a savings account.
What are the alternatives to a savings account?
He investigates these investments in addition to a savings account:
Term deposit
Canada Savings Bonds
Treasury bill
Yuri can compare these investments on the basis of the following: interest rate, access to his money, minimum deposit, security, and convenience. In this section, we will invest Yuri's $5000 over a two-year period in a savings account and in each of the above investments.

SAVINGS ACCOUNT

In an earlier section, we saw how interest is compounded in a savings account if there are no deposits or withdrawals.

YURI'S SAVINGS IN A SAVINGS ACCOUNT

$5000 is invested at 6.5%/a compounded quarterly for 2 a.

Since the interest is quarterly, there will be 8 interest periods in 2 a.

Interest Period	Principal P	Rate r	Time t	Interest I = Prt	Amount A = P + I
1	$5000.00	6.5%	0.25	$81.25	$5081.25
2	$5081.25	6.5%	0.25	$82.57	$5163.82
3	$5163.82	6.5%	0.25	$83.91	$5247.73
4	$5247.73	6.5%	0.25	$85.28	$5333.01
5	$5333.01	6.5%	0.25	$86.66	$5419.67
6	$5419.67	6.5%	0.25	$88.07	$5507.74
7	$5507.74	6.5%	0.25	$89.50	$5597.24
8	$5597.74	6.5%	0.25	$90.96	$5688.70

Press

5 0 0 0 × · 0 6 5 × · 2 5 = + 5 0 0 0 =

The display reads 5081.25

After 2 a in a savings account at 6.5%/a compounded quarterly, Yuri's $5000 will increase to $5688.70.

Some advantages of a savings account are that Yuri is able to make withdrawals and deposits at any time, which means that Yuri has free access to the account. But, in this case, Yuri would lose interest if he was to make any withdrawals.

TERM DEPOSIT

Most financial institutions offer term deposits. Yuri can invest money at a specific rate of interest for a specific period of time or term. The longer the term, the higher the interest. More money invested will also earn a higher interest rate.
A Guaranteed Investment Certificate (GIC) is very similar to a term deposit. A term deposit can be redeemed before the end of the term, but at a reduced interest rate. However, a GIC is usually non-redeemable. For this reason, Yuri should think about his future finances and read the contract carefully.

YURI'S SAVINGS IN A TERM DEPOSIT

$5000 is invested at 9.75% compounded annually for 2 a.

Since interest is compounded annually, there will be 2 interest periods.

Interest Period	Principal P	Rate r	Time t	Interest I	Amount A
1	$5000.00	9.75%	1.0	$487.50	$5487.50
2	$5487.50	9.75%	1.0	$535.03	$6022.53

After 2 a in a term deposit at a guaranteed rate of 9.75%, Yuri's $5000 amounts to $6022.53.

Example 1.
The table shows the term deposit interest rate for different periods.
Calculate the interest earned for
(a) $2000 invested for 30 d and reinvested for another 30 d
(b) $2000 invested for 60 d

Term Deposit Rates	
Number of Days	Rate Paid
30 – 59	6.5%/a
60 – 90	7.0%/a
90	7.5%/a
1 a – 5 a	9.75%

Solution:

(a) $I = Prt$

For the first 30 d,

$P = \$2000.00$, $r = 6.5\% = 0.065$, $t = \frac{30}{365}$ a

$I = \$2000.00 \times 0.065 \times \frac{30}{365}$

$\doteq \$10.68$

For the second 30 d,

$P = \$2000.00 + \$10.68 = \$2010.68$

$I = \$2010.68 \times 0.065 \times \frac{30}{365}$

$\doteq \$10.74$

Adding to find total interest,

$\$10.68 + \$10.74 = \$21.42$

Total interest earned is $21.42.

(b) $I = Prt$

For 60 d,

$P = \$2000.00$, $r = 7.0\% = 0.07$, $t = \frac{60}{365}$ a

$I = \$2000.00 \times 0.07 \times \frac{60}{365}$

$\doteq \$23.01$

The interest earned is $23.01.

Press

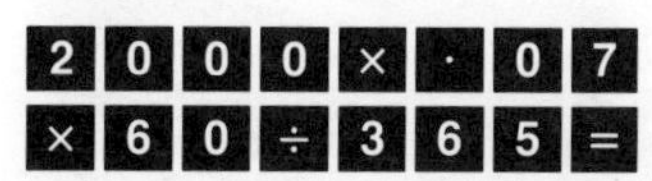

The display is 23.013699

What are some of the factors to consider when choosing the term of the deposit?

CANADA SAVINGS BONDS

Each year, Canada Savings Bonds (CSBs) are issued by the Government of Canada for a relatively short time. Yuri can purchase these from a financial institution or through a payroll deduction plan. A CSB can be cashed in, or redeemed, for its full face value at any time. If a CSB is redeemed within a few months after purchase, no interest is paid. Two types of CSBs are available:

Regular Interest Bond: pays simple interest by cheque or direct deposit into an account.

Compound Interest Bond: pays compound interest annually. The interest is added to the bond once a year and is compounded. The interest is paid when the bond reaches maturity or is redeemed.

CSBs are offered each year; each annual issue is given a series number. Interest rates, set by the federal government, can also change with each issue and CSB bonds are adjusted automatically to this new rate. The following chart shows the value of a Compound Interest Bond in several series.

Series	Maturity	Value of a $1000 Compound Interest Bond	
		Nov 1/87	Nov 1/88
S46			
S45			
S44			
S43 (1988)			
S42 (1987)	1994	$1000.00	$1090.00
S41 (1986)	1993	$1077.50	$1174.48
S40 (1985)	1992	$1174.48	$1280.18
S39 (1984)	1991	$1306.60	$1424.20
S38 (1983)	1990	$1432.91	$1561.87

Similar charts are available where bonds are sold.

To find the value of Yuri's $5000.00 use the table.

Read the value for $1000.00 from the table. $1174.48

Multiply by 5 to get the value for $5000.00. $5 \times \$1174.48$

Write the value of the bond. $5872.40

YURI'S SAVINGS IN AN S41 COMPOUND INTEREST BOND

The interest rate for each year can be different.
In the first year, S41 bonds paid 7.75%/a.
In the second year, S41 bonds paid 9.0%/a.
If Yuri invests $5000.00 in S41 bonds, the interest is

$$I = Prt$$

In the first year, $P = \$5000.00$, $r = 7.75\%$, $t = 1$ a

$$= 0.0775$$

$$I = \$5000 \times 0.0775 \times 1 = \$387.50$$

$$A = \$5000.00 + \$387.50 = \$5387.50$$

In the second year, $P = \$5387.50$, $r = 9\%$, $t = 1$ a

$$= 0.09$$

$$I = \$5387.50 \times 0.09 \times 1 \doteq \$484.88$$

$$A = \$5872.38$$

After 2 a in an S41 CSB at compounded interest, Yuri's $5000 amounts to $5872.38.

The variance of $0.02 is due to rounding.

TREASURY BILLS

Treasury bills are short term, that is one year or less, securities issued by the Government of Canada. They are sold at a discount to the purchaser. At maturity, the purchaser is paid the face value of the bill. The difference between the discounted price and the face value is the yield. You can think of yield as something like interest.

For example, Yuri could buy a $1000 treasury bill at a discounted price of $925. The bond matures at the end of one year.
What is the yield, expressed as a percent?

$$\text{Yield} = \frac{1000 - \text{price}}{\text{price}} \times \frac{365}{\text{term}} \times 100\%$$
$$= \frac{1000 - 925}{925} \times \frac{365}{365} \times 100\%$$
$$\doteq 8.1\%$$

$A = P + Prt$
$\frac{A - P}{P} \times \frac{1}{t} = r$; t in years
$\frac{A - P}{P} \times \frac{365}{t} \times 100\% = r$; t in days;
$100\% = 1$

The yield is 8.1%.

YURI'S SAVINGS IN A TREASURY BILL

The yield for treasury bills varies with each issue. Using past figures, Yuri estimates that the average yield for this year will be 11% and for the next year, 11.5%.

In the first year, a $5000 treasury bill costs

$P = A \div (1 + rt)$
$P = \$5000.00 \div (1 + 0.11 \times 1) = \$4504.50$

This means that Yuri paid $4504.50 for the bill and received $5000 a year later.
He could invest the balance in a savings account at 6.5%/a bank interest and receive interest:

$I = Prt$, $P = \$5000.00 - \$4504.50$, $r = 6.5\%$, $t = 1$ a
$= \$495.50$ $= 0.065$
$I = \$495.50 \times 0.065 \times 1 \doteq \$32.21$

$A = P + I$
$A = P + Prt$
$A = P(1 + rt)$
$P = A \div (1 + rt)$

The amount is $495.50 + $32.21 or $527.71.

Total at the end of one year:		
	Treasury bill	$5000.00
	Savings account	527.71
	Total	$5527.71

In the second year, he can invest $6000 in treasury bills.
$6000 of treasury bills at 11.5% yield would cost:

$P = A \div (1 + rt)$
$P = \$6000.00 \div 1.115 = \$5381.17$

This means that Yuri would pay $5381.17 for the bills and will receive $6000 a year later.
He could leave the balance in a savings account and receive interest:

$I = Prt$, $P = \$5527.71 - \$5381.17$, $r = 6.5\%$ (bank interest), $t = 1$ a
$= \$145.83$ $= 0.065$
$I = \$146.54 \times 0.065 \times 1 \doteq \$9.53$

The amount is $146.54 + $9.53 or $156.07.

Total at the end of two years:		
	Treasury bill	$6000.00
	Savings account	156.07
	Total	$6156.07

EXERCISE 8.7

A

1. Why would a term deposit pay a higher interest rate than a savings account?

2. How does a Government of Canada treasury bill differ from a Canada Savings Bond?

3. What is the difference between a CSB Compound Interest Bond and a CSB Regular Interest Bond?

4. When are Canada Savings Bonds offered for sale to the public?

B

5. Calculate the compound interest in each of the following savings accounts. Unless otherwise stated, all rates are per annum.
(a) a principal of $2000 on deposit for one year at 7% compounded semi-annually
(b) a principal of $650 on deposit for 6 months at 12% compounded monthly
(c) a principal of $4000 on deposit for 2 a at 6% compounded quarterly
(d) a principal of $5000 on deposit for 3 months at 8% compounded monthly
(e) a principal of $3000 on deposit for 6 months at 7.5% compounded semi-annually

6. Mike Boler has $10 000. He wants to keep the money in a savings account so he can make a down payment on some logging equipment in six months. Account A pays interest at a rate of 6%/a compounded monthly. Account B pays interest at a rate of 6.5%/a compounded semi-annually. Which account pays more interest over the six months?

7. Compare the four ways to deposit Yuri's savings in this section under the following headings.
(a) return on savings
(b) access
(c) convenience
(d) security

8. Calculate the interest compounded on each of the following term deposits. Unless otherwise stated, all interest rates are per annum.
(a) $2000 for 6 months at 8%
(b) $8000 for one year at 8.5%
(c) $1000 for 60 d at 12%
(d) $3000 for 100 d at 9%
(e) $10 000 for 2 a at 9%
(f) $20 000 for 90 d at 10%

9. The following table shows the term deposit rates on a certain date at the Upper Canada Bank.

TERM DEPOSIT RATES	
Length of Term	Interest Rate
30 – 59 d	6.5%/a
60 – 89 d	7.0%/a
90 – 120 d	8.0%/a
121 – 364 d	8.5%/a
1 a	9.0%/a
2 – 5 a	9.5%/a

Calculate the interest earned for each term deposit. (Assume compound interest.)
(a) $2000 for 60 d
(b) $5000 for 41 d
(c) $2500 for 200 d
(d) $4000 for 100 d
(e) $1000 for 2 a
(f) $3000 for 300 d
(g) $5000 for 4 a
(h) $5000 for 200 d
(i) $4000 for 1 a
(j) $2500 for 100 d

10. Which term deposit pays a greater return?
(a) $2000 on deposit for 2 a at a rate of 8%/a compounded annually
(b) $2000 on deposit for 2 a at a rate of 7.5%/a compounded semi-annually

11. Jackie has $2800 to put into a term deposit for 59 d at 7%/a. Then she puts the savings into another term deposit for 90 d at 8.5%/a.
(a) Calculate the interest earned after 59 d.
(b) How much money does Jackie have to reinvest for the next 90 d period?
(c) What is the amount after 149 d?

12. George has a $2000 Regular Interest Canada Savings Bond, that pays 8.75%/a for the first three years.
What is George's return on the bond if he cashes it in after 2 a?

13. The interest rate of a $500 Compound Interest Canada Savings Bond has been 9%/a for the past 5 a.
What was this bond worth after 3 a and 4 months?

14. Tara owns a $1000 Compound Interest Canada Savings Bond. For the first year, the interest rate is 9.5%/a, while for the second year, the rate is 10%/a.
(a) What is the interest earned in the first year?
(b) What is the amount after the first year?
(c) What is the interest earned in the second year?
(d) What is the total interest earned for the two years?

15. Geraldine has $9000 in savings to deposit for 3 a. She can buy Compound Interest Canada Savings Bonds that pay 9.5%/a compound interest each year, or she can put the money into a savings account that pays 9%/a with interest compounded annually.
Calculate the amount at the end of 3 a for each investment.

16. Lorraine has $10 000 in S41 Compound Interest Canada Savings Bonds. How much did she receive if she cashed the bonds in on November 1, 1988?

17. Bert MacDougall has $5000 in S40 Compound Interest Canada Savings Bonds. How much did Bert receive from the bonds if he cashed them in on November 1, 1987?

18. Calculate the cost of a $1000 treasury bill.

$P = A \div (1 + rt)$

(a) 360 d at 11% yield
(b) 180 d at 12.5% yield
(c) 90 d at 9.75% yield
(d) 90 d at 10.25% yield
(e) 182 d at 10.5% yield
(f) 181 d at 10.75% yield

19. Calculate the yield to the nearest tenth of a percent for each treasury bill.

	Value	Discounted Price	Term
(a)	$5 000	$4 911.19	60 d
(b)	$50 000	$47 538.42	180 d
(c)	$25 000	$23 260.26	260 d
(d)	$15 000	$14 769.36	60 d

C

20. Murray and Tiia have $45 000 on November 1. They would like to deposit the money into an account which pays good interest for two years, and then use the money to make a down payment on a house.
(a) Calculate the amount for a period of two years in each of the following.
  (i) savings account: 7%/a compounded quarterly
  (ii) term deposit: 11.5%/a compounded annually
  (iii) S41Compound Interest Canada Savings Bonds
(b) Which of the options in part (a) provides the greatest return?

MIND BENDER

If you could say one number every second, which of the following is the best estimate of how long you would take to count to one trillion?
1. 32 days
2. 32 years
3. 32 thousand years
4. 32 million years

8.8 BUYING AND OPERATING A CAR

Next to a house, a car is probably the most expensive purchase the average person will make. Selecting a car that is both suitable and affordable is an important decision. A car can be purchased outright, or it can be leased. The decision to lease or buy is usually determined by the person's ability to pay for the original purchase or lease, plus the upkeep. The purchase of a car is such a great expense that many people must borrow a portion of the cost or make lease payments. The money can be borrowed from the car dealer or from a financial institution.

When buying a car on an instalment plan, a person must ask two questions:

1. Can I afford the down payment (with or without a trade-in)?
2. Can I afford the monthly payments?

The table allows you to estimate the monthly payment.

Monthly Payment Factors			
15%		18%	
Number of Payments	Factor	Number of Payments	Factor
24	0.048 486 6	24	0.049 924 1
30	0.040 178 5	30	0.041 639 2
36	0.034 665 3	36	0.036 152 3
Monthly payment = (Amount to be financed) × (monthly payment factor)			

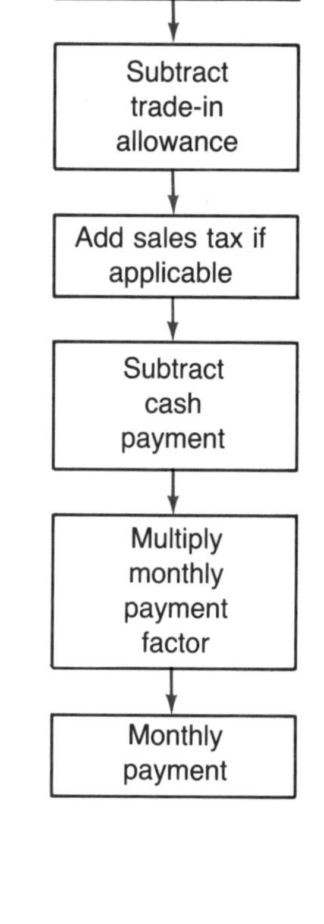

Example 1.
Carlos Rivera purchased a new Roadhandler for \$32 575. He was allowed \$12 500 on his GP300 trade-in, and financed the balance at 15%/a over 36 months. The sales tax is 8%. Calculate the instalment price.

Solution:

Price of car to be purchased	\$32 575.00
Trade-in allowance	12 500.00
Difference	\$20 075.00
Sales tax at 8% (0.08 × \$20 075.00)	1 606.00
Total cost (also 1.08 × \$20 075.00)	\$21 681.00
Cash payment	0.00
Amount to be financed	\$21 681.00
Monthly payment factor (15%, 36 months)	0.034 665 3
Approximate payment	\$751.58

∴ Carlos will make 36 payments of \$751.58 each.

The instalment price of the car is:

\$12 500.00 + (36 × \$751.58) = \$39 556.88

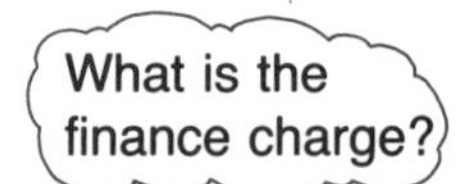

The trade-in value of a car decreases every year as the car gets older. This loss in trade-in value is called depreciation. Cars can depreciate at different rates, depending on such things as popularity and repair record. The following table gives the depreciation rate of an average new car.

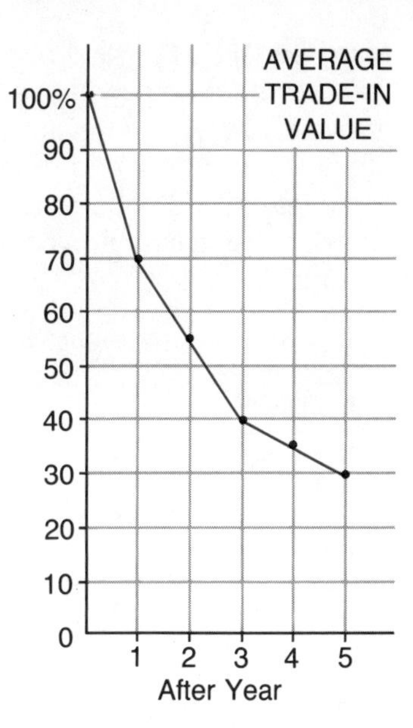

Average Automobile Depreciation and Trade-in Value					
Age of car in years	1	2	3	4	5
Depreciation (%)	30	45	60	65	70
Percent of selling price	70	55	40	35	30

The relation between depreciation and age can be displayed in a graph as shown.

Example 2.

Calculate the depreciation and the trade-in value of Carlos Rivera's Roadhandler after 3 a if the car cost \$32 575 when new.

Solution:

From the table or graph: after 3 a, the depreciation is 60% and the trade-in value is 40%.

Depreciation:		Trade-in value:	
Price of new car	\$32 575.00	Price of new car	\$32 575.00
Rate of depreciation: 60%	× 0.6	Rate of trade-in: 40%	× 0.4
Depreciation	\$19 545.00	Trade-in value	\$13 030.00

We can also find the trade-in value by subtracting \$32 575.00 (original cost)

− 19 545.00 (depreciation)

Trade-in value \$13 030.00

∴ After 3 a the value of the Roadhandler is \$13 030.

The costs of maintenance and operation depend on mileage and under what conditions a car is driven.

Example 3.

On July 1, the odometer reading on Celia's car was 31 328.2 and the gas tank was full. On July 15, Celia put 42.7 L of gasoline at a cost of \$32.25 into the tank. On July 31, the odometer reading was 32 416.7 and 55.3 L of gasoline at a cost of \$39.00 was needed to fill the tank.

Calculate (a) the fuel consumption in L/100 km and (b) the cost of gasoline per kilometre.

Solution:

(a) Distance travelled (km) = 32 416.7 km − 31 328.2 km = 1088.5 km

Gasoline (L) = 42.7 L + 55.3 L = 98.0 L

(b) Fuel consumption (L/100 km) = total litres ÷ (kilometres driven ÷ 100)

= 98.0 L ÷ (1088.5 km ÷ 100) ≐ 9.00 L/100 km

Cost = total cost ÷ kilometres driven

= (\$32.25 + \$39.00) ÷ 1088.5 ≐ 0.065

∴ The gasoline cost for the car was \$0.065/km.

Many people prefer to lease cars rather than buy them for a number of reasons: companies lease cars for employees; some self-employed people prefer to lease cars; and it is possible to avoid depreciation loss by leasing. The person who leases the car, the lessee, pays a monthly fee. At the end of the lease period, the lessee has the option to purchase the car at a predetermined price or to return the car to the leasing company. The predetermined price is called the lease end value (LEV). The following table shows several average monthly leasing fee factors.

Monthly Leasing Fee Factors			
Months of Lease	24	30	36
Lease factor @ 15%	0.048 888	0.040 683	0.035 237
LEV factor @ 15%	0.035 542	0.027 337	0.021 892
LEV	55%	46%	40%

Example 4.

Carlos Rivera would like to lease the Roadhandler, which costs \$32 575. The lease is financed at 15%/a and sales tax is 8%.
He wants to reduce the cost of the lease by trading in his old car for \$12 500.

(a) Calculate the LEV after 3 a.
(b) Calculate the monthly leasing fee and the total cost of leasing the car for 36 months.

Solution:

(a) After 3 a, the LEV is

$$\begin{aligned} 40\% \text{ of } \$32\,575.00 &= 0.4 \times \$32\,575 \\ &= \$13\,030.00 \end{aligned}$$

∴ After 3 a, Carlos has the option to purchase the car from the leasing company for \$13 030.

(b) Base monthly rental payment = (acquisition cost × lease factor) − (LEV × LEV factor)

Price of the car:	\$32 575.00
Less value of trade-in:	12 500.00
Acquisition cost of the car:	\$20 075.00

Acquisition cost × Lease factor:	\$20 075.00 × 0.035 237 =	\$707.38
LEV × LEV factor:	\$13 030.00 × 0.021 892 =	\$285.25
Base monthly rental payment:		\$422.13

Sales tax of 8% must be added so that each payment is

$$\$422.13 + 8\% \text{ of } \$422.13 = \$455.90$$

∴ The total cost of leasing, including tax, is 36 × \$455.90 or \$16 412.40.

Compare the costs of owning the Roadhandler after 3 a by
1. paying cash 2. buying with instalments
3. leasing and purchasing

EXERCISE 8.8

Unless otherwise stated, interest rates in this exercise are per annum.

B 1. Calculate the 8% sales tax on each of the following cars.

(a) Capello	\$18 655
(b) Scorcher	\$42 795
(c) Tracker	\$25 000
(d) Sumito	\$36 500
(e) Adio	\$49 600
(f) Arrow	\$54 000

2. Use the monthly payment factors to find an estimate of the monthly payment for the following car loans at 18%/a.

(a) \$45 850 for 24 months
(b) \$21 418 for 36 months
(c) \$34 500 for 30 months
(d) \$50 000 for 30 months
(e) \$35 500 for 24 months
(f) \$40 000 for 30 months

3. Find an estimate of the monthly payment for the following car loans at 15%/a.

(a) \$18 625 for 30 months
(b) \$28 350 for 24 months
(c) \$22 350 for 36 months
(d) \$25 675 for 30 months
(e) \$39 995 for 36 months
(f) \$36 500 for 30 months

4. Find an estimate of the monthly payment for the following car loans.

(a) \$24 260 at 18% for 24 months
(b) \$21 550 at 15% for 30 months
(c) \$25 800 at 15% for 36 months
(d) \$46 250 at 18% for 30 months
(e) \$18 480 at 15% for 36 months
(f) \$25 000 at 18% for 24 months

5. A car dealer has a special promotion: cars are sold for 10% down, and the balance is financed.
Calculate the balance to be financed on each of the following cars.

(a) Roadhandler	\$32 575
(b) Thundercloud	\$48 500
(c) Hawk	\$25 995

6. Find an estimate of the monthly payment for each car loan.

(a) a new X92 for \$29 500 with a trade-in allowance of \$13 550; sales tax is 8%, with financing at 15%/a for 36 months
(b) a used Vulcan costing \$19 500 with no trade-in; with a down payment of \$5000, and the rest is financed at 15%/a for 24 months; sales tax is 7%
(c) a new Starcruiser van costing \$41 850 with a Carsun trade-in worth \$23 150; the balance is financed at 18%/a for 36 months; sales tax is 8%

7. Erika Jensen buys a new Towncoup for \$52 500. She trades in a Hawk for \$18 500 and pays \$10 000 in cash, then finances the rest for 36 months at 18%. Sales tax is 8%.
Estimate the monthly payment.

8. Rene Rivard buys a new Skodar for \$38 500 and makes a down payment of \$16 500. He finances the balance at 15%/a over 30 months.
Estimate the monthly payment if sales tax is 8%.

9. Ruth Jarvis buys a new Quattro for \$44 500 and makes a trade-in worth \$16 000. She finances the balance with the dealer at 18%/a for 36 months. There is no sales tax.

(a) Find the monthly payment.
(b) What is the instalment price of the car?
(c) How much interest is paid in 36 months?

10. Tom Drew buys a new XM15 for \$49 400 and trades in his old car for \$15 000. A sales tax of 8% is paid. The balance is financed for 36 months at 18%/a.

(a) Find the monthly payment.
(b) What is the instalment price of the car?
(c) How much interest is paid in 36 months?

11. A new car is purchased for $37 850 with a trade-in allowance of $12 400. Sales tax of 8% is paid on the price difference. A cash payment of $5000 is made and the remainder is financed at 15%/a for 36 months.

(a) Find the monthly payment.
(b) What is the instalment price of the car?
(c) How much interest is paid on the loan?

12. Calculate the depreciation on each car.

(a) a $29 000 sedan after 2 a
(b) a $35 500 coupe after 4 a
(c) a $49 995 luxury car after 5 a
(d) a $39 500 imported car after 1 a

13. Calculate the trade-in value of each of the following cars.

(a) a $53 000 Roadster after 3 a
(b) a $27 500 Songbird after 2 a
(c) a $23 000 Devant after 1 a
(d) a $29 995 Quattro after 4 a

14. Margot bought a new car for $22 500 plus 8% sales tax. In the first year, she drove the car for 32 000 km. Complete the following table to determine Margot's cost of driving in the first year.

Item	Cost
Purchase of automobile	
Sales tax	
Regular service @ $29.95 every 8000 km	
Gasoline @ 52 cents/L 7 L/100 km	
Insurance $420.00/6 months	
14 car washes @ $5.50	
Total	

15. What is the cost of the following service and parts, plus 8% sales tax?

1 oil filter	$7.00
1 fuel filter	$4.95
1 air filter	$16.50
6 spark plugs	$15.90
Engine tune-up	$99.50
Change oil, lube service	$24.95

16. What is the total cost of the following service and parts, plus 7% sales tax?

1 upper steering rack @ $174
5 h labour @ $42/h
4 tires @ $125 each
wheel balance @ $5.50 each

17. What is the total cost of the following repair work if sales tax is 8%?

3 h @ $45.00/h remove and install new quarter panel

1 left front quarter panel	$275.00
1 L metallic paint	$22.50
1 L metallic paint primer	$11.25
miscellaneous shop materials	$10.00

18. Jessica Baer wants to lease a new Roadhandler, which costs $32 575, for a three-year period. The leasing company finances the car at 15%/a, and the provincial sales tax is 8%. There is no down payment.

(a) What is the LEV at the end of 3 a?
(b) What is the monthly leasing fee?
(c) What is the cost of leasing the car for 3 a?
(d) What is the total cost of the car if Jessica decides to buy the car at the end of the lease period?

19. Monty and Barbara trade in their Z28 for $10 800, which they use as a down payment towards a two-year lease on a new Stinger which costs $35 000. The leasing company finances the lease at 15%/a, and sales tax is 9%.

(a) What is the acquisition cost of the car?
(b) What is the LEV of the car after 2 a?

(c) Calculate the monthly leasing fee.
(d) What is the total cost of leasing the car for 2 a, including the value of the trade-in?

20. The price of a new car is $35 000 and sales tax is 9%.
(a) What is the cash price of the car, including sales tax?
(b) What is the monthly payment if the purchaser pays $15 000 down and finances the balance at 15% for 36 months?
(c) Calculate the LEV of the car after 3 a.
(d) What is the monthly leasing fee if the leasing company finances the car at 15% for 36 months? Assume a $15 000 down payment.
(e) Complete the following table using your answers to parts (a) to (d).

Cash price, including tax	
Instalment price of car	
Total cost of leasing + lease end purchase	

(f) What is the difference in cost, over a three-year period, between buying the car by payments and by leasing with the option to purchase at the end of the lease?
(g) What other important factors should be considered to determine whether leasing or purchasing is better?

21. The price of a new car is $47 500, and sales tax is 8%. Murray trades in a car that is worth $14 750.
(a) What is the cash difference that Murray pays to purchase the car, including sales tax?
(b) What is the monthly payment if Murray pays $5000 down plus the trade-in value, and finances the balance at 15%/a for 30 months?
(c) What is the monthly leasing fee if the leasing company finances the car at 15%/a for 30 months?

22. Juanita buys a car for $21 500 plus 8% sales tax. She makes a down payment of $6500. There is no trade-in.
(a) Calculate the finance charge at 15%/a over a 30 month period.
(b) Calculate the finance charge at 18%/a over a 30 month period.
(c) Compare the finance charges for interest rates of 15%/a and 18%/a over a period of 30 months.

23. Korman buys a car for $19 995 plus 9% sales tax. His trade-in is worth $4750.
(a) Calculate the finance charge at 15%/a over 24 months.
(b) Calculate the finance charge at 15%/a over 30 months.
(c) Calculate the finance charge at 15%/a over 36 months.
(d) Draw a graph to show how the finance charges in parts (a) to (c) vary with the time period.

MIND BENDER

It takes over eleven and one-half days for one million seconds to pass.
How long will it take for one billion seconds to pass?
How long will it take for one trillion seconds to pass?

8.9 PROBLEM SOLVING

1. A rectangular garden is surrounded by a ditch that is 3 m wide. You have two planks that are each 2.5 m long.
Can you arrange the planks in the following pattern to cross the ditch?

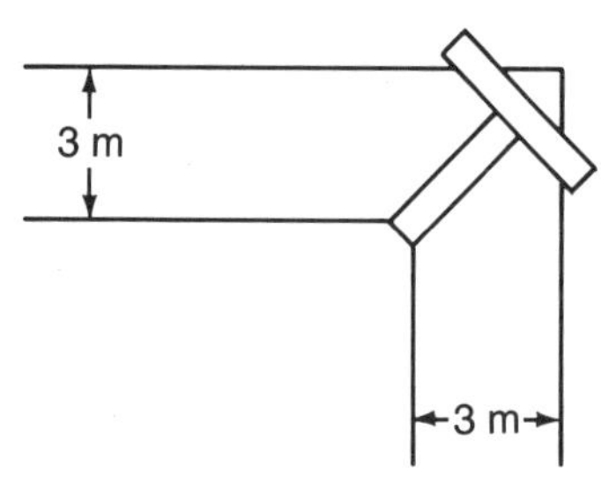

2. Water lilies double in area every 72 h. On June 1, there was exactly one water lily on the pond. At noon on August 1, the pond was completely covered with water lilies. On what day was the pond only half covered with water lilies?

3. Find the difference between the smallest possible six-digit number and the largest possible five-digit number if the digits in each number cannot be repeated.

4. Bill can get the tools and change the spark plugs in an eight-cylinder car in 21 min. He takes 17 min to do the same job for a six-cylinder car, and 13 min for a four-cylinder car.
How long does Bill take to get the tools?

5. The sum of three consecutive multiples of 5 is 390.
What are the numbers?

6. Which of the following wheels are turning clockwise?

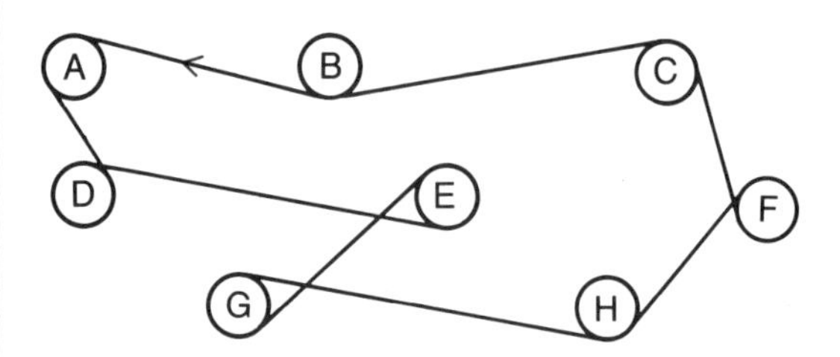

7. On what date would the last Friday of the month fall if the second Tuesday in the same month was the smallest possible two-digit odd number?

8. On what date in a leap year has one third of the year elapsed?

9. The tread on a new tire is 1 cm thick. The diameter of the tire is 63.5 cm. The tire is driven a distance of 97 656 km until all the tread is worn down.
State any assumptions you make to answer these questions. Use $\pi = 3.14$

(a) How far does the car travel during one complete rotation of the tire?
(b) How many times did the tire turn during the life of the tire?
(c) What is the average amount of tread which is left on the road from each turn of the tire?

10. There are 300 pages in a textbook. How many page numbers have the number 7 for at least one of the digits?

11. (a) How much does a ten-minute phone call to New York City cost during the most expensive time period?
(b) How much does a ten-minute phone call to New York City cost during the least expensive time period?

12. Place the numbers from 1 to 6 in the squares so that the product is 56 384.

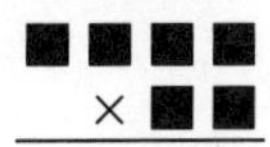

13. Barry, Elaine, Jane, Peter, and Richard work downtown in a large city. They want to meet on a street corner and then go for lunch together.

(a) At what corner should they meet in order to minimize the total distance they must travel?

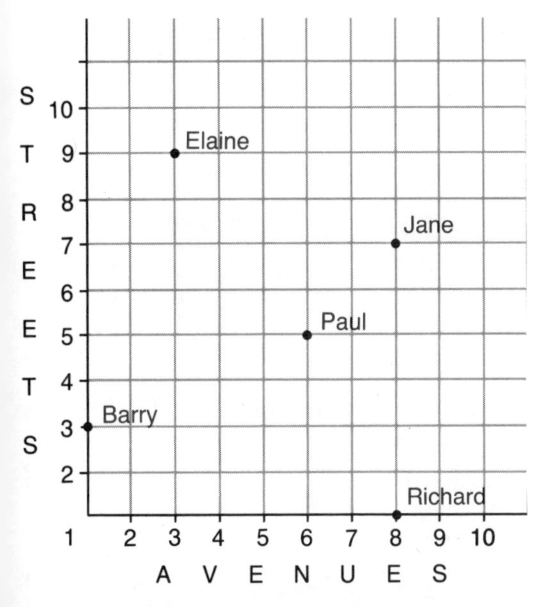

(b) At what corner should they meet in order to minimize the total distance travelled if they are to be joined by Heather who works at the corner of 2nd Avenue and 7th Street?

14. Steve took four subjects in one semester. His average mark for the four subjects was 78.5. His French mark was 72 and his mark in Mathematics was 84. His Science mark was 2 more than his English mark.
What were Steve's marks in English and Science?

CAREER

ENTREPRENEUR

An entrepreneur is a person who undertakes to start and run a business, assuming full responsibility for the risk associated with the business. In order to avoid bankruptcy, it is important for the entrepreneur to combine good ideas with a keen sense for business.

1. Larry Chin sells personal computers. For every 14 calls he makes, he sells one computer and printer.

(a) How many calls must he make in order to sell 15 computers?

(b) What is his net profit on one computer if he gives a gift worth \$5.50 with every call and makes \$544 profit on each computer he sells?

2. Marian operates a small business called The Housechecker. For \$5 plus \$1 per room, Marian will check your house twice per week while you are away and water your plants.
How much will Marian earn on the following calls?

Address	Name	Rooms
5 Elm	Jones	6
12 Ash	Scampi	8
96 Oak	Makie	7
88 Maple	Bowen	12
54 Birch	Robson	9

EXTRA

8.10 PUZZLES, PUZZLES, AND MORE PUZZLES

The following are examples of puzzles and problems that can be found in newspapers, magazines, aptitude tests, and interest tests.

EXERCISE

1. Find the missing number.

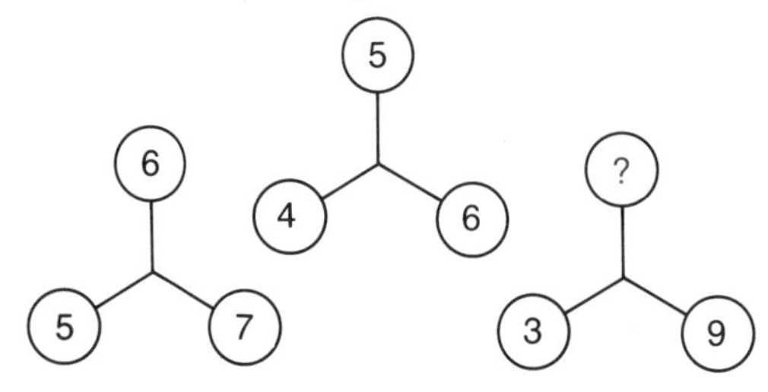

2. Which of the lines contains the scrambled letters of the name of a vegetable?

Q T B A T F R
E L M P T O O
C E T L U T E
I T I S A O N

3. Which word does not belong?

train sled bike van wagon

4. Which figure does not belong?

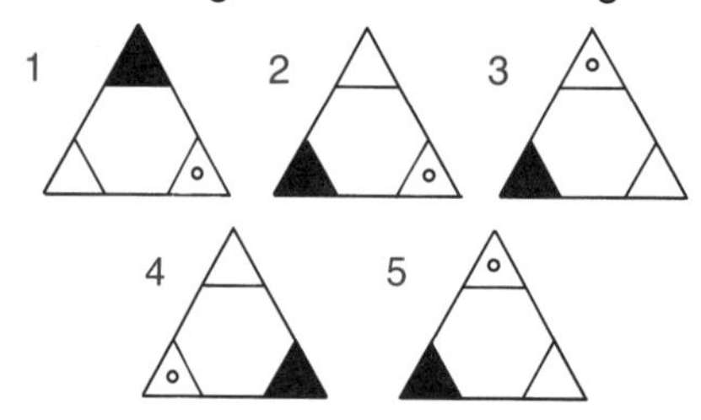

5. What letter is missing?

A D F I ■ N

6. Write the next two terms for each sequence.
(a) 20, 19, 17, 14, ■, ■
(b) 1, 3, 2, 4, 3, 5, ■, ■
(c) A, D, C, F, E, H, ■, ■

7. Find the missing letter.

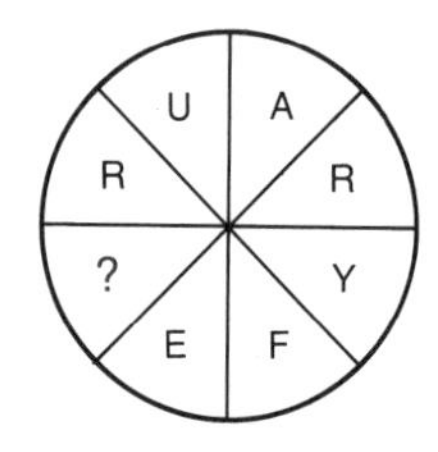

8. Find the missing number.

3	5	9	4
7	11	19	?

9. Which word does not belong?

TAIONOR
ELABATR
MRENVTO
OBMNITAA

10. Which figure does not belong?

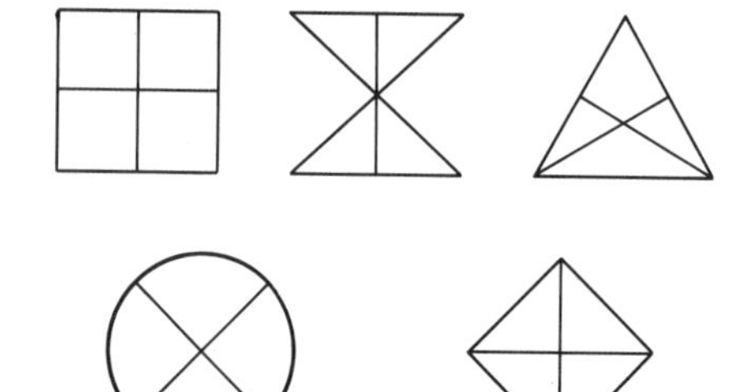

11. Which figure does not belong?

12. Find the missing number.

7	5	6
2	8	5
1	9	?

13. Which three faces do not belong?

14. Find the missing number.

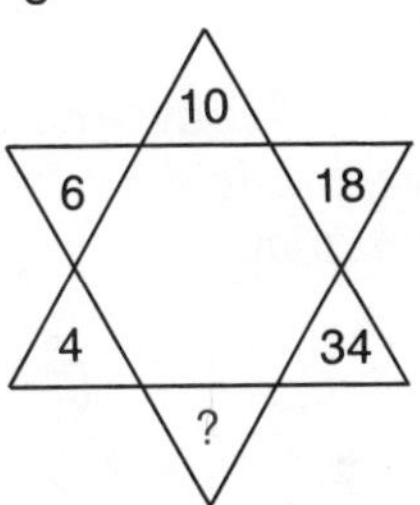

15. Find the missing letter.

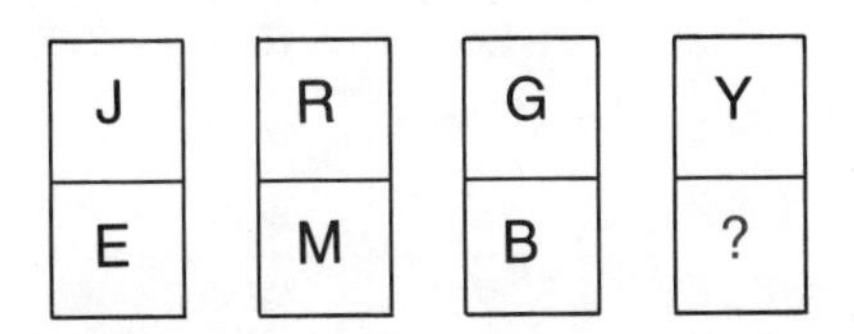

16. Which of the following is not a boy's name?

NPHEETS
LIAMESS
TOBRER
HICDRAR

17. Find the missing number.

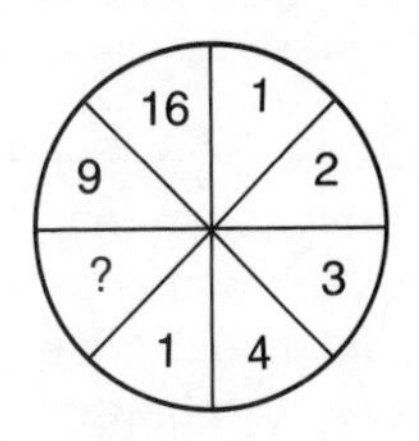

18. Which figure does not belong?

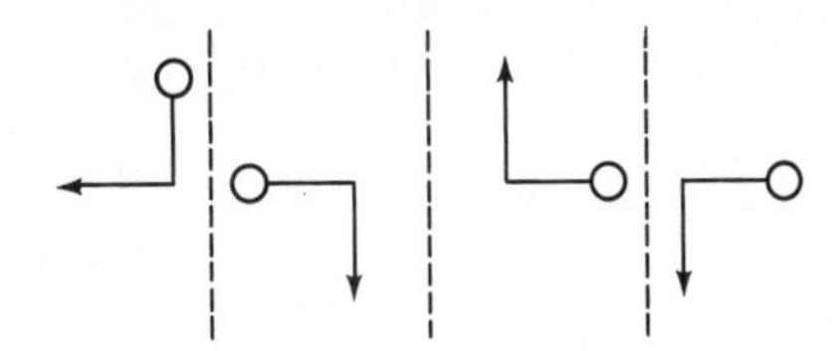

19. Ed and his brother Tim decided to meet at the park every Saturday at 08:00 to jog.
The first time Tim arrived at 08:10.
The second time Tim arrived at 08:20.
The third time he arrived at 08:40.
The fourth time he arrived at 09:20.
At what time did Tim arrive on the fifth Saturday?

20. Which two do not belong?

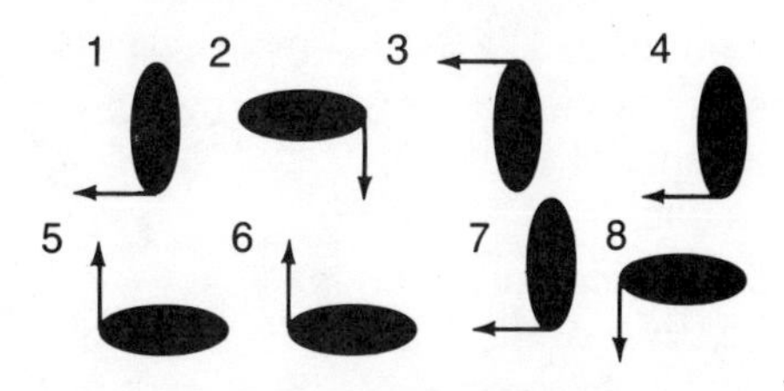

21. Find the missing number.

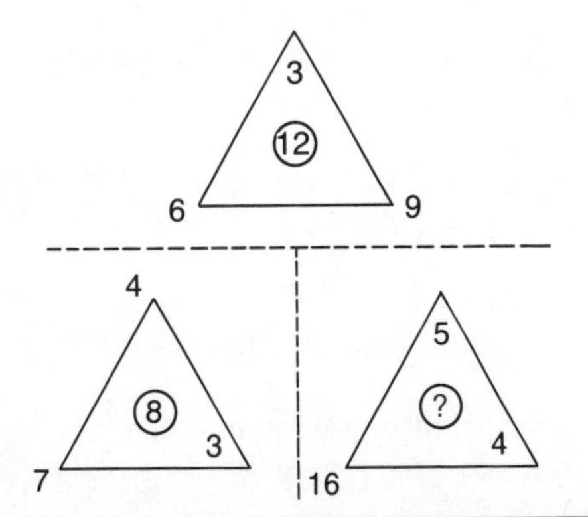

8.11 REVIEW EXERCISE

1. State definitions for the following terms.
(a) down payment
(b) principal
(c) instalment plan
(d) credit
(e) instalment price
(f) rate
(g) finance charge
(h) interest
(i) simple interest
(j) compound interest
(k) promissory note

2. Explain the difference between the following.
(a) savings account and term deposit
(b) Canada Savings Bonds and Treasury bills

3. What is the difference between financing a car and leasing a car?

4. Explain the meaning of the following terms related to operating a car.
(a) monthly payment factor
(b) lease end value
(c) depreciation
(d) acquisition cost
(e) upkeep
(f) insurance

5. A stereo costs $2890 plus 8% sales tax. Sonja buys the stereo for $1000 down and 12 instalments of $207.20.
Calculate each.
(a) the cash price, including tax
(b) the principal of the loan
(c) the total instalments
(d) the instalment price
(e) the finance charge

6. Use the table in section 8.2 to calculate the monthly payment.
(a) $2500 at 15% for 30 months
(b) $3600 at 17% for 12 months
(c) $4800 at 20% for 6 months
(d) $5000 at 18% for 36 months

7. An accordion costs $1200 plus 9% sales tax.
What is the monthly payment at 12% interest over 18 months, with a down payment of $300?

8. A VCR costs $456.95 cash or 30 monthly payments of $17.20 each. What is the effective annual interest rate?

9. A mini trail-bike costs $325 cash or 24 monthly payments of $18.30 each. What is the effective annual rate of interest?

10. (a) A loan of $2500 is paid at a rate of $85.00/month, which includes principal and interest. The loan bears interest at a rate of $\frac{3}{4}$%/month.
Draw up a schedule of payments for the first 5 months.
(b) A loan of $2500 is paid at a rate of $85.00/month on the principal, plus interest of $\frac{3}{4}$%/month on the unpaid balance.
Draw up a schedule of payments for the first 5 months.
(c) Compare the balances at the end of 5 months for each repayment schedule.

11. Fill in the missing quantities in the following simple interest table. Interest rates are per annum.

	I ($)	P ($)	r (%)	t
(a)	■	$500	6	30 d
(b)	■	$1250	12	2 months
(c)	$36.50	■	8	6 months
(d)	$5.20	■	10	30 d
(e)	$4.50	$1200	■	3 months
(f)	$39.45	$2400	■	60 d
(g)	$3.55	$3600	12	■ d
(h)	$4.25	$970	10	■ d
(i)	$168.75	$4500	9	■ months

12. Maurice deposits $10 000 into an account that pays 10%/a interest compounded semi-annually.
What is the amount after 2 a?

13. Ellen deposits $8500 into an account that pays 9%/a interest compounded quarterly.

(a) What is the amount after one year?
(b) How much interest is earned after one year?

14. Rose invests $4000 in a term deposit that pays $9.5%/a compounded semi-annually.
What is the amount at the end of 2 a?

15. Use the tables to find an estimate of the monthly payment for each car loan.

Monthly Payment Factor	
15%	
Number of Payments	Factor
24	0.048 486 6
30	0.040 178 5
36	0.034 665 3

Monthly Payment Factor	
18%	
Number of Payments	Factor
24	0.049 924 1
30	0.041 639 2
36	0.036 152 3

(a) $30 000 at 15% for 24 months
(b) $27 500 at 18% for 30 months
(c) $18 750 at 15% for 36 months
(d) $23 900 at 18% for 30 months
(e) $50 000 at 15% for 36 months

16. A new car is sold for $47 000. The customer's trade-in vehicle is worth $15 000 and the customer pays $5000 in cash. Sales tax is 8%, and the car is financed at 15%/a over 36 months.

(a) What is the principal of the loan?
(b) Calculate the monthly payment.
(c) What is the instalment price of the car?
(d) How much interest is paid in 36 months?

17. A new car is purchased for $27 500 with a trade-in allowance of $12 500. Sales tax, at a rate of 8%, is paid on the difference. A $3000 cash payment is made and the balance is financed at 15%/a for 30 months.

(a) Find the monthly payment.
(b) Find the total interest charged.

18. Calculate the depreciation on each of the following cars.

(a) a $27 500 sedan after 4 a
(b) a $33 499 coupe after 3 a

19. Munroe trades in his old car for $7800 which he uses as a down payment towards a three-year lease on his new Dart that costs $22 500. The leasing company finances the lease at 15%/a, and sales tax is 8%.

(a) What is the acquisition cost of the car?
(b) What is the lease end value of the car after 3 a?
(c) Calculate the monthly leasing fee.
(d) What is the cost of the car if Munroe decides to buy the car at the end of the lease period?

20. Joel buys a car for $29 995, plus 8% sales tax. He was given an allowance of $12 000 on his trade-in.

(a) Calculate the finance charge over 24 months at 15%/a interest.
(b) Calculate the finance charge over 30 months at 15%/a.
(c) Calculate the finance charge at 15%/a over a 36 month period.
(d) Draw a graph to show how the finance charge varies with time.

8.12 CHAPTER 8 TEST

1. Suzanne and Ronald purchase new furniture costing \$4320, including sales tax. They put \$1000 down, and pay the balance in 30 equal payments of \$133.40 each. Calculate the instalment price of the furniture.

2. Use the table at the right to calculate the monthly payment on a loan of \$2565.72 for 24 months at 15%/a.

Monthly Payments per \$100			
Annual Rate (%)	Number of Payments		
	18	24	30
12	6.098	4.707	3.875
13	6.145	4.754	3.922
14	6.192	4.801	3.970
15	6.238	4.849	4.019
16	6.286	4.898	4.066

3. The formula for the effective interest rate is

$$r = \frac{2NI}{P(n + 1)}$$

Find the effective interest rate on a loan of \$16 000, which is to be paid back in 30 monthly instalments of \$616 each.

4. What is the simple interest on a deposit of \$500 for 7 months at 18%/a?

5. Calculate the compound interest on a term deposit of \$5000 invested at 8.5%/a compounded semi-annually for 2 a.

6. Over a five-year period, a Roadster depreciates as follows.

Age of Car (a)	1	2	3	4	5
Depreciation (%)	30	45	60	65	70

What is the trade-in value of a three-year-old Roadster which cost \$18 600 when new?

7. Aurel Boivin buys a car for \$38 500 less his trade-in which was worth \$17 200. He finances the balance with the dealer at 15%/a over 24 months. Applicable sales tax is 8%. What is the monthly payment?

Monthly Payment Factor 15%	
Number of Payments	Factor
24	0.048 486 6
30	0.040 178 5
36	0.034 665 3

8.13 CUMULATIVE REVIEW FOR CHAPTERS 5 TO 8

1. Solve each pair of equations graphically.

(a) $y = 2x + 1$
$y = 3x - 2$

(b) $x + y = 4$
$x - y = 2$

2. Solve and check each pair of equations.

(a) $2a - b = 5$
$a - b = 3$

(b) $5x + 3y = -15$
$3x - 2y = 10$

(c) $9s + 7t = 21$
$s + 12t = 36$

(d) $2x + 3y = 7$
$x + 2y = 3$

(e) $4a + b = 0$
$6a - b = 5$

3. Solve each pair of equations by substitution.

(a) $x - y = 4$
$x - 2y = 1$

(b) $4a + b = 9$
$3a - 2b = 4$

4. The sum of two numbers is 40. Their difference is 12.
Find the numbers.

5. The Applewood Company makes baseball caps to sell as souvenirs at rock concerts. The cost of making the caps is described by the formula
$y = 4x + 30\,000$
where y is the total cost and x is the number of caps.
The income from selling the caps is given by the formula
$y = 10x$
where y is the income and x is the number of caps sold.
How many caps must the Applewood Company sell to break even?

6. The Haniford Company makes fireworks. The company makes 10 000 Red Star rockets each week. A sample of 50 rockets were tested and one was found to be defective.
How many of the week's production can you predict will be defective?

7. A national survey was conducted to determine the reasons why people change jobs. There were 200 people surveyed. The table gives the results of the survey.

Reason	Number
Lack of promotion	94
Inadequate salary	12
Bored	12
Lack of recognition	52
Unhappy with boss	30

Display this information on a graph.

8. The table gives the approximate number of Canadian geese feeding at Culver City over a one-week period during the migration season.

Day	S	M	T	W	T	F	S
Number	80	120	230	200	150	300	280

Display this information in a graph.

9. Determine the mean, median, and mode of the following sets of data.

(a) 10, 11, 12, 12, 14, 15, 13, 11, 10, 11, 14
(b) 20, 25, 25, 23, 22, 22, 21, 20, 22, 25, 23, 25

10. The following are the marks earned by students on a scuba diving test.

54, 87, 98, 63, 55, 64, 77, 63, 63, 65, 59, 96, 80, 72, 64, 61, 56, 88, 75, 62, 57, 72, 82, 95, 56, 60, 59, 75, 60

Display this information in a box-and-whisker plot.

11. Complete the following pay slip.

REGULAR EARNINGS:		\$256.80
OVERTIME EARNINGS:		78.52
GROSS EARNINGS:		(a)
DEDUCTIONS:		
C.P.P.:	\$5.95	
U.I.:	\$6.54	
INC. TAX:	\$54.00	
TOTAL:	(b)	
NET PAY:		(c)

12. The circle graph shows a budget for a student living away from home at college. The student has $950 for living expenses each month.
How much is spent on each item?

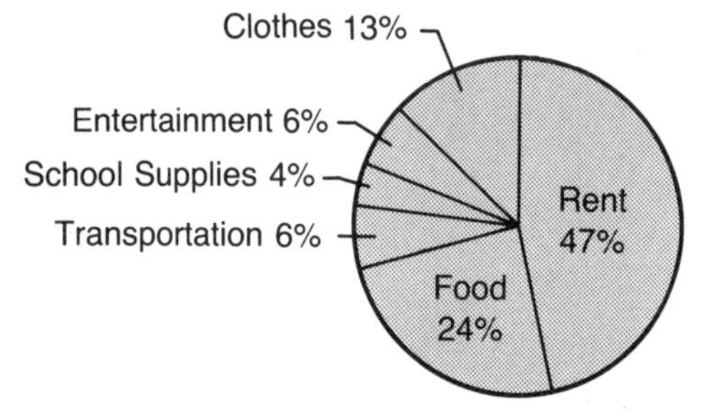

13. Jerome Pearcy's personal cheque book balance is $408.65. The closing balance on his monthly account statement is $433.89. Since the statement date, Jerome made a deposit of $350 and wrote cheques totalling $375.24. These cheques are not listed on the bank statement.
Prepare a reconciliation to determine if Jerome's balance agrees with the bank statement.

CLOSING BALANCE THIS STATEMENT	(a)
DEPOSITS MADE AFTER STATEMENT CLOSING DATE	(b)
SUB TOTAL	(c)
LESS OUTSTANDING CHEQUES	(d)
EQUALS	(e)
CHEQUE BOOK BALANCE	(f)
DIFFERENCE (IF ANY)	(g)

14. Fill in the missing quantities in the following simple interest table. Interest rates are per annum.

	I	P	r	t
(a)	$125.00	$2500.00	■	3 months
(b)	$425.21	$97 000	5%/a	■
(c)	$355.07	$24 000	■	90 d
(d)	■	$55 000	9%/a	1 a
(e)	■	$25 000	10%/a	6 months

15. Daniel deposits $2000 into an account that pays 9%/a interest compounded semi-annually.
What is the amount after 3 a?

16. Jessie puts $5000 into a term deposit that pays 12%/a compounded quarterly. What is the amount after 18 months?

17. Nancy buys $4000 of Compound Interest Canada Savings Bonds which pay 9.5%/a interest.

(a) What amount will Nancy receive if the bonds are cashed at the end of 3 a and the interest rate does not change?
(b) How much interest is earned?

18. Use the following chart to find an estimate of the monthly payment for each car loan.

Monthly Payment Factor	
15%	
Number of Payments	Factor
24	0.048 486 6
30	0.040 178 5
36	0.034 665 3

(a) $24 657 at 15%/a for 30 months
(b) $18 500 at 15%/a for 24 months
(c) $16 750 at 15%/a for 36 months
(d) $28 000 at 15%/a for 24 months
(e) $21 750 at 15%/a for 36 months
(f) $32 800 at 15%/a for 24 months
(g) $87 250 at 15%/a for 36 months
(h) $49 000 at 15%/a for 30 months

19. (a) Compute the monthly payments on the following car loans.
(i) $20 000 for 24 months at 15%/a
(ii) $15 000 for 30 months at 15%/a
(iii) $13 333 for 36 months at 15%/a

(b) Which is the lowest monthly payment?
(c) Which is the highest?

RATIO, RATE, PROPORTION, AND SCALE

REVIEW AND PREVIEW TO CHAPTER 

THE RULE OF THREE

The Rule of Three is a series of three steps that can be used to solve a familiar type of problem.

Example 1.
The Ambassador Bridge, the world's longest international suspension bridge, was opened in 1929 and joins Windsor to Detroit. As part of a government survey, Jill counts the number of trucks that cross the Ambassador Bridge from Detroit to Windsor in three hours. She counts 273.
(a) At this rate, how many trucks would cross in one day?
(b) How many would cross in one year?

Solution:
(a) The steps in the Rule of Three are outlined below.

Statement of Fact: In 3 h there are 273 trucks.

Reduce to 1: In 1 h there are $\frac{273}{3}$ or 91 trucks.

Multiply: In 24 h there would be 24×91 or 2184 trucks.

At this rate, 2184 trucks will cross the bridge in a day.

(b) In one year there would be 365×2184 or approximately 800 000 trucks crossing the bridge.

In some problems you will find it convenient, when reducing to 1, to leave the answer as a fraction.

Example 2.
A bicycle race had 18 laps. Theresa completed the first 4 laps in 51 min. At this rate how long will she take to complete the race?

Solution:

Statement of Fact: 4 laps take 51 min.

Reduce to 1: 1 lap takes $\frac{51}{4}$ min.

Multiply: 18 laps take $18 \times \frac{51}{4} = 229.5$ min.

Theresa will take 3 h and 49.5 min to complete the race.

EXERCISE 1

1. An express train travels 475 km in 5 h.
How far will the train travel in 8 h at the same rate?

2. Noreen bought 200 shares of Gold Dust stock and earned a total of $60 in dividends.
How much could she have earned if she had bought 350 shares of Gold Dust stock?

3. City planners predict that in a new housing development there will be seven school-aged children in every ten new homes.
How many school-aged children can be expected in a development with 280 new homes?

4. A 40 cm long gold chain costs $102.40.
How much will a 35 cm gold chain cost?

5. It takes Francine 3 h to cross-country ski 25.5 km.
How far will she ski in 5 h at the same rate?

6. A dozen tournament tennis balls cost $31.20.
How much would fifty balls cost?

7. If you run for two hours, you will use the same amount of energy needed to boil water for 123 cups of coffee.
How much water could you boil with the energy used to run for twenty minutes?

8. While monitoring traffic at the Main and Park intersection during rush hour, Bianca counted 220 vehicles passing through the intersection in four minutes.
How many vehicles will pass through the intersection in 90 min?

9. A 23 kg watermelon costs $5.75.
How much does a 57 kg watermelon cost?

10. Alfredo got 5 hits in the first 9 games of the baseball season.
At this rate, how many hits will he get in 162 games?

11. During the first three days of the World's Fair there were 186 000 visitors.
If the fair lasts 14 d, how many visitors can be expected?

MENTAL MATH

EXERCISE 2

Find the value of the unknown.

1. $x + 7 = 15$
2. $y + 9 = 21$
3. $s - 4 = 11$
4. $t - 2 = 32$
5. $4m = 36$
6. $7w = 56$
7. $\frac{x}{2} = 5$
8. $\frac{x}{3} = 4$
9. $x + 1.4 = 8$
10. $a - 2.2 = 5$
11. $8d = 40$
12. $t + 13 = 88$
13. $5y = 85$
14. $m - 63 = 100$
15. $\frac{1}{2}t = 10$
16. $\frac{m}{4} = 1$
17. $7t = 140$
18. $t + 15 = 65$
19. $r - 3.6 = 4$
20. $x + 45 = 80$
21. $x - 17 = 20$
22. $8t = 4$
23. $\frac{t}{4} = 2$
24. $\frac{m}{5} = 3$
25. $9 + k = 30$
26. $20k = 5$
27. $w + 3.5 = 7$
28. $s - 11 = 90$

9.1 RATIOS

The Quigley family has a flower stall at the market. Every spring they sell packages of tulip bulbs. The most popular package is one that contains 4 red tulip bulbs and 2 yellow tulip bulbs. We can describe this package mathematically as

$$\frac{\text{Number of red bulbs}}{\text{Number of yellow bulbs}} = \frac{4}{2}$$

We say that the number of red bulbs and the number of yellow bulbs are in a ratio of 4 to 2, or in colon form 4 : 2.

We can write the ratio of red bulbs to yellow bulbs in four ways.

Word form	4 to 2	Colon form	4 : 2	Fraction form	$\frac{4}{2}$	Division form	$4 \div 2$

A ratio is the comparison, by division, of two numbers with the same units.

We express a ratio in lowest terms in the same way we express fractions in lowest terms.

$$\frac{4}{2} = \frac{4 \div 2}{2 \div 2} = \frac{2}{1} \quad \text{or} \quad 2:1$$

To find equal, or equivalent ratios, multiply or divide each term by the same number.

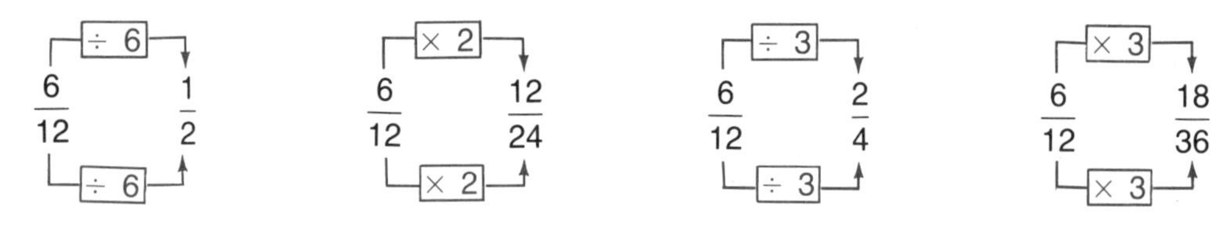

$$\frac{6}{12} \xrightarrow{\div 6} \frac{1}{2} \qquad \frac{6}{12} \xrightarrow{\times 2} \frac{12}{24} \qquad \frac{6}{12} \xrightarrow{\div 3} \frac{2}{4} \qquad \frac{6}{12} \xrightarrow{\times 3} \frac{18}{36}$$

Example.
Dalton won 7 games and lost 5. Marjory won 11 games and lost 8.
Who had the better wins to losses ratio?

Solution:
To compare, express each ratio with the same denominator.

$$\frac{7}{5} = \frac{7 \times 8}{5 \times 8} = \frac{56}{40} \qquad \frac{11}{8} = \frac{11 \times 5}{8 \times 5} = \frac{55}{40}$$

Since $\frac{56}{40} > \frac{55}{40}$,

Dalton had the best wins to losses ratio.

THREE-TERM RATIOS

A bill was passed by the members of parliament in the House of Commons. For every 5 yes votes there were 3 no votes and 2 abstentions. We can write this as a three-term ratio as follows.

5 : 3 : 2

We write three-term ratios only in colon form.

EXERCISE 9.1

A 1. State ratios for the following.

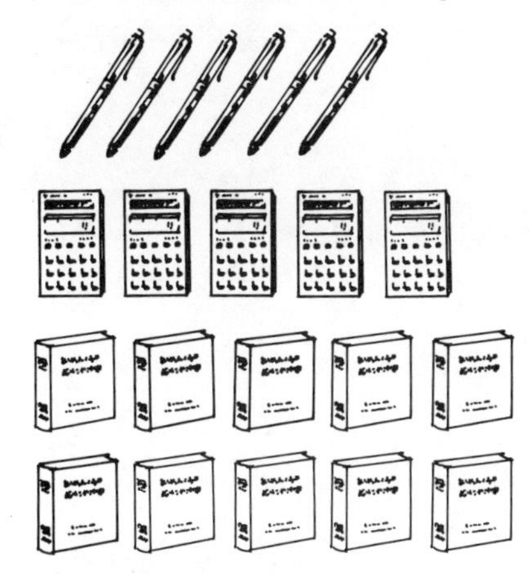

(a) calculators to pens
(b) pens to calculators
(c) books to pens
(d) books to calculators
(e) pens to books
(f) pens to calculators to books
(g) calculators to books to pens
(h) calculators to pens to books

2. Express the following ratios in their simplest form.

(a) 4 : 2 (b) 18 : 3
(c) 5 : 35 (d) 12 : 9
(e) 15 : 6 (f) 27 : 18
(g) 48 : 16 (h) 100 : 40

B 3. Express the following ratios in simplest form.

(a) 36 : 18 : 9 (b) 8 : 4 : 4
(c) 13 : 26 : 52 (d) 48 : 16 : 24
(e) 42 : 14 : 28 (f) 18 : 45 : 72
(g) 8 : 8 : 20 (h) 76 : 38 : 19
(i) $\frac{4}{6}$ (j) $\frac{10}{12}$ (k) $\frac{15}{12}$ (l) $\frac{21}{35}$
(m) $\frac{14}{7}$ (n) $\frac{56}{16}$ (o) $\frac{15}{90}$ (p) $\frac{5}{25}$

4. Write two equivalent ratios for each.

(a) 3 : 5 (b) 1 : 3 (c) 4 : 3
(d) 6 : 7 (e) 2 : 5 (f) 2 : 3
(g) 1 : 3 : 2 (h) 2 : 4 : 5 (i) 3 : 8 : 2

5. Find the missing term in each ratio.

(a) 4 : ■ = 5 : 10
(b) $\frac{3}{5} = \frac{18}{\blacksquare}$ (c) $\frac{2}{3} = \frac{\blacksquare}{15}$
(d) 3 : 7 = ■ : 35 (e) ■ : 20 = 2 : 5
(f) 1 : 2 : 3 = 5 : ■ : ■
(g) 3 : ■ : 7 = 12 : 20 : ■

6. State which ratio is greater and explain your choice.

(a) 2 : 3 or 5 : 6 (b) 3 : 4 or 5 : 8
(c) 4 : 5 or 3 : 4 (d) 2 : 5 or 3 : 4
(e) $\frac{2}{7}$ or $\frac{3}{10}$ (f) $\frac{3}{8}$ or $\frac{2}{9}$
(g) 5 : 7 or 6 : 8 (h) 2 : 5 or 3 : 7

7. A bag of nuts contains 5 parts peanuts, 2 parts cashews, and 3 parts pecans.

(a) What is the ratio of cashews to peanuts to pecans?
(b) What fraction of the mixture is pecans?
(c) What fraction of the mixture is peanuts?
(d) What fraction of the mixture is cashews?

C 8. When comparing quantities in a ratio, use the same units.
Express each of the following as a ratio.

(a) 15 cm to 2.5 m (b) 3 L to 400 mL
(c) 5 m to 23 cm (d) 3 kg to 120 g
(e) 48 cm^2 to 3000 mm^2
(f) 90 000 m^2 to 4 ha

MIND BENDER

Identify the pattern and complete the table.

7	7	9	■	■	■
4	5	6	6	7	5
3	2	■	1	■	4
1	3	■	■	4	■

9.2 USING RATIOS

Brenda and Larry want to buy a canoe for the summer. The canoe costs \$3500. Brenda will use the canoe 4 d/week while Larry will have it 3 d/week. They divide the cost of the canoe between them in a ratio of 4 : 3.
How much should each of them pay?

We can look at this problem in two ways.

Since the cost will be divided in a ratio of 4 : 3, we can think of the cost as a total of $4 + 3$ or 7 shares.

Brenda's share is $\frac{4}{7}$ of (\$3500) $= \frac{4}{7} \times \$3500 = \$2000$

Larry's share is $\frac{3}{7}$ of (\$3500) $= \frac{3}{7} \times \$3500 = \$1500$

The second way is to use equivalent ratios. The ratio 4 : 3 is equivalent to $4x : 3x$

If we let $4x$ be Brenda's share and $3x$ be Larry's share, then

$$4x + 3x = 3500$$

and

$$7x = 3500$$
$$x = 500$$

Brenda's share is $4x$ or $4 \times \$500 = \$2000$
Larry's share is $3x$ or $3 \times \$500 = \$1500$

Example.
Franco, Jennifer, and Lola bought one pair of season's tickets for the Blue Jays games. They divided the cost in a ratio of 5 : 7 : 6, according to the number of games each plans to attend. The total cost of the tickets is \$3060.
How much should each pay?

Solution:
METHOD I

Think of the cost as a total of $5 + 7 + 6$ or 18 shares.
Then each share is worth

$\frac{\$3060}{18}$ or \$170

Franco has 5 shares: $5 \times \$170 = \$850$
Jennifer has 7 shares: $7 \times \$170 = \$1190$
Lola has 6 shares: $6 \times \$170 = \$1020$

METHOD II

Use equivalent ratios.
$5 : 7 : 6 = 5x : 7x : 6x$
Then

$$5x + 7x + 6x = 3060$$
$$18x = 3060$$
$$x = 170$$

Franco's share: $5x$ or $5(170) = \$850$
Jennifer's share: $7x$ or $7(170) = \$1190$
Lola's share: $6x$ or $6(170) = \$1020$

EXERCISE 9.2

B

1. Ivan and Ronald were hired for the summer to patrol Bass Lake. Ivan worked 4 d/week and Ronald worked 3. The total pay for eight weeks of work is $6020.
How much should each of them receive?

2. A pipeline is 45 km long. It is to be cut into two pieces in a ratio of 4 : 1.
How long is each piece?

3. Olivia and Carmella bought the Action Fitness Centre. The cost of the business was $880 000. Olivia and Carmella shared the cost in a ratio of 6 : 5.
How much did each of them pay?

4. There are three partners in The Treasure Find company. They share profits in a ratio of 2 : 3 : 5. Last year the profits were $256 000.
How much should each partner receive?

5. A batting average in baseball is really a ratio of the number of hits to the number of times at bat. This ratio is written as a decimal to the nearest thousandth.

(a) In 1927, the year he hit 60 home runs, Babe Ruth got 192 hits in 540 times at bat.
What was his batting average?

(b) In major league baseball the average batting average has been 0.260 since the beginning of the modern era.
What percent of the time will a player batting 0.260 get a hit?

(c) The last player to hit over 0.400 was Ted Williams. In 1941 he got 185 hits in 456 times at bat.
What was his batting average?

6. An eccentric millionaire divided his wealth among his four children in a ratio according to their ages.
If they are 21, 23, 26, and 30 a old, and his fortune is $156 000 000, how much did each receive?

7. Four friends decided to buy lottery tickets and share the winnings according to the number of tickets each bought. Ho bought 3 tickets, Marvin bought 4, Al bought 5, and Margaret 2. One of the tickets won $1 000 000.
How much should each receive?

EXTRA

GOLD IN RATIOS

The amount of gold in jewellery is given using carats (K). For example, 24 K gold coin means that the coin is pure gold. However, gold is a very soft metal so that jewellery is rarely made out of pure gold. Gold is alloyed with other metals to make it stronger.

A necklace which is 12 K gold has 12 parts gold and 12 parts of another metal. We can also describe 12 K as a ratio,

$$\frac{\text{mass of gold in necklace}}{\text{mass of necklace}} = \frac{12}{24} = \frac{1}{2}$$

1. A bracelet is marked 18 K gold. The mass of the bracelet is 100 g.
What is the mass of the gold in the bracelet?
2. A coin is marked as 10 K gold. The mass of the coin is 75 g.
How much gold is in the coin?

9.3 PROPORTIONS AND TWO-TERM RATIOS

Last year John Evans entered 12 riding events and won prize money in 9 of them. This year he entered 20 events and won money in 15 of them. John's ratio of wins to events last year was 9 : 12. This year his ratio of wins to events is 15 : 20. When these two ratios are written in simplest form, they are identical or equivalent.

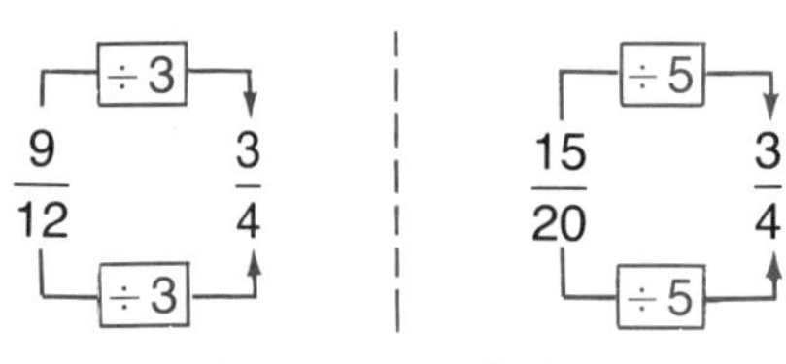

We can say that

$$\frac{9}{12} = \frac{15}{20}$$

This statement, showing that two ratios are equal, is called a proportion. In this case, the ratio of wins to events is said to be proportional.

We showed that the two ratios $\frac{9}{12}$ and $\frac{15}{20}$ are equal

by using equivalent fractions, or ratios. Another way to check that ratios are equal is to use cross products.

$\frac{9}{12} \, ? \, \frac{15}{20}$ $\quad$ $\frac{9}{12} \times \frac{15}{20}$ $\quad$ and $\quad$ $9 \times 20 = 15 \times 12$

$$180 = 180$$

Since the cross products are equal, the ratios are also equal.

We use equivalent fractions and cross products to solve many problems involving proportions.

Example 1.

A survey found that 3 out of 4 students bought yearbooks last year. This year there are 640 students in the school.
How many yearbooks should be sold this year?

Solution:

Write a proportion.

$$\frac{\text{sales last year}}{\text{students last year}} = \frac{\text{sales this year}}{\text{students this year}}$$

Let n represent the sales this year and solve for n.

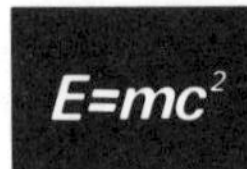

$$3 : 4 = n : 640 \quad \text{or} \quad \frac{3}{4} = \frac{n}{640}$$

METHOD I: Equivalent Fractions

The lowest common denominator of 4 and 640 is 160.

$$\frac{3}{4} = \frac{3 \times 160}{4 \times 160} = \frac{480}{640}$$

$$\frac{480}{640} = \frac{n}{640}$$

$$n = 480$$

METHOD II: Rules for Equations

Multiply both sides by 640.

$$\frac{3}{4} = \frac{n}{640}$$

$$640 \times \left(\frac{3}{4}\right) = \left(\frac{n}{640}\right) \times 640$$

$$480 = n$$

$$n = 480$$

METHOD III: Cross Products

$$\frac{3}{4} \times \frac{n}{640}$$

$$1920 = 4n$$

$$n = 480$$

There should be 480 yearbooks sold.

Example 2.

To determine the trout population of a lake, 300 trout were tagged and then released. A month later, a sample of trout were caught, and 25 out of the 200 in the sample had the tag. How many trout were in the lake?

Solution:

Write a proportion. Let n be the number of trout in the lake.

$$\frac{\text{number of trout in the lake}}{\text{number of tagged trout in the lake}} = \frac{\text{number of trout in the sample}}{\text{number of tagged trout in the sample}}$$

E=mc²

$$n : 300 = 200 : 25 \quad \text{or} \quad \frac{n}{300} = \frac{200}{25}$$

METHOD I: Equivalent Fractions

The lowest common denominator of 25 and 300 is 300.

$$\frac{200}{25} = \frac{200 \times 12}{25 \times 12} = \frac{2400}{300}$$

$$\frac{n}{300} = \frac{2400}{300}$$

$$n = 2400$$

METHOD II: Rules for Equations

$$\frac{n}{300} = \frac{200}{25}$$

Multiply both sides by 300.

$$300 \times \left(\frac{n}{300}\right) = \left(\frac{200}{25}\right) \times 300$$

$$n = 2400$$

METHOD III: Cross Products

$$\frac{n}{300} \times \frac{200}{25}$$

$$25n = 60\,000$$

$$n = 2400$$

There are approximately 2400 trout in the lake.

EXERCISE 9.3

A 1. Which pairs of ratios are equal?

(a) 3 : 5, 12 : 20 (b) 3 : 2, 6 : 5

(c) $\frac{8}{6}, \frac{20}{15}$ (d) $\frac{10}{12}, \frac{15}{18}$

(e) 14 : 4, 21 : 8 (f) 8 : 12, 12 : 18

(g) $\frac{21}{28}, \frac{27}{36}$ (h) $\frac{12}{20}, \frac{16}{28}$

(i) 28 : 35, 32 : 40 (j) 5 : 25, 6 : 30

B 2. Find the missing term.

(a) $\frac{x}{9} = \frac{5}{3}$ (b) $\frac{4}{7} = \frac{n}{21}$

(c) $\frac{5}{3} = \frac{n}{6}$ (d) $\frac{4}{t} = \frac{1}{3}$

(e) $\frac{4}{9} = \frac{8}{m}$ (f) $\frac{6}{5} = \frac{t}{10}$

(g) x : 4 = 6 : 12 (h) 10 : 3 = m : 6

(i) 2 : 11 = 4 : y (j) 8 : 5 = 20 : r

3. Ibrahim's punch recipe calls for 8 L of ginger ale to be mixed with 3 L of orange juice.
How much orange juice should be mixed with 12 L of ginger ale?

4. Shandra sold 8 cars in 15 d.
At this rate, how many cars could she expect to sell in 210 d?

5. Three buses are needed to take 8 softball teams from a hotel to the tournament diamonds.
How many buses are needed for 13 teams?

6. Seven out of ten college students prefer cross-trainer shoes.
If there are 840 students, how many will prefer cross-trainers?

7. Ella's ratio of hits to times at bat is 7 : 25.
If she has 84 hits so far, how many times has she been at bat?

8. Giovanni teaches tourists how to pan for gold. The ratio of gold nuggets to fool's gold nuggets is 2 : 5. On one day the tourists found 40 gold nuggets.

(a) How many fool's gold nuggets were found?

(b) What percent of the nuggets found were gold?

9. Three large pizzas will serve 10 people. How many pizzas are needed to serve 24 people?

10. In a sports store the school discount on a pair of $120 basketball shoes is $9.60. What is the school discount on a $90 purchase from this store?

C 11. Mustard and relish are mixed to make a special sausage sauce. The ratio of mustard to relish is 5 : 8.

(a) If 25 L of mustard are used to make the sauce, how much relish is needed?

(b) If 44 L of relish are used, how much mustard is needed?

(c) In 91 L of the sauce, how much is mustard?

(d) How much relish is needed to make 104 L of the mixture?

MIND BENDER

In the figure below there can be found a special square. The corners of the square are dots. Exactly two of the dots are in the rectangle and exactly two of the dots are in the triangle. One dot is in the ellipse.
Find the square.

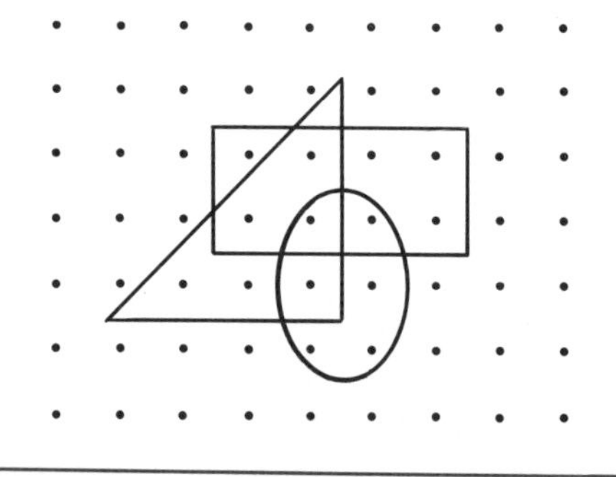

9.4 PROPORTIONS AND THREE-TERM RATIOS

Each year the Klondike Days Fair has a prospector's baked bean competition. Beans must be cooked over an open fire. Badland Jack's recipe calls for navy beans, molasses, and brown sugar in a ratio of 16 : 3 : 2. To serve all the judges and other contestants Jack figures he will need 40 L of beans.
How much molasses and brown sugar will he need?

We can write a proportion using m for the amount of molasses and s for the amount of brown sugar needed.

$16 : 3 : 2 = 40 : m : s$

Step 1. To find m, solve the proportion

$16 : 3 = 40 : m$ or $\frac{16}{3} = \frac{40}{m}$

$16m = 120$

$m = 7.5$

Step 2. To find s, solve the proportion

$16 : 2 = 40 : s$ or $\frac{16}{2} = \frac{40}{s}$

$8 = \frac{40}{s}$

$8s = 40$

$s = 5$

Jack will need 7.5 L of molasses and 5 L of brown sugar.

EXERCISE 9.4

A 1. How are the first and second ratios related?

(a) $2 : 5 : 9 = 10 : 25 : 45$
(b) $14 : 16 : 8 = 7 : 8 : 4$
(c) $3 : 1 : 2 = 9 : 3 : 6$
(d) $20 : 40 : 80 = 2 : 4 : 8$
(e) $8 : 10 : 6 = 12 : 15 : 9$

2. Which pairs of three-term ratios are equal?

(a) 3 : 2 : 1, 6 : 4 : 2
(b) 4 : 1 : 5, 12 : 4 : 15
(c) 10 : 8 : 6, 5 : 4 : 3
(d) 20 : 25 : 40, 4 : 6 : 8
(e) 18 : 9 : 12, 6 : 3 : 4
(f) 40 : 60 : 10, 5 : 6 : 1

B 3. Find the missing terms.

(a) $3 : 1 : 2 = 12 : x : y$
(b) $4 : x : 9 = y : 14 : 18$
(c) $8 : 9 : m = 12 : n : 24$
(d) $r : s : 24 = 5 : 4 : 6$
(e) $8 : 6 : 10 = 12 : a : b$
(f) $30 : s : 80 = 6 : 12 : t$
(g) $15 : x : 30 = 6 : 8 : y$

4. Sand, gravel, and cement are mixed in a ratio of 5 : 10 : 3 to make concrete.
If you used 35 shovels full of sand for making a concrete pad of a garage, how much cement and gravel would you also use?

5. The Theatre Aquarius sold adult, student, and child tickets in the ratio of 10 : 5 : 2.

(a) If 350 adult tickets were sold, how many student tickets were also sold?
(b) If tickets sell for $35, $12, and $8 respectively, how much money was collected?

6. The ratio of ginger ale to orange juice to grapefruit juice in a punch is 4 : 2 : 1.

(a) If 6 L of grapefruit juice is used, how much orange juice and ginger ale are needed?
(b) If 32 L of ginger ale is used, how much grapefruit juice and orange juice are needed?

9.5 RATE

While attending the Olympics, Kate rented a video camera. She paid \$78 to rent it for six days. Ross paid \$112 to rent the same model of camera for eight days.
Who rented the camera at the greater rate?

In these examples we are "comparing dollars to days." A comparison of two quantities with different units is called a rate. Kate's rental rate was \$78/6 d. Ross's rental rate was \$112/8 d.

To compare the rates we should find the rental for one day. This is called the unit rate.

Kate's rate: $\frac{78}{6} = \frac{c}{1}$

$78 = 6c$

$13 = c$

Kate's unit rate was \$13/d.

Ross's rate: $\frac{112}{8} = \frac{f}{1}$

$112 = 8f$

$14 = f$

Ross's unit rate was \$14/d.

We could also use the first two steps in the Rule of Three to determine the unit rate.

For Kate:

6 d cost \$78

1 d costs .... $\frac{\$78}{6}$

$= \$13$

For Ross:

8 d cost \$112

1 d costs .... $\frac{\$112}{8}$

$= \$14$

Ross rented the camera at the greater rate.

There are many examples of rate. One is speed, which compares distance to time.

$$\text{Speed} = \frac{\text{Distance}}{\text{Time}}$$

Example 1.

READ

Lydia rode her bicycle 147 km in 6 h.
What was her average speed?

Solution:

PLAN

$$\text{Speed} = \frac{\text{Distance}}{\text{Time}}$$

SOLVE

$$= \frac{147}{6}$$

$$= 24.5$$

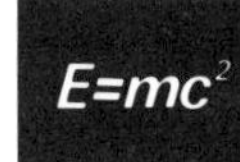

ANSWER

Lydia's average speed was 24.5 km/h.

When shopping, some people choose from similar products by comparing prices for a common unit, for example, 100 g of the product. Unit pricing is another example of a rate.

$$\text{Unit price} = \frac{\text{Cost}}{\text{Amount}}$$

Example 2.
The Apple Works sells apple cider at $6.80 for a 5 L jug. The Juice Factory sells the same cider at $5.40 for a 4 L jug. Which company offers the better value?

Solution:
Calculate the unit price per litre for each.

Apple Works:

$$\text{Unit price} = \frac{\text{Cost}}{\text{Amount}} = \frac{\$6.80}{5\text{ L}} = \$1.36/\text{L}$$

Juice Factory:

$$\text{Unit price} = \frac{\text{Cost}}{\text{Amount}} = \frac{\$5.40}{4\text{ L}} = \$1.35/\text{L}$$

The Juice Factory offers the better value.

EXERCISE 9.5

B

1. Calculate the unit rate for each of the following.
(a) 315 words typed in 5 min
(b) $59.15 for 7 h of work
(c) 8 kg of cheese for $38
(d) 300 km driven in 6 h
(e) $142 for 4 concert tickets
(f) 231 points in 7 basketball games
(g) $17.85 for 3 video cassettes

2. Calculate the unit rate for each of the following.
(a) 245 km driven in 3.5 h
(b) $12.15 for 2.7 L
(c) $72 for 7.5 h of work

3. Leonard drove 225 km in 2.5 h.
(a) What was his speed in kilometres per hour?
(b) How long would it take him to drive 495 km at this rate?

4. Peggy earned $47.85 for 5.5 h of work.
(a) What was her rate of pay?
(b) How much would she earn for 40 h at this rate?

5. Vince drove 380 km in 4 h. Sandra drove 558 km in 6 h.
Who drove at the faster speed?

6. Which is the better buy: 10 kg of beef for $89 or 15 kg of beef for $130.50?

7. Which is the better buy: 12 L of detergent for $11.28 or 5 L of detergent for $4.85?

8. A 10 kg bag of firewood costs $27.50.
(a) How much would you pay for a 25 kg bag?
(b) How much could you buy for $55?

9. The ad says that 1.46 kg of margarine costs $3.40.
Approximately how much does 1 kg of this margarine cost?

10. The Clearwater Boat Club will rent you a sailboat for 8 d at $624. Sunset Boats will rent you the same kind of boat for 6 d at $474.
Which company offers the better rate and by how much?

11. Which is the better buy: 1 L of oil for $8.80 or 750 mL of the same oil for $6.45?

C

12. A car travels 550 km on one tank of gas. The tank refill is 38.5 L.
What is the rate of gas consumption in litres per 100 kilometres?

9.6 SCALE DRAWINGS AND MAPS

The illustration of the jumbo jet is a scale drawing. The artist used a scale or ratio to draw the jet in a smaller, but proportional, size. The scale the artist used is 1 : 1000, which means that 1 cm measured along the jet in the drawing represents 1000 cm or 10 m of the actual plane.

What is the actual length of a jumbo jet?

$$\text{Scale} = \begin{pmatrix}\text{the length on the} \\ \text{scale diagram}\end{pmatrix} : \begin{pmatrix}\text{the length on the} \\ \text{actual object}\end{pmatrix}$$

Example.

The map at the right is drawn to a scale of 3.5 cm to 120 km. The scale at the bottom of the map gives us this information.

What is the approximate flying distance between North Bay and Toronto?

Solution:

The measured distance on the map between North Bay and Toronto is 9.2 cm. We can use a proportion to find the flying distance.

Let x be the distance between Toronto and North Bay in kilometres. Then,

3.5 cm : 120 km = 9.2 cm : x km

$$\frac{3.5}{120} = \frac{9.2}{x}$$

$$3.5x = 1104$$

$$x \doteq 315$$

The flying distance between Toronto and North Bay is approximately 315 km.

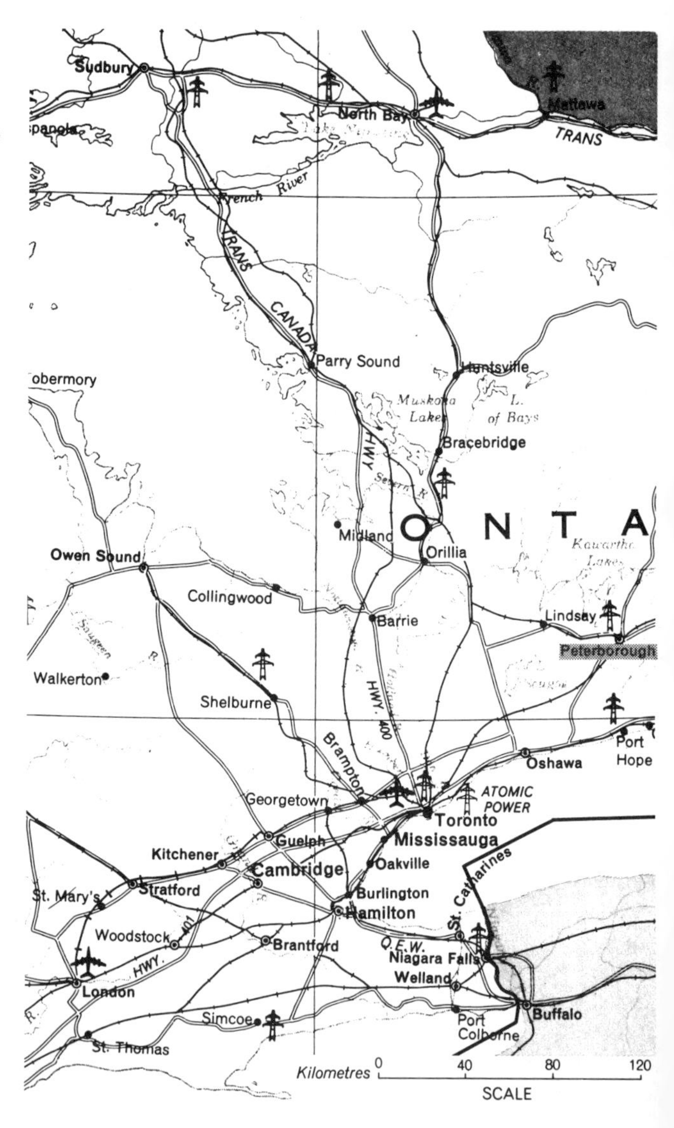

EXERCISE 9.6

B

1. The floor plan of a cottage is shown below. It is drawn to a scale of 1 : 100.

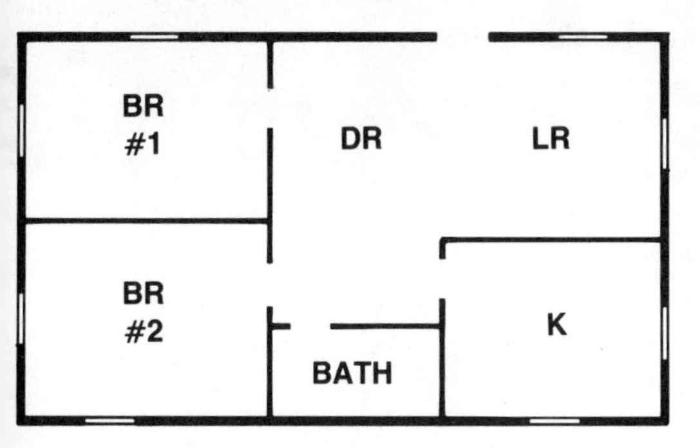

(a) What are the outside dimensions of the cottage in metres?
(b) What are the inside dimensions of the kitchen in metres?
(c) What is the approximate total floor area of the cottage?

2. Use the map on the left page to determine:
(a) the approximate flying distance between Niagara Falls and Sudbury
(b) the approximate driving distance between Collingwood and St. Thomas
(c) the approximate flying distance between London and Peterborough
(d) the approximate driving distance between Sudbury and Toronto

3. The blue whale was drawn using a scale of 1 : 600.
What is the actual length of the whale?

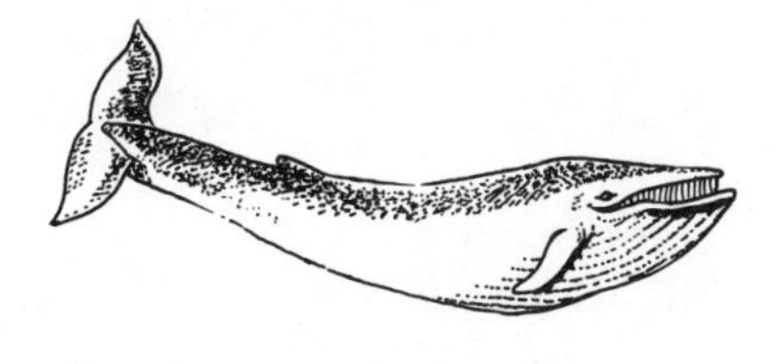

4. The scale on the map is 3.5 cm to 120 km.
(a) What is the approximate driving distance from Halifax to Moncton to Chatham?
(b) What is the approximate flying distance from Halifax to Summerside?
(c) What is the approximate flying distance from Moncton to Charlottetown?
(d) What is the driving distance from Truro to Sussex?
(e) What is the driving distance from Sussex to Liverpool?

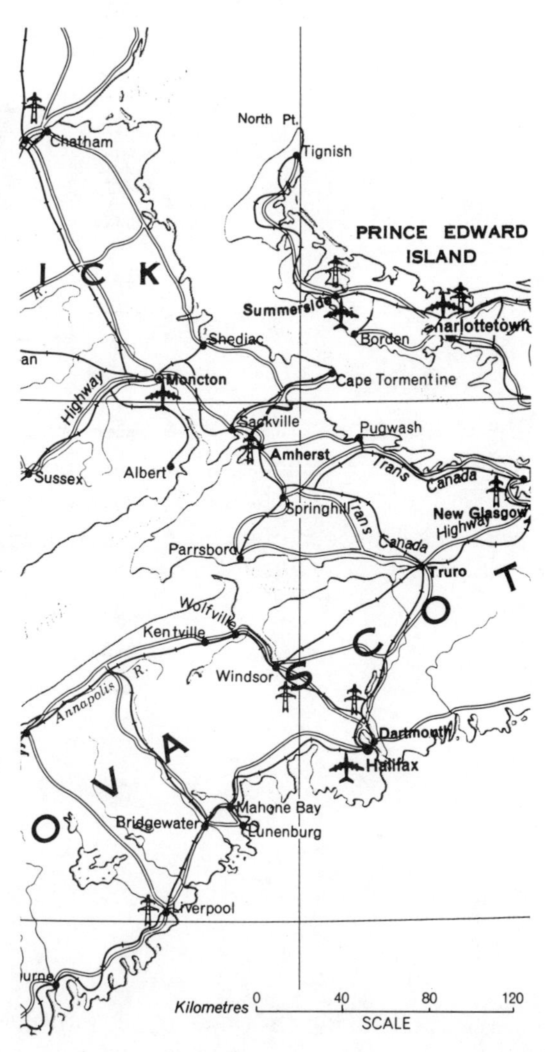

9.7 DATA AND PROPORTIONS IN TABLES

A restaurant charges $2500 for a banquet serving 100 people and $4500 for 200 people. What should the school band pay for a banquet serving 125 people?

There are many occasions when the information you need is given, but you cannot use it directly. In many cases this kind of information is concealed in tables

The table at the right is used in accident investigations. Given the length of a car's skidding distance, which is measured along the rubber marks on the road, the accident investigator can determine the speed of the car immediately before the brakes were applied. For example, a skid distance of 25 m indicates a speed of 70 km/h.

Skid Distance (m)	Speed (km/h)
12	50
18	60
25	70
33	80

What was the speed of a car if the skid distance was 14 m? In the table, 14 m is between 12 m and 18 m. Finding the differences, we have

Distance: 12, 14, 18 (12 to 14: 2; 12 to 18: 6)

Speed: 50, ?, 60 (50 to ?: x; 50 to 60: 10)

We now write a proportion. $\frac{6}{2} = \frac{10}{x}$

and $6x = 20$

$x \doteq 3.3$

The speed of the car before the brakes were applied was (50 + 3.3) km/h or about 53 km/h.

Since 14 m is in the table, we call the above calculation interpolation. When the values lie outside the table, we call the calculation extrapolation.

What was the speed of a car, if the skid distance is 44 m? This value is not listed in the table. Finding differences, we have

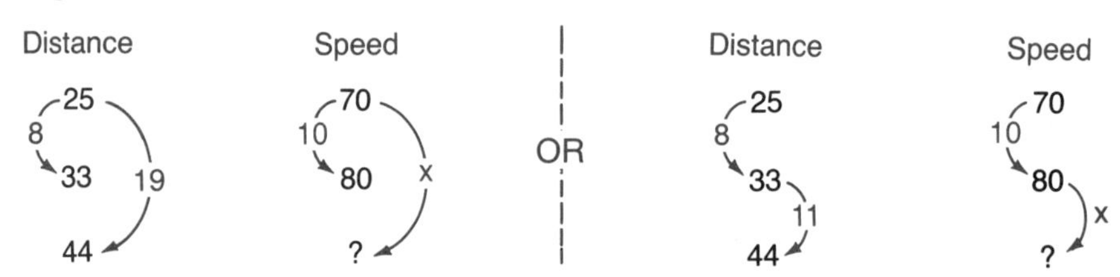

Set up a proportion.

$\frac{8}{19} = \frac{10}{x}$

$8x = 190$

$x \doteq 23.8$

Set up a proportion.

$\frac{8}{11} = \frac{10}{x}$

$8x = 110$

$x \doteq 13.8$

$\therefore$ The speed of the car was (70 + 23.8) km/h, or (80 + 13.8) km/h, or about 94 km/h.

EXERCISE 9.7

A

1. Carl Edwards refinishes pool decks. The table gives the cost of refinishing a deck according to area.

Area (m^2)	20	30	40	50	60	70
Cost ($)	120	160	200	240	280	310

(a) How much would it cost to have a deck with an area of 35 m^2 refinished?
(b) How much would it cost to have a deck with an area of 80 m^2 refinished?
(c) Nuzhat's deck has an area of 15 m^2. How much would it cost to have her deck refinished?

B

2. The Sunburst Airline delivers packages to the headwaters of the Amazon River. The cost varies according to the mass of the package.

Mass (kg)	50	100	150	200
Cost ($)	1900	3000	4100	5200

(a) How much would it cost to have a 175 kg package delivered?
(b) Poindexter's stereo system has a mass of 57 kg.
What is the cost to have it delivered?
(c) The Waterford Company needs 260 kg of gold delivered to pay salaries and operating expenses.
How much will this delivery cost?
(d) Stephanie's camera has a mass of 3 kg. How much will it cost to deliver the camera?

3. An experimental weather rocket was launched in the desert. The table gives the height of the rocket in kilometres according to the number of seconds after launch.

Time (s)	0	5	10	15	20	25	30
Height (km)	0	125	200	225	200	125	0

(a) What was the maximum height reached by the rocket?
(b) How long was the rocket's flight?
(c) How high was the rocket after 7 s?
(d) How high was the rocket after 14 s?
(e) How high was it after 23 s?
(f) When did the rocket reach a height of 100 km?

4. The following table gives the time, in hours, required to clean a major league baseball stadium according to the number of cleaners.

Number of Cleaners	10	15	20	25	30
Time to Clean (h)	30	20	15	12	10

(a) How long will 17 people take to clean the stadium?
(b) How long will 8 people take to clean the stadium?
(c) How many people must be hired to clean the stadium in 8 h?
(d) If cleaners get paid $12/h, is it cheaper to hire 25 people or 10 people?

9.8 PROBLEM SOLVING

1. In 1987 Garry Sowerby drove a light truck from Tierra Del Fuego to Prudhoe Bay. The total distance was 24 086 km. Garry made the trip in a record time of 23 d, 22 h, and 43 min.

(a) What was Garry's average speed per hour?
(b) Where is Prudhoe Bay?
(c) Where is Tierra Del Fuego?
(d) How long would it take a commercial airliner flying at 800 km/h to make the same trip?
(e) How many countries did Garry pass through?

2. How many squares can you draw on a grid of nine equally spaced dots so that each vertex of each square is on a dot?

. . .

. . .

. . .

3. Your plane leaves for New York at 20:30. You have to be at the airport at least one hour before take-off to go through customs. Your boss wants a half-hour meeting with you at the airport before you go through customs. The cab ride to the airport takes forty minutes. You will telephone a cab to take you to the airport and the cab takes fifteen minutes to get to your place after you call.
At what time should you call for the cab?

4. Paulo has an 8 L container and a 3 L container.
How can he use these containers to measure exactly 4 L of water?

5. A stamp has four edges. When two stamps are attached, we say that each stamp has three free edges.

Suppose you have a sheet of forty stamps. The stamps are attached in eight rows of five stamps each.
(a) How many stamps have one free edge?
(b) How many stamps have two free edges?
(c) How many stamps have no free edges?

6. Dee bowled 152 in her first game and 153 in the second game.
What will she have to bowl in the third game to have an average of 164?

7. The box will hold one die. The die is marked as shown.
In how many different ways can the die be placed in the box?

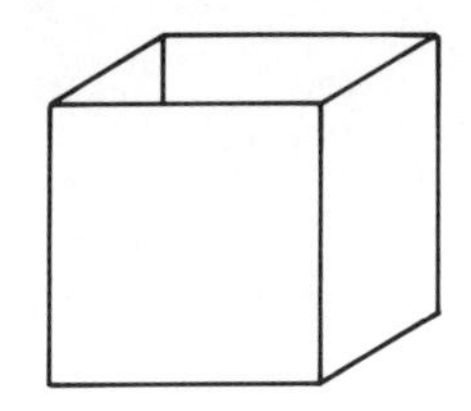

8. Find three consecutive integers whose product is 13 800.

9. Riça has a total of 14 dimes and nickels. Their total value is one dollar.
How many dimes does she have?

10. After a basketball game each of the 10 members of the winning team shakes hands with each of the 10 members of the losing team.
How many handshakes take place?

11. A room measures 4 m by 4 m. The floor is tiled with square tiles that measure 20 cm by 20 cm each.
How many tiles are needed?

12. The cost of a federal election works out to be about $5.25 per Canadian taxpayer.
(a) How much does it cost to run a federal election?
(b) For what is this money used?

13. Name five songs whose titles are questions.

14. Walt Disney's movie Snow White and the Seven Dwarfs has been released in many languages. In French, the title of Snow White is Blanche Neige, and the dwarfs' names were changed to the seven names below.
Identify the English name of the French version.

(a) Prof
(b) Simplet
(c) Dormeur
(d) Timide
(e) Grincheux
(f) Joyeux
(g) Atchoum!

CAREER

CAR MECHANIC

The power of a car's engine is transmitted to the wheels via the transmission and drive shaft.

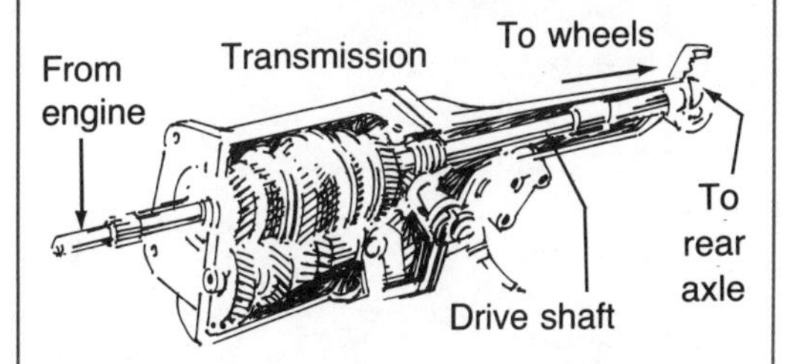

Because the transmission, which has a series of gears, changes the speed for specific driving conditions, the speed of the engine, called the tach and measured with the tachometer, is not necessarily the same as the drive-shaft speed. The transmission ratio compares the engine's speed to the drive-shaft speed. Both speeds are expressed in revolutions per minute (rpm), to the nearest thousandth.

$$\text{Transmission ratio} = \frac{\text{Engine speed (rpm)}}{\text{Drive-shaft speed (rpm)}}$$

A car is travelling in fourth gear. The tach reads 4000 rpm. The transmission ratio in this situation is 1.700 to 1.000.
What is the speed of the drive-shaft?

Let s be the speed of the drive shaft and write a proportion.

$$\frac{1.700}{1.000} = \frac{4000}{s} \quad \text{and} \quad s \doteq 2353$$

1. The tach reads 2500 rpm in third gear. The transmission ratio in this case is 3 : 1.
What is the drive-shaft speed?

2. The tach shows an engine speed of 2800 rpm. At the same time the drive-shaft speed is 800 rpm.
What is the transmission ratio?

EXTRA

9.9 NIAGARA FALLS — A WONDER OF THE WORLD

The waterfalls at Niagara are one of the natural wonders of the world. The Falls are located on the Niagara River, which joins Lake Erie and Lake Ontario.
The waterfalls is actually two waterfalls: the American Falls and the Horseshoe Falls. Goat Island lies between the two.

The American Falls are 64 m high and 305 m wide and the Horseshoe Falls are 54 m high and 673 m wide.

The water flow in the Niagara River varies, but the average rate is about 6000 m^3/s. Seven percent of the water flows over the American Falls and the remainder over the Horseshoe Falls.

The great volume of water passing over the Horseshoe Falls has caused the underlying rock to erode and the falls to shift toward Lake Erie. From 1764 to 1949 the falls moved 264 m.
In 1949 river channels on both sides of the horseshoe were artificially deepened to even out the flow of water and stop the erosion.

LAKE ONTARIO
NIAGARA ON-THE-LAKE
YOUNGSTOWN
Niagara River
QUEENSTON
LEWISTON
Goat Island
Whirlpool
Whirlpool Rapids
American Falls
Horseshoe Falls
direction of water flow
Niagara River
GRAND ISLAND
N
FORT ERIE
BUFFALO
LAKE ERIE

WHIRLPOOL RAPIDS

Beyond the Falls, the Niagara River flows into a canyon-like gorge that extends to Queenston. The Whirlpool Rapids are located 5 km below the falls in this gorge, which is 76 m wide at the rapids. The waves reach a height of 6.1 m. Huge blocks of limestone have choked the river so that the depth at this point is only 12 m. The rapids are about 200 m in length, and the water flows at a rate of 50 km/h.

THE WHIRLPOOL

Since 1916, millions of tourists have been carried safely across the swirling whirlpool in the Spanish Aero Car that travels a distance of 549 m. The car makes 45 trips per day with a maximum of 45 people per trip and it operates 192 days a year.

EXERCISE 9.9

1. (a) How much water flows over both falls in one week?
(b) How much of this water flows over the American Falls?
(c) How much flows over the Horseshoe Falls?
(d) A large oil tanker can hold about 670 000 m^3. It would take a fast-running kitchen tap 740 d to fill the tanker.
How long would it take the water from the Niagara River to fill the tanker?

2. (a) What is the approximate surface area of the American Falls?
(b) What is the approximate surface area of the Horseshoe Falls?

3. (a) How many metres per year did the Horseshoe Falls recede between 1764 and 1949?
(b) At this rate, how long would it take the falls to move the length of your classroom?

4. The Floral Clock was built in 1902. The face of the clock is 12.2 m in diameter.
(a) What is the area of the clock?
(b) What is the circumference?
(c) How long ago was the clock built?

5. (a) If the Spanish Aero Car operates at an average of 80% capacity, how many people make the trip in a year?
(b) If the cost per ticket is $3.50, what is the revenue for one year?

6. The Niagara River can be described by another geographic term.
What is the name used for a body of water that joins two lakes?

7. At one time, Sir Harry Oakes lived in the city of Niagara Falls.
Who was Sir Harry Oakes and how did he make his fortune?

9.10 REVIEW EXERCISE

1. State the ratios for the following.

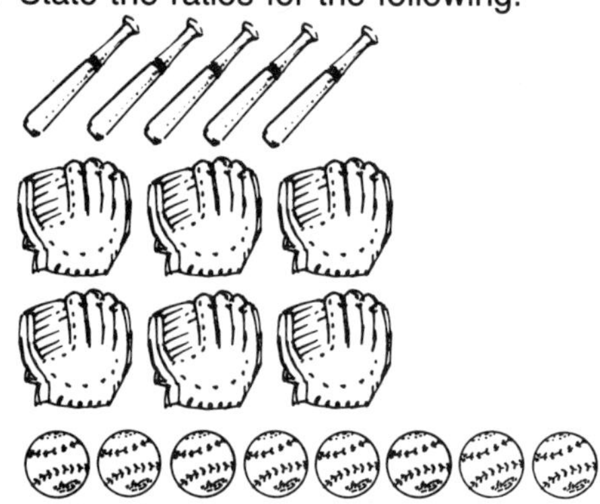

(a) bats to balls
(b) gloves to bats
(c) balls to gloves
(d) gloves to balls
(e) bats to gloves to balls
(f) balls to bats to gloves

2. Express the following ratios in simplest form.

(a) 10 : 5 (b) 4 : 12
(c) 15 : 20 (d) 6 : 4
(e) 4 : 6 : 2 (f) 15 : 10 : 25

3. Find two equivalent ratios for each.

(a) 2 : 3 (b) 4 : 1
(c) 3 : 5 (d) 2 : 5 : 3

4. From the following list, pick out four pairs of equivalent ratios.

7 : 28, 14 : 2, 9 : 6, 4 : 16,
8 : 10, 56 : 8, 27 : 18, 12 : 6
7 : 1, 32 : 40

5. Find the missing term in each.

(a) x : 12 = 3 : 4
(b) a : 27 = 2 : 3
(c) 5 : 6 = b : 42
(d) 3 : m = 9 : 15
(e) $\frac{x}{2} = \frac{4}{8}$
(f) $\frac{m}{7} = \frac{5}{21}$
(g) 3 : 2 : 1 = 6 : x : y
(h) 5 : a : 4 = 20 : 24 : b

6. Esther and Helga bought a sailboat. The cost of the boat was $24 000, and they shared the cost in a ratio of 5 : 3.
How much did each of them pay?

7. The three Smith brothers, Adam, Aaron, and Aden buy a $50 charity lottery ticket. Adam contributes $15, Aaron $10, and Aden $25. Their ticket won first prize, which is $250 000.
What share does each receive?

8. Find the missing term.

(a) $\frac{x}{9} = \frac{8}{3}$ (b) $\frac{4}{t} = \frac{2}{7}$
(c) $\frac{3}{5} = \frac{m}{10}$ (d) $\frac{4}{9} = \frac{5}{s}$
(e) $\frac{x}{7} = \frac{7}{10}$ (f) $\frac{5}{4} = \frac{4}{y}$

9. Three out of five high school students prefer leather jackets.
If there are 455 students, how many prefer leather jackets?

10. The most popular hot dog topping at the ball park is a mixture of tomatoes and chopped onions in the ratio of 5 : 3.

(a) How many kilograms of onions are mixed with 60 kg of tomatoes?
(b) How many kilograms of tomatoes are mixed with 75 kg of onions?
(c) How many kilograms of tomatoes are in 160 kg of the mixture?
(d) How many kilograms of onions are in 140 kg of the mixture?

11. Find the missing terms.

(a) 1 : 2 : 3 = 5 : x : y
(b) 4 : 5 : 7 = x : 25 : y
(c) 20 : x : 6 = 5 : 3 : y

12. The bus company sells bus passes for adults, students, and children. The ratio of sales is 3 : 10 : 2.
If 450 adult passes were sold, then how many student and children passes were sold?

13. In a reforestation project, maple, elm, and birch trees were planted in a ratio of 6 : 2 : 1.
If there were 1170 maples planted, how many elm and birch trees were also planted?

14. Calculate the unit rate for each of the following.

(a) $105.75 for nine hours of work
(b) $4.45 for five litres of apple cider
(c) 511 km driven in 7 h
(d) $3.65 for five loaves of bread
(e) 360 km driven in 4.5 h

15. Which is the better buy: 7 kg of bananas for $8.75 or 9 kg of bananas for $11.70?

16. Below is a sketch of the floor plan for a clothing store.
Make a scale drawing of the floor plan.

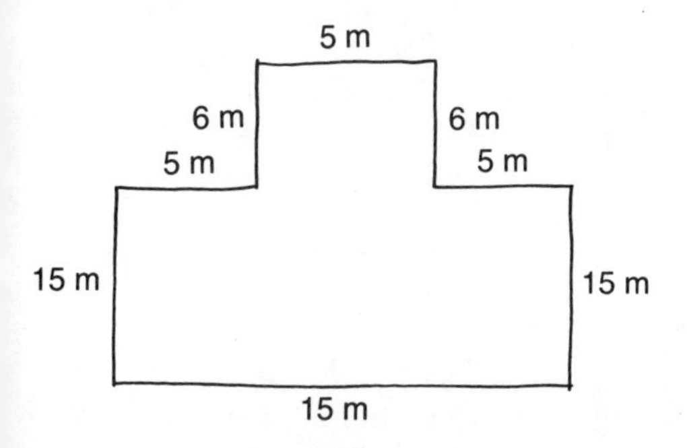

17. The black bear is drawn to a scale of 1 : 25.
What is the bear's length?

18. The Elite Company caters meals for parties. The table gives the total cost for a specific number of people.

Number	20	40	60	80
Cost ($)	575	1075	1575	2075

(a) How much does a party of 50 people cost?
(b) How much does a party of 105 people cost?
(c) How much does a party of 63 people cost?
(d) How many people could you have at a party if you had $950 to spend?

19. A test driver drove a new car from North Bend to Oxbow several times at constant speeds of 70 km/h, 80 km/h, 90 km/h, and 100 km/h for each respective trip. The cost of each trip is given in the table.

Speed (km/h)	70	80	90	100
Cost ($)	46	56	70	80

(a) How much would a trip cost if you drove at 95 km/h?
(b) How much would a trip cost if you drove at 65 km/h?
(c) How fast can you drive if you have $50 for the trip?

MIND BENDER

There are thirteen high school teams in a soccer league. Each team played each other once. At the end of the season each team had won six games and lost six games.
How many games were played?

9.11 CHAPTER 9 TEST

1. Express the following ratios in simplest form.

(a) 15 : 5 (b) 4 : 24 (c) $\frac{18}{12}$

2. Write two equivalent ratios for each.

(a) 2 : 5 (b) 4 : 3 (c) $\frac{3}{4}$

3. Find the missing term in each.

(a) x : 5 = 12 : 15 (b) 36 : a = 9 : 24

(c) $\frac{x}{8} = \frac{7}{2}$ (d) $\frac{3}{5} = \frac{x}{20}$ (e) $\frac{21}{9} = \frac{7}{x}$

4. Robyn and Emil bought a sailboat for $21 000. They shared the cost in a ratio of 5 : 2.
How much did each of them pay?

5. The ratio of tonic water to grapefruit juice to pineapple juice in a punch is 4 : 3 : 1.
In 80 L of punch, how many litres of grapefruit juice is there?

6. Calculate the unit rate for each of the following.
(a) $78 for eight hours of work
(b) $18.20 for 7 L of windshield washer fluid

7. Which is the better buy: 5 kg of apples for $7.80 or 4 kg of apples for $6.08?

8. The scale on a drawing of a building is 1 : 40.
If the height of the building is 30 cm on the drawing, how tall is the building?

9. The Fast Lane courier service delivers large packages anywhere in the country for a fee according to the following rate table.

Mass of Package (kg)	10	20	30	40
Cost ($)	25	45	65	85

(a) How much would a 35 kg package cost to deliver?
(b) How much would a 55 kg package cost to deliver?

GEOMETRY AND MEASUREMENT

REVIEW AND PREVIEW TO CHAPTER 

BASIC TERMS IN GEOMETRY

The chart summarizes the basic terms in geometry.

Term	Labelled Diagram	Description	Symbol
Point	A •	Point A	A
Line	A B ℓ	Line $\overleftrightarrow{AB}$ or $\overleftrightarrow{BA}$ or line ℓ	$\overleftrightarrow{AB}$ or $\overleftrightarrow{BA}$ or ℓ
Line Segment	A B	Line segment AB, or BA, where A and B are the end points of the line segment	AB or BA
Ray	A B	Ray $\overrightarrow{AB}$	$\overrightarrow{AB}$
Angle	A B C A B C	Angle ABC or angle CBA, where BA and BC are the rays (arms) of the angle and B is the vertex of the angle	∠ABC or ∠CBA or ∠B
Triangle	A B C	Triangle ABC, where AB, BC, and AC are sides of the triangle and A, B, and C are vertices	△ABC
Quadrilateral	W Z X Y	Quadrilateral WXYZ, where WX, XY, YZ, and WZ are sides and W, X, Y, and Z are vertices	WXYZ

EXERCISE 1

1. In the diagram, name the following.
(a) 3 points (b) 3 lines
(c) 3 line segments (d) 3 rays
(e) 6 angles (f) a triangle

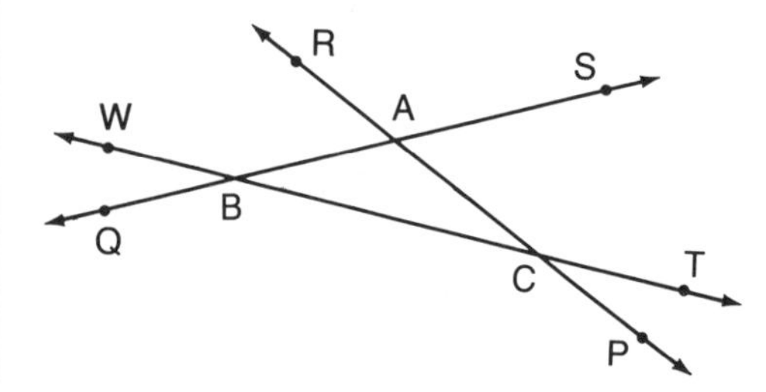

2. In the diagram, name the following.
(a) 6 angles (b) 6 triangles
(c) 4 quadrilaterals (d) 6 line segments

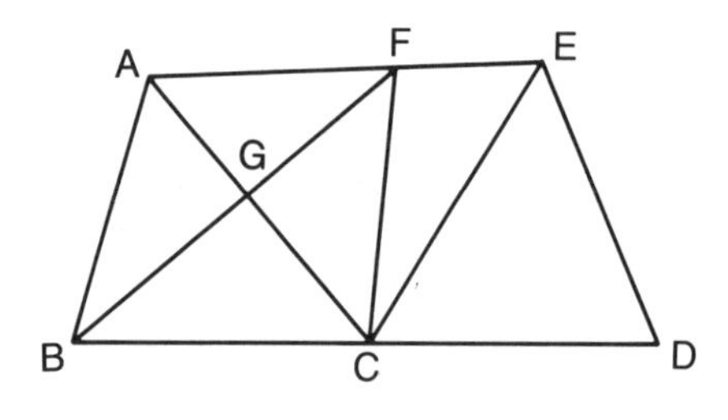

ANGLES

An angle is a figure formed by two rays or line segments. Angles are measured in degrees and we use a protractor to measure an angle. The angle at the right measures 54°.

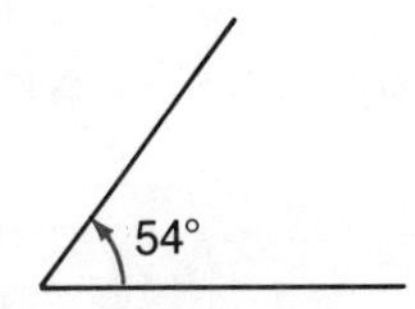

EXERCISE 2

1. State the measure of each angle in the diagram.

(a) ∠DBA (b) ∠ABE
(c) ∠FBA (d) ∠CBG
(e) ∠CBH (f) ∠CBE
(g) ∠CBD (h) ∠GBA
(i) ∠EBC (j) ∠ABH

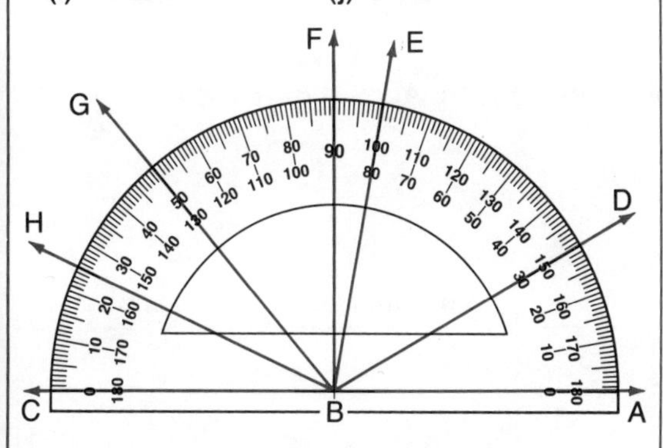

2. State the measure of each angle in the diagram.

(a) ∠TXZ (b) ∠RXZ
(c) ∠ZXS (d) ∠MXZ
(e) ∠YXN (f) ∠MXY
(g) ∠RXY (h) ∠NXZ
(i) ∠YXT (j) ∠ZXY

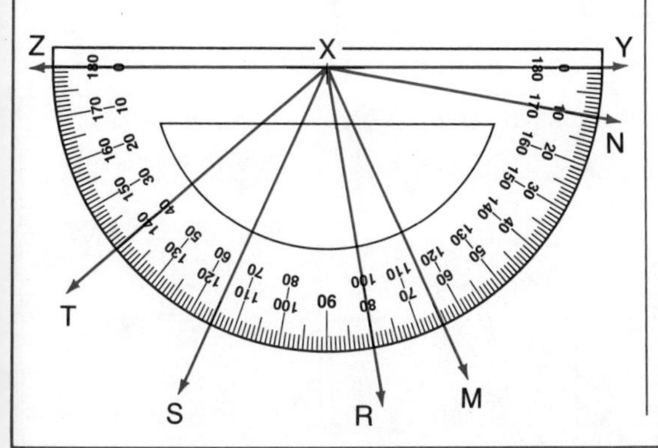

3. Use a protractor to construct each angle.

(a) ∠ABC = 30° (b) ∠DEF = 47°
(c) ∠RST = 78° (d) ∠XYZ = 90°
(e) ∠MNO = 180° (f) ∠PQR = 120°
(g) ∠JKL = 155° (h) ∠GHI = 177°

4. What is the angle formed by the hands of a clock when the time is

(a) three o'clock?
(b) six o'clock?
(c) two o'clock?
(d) eleven o'clock?

5. Calculate the number of degrees through which the minute hand of a clock moves in

(a) one hour (b) 30 min
(c) 15 min (d) 45 min
(e) 10 min (f) 5 min
(g) 35 min (h) 55 min

6. Estimate the size of each angle.

(a) (b)

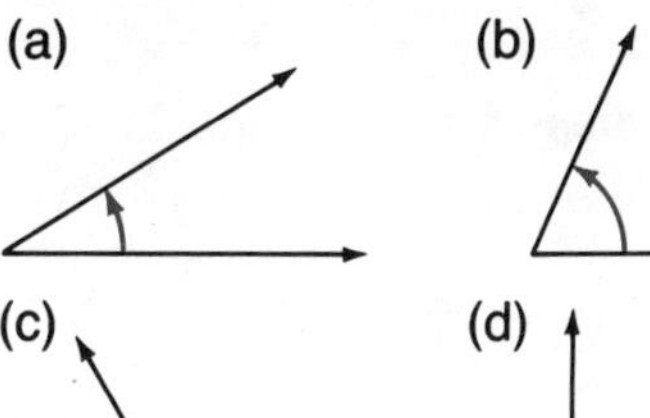

(c) (d)

7. Using only a ruler, and no protractor, draw each of the following angles. Check the accuracy of your work with a protractor.

(a) 60° (b) 30°
(c) 75° (d) 45°

10.1 ANGLES AND TRIANGLES

Geometry is illustrated in buildings, art, navigation, computer design, and flight.

Like algebra, geometry is a language of mathematics and science. Learning this language will help you solve problems, and work in the modern world.

We will start by reviewing geometrical terms.

You can identify a number of angles in the above photographs. Angles are classified according to size.

CLASSIFYING ANGLES

Acute Angle	Right Angle	Obtuse Angle	Straight Angle	Reflex Angle
Less than 90°	Equal to 90°	Greater than 90° but less than 180°	Equal to 180°	Greater than 180° but less than 360°

Between some angles there are special relationships, which are useful tools for solving problems.

ANGLE RELATIONSHIPS

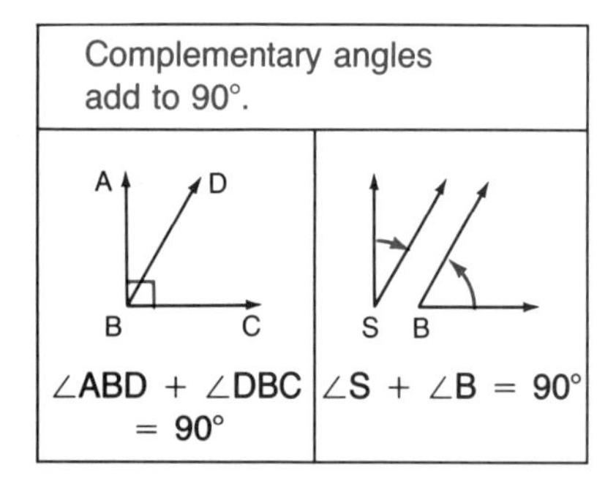

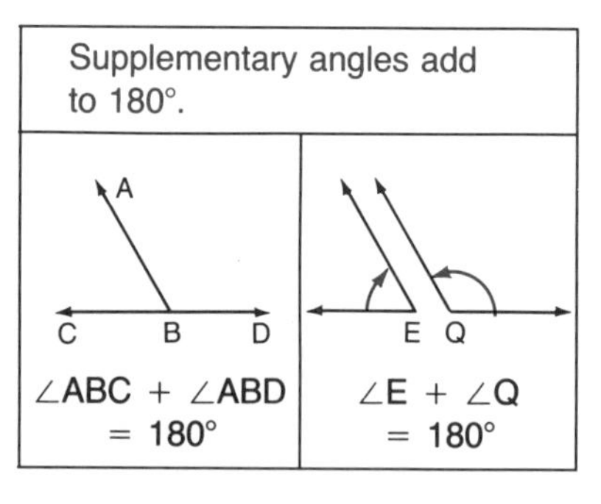

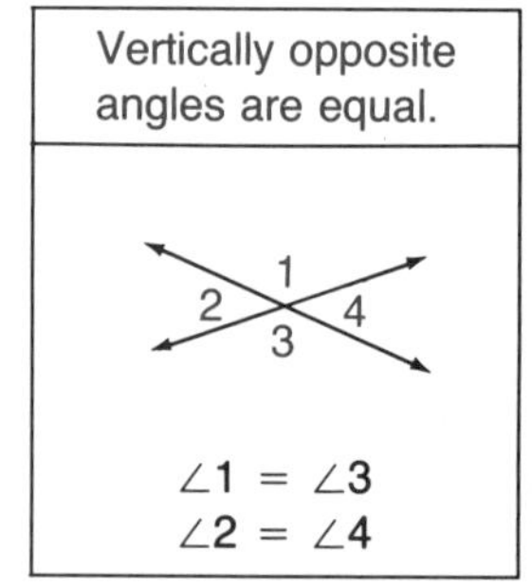

The triangle is the most common shape in any construction.
Triangles are classified in two ways:

(a) by how the lengths of the sides compare;
(b) by how the measures of the angles compare.

CLASSIFYING TRIANGLES BY SIDES

Scalene Triangle	Isosceles Triangle	Equilateral Triangle
Each side has a different length.	Two sides have the same length.	All sides have the same length.

CLASSIFYING TRIANGLES BY ANGLES

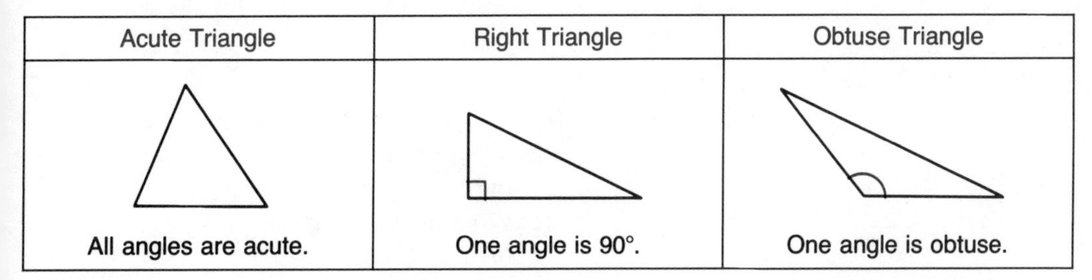

Acute Triangle	Right Triangle	Obtuse Triangle
All angles are acute.	One angle is 90°.	One angle is obtuse.

There are angle relationships in triangles, too. The following are important ones.

ANGLE RELATIONSHIPS FOR TRIANGLES

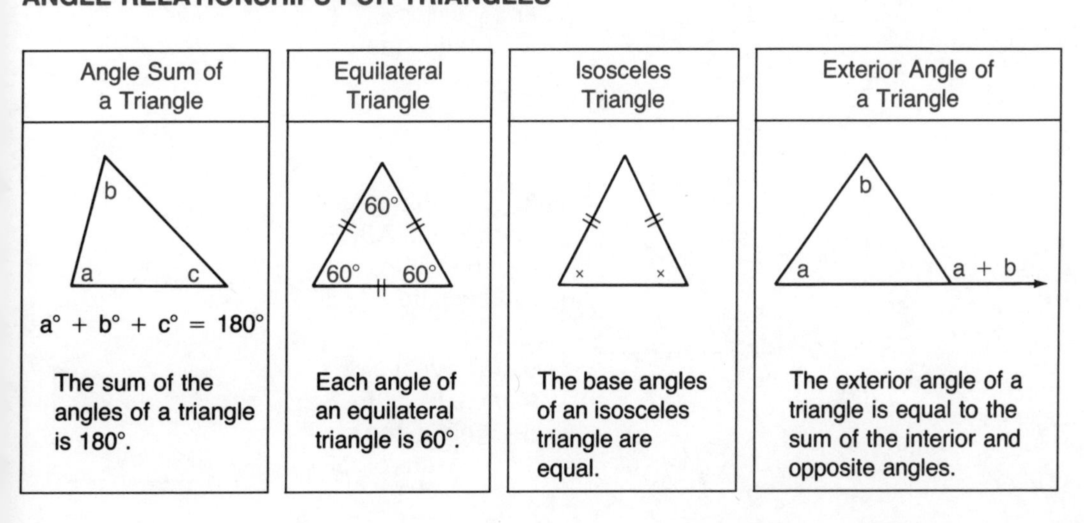

Angle Sum of a Triangle	Equilateral Triangle	Isosceles Triangle	Exterior Angle of a Triangle
$a° + b° + c° = 180°$ The sum of the angles of a triangle is 180°.	Each angle of an equilateral triangle is 60°.	The base angles of an isosceles triangle are equal.	The exterior angle of a triangle is equal to the sum of the interior and opposite angles.

READ

Example.
Find the size of each angle in the triangle.

PLAN

Solution:
The sum of the angles in a triangle is 180°.

SOLVE

$$2x + 3x + 4x = 180$$
$$9x = 180$$
$$x = 20$$

Check: $40 + 60 + 80 = 180$

ANSWER

Then $2x = 40$, $3x = 60$, and $4x = 80$.
So $\angle A = 40°$, $\angle B = 60°$, and $\angle C = 80°$.

EXERCISE 10.1

A 1. Classify each angle.

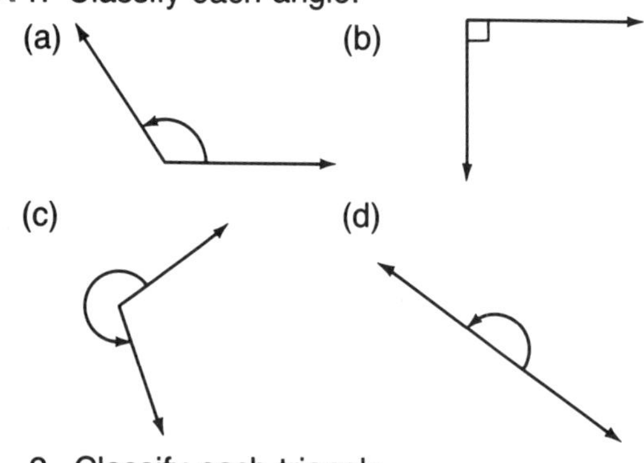

2. Classify each triangle.

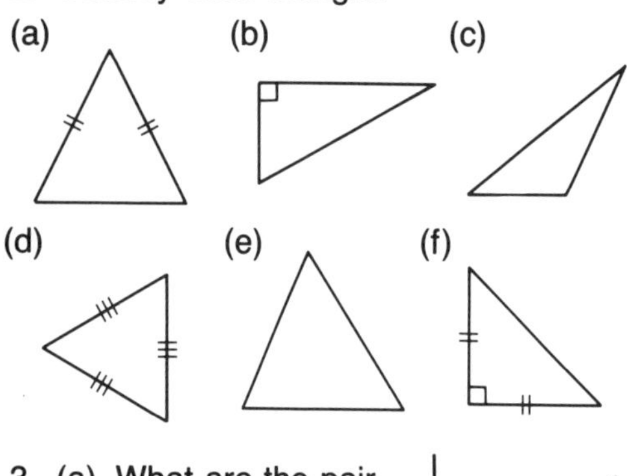

3. (a) What are the pair of angles called?
(b) What is the measure of $\angle x$?

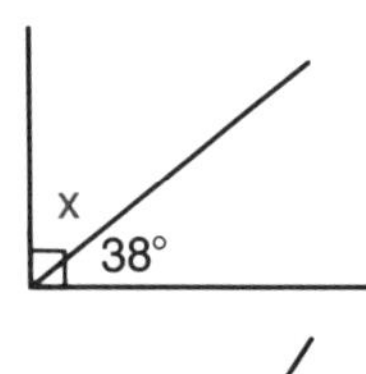

4. (a) What are the pair of angles called?
(b) What is the measure of $\angle x$?

5. $\angle DEF + \angle ABC = 90°$
(a) What are the pair of angles called?
(b) What is the measure of $\angle ABC$?

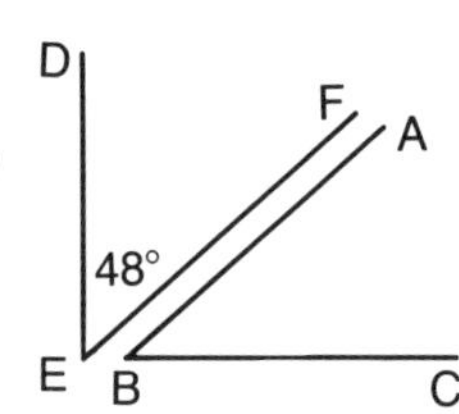

6. $\angle RST + \angle WXY = 180°$
(a) What are the pair of angles called?
(b) What is the measure of $\angle RST$?

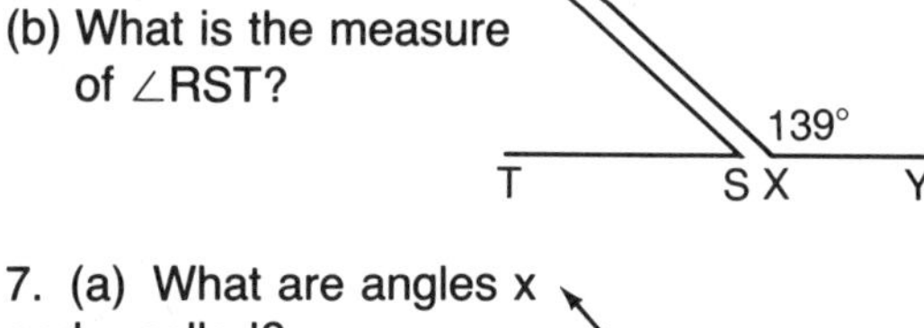

7. (a) What are angles x and y called?
(b) What is the measure of $\angle z$?

8. (a) What type of triangle is $\triangle ABC$?
(b) What is the measure of $\angle B$?

9. (a) What type of triangle is $\triangle DEF$?
(b) What is the measure of $\angle F$?

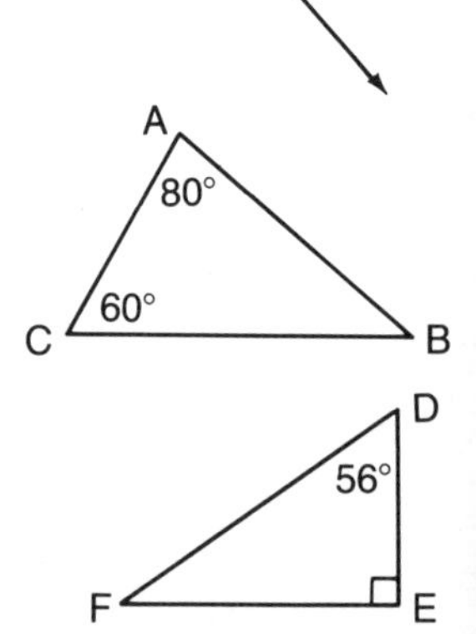

10. (a) What type of triangle is $\triangle RST$?
(b) What is the measure of $\angle T$?

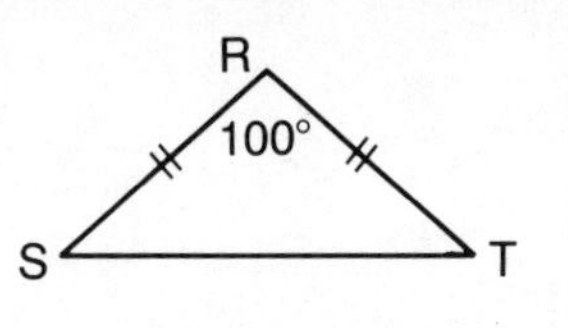

11. (a) What type of triangle is $\triangle JKL$?
(b) What are the measures of $\angle J$, $\angle K$, and $\angle L$?

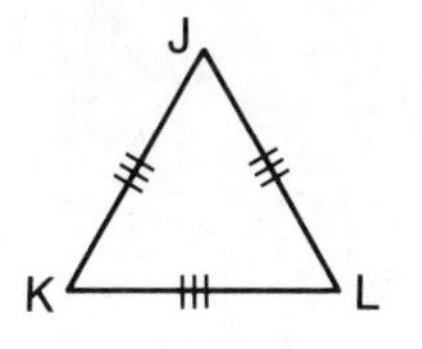

B 12. Find the measures of the unknown angles.

(a)

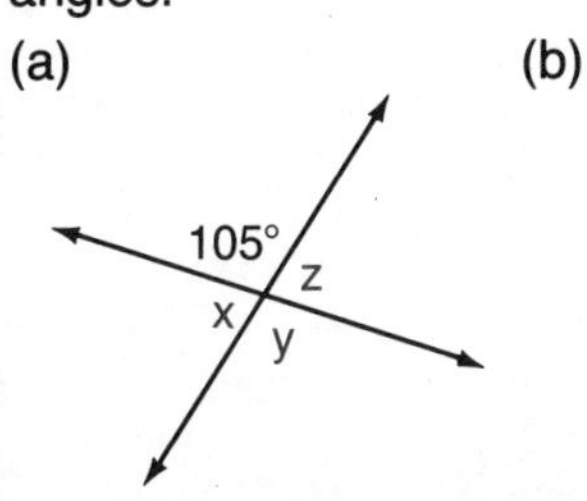

(b)

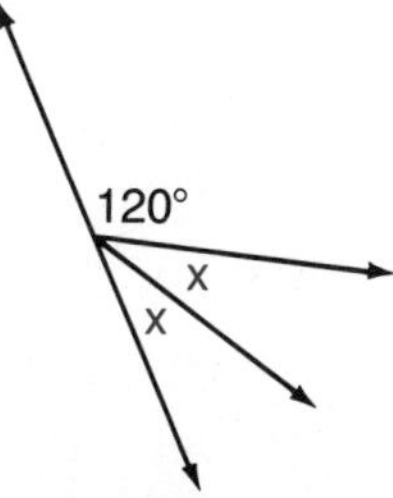

13. Find the measures of the unknown angles.

(a)

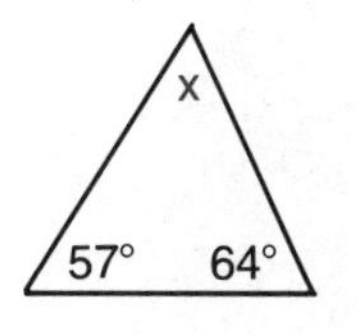

(b)

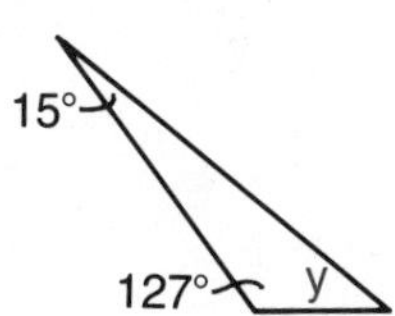

(c)

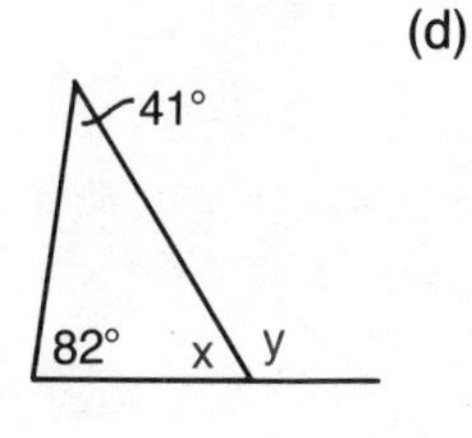

(d)

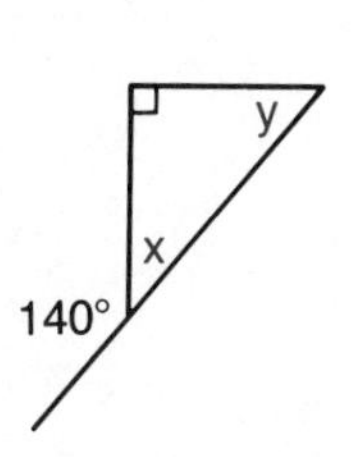

14. Find the measures of the unknown angles.

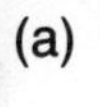

(a)

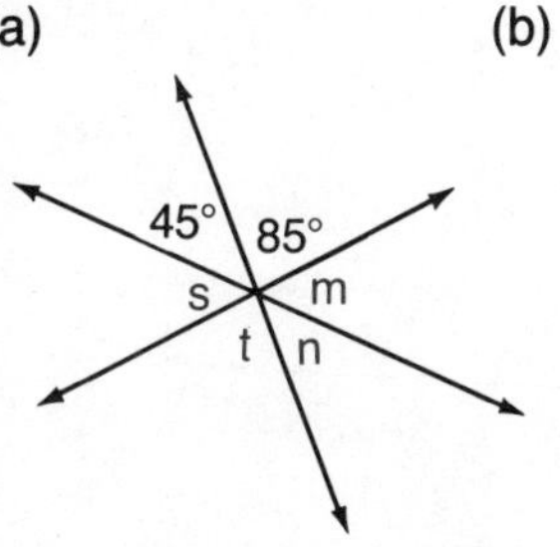

(b)

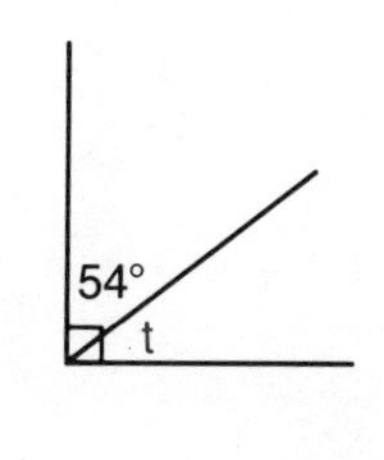

(c) (d)

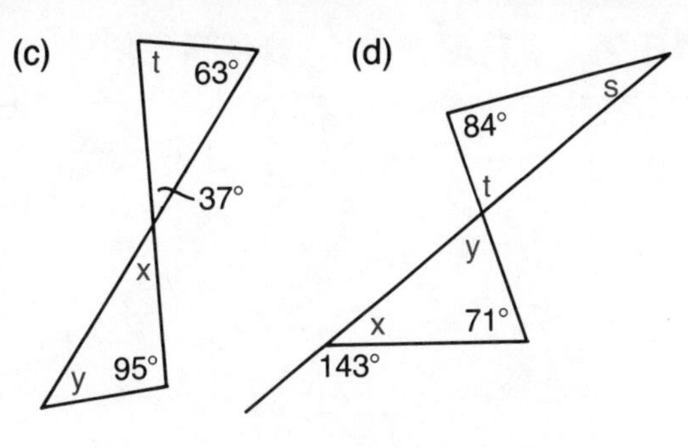

15. Find the measures of the unknown angles.

(a)

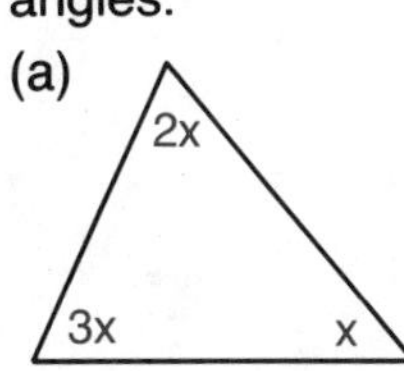

(b)

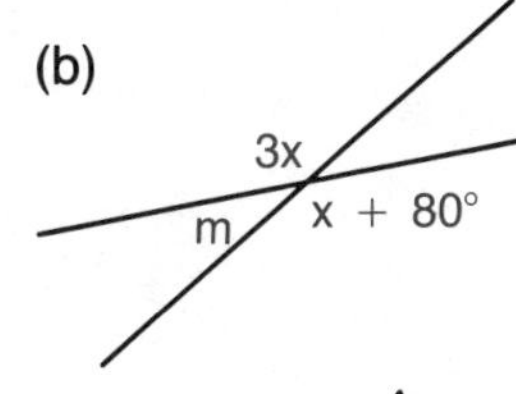

(c)

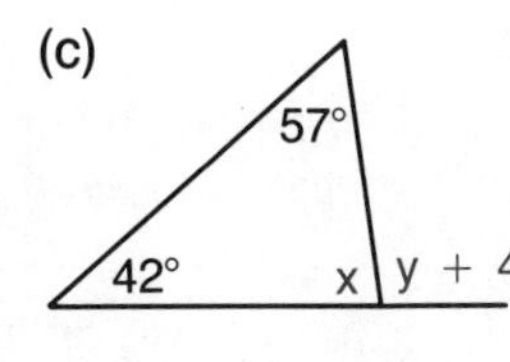

(d)

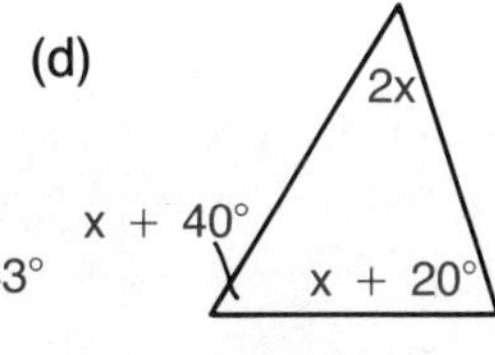

16. Find the measures of the unknown angles.

(a)

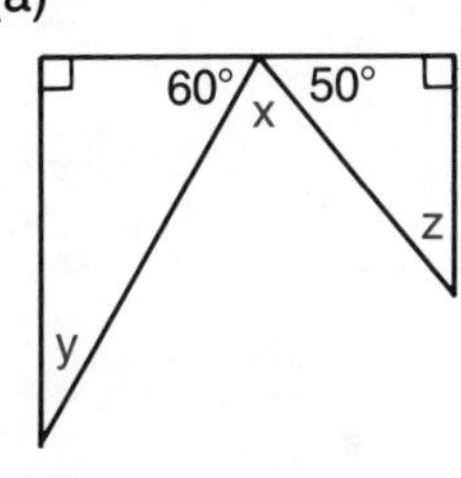

(b)

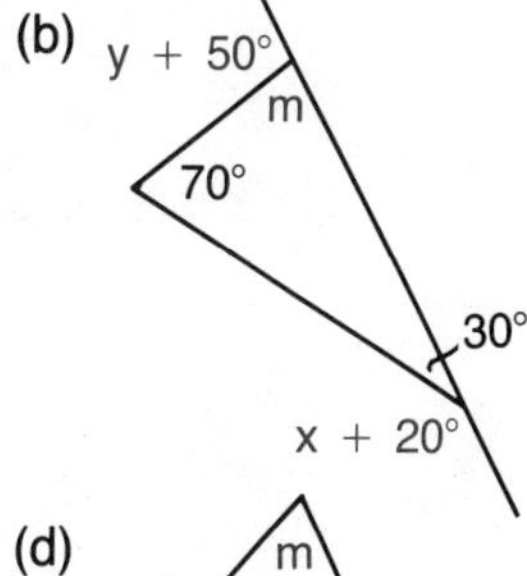

(c)

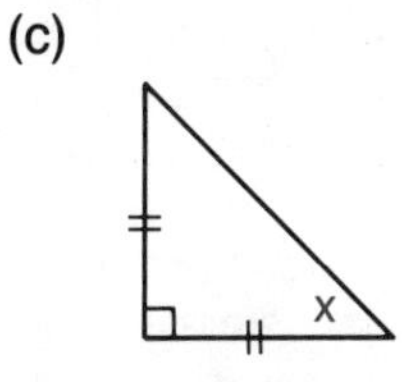

(d)

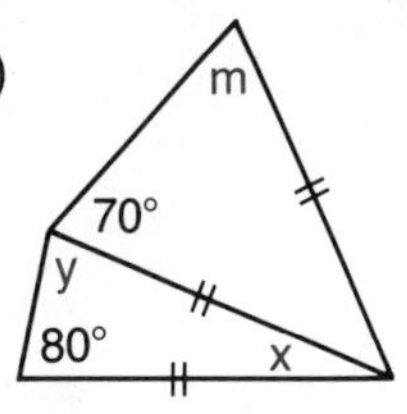

(e)

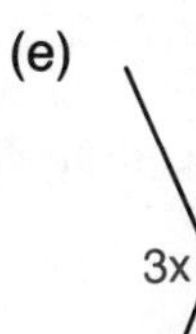

(f)

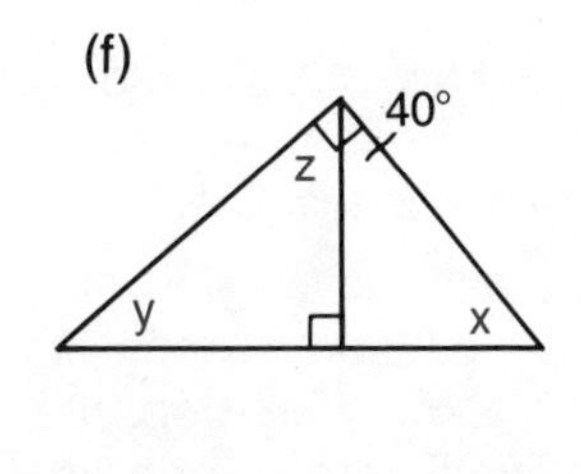

10.2 PARALLEL LINES

Parallel lines are in the same plane but never meet. You can find many examples of parallel lines around you.

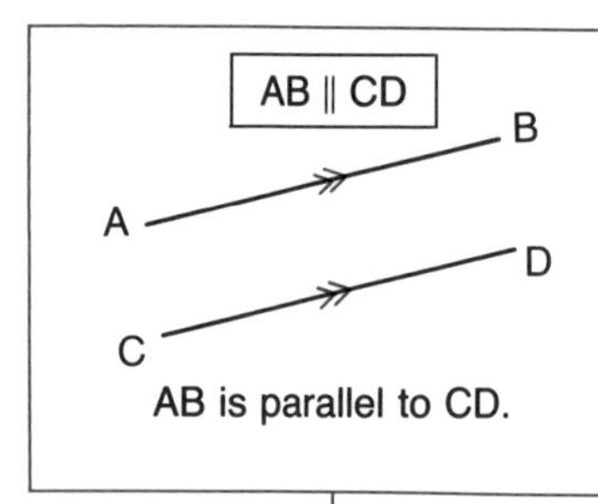

AB is parallel to CD.

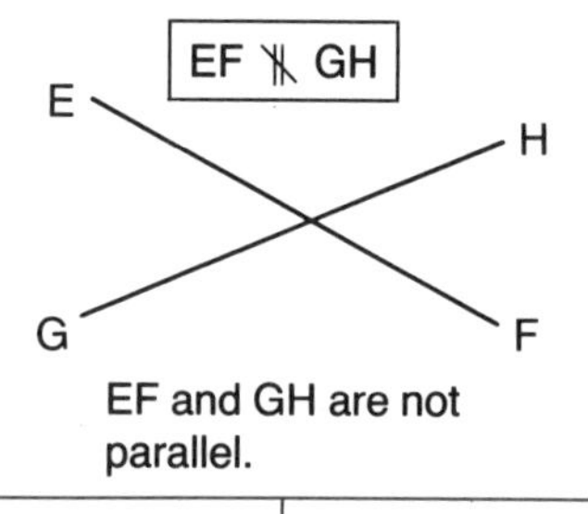

EF and GH are not parallel.

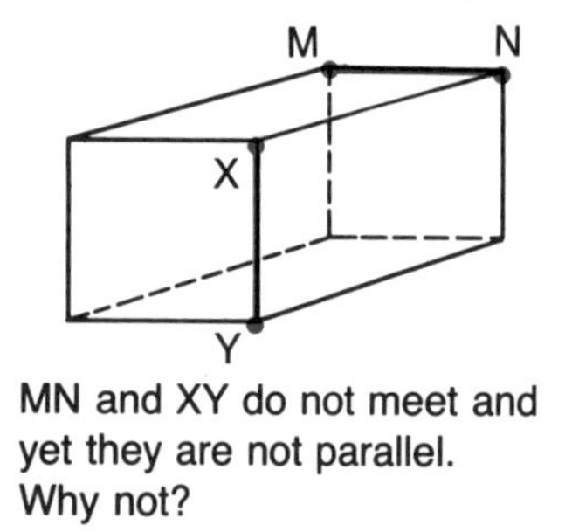

MN and XY do not meet and yet they are not parallel. Why not?

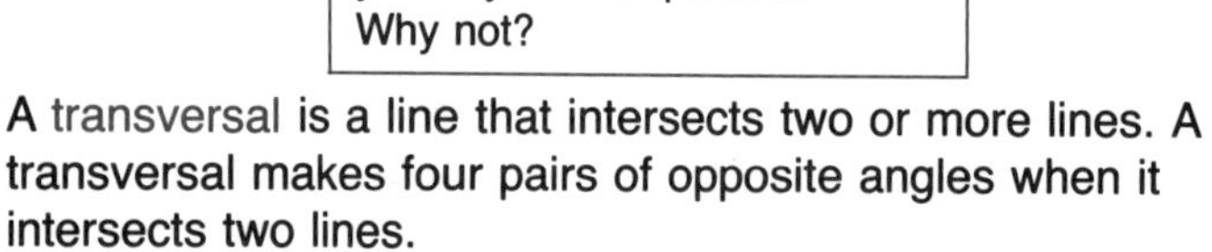

A transversal is a line that intersects two or more lines. A transversal makes four pairs of opposite angles when it intersects two lines.

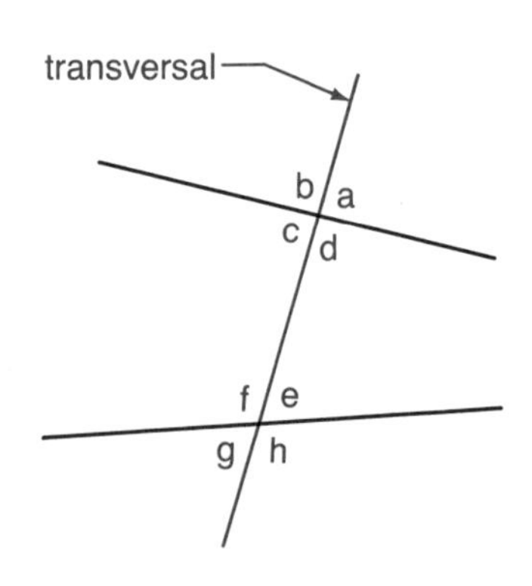

∠a and ∠c are opposite.
∠b and ∠d are opposite.
∠e and ∠g are opposite.
∠f and ∠h are opposite.

The diagram at the right shows two parallel lines cut by a transversal. The angles formed by the transversal and the lines can be paired in different ways and are given special names.

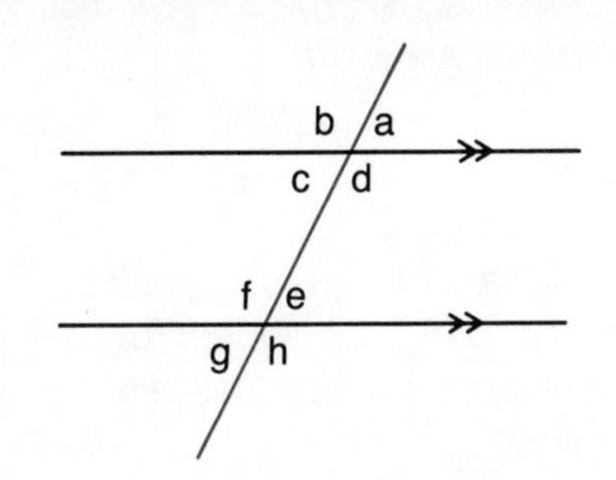

Alternate angles	∠c and ∠e ∠d and ∠f	These angles form a Z pattern.
Corresponding angles	∠a and ∠e ∠b and ∠f ∠c and ∠g ∠d and ∠h	These angles form an F pattern.
Interior angles on the same side of the transversal	∠c and ∠f ∠d and ∠e	These angles form a C pattern.

EXERCISE 10.2

A 1. For each diagram list all pairs of:
(a) alternate angles
(b) corresponding angles
(c) interior angles on the same side of the transversal

(i)

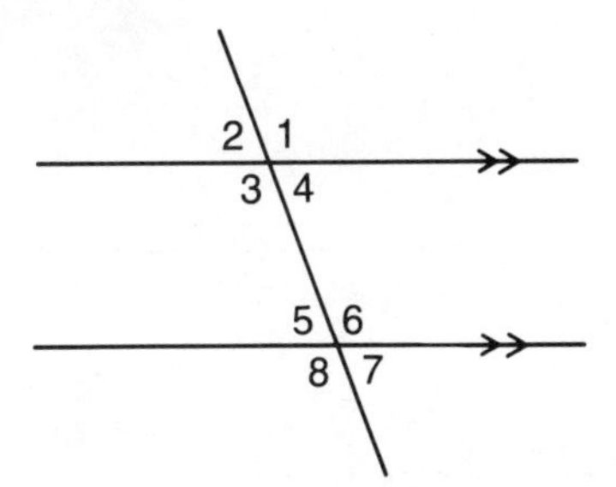

(ii)

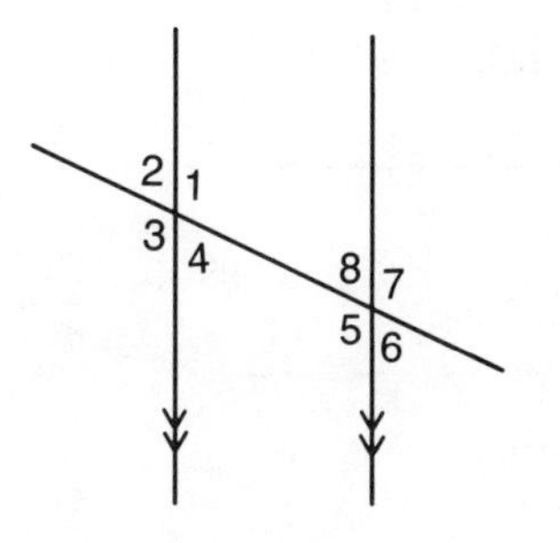

(iii)

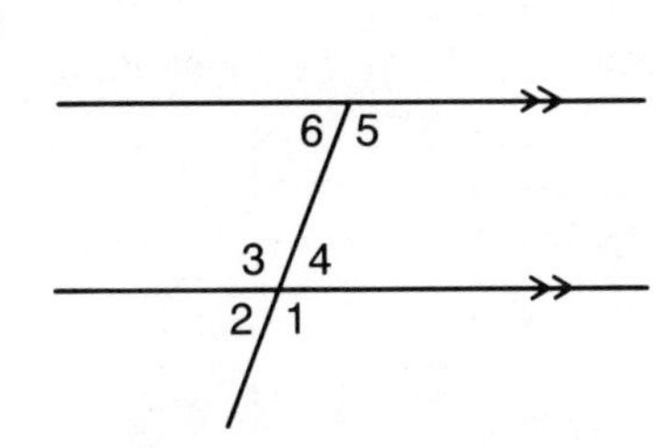

(iv)

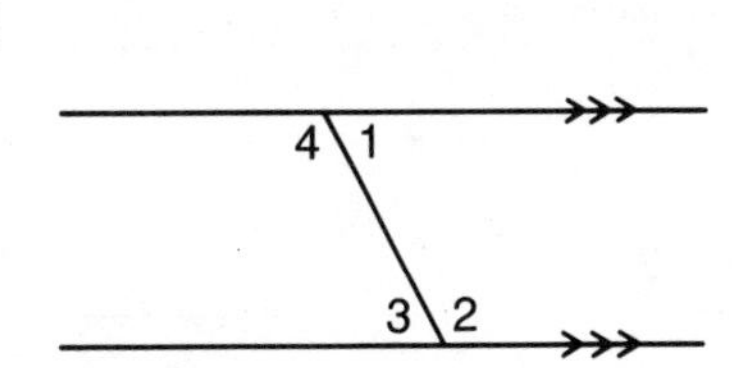

(v)

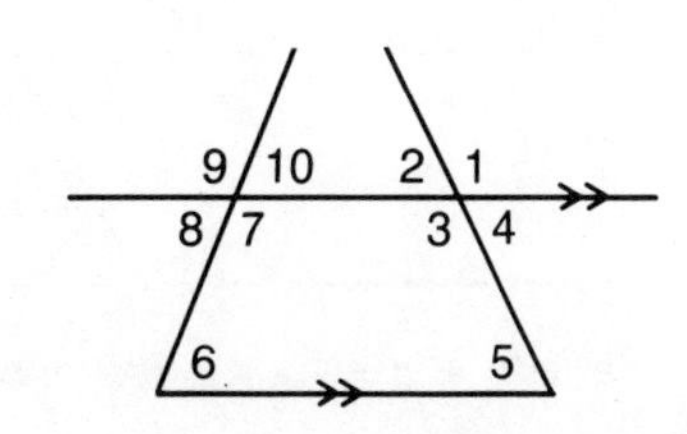

INVESTIGATION

Follow the steps to investigate the relationships between angles formed by a transversal intersecting two parallel lines.

1. Draw parallel lines RS and PQ. One way is to place your ruler down on the page and draw a line on each side.
2. Draw a transversal to intersect the parallel lines. Label the angles a, b, c, d, e, f, g, and h as shown.

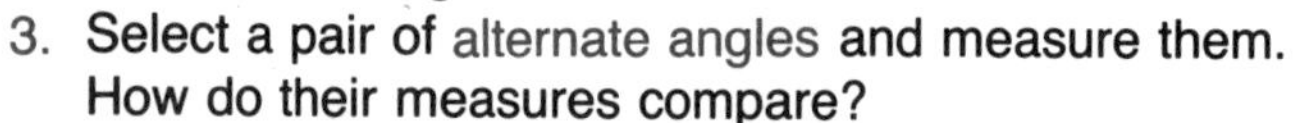

3. Select a pair of alternate angles and measure them. How do their measures compare?
4. Select another pair of alternate angles. Measure them. How do their measures compare?
5. Select a pair of corresponding angles and measure them. How do their measures compare?
6. Repeat step 4 for 3 other pairs of corresponding angles.
7. Select a pair of interior angles and measure them. How do their measures compare?
8. Repeat step 6 for another pair of interior angles.

Compare your results of the investigation to the chart below.

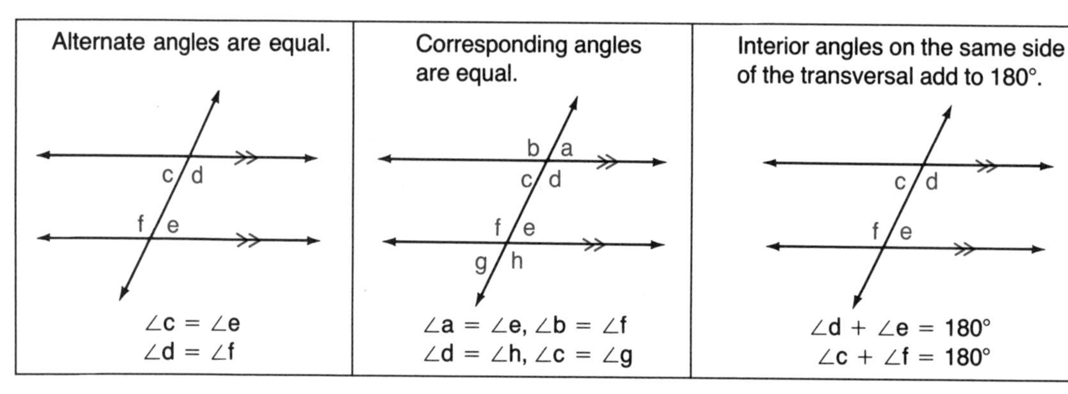

You can remember the angle pairs by thinking of the Z, F, and C patterns.

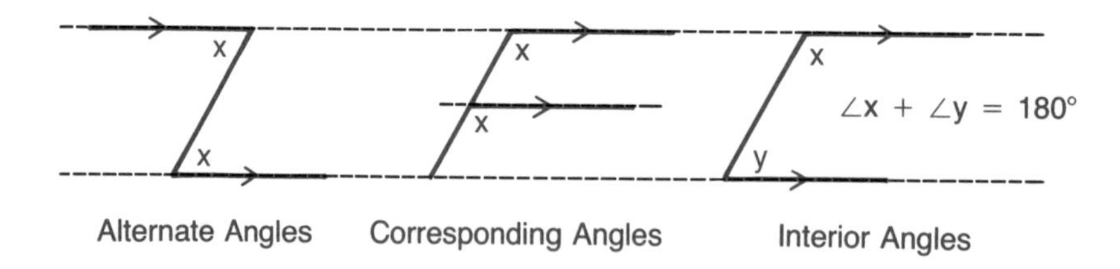

EXERCISE 10.2 (continued)

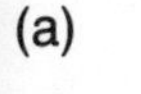

2. Find the measure of each missing angle.

(a) (b)

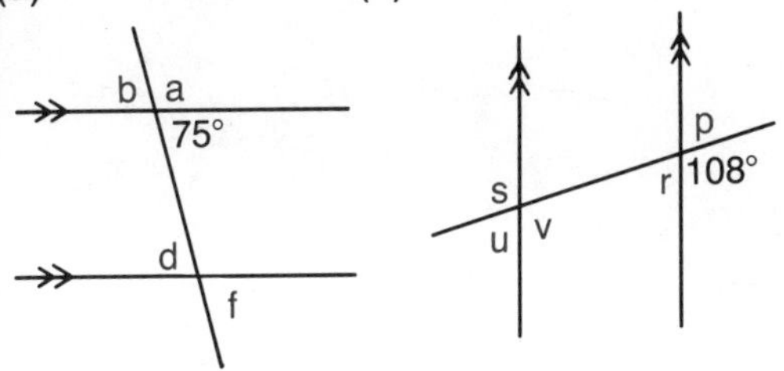

(c) (d)

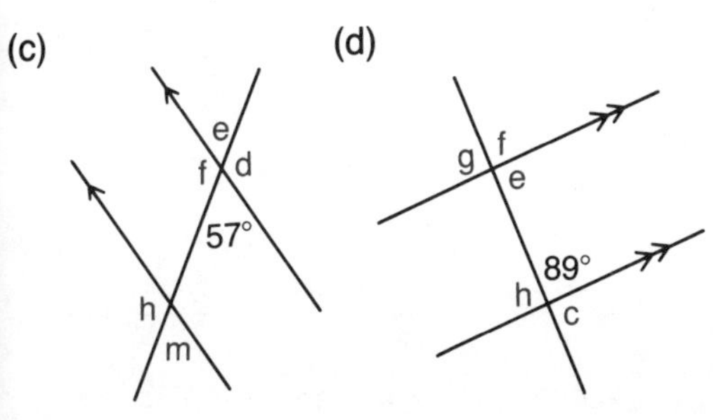

3. Find the measure of each missing angle.

(a) (b)

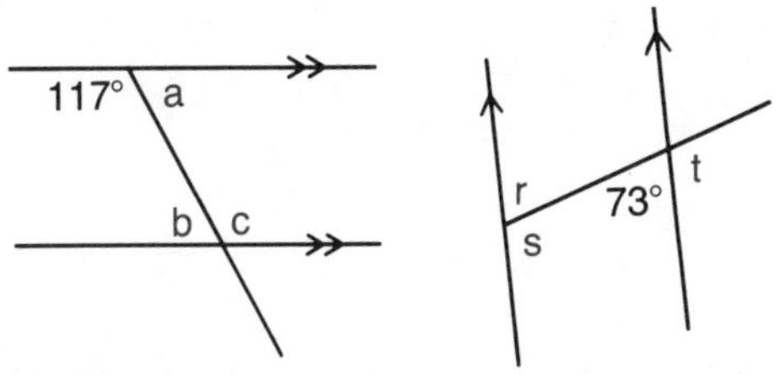

4. Find the measure of each missing angle.

(a) (b)

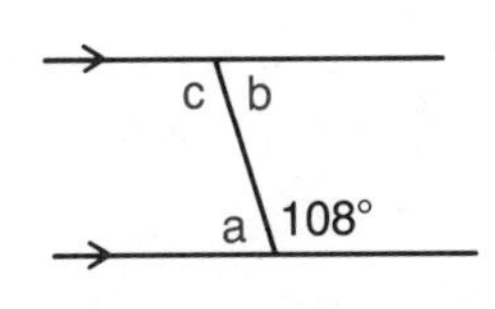

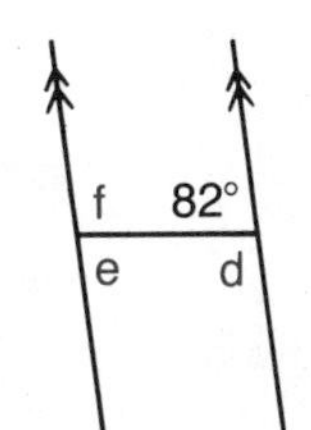

(c) (d)

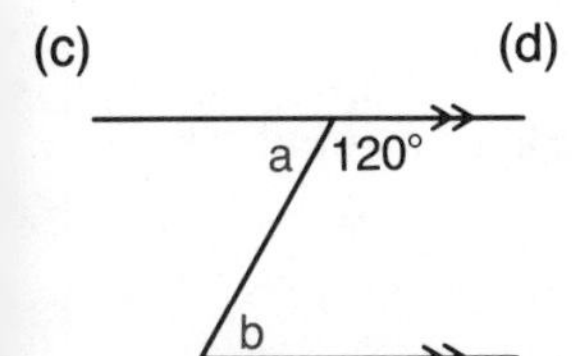

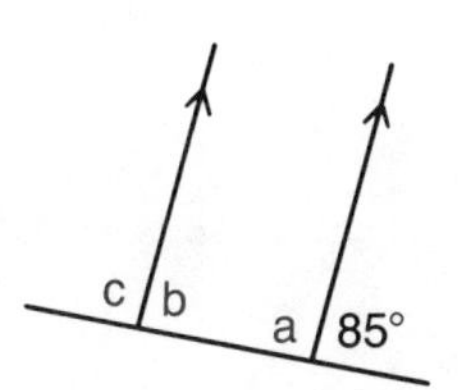

5. Find the measure of each missing angle.

(a) (b)

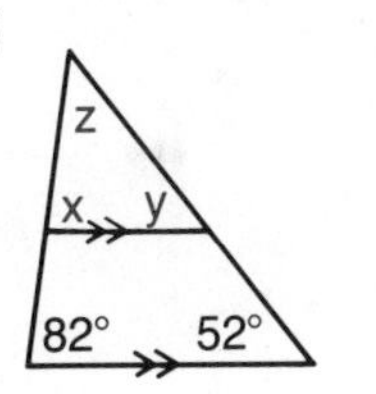

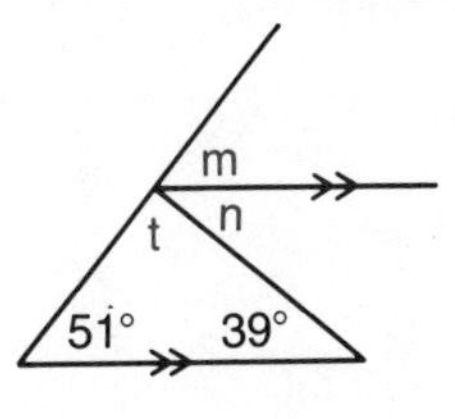

(c) (d)

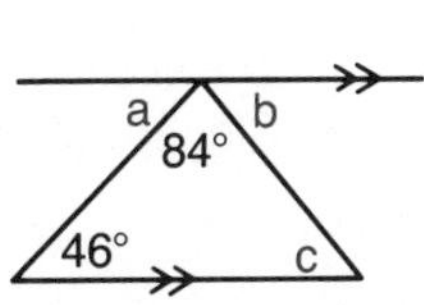

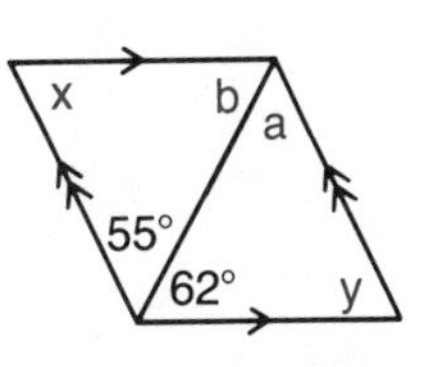

6. The captain of a submarine uses a periscope to look outside. Why is the ray of light entering the periscope parallel to the ray of light entering the viewer's eye?

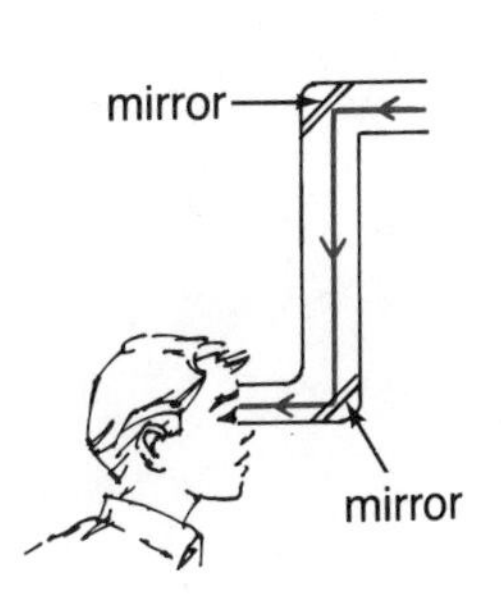

7. How would you build an adjustable ironing board so that the top is always parallel to the floor?

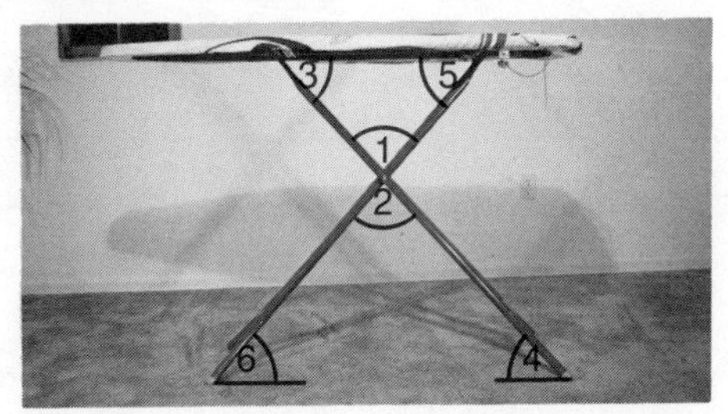

MIND BENDER

Suppose a used car salesperson sold a car one day for $2000 and made a 10% profit. The following week he sold another car for $2000 at a 10% loss. Compare the two deals: did he make or lose money?

10.3 QUADRILATERALS

A quadrilateral has four sides. We use these terms to describe quadrilaterals.

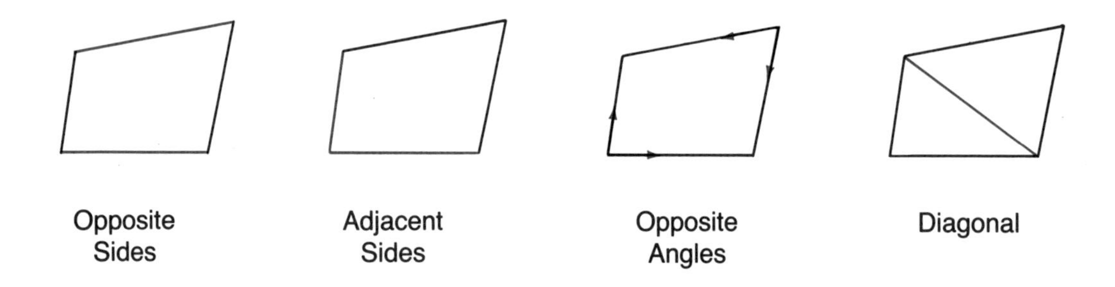

Quadrilaterals have different shapes.

Quadrilateral	Diagram	Definition
Trapezoid		A quadrilateral with one pair of parallel sides.
Parallelogram		A quadrilateral with both pairs of opposite sides parallel.
Rectangle		A parallelogram with one right angle.
Rhombus		A parallelogram with all sides equal.
Square		A quadrilateral with all sides equal and four right angles.

READ

PLAN

SOLVE

ANSWER

Example.
Calculate the sum of the interior angles of quadrilateral ABCD.

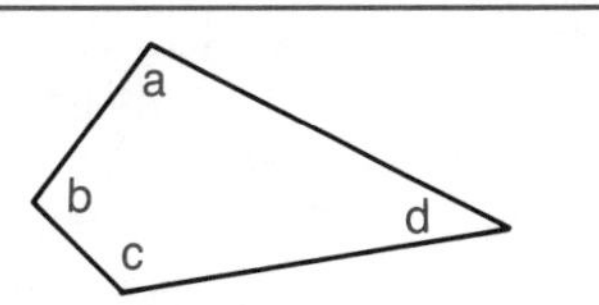

Solution:
Draw diagonal BD.
Since the sum of the interior angles of a triangle is 180°, then

In $\triangle ABD$, $\angle A + \angle ABD + \angle ADB = 180°$
In $\triangle BCD$, $\angle C + \angle CBD + \angle CDB = 180°$

Adding

$\angle A + \angle C + (\angle ABD + \angle CBD) + (\angle ADB + \angle CDB) = 360°$
$\angle A + \angle C + \angle B + \angle D = 360°$

You can treat any quadrilateral in this way.

The sum of the interior angles of a quadrilateral is 360°.
$\angle a + \angle b + \angle c + \angle d = 360°$

EXERCISE 10.3

A 1. Identify the following figures.

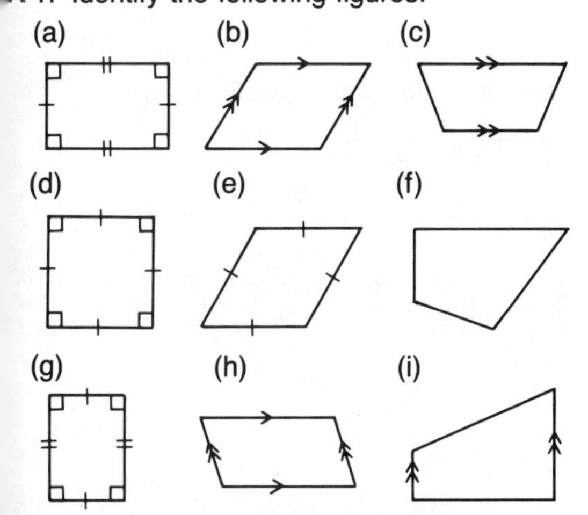

B 2. Find the measure of the missing angle.

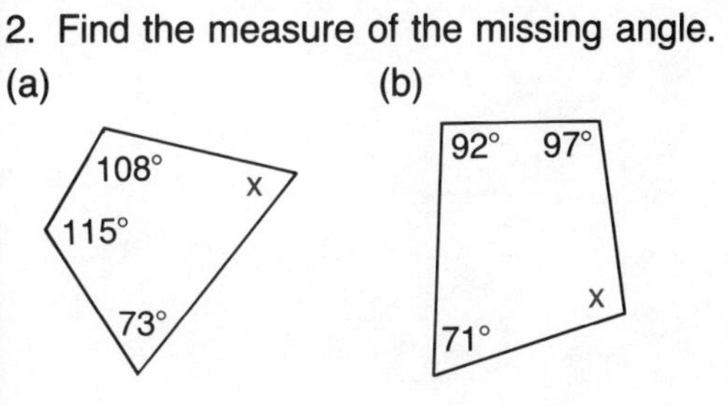

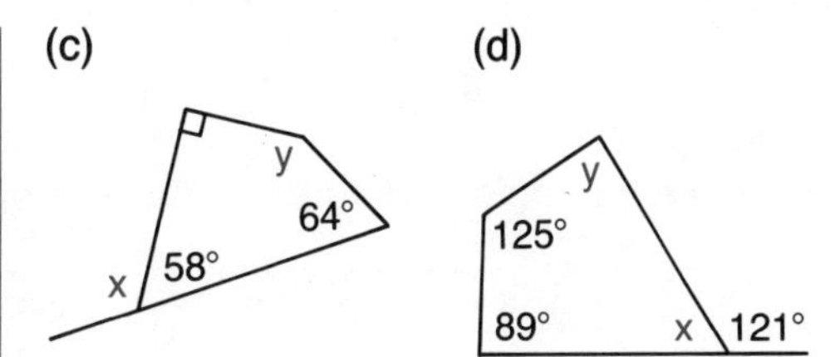

3. (a) Calculate the measures of the missing angles in each parallelogram.

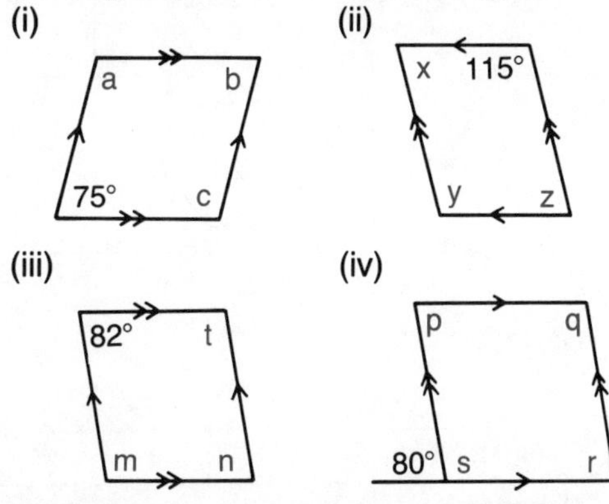

(b) What conclusion can you draw about the opposite angles of a parallelogram?

4. Find the unknown lengths and measures of the indicated angles.

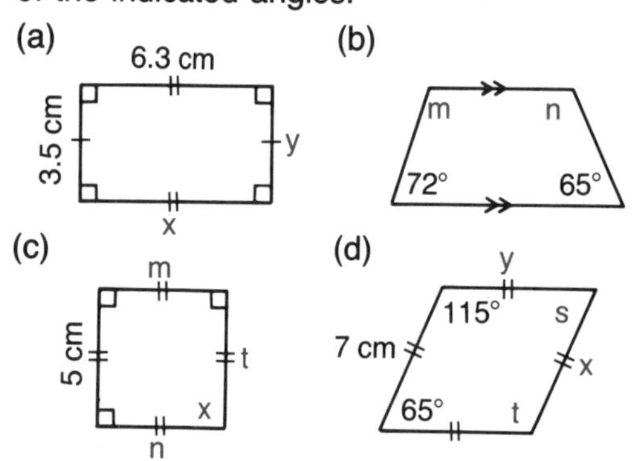

5. Find the measures of the missing angles.

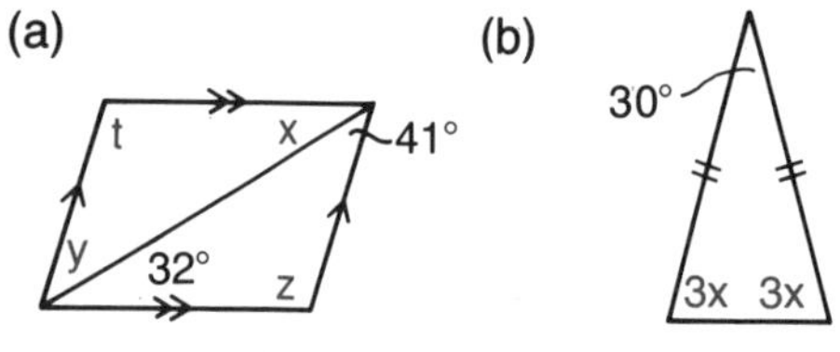

(c)

r
m
x
y
81°

(d)

x
x
62°
y
67°

6. Find the measures of the missing angles.

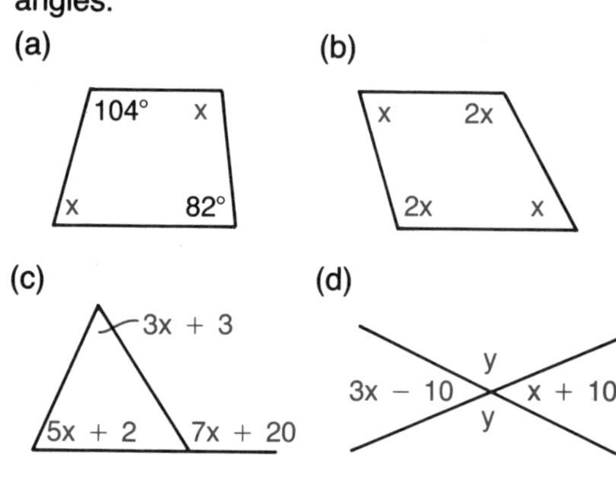

INVESTIGATION

7. Complete the table to determine the sum of the interior angles in the given polygons.

Polygon	Number of Sides	Number of Triangles	Sum of Interior Angles
Pentagon	5	3	$180° \times 3 = 540°$
Hexagon	6	4	
Septagon	7		
Octagon	8		
Nonagon	9		
Decagon	10		

10.4 RULER AND COMPASS CONSTRUCTIONS

In this section six basic ruler and compass constructions are reviewed.

BISECTING AN ANGLE

Bisect ∠RST.

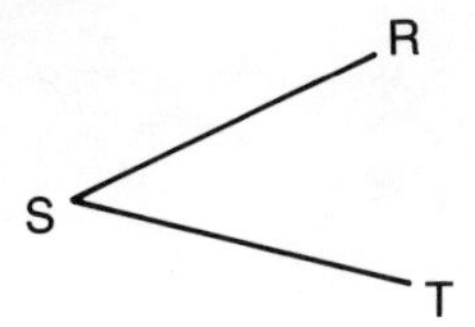

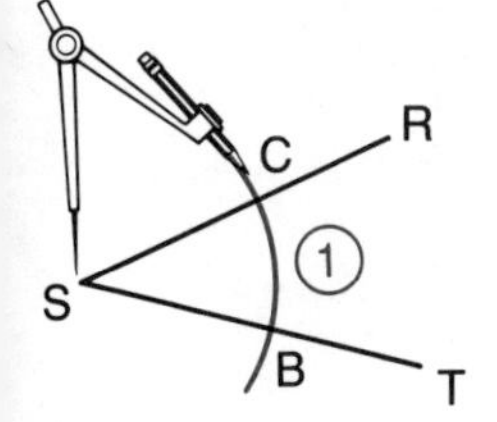

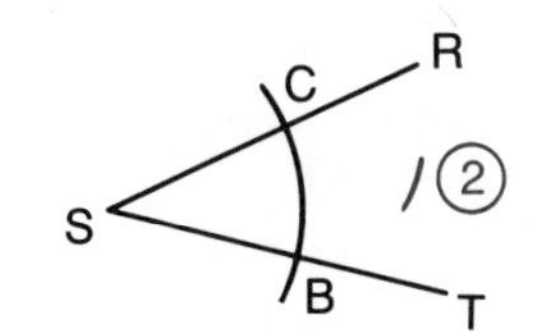

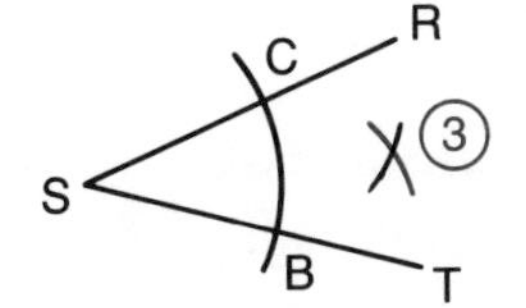

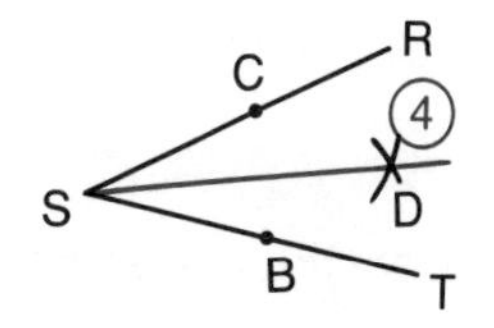

CONSTRUCTING A RIGHT BISECTOR OF A LINE SEGMENT

Construct the right bisector of AB.

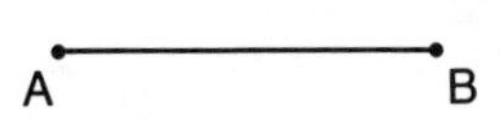

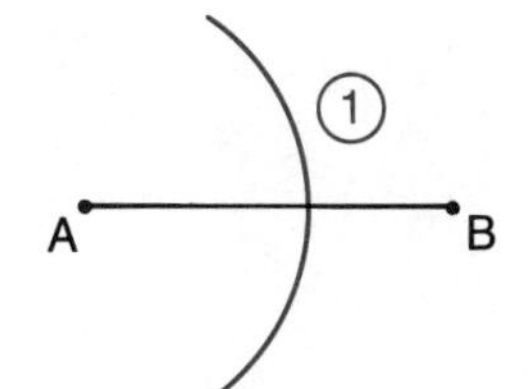

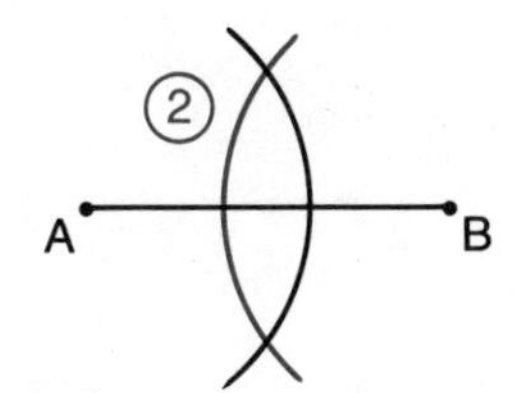

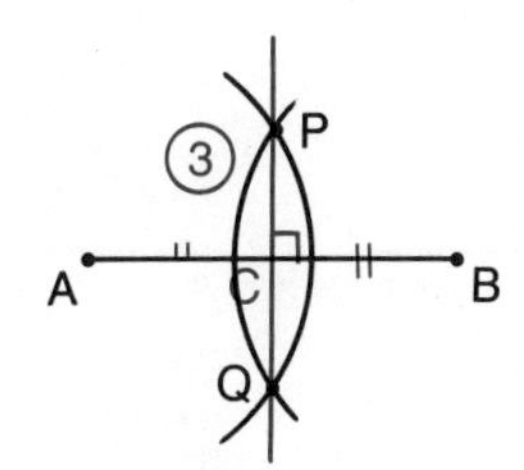

CONSTRUCTING A PERPENDICULAR TO A LINE FROM A POINT

Construct a perpendicular from P to AB. • P

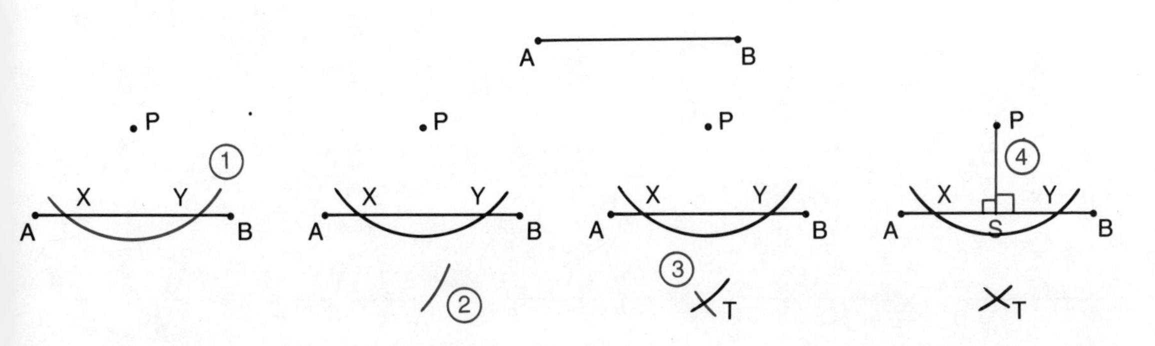

CONSTRUCTING A PERPENDICULAR TO A LINE AT A POINT

Construct a perpendicular to AB at Q.

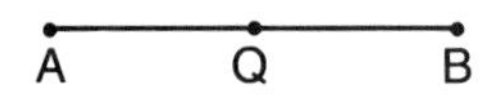

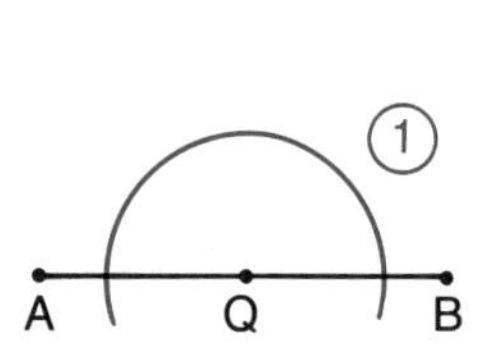

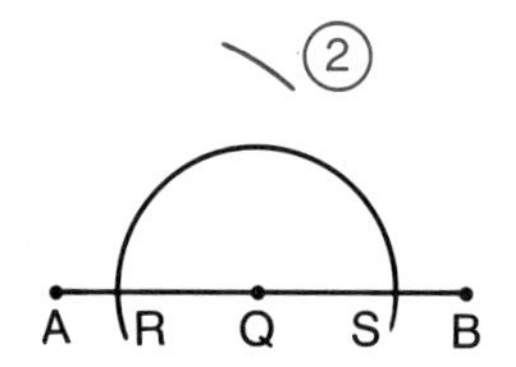

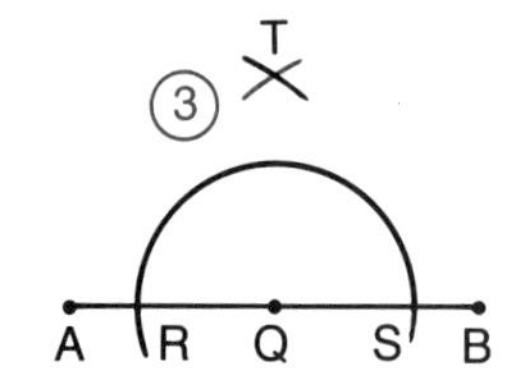

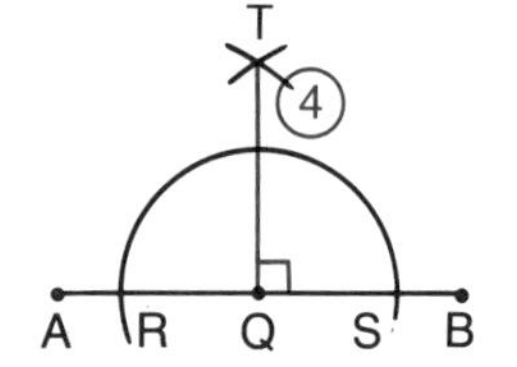

CONSTRUCTING AN ANGLE EQUAL TO A GIVEN ANGLE

Construct an angle equal to ∠DEF.

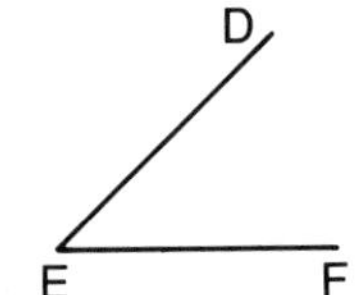

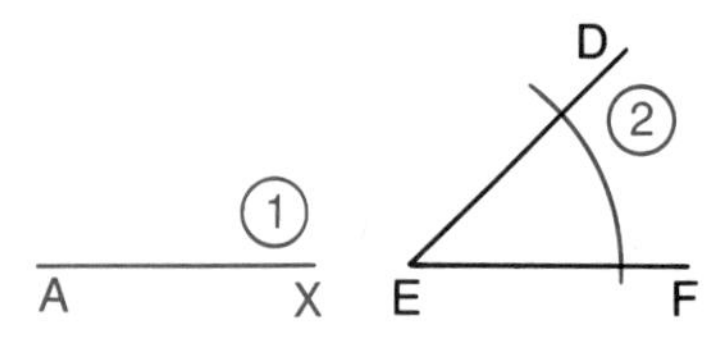

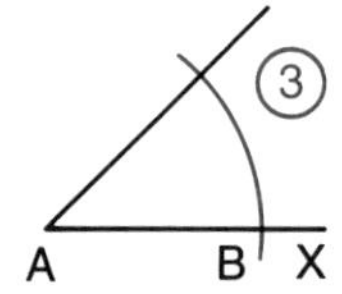

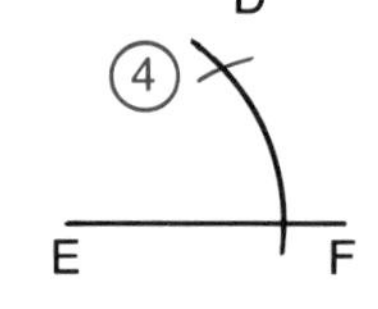

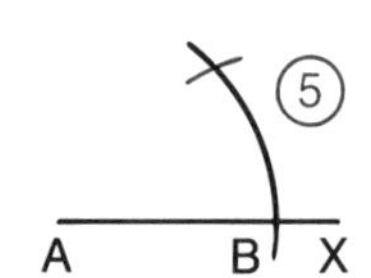

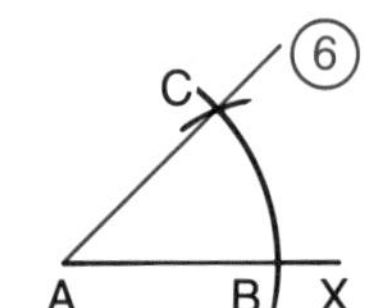

CONSTRUCTING A LINE THROUGH A POINT PARALLEL TO A LINE

Construct a line through P and parallel to AB.

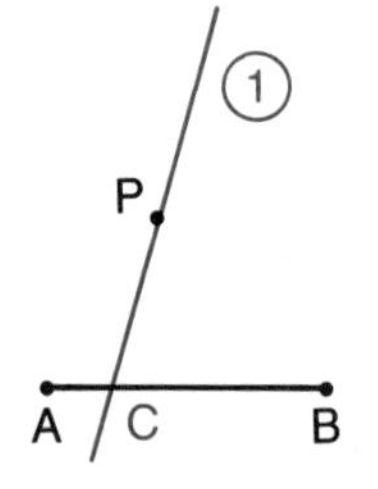

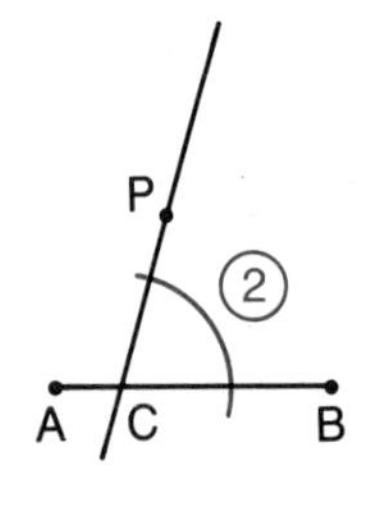

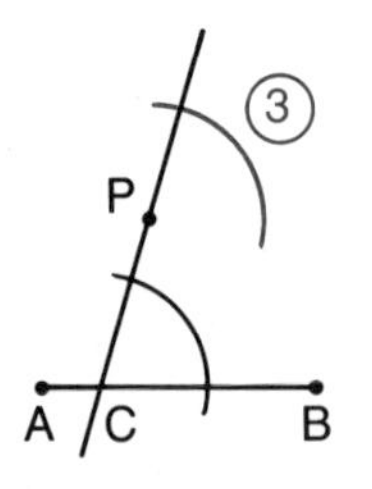

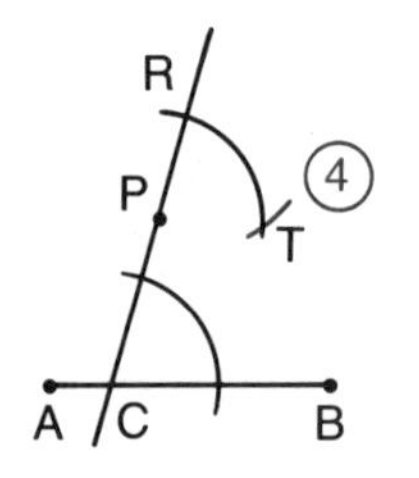

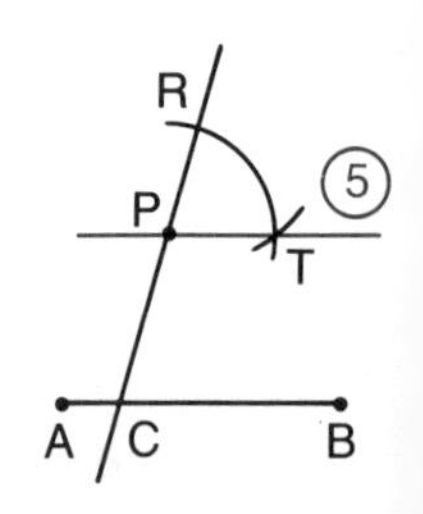

EXERCISE 10.4

B 1. Draw an angle similar to the one shown. Bisect each angle.

(a)

A

B C

(b)

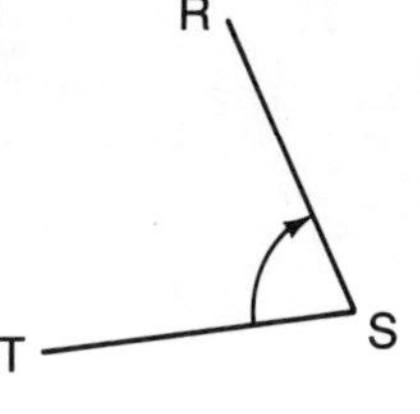

(c)

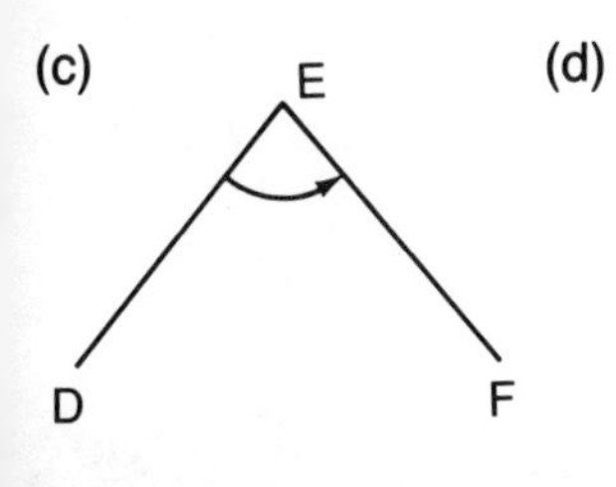

(d)

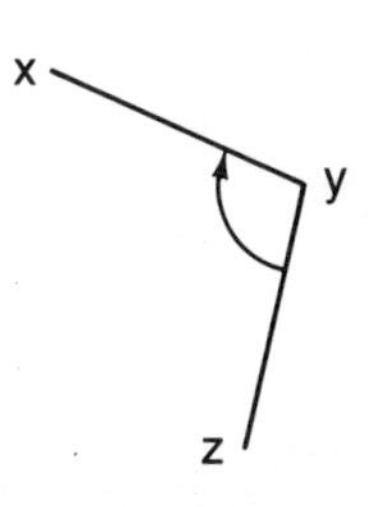

2. Draw a line segment similar to the one shown.
Construct a right bisector.

(a)

A B

(b)

3. Draw a diagram similar to the one shown.
Construct a perpendicular from P to the line segment.

(a) (b) (c) (d)

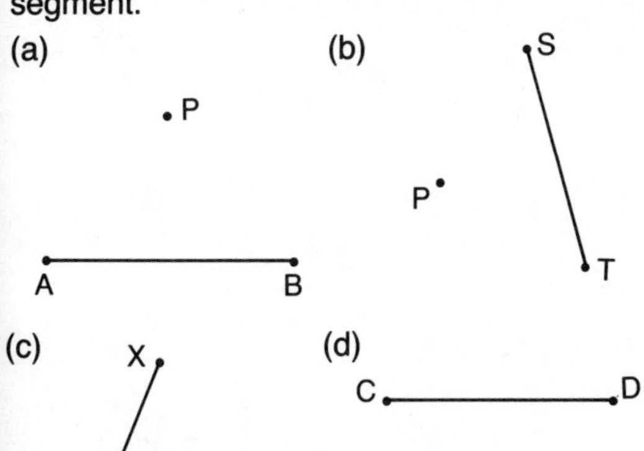

4. Draw a diagram similar to the one shown.
Construct the perpendicular to the line segment at P.

(a) (b)

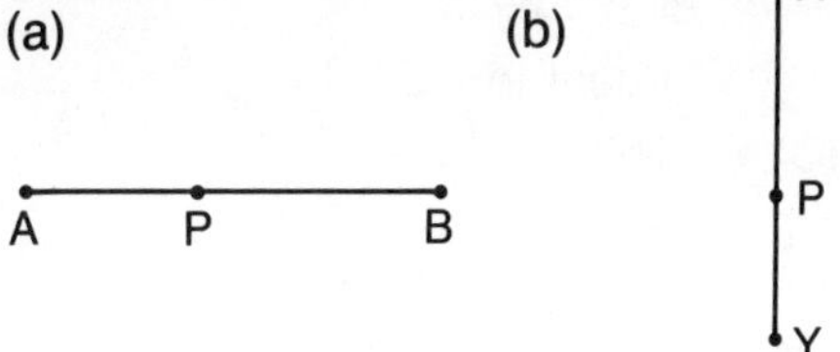

5. Draw an angle similar to the one shown.
Construct an angle equal to each.

(a) (b)

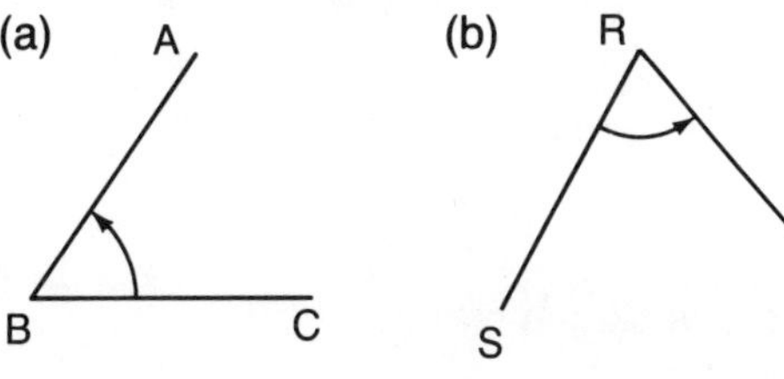

(c) (d)

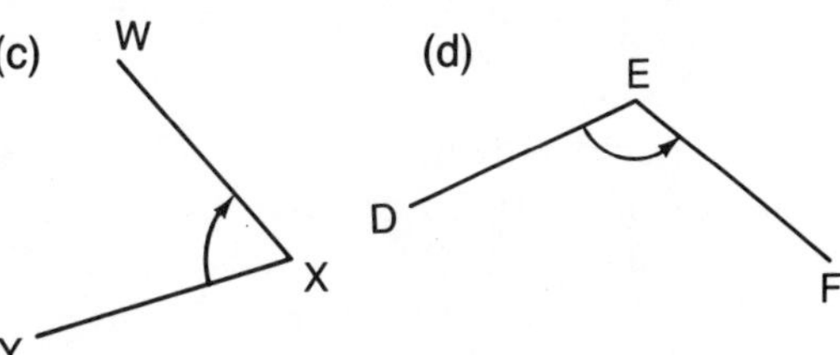

6. Draw a diagram similar to the one shown.
Construct a line through P parallel to each line segment.

(a) (b)

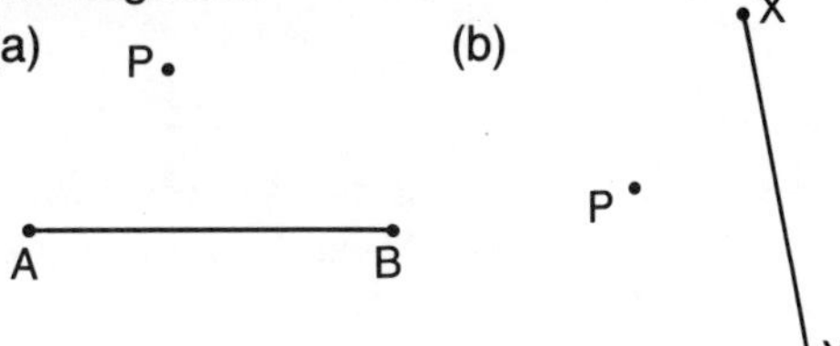

(c) (d)

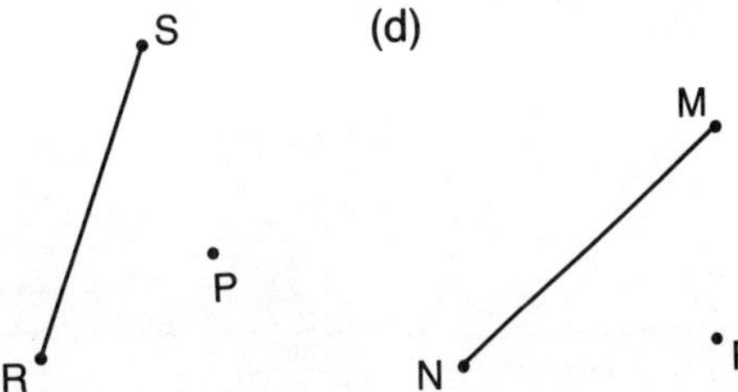

10.5 INVESTIGATING CONGRUENCE

Manufacturers could make each souvenir pennant a different shape, but they don't.

The pennants at the right are identical, or congruent. They are equal in all respects.

△ABC is congruent to △DEF.

$\triangle ABC \cong \triangle DEF$

AB = DE	and	∠A = ∠D
BC = EF		∠B = ∠E
AC = DF		∠C = ∠F

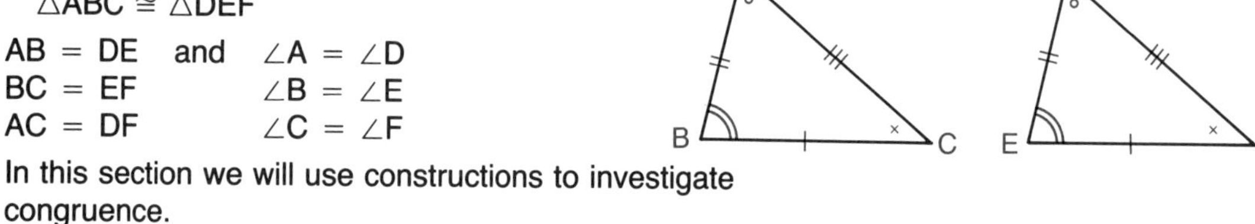

In this section we will use constructions to investigate congruence.

Example.
Construct △RST where, RS = 3.8 cm, ST = 5.0 cm, and RT = 4.4 cm.

Solution:
1. Make a sketch of the triangle.
2. Draw ray $\overrightarrow{SM}$ and mark ST = 5.0 cm.
3. With centre S and radius 3.8 cm draw an arc.
4. With centre T and radius 4.4 cm, draw an arc to intersect the first arc at R.
5. Join RS and RT.

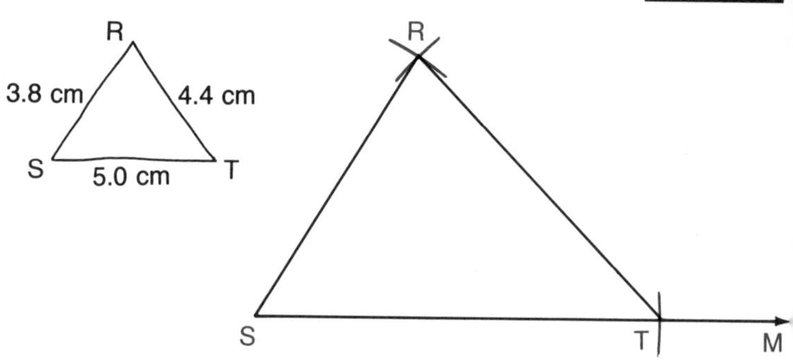

EXERCISE 10.5

B 1. Use the information in the sketches to construct the triangles. Measure and record all dimensions.

(a)

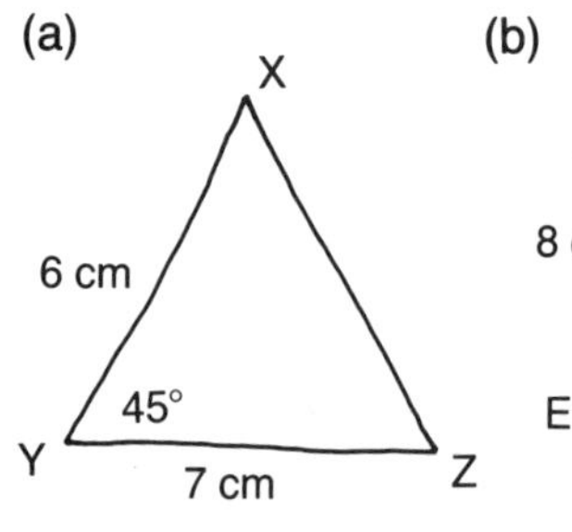

(b)

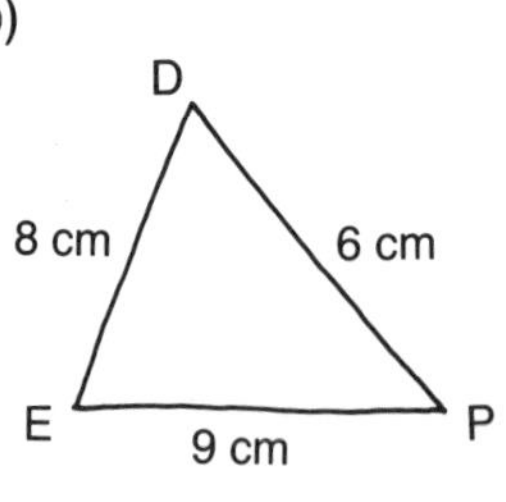

(c) (d)

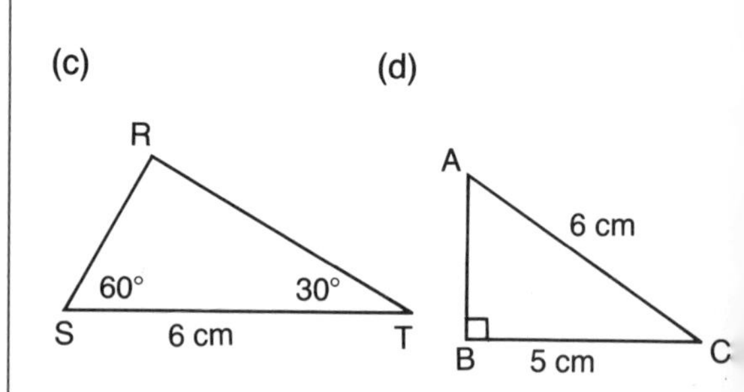

INVESTIGATIONS

2. (a) Construct $\triangle PQR$ where
   PQ = 6.5 cm
   QR = 7 cm
   PR = 9 cm
   Measure and record the missing parts.
   (b) Construct $\triangle ABC$ where
   BC = 7 cm
   AC = 6.5 cm
   AB = 9 cm
   Measure and record the missing parts.
   (c) How do $\triangle PQR$ and $\triangle ABC$ compare?
   (d) Compare your triangles with the triangles of other students in the class. Are the triangles congruent?

3. (a) Construct $\triangle RST$ where
   RS = 5.5 cm
   $\angle S = 60°$
   ST = 6 cm
   Measure and record the missing parts.
   (b) Construct $\triangle MNL$ where
   $\angle L = 60°$
   NL = 6 cm
   ML = 5.5 cm
   Measure and record the missing parts.
   (c) How do $\triangle RST$ and $\triangle MNL$ compare?
   (d) Compare your triangles with the triangles of other students in the class. Are the triangles congruent?

4. (a) Construct $\triangle XYZ$ where
   YZ = 6.2 cm
   $\angle Y = 45°$
   $\angle Z = 30°$
   Measure and record the missing parts.
   (b) Construct $\triangle GHI$ where
   $\angle H = 30°$
   $\angle I = 45°$
   HI = 6.2 cm
   Measure and record the missing parts.
   (c) How do $\triangle XYZ$ and $\triangle GHI$ compare?
   (d) Compare your triangles with the triangles of other students in the class. Are the triangles congruent?

5. (a) Construct $\triangle JKL$ where
   $\angle K = 90°$
   KL = 6.7 cm
   JL = 7.8 cm
   Measure and record the missing parts.
   (b) Construct $\triangle UVW$ where
   WU = 7.8 cm
   VW = 6.7 cm
   $\angle V = 90°$
   Measure and record the other dimensions.
   (c) How do $\triangle JKL$ and $\triangle UVW$ compare?
   (d) Compare your triangles with the triangles of other students in the class. Are the triangles congruent?

6. It is not necessary to know the measures of all three lengths or all three sides of a triangle before you can construct a congruent triangle. In the investigations you used combinations of angles and lengths to construct triangles.
   List four combinations of three lengths and angles that would enable you to construct a congruent triangle.

7. Identify pairs of congruent triangles.

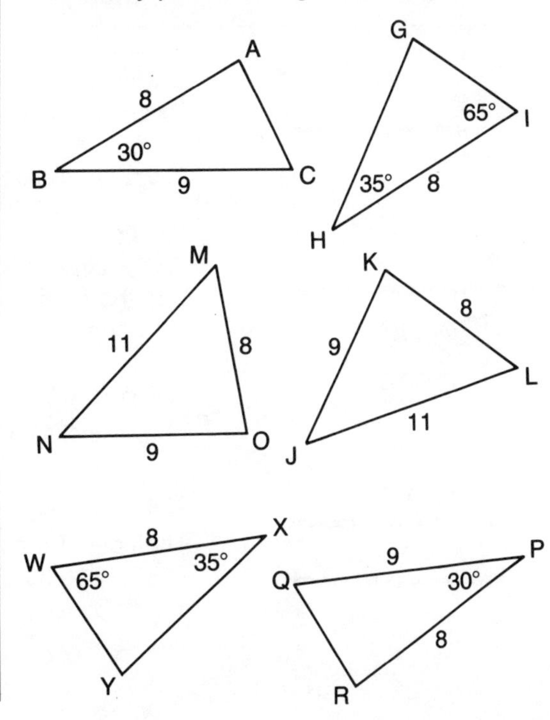

10.6 APPLYING PERIMETER

The distance around a figure is called the perimeter.

The Southside Athletic Department is sponsoring a 3 km run for charity. To calculate how many laps of the field will add to 3 km, you need to know the distance around the field or the perimeter.

The perimeter of the field is

$$175 \text{ m} + 89 \text{ m} + 155 \text{ m} + 81 \text{ m} = 500 \text{ m}$$

Since 3 km = 3000 m, then there are 6 laps of the field in a 3 km run.

The table gives the formulas for the perimeter of several figures. The perimeter of a circle is called the circumference.

Triangle	Quadrilateral	Square	Rectangle	Circle
a, b, c	a, b, c, d	s, s, s, s	ℓ, w	r
$P = a + b + c$	$P = a + b + c + d$	$P = 4s$	$P = 2(\ell + w)$ $P = 2\ell + 2w$	$C = 2\pi r$ $C = \pi d$

READ

Example.
Calculate the cost of fencing the pool area if fencing costs $15.50/m to install.

PLAN

Solution:
The pool area is a rectangle. Use the formula to calculate the perimeter.

$$P = 2(\ell + w)$$
$$P = 2(72 + 46)$$
$$= 2(118)$$
$$= 236$$

SOLVE

The perimeter is 236 m.

ANSWER

The cost of fencing is
236 × $15.50 or $3658.00

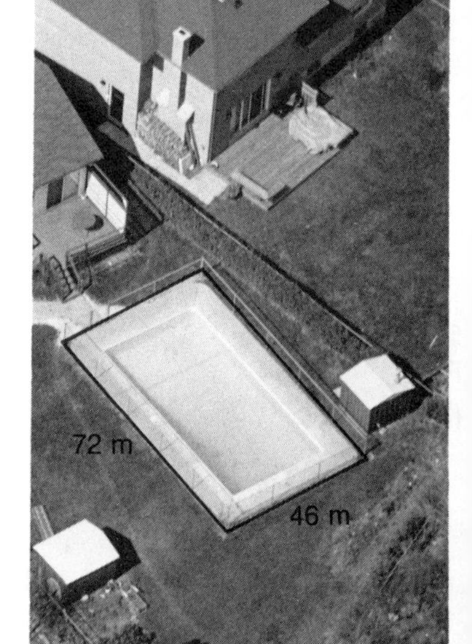

EXERCISE 10.6

B 1. Calculate the perimeter of each figure.

(a)

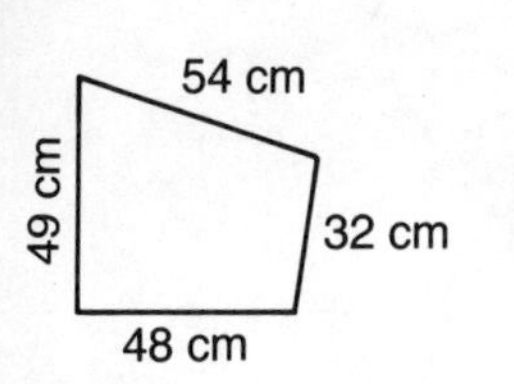

(b)

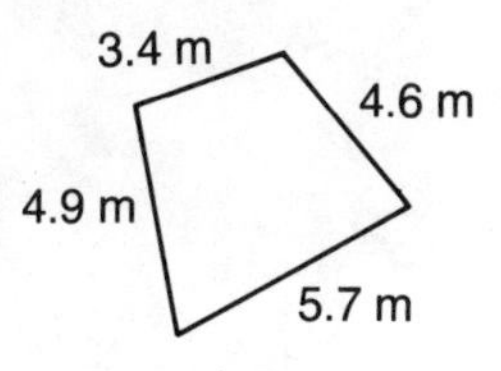

2. Calculate the perimeter of each patio.

(a)

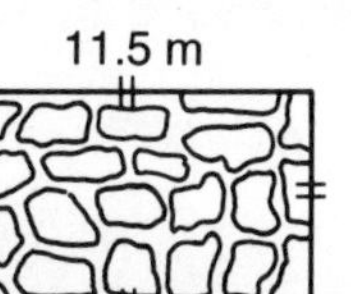

(b)

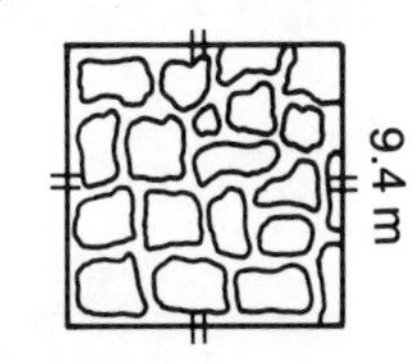

3. Calculate the circumference to the nearest tenth. (Use $\pi = 3.14$)

(a)

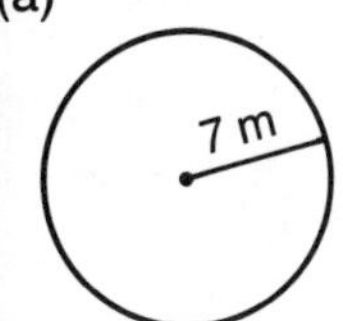

(b)

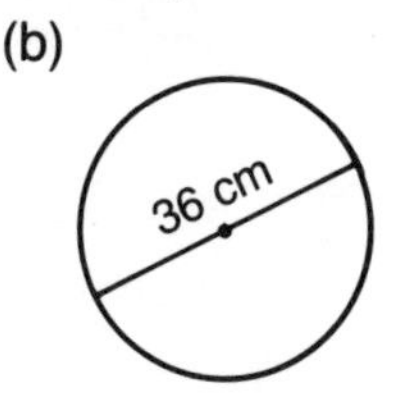

4. Calculate the perimeter to the nearest tenth.

(a)

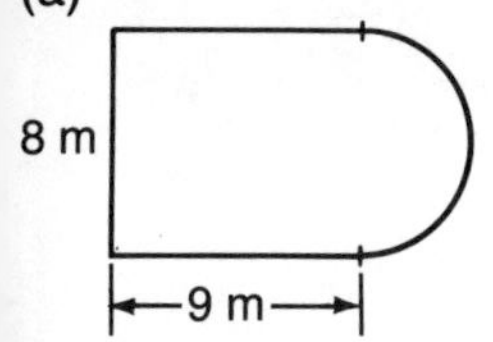

(b)

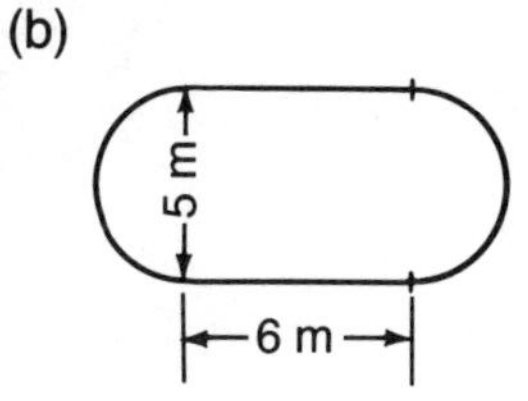

5. The perimeter of each figure is given. Find the missing dimension.

(a)

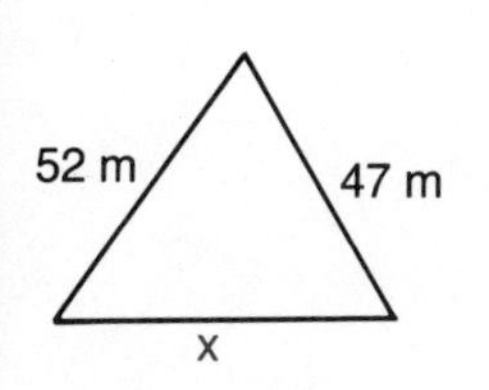

P = 159 m

(b)

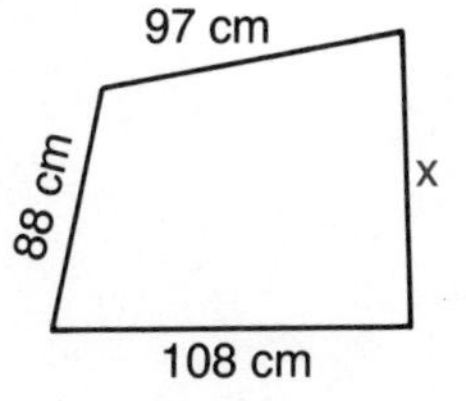

P = 406 cm

6. Find the perimeter of a rectangular field with a length of 28.4 m and a width of 17.9 m.

7. Find the perimeter of a square with each side measuring 23.6 cm.

8. Find the circumference, to the nearest tenth, of a circle with a radius of 13 m.

9. Rugby is played on a field that measures 146.3 m by 68.62 m. What is the perimeter of the field?

10. A soccer field measures 100 m by 73 m. To keep the fans off the field a fence is installed 10 m from each side of the field.

(a) Calculate the amount of fencing needed.

(b) Calculate the cost of the fence if fencing costs $16.75/m to install.

11. The radius of a basketball hoop is 28 cm.
Calculate the circumference of the hoop to the nearest tenth.

12. Tom has been hired to put aluminum trim around a factory's windows. There are 106 windows. Each window measures 1.2 m by 2.7 m.
How much aluminum trim does he need?

13. The diagram shows the distances travelled by the park rangers on a night patrol.

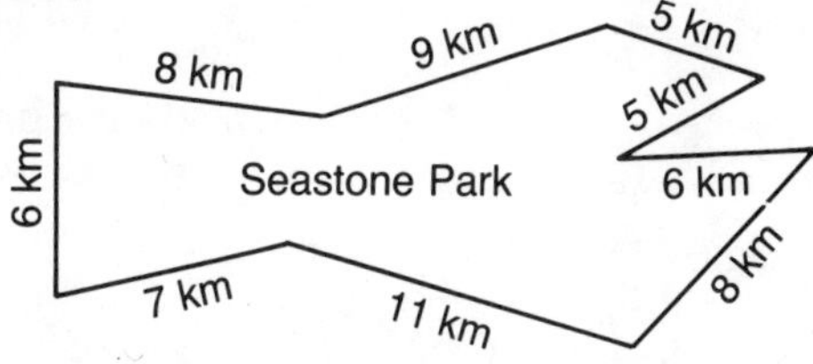

(a) What is the total distance they travel?

(b) Rangers travel in jeeps and drive at 10 km/h.
How long do they take to travel the route once?

14. Calculate the circumference of a quarter.

10.7 APPLYING AREA

The area of a figure is the measure of the amount of surface.
What is the area of the concert stage if each block is 1 m^2?
The dimensions of the stage are 6 m by 5 m.

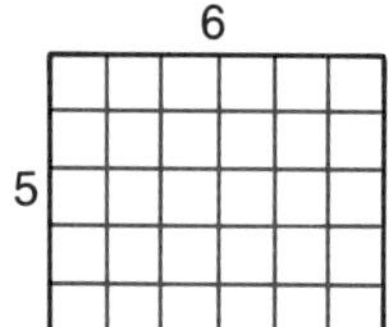

The area of the stage is 30 m^2.

The table gives the formulas for the area of different figures.

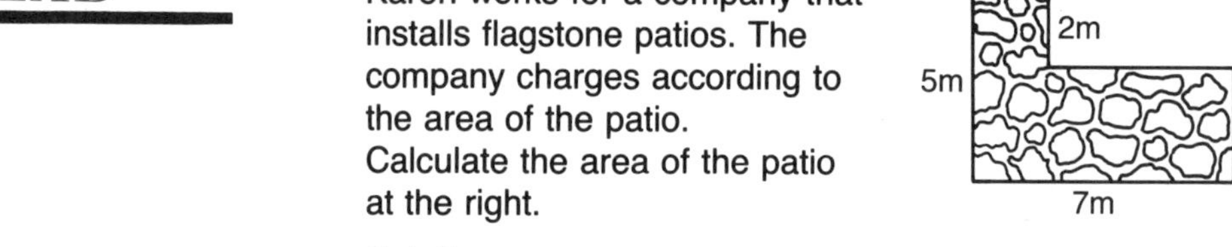

Square	Rectangle	Triangle	Parallelogram	Trapezoid	Circle
s, s	ℓ, w	h, b	h, b	a, h, b	r
$A = s^2$	$A = \ell \times w$	$A = \frac{1}{2} \times b \times h$	$A = b \times h$	$A = \frac{1}{2} \times h(a + b)$	$A = \pi r^2$

READ

Example.
Karen works for a company that installs flagstone patios. The company charges according to the area of the patio.
Calculate the area of the patio at the right.

2m
2m
5m
7m

PLAN

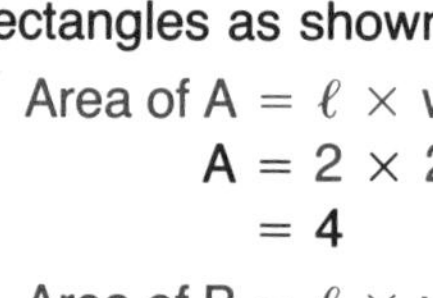

Solution:
Divide the patio into two rectangles as shown.

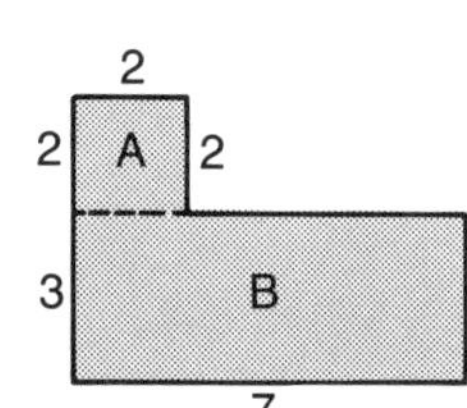

SOLVE

Area of A $= \ell \times w$
$A = 2 \times 2$
$= 4$

Area of B $= \ell \times w$
$B = 7 \times 3$
$= 21$

Total Area $= A + B$
$= 4 + 21$
$= 25$

ANSWER

The area of the patio is 25 m^2.

EXERCISE 10.7

1. State the area of each figure.

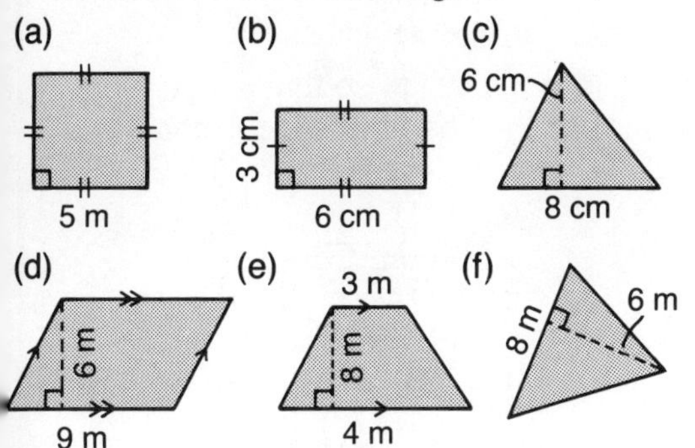

2. Calculate the area of each figure.

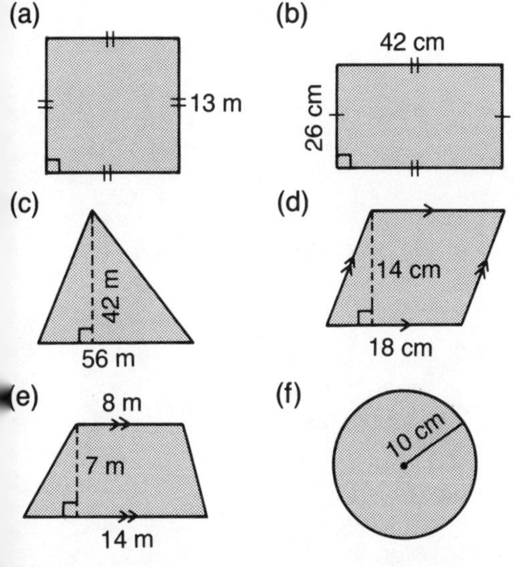

3. Calculate the area of each patio to the nearest tenth.

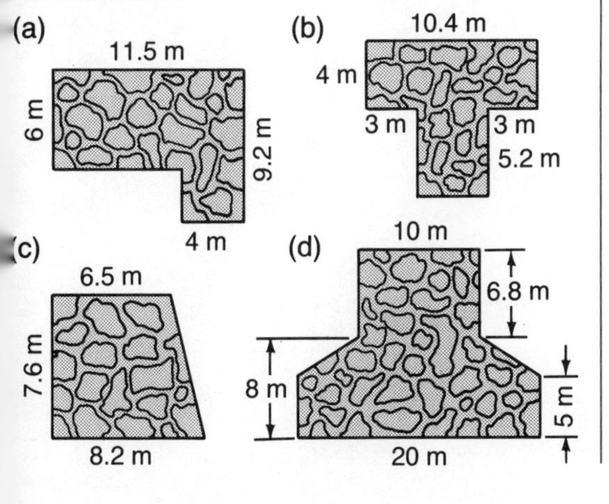

4. A rectangular park has a length of 134.5 m and a width of 77.4 m.
Calculate the area to the nearest tenth.

5. A square-shaped window pane has sides of 42 cm.
Calculate the area.

6. A campsite is in the shape of a parallelogram. The height of the parallelogram is 10.4 m and the base is 21.7 m.
Calculate the area to the nearest tenth.

7. A basketball court measures 26 m by 14 m. A soccer field measures 100 m by 73 m. One person can stand comfortably in a 40 cm by 40 cm square.

(a) How many people could stand on a basketball court?

(b) How many people could stand on a soccer field?

8. Calculate the cost of the glass tabletop if glass costs $12.70/m^2.

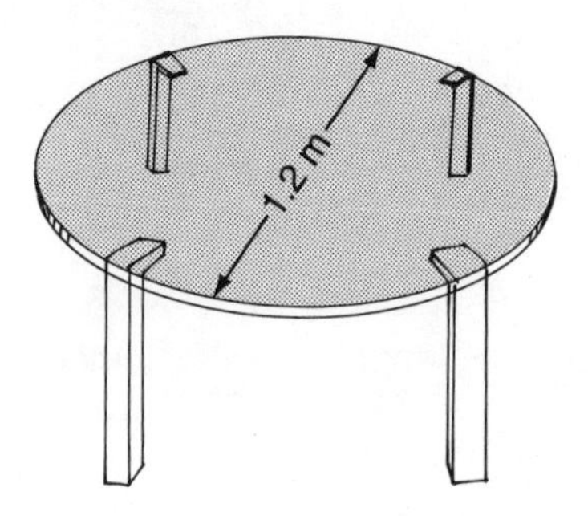

9. Each square in the grids has sides of 1 cm.
Estimate and then calculate the area of each circle.

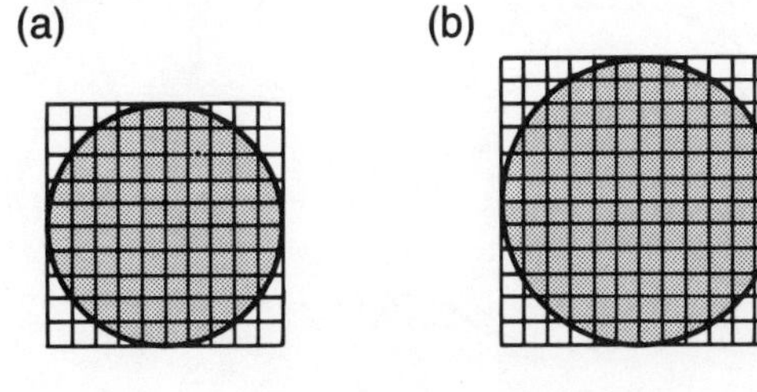

10.8 APPLYING VOLUME

The volume of an object is the space it occupies.
Volume is usually measured in cubic metres (m^3) or cubic centimetres (cm^3).

The base of the rectangular solid below has an area of 6×4 or $24\ cm^2$. It takes $24\ cm^2$ to make the bottom layer.

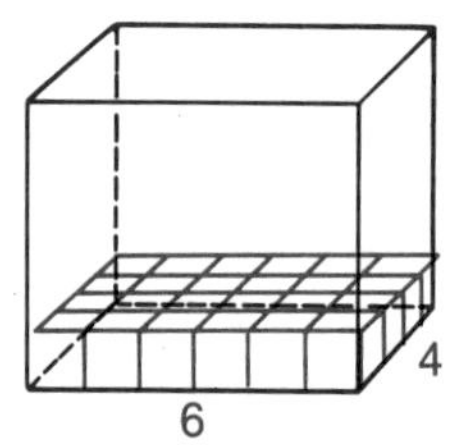

The solid is 5 cm high so that 5 layers of $24\ cm^2$ each will be needed to fill the solid.

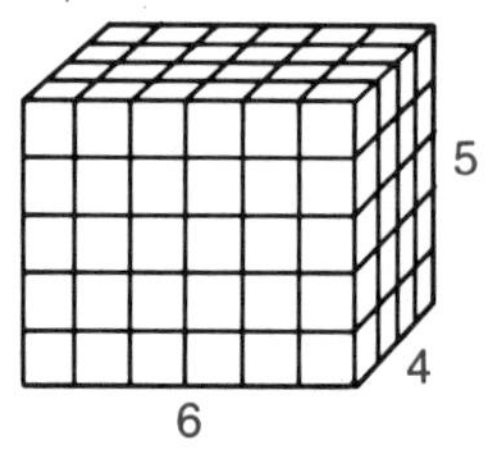

The volume of the solid is $5\ cm \times 24\ cm^2$ or $120\ cm^3$.

$$\text{Volume} = \overbrace{6 \times 4}^{\text{cubes in 1 layer}} \times \underbrace{5}_{\text{5 layers}} = 120\ cm^3$$

(6 = length, 4 = width, 5 = height)

The volume can be calculated using the formula

Volume = length × width × height $V = \ell \times w \times h$

Example 1.
Find the volume of the suitcase.

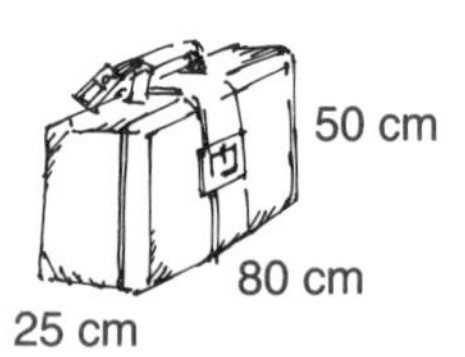

Solution:
$V = \ell \times w \times h$
$V = 80 \times 25 \times 50$
$= 100\ 000$
The volume is $100\ 000\ cm^3$.

To find the volume of other objects modify the formula for volume.

$$\text{Volume} = \underbrace{\text{length} \times \text{width}}_{\text{area of base}} \times \text{height}$$
$$= (\text{area of base}) \times \text{height}$$

$V = B \times h$

Example 2.
Calculate the volume of each solid.

(a)

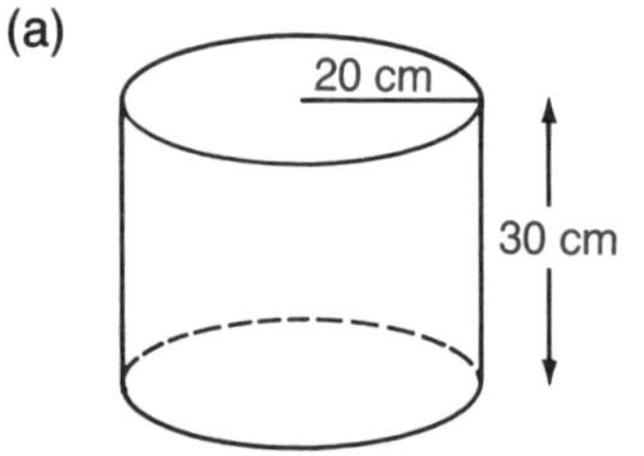

(b)

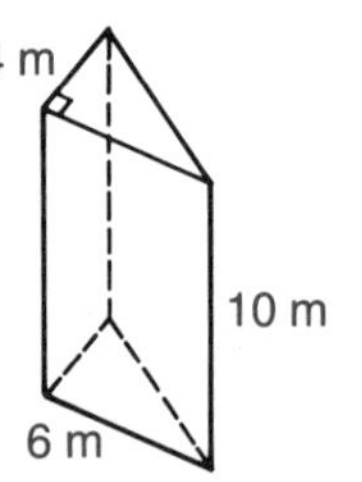

Solution:

(a) Find the area of the cylinder's base.

$$\begin{aligned}\text{Area of base} &= \pi r^2 \\ &= 3.14 \times 20^2 \\ &= 3.14 \times 400 \\ &= 1256\end{aligned}$$

Then,

$$\begin{aligned}V &= B \times h \\ V &= 1256 \times 30 \\ &= 37\,680\end{aligned}$$

The volume is 37 680 cm^3.

(b) Find the area of the prism's base.

$$\begin{aligned}\text{Area of base} &= \tfrac{1}{2} \times b \times h \\ &= \tfrac{1}{2} \times 6 \times 4 \\ &= 12\end{aligned}$$

Then,

$$\begin{aligned}V &= B \times h \\ V &= 12 \times 10 \\ &= 120\end{aligned}$$

The volume is 120 m^3.

EXERCISE 10.8

1. State the volume of each solid.

(a)

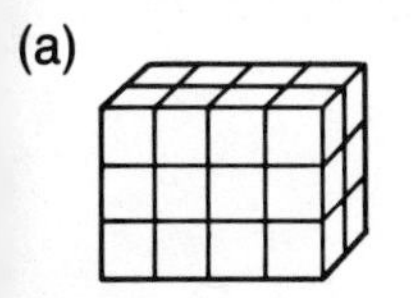

(b)

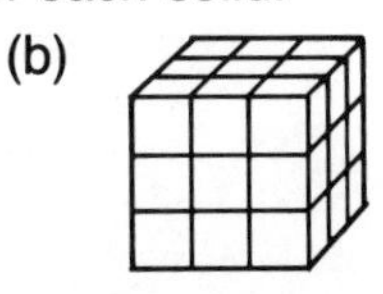

(c)

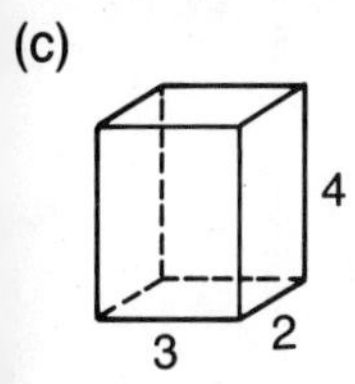

(d)

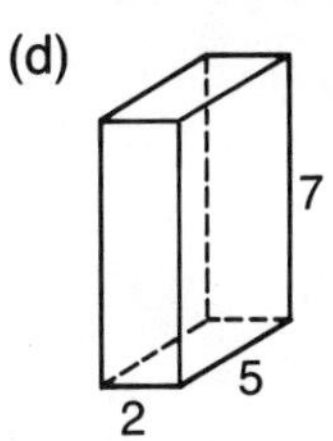

2. Calculate the volume of each solid.

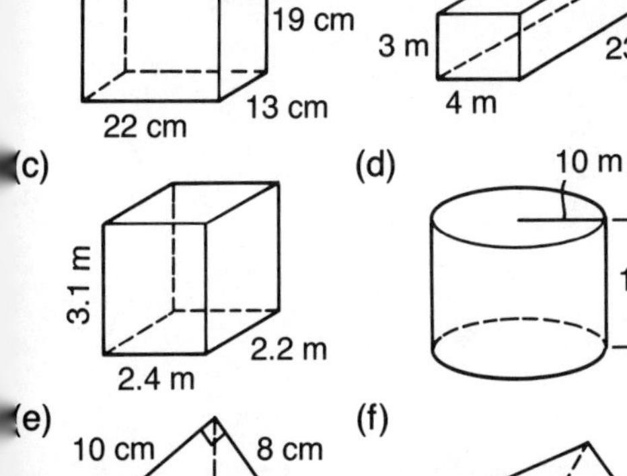

(e) 10 cm, 8 cm, 12 cm

(f) 6 m, 8.4 m, 14 m

3. Calculate the volumes of the solids.

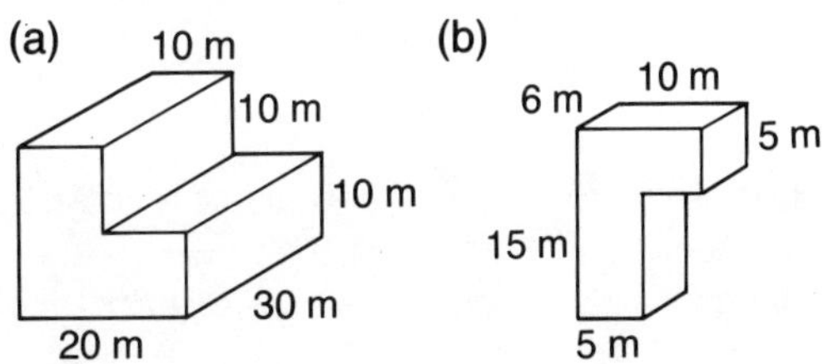

4. An Olympic-sized training pool has a length of 50 m and a width of 21 m. The floor of the pool is movable so that the depth of the water can be adjusted according to how the pool is to be used.
   (a) For adult non-swimmer lessons the depth is 1.5 m. What is the volume of the pool?
   (b) For waterpolo the required depth is 4 m. What is the volume of the pool?

5. Below is an illustration of an archaeological dig.

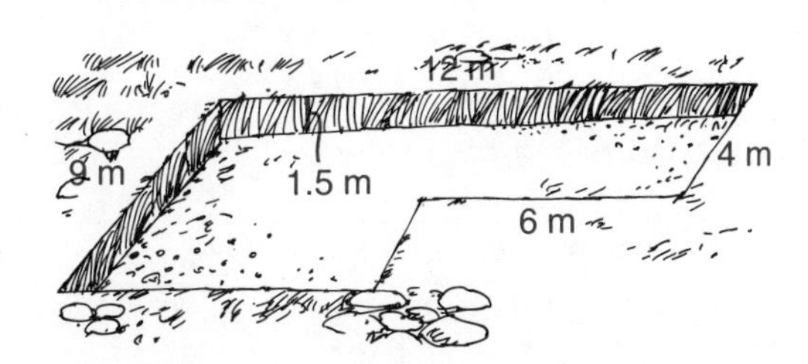

   (a) What is the area of the dig?
   (b) What is the volume of soil that has been removed from the site?

10.9 THE RIGHT TRIANGLE AND THE PYTHAGOREAN THEOREM

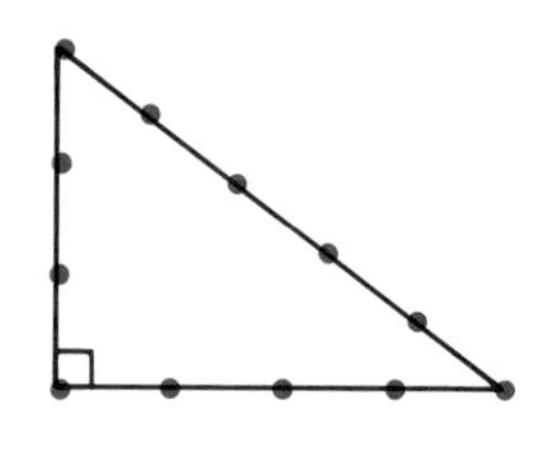

A triangle containing a right angle, 90°, is called a right triangle. The side opposite the right angle is the hypotenuse.

The ancient Egyptians used a right triangle to survey their lands. Surveyors called "rope stretchers" used ropes with knots to mark boundaries. One kind of rope had 12 knots and was stretched between three stakes. The result was a right triangle with sides in the ratio 3 : 4 : 5.

While Pythagoras, a Greek mathematician who lived in 500 B.C., may not have been the first to discover this ratio, he is credited with proving the theorem that bears his name.

INVESTIGATION

1. (a) Using ruler and compass only, construct a triangle with sides 3 cm, 4 cm, and 5 cm.
(b) Measure and record the size of each angle to the nearest degree.
(c) Is the triangle a right triangle?

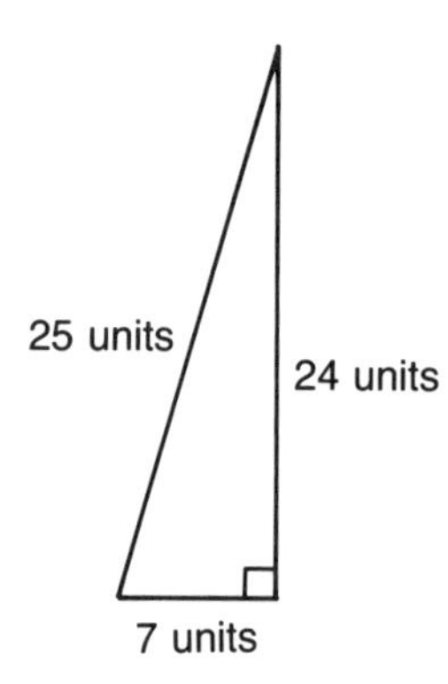

7 : 24 : 25

2. The triangle in question 1 part (a) is sometimes referred to as a "3 : 4 : 5 triangle."
Construct a 6 : 8 : 10 triangle using 1 cm as a unit.
(a) Is the 6 : 8 : 10 triangle also a right triangle?
(b) Suggest possible dimensions for three other right triangles.

3. (a) Using suitable units and scale, construct the following triangles and determine which are right triangles.
(i) 9 : 12 : 15 (ii) 2 : 3 : 4 (iii) 5 : 12 : 13 (iv) 8 : 15 : 17
(b) Which triangles have sides that are multiples of the right triangles in part (a)?
(i) 10 : 24 : 26 (ii) 16 : 30 : 34 (iii) 12 : 16 : 20
(iv) 15 : 20 : 25 (v) 8 : 14 : 15

4. (a) Complete the following table.

Squares of the Sides of a Right Triangle								
Shortest Side Squared	$3^2 = 9$	$6^2 = 36$	9^2	12^2	15^2	5^2	8^2	7^2
Second Side Squared	$4^2 = 16$	$8^2 = 64$						
Hypotenuse Squared	$5^2 = 25$	$10^2 = 100$						

(b) In the first row of the table in part a, $3^2 = 9$, $4^2 = 16$, and $5^2 = 25$.
How are the numbers 9, 16, and 25 related?
Test this relationship for all the entries in the table.

(c) From your observations above, tell how the squares of the sides of a right triangle are related.

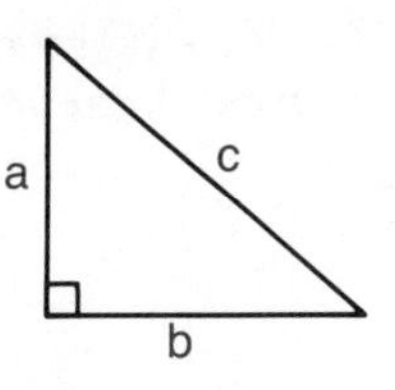

In any right triangle

$$a^2 + b^2 = c^2$$

where c is the length of the hypotenuse and a and b are the lengths of the other two sides.

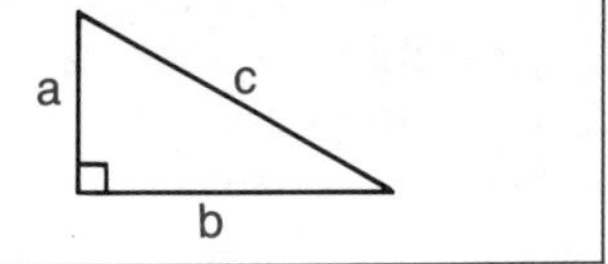

EXERCISE 10.9

A

1. Use the formula $a^2 + b^2 = c^2$ to determine whether each set of numbers represents the lengths of the sides in a right triangle.

(a) (30, 40, 50)
(b) $(1, 2, \sqrt{5})$
(c) $(2, 4, 2\sqrt{5})$
(d) (1, 1, 2)
(e) $(3, 4, \sqrt{7})$
(f) $(1, \sqrt{3}, 2)$
(g) $(1, \frac{3}{4}, \frac{5}{4})$
(h) $(1, 1, \sqrt{2})$

2. If a, b, and c are the lengths of the sides of a right triangle, and c is the hypotenuse, then complete each table.

(a)

a	b	c
4	3	
4	4	
12	5	
6	8	

(b)

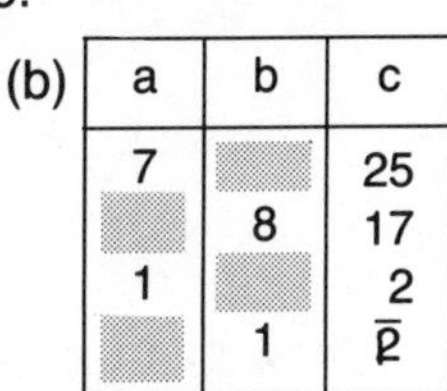

a	b	c
7		25
	8	17
1		2
	1	$\sqrt{2}$

Why is it not necessary to state the units used for the measures in questions 1 and 2?

3. In each case indicate which two whole numbers are nearest the given number.

(a) If c^2 is 18, then c is between ■ and ■.
(b) If c^2 is 61, then c is between ■ and ■.
(c) If a^2 is 91, then a is between ■ and ■.
(d) If b^2 is 157, then b is between ■ and ■.
(e) If c^2 is 600, then c is between ■ and ■.
(f) If a^2 is 175, then a is between ■ and ■.

B

4. Use the Pythagorean Theorem to find c^2 for each right triangle. Then use a calculator to find c.

(a)

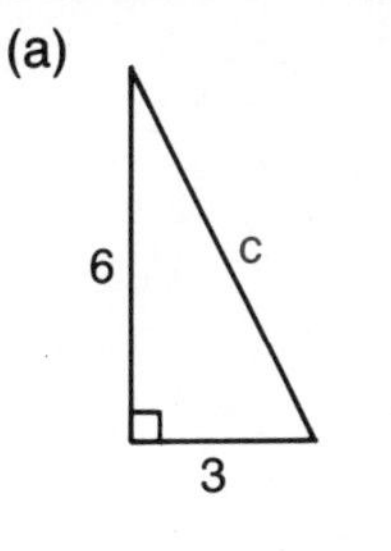

(b)

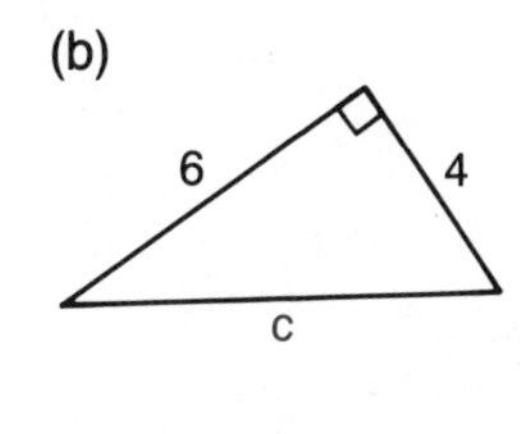

(c)

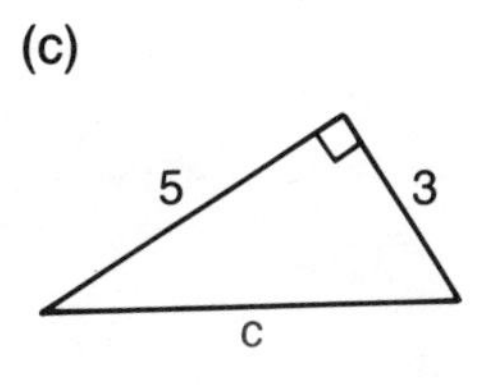

(d)

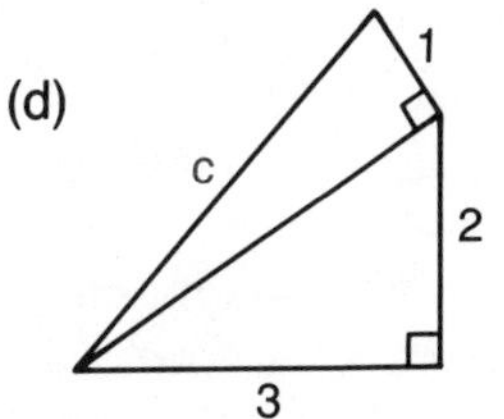

(e)

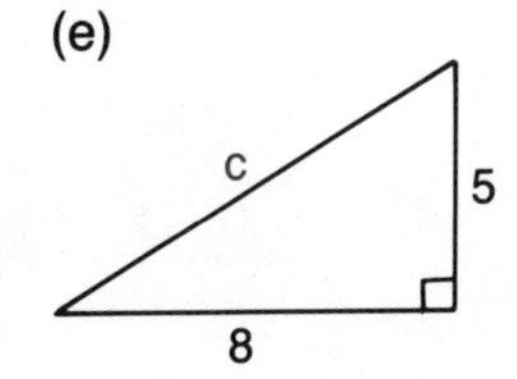

(f)

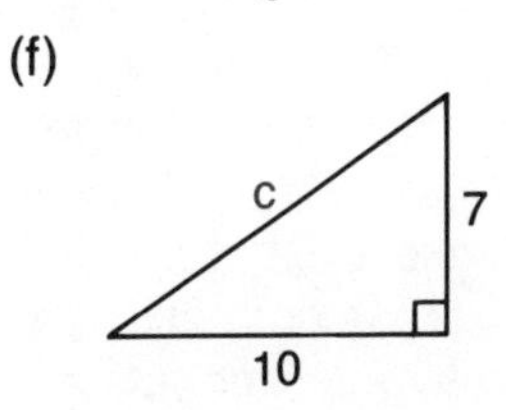

5. A rectangular piece of steel plate is 20 cm long and 7 cm wide.
Calculate the length of the diagonal to the nearest tenth.

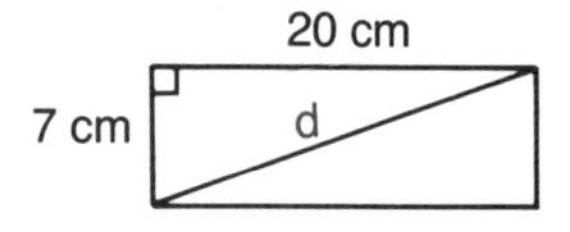

6. How many metres of cable will be required to fasten a hook at point A, 27 m up on a tower, to a hook at point C on the ground 36 m from the base B of the pole?

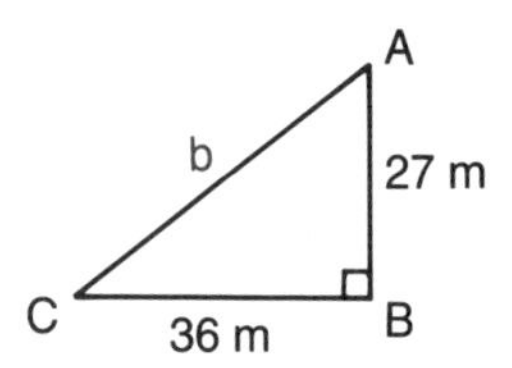

7. A pipe is used to make a diagonal brace for a gate 4 m high and 8 m long. How long is the pipe?

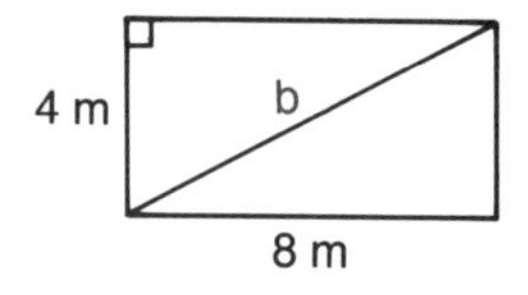

8. A board is used to make a diagonal brace for a garage door 3 m by 2.5 m. How long is the board?

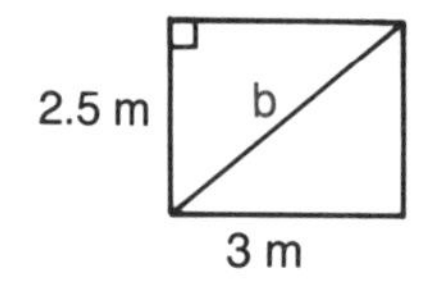

9. (a) How high up on a building will a 6 m ladder reach if its lower end must be 2 m from the building?
(b) How far down the wall will the top of the ladder slide if the lower end moves another metre away from the wall?
Make a diagram and express both answers to two decimal places.

10. Leaving Halifax, the aircraft carrier *Birdsnest* sails 300 km east, and then sails 500 km south.
How far is the ship from Halifax?
Make a diagram and express your answer to the nearest kilometre.

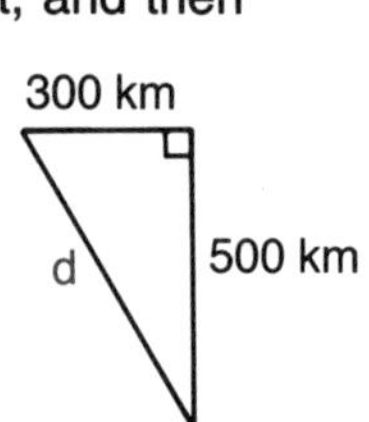

11. The given diagram consists of three pipes with diameters 3 cm, 4 cm, and 5 cm.

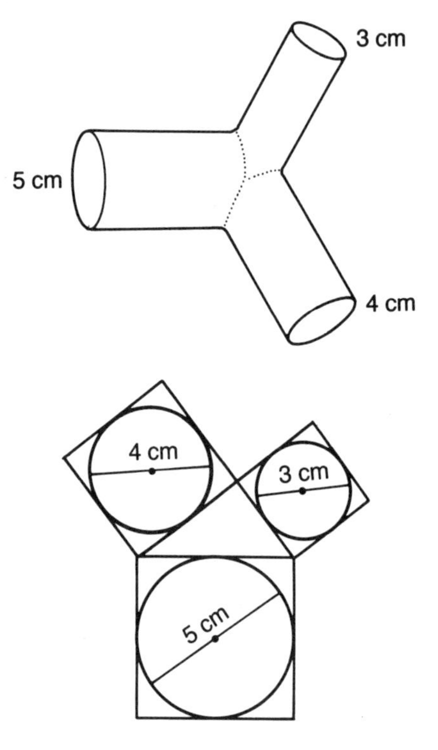

(a) Show that the same amount of water will flow through the 3 cm and 4 cm pipes as will flow through the 5 cm pipe, provided the pressure is exactly the same in all the pipes.
(b) Find the size of pipe that will carry the same amount of water as a 12 cm pipe and a 5 cm pipe.

12. What is the longest rod, to the nearest 0.1 cm, that can be placed inside a trunk 30 cm wide, 30 cm deep, and 50 cm long?

13. Draw △ABC, with ∠ACB = 90° and AC = BC = 6 cm. Divide AC in 1 cm intervals at D, E, F, G, and H. Then divide CB in 1 cm intervals at J, K, L, M, and N. Join and calculate the lengths of AJ, DK, EL, FM, GN, and HB.

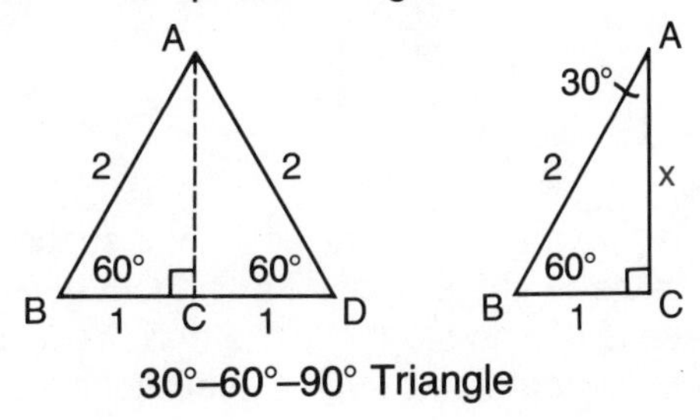

14. Find the length of the diagonal, AG, of the rectangular prism pictured below. Make diagrams to show the position of the diagonal in the face ABCD and in the face BCGH.

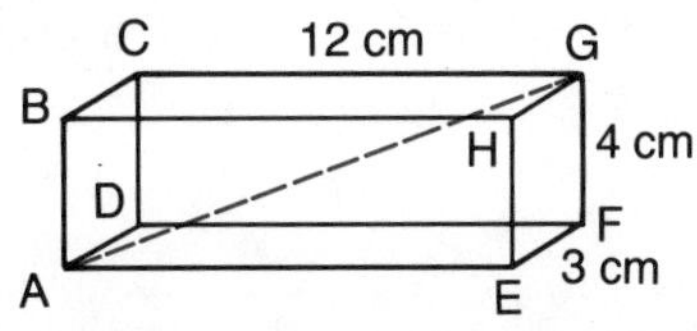

15. A doorway is 1 m by 2 m. What is the maximum width of a sheet of plywood that will fit through the doorway?

16. A corner lot is 30 m by 60 m. How much shorter is it for you to cut across the corner?

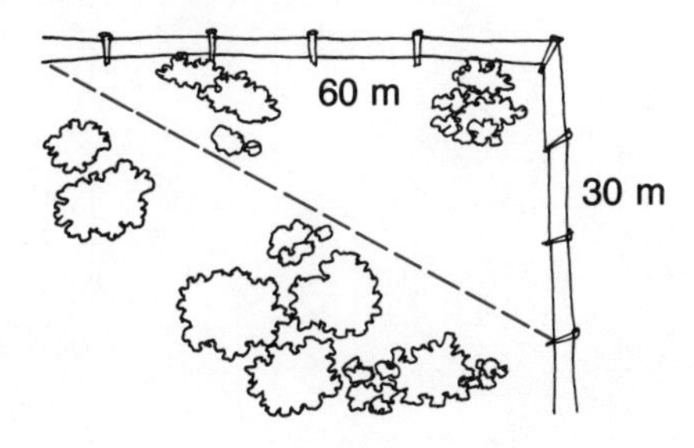

C 17. Two special triangles:

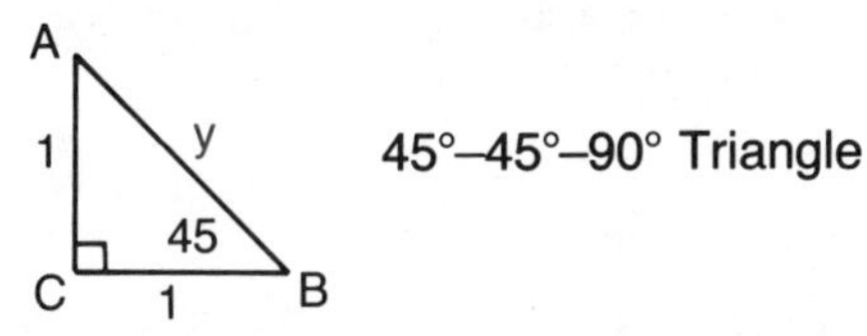

A
1
y
45
C
1
B
45°–45°–90° Triangle

(a) The 30°–60°–90° triangle is found by taking an equilateral triangle, ABD, with each side 2 units, and drawing the perpendicular AC to form the required triangle, ABC. In △ABC, AB = 2 units and BC = 1 unit.
Make a diagram and calculate the length of AC, and express your answer as a Pythagorean triple, 1, x, 2.
Repeat this process beginning with the equilateral triangle, ABD, having sides equal to 10 units.

(b) The 45°–45°–90° triangle is an isosceles triangle, ABC, with ∠C = 90° and AC = BC = 1 unit.
Make a diagram, calculate the length of AB, and express your answer as a Pythagorean triple, 1, 1, y.
Repeat this process beginning with the isosceles triangle, ABC, having both sides AC and BC equal to 5 units.

MIND BENDER

The numbers 3, 4, and 5 are called a Pythagorean triple because

$$3^2 + 4^2 = 5^2$$

Find a Pythagorean triple where each number is less than 100 and all three numbers start with the same digit.

10.10 CREATING ESCHER-TYPE DRAWINGS

"Imagine that you have an infinite supply of jigsaw puzzle pieces, all identical. If it is possible to fit them together without gaps or overlaps to cover the entire plane, the piece is said to tile the plane and the resulting pattern is called a tessellation. From the most ancient times such tessellations have been used throughout the world for floor and wall coverings and as patterns for furniture, rugs, tapestries, quilts, clothing, and other objects."

One very popular artist to use repetitive patterns and tessellations was M. C. Escher, a Dutch artist born in 1898.

Plane tessellations are the basis for Escher-type drawings. A plane tessellation is a complete covering of a plane, or surface, by one or more figures in a repeating pattern, with no overlapping of figures.

Any triangle will tessellate the plane.

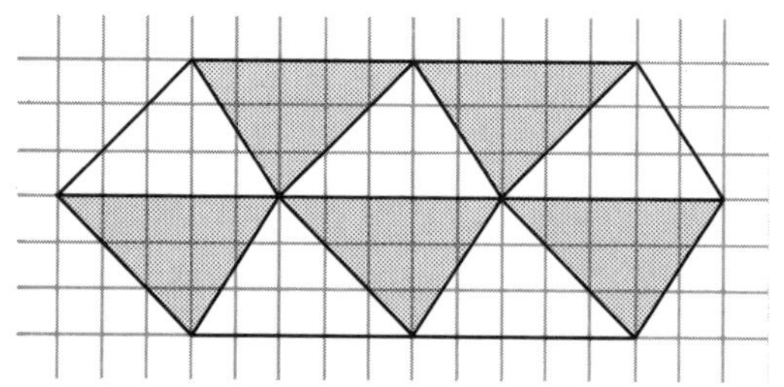

Any quadrilateral will tessellate the plane.

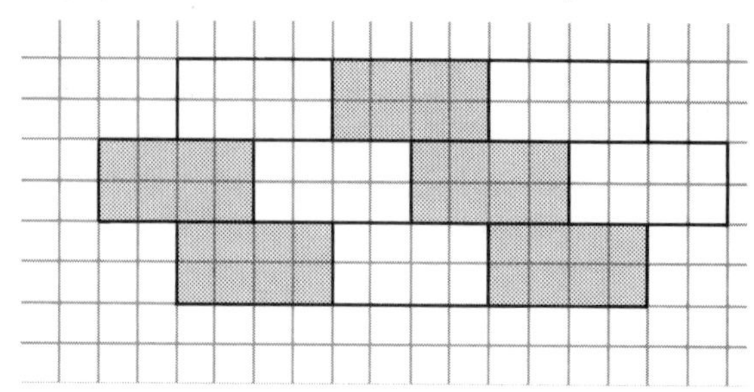

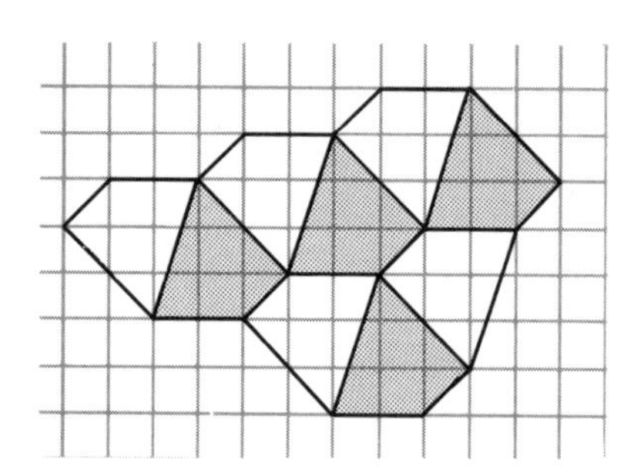

You can combine two or more polygons and tessellate the plane.

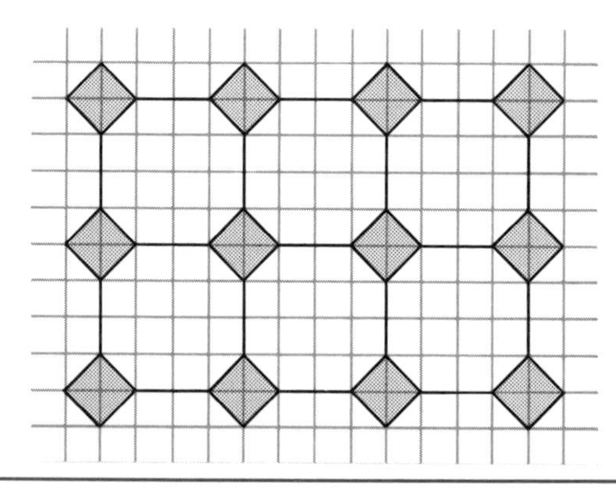

You can also tessellate the plane using figures that are not polygons.

You can start with one figure and alter it to form a new figure that will tessellate a plane. The changes made to one side must be made in an opposite mode to another side. The area of the new figure must be equal to the area of the original figure.

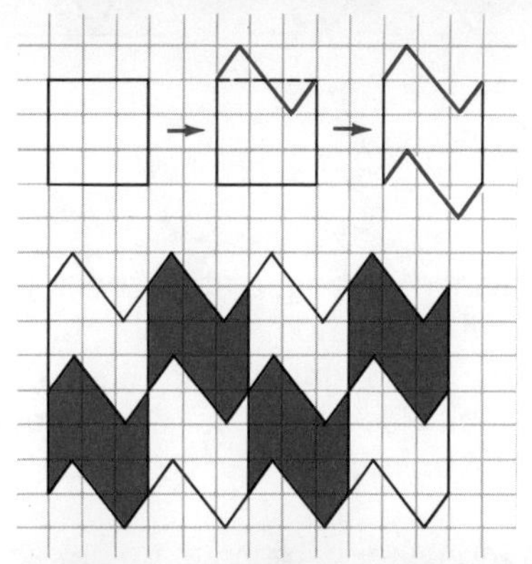

Use your imagination to draw many interesting designs.

EXERCISE 10.10

1. Alter these basic shapes to create tessellation drawings.

(a)

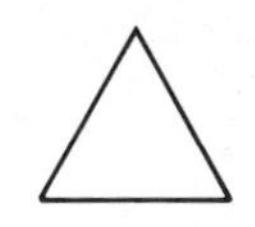

(b)

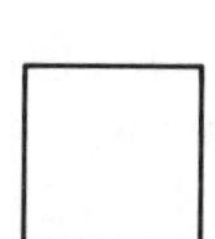

(c)

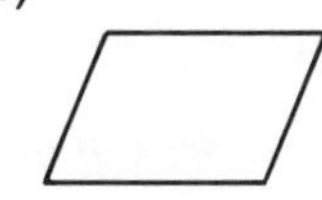

(d)

(e)

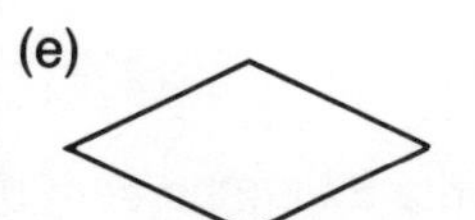

(f)

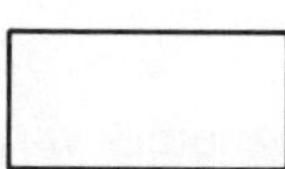

2. Below is an example of Escher's impossible staircase.

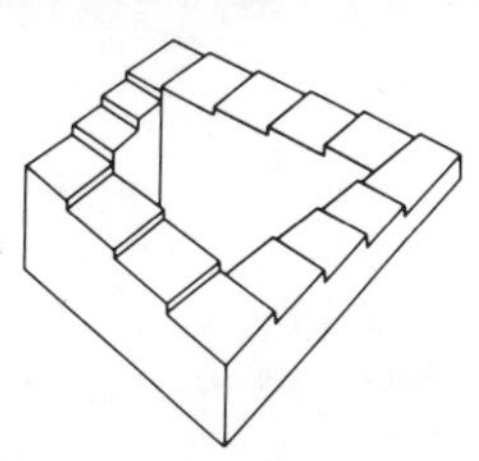

(a) Why is it called impossible?
(b) In which of his drawings does the staircase appear?
(c) He also used an impossible triangle. Sketch an impossible triangle.

10.11 PROBLEM SOLVING

1. (a) Each figure is made of toothpicks. How many toothpicks make up the 12th figure?

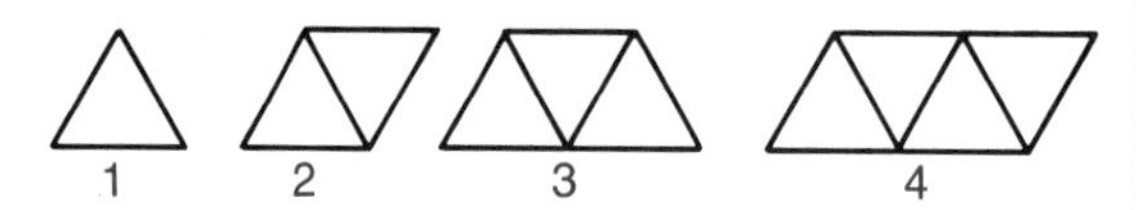

(b) How many toothpicks make up the 26th figure?

2. Glynis cashed a cheque at the bank for \$284.67. On the way home she bought groceries for \$56.83, gasoline for \$26.50, and three blank video tapes for \$7.20 each. When she got home, she had \$275.72 left in her wallet.
How much money did she have before she cashed the cheque?

3. Every morning at eight o'clock a train leaves Edmonton for Quebec City. At the same time the train leaves Edmonton, another train leaves Quebec City for Edmonton. It takes three days to get from one city to the other.
If you boarded a train in Edmonton to go to Quebec City, how many trains going to Edmonton would you pass?

4. To some, the depression year 1930 was unlucky because the sum of the digits is 13. What is the next year in which the sum of the digits will be 13?

5. It takes Carla 2 h to cut a rectangular park lawn that measures 10 m by 30 m. How long should she take to cut a similar park lawn that is twice as wide and three times as long?

6. The Oceanview Souvenir Shop sells sea shells in bags of 10 and 20. Frank bought 9 packages for the school aquariums and got 150 shells.
How many bags of each did he buy?

7. Each circle has a radius of 6 cm. What is the perimeter and area of the square ABCD?

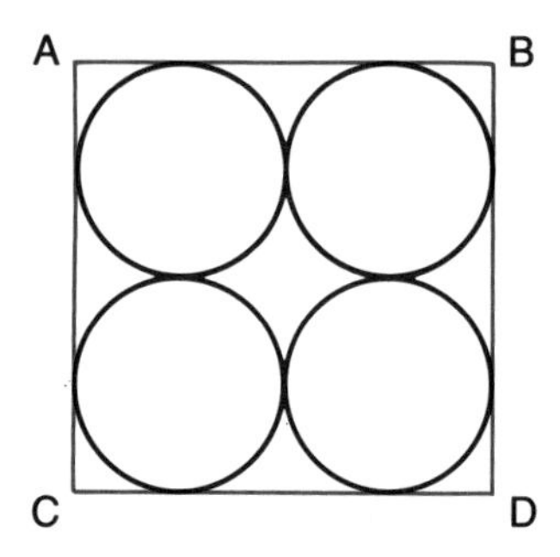

8. How many squares are in the figure?

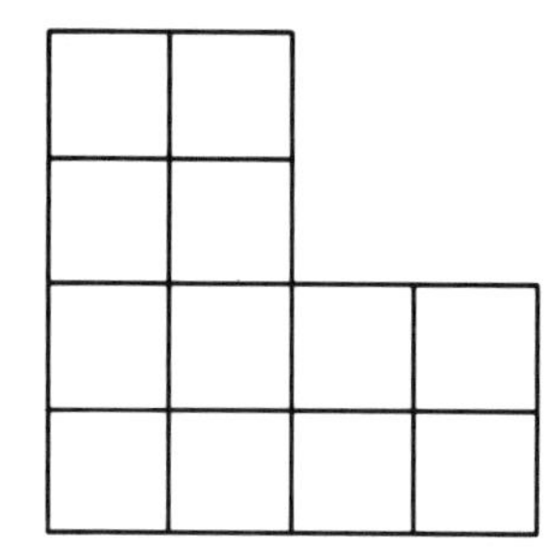

9. List the names of provinces in which the following words are found. The letters of each word must remain in the given order.

(a) it (b) ask
(c) run (d) cot
(e) is (f) tar
(g) be (h) in

10. The time is now 09:13.
What time was it fourteen hours and fifty-two minutes ago?

11.

57

The number 57 can be written as the sum of consecutive whole numbers. For example,

$28 + 29 = 57$

Find another way to write 57 as the sum of consecutive whole numbers.

12. Yolanda paid $7.29 for a new pen. She gave the clerk a twenty-dollar bill. The clerk gave Yolanda change using the least number of coins and bills possible.
What coins and bills did she give Yolanda?

13. Use your calculator to find each product.

$9 \times 9 = ■$
$99 \times 99 = ■$
$999 \times 999 = ■$

Determine the pattern and then predict the following products.

$9999 \times 9999 = ■$
$99999 \times 99999 = ■$

14. Every year a car depreciates by 20%. This year a car is worth $20 000.
What was it worth last year?

15. One side of a field is bounded by a lake.

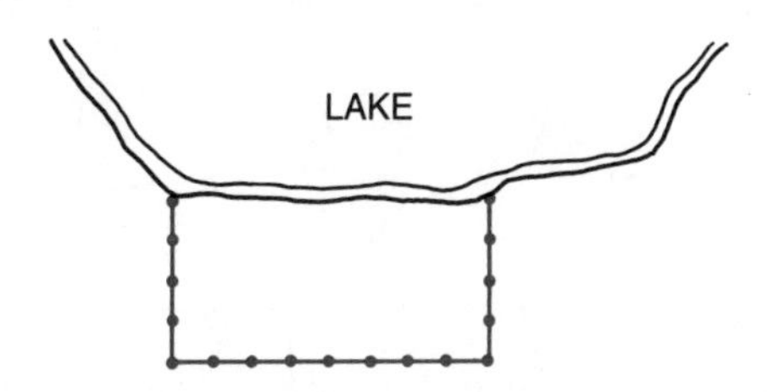

Erica wants to fence a rectangular riding area with the lake as one side. Erica has 200 m of fencing.
What are the dimensions of the largest area that can be enclosed?

16. Which container has the greater volume?

(a) (b)

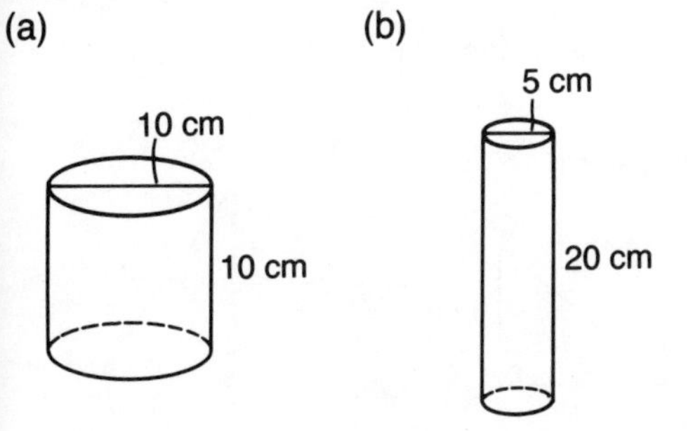

CAREER

MARINE BIOLOGIST

A marine biologist studies things that live in the sea.
One of the more interesting creatures is the squid which swims, in North America, in waters from Nova Scotia to Florida. Ten arms, or tentacles, with sucking discs, extend from the head of a squid.

Squids usually measure from 15 cm to 45 cm in length. One third of the length is the mantle and two thirds is the length of the arms.

Several giant squids have washed up on the shore of Newfoundland, near Cape Bonavista. One squid measured 10 m in length and the diameter of its suction pod was 5 cm.

1. Whales have been found with suction pod scars around their heads from battles with giant squids. The diameter of each of these pods was 25 cm.
About how big were the squids that the whales were fighting?
2. How long would be the arms on a squid that measures 10 m?
3. Locate Cape Bonavista on a map.
4. Identify the body parts of a squid.

10.12 REVIEW EXERCISE

1. Classify each triangle.

(a)

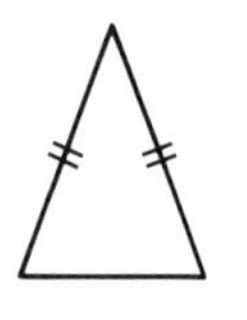

(b)

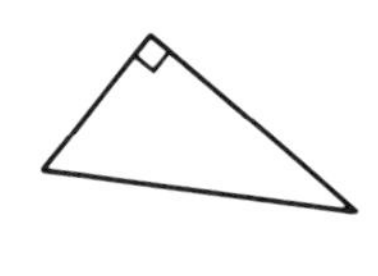

(c)

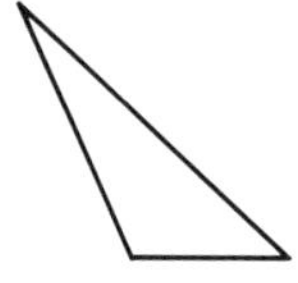

(d)

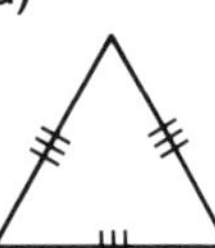

(e)

(f)

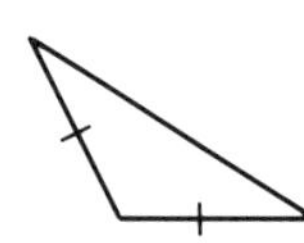

2. Find the measures of the unknown angles.

(a)

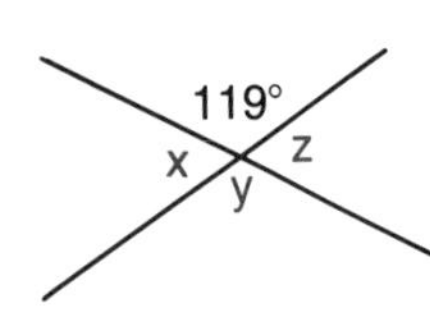

(b)

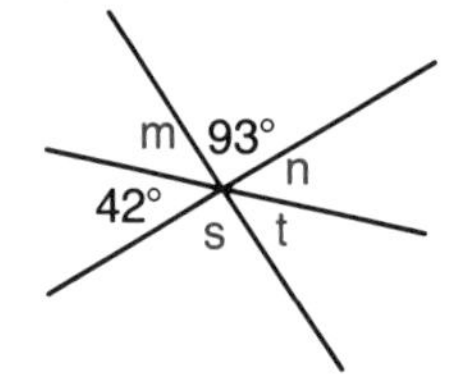

(c)

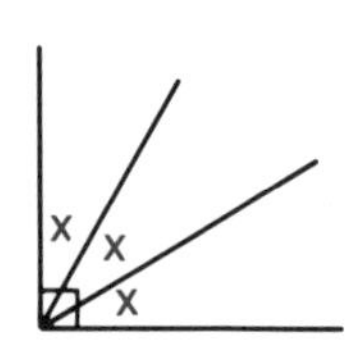

(d)

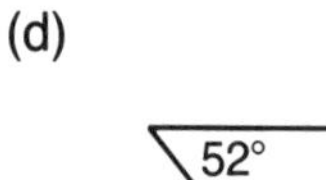

(e)

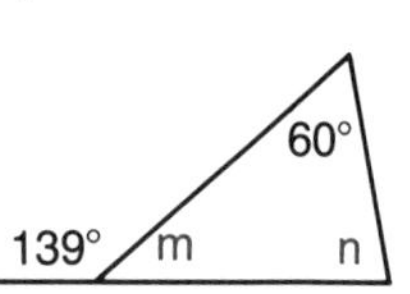

(f)

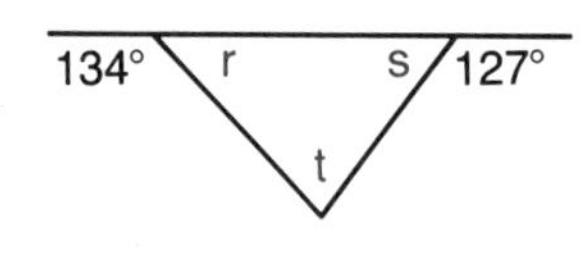

(g)

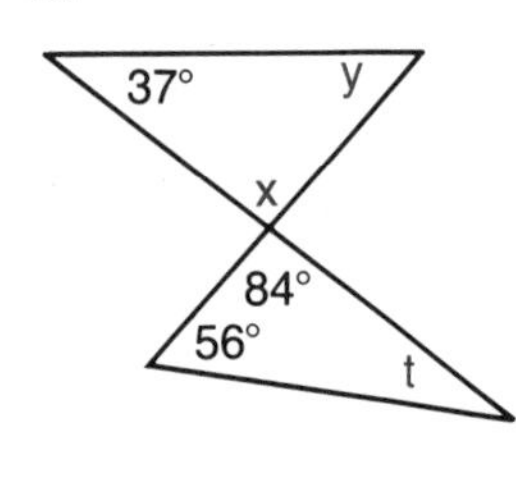

(h)

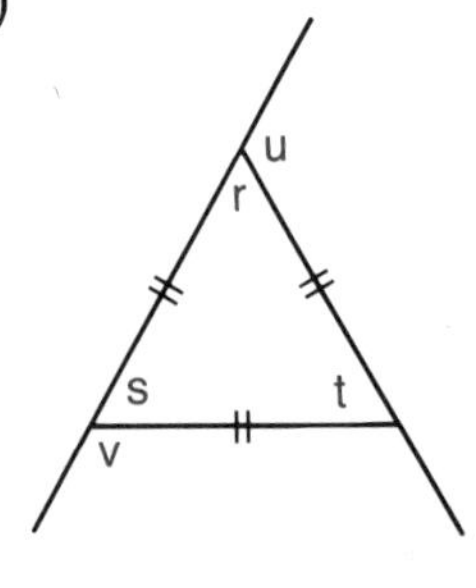

3. Find the measure of each missing angle.

(a)

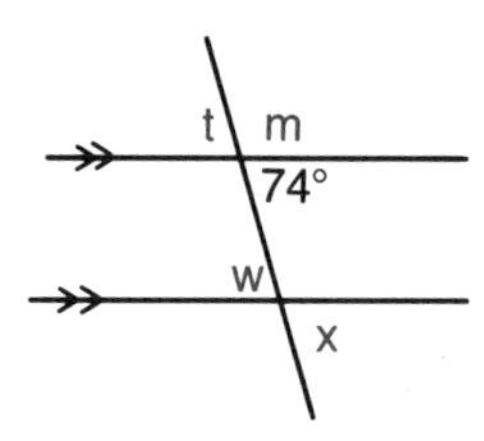

(b)

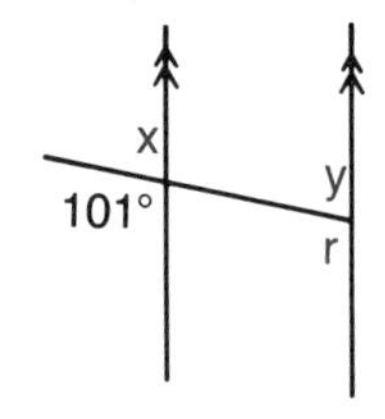

(c)

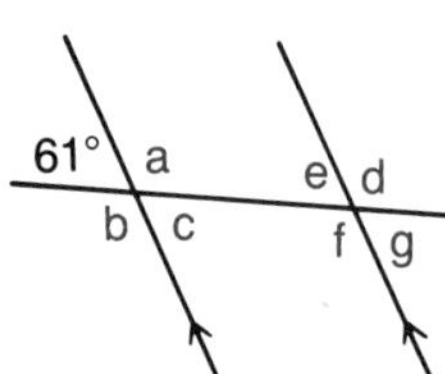

(d)

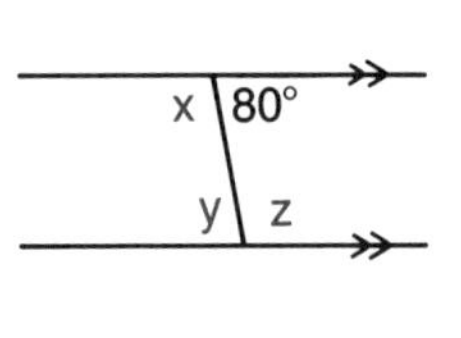

4. Find the unknown angles and sides..

(a)

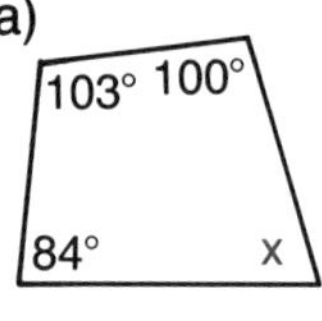

(b)

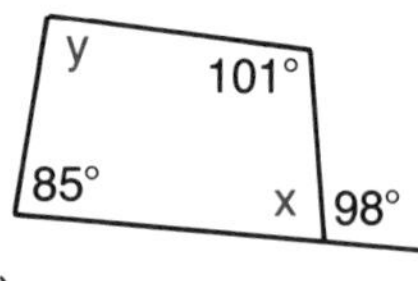

(c)

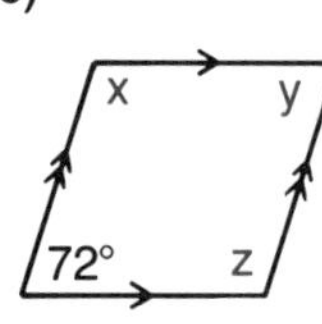

(d)

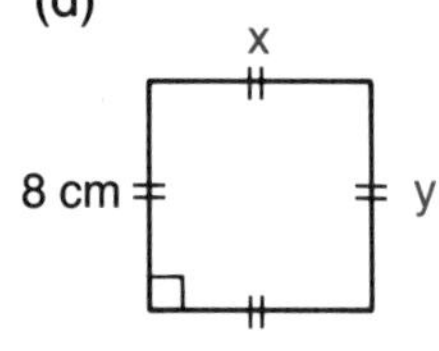

(e)

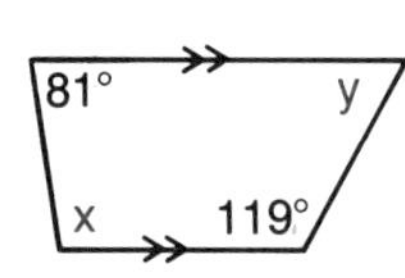

(f)

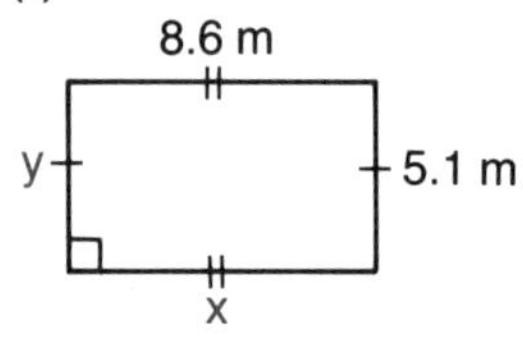

5. Calculate the length of the missing side to the nearest tenth.

(a)

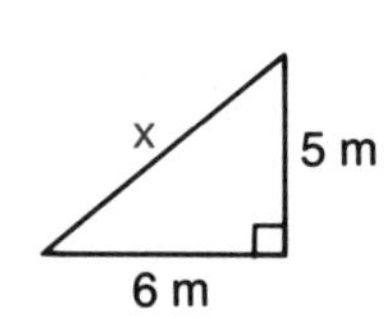

(b)

x

12 m

7 m

6. Draw an angle similar to the one shown. Bisect the angle.

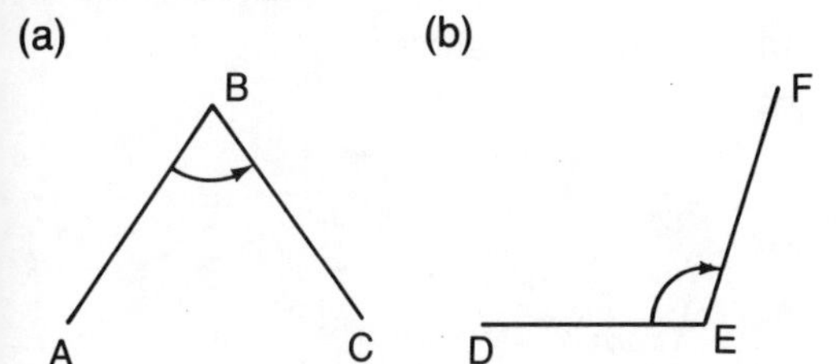

7. Draw a line segment similar to the one shown.
Construct the right bisector.

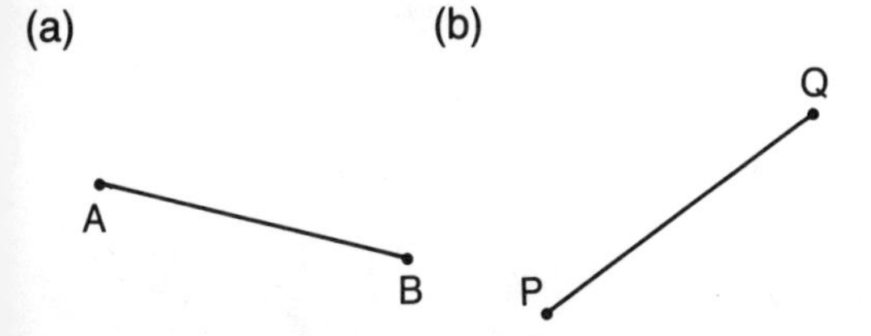

8. Construct the perpendicular from P to the line.

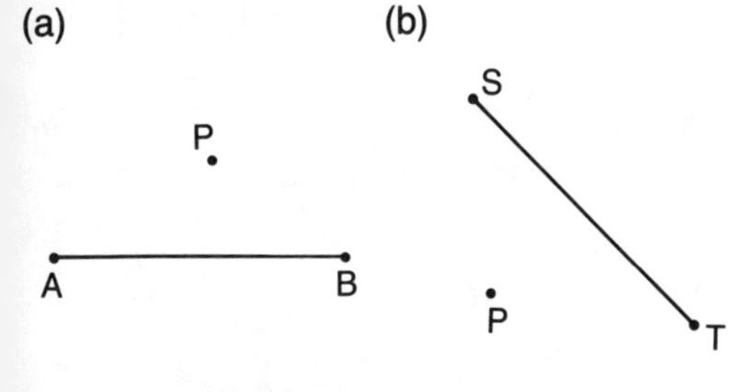

9. Construct the perpendicular to the line at P.

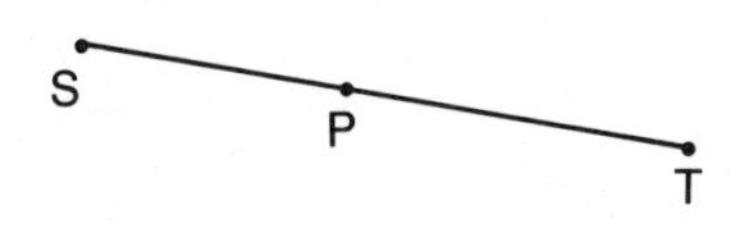

10. Draw an angle similar to the one shown. Construct an equal angle.

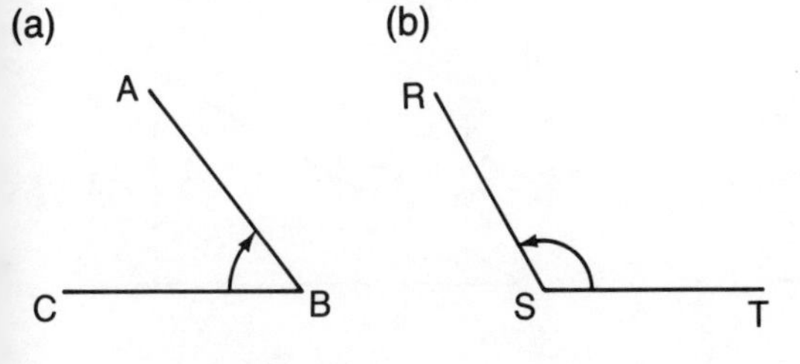

11. Construct a line through P parallel to each given line.

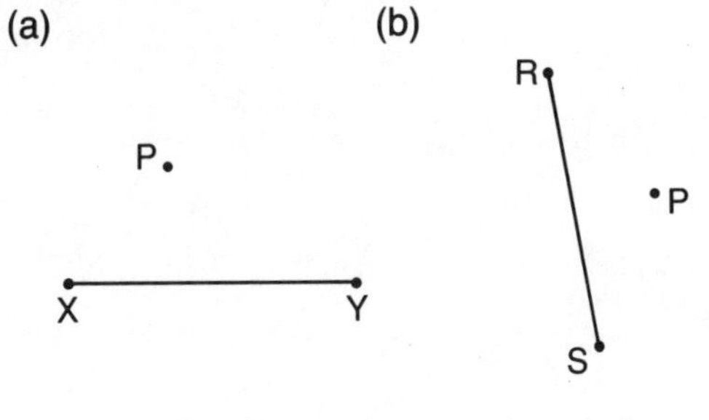

12. Calculate the perimeter of each figure.

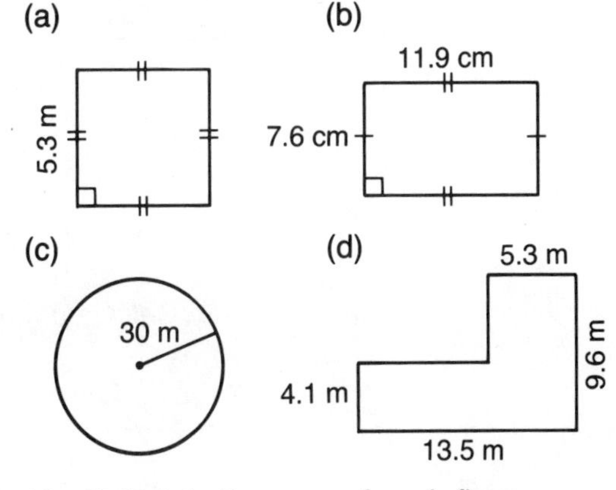

13. Calculate the area of each figure.

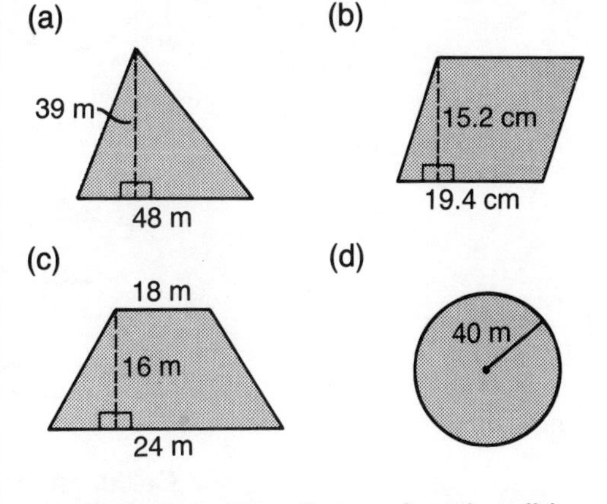

14. Calculate the volume of each solid.

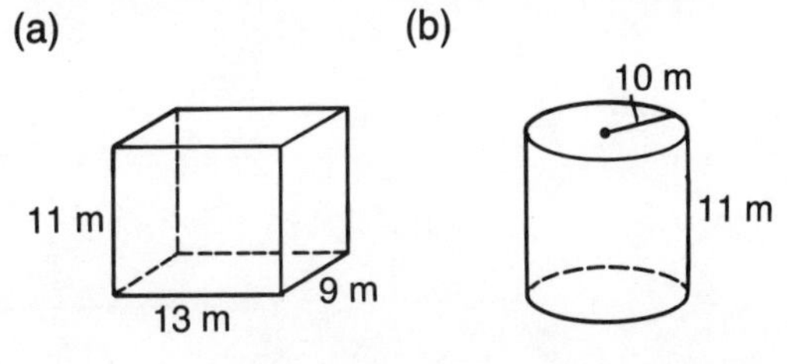

10.13 CHAPTER 10 TEST

1. Find the measure of each missing angle.

(a)

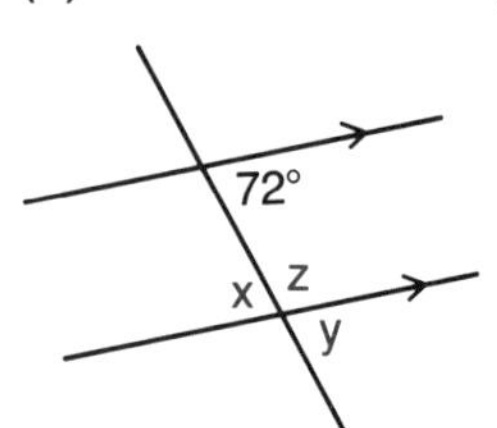

(b)

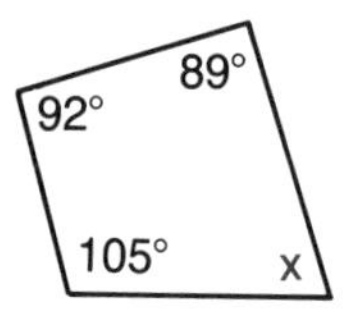

(c)

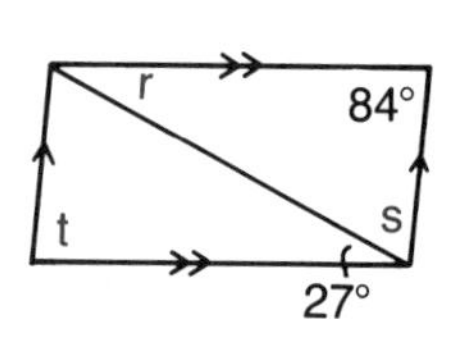

2. (a) Bisect ∠ABC.

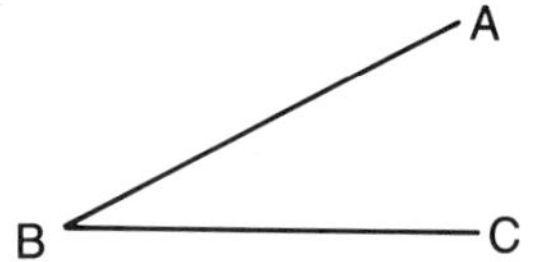

(b) Construct the perpendicular from P to AB.

• P

3. Calculate the perimeter and area of each figure.

(a)

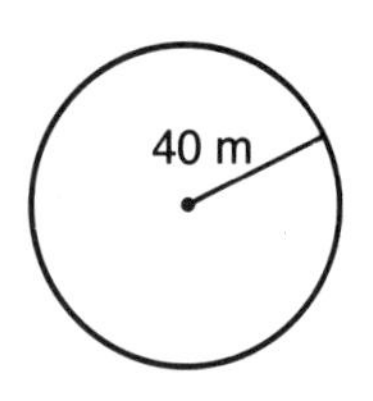

(b)

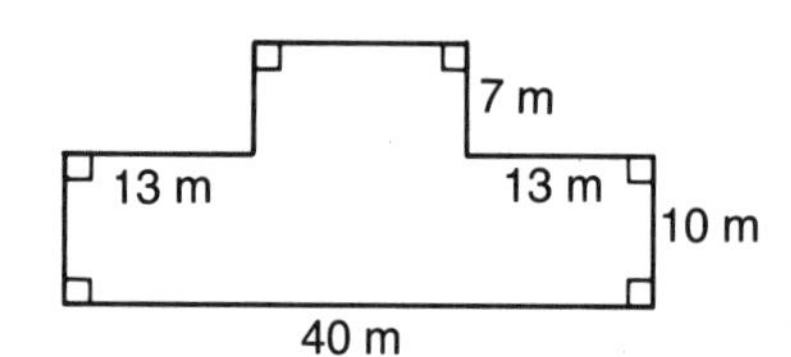

4. Calculate the volume of each solid.

(a)

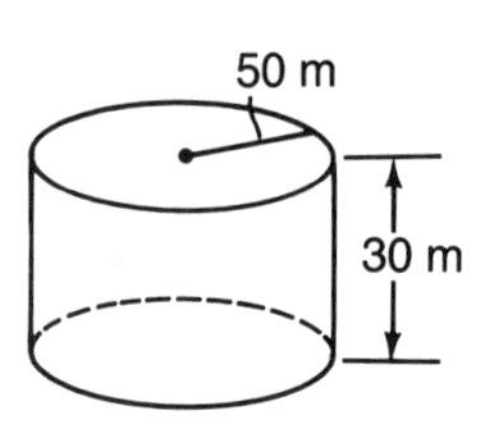

(b)

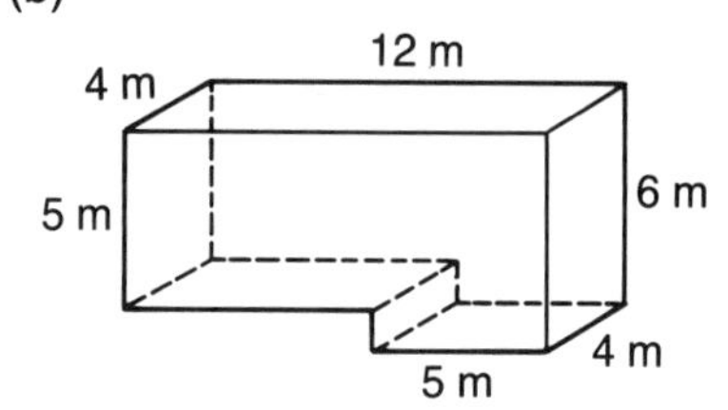

5. Calculate the length of the missing side to the nearest tenth.

(a)

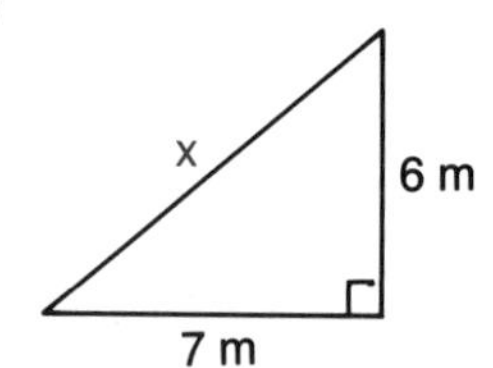

(b)

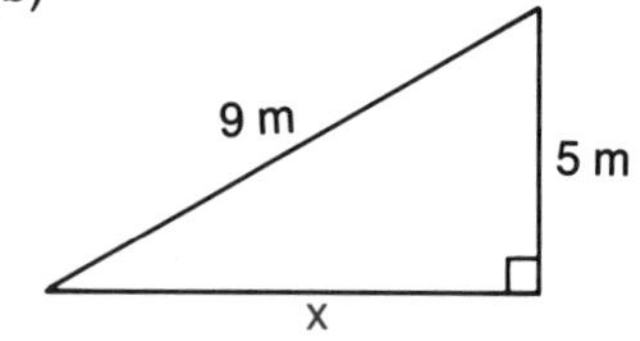

TRIGONOMETRY

REVIEW AND PREVIEW TO CHAPTER 

RATIO AND PROPORTION

If $\frac{a}{b} = \frac{c}{d}$

Then $ad = bc$

EXERCISE 1

1. Simplify to the nearest tenth.

(a) $\frac{5}{13}$ (b) $\frac{7}{24}$

(c) $\frac{7}{25}$ (d) $\frac{28}{4.5}$

(e) $\frac{5.5}{8}$ (f) 3.5×2.7

(g) 6.8×4.6 (h) 3.8^2

(i) 4.2×8.1 (j) 1.73×1.41

2. Solve each equation to the nearest tenth.

Example. $\frac{x}{3.5} = \frac{4.7}{2.8}$

$$x = \frac{3.5 \times 4.7}{2.8}$$

$$\doteq 5.9$$

Press

3 · 5 × 4 · 7 ÷ 2 · 8 =

The display is 5.875

(a) $\frac{a}{5.7} = \frac{8.2}{12.5}$ (b) $\frac{h}{2.7} = \frac{6.5}{2.3}$

(c) $\frac{h}{6.2} = \frac{37}{44}$ (d) $\frac{d}{140} = \frac{4}{3}$

(e) $\frac{x}{53} = \frac{2.7}{4.1}$ (f) $\frac{m}{6.5} = 0.337$

(g) $\frac{n}{58} = 0.246$ (h) $\frac{p}{62} = 0.331$

(i) $\frac{h}{135} = 1.73$ (j) $\frac{h}{2.6} = 0.866$

PYTHAGOREAN THEOREM

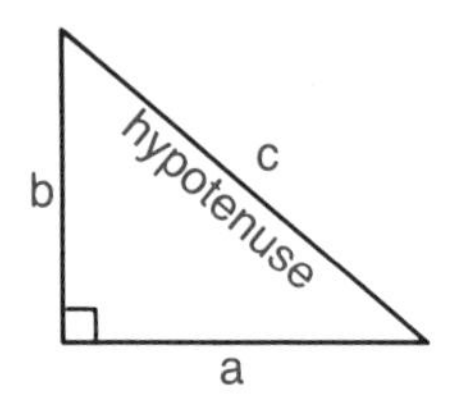

$$c^2 = a^2 + b^2$$
$$a^2 = c^2 - b^2$$
$$b^2 = c^2 - a^2$$

EXERCISE 2

1. Use the Pythagorean Theorem to find the hypotenuse in each of the following triangles, to the nearest tenth.

(a)

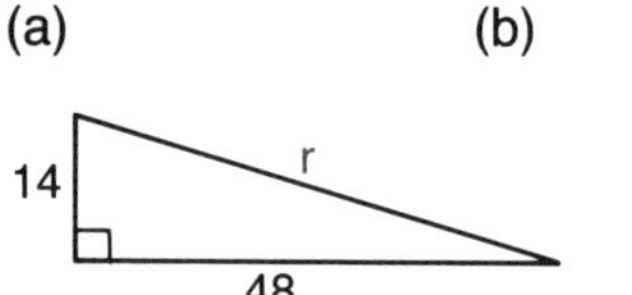

(b)

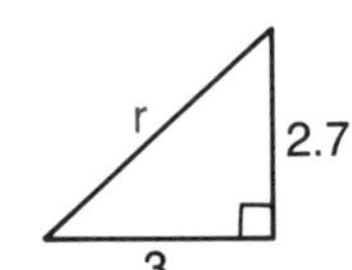

(c)

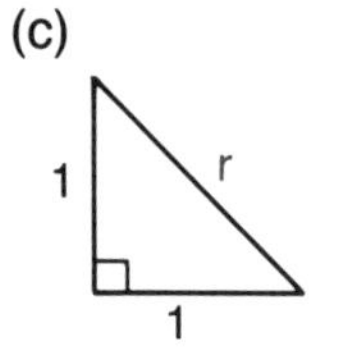

(d)

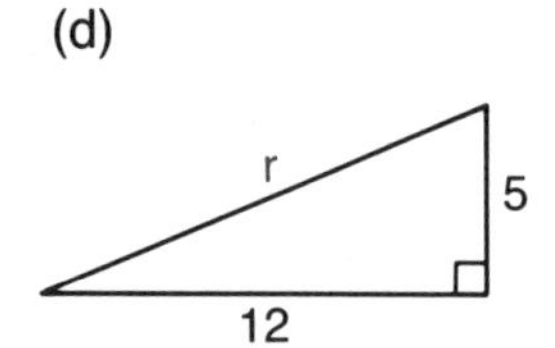

2. Find the length of the indicated side to the nearest tenth.

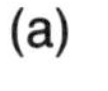

(a)

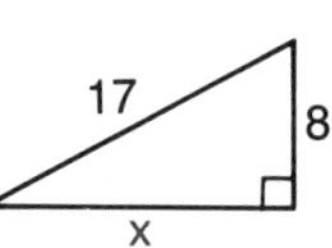

(b)

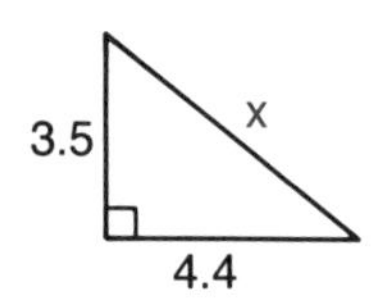

(c)

x x 2.82

(d)

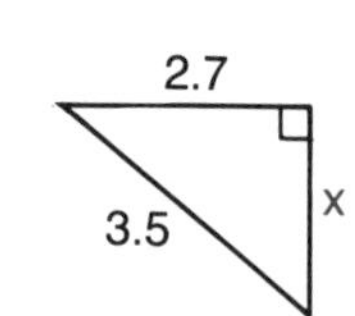

ANGLE CALCULATIONS

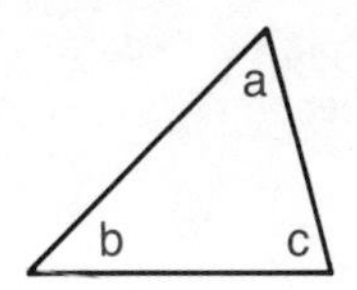

$a + b + c = 180°$

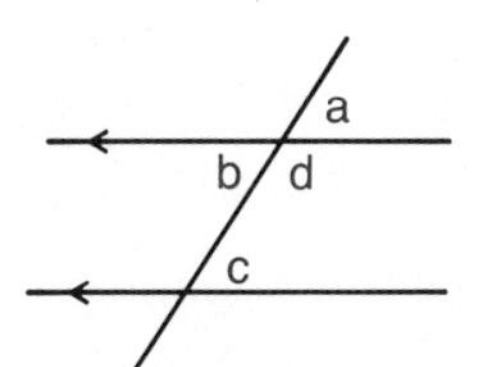

$a = c$
$b = c$
$c + d = 180°$

EXERCISE 3

1. Calculate the measures of the unknown angles.

(a)

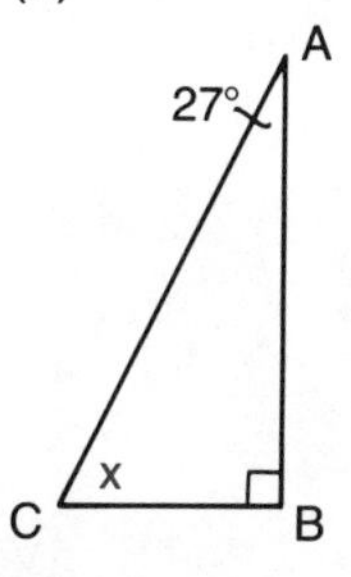

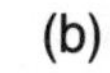

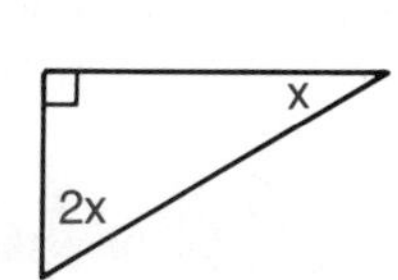

(c)

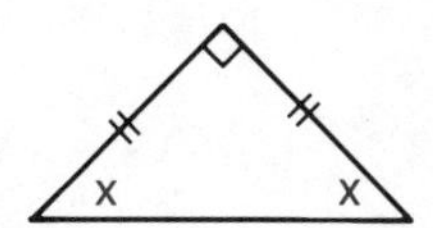

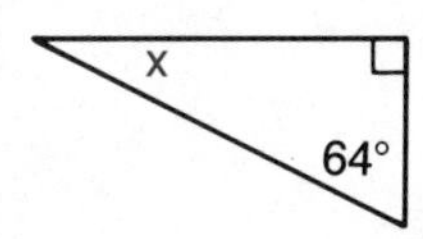

2. Calculate the measures of the indicated angles.

(a)

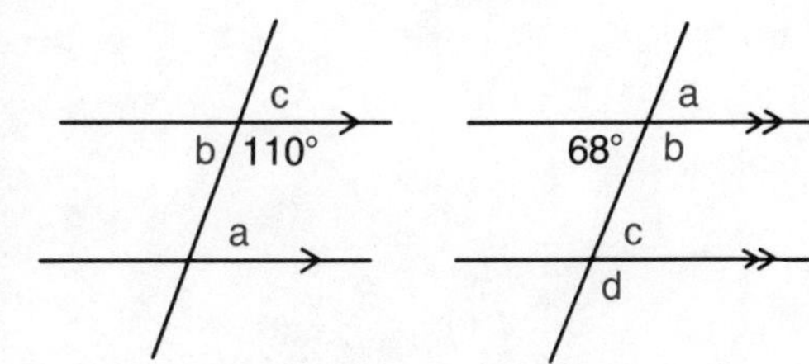

MENTAL MATH

Solve.

1. $5x = 20$
2. $5 + x = 20$
3. $4x = 20$
4. $4 + x = 20$
5. $2x + 1 = 7$
6. $x + 7 = 10$
7. $7x = 35$
8. $7x + 2 = 37$
9. $x - 5 = 3$
10. $2x = -6$
11. $2x - 1 = 11$
12. $3x - 1 = 14$
13. $5 - x = 7$
14. $7 - x = 4$
15. $5x + 0 = 15$
16. $4x - 8 = 0$
17. $3y + 5 = 11$
18. $2t + 3 = 11$
19. $z + z = 16$
20. $2r + 3r = 30$
21. $\frac{x}{7} = \frac{15}{21}$
22. $\frac{y}{4} = \frac{12}{8}$
23. $\frac{a}{3} = \frac{10}{15}$
24. $\frac{a}{12} = \frac{7}{3}$
25. $\frac{4}{b} = \frac{16}{20}$
26. $\frac{12}{b} = \frac{3}{2}$
27. $\frac{3}{5} = \frac{c}{20}$
28. $\frac{5}{12} = \frac{c}{36}$
29. $\frac{3}{5} = \frac{9}{d}$
30. $\frac{16}{10} = \frac{32}{d}$
31. $\frac{a}{16} = \frac{50}{8}$
32. $\frac{12}{b} = \frac{24}{14}$
33. $\frac{6}{9} = \frac{c}{27}$
34. $\frac{15}{e} = \frac{5}{7}$
35. $\frac{4}{d} = \frac{40}{3}$
36. $\frac{50}{9} = \frac{f}{18}$
37. $\frac{3}{4} = \frac{12}{x}$
38. $\frac{y}{9} = \frac{6}{27}$

11.1 SIMILAR TRIANGLES

Similar figures have the same shape but are different in size. In computer-assisted design (CAD), the computer operator works with similar figures to make enlargements and reductions on the monitor screen.

Scale drawings, blueprints, and map making are applications of similar figures. Architects use scale models to show how finished buildings will look.

In this section we will study and compare the measurements of similar triangles, in particular, similar right triangles. This branch of mathematics is called trigonometry.

Astronomers, navigators, and surveyors first used trigonometry more than 2000 a ago. Today, modern technologists and people working in robotics continue to use trigonometry in their day-to-day work.

In the diagram at the right, we see that in $\triangle ABC$, $\triangle ADE$, and $\triangle AFG$

$\angle A$ is common to all three triangles and

$\angle ABC = \angle ADE = \angle AFG = 90°$

From our earlier work in geometry,

$\angle ACB = \angle AED = \angle AGF$

Since the corresponding angles of these triangles are equal, the triangles have the same shape and they are similar.

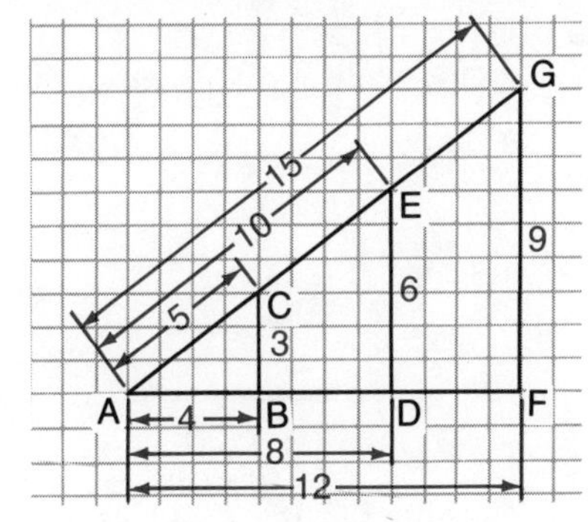

We can measure the side lengths of the triangles in grid units. In the table below, we investigate the relationship between the side lengths of these similar triangles.

$\triangle ABC$	$\triangle ADE$	$\triangle AFG$
AB = 4 BC = 3 AC = 5	AD = 8 DE = 6 AE = 10	AF = 12 FG = 9 AG = 15

To find the hypotenuse use the Pythagorean Theorem: $a^2 + b^2 = c^2$

We can compare any two similar triangles by finding the ratios of corresponding sides.

In $\triangle ABC$ and $\triangle ADE$,

$$\frac{AB}{AD} = \frac{4}{8} = \frac{1}{2} \qquad \frac{BC}{DE} = \frac{3}{6} = \frac{1}{2} \qquad \frac{AC}{AE} = \frac{5}{10} = \frac{1}{2} \qquad \text{and} \qquad \frac{AB}{AD} = \frac{BC}{DE} = \frac{AC}{AE}$$

The ratios are equal.

In $\triangle ABC$ and $\triangle AFG$,

$$\frac{AB}{AF} = \frac{4}{12} = \frac{1}{3} \qquad \frac{BC}{FG} = \frac{3}{9} = \frac{1}{3} \qquad \frac{AC}{AG} = \frac{5}{15} = \frac{1}{3} \qquad \text{and} \qquad \frac{AB}{AF} = \frac{BC}{FG} = \frac{AC}{AG}$$

The conditions for similar triangles are summarized:

In similar triangles, $\triangle ABC \sim \triangle DEF$

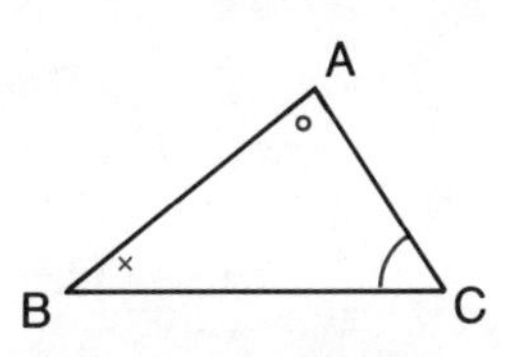

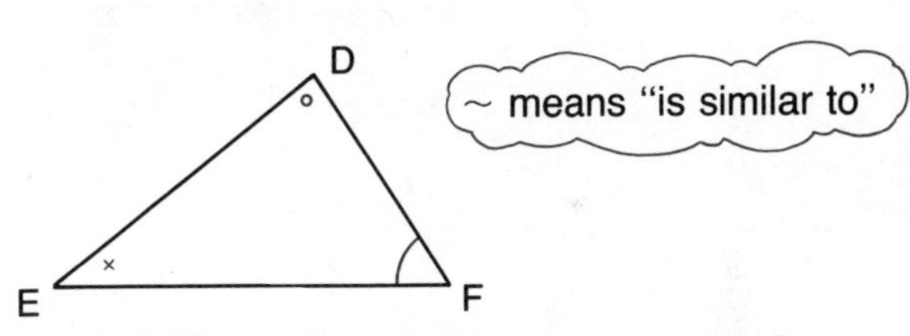

1. Corresponding pairs of angles are equal.
   $\angle A = \angle D$
   $\angle B = \angle E$
   $\angle C = \angle F$

2. Corresponding pairs of sides are in proportion.
   $$\frac{AB}{DE} = \frac{BC}{EF} = \frac{AC}{DF} \quad \text{and} \quad \frac{DE}{AB} = \frac{EF}{BC} = \frac{DF}{AC}$$

To solve problems using similar triangles:

1. Make neat, well-labelled diagrams.
2. Identify the pairs of corresponding sides.
3. Write a proportion.
4. Solve for the unknown value using algebra.

Example 1.
If $\triangle ABC \sim \triangle DEF$, find the lengths of the unknown sides to the nearest tenth.

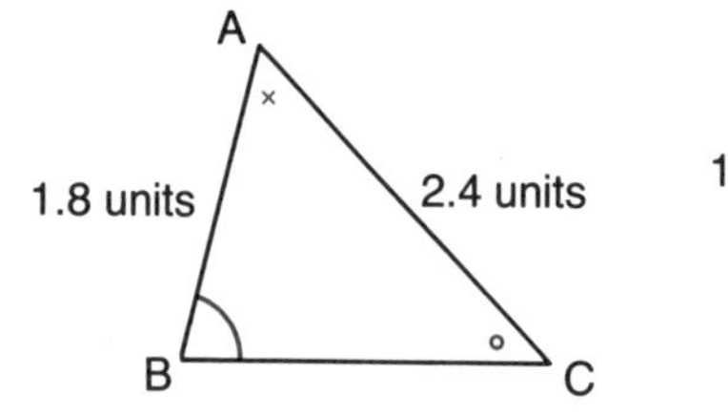

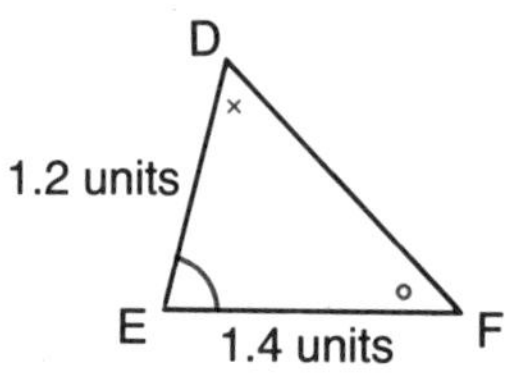

Solution:
The unknown sides are BC and DF.
Since the triangles are similar, find BC by using

$$\frac{BC}{EF} = \frac{AB}{DE}$$

and EF = 1.4, AB = 1.8, DE = 1.2

$$\frac{BC}{1.4} = \frac{1.8}{1.2}$$

$$BC = \frac{1.4 \times 1.8}{1.2}$$

$$= 2.1$$

Write the unknown in the numerator.

Find DF by using

$$\frac{DF}{AC} = \frac{DE}{AB}$$

and AC = 2.4, DE = 1.2, AB = 1.8

$$\frac{DF}{2.4} = \frac{1.2}{1.8}$$

$$DF = \frac{2.4 \times 1.2}{1.8}$$

$$= 1.6$$

Instruments, such as rulers and tape measures, can be used to directly measure distances. We can indirectly, that is without actually measuring, find distances and dimensions using similar triangles.

READ

Example 2.
A tree casts a shadow 18 m long at the same time a metre stick, positioned vertically, casts a shadow 0.75 m long. Find the height of the tree in metres.

PLAN

Solution:
We show the given information in a labelled diagram.
Let the height of the tree be h.

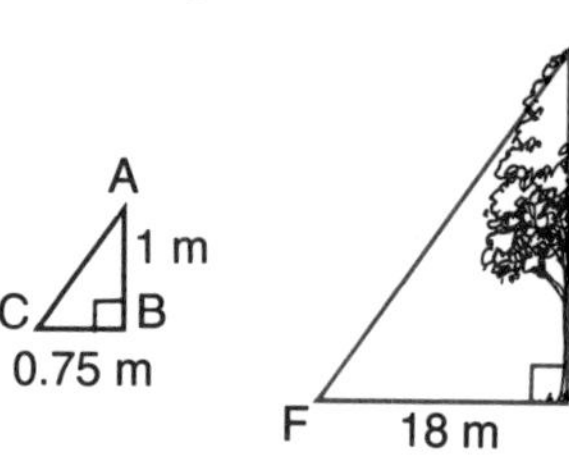

Since $\angle B = \angle E = 90°$
$\angle C = \angle F$
$\triangle ABC \sim \triangle DEF$

$$\therefore \frac{DE}{AB} = \frac{EF}{BC} = \frac{DF}{AC}$$

SOLVE

From the similar triangles,

$$\frac{h}{1} = \frac{18}{0.75}$$

$$h = \frac{1 \times 18}{0.75}$$

$$= 24$$

ANSWER

$\therefore$ The height of the tree is 24 m.

Example 3.
The width of a river can be calculated using only measurements taken on one side of the river as shown in the diagram.
(a) Why are the triangles similar?
(b) Find the width of the river to the nearest metre.

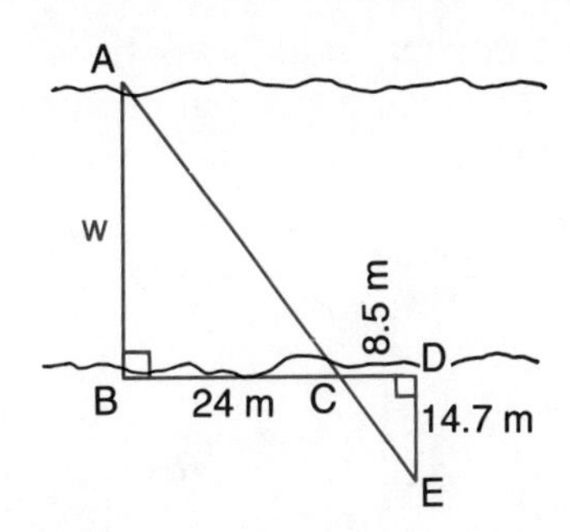

Solution:
(a) In $\triangle ABC$ and $\triangle EDC$,
$\angle ABC = \angle EDC = 90°$
$\angle ACB = \angle ECD$ (opposite angles)
$\therefore \angle BAC = \angle DEC$
Thus, corresponding angles are equal, and the triangles are similar.
Since $\triangle ABC \sim \triangle EDC$,
we can use the proportion
$$\frac{AB}{ED} = \frac{BC}{DC}$$
to find AB.

(b) Let the width of the river be w.
$$\frac{w}{14.7} = \frac{24}{8.5}$$
$$w = \frac{14.7 \times 24}{8.5}$$
$$w \doteq 41.5$$
The width of a river is about 42 m.

Press

The display is 41.505882

EXERCISE 11.1

A 1. (a) Name the similar triangles.
(b) State why the triangles are similar.

(i)

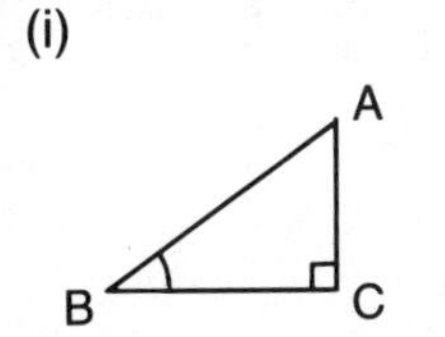

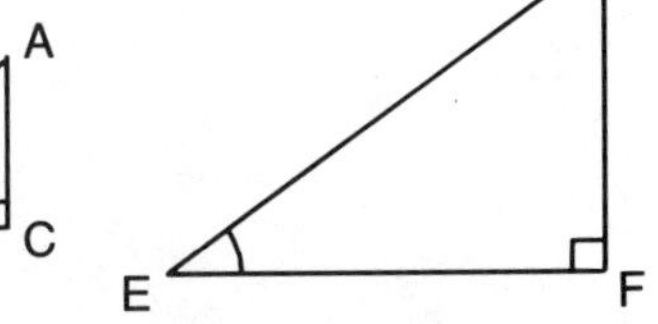

(ii)

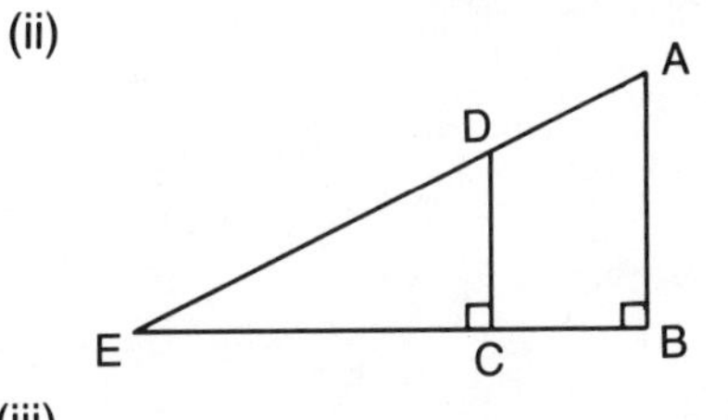

(iii)

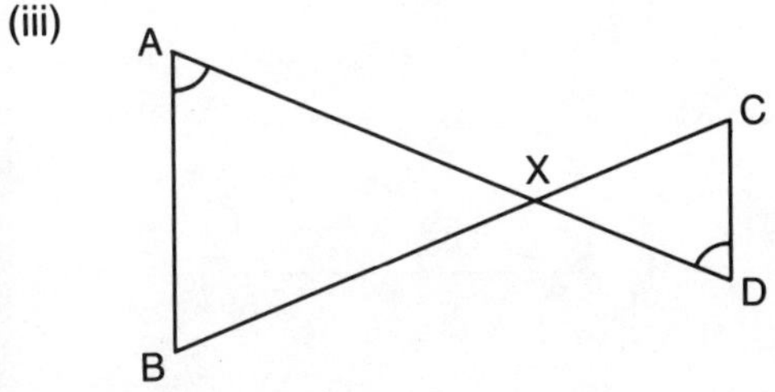

2. In each of the following, name a triangle which is similar to $\triangle ABC$. Give reasons for your answer.

(a)

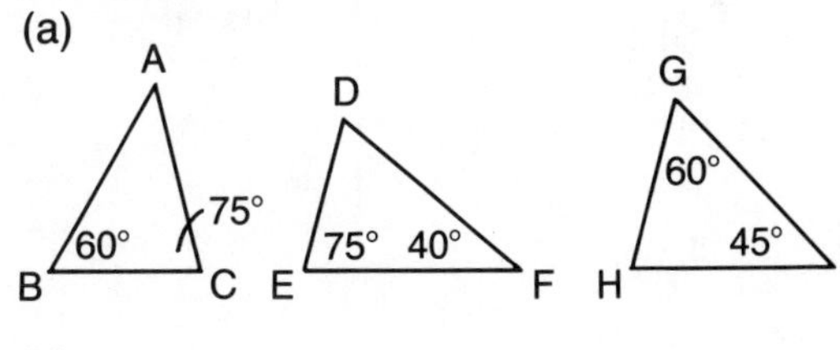

(b)

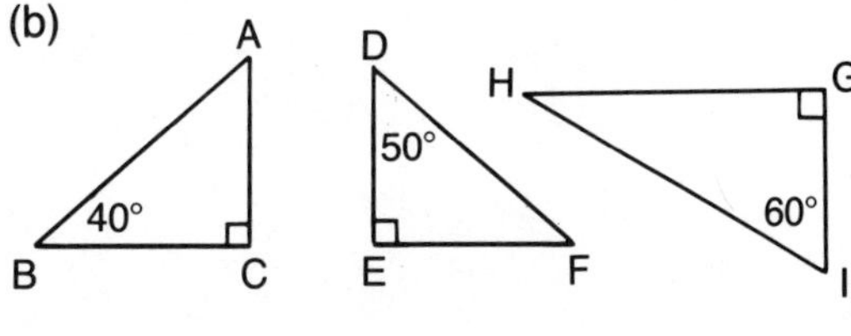

(c)

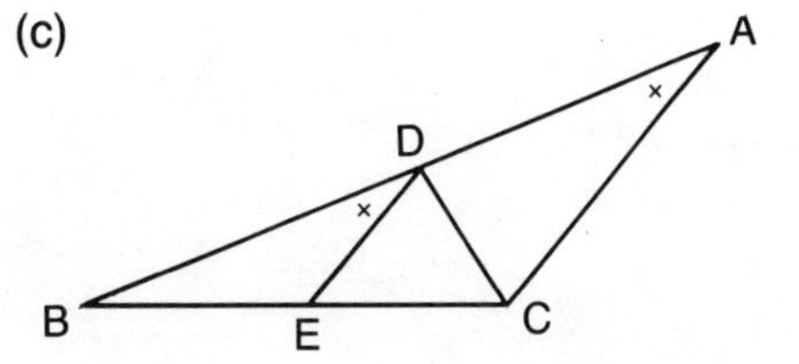

3. State the proportions between corresponding sides in each pair of similar triangles.

(a)

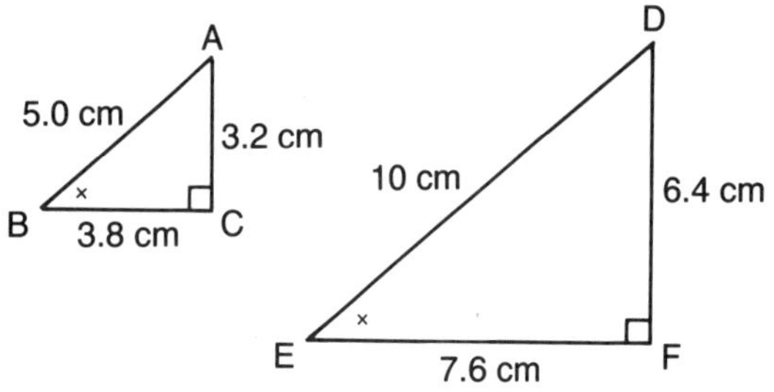

A, B, C; 5.0 cm, 3.2 cm, 3.8 cm

D, E, F; 10 cm, 6.4 cm, 7.6 cm

(b)

P, Q, R; r, q, p

A, B, C; c, b, a

(c)

A, B, 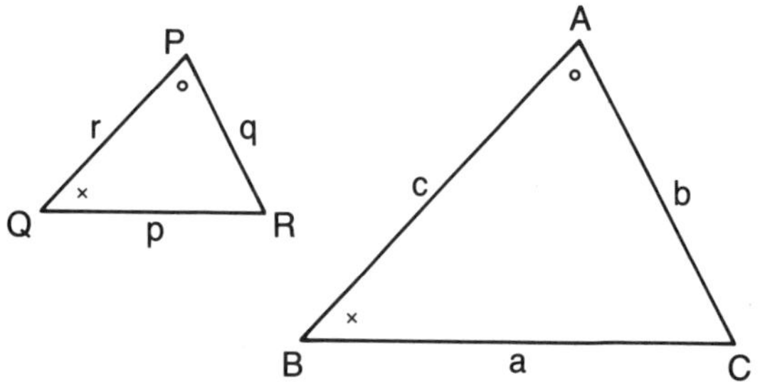 C, 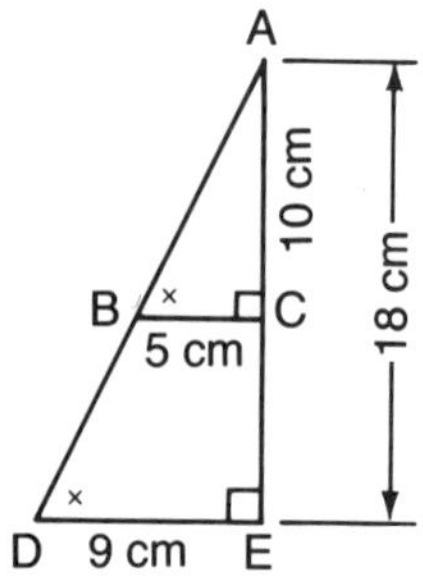 D, 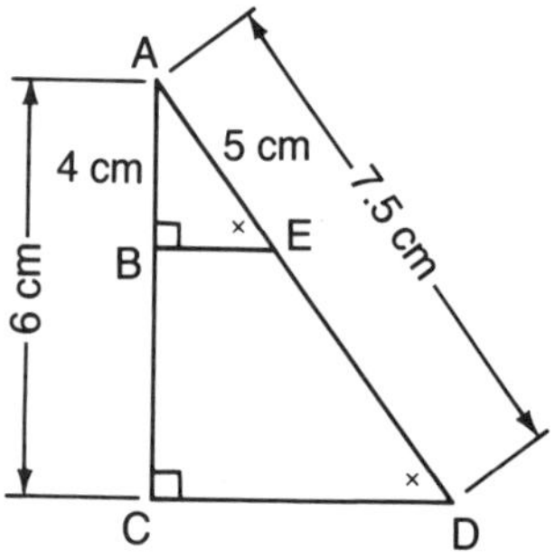 E; 10 cm, 18 cm, 5 cm, 9 cm

(d)

A, B, C, D, E; 4 cm, 6 cm, 5 cm, 7.5 cm

(e)

A, B, C, D, E; 5 cm, 4 cm, 2 cm, 1.6 cm

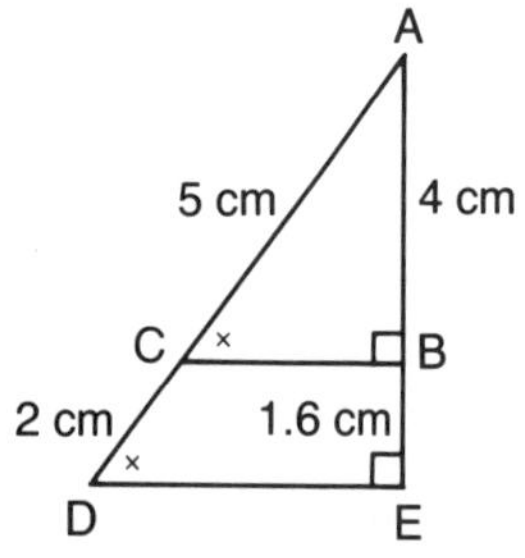

(f)

A, B, C, D, E; c, d, a, b

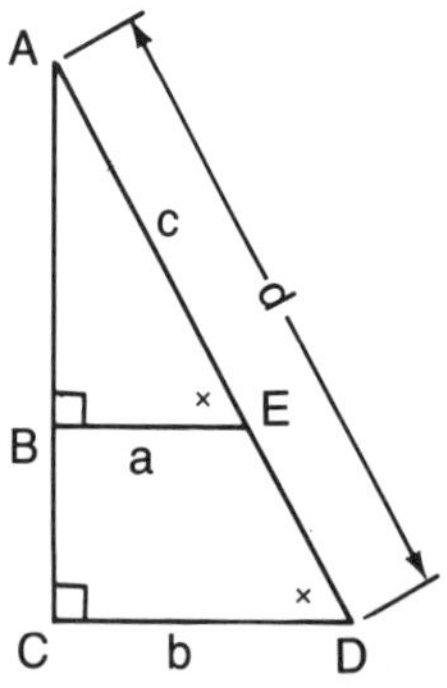

B 4. Each pair of triangles is similar. Calculate the lengths of the unknown sides.

(a)

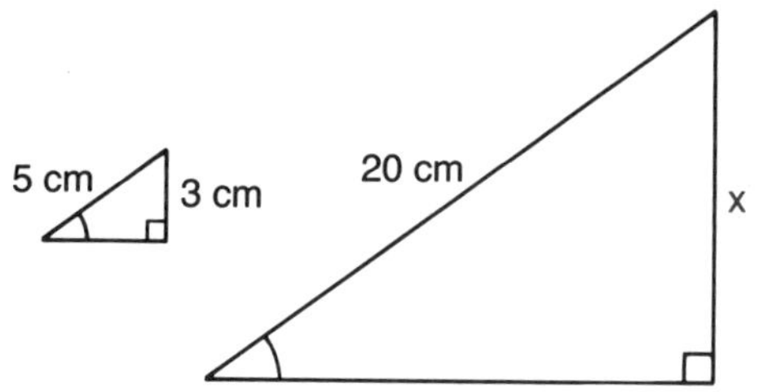

5 cm, 3 cm, 20 cm, x

(b)

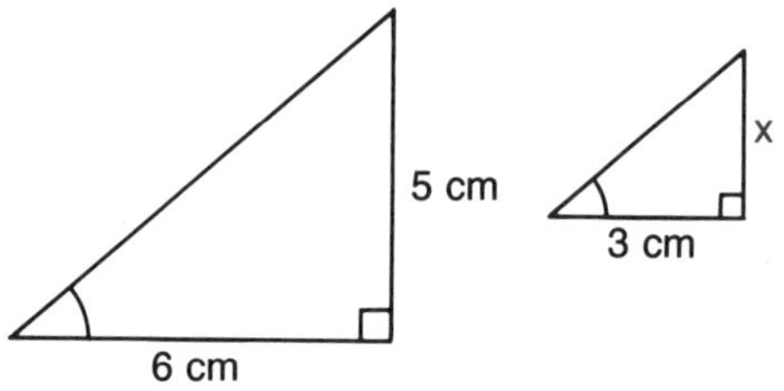

5 cm, 6 cm, x, 3 cm

(c)

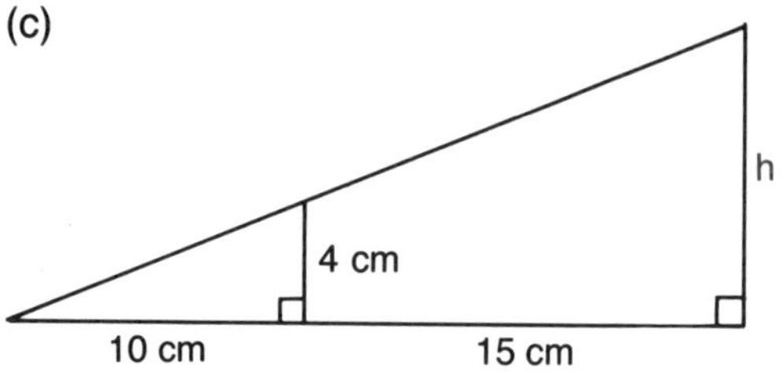

h, 4 cm, 10 cm, 15 cm

(d)

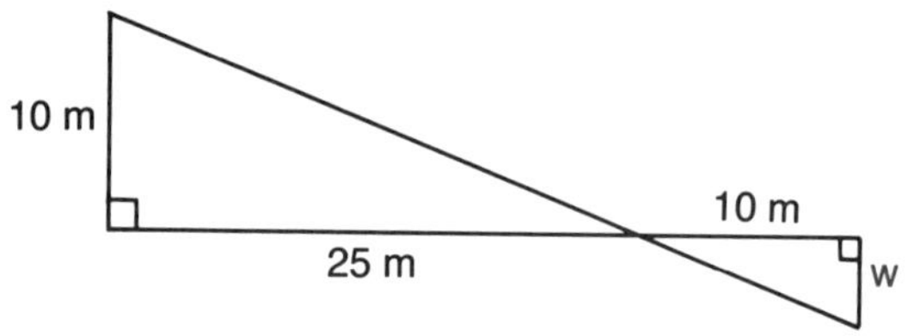

10 m, 10 m, 25 m, w

(e)

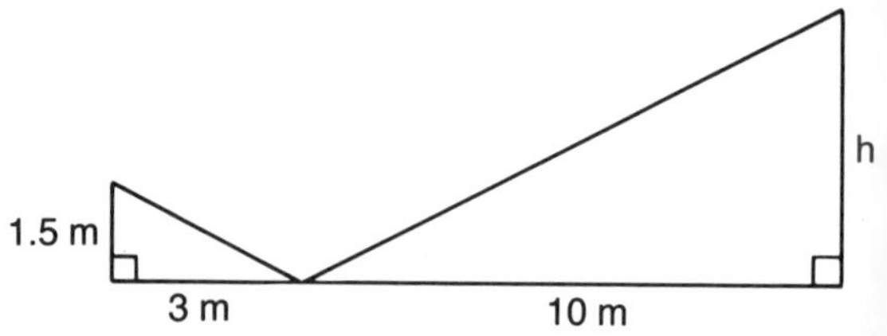

h, 1.5 m, 3 m, 10 m

5. In the diagrams below, △ABC, △DEF, and △IGH are similar.
Find the lengths of DE and GH to the nearest tenth.

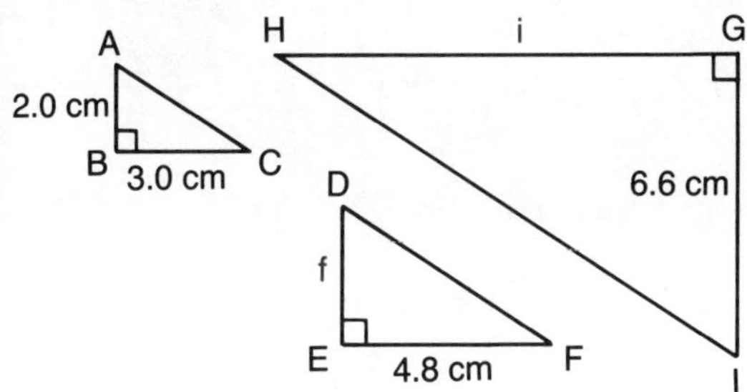

6. (a) State why the three triangles are similar.
(b) Calculate the values of a and b to the nearest tenth.

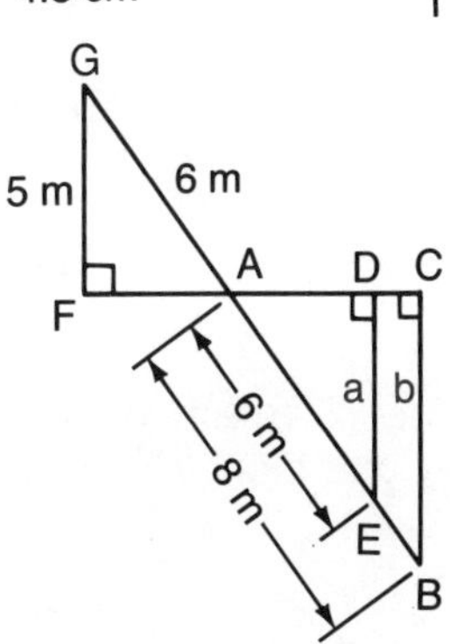

7. A flagpole casts a shadow 14.5 m at the same time that a person 1.8 m tall casts a shadow 2.5 m long.
Calculate the height of the flagpole.

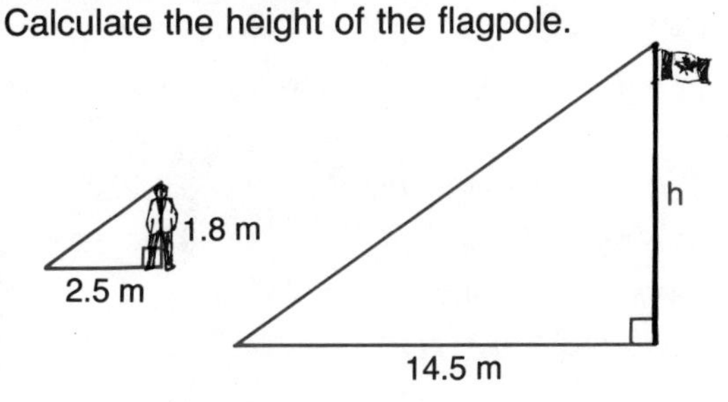

8. Calculate the distance across the river.

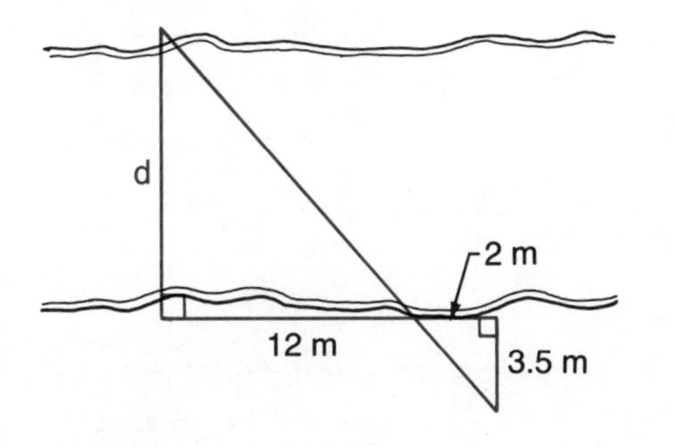

9. Calculate the distance across the river.

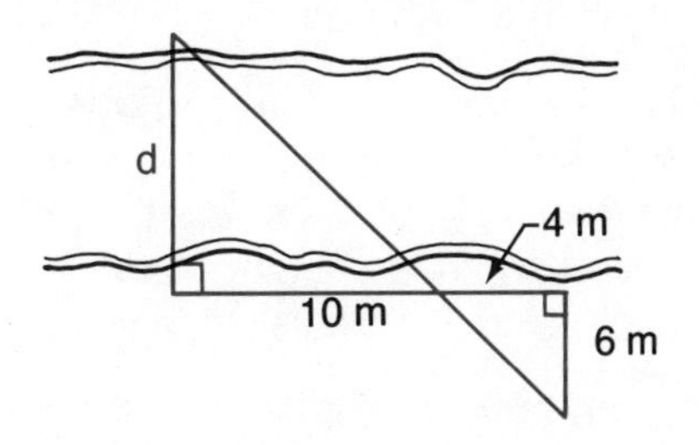

10. Calculate the distance across the pond.

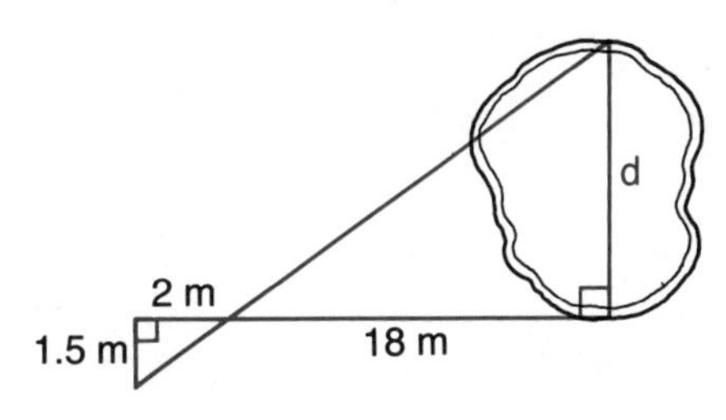

INVESTIGATION

You can calculate height by using a measuring tape and a mirror.
(i) Pick out an object that is much taller than you. Place a mirror on the ground between you and the object.
(ii) Position yourself so you can see the reflection of the top of the object in the mirror.
(iii) Measure the distance from yourself to the mirror and from the mirror to the object.
(iv) Measure the height of your eyes above the ground.

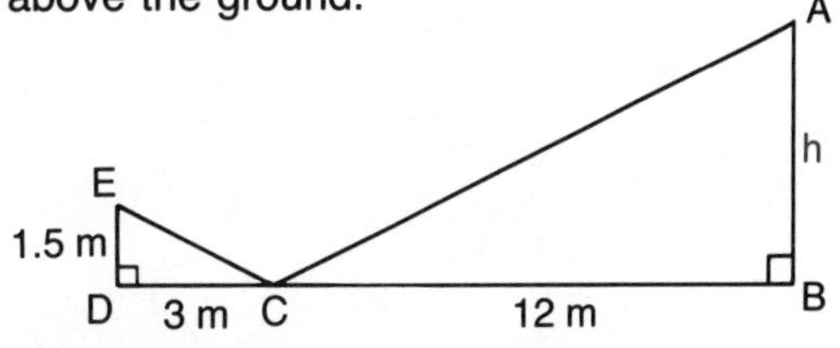

1. Why is ∠ECD = ∠ACB?
2. Name a pair of similar triangles.
3. Calculate the height of the object in the diagram.

11.2 THE TANGENT RATIO

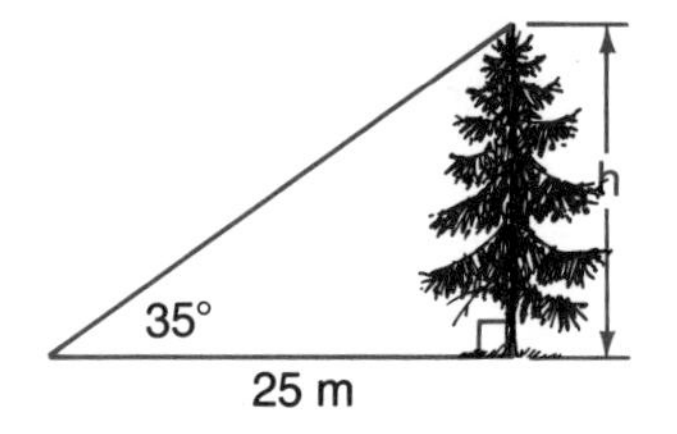

A tree casts a shadow 25 m long when the angle of elevation of the sun is 35°. We would like to use this information to calculate the height of the tree. In the previous section, we calculated a length, height, or distance by setting up a proportion, using two similar triangles.

We draw and measure △DEF so it is similar to the given triangle.

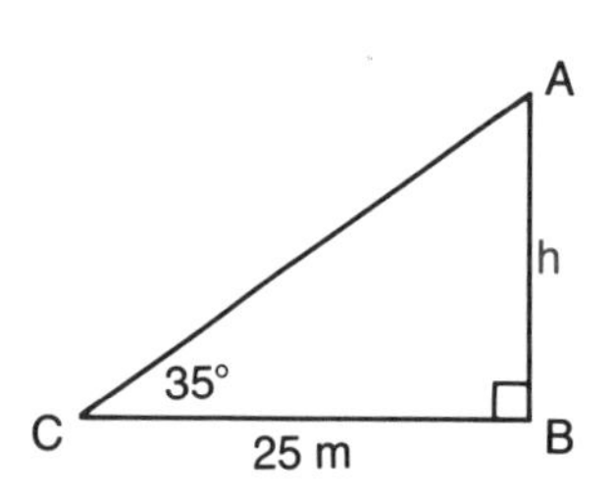

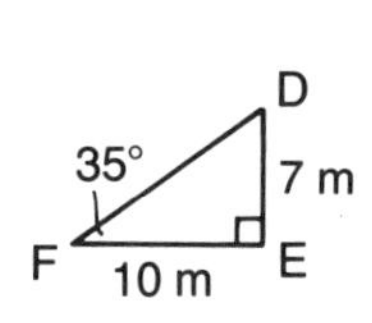

Using similar triangles,

$$\frac{h}{7} = \frac{25}{10}$$

$$h = \frac{7 \times 25}{10}$$

$$= 17.5$$

∴ The height of the tree is 17.5 m.

How would you solve this problem when you are only given one triangle? Trigonometry provides a method to indirectly measure, using only one triangle.

In △ABC at the right,
BC is the side opposite to ∠A
AC is the side adjacent to ∠A
AB is the hypotenuse.

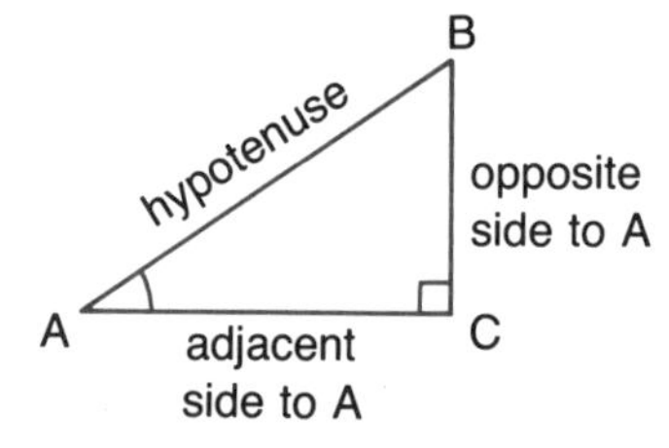

In trigonometry, we use the words in the diagram to define special ratios. One of these special ratios is called the tangent ratio and is:

$$\text{tangent of } \angle A = \frac{\text{measure of side opposite } \angle A}{\text{measure of side adjacent to } \angle A}$$

For convenience, we write $\tan A = \frac{\text{opposite}}{\text{adjacent}}$

The following three triangles are similar because the corresponding angles are equal. We can compare the tangent ratio for the 35° angle in each triangle.

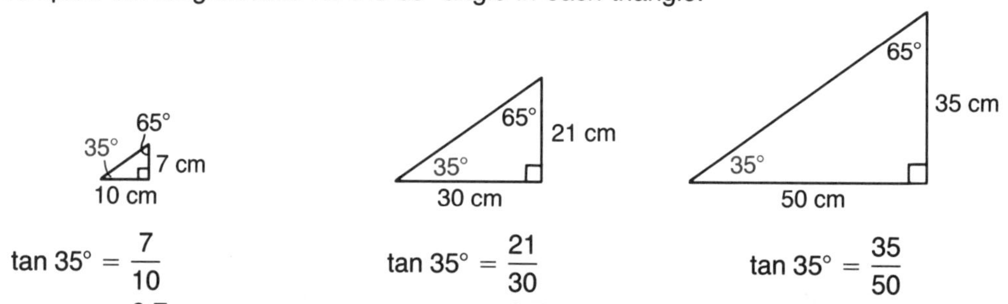

$$\tan 35° = \frac{7}{10} = 0.7$$

$$\tan 35° = \frac{21}{30} = 0.7$$

$$\tan 35° = \frac{35}{50} = 0.7$$

From the above results, we note the tangent ratio for 35° is constant. The tangent ratio for 35 ° in each different triangle is a constant number, and we can use this fact to calculate.

For convenience, tables of tangent ratio values to four decimal places are given on pages 339 and 406.
From the partial tangent table at the right, we read
tan 35° = 0.7002

θ°	tan θ
0	0.0000
1	0.0175
34	0.6745
35	0.7002
36	0.7265
74	3.4874

To find the value of tan 35° on a calculator,

press **3** **5** **tan**

and the display is 0.7002075

Note that it is not necessary to press the **=** key to find a tangent ratio.

Example 1.
Calculate the value of the tangent ratio of the indicated angle.

(a)

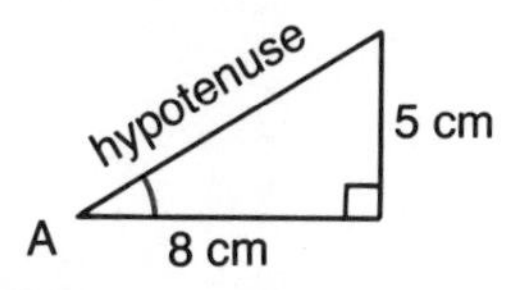

(b)

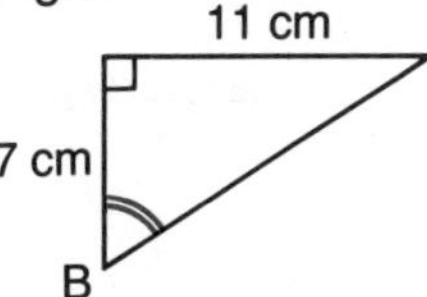

Solution:

(a) The length of the side opposite to ∠A is 5 cm. The length of the side adjacent to ∠A is 8 cm.

$$\text{tangent ratio} = \frac{\text{opposite}}{\text{adjacent}}$$
$$\tan A = \frac{5}{8}$$
$$= 0.6250$$

(b) The length of the side opposite to ∠B is 11 cm. The length of the side adjacent to ∠B is 7 cm.

$$\text{tangent ratio} = \frac{\text{opposite}}{\text{adjacent}}$$
$$\tan B = \frac{11}{7}$$
$$\doteq 1.5714$$

Example 2.
Calculate.
(a) tan A
(b) tan C
(c) the measure of ∠A

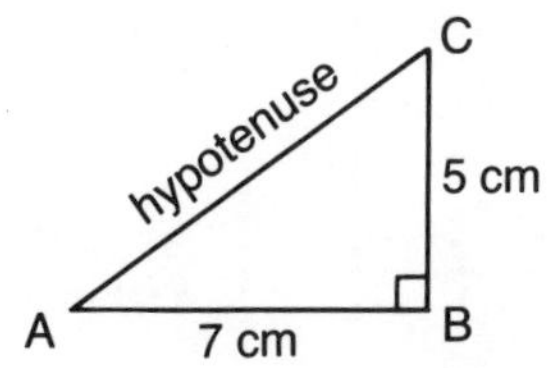

Solution:

(a) $\tan A = \frac{BC}{AB} = \frac{5}{7}$
$\doteq 0.7143$

(b) $\tan C = \frac{AB}{BC} = \frac{7}{5}$
$= 1.4000$

(c) To find the measure of ∠A:
Using a calculator, press

5 **÷** **7** **=** **inv** **tan**

The display is 35.537678
∠A = 36°, to the nearest degree.

Using the table, we find that
0.7265 = tan 36°
Hence,
∠A = 36°, to the nearest degree

The tangent ratio depends only on the measure of the angle and not on the size of the right triangle. This fact permits us to solve some interesting problems.

Example 3.
Calculate the height of a flagpole which casts a 26.5 m shadow when the angle of elevation to the sun is 32°.

Solution:
Let the height of the flagpole in metres be h.

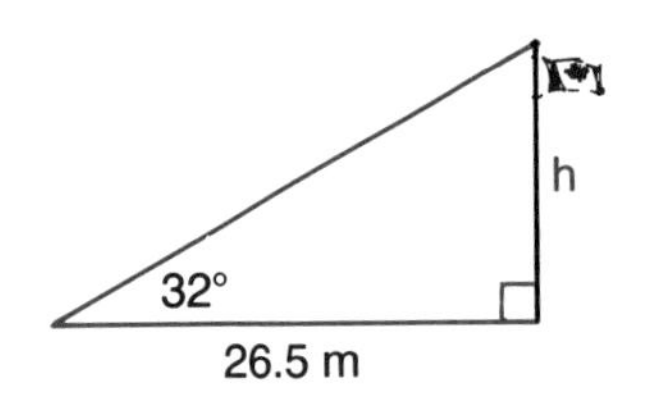

$$\frac{h}{26.5} = \tan 32°$$
$$h = 26.5 \times \tan 32°$$
$$= 26.5 \times 0.6249$$
$$\doteq 16.5590$$

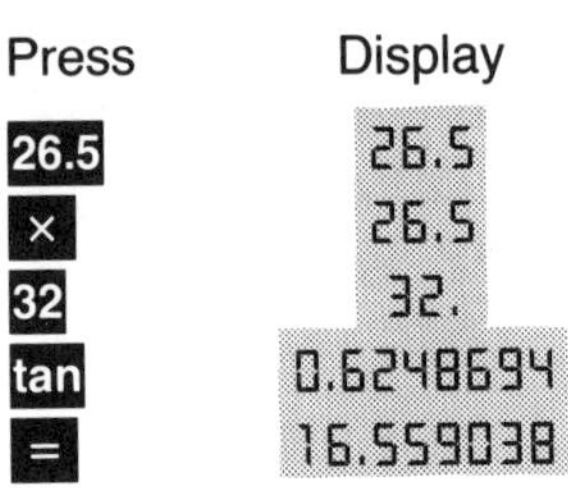

The height of the flagpole is 16.6 m to the nearest tenth.

EXERCISE 11.2

A 1. Identify the hypotenuse and the sides opposite and adjacent to the indicated angle in each triangle.

θ is the Greek letter theta, which is often used to denote the measure of an angle.

(a)

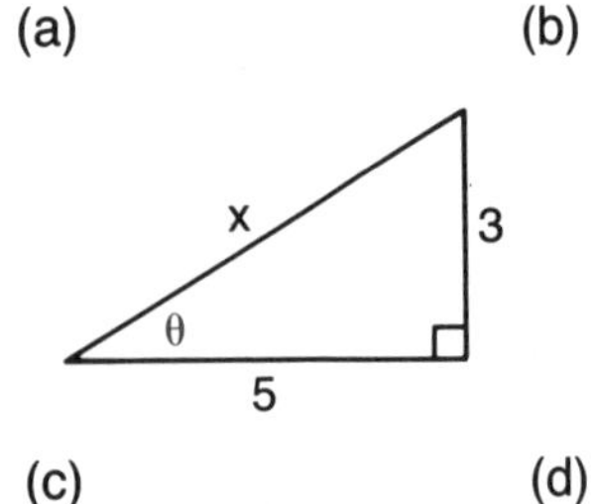

(b)

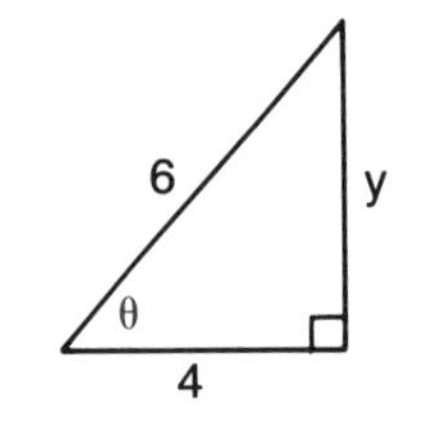

(c)

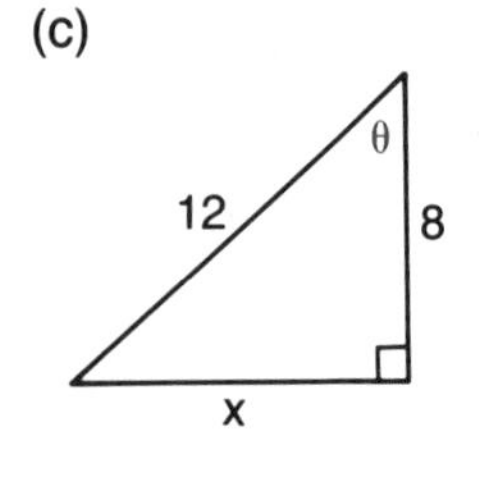

(d)

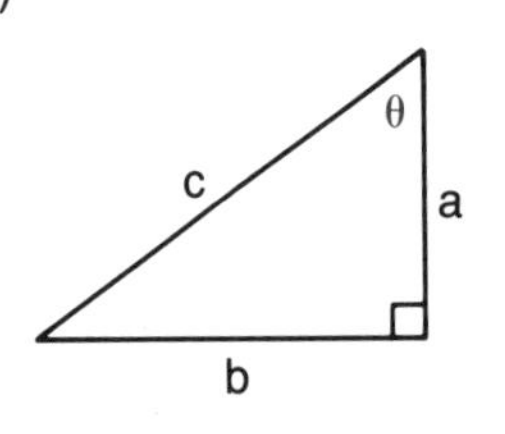

(e)

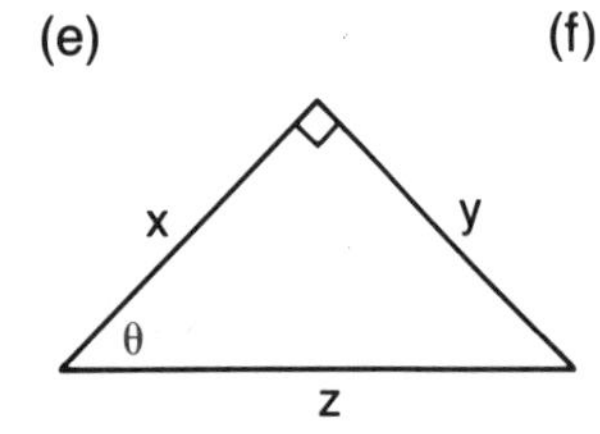

(f)

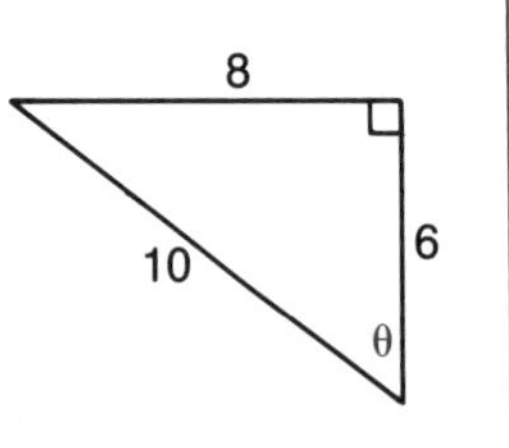

2. State the ratio tan θ in fraction form for each of the following.

(a)

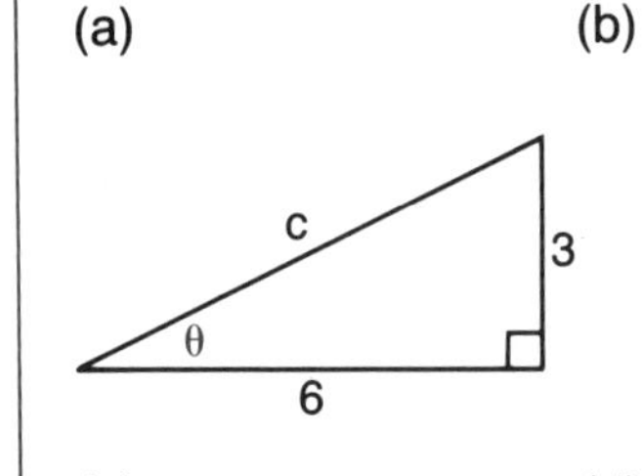

(b)

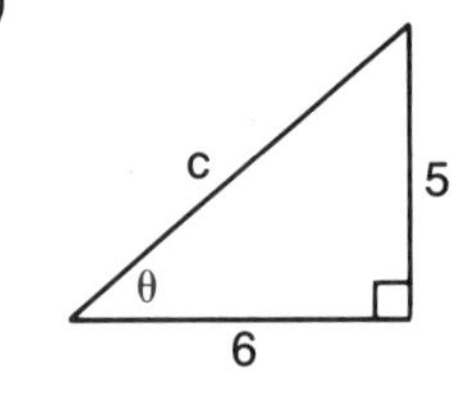

(c)

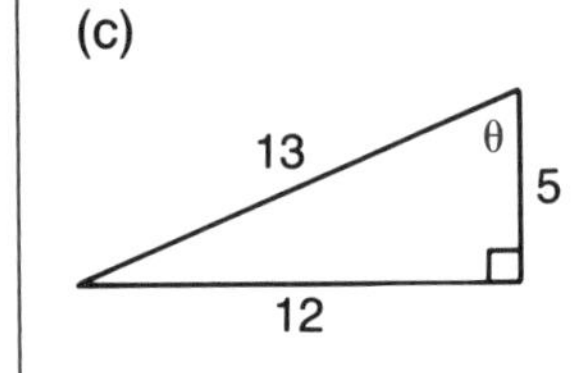

(d)

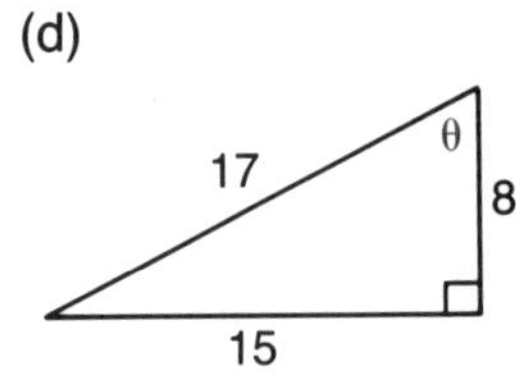

(e)

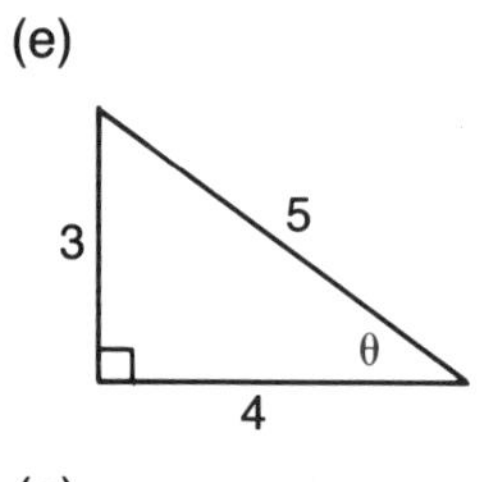

(f)

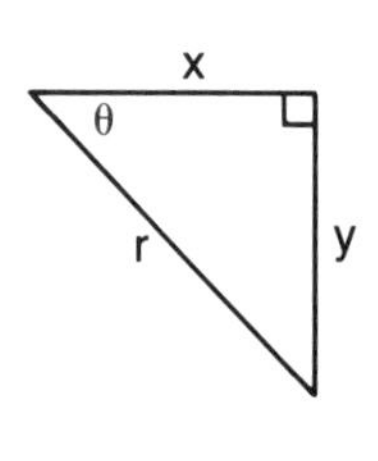

(g)

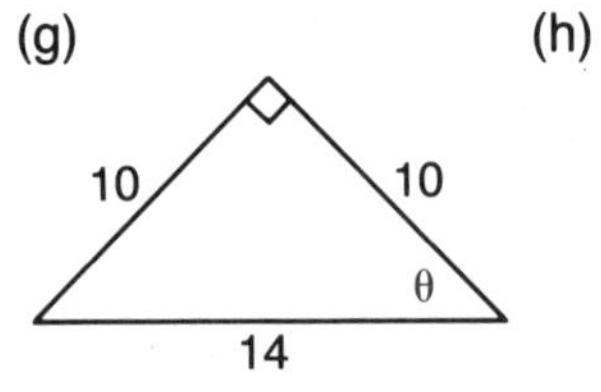

(h)

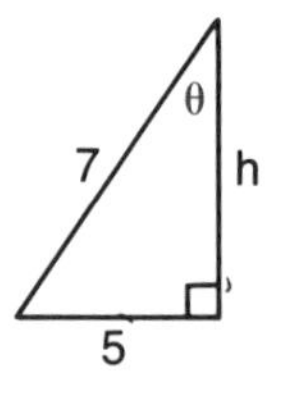

B Questions 3, 4, and 5 can be answered using a calculator or the table at the right.

3. Find a value for each of the following tangent ratios.

(a) tan 18° (b) tan 36°
(c) tan 52° (d) tan 26°
(e) tan 45° (f) tan 25°
(g) tan 40° (h) tan 80°
(i) tan 10° (j) tan 20°
(k) tan 28° (l) tan 56°
(m) tan 30° (n) tan 60°

4. Find the value of θ to the nearest degree.

(a) $\tan \theta = 0.2126$ (b) $\tan \theta = 0.5543$
(c) $\tan \theta = 0.7265$ (d) $\tan \theta = 1.0000$
(e) $\tan \theta = 0.5774$ (f) $\tan \theta = 3.7321$
(g) $\tan \theta = 1.7321$ (h) $\tan \theta = 0.8391$
(i) $\tan \theta = 1.2645$ (j) $\tan \theta = 0.5555$
(k) $\tan \theta = 2.5287$ (l) $\tan \theta = 9.5000$
(m) $\tan \theta = 2.5000$ (n) $\tan \theta = 0.9999$

5. Find the measure of the angle θ to the nearest degree in each triangle.

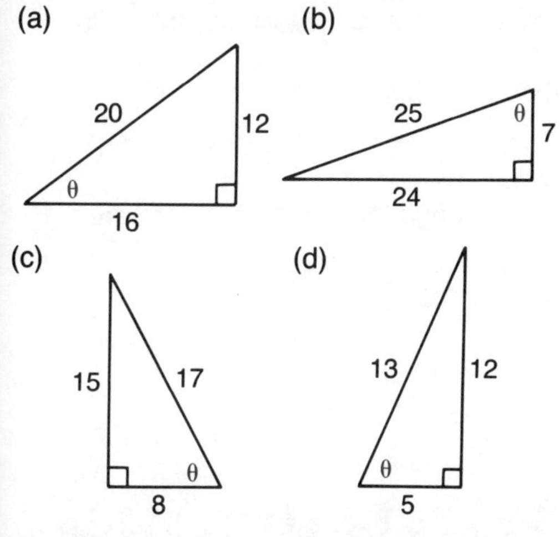

The table at the right gives the values of tangent ratios to four decimal places for angles from 0° to 90°. To find a tangent value, first look in the angle column, then find the corresponding value in the tangent column. To find the measure of an angle to the nearest degree, first look in the tangent column, then find the corresponding value in the angle column.

θ°	tan θ	θ°	tan θ
0	0.0000	46	1.0355
1	0.0175	47	1.0724
2	0.0349	48	1.1106
3	0.0524	49	1.1504
4	0.0699	50	1.1917
5	0.0875	51	1.2349
6	0.1051	52	1.2799
7	0.1228	53	1.3270
8	0.1405	54	1.3764
9	0.1584	55	1.4281
10	0.1763	56	1.4826
11	0.1944	57	1.5399
12	0.2126	58	1.6003
13	0.2309	59	1.6643
14	0.2493	60	1.7320
15	0.2680	61	1.8040
16	0.2867	62	1.8807
17	0.3057	63	1.9626
18	0.3249	64	2.0503
19	0.3443	65	2.1445
20	0.3640	66	2.2460
21	0.3839	67	2.3558
22	0.4040	68	2.4751
23	0.4245	69	2.6051
24	0.4452	70	2.7475
25	0.4663	71	2.9042
26	0.4877	72	3.0777
27	0.5095	73	3.2708
28	0.5317	74	3.4874
29	0.5543	75	3.7320
30	0.5774	76	4.0108
31	0.6009	77	4.3315
32	0.6249	78	4.7046
33	0.6494	79	5.1445
34	0.6745	80	5.6713
35	0.7002	81	6.3137
36	0.7265	82	7.1154
37	0.7536	83	8.1443
38	0.7813	84	9.5144
39	0.8098	85	11.430
40	0.8391	86	14.301
41	0.8693	87	19.081
42	0.9004	88	28.636
43	0.9325	89	57.290
44	0.9657	90	—
45	1.0000		

6. Calculate the lengths of the indicated sides to the nearest tenth using a tangent ratio.

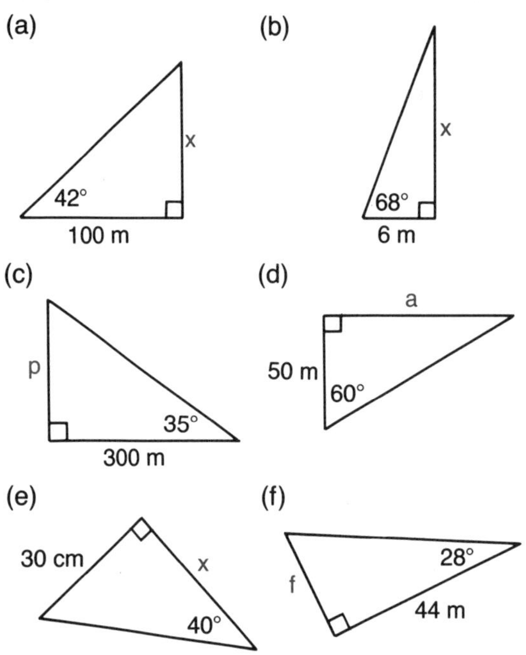

7. Calculate the values of the variables.

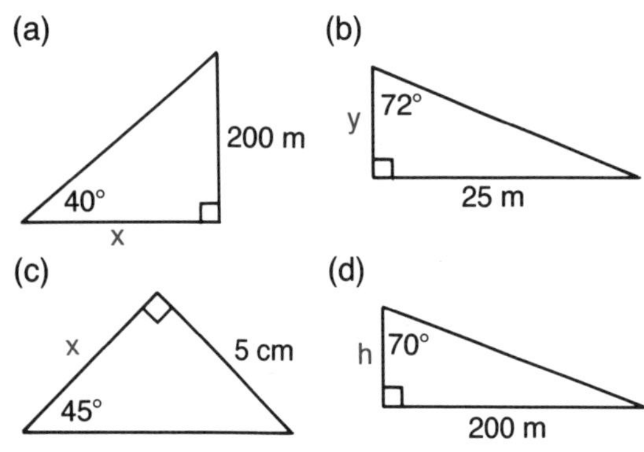

8. A tree casts a shadow 20 m long when the angle of elevation to the sun is 35°. Find the height of the tree.

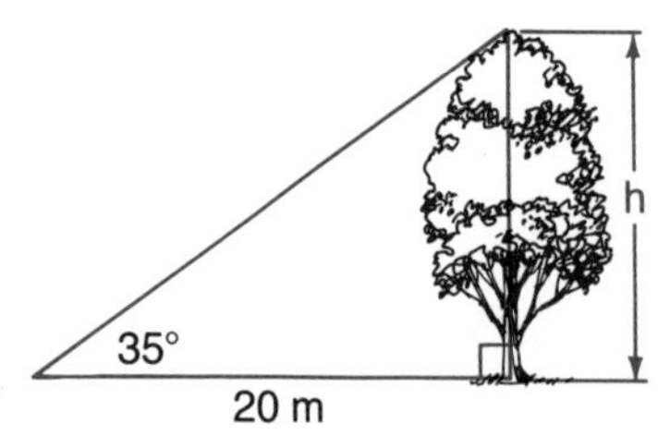

9. Calculate the height of a flagpole which casts a shadow of 12 m when the angle of elevation to the sun is 62°.

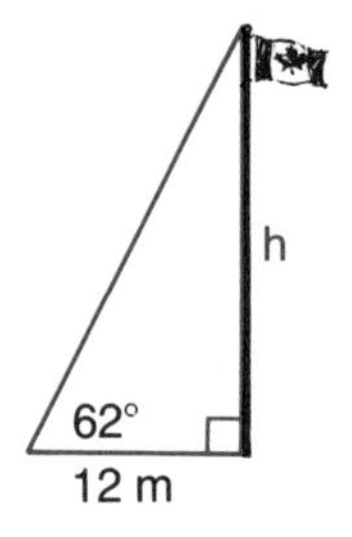

10. A ladder which rests against a wall makes an angle of 65° with the ground. The base of the ladder is 2.35 m from the wall.
How high up the wall is the ladder?

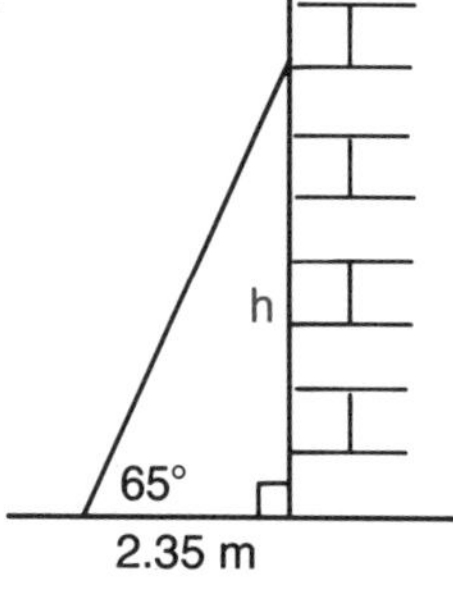

11. The sun's rays make an angle of 25° with the ground. A tree is 21.5 m high. How long is the tree's shadow?

12. A hot air balloon rises vertically. From a spot 1000 m away from where the balloon rose, the angle of elevation to the balloon is 35°.
What is the height of the balloon?

13. A building which is 24 m high casts a shadow 15 m long.
Find the angle of elevation to the sun.

MIND BENDER

Jane and Alice ran the 100 m dash. When Jane crossed the finish line, Alice still had 10 m to go. They raced again with Jane starting 10 m behind the starting line. They both ran the second race at the same speed as the first. Who won the second race?

11.3 THE SINE RATIO

A guy rope stretches 40 m from the top of the circus tent side to the ground. The rope makes an angle of 30° with the ground.
How high is the side of the tent?

40 m
h
30°

The tangent ratio compares the opposite side to the adjacent side. $\frac{\text{opposite side}}{\text{adjacent side}}$

However, to solve this problem, we want to compare the opposite side to the hypotenuse. $\frac{\text{opposite side}}{\text{hypotenuse}}$

We can calculate this ratio by constructing similar triangles.

$\frac{\text{opposite side}}{\text{hypotenuse}}$

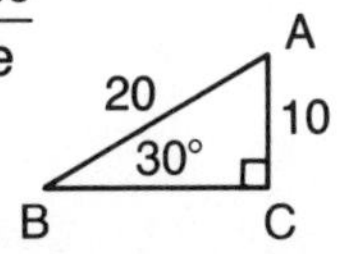

$\frac{AC}{AB} = \frac{10}{20} = 0.5$

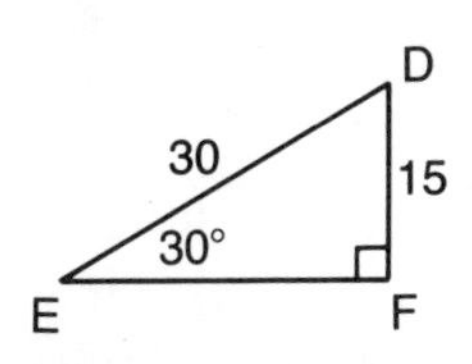

$\frac{DF}{DE} = \frac{15}{30} = 0.5$

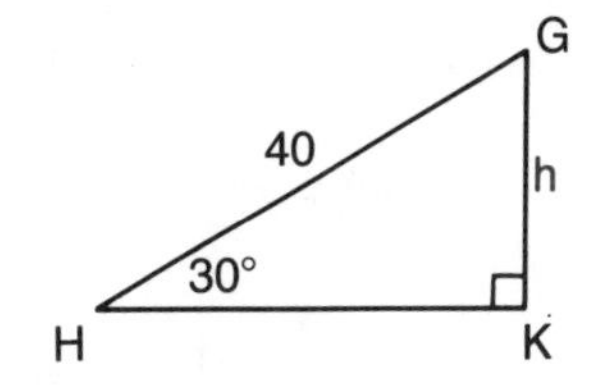

$\frac{GK}{GH} = \frac{h}{40} = 0.5 \quad h = 20$

∴ The height of the side of the tent is 20 m.

The ratios are constant for an angle of 30°. Each angle has a different sine value.

The ratio $\frac{\text{opposite side}}{\text{hypotenuse}}$ is called the sine ratio. The sine of ∠A is written sin A.

Values for the sine ratio are given in tables on pages 342 and 406.

Example 1.
State sin θ.

5 cm
3 cm
θ
4 cm

Solution:

$\sin \theta = \frac{opp}{hyp}$

$\sin \theta = \frac{3}{5}$

Example 2.
Calculate the measure of b, to the nearest centimetre.

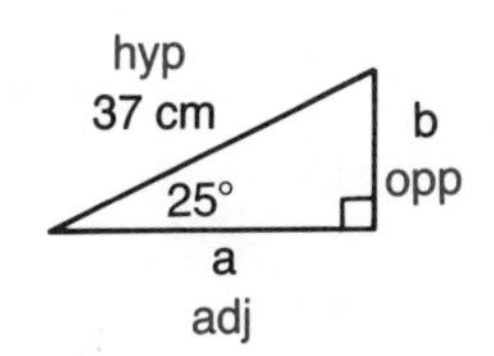

Solution:

$\frac{b}{37} = \frac{opp}{hyp}$

Write an equation using a trigonometric ratio.

$\frac{b}{37} = \sin 25°$

$b = 37 \times \sin 25°$

$\doteq 37 \times 0.4226$

$\doteq 15.6362$

From the table on the next page, sin 25° = 0.4226

Press	Display
3 7	37.
×	37.
2 5	25.
sin	0.4226182
=	15.636876

∴ In the given triangle, b is 16 cm to the nearest centimetre.

EXERCISE 11.3

A

1. State sin θ for each triangle.

(a) (b)

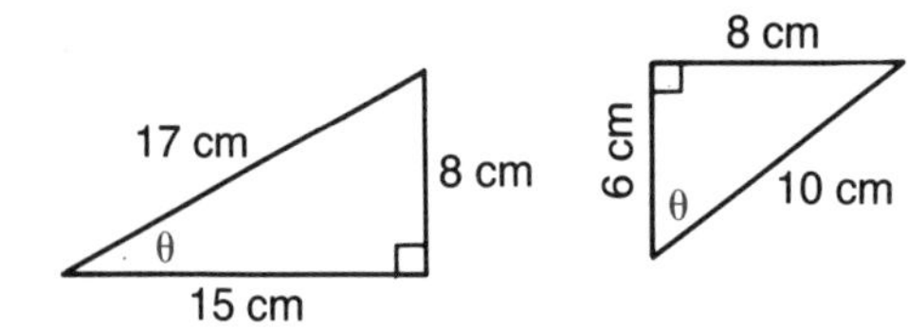

(c)

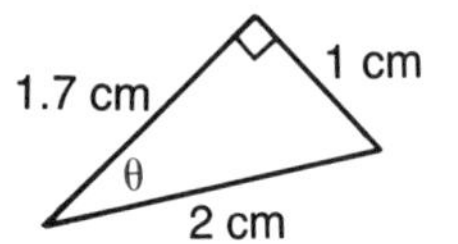

(d)

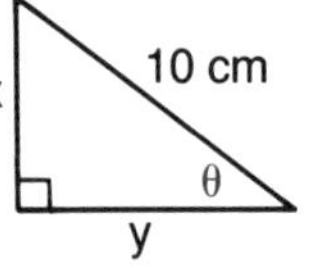

(e)

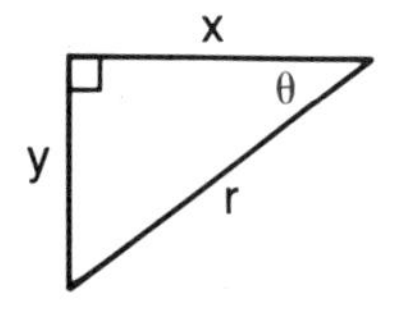

(f)

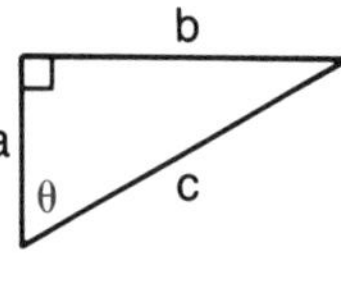

2. State the equation you would use to find the indicated side.

(a)

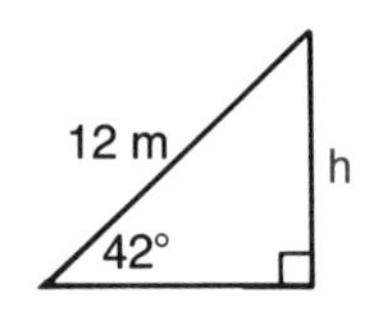

(b)

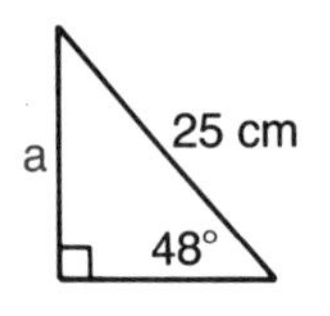

(c)

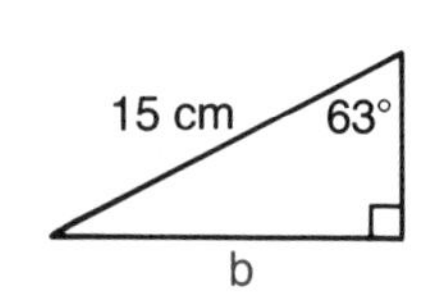

(d)

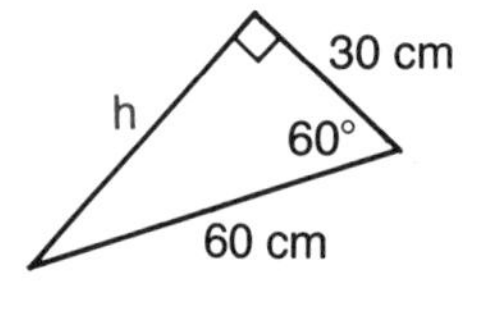

(e)

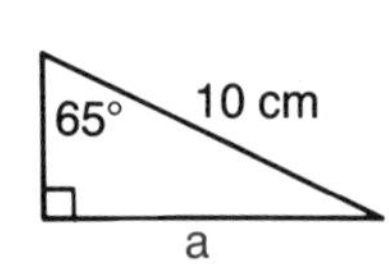

(f)

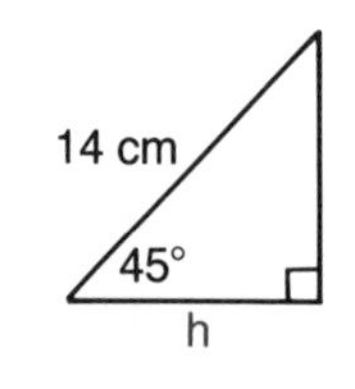

θ°	sin θ	θ°	sin θ
0	0.0000	**46**	0.7193
1	0.0175	**47**	0.7314
2	0.0349	**48**	0.7431
3	0.0523	**49**	0.7547
4	0.0698	**50**	0.7660
5	0.0872	**51**	0.7772
6	0.1045	**52**	0.7880
7	0.1219	**53**	0.7986
8	0.1392	**54**	0.8090
9	0.1564	**55**	0.8192
10	0.1737	**56**	0.8290
11	0.1908	**57**	0.8387
12	0.2079	**58**	0.8481
13	0.2250	**59**	0.8572
14	0.2419	**60**	0.8660
15	0.2588	**61**	0.8746
16	0.2756	**62**	0.8830
17	0.2924	**63**	0.8910
18	0.3090	**64**	0.8988
19	0.3256	**65**	0.9063
20	0.3420	**66**	0.9136
21	0.3584	**67**	0.9205
22	0.3746	**68**	0.9272
23	0.3907	**69**	0.9336
24	0.4067	**70**	0.9397
25	0.4226	**71**	0.9455
26	0.4384	**72**	0.9511
27	0.4540	**73**	0.9563
28	0.4695	**74**	0.9613
29	0.4848	**75**	0.9659
30	0.5000	**76**	0.9703
31	0.5150	**77**	0.9744
32	0.5299	**78**	0.9782
33	0.5446	**79**	0.9816
34	0.5592	**80**	0.9848
35	0.5736	**81**	0.9877
36	0.5878	**82**	0.9903
37	0.6018	**83**	0.9926
38	0.6157	**84**	0.9945
39	0.6293	**85**	0.9962
40	0.6428	**86**	0.9976
41	0.6561	**87**	0.9986
42	0.6691	**88**	0.9994
43	0.6820	**89**	0.9999
44	0.6947	**90**	1.0000
45	0.7071		

B 3. Find the length of the indicated side to the nearest tenth.

(a)

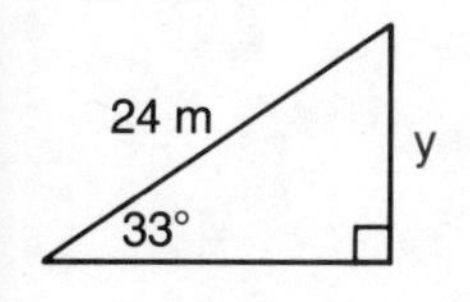

(b)

(c)

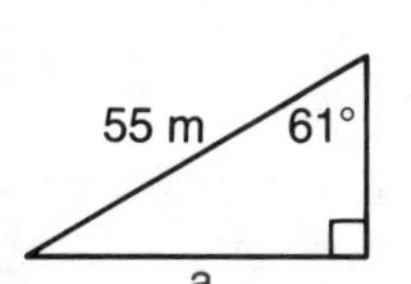

(d)

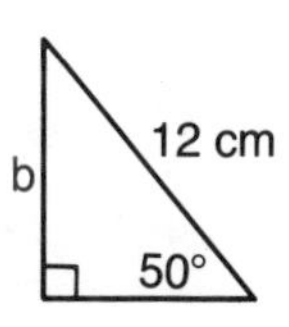

(e)

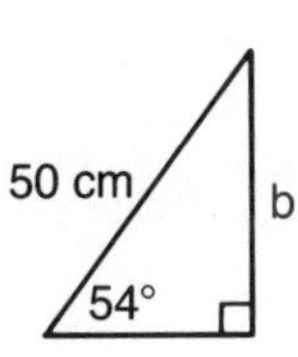

(f)

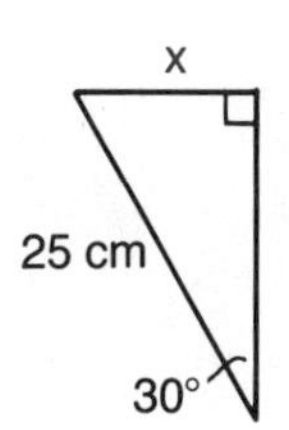

4. Find the measure of the indicated angle to the nearest degree.

(a)

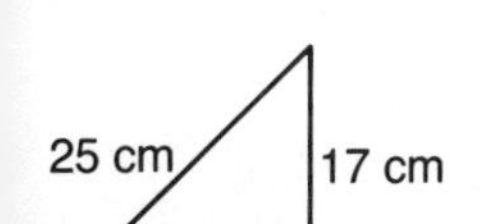

(b)

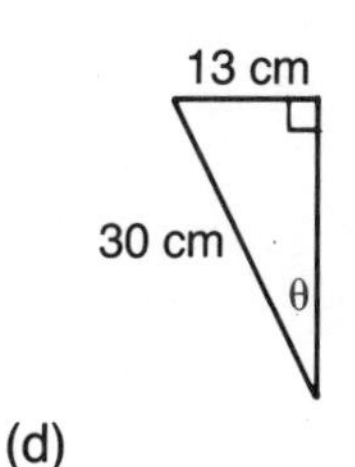

(c)

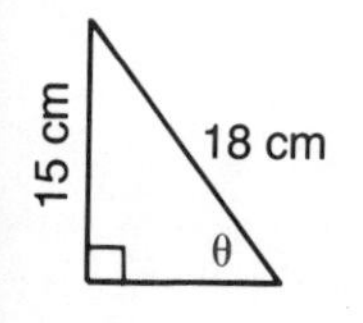

(d)

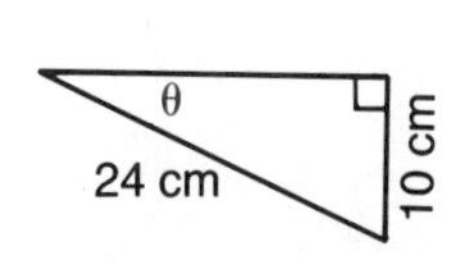

(e)

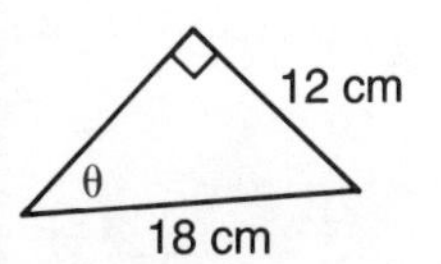

(f)

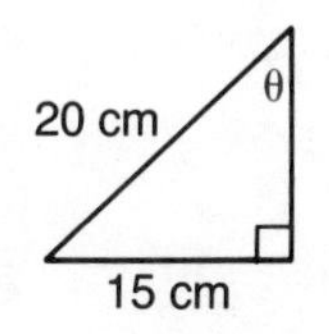

5. A 4.7 m ladder leans against a wall. The foot of the ladder makes an angle of 75° with the ground.
How high up the wall is the ladder?

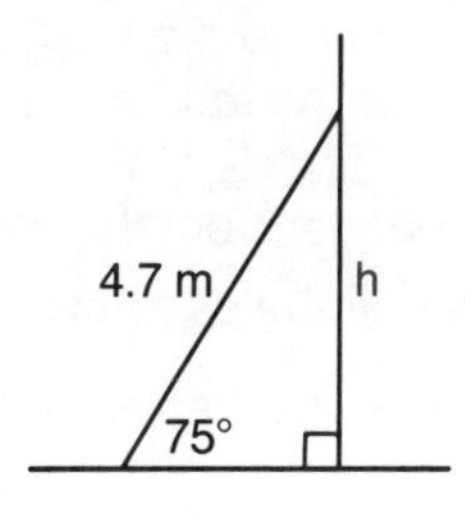

6. A ladder that is 5.1 m long leans against a wall and reaches a window 5.0 m above the ground.
Find the angle between the ladder and the ground.

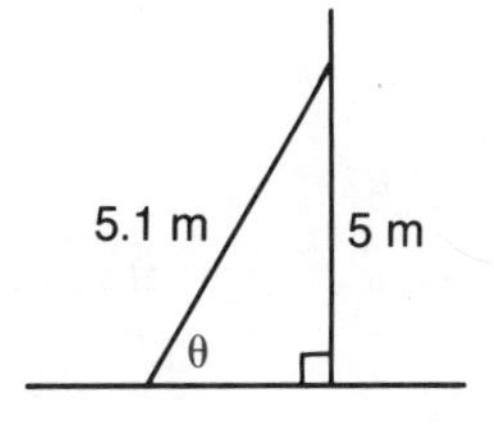

7. Tommy lets out 350 m of kite string. He estimates that the angle between the string and the ground is 30°.
What is the vertical height of the kite?

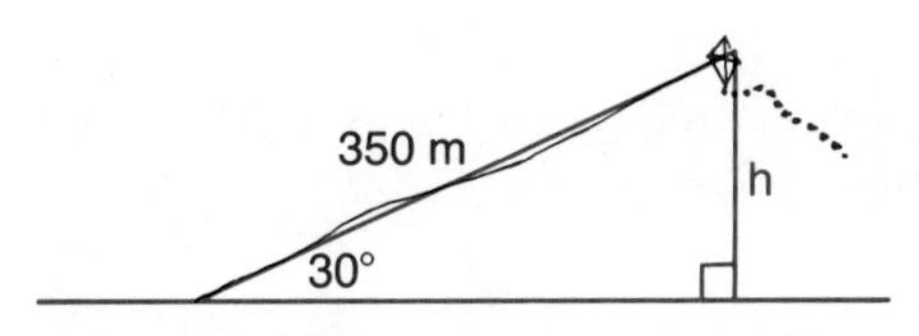

C 8. Along a distance of 1000 m up a hill, the elevation changes from 3245 m to 3310 m. How steep is the hill? (In other words, what is the angle of inclination?)

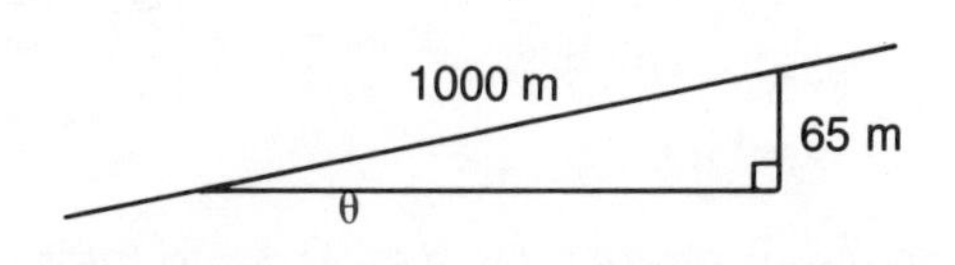

11.4 THE COSINE RATIO

Ross has a summer job painting houses. To use a ladder safely, Ross must ensure the angle the ladder makes with the ground is not more than 65°. He places a 4.5 m ladder so that the foot of the ladder makes an angle of 65° with the wall.
How far is the ladder from the foot of the wall?

Drawing a diagram helps to organize the information.

From the diagram, we can write the ratio $\frac{\text{adjacent side}}{\text{hypotenuse}}$

This ratio is called the cosine ratio.
The cosine of $\angle B$ is written cos B.
Values for the cosine ratio are given in tables on pages 345 and 406.

From the diagram,

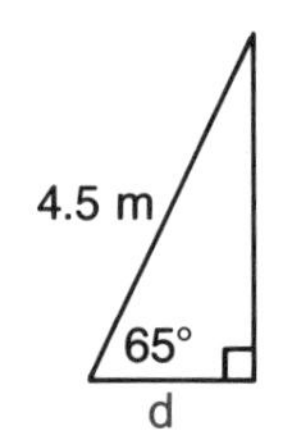

$$\frac{d}{4.5} = \cos 65°$$
$$d = 4.5 \times \cos 65°$$
$$\doteq 4.5 \times 0.4226$$
$$\doteq 1.9018$$

The foot of the ladder is 1.9 m from the foot of the wall.

Example 1.
State $\cos \theta$.

Solution:

$$\cos \theta = \frac{\text{adjacent side}}{\text{hypotenuse}}$$
$$\cos \theta = \frac{4}{5}$$

Example 2.
Calculate the measure from a ground anchor to a TV tower, to the nearest metre.

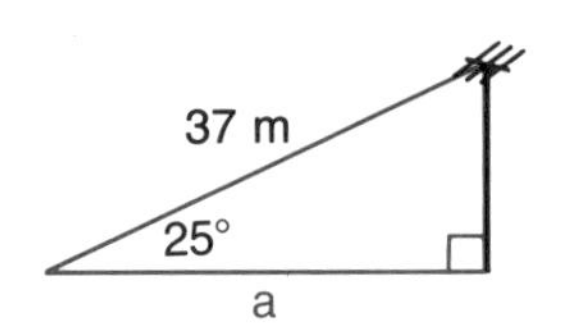

Solution:
Write an equation using a trigonometric ratio.

$$\frac{a}{37} = \frac{adj}{hyp}$$
$$\frac{a}{37} = \cos 25°$$
$$a = 37 \times \cos 25°$$
$$\doteq 37 \times 0.9063$$
$$\doteq 33.53$$

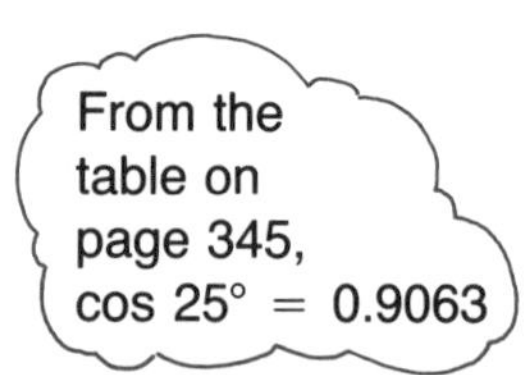

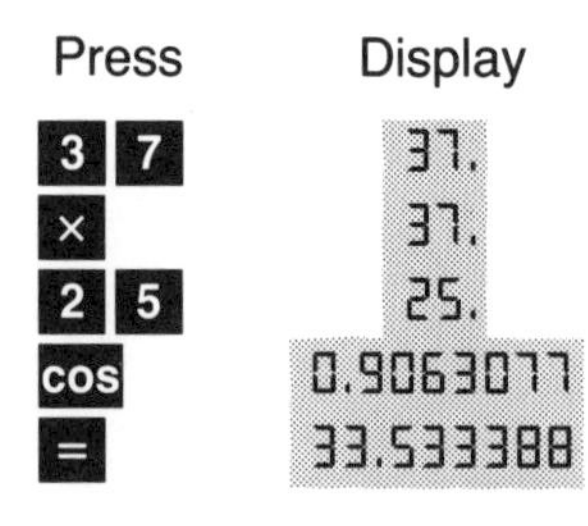

$\therefore$ In the given triangle, a is 34 m, to the nearest metre.

EXERCISE 11.4

1. State cos θ.

(a)

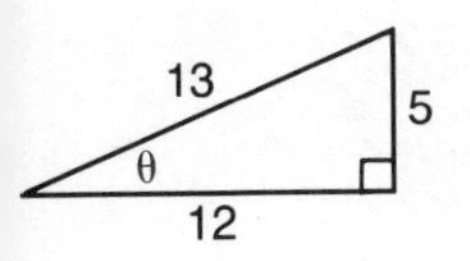

(b)

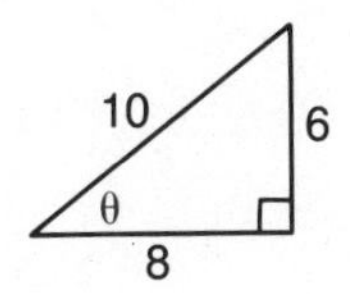

(c)

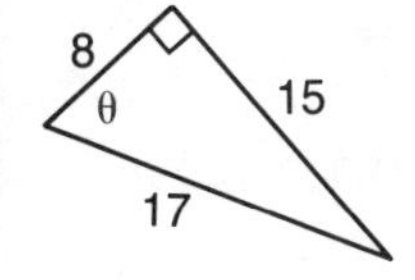

(d)

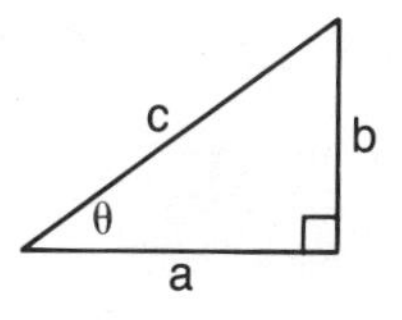

(e)

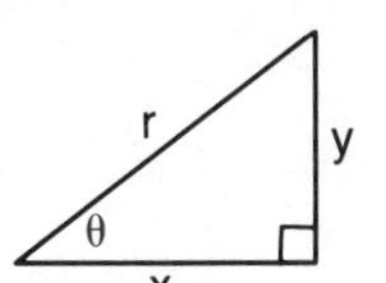

(f)

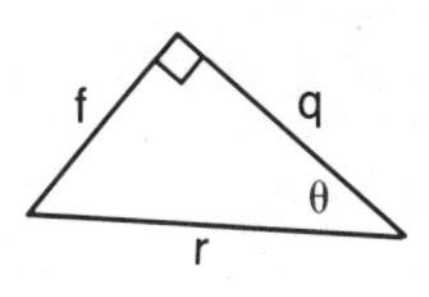

2. State the equation you would use to find the indicated side.

(a)

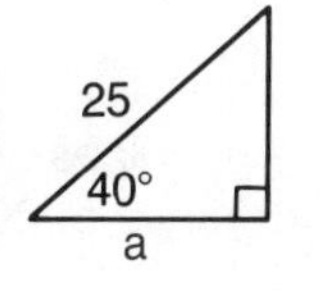

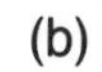

(b)

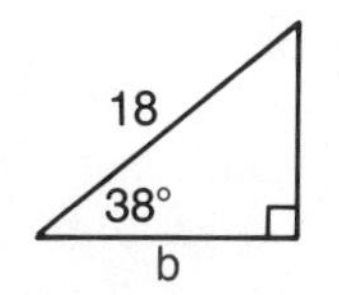

(c)

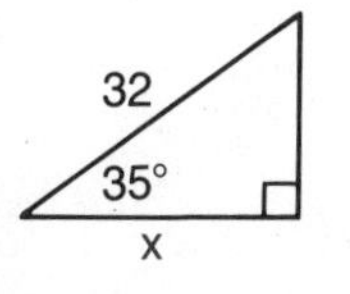

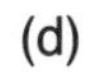

(d)

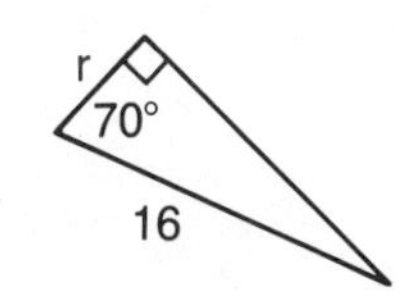

(e)

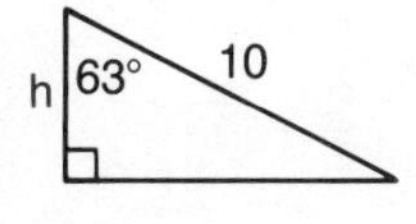

(f)

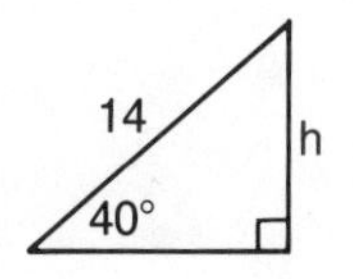

0°	cos θ	0°	cos θ
0	1.0000	46	0.6947
1	0.9999	47	0.6820
2	0.9994	48	0.6691
3	0.9986	49	0.6561
4	0.9976	50	0.6428
5	0.9962	51	0.6293
6	0.9945	52	0.6157
7	0.9926	53	0.6018
8	0.9903	54	0.5878
9	0.9877	55	0.5736
10	0.9848	56	0.5592
11	0.9816	57	0.5446
12	0.9782	58	0.5299
13	0.9744	59	0.5150
14	0.9703	60	0.5000
15	0.9659	61	0.4848
16	0.9613	62	0.4695
17	0.9563	63	0.4540
18	0.9511	64	0.4384
19	0.9455	65	0.4226
20	0.9397	66	0.4067
21	0.9336	67	0.3907
22	0.9272	68	0.3746
23	0.9025	69	0.3584
24	0.9136	70	0.3420
25	0.9063	71	0.3256
26	0.8988	72	0.3090
27	0.8910	73	0.2924
28	0.8830	74	0.2756
29	0.8746	75	0.2588
30	0.8660	76	0.2419
31	0.8572	77	0.2250
32	0.8481	78	0.2079
33	0.8387	79	0.1908
34	0.8290	80	0.1737
35	0.8192	81	0.1564
36	0.8090	82	0.1392
37	0.7986	83	0.1219
38	0.7880	84	0.1045
39	0.7772	85	0.0872
40	0.7660	86	0.0698
41	0.7547	87	9.0523
42	0.7431	88	0.0349
43	0.7314	89	0.0175
44	0.7193	90	0.0000
45	0.7071		

B 3. Find the length of the indicated side to the nearest tenth.

(a)

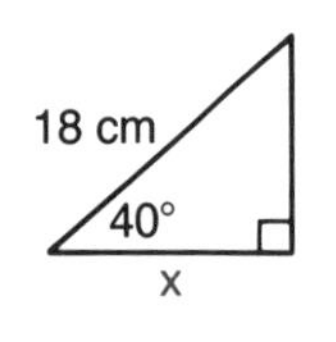

(b)

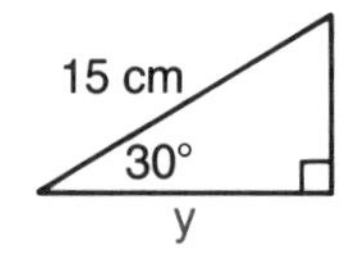

(c)

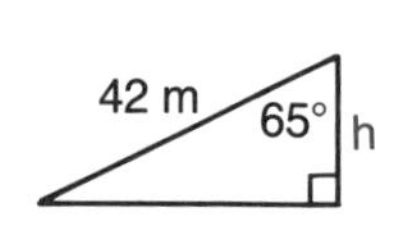

(d)

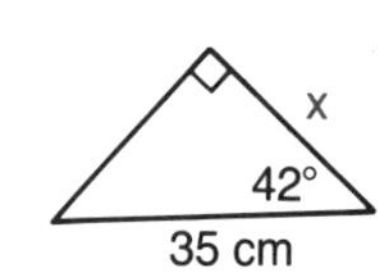

(e)

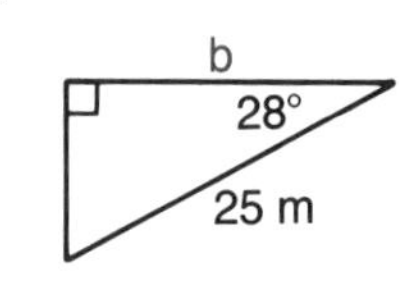

(f)

4. Calculate the measure of the indicated angle.

(a)

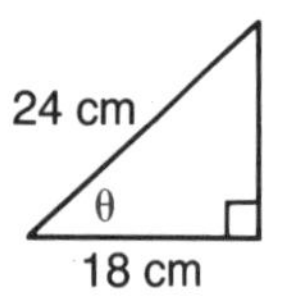

(b)

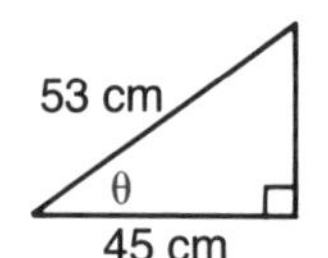

(c)

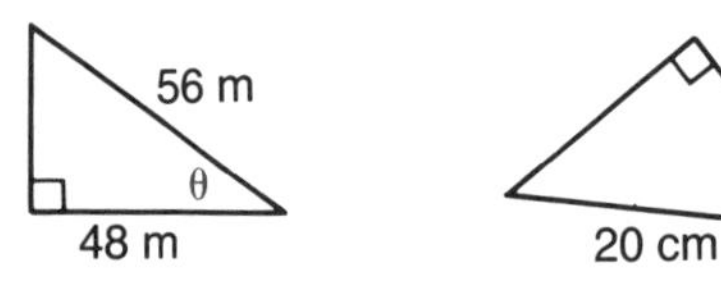

(d)

(e)

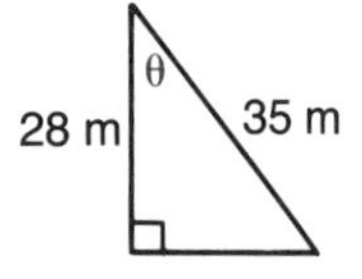

(f)

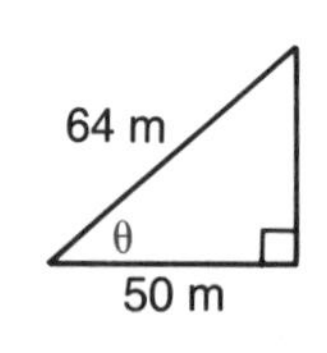

5. A ladder 5.3 m long leans against a wall. The foot of the ladder is 1.3 m from the wall.
Calculate the angle that the ladder makes with the ground.

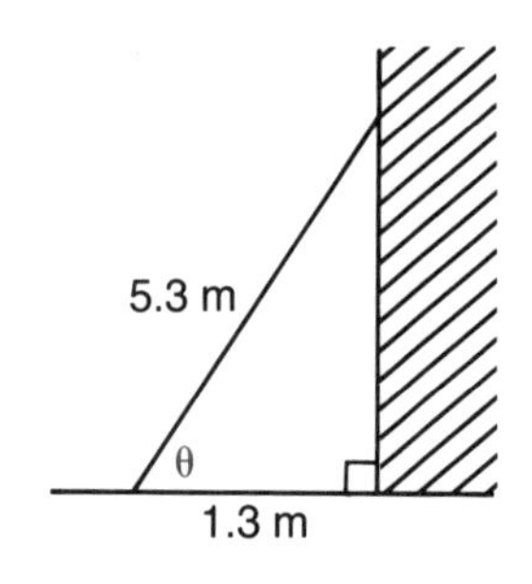

6. Find the distance, d, from the ship to the shoreline.

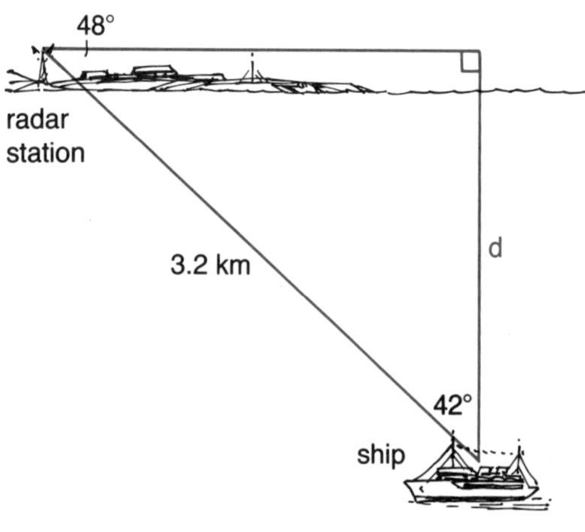

7. A radio tower is secured by guy lines 100 m long. The lines make an angle of 20° with the top of the tower.
How far up the tower are the lines fastened?

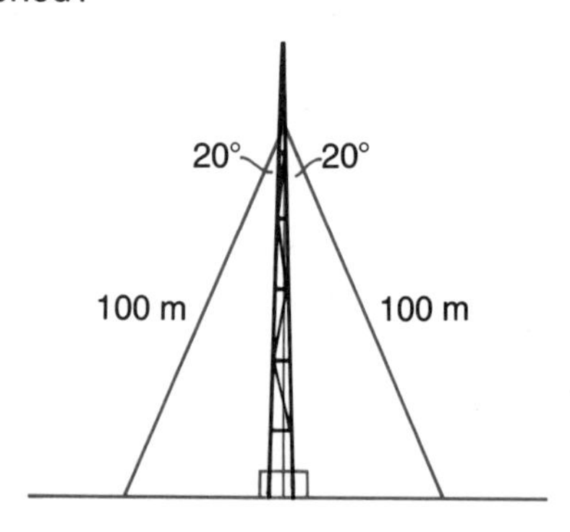

11.5 INVESTIGATING GRAPHS IN TRIGONOMETRY

INVESTIGATION

1. The graph of sin θ:

(a) Using a calculator or trigonometric tables, copy and complete the table.

θ	0°	5°	10°	15°	20°	25°	60°	65°	70°	75°	80°	85°	90°
sin θ													

(b) Using θ and sin θ axes, plot on graph paper the values of sin θ from the table and draw a smooth curve.

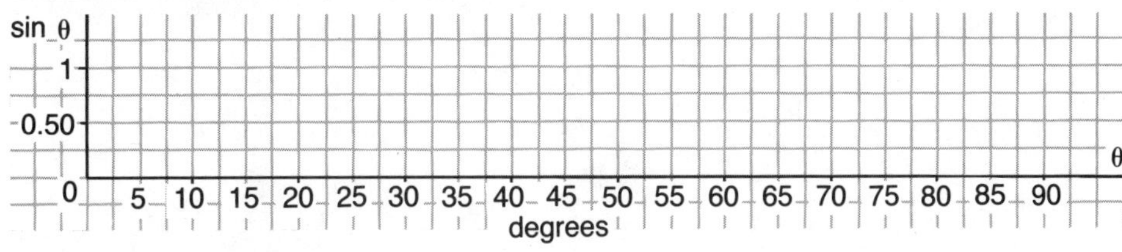

2. The graph of cos θ:

(a) Using a calculator or trigonometric tables, copy and complete the table.

θ	0°	5°	10°	15°	20°	25°	60°	65°	70°	75°	80°	85°	90°
cos θ													

(b) Using θ and cos θ axes, plot on graph paper the values of cos θ from the table and draw a smooth curve.

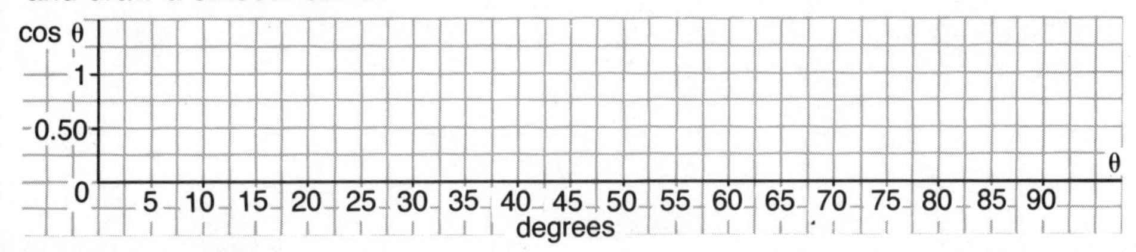

3. The graph of tan θ:

(a) Using a calculator or trigonometric tables, copy and complete the table.

θ	0°	5°	10°	15°	20°	25°	60°	65°	70°	75°	80°	85°	90°
tan θ													

(b) Using θ and tan θ axes, plot on graph paper the values of tan θ from the table and draw a smooth curve.

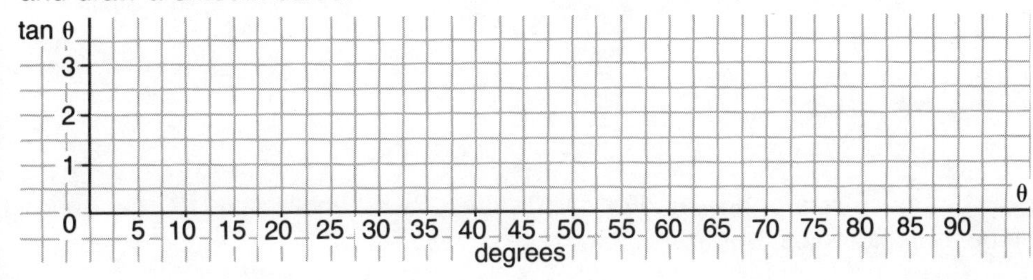

11.6 FINDING TRIGONOMETRIC RATIOS

Tables of trigonometric ratios are provided in the three previous sections and on page 406. We can also find the values of the trigonometric ratios efficiently using a scientific calculator. Since not all calculators operate in the same way, the following instructions are general. You should get specific instructions from your calculator manual.

Since we use degree measure for our calculations with angles, we must first be sure that the calculator is set to degree measure. This is called the angular mode, and you should set the calculator at DEG using the DRG key. Many calculators are automatically in the degree mode when they are turned on.

When you press one of the trigonometric function keys, sin, cos, or tan, it calculates the ratio of the angle in the display.

Instruction	Press	Display
To find the sine of 25°	2 5 sin	25. 0.4226182
To find the cosine of 55°	5 5 cos	55. 0.5735764
To find the tangent of 40°	4 0 tan	40. 0.8390996

We use an inverse operation to find an angle, given a trigonometric ratio. An inverse trigonometric function calculates the smallest angle.

Instruction	Press	Display
To find the smallest angle with sine ratio 0.25	· 2 5 arc sin or 2nd $\sin^{-1}$	0.25 14.477512
To find the smallest angle with cosine ratio 0.72	· 7 2 arc cos or 2nd $\cos^{-1}$	0.72 43.94552
To find the smallest angle with tangent ratio 1.15	1 · 1 5 arc tan or 2nd $\tan^{-1}$	1.15 48.990913

We give 8 digits in the calculator display. For calculations, round angle values to the nearest degree and trigonometric ratios to four decimal places.

The BASIC computer program at the right generates a table of trigonometric ratios for angles from 0° to 90°. For an angle of 10°, the program gives these values:

$\sin 10° = 0.1736482$
$\cos 10° = 0.9848078$
$\tan 10° = 0.176327$

```
10 PRINT"TRIGONOMETRIC TABLES"
20 PRINT"ANGLE","SINE","COSINE",
   "TANGENT"
30 FOR A = 1 TO 90
40 R = A*.017453293#
50 PRINT A,SIN(R),COS(R),TAN(R)
60 NEXT A
70 END
```

We round the computer values to four decimal places.

The following is a partial printout of the above program. Note that for very small angles, the values for the sine and tangent ratios are given in scientific notation.

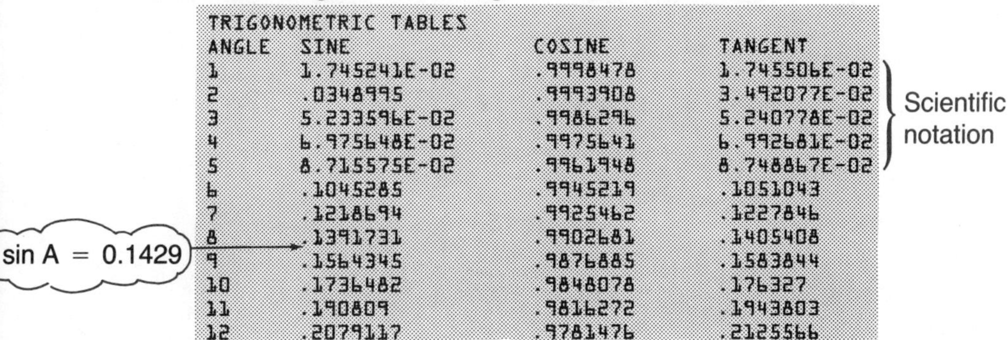

TRIGONOMETRIC TABLES

ANGLE	SINE	COSINE	TANGENT
1	1.745241E-02	.9998478	1.745506E-02
2	.0348995	.9993908	3.492077E-02
3	5.233596E-02	.9986296	5.240778E-02
4	6.975648E-02	.9975641	6.992681E-02
5	8.715575E-02	.9961948	8.748867E-02
6	.1045285	.9945219	.1051043
7	.1218694	.9925462	.1227846
8	.1391731	.9902681	.1405408
9	.1564345	.9876885	.1583844
10	.1736482	.9848078	.176327
11	.190809	.9816272	.1943803
12	.2079117	.9781476	.2125566

In $\triangle ABC$, $\sin A = \frac{1}{7}$.

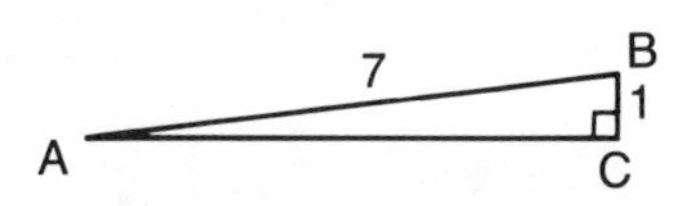

We can find the measure of $\angle A$ using a calculator or the trigonometric tables.

Using a calculator:

Press [1] [÷] [7] [=] [2nd] [inv] [sin]

The display is 8.2132108

$\therefore \angle A \doteq 8°$

Using the table on page 342:

$\sin A = \frac{1}{7} \doteq 0.1429$

Look down the sine column until you see a number close to 0.1429. This value lies between 0.1392 and 0.1564. Since $\sin A$ is closer to 0.1392, then $\angle A \doteq 8°$

EXERCISE 11.6

1. Read the following values from a calculator display or from the tables.

(a) sin 47° (b) tan 70°
(c) cos 35° (d) tan 42°
(e) sin 25° (f) tan 80°
(g) sin 75° (h) cos 20°
(i) cos 50° (j) sin 65°
(k) tan 34° (l) cos 40°
(m) sin 85° (n) cos 15°
(o) tan 54° (p) tan 18°
(q) sin 35° (r) cos 35°

B 2. Find the measure of each angle to the nearest degree.

(a) $\sin A = 0.2556$ (b) $\tan B = 0.3617$
(c) $\cos C = 0.2556$ (d) $\tan D = 1.1416$
(e) $\cos E = 0.8675$ (f) $\sin F = 0.5000$
(g) $\tan G = 0.8465$ (h) $\cos H = 0.4575$
(i) $\sin I = 0.3285$ (j) $\tan J = 0.0535$
(k) $\cos K = 0.8362$ (l) $\tan L = 0.5$
(m) $\sin M = 0.75$ (n) $\cos N = 0.25$
(o) $\sin O = 0.5257$ (p) $\cos P = 0.7007$
(q) $\sin Q = 1.0$ (r) $\tan R = 1.0$

11.7 SOLVING RIGHT TRIANGLES

In this section we will apply the trigonometric ratios sine, cosine, and tangent along with the Pythagorean Theorem to solve problems.

We have defined sine, cosine, and tangent as the ratios of sides in a right triangle. In $\triangle ABC$, the side opposite to $\angle A$ is a units long, the side opposite to $\angle B$ is b units long, and the side opposite to $\angle C$ is c units long. The letter A is the measure of the angle at the vertex.

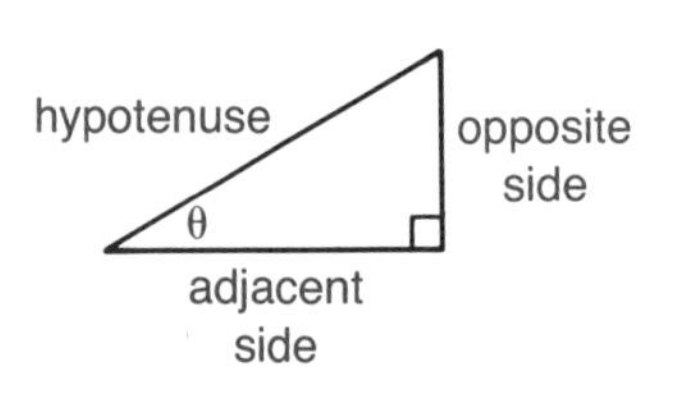

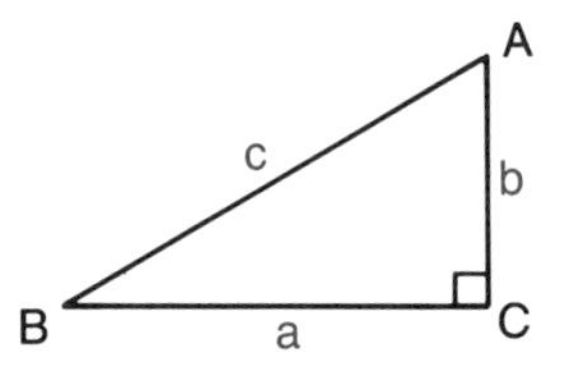

From the diagram,

$$\sin\theta = \frac{opp}{hyp} \qquad \sin A = \frac{a}{c} \qquad \sin B = \frac{b}{c}$$

$$\cos\theta = \frac{adj}{hyp} \qquad \cos A = \frac{b}{c} \qquad \cos B = \frac{a}{c}$$

$$\tan\theta = \frac{opp}{adj} \qquad \tan A = \frac{a}{b} \qquad \tan B = \frac{b}{a}$$

Example 1.
Calculate a and b to one decimal place.

Solution:

$\frac{a}{30} = \sin 24°$

$a = 30 \times \sin 24°$
$\doteq 30 \times 0.4067$
$\doteq 12.2$

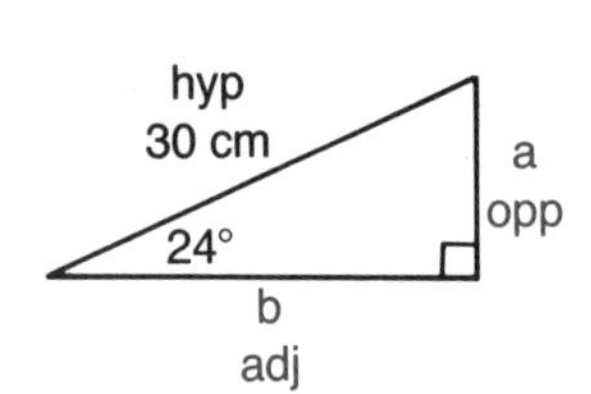

$\frac{b}{30} = \cos 24°$

$b = 30 \times \cos 24°$
$\doteq 30 \times 0.9135$
$\doteq 27.4$

Press 3 0 × 2 4 sin =

The display is 12.202099

Press 3 0 × 2 4 cos =

The display is 27.406364

Apply the problem solving model outlined in the table to solve trigonometric ratio problems.

READ	PLAN	SOLVE	ANSWER
Identify the given information and what is asked.	Choose a strategy. Diagram, Formula ($E=mc^2$), Estimate	Substitute known angles and sides into the formula. Solve the equation.	Check the estimate. Make a final statement.

Example 2.

In $\triangle ABC$, $\angle A = 90°$, $b = 7.0$ cm, and $c = 9.0$ cm.

Find (a) the measure of $\angle B$.
(b) the measure of $\angle C$.
(c) the length of a.

Solution:

Make a neat diagram.
Mark all the given information on the diagram.
Indicate, also on the diagram, the dimensions to be calculated.

(a) To find $\angle B$:

$$\tan B = \frac{b}{c}$$

$$\tan B = \frac{7.0}{9.0}$$

$$\doteq 0.7778$$

$$\angle B \doteq 38°$$

Press

The display is

37.874984

(b) To find $\angle C$:

Using the results of part (a)

$$\angle B = 38°$$

$$\angle C = 90° - 38°$$

$$= 52°$$

(c) To find a:

$$\frac{b}{a} = \sin B$$

$$\frac{7.0}{a} = \sin 38°$$

$$7.0 = a \times \sin 38°$$

$$\frac{7.0}{\sin 38°} = a$$

$$\frac{7.0}{0.6157} \doteq a$$

$$a \doteq 11.4$$

Press

The display is

11.369885

$\therefore \angle B \doteq 38°$, $\angle C \doteq 52°$, and $a \doteq 11.4$ cm.

Example 3.

Calculate the height of the cliff in metres.

Solution:

Let h represent the height of the cliff.
From the diagram,

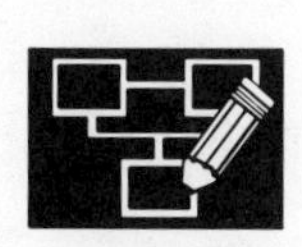

$$\frac{AB}{BC} = \tan C$$

$$\frac{h}{180} = \tan 48°$$

$$h = 180 \times \tan 48°$$

$$\doteq 180 \times 1.1106$$

$$\doteq 200$$

Press

The display is

199.91025

$\therefore$ The height of the cliff is 200 m.

EXERCISE 11.7

A 1. State a trigonometric equation you would use to find the length of the side marked x.

(a) (b)

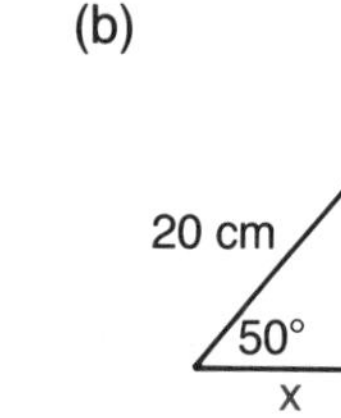

(c) (d)

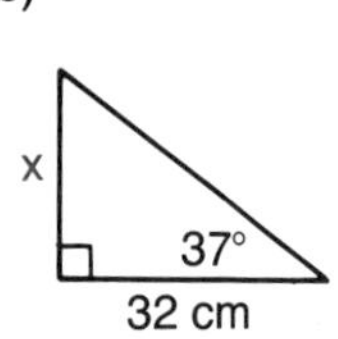

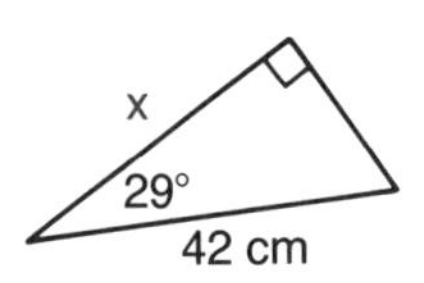

2. State a trigonometric equation you would use to find the measure of the indicated angle.

(a) (b)

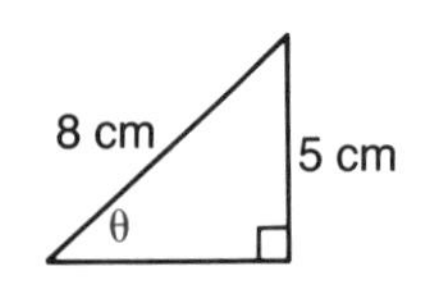

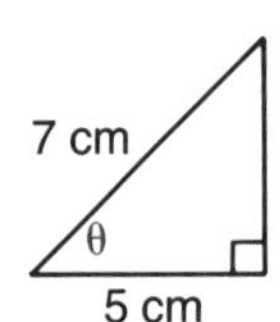

(c) (d)

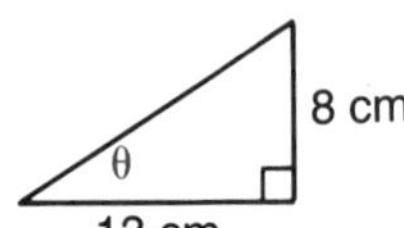

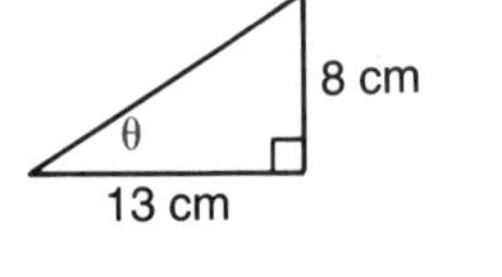

(e) (f)

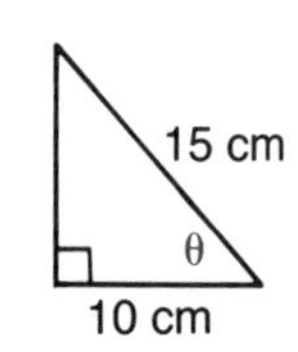

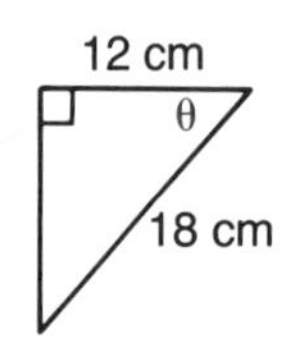

B 3. In △ABC, ∠A = 90°, b = 11.5 cm, and a = 15.0 cm.
Find:

(a) the measure of ∠B.
(b) the measure of ∠C.
(c) the length of c.

4. In △ABC, ∠A = 90°, ∠B = 38°, and a = 25 cm.
Find:

(a) the measure of ∠C.
(b) the length of b.
(c) the length of c.

5. A flagpole casts a shadow 24.5 m long. At the same time, the sun's rays strike the ground at an angle of 32°.
Calculate the height of the flagpole.

6. A rectangular playing field is 60 m by 120 m.
What is the length of a diagonal running between two corners of this field?

7. A ladder that is 8.5 m long leans against the wall. The ladder just reaches a window that is 7.8 m above the ground.

(a) What is the measure of the angle that the ladder makes with the ground?
(b) How far is the foot of the ladder from the wall?

8. A ship sails 510 km in a northerly direction, then changes course and sails 420 km in an easterly direction.

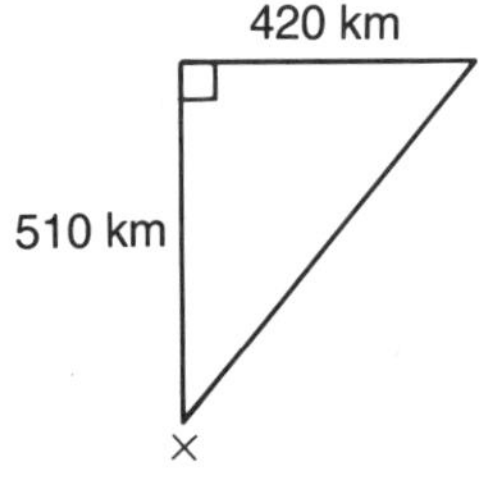

(a) In what direction should a second ship sail from X to go directly to the location of the first ship?
(b) How far is the first ship from X?

11.8 APPLICATIONS: NAVIGATION

Wind will alter the speed and direction of an airplane in flight. To fly from A to B in a crosswind as shown by the arrow, w, you have to set a course toward M. If you set a course toward B, then you will reach N.

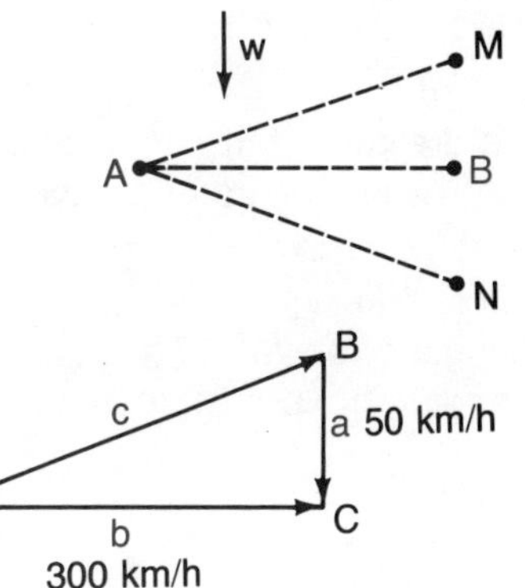

Example 1.
A pilot intends to fly her aircraft at 300 km/h in an easterly direction. What course should the pilot set in order to overcome a 50 km/h north wind?

Solution:
From the diagram, we see that the pilot should set a course towards B.

E=mc²

To find the air speed of the aircraft:
Using the Pythagorean Theorem,

$c^2 = a^2 + b^2$
$c^2 = 50^2 + 300^2$
$= 2500 + 90\ 000$
$= 92\ 500$
$\doteq 304$

Press

5 0 x² + 3 0 0 x² = √

The display is 304.13813

To find the direction:

$\tan A = \frac{a}{b}$

$\tan A = \frac{50}{300}$

$\doteq 0.1667$

$\angle A \doteq 9°$

Press

5 0 ÷ 3 0 0 = inv tan

The display is 9.4623224

∴ The pilot should set a course 9° north of east at 304 km/h.

EXERCISE 11.8

1. A pilot wishes to fly at 210 km/h toward the west. There is a 30 km/h wind from the south.
Find the course and the air speed that the pilot must set.

2. A small boat can travel at 20 km/h in still water. A small river has a current of 4 km/h. Find the direction the boat must take in order to travel directly across the river.

3. What is the true air speed of an airplane flying north at 450 km/h if there is a 50 km/h west wind?

4. In what direction would you launch a motorboat that travels at 50 km/h to cross a river with a current of 5 km/h?

5. Find the number of degrees off course an airplane would be if it travels south at 400 km/h and the pilot did not account for a 40 km/h east wind.

6. A small aircraft flies at 250 km/h in a northerly direction. There is a 55 km/h west wind.
(a) Find the ground speed of the aircraft.
(b) In what direction will the aircraft travel relative to the ground?

11.9 APPLICATIONS: ANGLES OF DEPRESSION AND ELEVATION

The diagram at the right illustrates some of the terminology used by surveyors and navigators:

line of sight
horizontal
angle of elevation
angle of depression

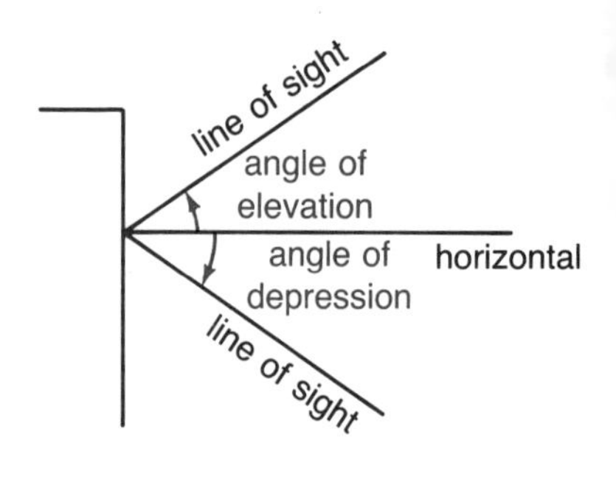

Example 1.
The angle of elevation to the top of a school is 32°.
The distance to the school is 19.7 m.
What is the height of the school?

Solution:
Let h represent the height of the school in metres.
From the diagram,

$$\frac{h}{19.7} = \tan 32°$$

$$h = 19.7 \times \tan 32°$$
$$\doteq 19.7 \times 0.6249$$
$$\doteq 12.3$$

Press

The display is 12.309926

∴ The height of the school is approximately 12.3 m.

Example 2.
From the top of a cliff 115 m high to a boat on the water the angle of depression is 16°.
How far is the boat from the cliff?

Solution:
From △ABC in the diagram,
$\angle A = 74°$ (complementary angles)
Let a represent the distance from the cliff in metres.

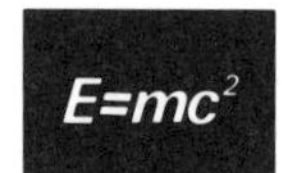

$$\frac{a}{c} = \tan A$$

$$\frac{a}{115} = \tan 74°$$

$$a = 115 \times \tan 74°$$
$$\doteq 115 \times 3.4874$$
$$\doteq 401$$

Press

The display is 401.05267

∴ The boat is approximately 401 m from the cliff.

EXERCISE 11.9

1. Find the angle of elevation, θ.

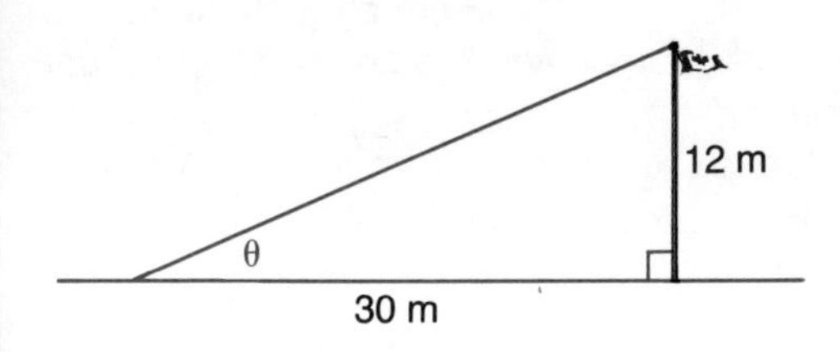

2. What is the height of the building?

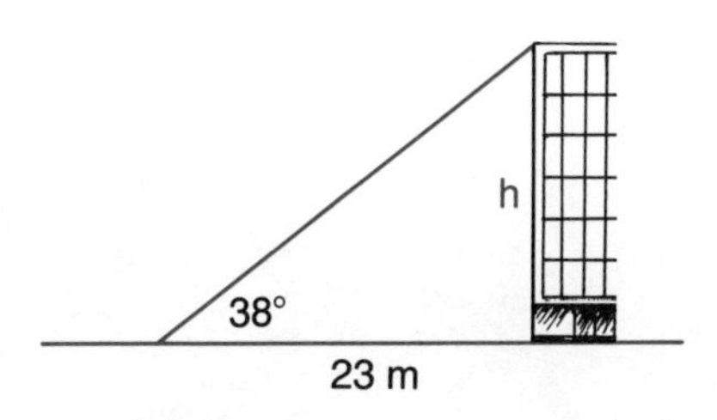

3. How far is the fire hydrant from the building?

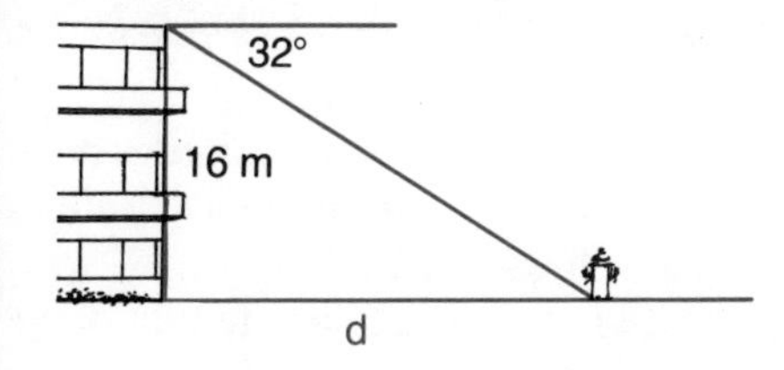

4. What is the angle of depression, θ?

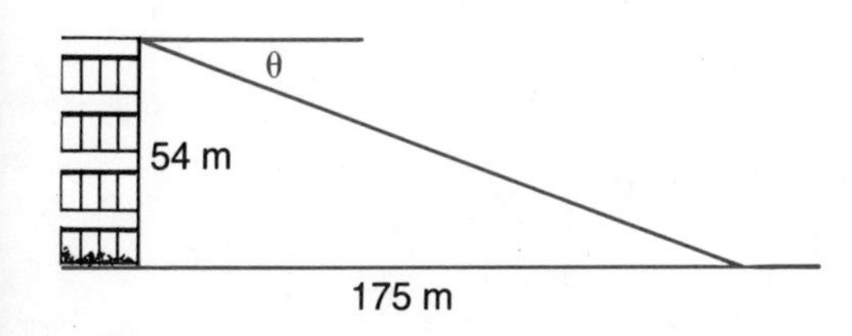

5. From a point on the edge of a cliff 142 m above the water, the angle of depression to a boat is 18°.
How far is the boat from the foot of the cliff?

6. A radio tower 100 m high casts a shadow that is 75 m long.
What is the angle of elevation to the sun?

7. The ends of a steel cable 125 m long are attached to the top of a radio tower and to the ground. The angle of elevation of the wire at the ground is 68°.
What is the height of the tower?

8. A fire hydrant is 312 m from the foot of a building. The angle of depression from the top of the building to the hydrant is 72°.
What is the height of the building?

9. A boat is 1000 m from the foot of a cliff which is 31 m high.
What is the angle of depression of the boat to the top of the cliff?

10. Two buildings are separated by a driveway that is 6.5 m wide. From a window in one building, the angle of depression to the foot of the other building is 34° and the angle of elevation to the top is 58°.
Find the height of the building on the right.

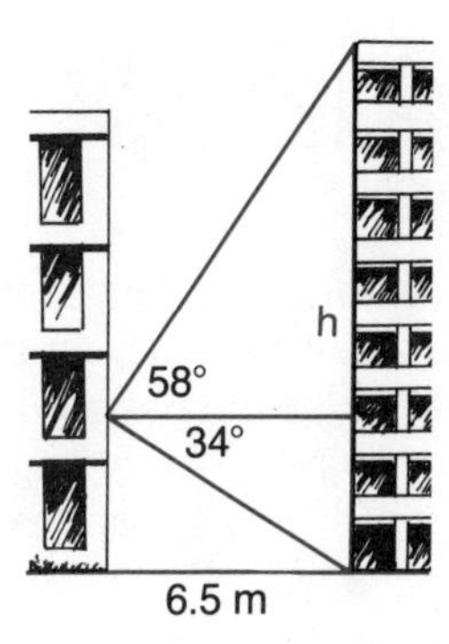

C 11. From a point 58.5 m away from the foot of a building, the angle of elevation to the top of the building is 63°.
What is the angle of elevation from a point 40 m away from the building to the top of the building?

12. From a window 35 m above the street, the angle of depression to the bottom of a building across the street is 32° and the angle of elevation to the top of the building is 48°.
What is the height of the building across the street?

11.10 PROBLEM SOLVING

1. (a) Which of these shapes can you draw in a single stroke without lifting your pencil from the paper and without going over any line twice?

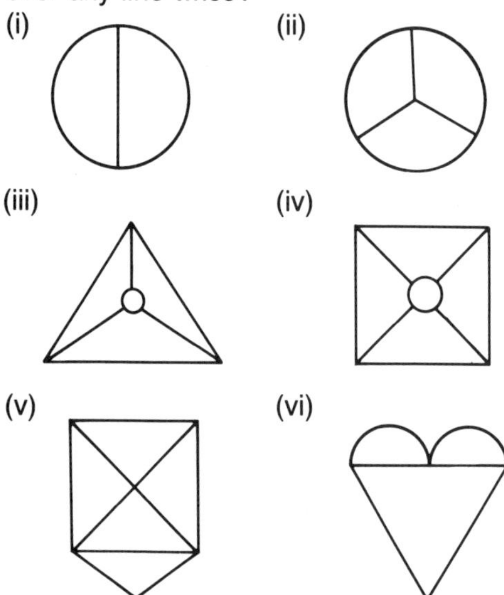

(b) What similarities are there among the shapes that can be drawn this way?
(c) Draw a shape that can be drawn this way.
(d) Draw a shape that cannot be drawn this way.

2. Copy the following shape onto stiff paper.

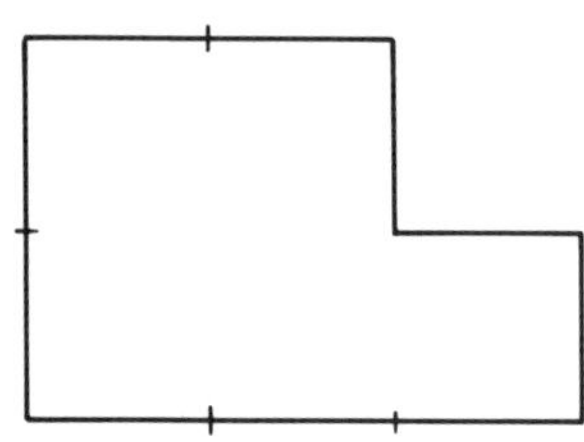

Cut the shape into three pieces using only two straight cuts so that you can rearrange the pieces to form a square.

3. A two-digit number meets the following two conditions:
it is a perfect square, and
it has exactly nine factors.
What is the number?

4. A square sheet of paper measures 19 cm by 19 cm. A 3 cm by 3 cm square is placed directly over the centre of the larger square.
What is the distance from a corner of the smaller square to a corresponding corner of the larger square?

5. Research to find the missing information and solve the following problem.
How many railway ties are there in 250 km of track?

6. In a single round-robin tournament, each team plays every other team once. There are 6 teams in the tournament.
How many games are required to complete the round robin?

7. In a single knockout tournament, each team plays until it loses. The schedule for four teams in a knockout tournament is given.

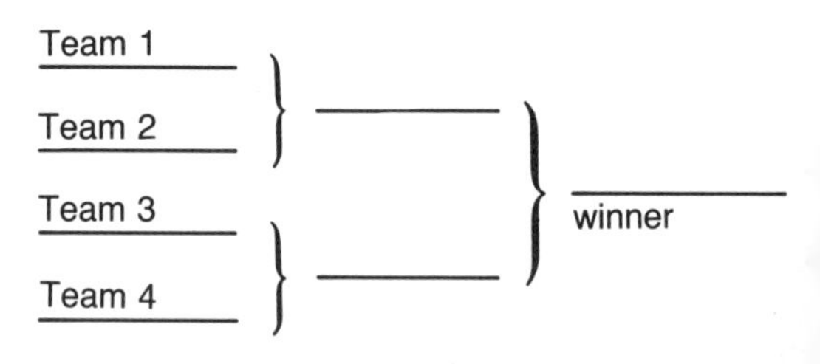

How many games are required to complete a single knockout tournament for 16 teams?

8. There are some chickens and rabbits in a farmyard. Mary counted them and said "There are 140 feet." Jim counted and replied "There are only 50 heads."
How many chickens and rabbits are there?

9. A bank teller has 100 bills in tens and twenties. The total value of the bills is $1630.
How many tens does the teller have?

10. An assembly-line worker takes 3 min to install a computer chip. When the production line is turned on in the morning, there are 2 chips ready for the worker to install. The first new chip arrives 2 min after the line is turned on. After that the next chip arrives every 6 min.
How long after the assembly line is turned on will the first "no-wait" chip arrive?

11. In the Directors Road Run, there are 10 checkpoints 4.5 km apart. The first checkpoint is 3 km after the start and the last checkpoint is 2 km from the finish.
How long is the race?

12. This set of numbers is called Pascal's Triangle. Find the next five lines of numbers in the pattern.

1
1 1
1 2 1
1 3 3 1
1 4 6 4 1

13. In a restaurant, the jobs of cook, manager, and waiter are held by Arnie, Sal, and Kenneth, but not necessarily in that order. The waiter is an only child and earns the least. Kenneth is married to Arnie's sister and earns more than the manager.
Who has which job?

CAREER

LAND SURVEYOR

A land surveyor takes readings of property boundaries and roadways to make drawings. Using a theodolite and steel measuring tape, the surveyor records readings on site and returns to the office to prepare the drawings.

A surveyor drew these sketches on site.
For each sketch
(a) select a suitable scale,
(b) prepare a scale drawing, and
(c) find the missing dimension.

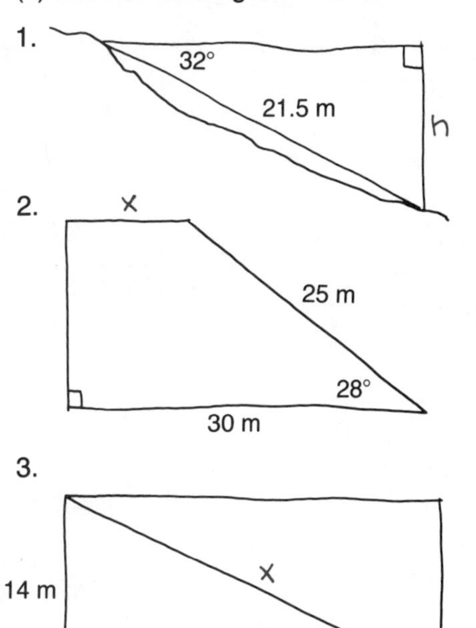

EXTRA

11.11 WAYNE GRETZKY AS AN EDMONTON OILER

On August 9, 1988, the Edmonton Oilers hockey club announced that Wayne Gretzky, the greatest player in the history of the National Hockey League (NHL), had been traded to the Los Angeles Kings.

The trade was reported as one of the biggest in the history of sports as there were several players and draft choices traded. The Los Angeles Kings received Wayne, Marty McSorley, and Mike Krushelnyski from the Oilers. In return the Oilers received Jimmy Carson, Martin Gelinas, three first-round draft choices, and cash.

Wayne was born on January 26, 1961, in Brantford, Ontario. He received his first taste of national exposure in the 1971/72 season when he was ten years old. He scored 378 goals and had 116 assists in 69 games as a novice player in Brantford.

On November 2, 1978, Wayne Gretzky's contract was sold to the Edmonton Oilers.

The table below gives Wayne's record as an Oiler, both in the regular season and the playoffs.

In what regular season did he score the most goals?

In what playoff season did he score the most goals?

What was the most number of games he played during the regular season and playoffs?

In what regular season did he have the most penalty minutes?

Years	Regular Season					Playoffs				
	GP	G	A	P	PIM	GP	G	A	P	PIM
1979–80	79	51	86	137	21	3	2	1	3	0
1980–81	80	55	109	164	28	9	7	14	21	4
1981–82	80	92	120	212	26	5	5	7	12	8
1982–83	80	71	125	196	59	16	12	26	38	4
1983–84	74	87	118	205	39	9	13	22	35	12
1984–85	80	73	135	208	52	18	17	30	47	4
1985–86	80	52	163	215	46	10	8	11	19	2
1986–87	79	62	121	183	28	21	5	29	34	6
1987–88	64	40	109	149	24	19	12	31	43	16
TOTAL	696	583	1086	1669	323	120	81	171	252	56

During Wayne's nine years with the Oilers, the team won the Stanley Cup four times. Can you tell from the table in which years the Oilers won the Cup?

EXERCISE 11.11

1. The Oilers' office received an average of 49 letters per day during Gretzky's nine seasons.
(a) How many letters would the office receive in one year?
(b) How many letters were received in the nine seasons?

2. Wayne was the NHL scoring leader for seven consecutive seasons, from 1980/81 to 1986/87.
(a) How many regular season goals did he score during these years?
(b) How many regular season assists did he have?
(c) What was his average number of goals per regular season?

3. Wayne was named the NHL's most valuable player in eight of his nine seasons with the Oilers.
What is the name of the trophy you receive for this award?

4. In a regular season each hockey team plays 80 games.
How many games did Wayne miss while playing for the Oilers?

5. During the nine years, the Oilers made the Stanley Cup finals five times — 1983, 1984, 1985, 1987, and 1988 — losing only in 1983.
(a) How many playoff goals did Wayne score in these years?
(b) How many assists did he have?
(c) How many goals did he score in the playoff when the Oilers won?

6. On July 1, 1984, Wayne Gretzky was awarded the Order of Canada.
What is the Order of Canada?

7. During his last season with the Oilers (1987/88), the team scored a total of 363 goals. The circle graph shows how Wayne contributed.

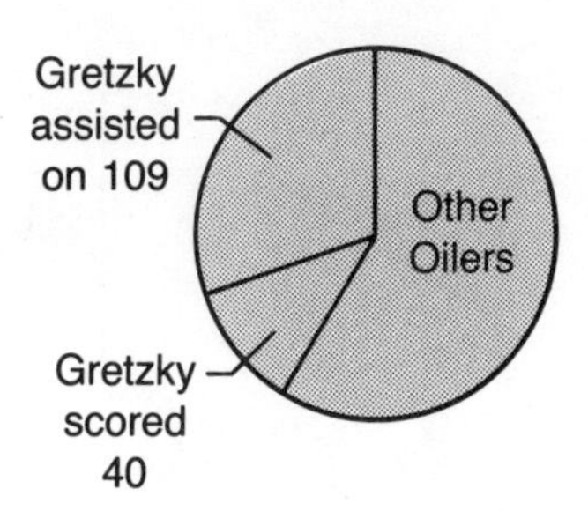

During his nine years with the Oilers the team scored 3478 goals. Wayne scored 583 of them and assisted on 1086 others.
Display this information on a circle graph similar to the one above.

8. The table gives Wayne's record as a junior hockey and professional player before signing with the Oilers.

Season	Club	Lea	GP	G	A	PTS	PIM
'76/77	Peter-bor-ough	OHA	3	0	3	3	0
'77/78	S.S. Marie	OHA	64	70	112	182	14
'78/79	Indi-anap-olis	WHA	8	3	3	6	0
	Ed-mon-ton	WHA	72	43	61	104	19

(a) What was his goals per game average, to the nearest tenth, as a junior?
(b) What was his goals per game average in the WHA?
(c) What was the name of the Sault Ste. Marie hockey team?
(d) What was the nickname of the Indianapolis team?

11.12 REVIEW EXERCISE

1. Calculate the lengths of the indicated sides using similar triangles.

(a)

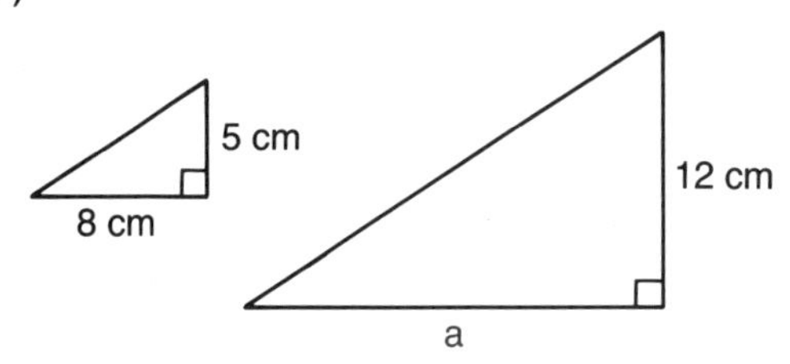

(b)

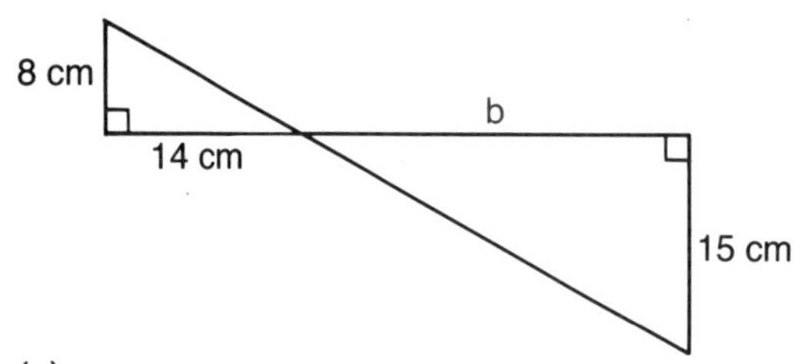

(c)

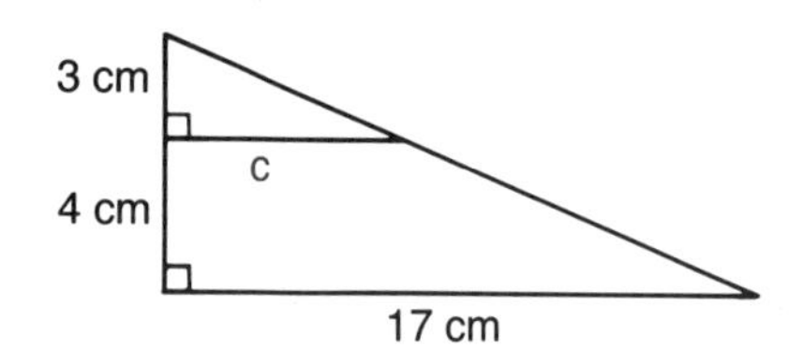

2. State the sine, cosine, and tangent ratios for each angle.

(a)

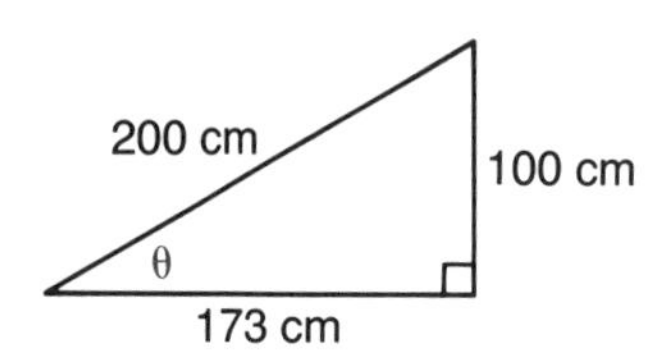

(b)

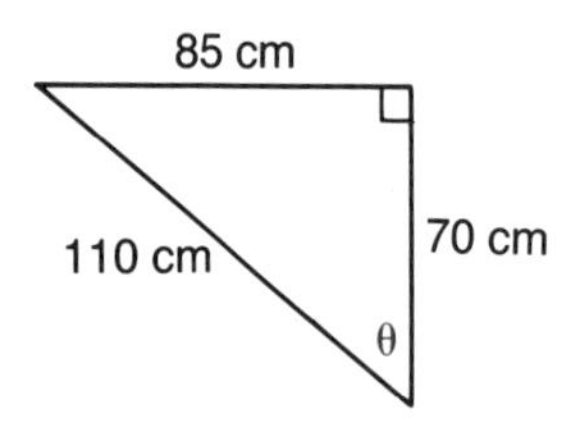

(c)

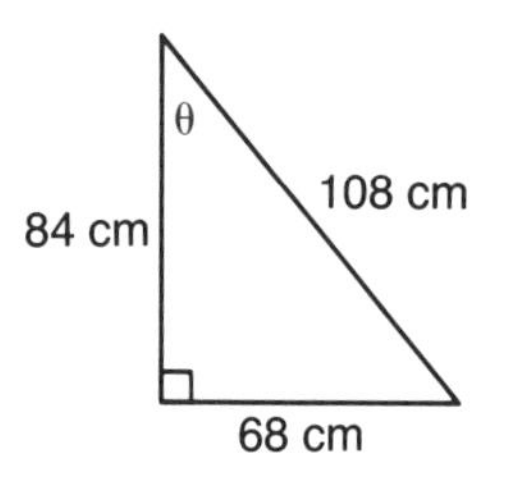

3. Solve for b using the tangent ratio. Solve for c using the Pythagorean Theorem.

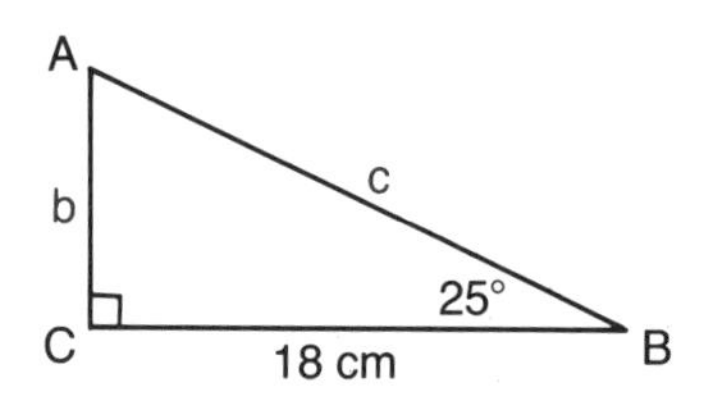

4. Solve for the variable in each triangle.

(a)

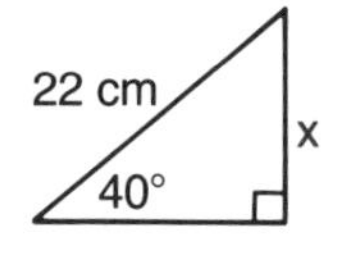

(b)

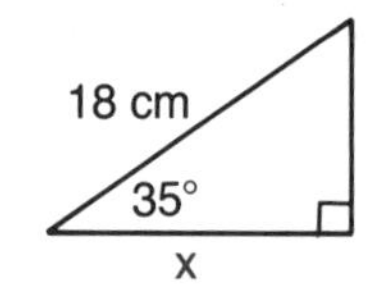

(e)

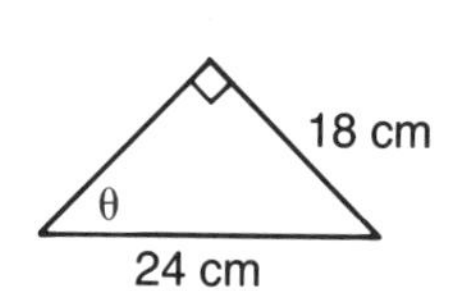

(f)

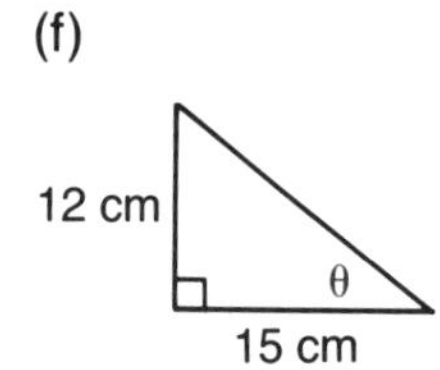

(c)

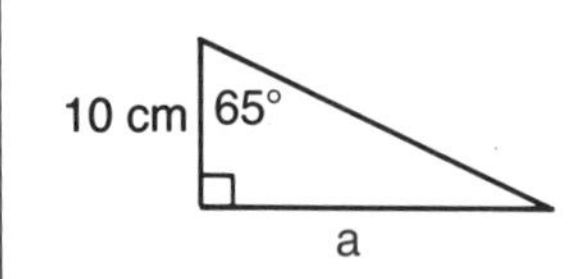

(d)

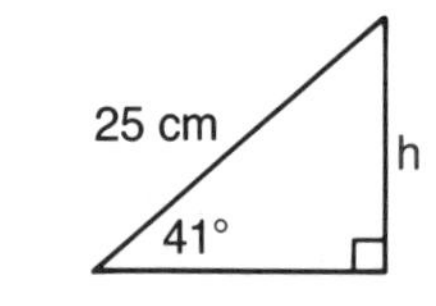

5. From the top of a cliff 256 m high to a boat on the water the angle of depression is 21°.
Calculate the distance from the boat to the foot of the cliff.

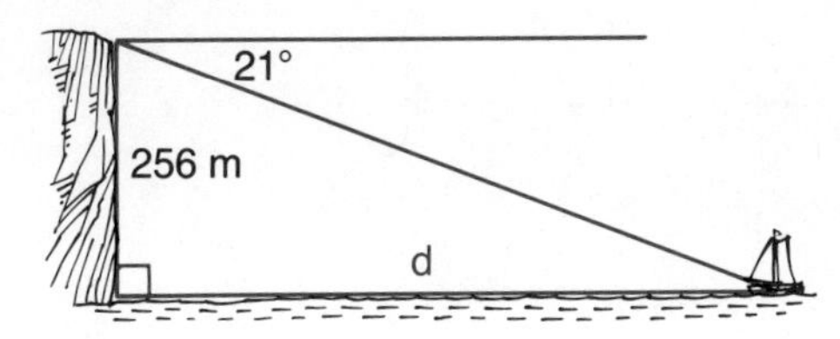

6. The shadow of a tree is 15.6 m long when the angle of elevation to the sun is 62°.
Calculate the height of the tree.

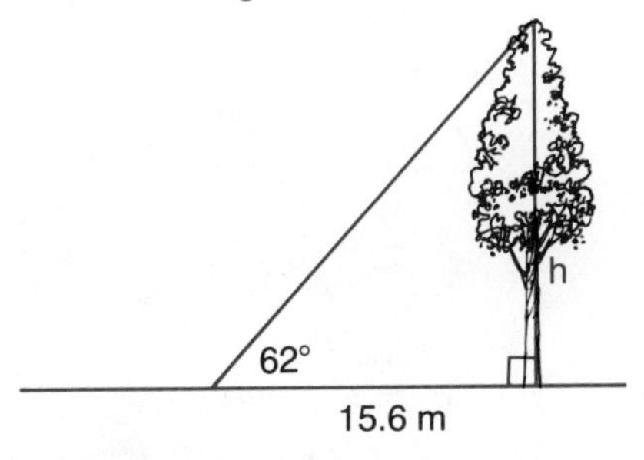

7. A flagpole 12 m long casts a shadow 8.2 m long.
What is the angle of elevation to the sun?

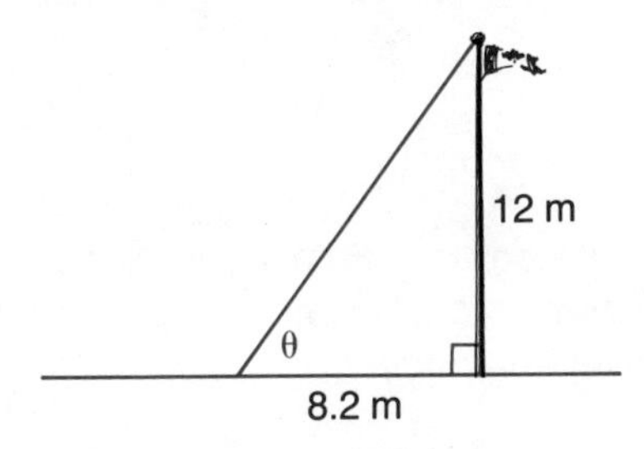

8. A boat is 1200 m from the foot of a cliff that is 115 m high.
Calculate the angle of depression to the boat from the top of the cliff.

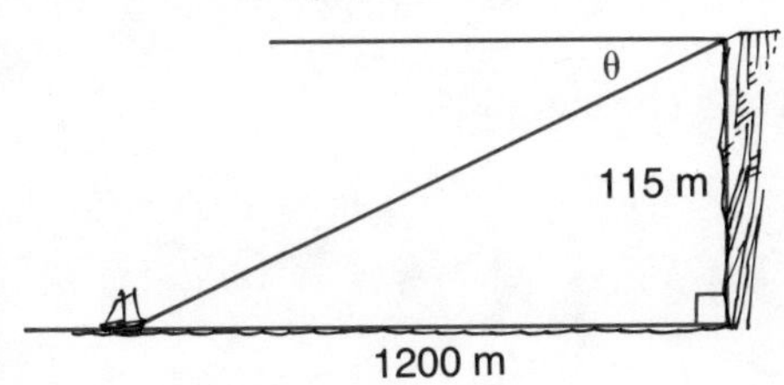

9. A steel plate has two holes in it as shown below.
Calculate the length of the slot from the centre of one hole to the centre of the other.

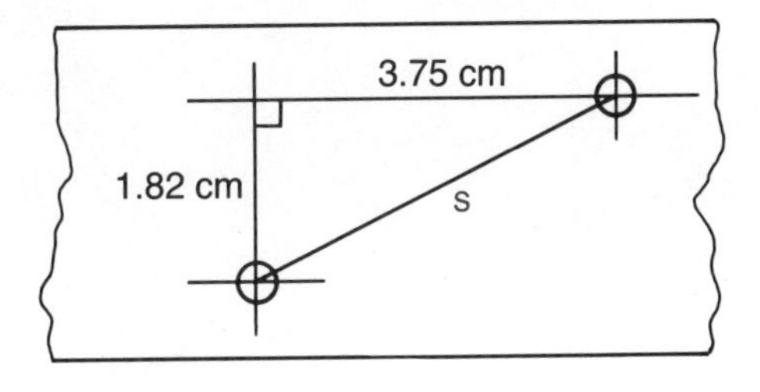

10. A 5 m ladder is to lean against a wall so that the foot of the ladder makes an angle of 75° with the ground.
How far should the foot of the ladder be placed from the wall?

11. Nine holes are to be drilled in a circular pattern with a diameter of 28 cm. The holes are to be equally spaced.
Calculate the distance, d, between the centres of any two holes.

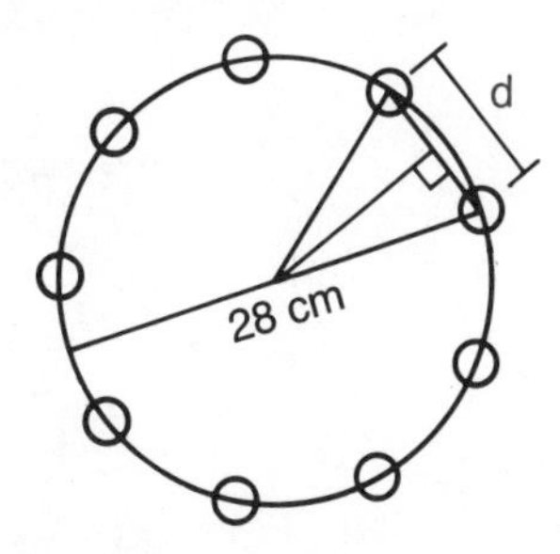

12. A ship sails directly north for 250 km, then turns 90° to the right and sails 125 km.
(a) How far is the ship from its starting point?
(b) In what direction should a second ship sail in order to go from the starting point directly to the ship?

13. A flagpole casts a shadow 16 m long. The angle of elevation to the sun is 63°. What is the height of the flagpole?

11.13 CHAPTER 11 TEST

1. Calculate the indicated length using similar triangles.

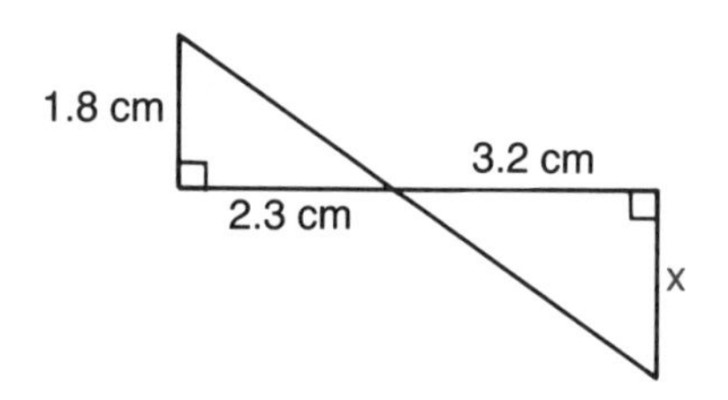

2. State the trigonometric ratios sin θ, cos θ, and tan θ.

(a)

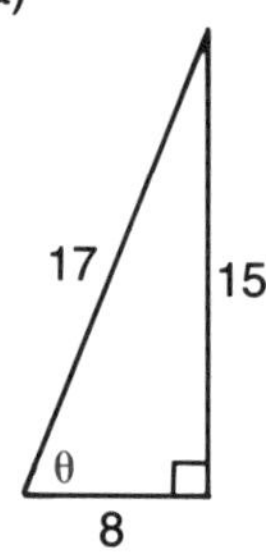

(b)

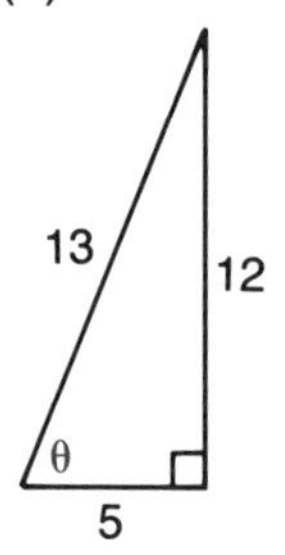

3. Solve for the indicated variable in each diagram.

(a)

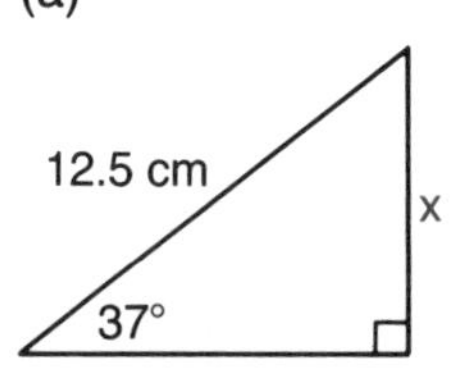

(b)

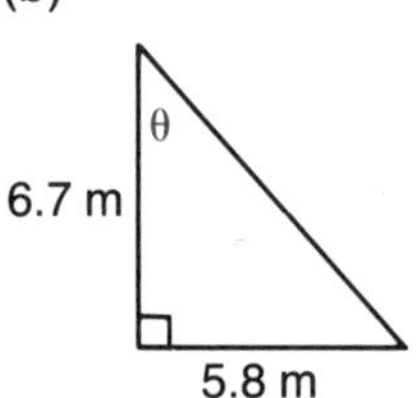

(c)

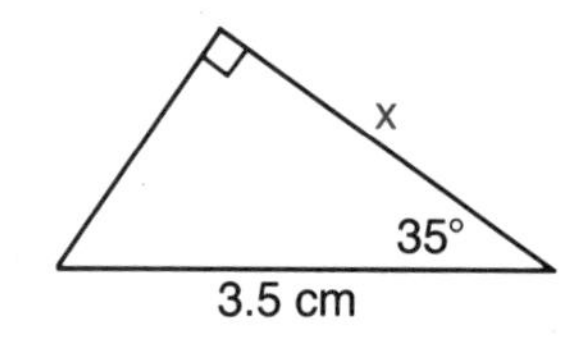

4. The shadow of a telephone pole is 7.5 m long. At the same time, the angle of elevation to the sun is 62°.
What is the height of the telephone pole?

5. From the top of a building 124 m high to a sewer cover, the angle of depression is 57°.
How far is the sewer cover from the foot of the building?

6. A tree that is known to be 97.5 m high casts a shadow of 70.4 m.
What is the angle of elevation to the sun?

COORDINATE GEOMETRY

REVIEW AND PREVIEW TO CHAPTER 

GRAPHING

EXERCISE 1

1. State the coordinates of each point on the coordinate grid.

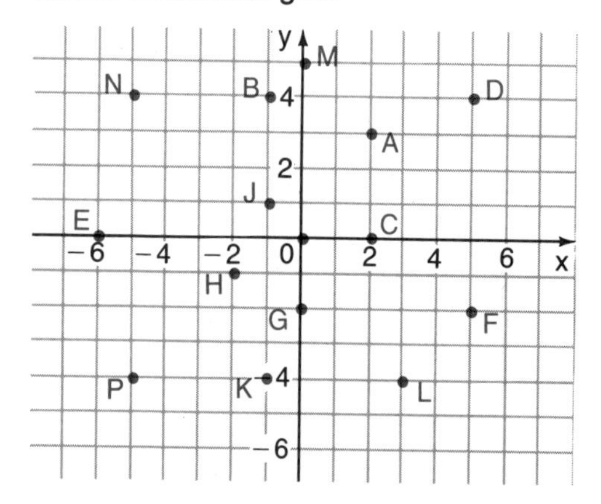

2. Plot and join the following points on a coordinate grid.
(−3, 0), (−4.5, 0.5), (−4, 1), (−3, 1.5), (−2, 2), (−1, 2.5), (0, 3.5), (2, 4), (3, 4), (2, 3), (1, 2), (2, 1.5), (3, 1), (6, 1.5), (5, 1), (4, 0.5), (3, 0), (4, −1.5), (3, −1), (2, −0.5), (1, −1), (0, −1.5), (1, −3), (0, −2.5), (−1, −2), (−2, −2.5), (−3, −2), (−4, −1), (−4.5, −0.5), (−3, 0).

3. Determine which ordered pairs satisfy the given equation.
(a) $y = 2x - 3$; A(1, −2), B(0, 3), C(4, 5)
(b) $3x + 2y = 12$; P(1, 4), Q(4, 0), R(−2, 9)

4. Rearrange the formulas to solve for the indicated variable.
(a) $d = vt$; v
(b) $c = 2\pi r$; r
(c) $P = a + b + c$; b
(d) $I = Prt$; P

THE PYTHAGOREAN THEOREM

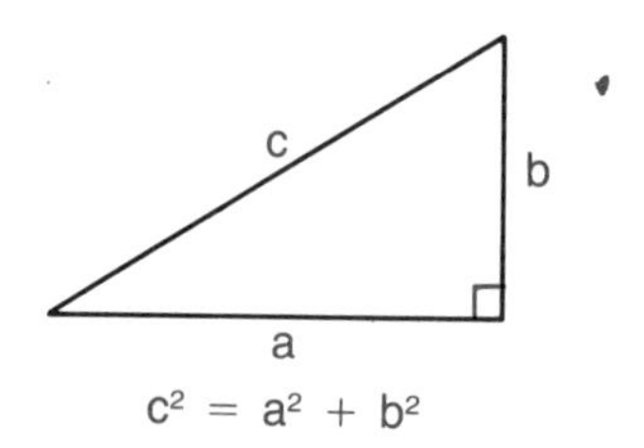

$$c^2 = a^2 + b^2$$

EXERCISE 2

1. Calculate the missing dimension to the nearest tenth.

(a)

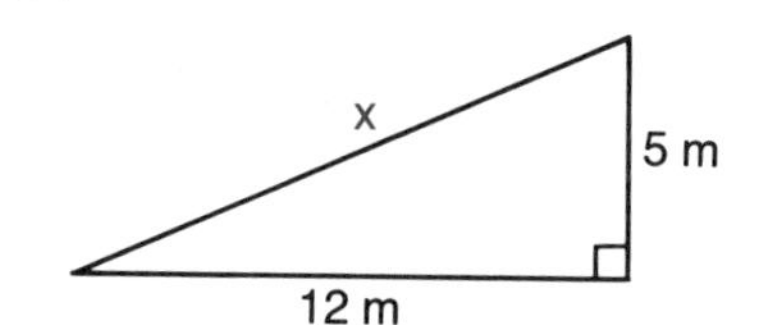

(b)

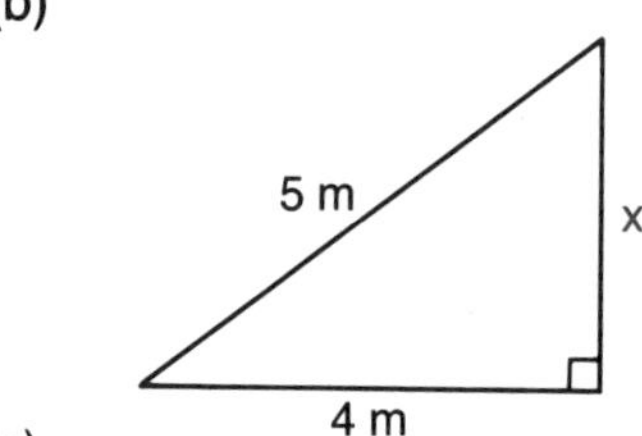

(c)

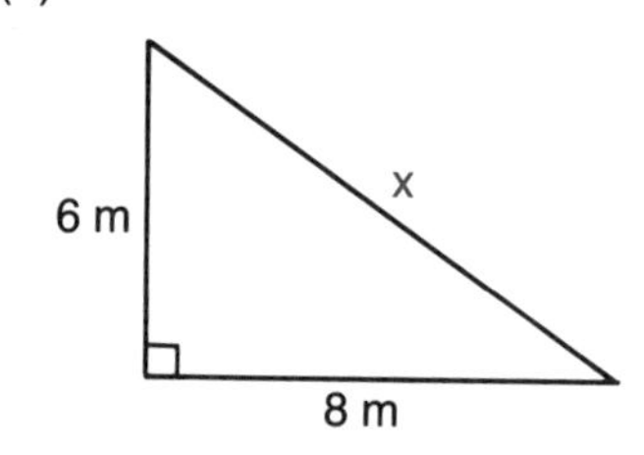

(d)

16 m

x

20 m

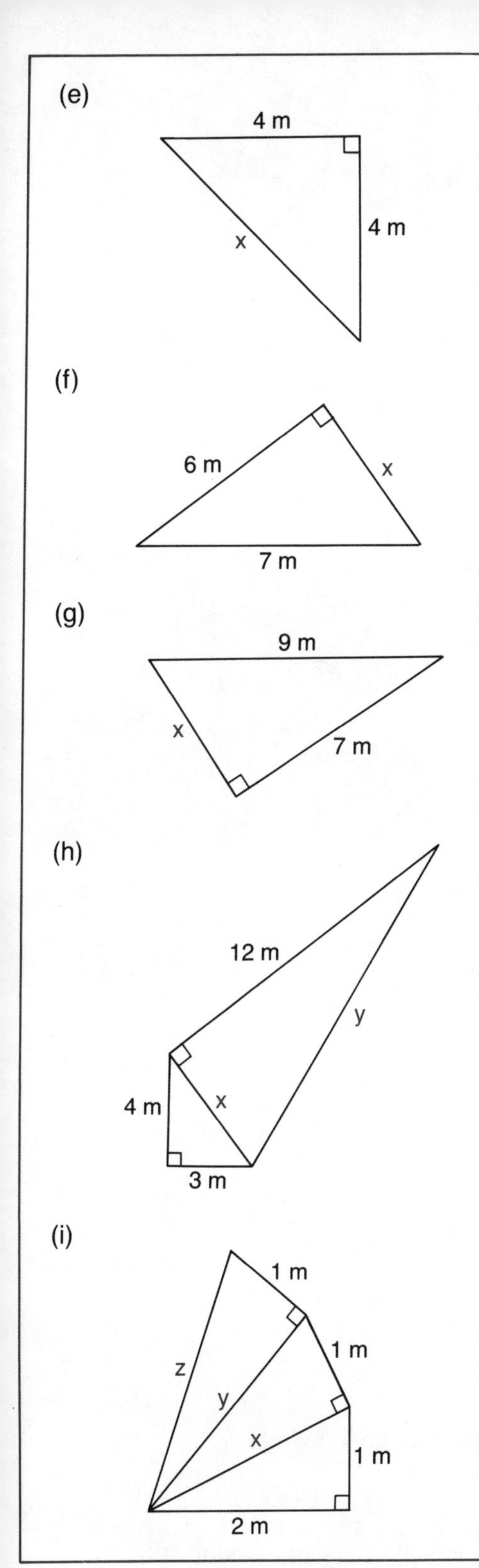

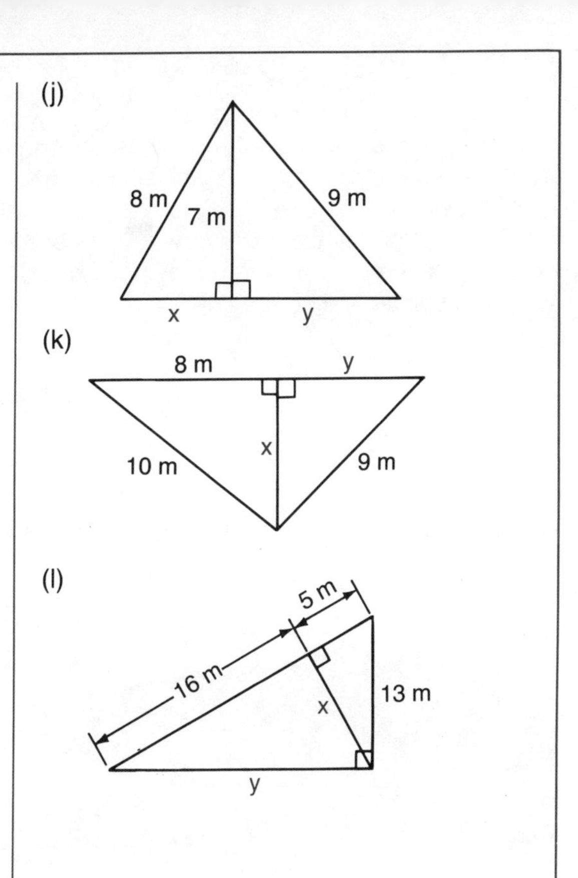

n	$\sqrt{n}$	n	$\sqrt{n}$	n	$\sqrt{n}$
1	1.000	21	4.582	41	6.403
2	1.414	22	4.690	42	6.481
3	1.732	23	4.795	43	6.557
4	2.000	24	4.898	44	6.633
5	2.236	25	5.000	45	6.708
6	2.449	26	5.099	46	6.782
7	2.645	27	5.196	47	6.856
8	2.828	28	5.291	48	6.928
9	3.000	29	5.385	49	7.000
10	3.162	30	5.477	50	7.071
11	3.316	31	5.567	51	7.141
12	3.464	32	5.656	52	7.211
13	3.605	33	5.744	53	7.280
14	3.741	34	5.830	54	7.348
15	3.872	35	5.916	55	7.416
16	4.000	36	6.000	56	7.483
17	4.123	37	6.082	57	7.550
18	4.242	38	6.164	58	7.616
19	4.358	39	6.244	59	7.681
20	4.472	40	6.324	60	7.746

12.1 SLOPE

Where a road is steeply graded, signs are posted to warn drivers, helping them to gauge their speed. The measure of a road's steepness is called the grade. A 5% grade means that for every 100 m horizontal distance, called the run, there is 5 m change in vertical distance, called the rise.

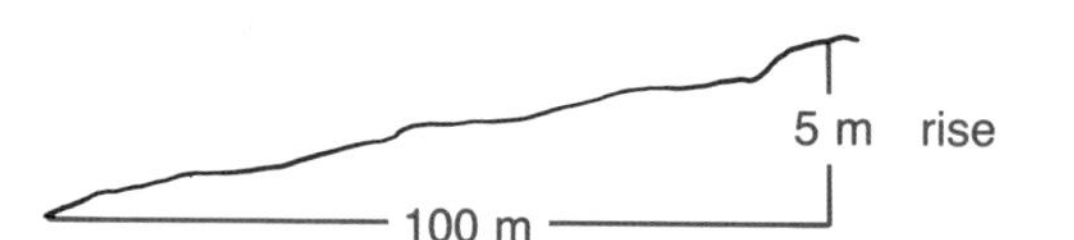

In mathematics we call this ratio of rise to run the slope.

$$\text{slope} = \frac{\text{rise}}{\text{run}}$$

Slope is illustrated in many situations, besides roads, and some examples are:

We use slope not only to describe steepness but also to compare the steepness of objects.

Example 1.
Calculate and compare the slopes of these ski hills.

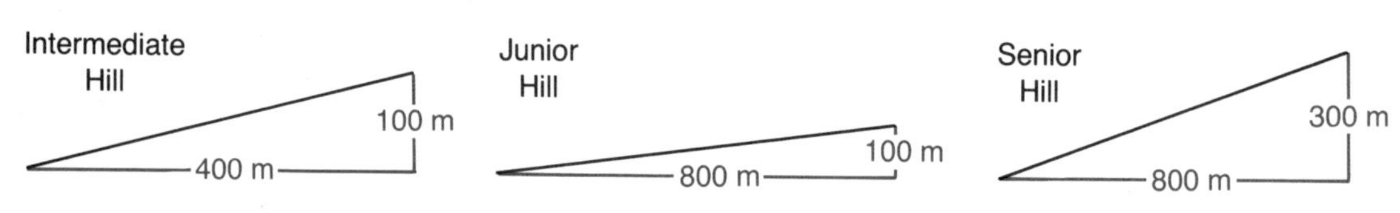

Solution:

$$\text{slope (intermediate)} = \frac{\text{rise}}{\text{run}} = \frac{100}{400} = \frac{1}{4} = \frac{2}{8}$$

$$\text{slope (junior)} = \frac{\text{rise}}{\text{run}} = \frac{100}{800} = \frac{1}{8}$$

$$\text{slope (senior)} = \frac{\text{rise}}{\text{run}} = \frac{300}{800} = \frac{3}{8}$$

To compare steepness, we express the slopes with the same denominator.

$$\text{slope (intermediate)} = \frac{2}{8}, \quad \text{slope (junior)} = \frac{1}{8}, \quad \text{slope (senior)} = \frac{3}{8}$$

Place the hills in order, beginning with the steepest: senior, intermediate, and junior.

Example 2.
Calculate the slope of AB.

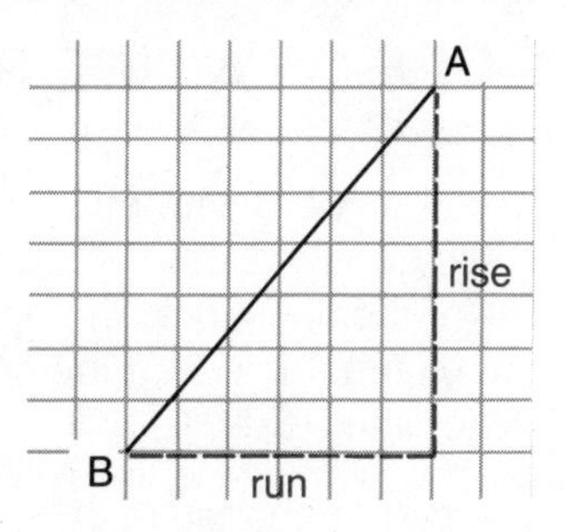

Solution:
By counting, rise $=$ 7 units and run $=$ 6 units.

$$\text{slope} = \frac{\text{rise}}{\text{run}} = \frac{7}{6}$$

EXERCISE 12.1

B 1. Calculate the slope of each ramp.

(a)

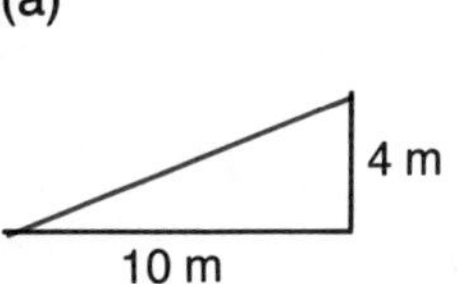

(b)

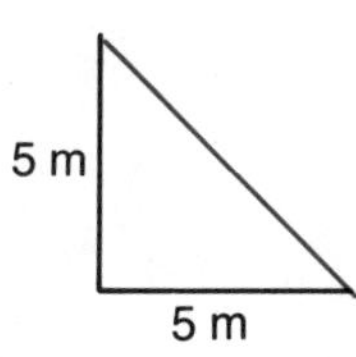

(c)

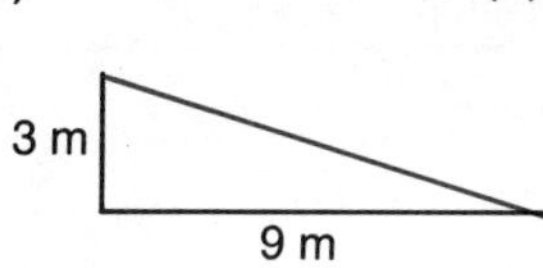

(d)

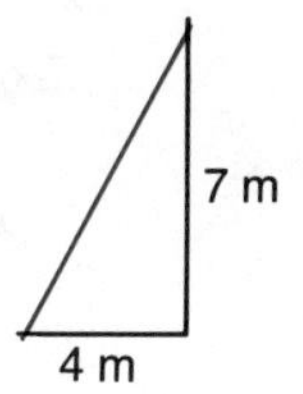

(e)

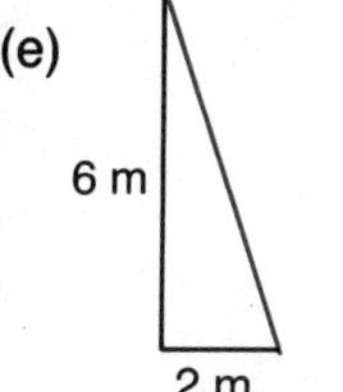

(f)

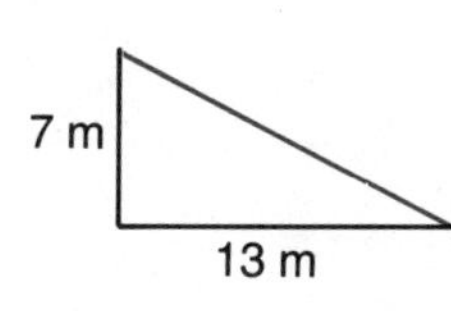

2. Calculate the slope of each set of stairs.

(a)

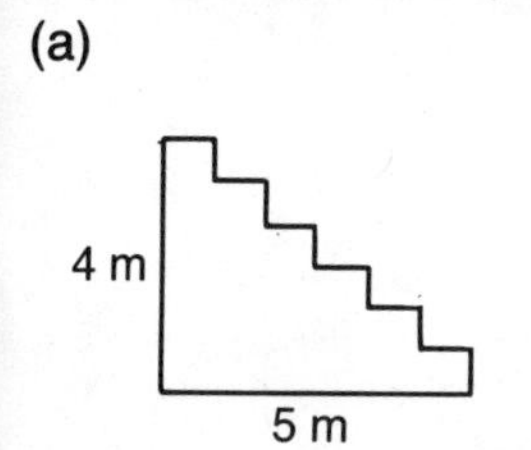

(b)

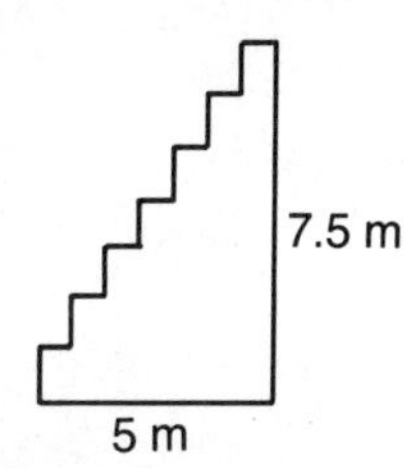

3. Calculate the slope of each roof.

(a)

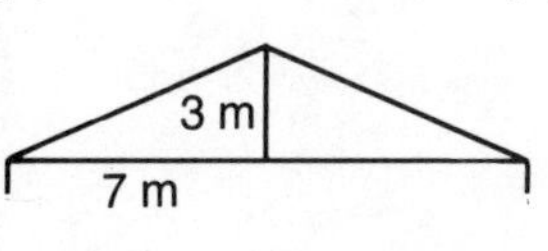

(b)

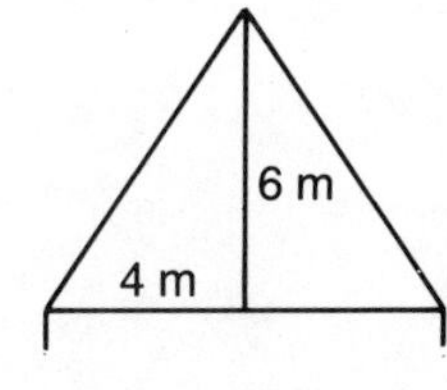

4. Find the slope of each line segment.

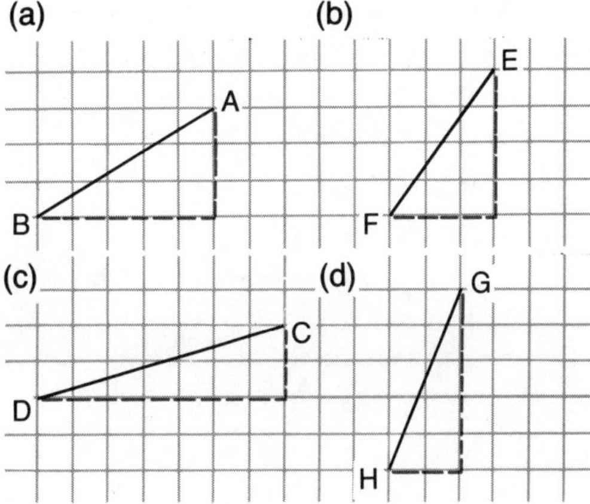

5. Find the slope of each line segment.

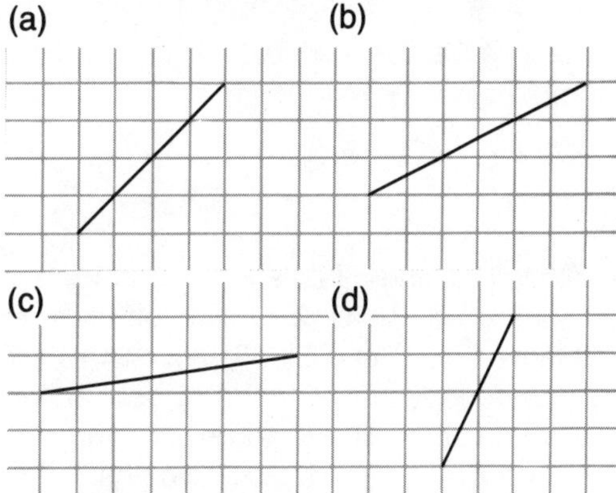

6. Draw a line with a slope of $\frac{1}{2}$.

7. Draw a line with a slope of $\frac{1}{4}$.

8. Draw a line with a slope of $\frac{6}{5}$.

9. Draw a line with a slope of $\frac{5}{8}$.

12.2 SLOPE OF A LINE SEGMENT

Lines drawn on a coordinate grid have slope. The rise, or vertical change, is called the change in y. The change in y for the line AB at the right is 4. The run, or horizontal change, is called the change in x. The change in x for line AB is 5.

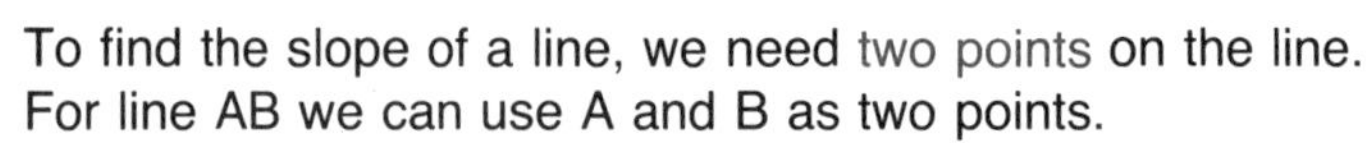

$$\text{slope} = \frac{\text{rise}}{\text{run}} = \frac{\text{change in y}}{\text{change in x}}$$

To find the slope of a line, we need two points on the line. For line AB we can use A and B as two points.

In the first quadrant: when we move from point A to B, the change in x for the line is 4 and the corresponding change in y is 3. The slope is $\frac{3}{4}$.

In the third quadrant: if we start at D and move to E, we see that the change in x is -4, and the corresponding change in y is -3.

The slope of the line is $\frac{-3}{-4}$ or still $\frac{3}{4}$.

We say that slope AB and slope DE are positive.

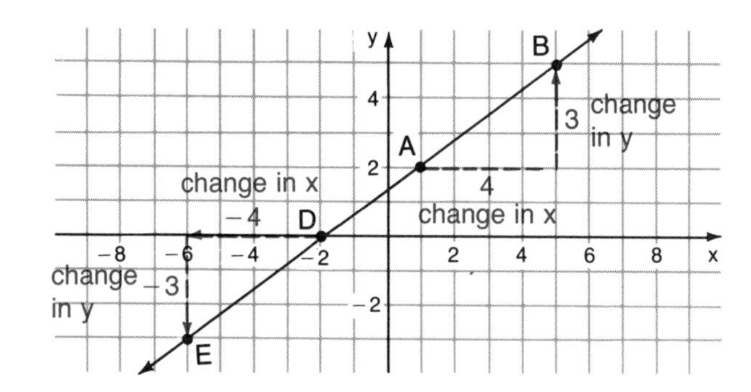

The slope of a line can also be negative.

For this line we use the points P and Q. From P to Q the change in x is -3 and the change in y is 2. Therefore, the slope is $\frac{2}{-3}$ or $-\frac{2}{3}$.

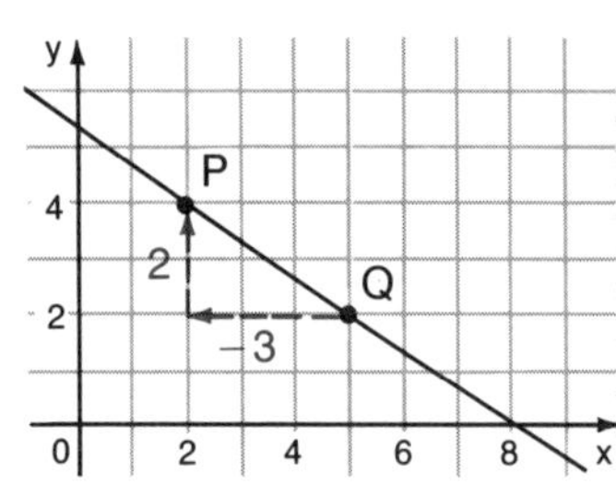

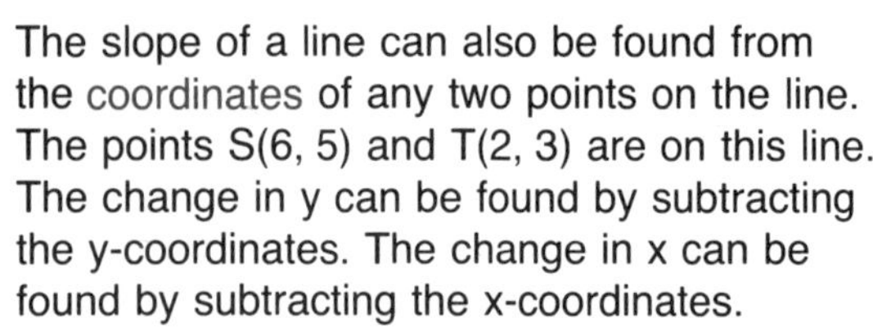

The slope of a line can also be found from the coordinates of any two points on the line. The points S(6, 5) and T(2, 3) are on this line. The change in y can be found by subtracting the y-coordinates. The change in x can be found by subtracting the x-coordinates.

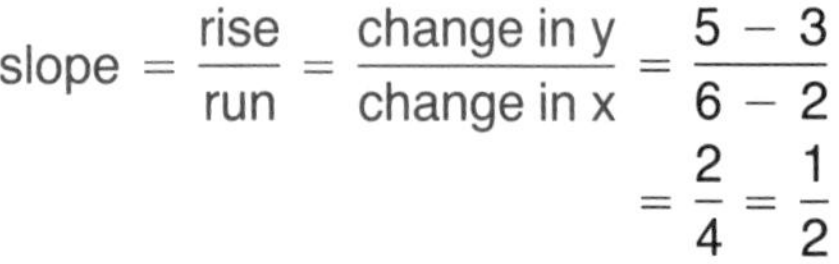

$$\text{slope} = \frac{\text{rise}}{\text{run}} = \frac{\text{change in y}}{\text{change in x}} = \frac{5-3}{6-2}$$
$$= \frac{2}{4} = \frac{1}{2}$$

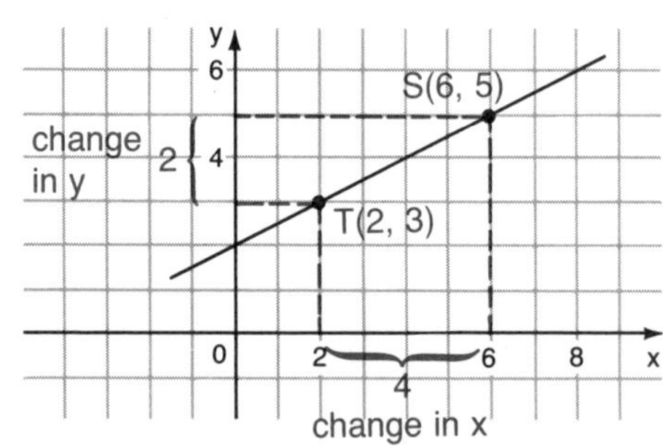

We can also calculate the slope in this way:

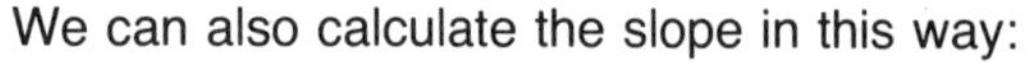

$$\text{slope} = \frac{\text{change in y}}{\text{change in x}} = \frac{3-5}{2-6} = \frac{-2}{-4} = \frac{1}{2}$$

The letter m is often used to mean slope.

The slope of a line, m, through any two points, $P(x_1, y_1)$ and $Q(x_2, y_2)$, is

$$m = \frac{\text{rise}}{\text{run}} = \frac{y_1 - y_2}{x_1 - x_2}$$

$$\text{or } \frac{y_2 - y_1}{x_2 - x_1}$$

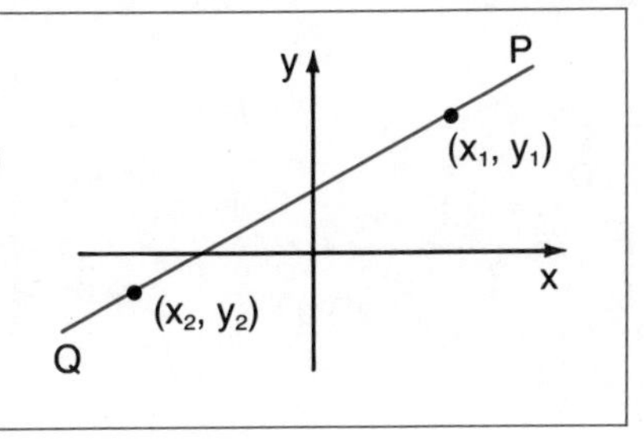

When we write x_1, the 1 is called a subscript.

Example 1.

Calculate the slope of the line that passes through each pair of points. Sketch each line.

(a) A(1, −1) and B(5, 2) (b) P(0, 0) and Q(−4, 2)

Solution:

(a) slope of AB $= \frac{\text{rise}}{\text{run}}$

$= \frac{-1 - 2}{1 - 5}$

$= \frac{-3}{-4}$

$= \frac{3}{4}$

(b) slope of PQ $= \frac{\text{rise}}{\text{run}}$

$= \frac{0 - 2}{0 - (-4)}$

$= \frac{-2}{4}$

$= -\frac{1}{2}$

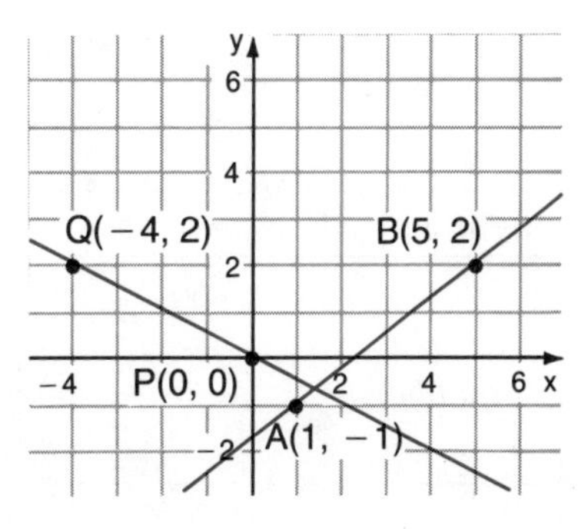

Example 2.

READ

The grade of a road is 5%. Calculate the rise, in metres, of a section of the road that is 2.5 km (or 2500 m) long.

Solution:

PLAN

If the grade is 5%, the slope is $\frac{5}{100}$.

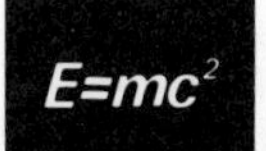

The run of the road is 2.5 km.

Let the rise of the road be x.

SOLVE

$$\text{slope} = \frac{\text{rise}}{\text{run}} = \frac{5}{100} = \frac{x}{2500}$$

$$\frac{5}{100} = \frac{x}{2500}$$

$$12\,500 = 100x$$

$$125 = x$$

ANSWER

∴ The rise is 125 m.

Example 3.

Find the slope of the line passing through A(−3, 4) and B(2, 4), and through C(5, 4) and D(5, −3).

Solution:

$$\text{slope of AB} = \frac{\text{rise}}{\text{run}} = \frac{4-4}{-3-2} = \frac{0}{-5} = 0$$

The slope of a horizontal line is 0.

$$\text{slope of CD} = \frac{\text{rise}}{\text{run}} = \frac{4-(-3)}{5-5} = \frac{7}{0}$$

Since division by 0 is undefined, the vertical line has no slope.

Description	Sketch	Slope
rising to the right		positive
horizontal		zero

Description	Sketch	Slope
falling to the right		negative
vertical		undefined

EXERCISE 12.2

A 1. Name the line segments with:

(a) positive slope (b) negative slope
(c) zero slope (d) no slope

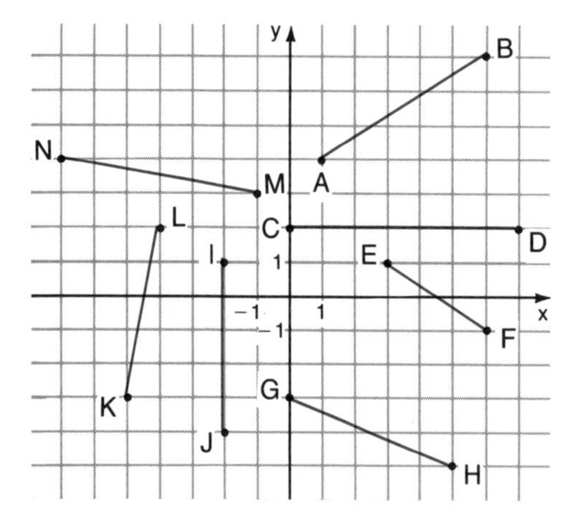

B 2. Find the slope of each line segment.

(a)

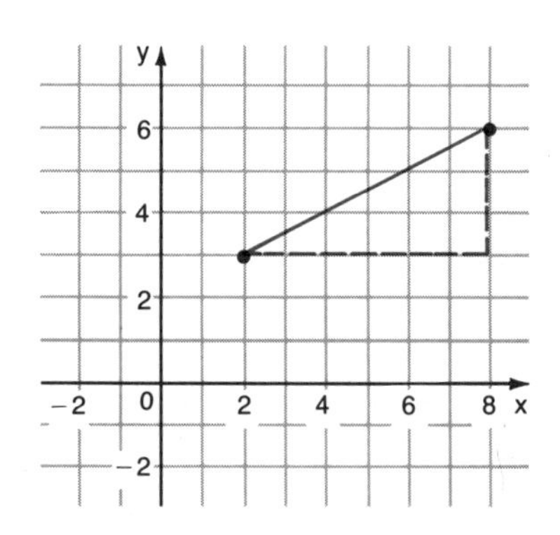

(b)

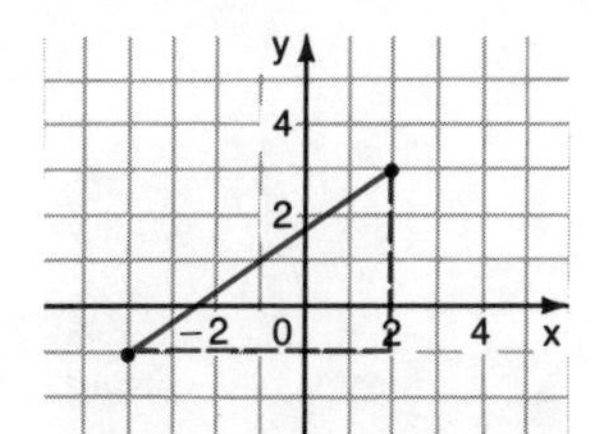

(c)

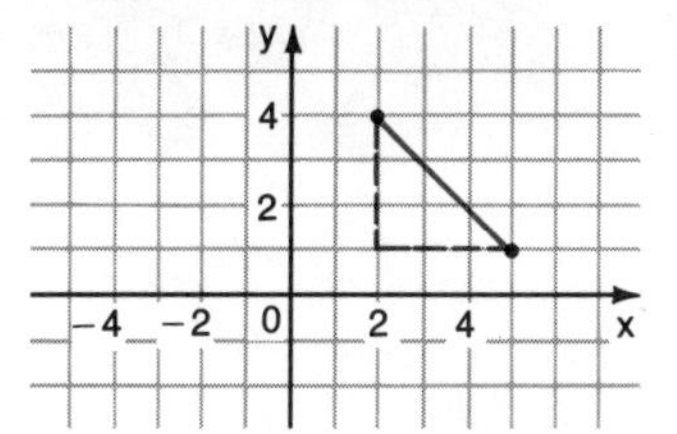

(d)

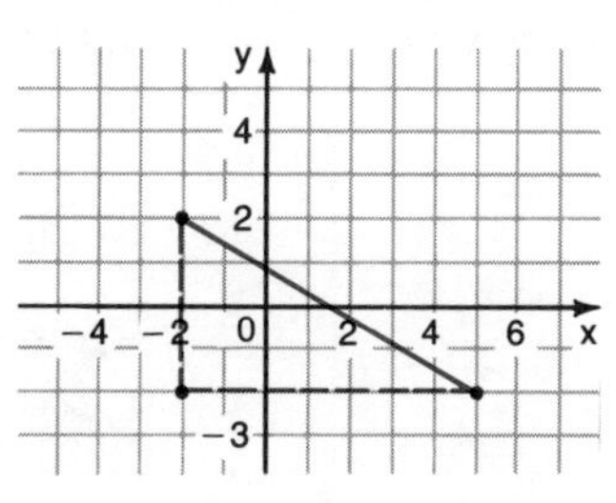

3. Find the slope of each line segment.

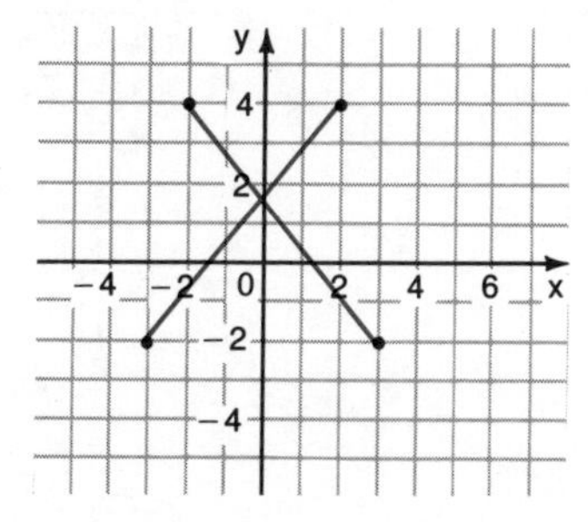

4. Calculate the slope of each line segment.

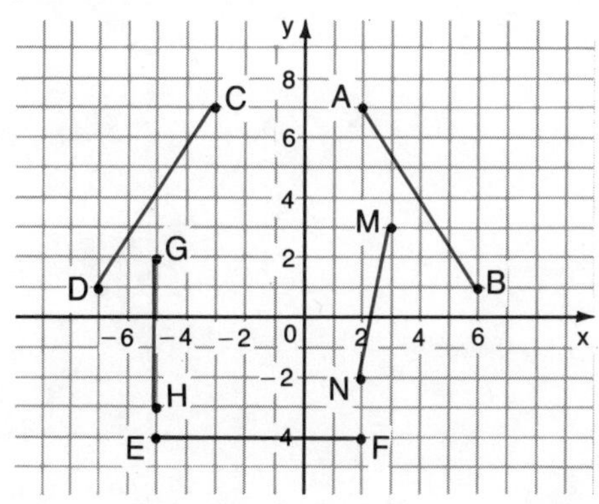

5. Calculate the slope of the line passing through each pair of points.

(a) A(2, 3), B(5, 9)
(b) P(4, 0), Q(6, 4)
(c) M(0, 0), N(4, 3)
(d) C(2, 5), D(4, 5)
(e) T(4, −3), S(5, 2)
(f) A(−1, −3), B(5, 4)
(g) H(−3, 1), K(0, 2)
(h) J(−4, −3), K(−1, 5)
(i) A(−6, 0), B(4, −5)
(j) P(1, −5), Q(−3, −2)
(k) M(1.5, 2), N(3, 3)
(l) E(0.3, 2), F(0.7, 1)
(m) F(4, 3.6), G(6, 5)
(n) H(2.5, 3.2), J(2.4, 3.3)
(o) A(10, 4.2), B(8.5, 4.8)
(p) S(−1.6, 1.8), T(2, −1.8)

6. On a grid draw a line through the point (3, 5) with a slope of $\frac{3}{4}$.

7. Draw a line through the point (4, 1) with a slope of $-\frac{1}{2}$.

8. Draw a line through (−3, −2) with a slope of $\frac{4}{3}$.

9. Draw a line through (−4, 3) with a slope of $-\frac{3}{2}$.

10. The grade of a road is 3%.

(a) Calculate the vertical climb, in metres, for a horizontal distance of 1.5 km.
(b) Calculate the run, in kilometres, necessary to reach the top of an 81 m high hill.

11. Two Olympic ski runs are compared. One has a vertical drop of 450 m for a horizontal run of 1.2 km, and the other a drop of 625 m for a run of 1.8 km. Which hill has the greater slope?

12.3 INVESTIGATING PARALLEL LINES

We can think of the two ramps that are used to load cars onto the platform as parallel lines. Parallel lines in the same plane never meet. In this section we will investigate how the slopes of parallel lines are related.

INVESTIGATION

1. (a) The lines AB, CD, EF, and GH on the coordinate grid are parallel.

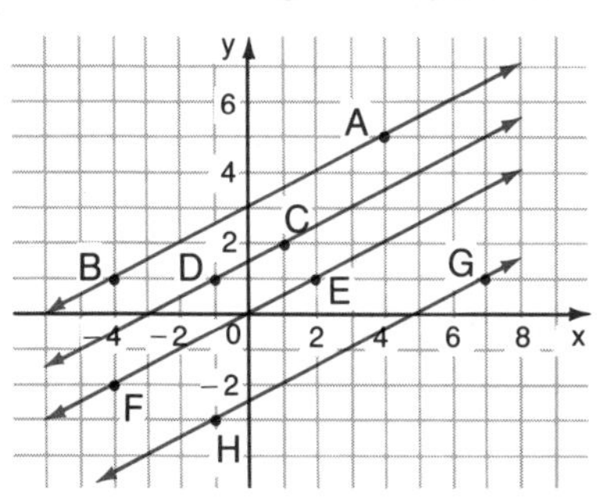

Calculate the slope of each line and complete the table.

Line	AB	CD	EF	GH
Slope				

(b) How are the slopes of the lines related?

The slopes of parallel lines are equal.

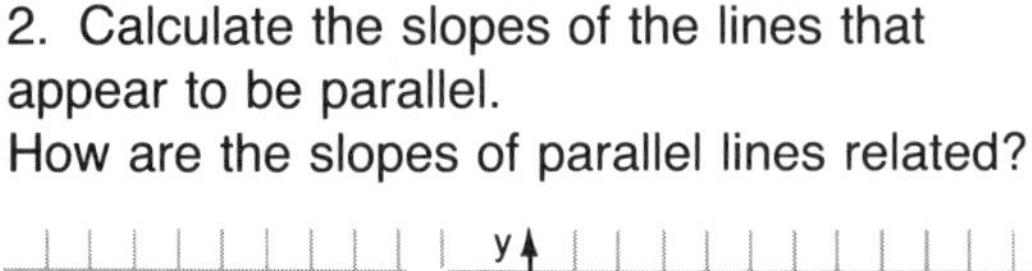

2. Calculate the slopes of the lines that appear to be parallel.
How are the slopes of parallel lines related?

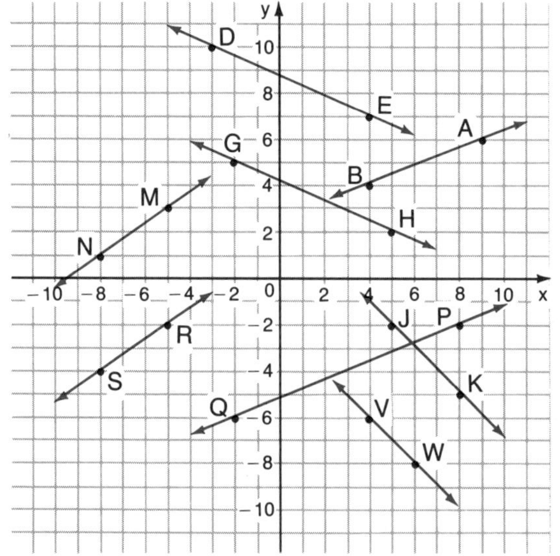

Example 1.
Show that the line through A(−3, 5) and B(1, 0) is parallel to the line through P(−1, −1) and Q(−5, 4).

Solution:

$$\text{slope of AB} = \frac{5-0}{-3-1} = \frac{5}{-4} = -\frac{5}{4}$$

$$\text{slope of PQ} = \frac{-1-4}{-1-(-5)} = \frac{-5}{4} = -\frac{5}{4}$$

Since the slopes are equal, the lines are parallel.

Example 2.

If the line through M(2, 1) and N(5, 3) is parallel to the line through P(x, − 1) and Q(4, 7), calculate the value of x.

Solution:

MN ∥ PQ

∴ slope of MN = slope of PQ

$$\frac{1-3}{2-5} = \frac{-1-7}{x-4}$$

$$\frac{-2}{-3} = \frac{-8}{x-4}$$

$$\therefore -2x + 8 = 24$$

$$-2x = 16$$

$$x = -8$$

MN ∥ PQ means MN is parallel to PQ.

EXERCISE 12.3

B

1. (a) Calculate the slope of the line joining A(4, 8) and B(2, 5).
(b) Calculate the slope of the line joining C(3, 1) and D(7, 7).
(c) Are the lines parallel?

2. (a) Calculate the slope of the line joining R(−6, 9) and T(−5, 7).
(b) Calculate the slope of the line joining P(−6, 5) and Q(−3, 1).
(c) Are the lines parallel?

3. Determine which pairs of line segments are parallel.
(a) A(−1, 2), B(0, 1) and C(2, 4), D(0, 6)
(b) R(−1, 1), P(4, 3) and S(1, −2), T(8, 0)
(c) P(−2, 1), Q(2, −1) and S(4, 3), T(7, 0)
(d) M(5, 9), N(8, 1) and O(0, 0), P(3, −8)
(e) A(−4, 0), B(−2, 6) and C(−9, 2), D(−7, 7)
(f) H(−5, −10), J(−3, −14) and K(−9, −8), L(−7, −12)
(g) Q(4, 1), R(5, 3) and S(8, 3), T(12, −5)

4. (a) Determine the slopes of the lines through P(−2, 4), Q(3, 4) and M(7, 0), N(2, 0).
(b) How are these two lines related?
(c) What is the slope of a line parallel to the x-axis?

5. (a) Describe the slopes of the lines through A(4, 6), B(4, −2) and C(0, 9), D(0, 1).
(b) How are these two lines related?
(c) Describe the slope of a line parallel to the y-axis.

6. Use slope to determine whether ABCD is a parallelogram.

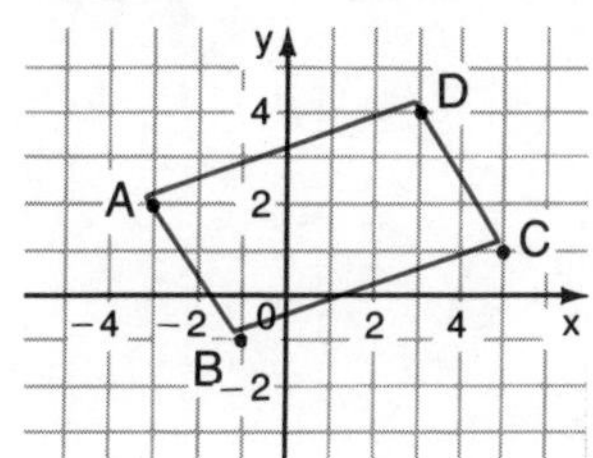

C

7. If the lines through each pair of points are parallel, find the value of the unknown coordinate.
(a) A(2, 5), B(4, y) and C(0, 2), D(2, 3)
(b) P(4, −2), Q(5, 1) and S(2, y), T(1, 0)
(c) M(5, 4), N(x, 8) and R(0, 0), S(1, −4)
(d) A(4, 3), M(1, 5) and P(x, 0), Q(0, 6)
(e) F(−1, −3), G(5, 2) and H(−3, y), K(3, 4)
(f) A(x, 9), B(5, 0) and W(−2, −2), T(8, 7)

12.4 INVESTIGATING PERPENDICULAR LINES

Rock climbers use a technique called rappelling to make a difficult descent. A climber backs to the edge of a rock face and springs off, holding on to a double rope for support. The climber swings down and towards the cliff face in a circular motion. In order not to slip on the face, the climber should land on the face so that his legs are perpendicular to the side of the cliff.

Perpendicular lines intersect at right angles. Since the lines are not parallel, their slopes cannot be equal. However, the slopes of perpendicular lines are related.

INVESTIGATION

1. (a) The lines AB and CD are perpendicular.
The lines EF and GH are perpendicular.
The lines PQ and RQ are perpendicular.

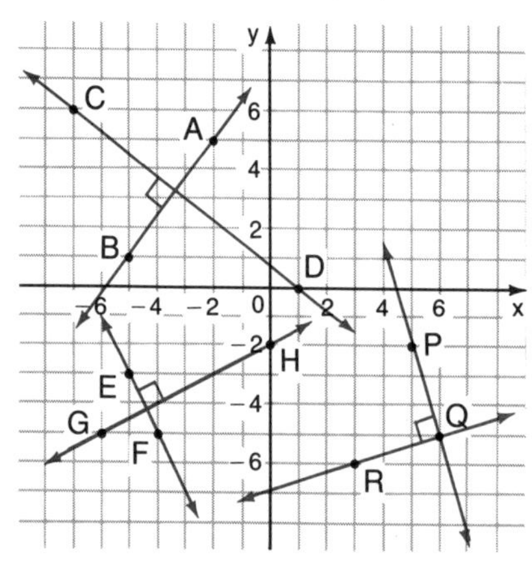

Calculate the slope of each line and complete the tables.

Line	Slope	Slope of AB × Slope of CD
AB		
CD		

Line	Slope	Slope of EF × Slope of GH
EF		
GH		

Line	Slope	Slope of PQ × Slope of RQ
PQ		
RQ		

(b) How are the slopes of perpendicular lines related?

From the investigation we conclude that:

Two lines are perpendicular if the product of their slopes is -1.

Example.
Show that the line through A(0, 2) and B(−3, −4) is perpendicular to the line through P(2, −4) and Q(−8, 1).

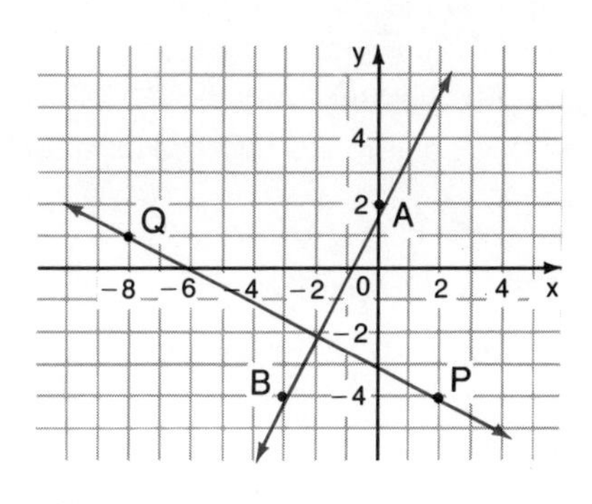

Solution:

$$\text{Slope of AB} = \frac{2-(-4)}{0-(-3)} = \frac{6}{3} = 2 \qquad \text{Slope of PQ} = \frac{-4-1}{2-(-8)} = \frac{-5}{10} = -\frac{1}{2}$$

Since $2 \times \left(-\frac{1}{2}\right) = -1$, AB ⊥ PQ

AB ⊥ PQ means
AB is perpendicular to PQ.

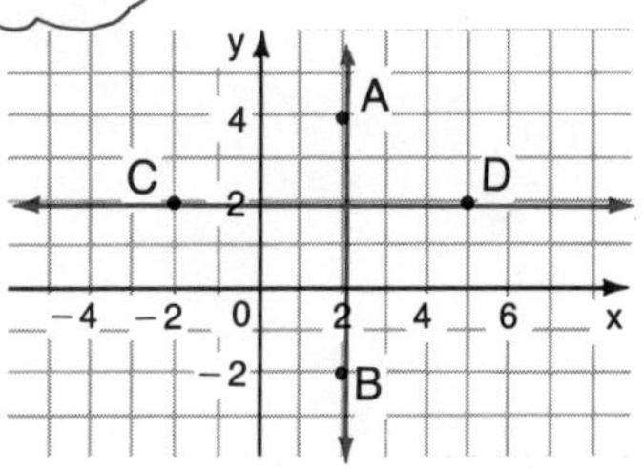

The lines illustrated at the right are perpendicular, but the slope of AB is undefined and the slope of CD is zero. The product of these slopes cannot be calculated.

EXERCISE 12.4

1. A line has the given slope. What is the slope of a line which is perpendicular to this line?

(a) $\frac{1}{2}$ (b) 3 (c) $\frac{2}{5}$ (d) $-\frac{3}{4}$

(e) -4 (f) $-\frac{6}{5}$ (g) 0 (h) -1

2. Determine the slope of the line perpendicular to the line through the given points.
(a) A(5, 7), B(4, 9)
(b) P(−1, 5), Q(2, 8)
(c) C(0, 4), D(5, 2)
(d) M(4, −2), N(−3, 1)
(e) R(2, 3), T(−2, 0)
(f) S(−5, 2), Q(−3, −3)

3. (a) Calculate the slope of the line joining A(−5, 4) and B(1, −4).
(b) Calculate the slope of the line joining C(−6, −3) and D(2, 3).
(c) Are the lines perpendicular?

4. (a) Calculate the slope of the line joining R(3, −1) and T(−2, 3).
(b) Calculate the slope of the line joining P(−1, −1) and Q(3, 4).
(c) Are the lines perpendicular?

5. Using slopes, show that the triangle formed by the three points is a right-angled triangle. Name the right angle.
(a) P(3, 1), Q(1, 7), R(5, 5)
(b) A(−3, 4), B(−6, 0), C(5, −2)
(c) M(0, 1), N(6, 1), O(6, −2)

12.5 DISTANCE BETWEEN TWO POINTS

We have used the coordinate system of graphing to:

(i) graph equations of lines
(ii) find the point of intersection of lines
(iii) find the slope of lines
(iv) investigate parallel and perpendicular lines

In this section we find the length of line segments from a graph.

VERTICAL LINES

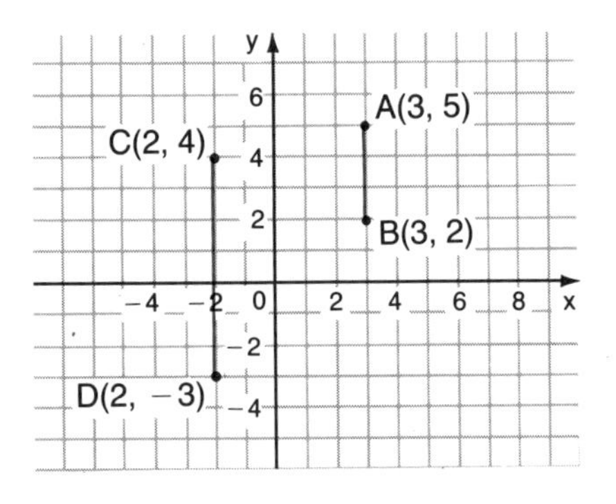

length of AB = difference of y-coordinates
$= 5 - 2$
$= 3$

length of CD = difference of y-coordinates
$= 4 - (-3)$
$= 7$

For the line segment CD, subtracting the y-coordinates in the opposite order gives

$-3 - 4 = -7$

Since length cannot be negative, CD is 7 units long.

HORIZONTAL LINES

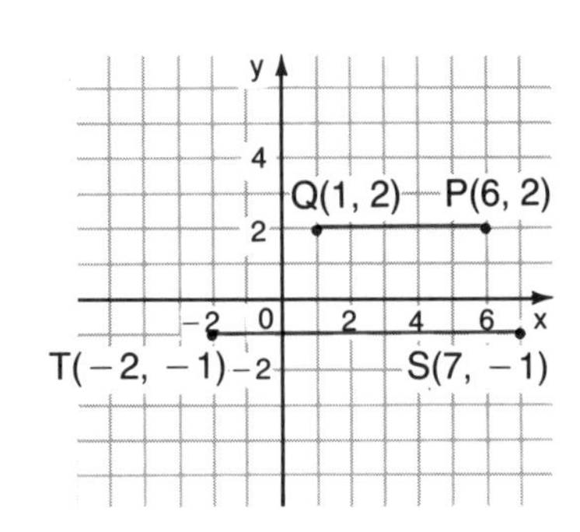

length of PQ = difference of x-coordinates
$= 6 - 1$
$= 5$

length of ST = difference of x-coordinates
$= 7 - (-2)$
$= 9$

To find the length of a line segment that is neither horizontal nor vertical, we use the Pythagorean Theorem.

Example 1.
Calculate the length of the line segment joining A(6, 4) and B(2, 1).

Solution:
Draw the line segment AB.
Construct the right triangle ABC.

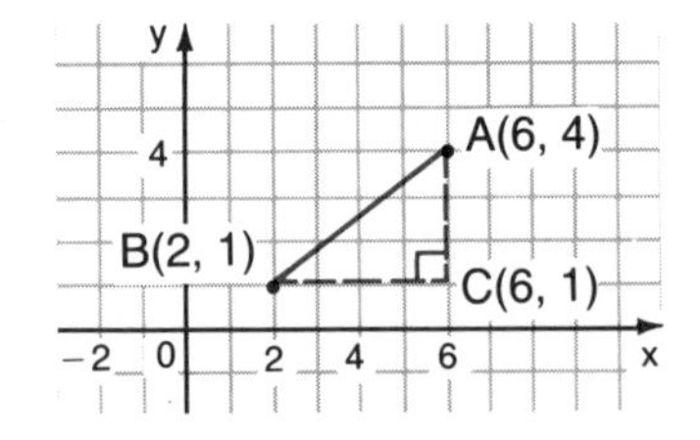

length of AC $= 4 - 1$
$= 3$

length of BC $= 6 - 2$
$= 4$

Using the Pythagorean Theorem:

$$\begin{aligned}(AB)^2 &= (AC)^2 + (BC)^2\\ &= 3^2 + 4^2\\ &= 9 + 16\\ &= 25\\ (AB)^2 &= 25\\ AB &= 5\end{aligned}$$

In general, $(AB)^2 = (\text{change in y})^2 + (\text{change in x})^2$

$$\begin{aligned}(AB)^2 &= (y_2 - y_1)^2 + (x_2 - x_1)^2\\ AB &= \sqrt{(y_2 - y_1)^2 + (x_2 - x_1)^2}\\ \text{or } AB &= \sqrt{(\text{rise})^2 + (\text{run})^2}\end{aligned}$$

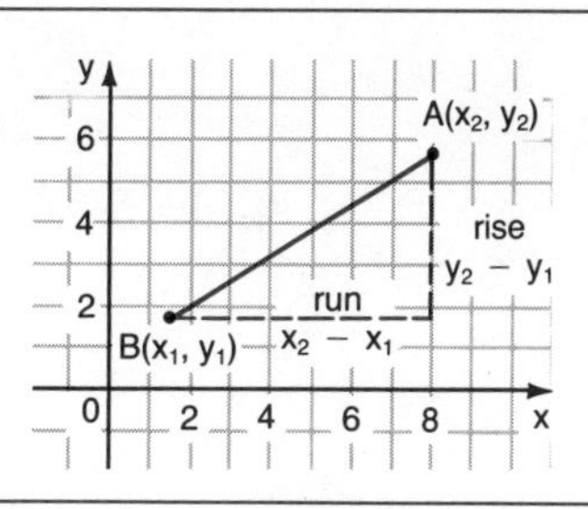

Example 2.
Calculate the distance between P(4, 5) and Q(−3, −1) to the nearest tenth.

Solution:
Draw the right triangle PQT.

$$\begin{aligned}PT &= 6, QT = 7\\ PQ &= \sqrt{(\text{rise})^2 + (\text{run})^2}\\ &= \sqrt{6^2 + 7^2}\\ &= \sqrt{36 + 49}\\ PQ &= \sqrt{85}\\ PQ &\doteq 9.2, \quad \text{to the nearest tenth}\end{aligned}$$

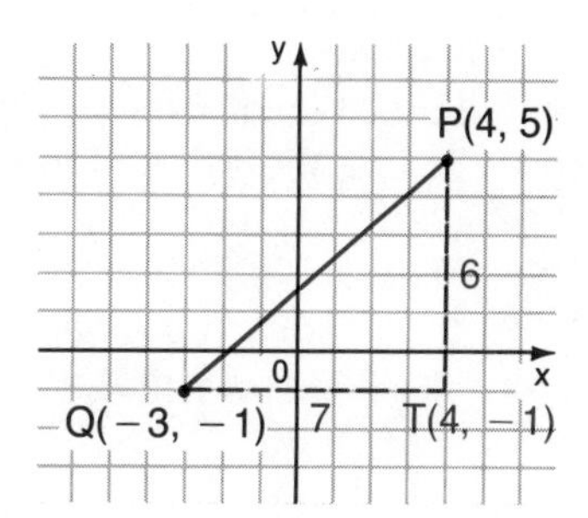

EXERCISE 12.5

1. State the length of the line segments.

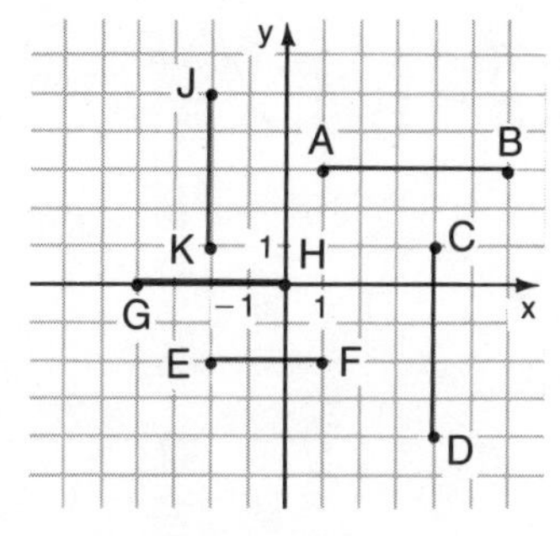

2. Calculate the distance between each pair of points to the nearest tenth.

(a) A(3, 1), B(6, 5)
(b) C(0, 0), D(−6, 4)
(c) P(3, 4), Q(5, 4)
(d) M(2, 3), N(1, 1)
(e) E(−3, 5), F(2, 1)
(f) T(0, −4), W(−7, 0)

3. Calculate the lengths of the sides in parallelogram ABCD.

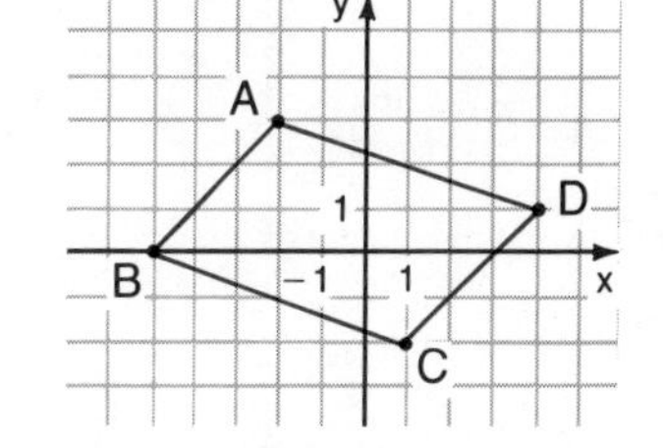

4. To show that M(1, 2) is the midpoint of PQ, calculate the lengths of PM and QM.

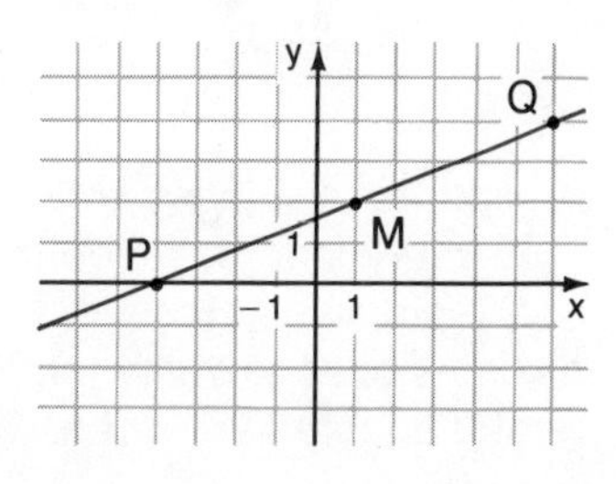

12.6 PROPERTIES OF PLANE FIGURES

In this section we will use coordinate geometry to investigate and verify some properties of squares, rectangles, parallelograms, and triangles.

INVESTIGATION

1. The vertices of the square are A(2, 5), B(−1, 2), C(2, −1), and D(5, 2).

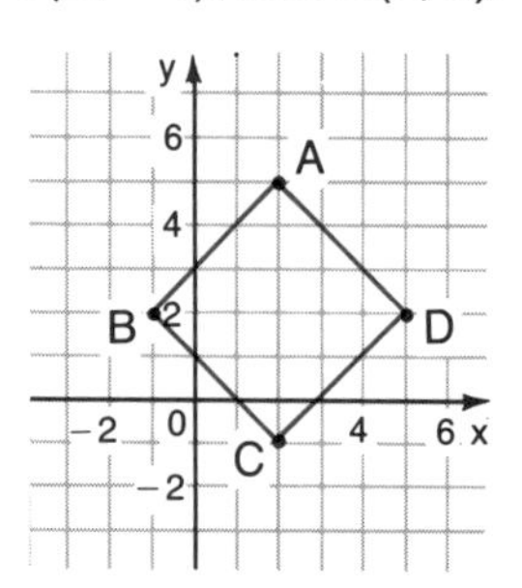

(a) Find the distances between the points to verify that the sides of the square are equal.
(b) Use slopes to verify that the opposite sides of the square are parallel.
(c) Use slopes to verify that each angle of the square is 90°.

2. The vertices of the square are P(4, 2), Q(−1, 2), R(−1, −3), and S(4, −3).

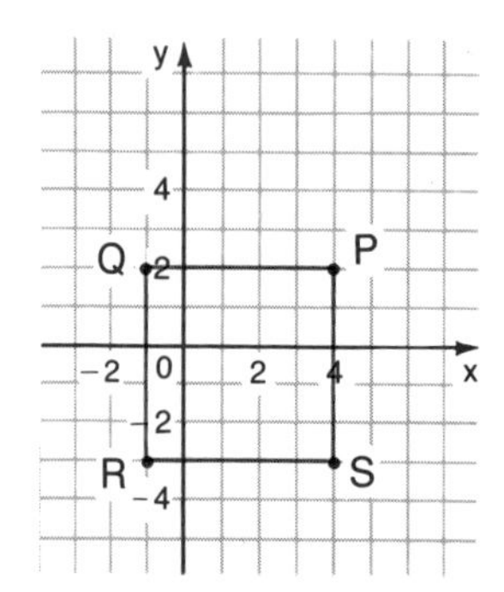

(a) Find the distances between the points to verify that the diagonals of a square are equal.
(b) Use slopes to verify that the diagonals of a square intersect at 90°.

3. The vertices of the rectangle are D(4, −1), E(−2, 5), F(−5, 2), and G(1, −4).

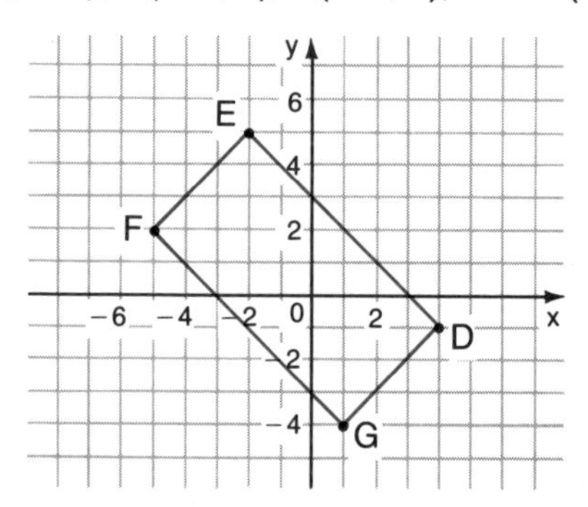

(a) Find the distances between the points to verify that the opposite sides of the rectangle are equal.
(b) Use slopes to verify that the opposite sides of the rectangle are parallel.
(c) Use slopes to verify that the angles of the rectangle are 90°.
(d) Verify that the diagonals of a rectangle are equal.

4. The vertices of the parallelogram are A(5, 4), B(−2, 4), C(−5, −1), and D(2, −1).

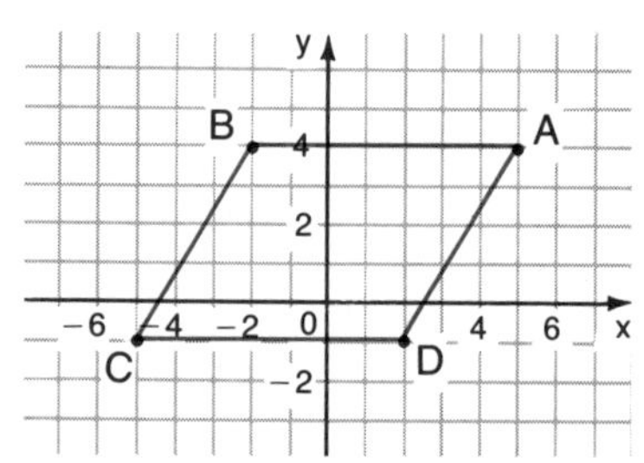

(a) Verify that the opposite sides of the parallelogram are equal.
(b) Verify that the opposite sides of the parallelogram are parallel.

5. The vertices of the triangle are A(4, 3), B(−1, −1), and C(8, −2).

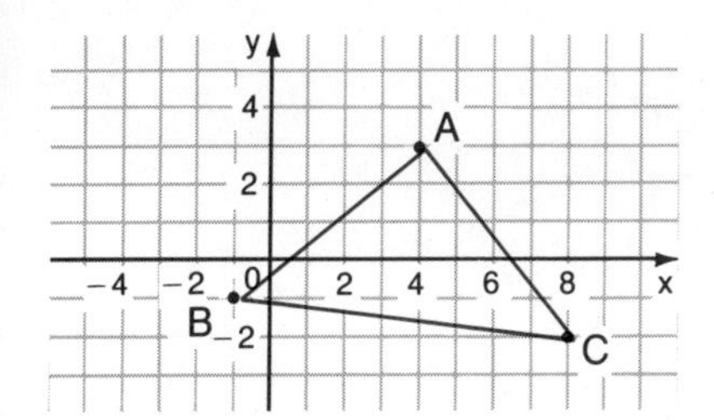

(a) Verify that the triangle is right angled.
(b) Verify that the triangle is isosceles.
(c) Verify that the square of the length of the hypotenuse is equal to the sum of the squares of the lengths of the other two sides.

6. The vertices of the triangle are D(4, −2), E(−4, 4), and F(−8, −6). Points M(−6, −1) and N(−2, −4) are midpoints.

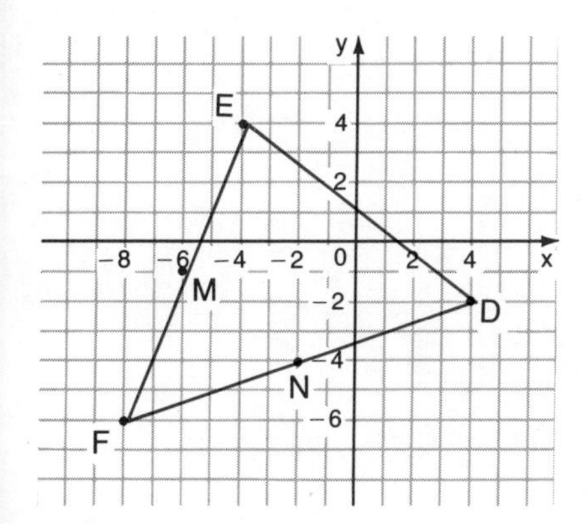

(a) Verify that M is the midpoint of EF.
(b) Verify that N is the midpoint of DF.
(c) Verify that the length of MN is half the length of DE.
(d) Verify that MN is parallel to DE.

7. The vertices of a rectangle are A(2, 1), B(2, 5), C(7, 1), and D(7, 5). Find the midpoint of each side.

8. The vertices of the quadrilateral are A(4, 4), B(−6, 2), C(−4, −6), and D(8, −2). E, F, G, and H are midpoints of the sides.

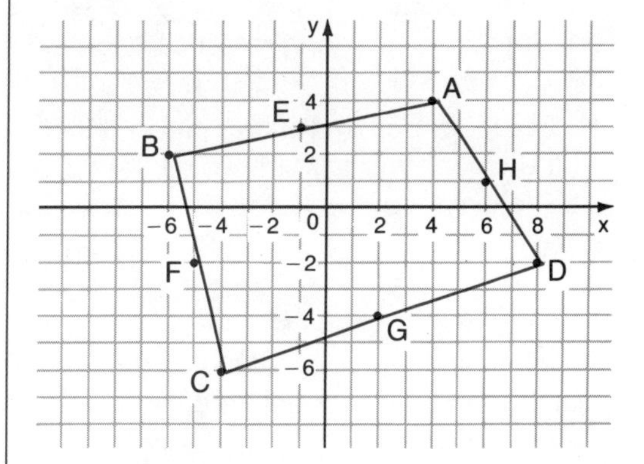

(a) Find the coordinates of E, F, G, and H.
(b) Verify that E, F, G, and H are the midpoints of the sides.
(c) Verify that the length of EH is equal to the length of FG.
(d) Verify that EF equals HG.
(e) Verify that EF is parallel to HG.
(f) Verify that EH is parallel to FG.

9. The vertices of a triangle are A(−3, 0), B(1, 8), and C(5, 1).
(a) Verify that the triangle is isosceles.
(b) Is the triangle a right triangle?

MIND BENDER

Each of the following number and letter combinations is commonly used to measure something.
What does each measure represent?

24K 64K 10K 273K

EXTRA

12.7 TELEPHONING LONG DISTANCE

When you make a long-distance telephone call, the cost of the call depends on two things:

1. The length of time of the call in minutes: the longer you talk, the more the call will cost.
2. The distance: the farther away, the more the call will cost.

For example, suppose a call from Toronto to Chicago, a distance of 688 km, costs $0.69 for the first minute and $0.58 for each additional minute. A call from Toronto to Los Angeles, a distance of 3480 km, would cost $1.00 for the first minute and $0.89 for each additional minute.

The length of a call is timed using a computer.

To determine the distance between two stations, a large grid is placed on a map of North America, as shown in the diagram.

The coordinates (0, 0) are located in the top right corner. The grid lines are numbered horizontally and vertically. Positions on the grid are labelled as ordered pairs.

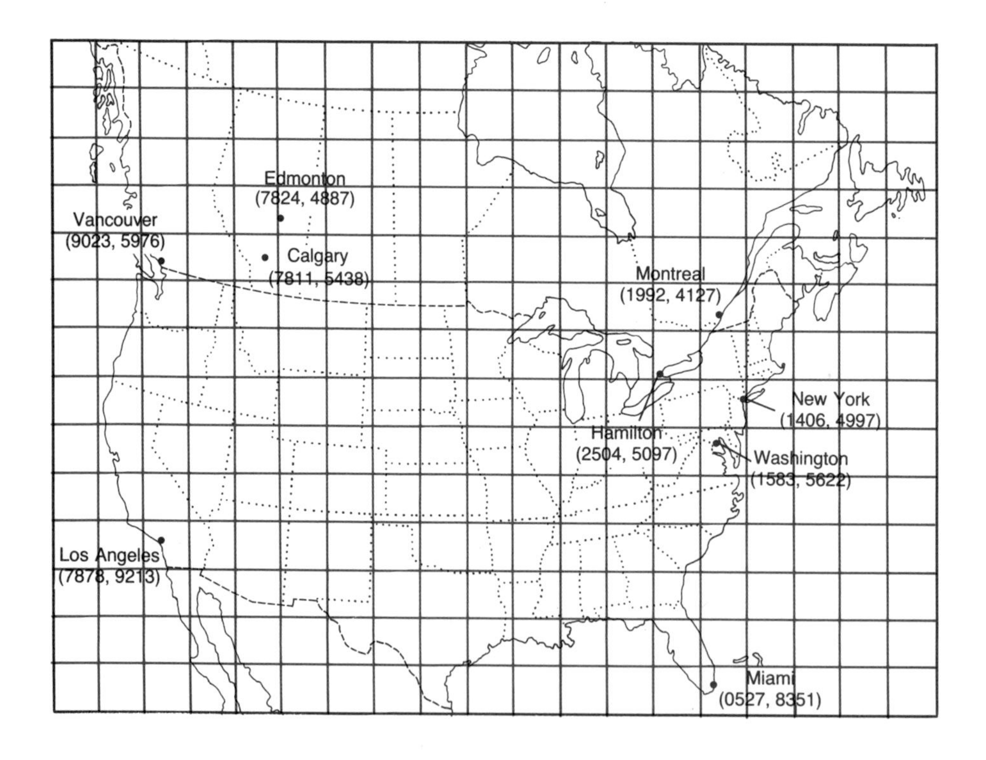

The coordinates of Montreal are (1992, 4127). This means that Montreal is 1992 units horizontally from (0, 0) and 4127 units vertically from (0, 0). The coordinates of Hamilton are (2504, 5097).

To find the distance between Hamilton and Montreal, we use the formula for the distance between two points.

$$d = \sqrt{(x_2 - x_1)^2 + (y_2 - y_1)^2}$$

$$d = \sqrt{(2504 - 1992)^2 + (5097 - 4127)^2}$$

$$= \sqrt{512^2 + 970^2}$$

$$= \sqrt{262\,144 + 940\,900}$$

$$= \sqrt{1\,203\,044}$$

$$\doteq 1097$$

The distance between Hamilton and Montreal is 1097 units. Each grid unit represents 0.51 km so that the distance between the cities is

$$1097 \times 0.51 \doteq 559 \text{ km}$$

EXERCISE 12.7

1. Calculate the distance between each pair of cities.
(a) Calgary and New York City
(b) Washington and Vancouver
(c) Miami and Edmonton
(d) Hamilton and Los Angeles

2. Why would it be impractical to calculate the distance between each and every city and town in North America?

3. What factors would determine the size of the squares on the telephone grid?

4. The town of Cassey Corners has coordinates (2002, 4521). The village of Fisher's Falls has coordinates (5283, 7812). The following is a partial rate chart.

Distance (km)	First Minute ($)	Each Additional Minute
1956–2157	2.09	0.56
2158–2357	2.24	0.61
2358–2557	2.39	0.66
2558–2757	2.41	0.71

(a) How much does a one minute call from Cassey Corners to Fisher's Falls cost?
(b) How much does a 9 min call from Fisher's Falls to Cassey Corners cost?

12.8 PROBLEM SOLVING

1. Two sides of an isosceles triangle measure 13 cm. The base is 10 cm.

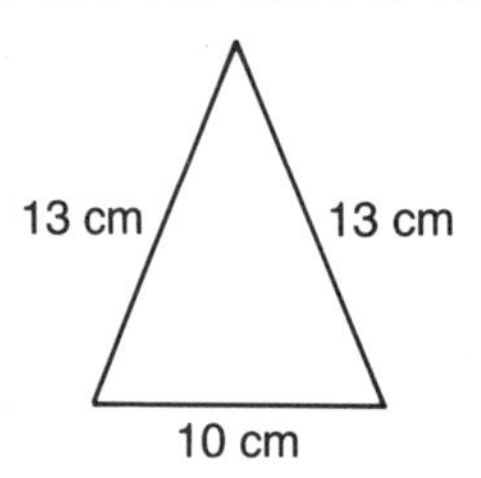

(a) Calculate the area of the triangle.
(b) The two equal sides of another isosceles triangle measure 13 cm, but the base of this triangle is not 10 cm. If this triangle has the same area as the one in part (a), find the base.

2. The digits from 0 to 7 have been classified as either A, B, or C.

A	1	4	7
B	2	5	
C	0	3	6

How would you classify the digit 8?

3. If July 1 falls on a Wednesday, how many more Wednesdays are there in the year?

4. Carol owns a miniature golf course. The course is in a rectangular lot that measures 45 m by 21 m.

(a) Carol wants to fence the course. Approximately how much fencing will she have to buy?
(b) If fencing is \$16.60/m, how much will the fencing around the golf course cost?
(c) There is to be a post every 3 m around the course.
How many posts are needed?
(d) Posts cost \$12.75 each.
How much does Carol pay for the posts?
(e) The costs of labour and other materials for the job are \$600.
What is the total cost of installing the fence?

5. There are four students on the executive of the school golf team. They can only appoint three executives to go to the provincial tournament organization meeting. How many different groups of three can there be?

6. Use only the digits 3, 4, 5, 6, and 7, and each only once, to write two whole numbers whose product is as large as possible.

7. At 07:00 Wilf and his crew left for Skull Island in a power boat. They travelled at 10 km/h.
At 08:30 Annette and her crew left for Skull Island in their power boat. They travelled at 12 km/h.
How far apart will they be at 11:30?

8. The number of people at the County Fair doubles every hour from 08:00 to noon.
If there were 12 000 people there at noon, how many were there at 08:00?

9. Approximately how thick is a page in this math book?

10. When Liz started her holidays, the odometer on her car read 12 346.8 km. When she returned, the odometer read 15 696.8 km.

(a) How many kilometres did she travel?
(b) If her car consumes fuel at the rate of 14L/100 km, approximately how many litres of fuel did the car use?

11. Determine the pattern and then find the next three numbers.

1, 4, 2, 6, 3, 8, 4, 10, ■, ■, ■

12. There are twelve basketball teams entered in a tournament. Each team plays until it loses.
If there are no ties, how many games must be scheduled?

13. The bus arrived at the Glendale station where 8 passengers got on and 11 passengers got off. At the Churchill station 9 passengers got on and 4 got off. At the Delta station 15 passengers got on and 6 got off. At the Westdale station 5 got on and 9 got off. When the bus left the Westdale station, there were 41 passengers on the bus.
How many passengers were on the bus when it originally arrived at the Glendale station?

14. Determine the pattern and then find the missing number.

3	11	9	5
9	2	5	6
5	7	8	4
7	6	?	8

15. There are eleven people at a birthday party. There is one round cake to divide among them.
How can you cut the cake into eleven parts, not necessarily of equal size, by making only four straight line cuts?

16. Eight cars were lined up bumper to bumper in a dealer's lot.
How many bumpers were actually touching each other?

CAREER

TRAVEL AGENT

Tourism is now the world's largest business. Canadians alone spend about 4 billion dollars on travel outside of Canada. Canadians are avid travellers: each year we take an average of six trips for every man, woman, and child in Canada.

1. If a small plane costs one million dollars, how many of them could you buy for 4 billion dollars?

2. How many trips do Canadians take every year?

3. When a travel agent sells an airline ticket, she receives a commission on the cost of the ticket.
If the rate of commission is 8%, what commission would an agent receive on a ticket costing $345.50?

4. Cecilia Chong works as a travel agent. She earns a weekly salary of $400 plus a 7% commission on all airline ticket sales.
How much will she earn in a month if she has ticket sales of $21 500?

5. It takes five hours to fly from Toronto to Los Angeles.
If a traveller leaves Toronto at 11:00, at what local time will she arrive in Los Angeles?

12.9 REVIEW EXERCISE

1. Find the slope of each ramp.

(a)

(b)

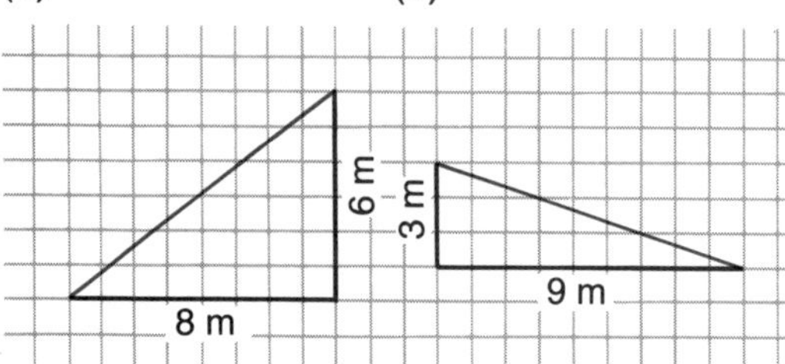

(c) (d)

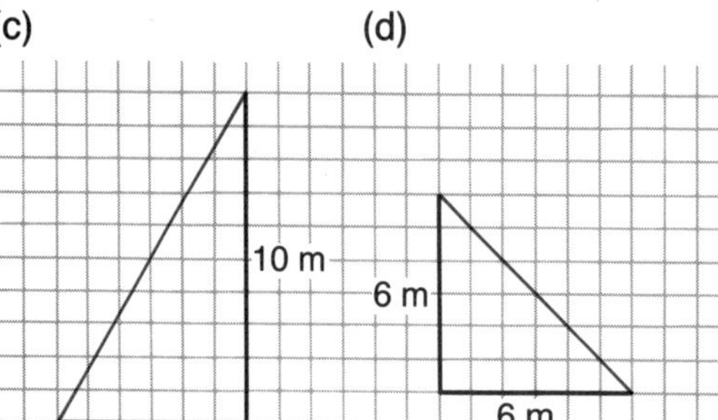

2. Find the slope of each line segment.

(a)

(b)

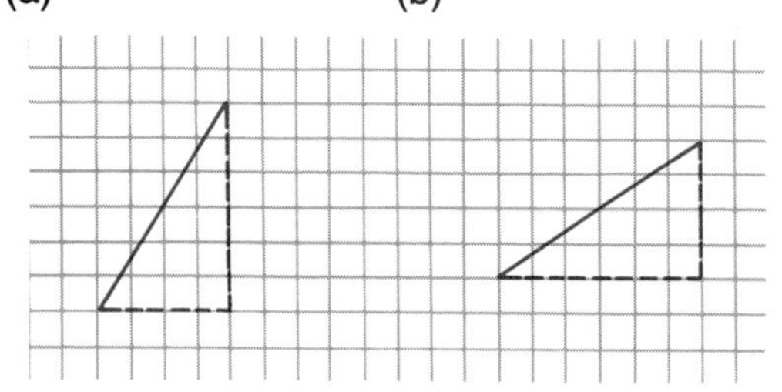

3. Draw a line with a slope of $\frac{1}{3}$.

4. Draw a line with a slope of $\frac{3}{4}$.

5. Find the slope of each line segment.

(a)

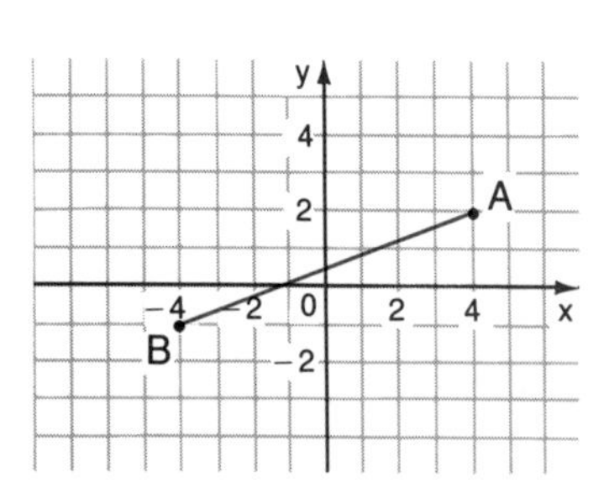

(b)

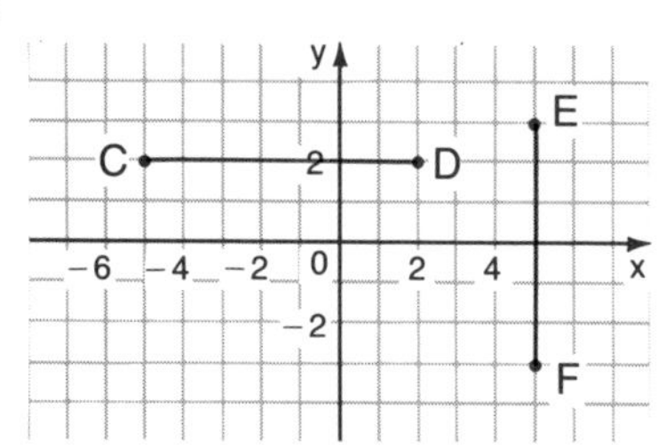

(c)

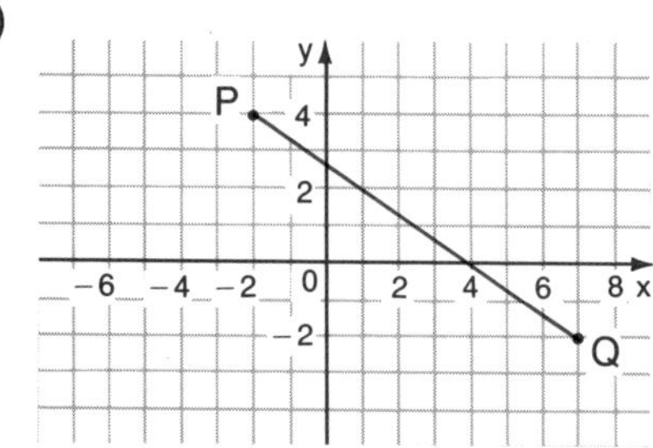

6. Calculate the slope of the line passing through each pair of points.

(a) (5, 3) and (8, 9)
(b) (3, −1) and (7, 3)
(c) (4, 2) and (3, 5)
(d) (8, 6) and (−2, 5)
(e) (−1, 0) and (4, −2)
(f) (5, −1) and (5, 3)
(g) (0, 2) and (0, −4)
(h) (8, −3) and (4, −3)

7. On a grid draw a line through the point (3, 2) and with slope $\frac{1}{2}$.

8. On a grid draw a line through the point (−2, 4) and with slope $-\frac{4}{3}$.

9. (a) Calculate the slope of the line joining A(5, 5) and B(1, 2).
(b) Calculate the slope of the line joining C(−2, 3) and D(4, −5).
(c) Are the lines parallel or perpendicular?

10. (a) Calculate the slope of the line joining S(0, 6) and T(−4, −6).
(b) Calculate the slope of the line joining P(1, 1) and Q(2, 4).
(c) Are the lines parallel or perpendicular?

11. Each pair of points represents a line segment. Determine if the line segments are parallel or perpendicular.

(a) A(1, 0), B(5, 3) and C(−1, 3), D(3, 6)
(b) P(−4, −1), Q(2, −3) and T(2, −1), V(3, 2)
(c) M(−1, 1), N(5, 4) and R(3, 3), S(2, 5)
(d) X(0, 0), Y(6, −2) and T(3, 1), W(9, −1)
(e) A(0, −1), C(−4, −3) and D(2, −3), E(10, 1)

12. The slope of a ramp for moving refrigerators must be $\frac{1}{4}$.

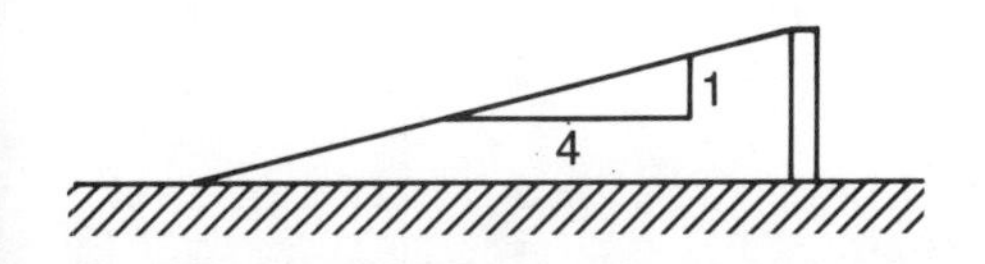

(a) How far back must a ramp be placed to move a refrigerator to a height of 3 m?
(b) To what height can you move the refrigerator from a distance of 32 m?

13. Calculate the length of each line segment.

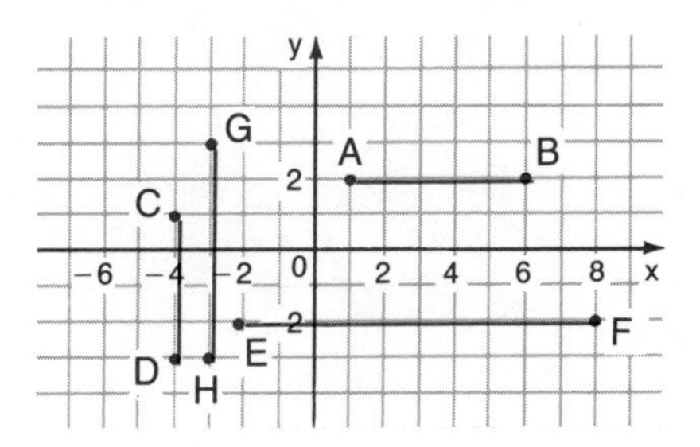

14. Calculate the length of each line segment.

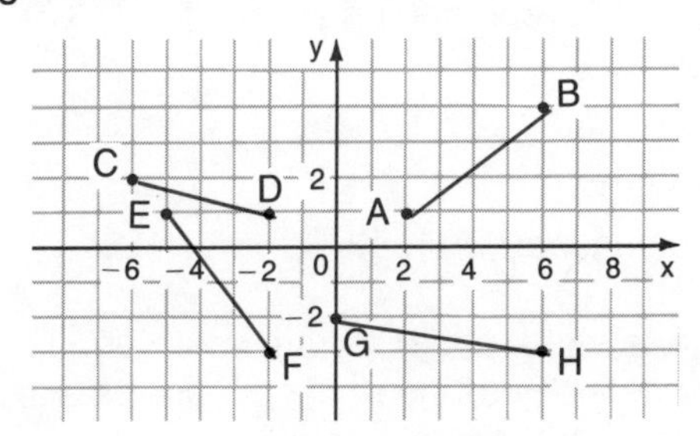

15. Calculate the distance between each pair of points.

(a) A(5, 7), B(9, 10)
(b) C(3, 12), D(8, 0)
(c) E(4, 3), F(4, −7)
(d) G(−2, 5), H(8, 5)
(e) P(−1, 1), Q(0, 2)
(f) R(4, −2), S(1, 5)
(g) X(−5, 0), Y(1, 3)

16. The vertices of the triangle are P(5, 6), Q(−3, 4), and R(3, −2).

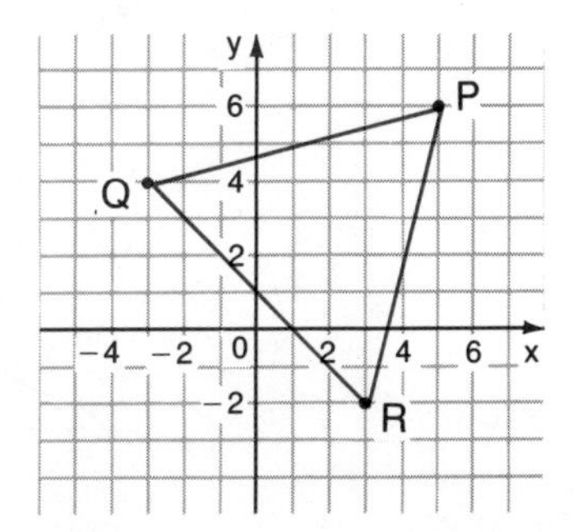

Verify that the triangle is isoceles.

17. The vertices of the parallelogram are A(6, 4), B(−2, 4), C(−4, −2), and D(4, −2).

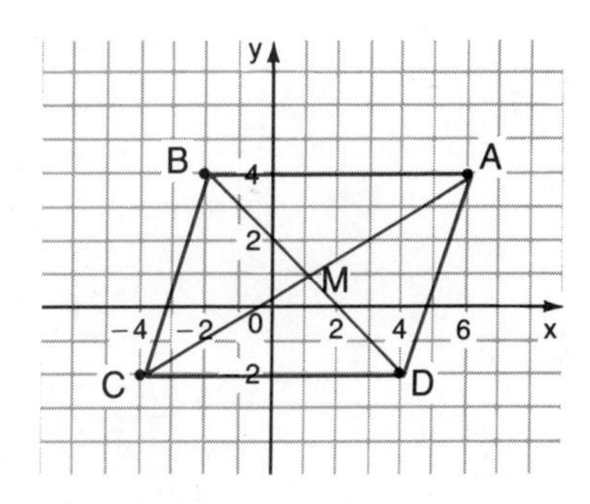

(a) Verify that the opposite sides of the parallelogram are parallel.
(b) Verify that the opposite sides of the parallelogram are equal.
(c) The diagonals of the parallelogram intersect at M(1, 1).

Verify that M is the midpoint of AC, and also the midpoint of BD.

12.10 CHAPTER 12 TEST

1. Find the slope of each ramp.

(a)

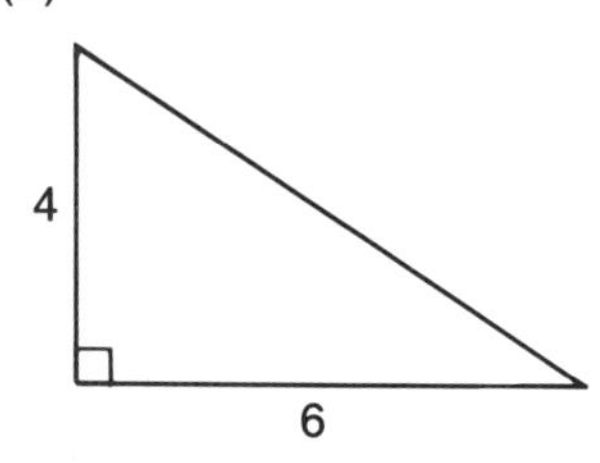

(b)

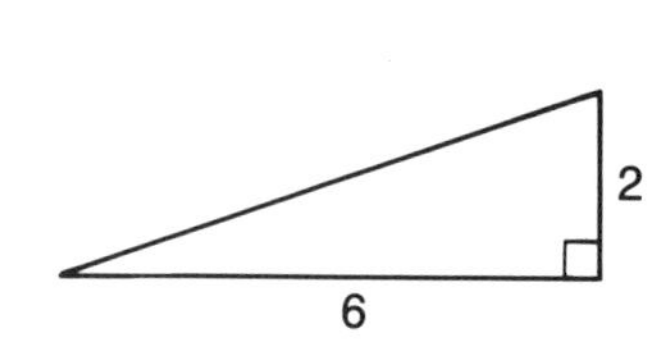

2. Find the slope of each line segment.

(a) (b)

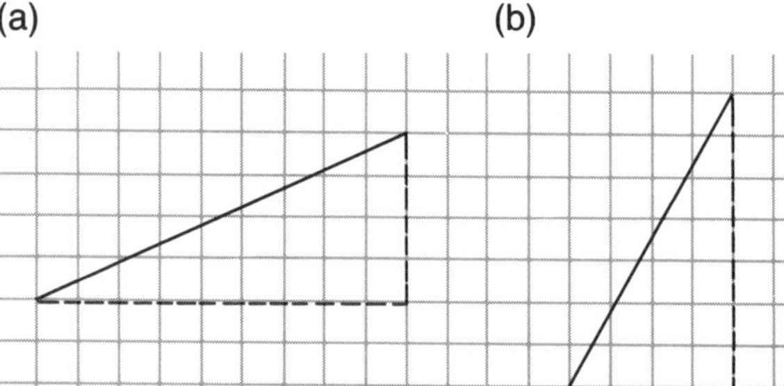

3. Calculate the slope of the line passing through each pair of points.

(a) A(8, 7) and B(1, 3)
(b) C(−6, 5) and D(−1, −2)
(c) E(−4, 5) and F(3, 5)

4. (a) Calculate the slope of the line segment joining each pair of points.

(i) A(4, 2) and B(1, −2)
(ii) M(5, 1) and N(3, 5)
(iii) C(−1, 5) and D(−7, −3)
(iv) G(−2, 1) and H(2, −2)

(b) Which line segments are parallel in part (a)?
(c) Which line segments are perpendicular in part (a)?

5. Calculate the distance between each pair of points.

(a) A(5, 6) and B(5, −3)
(b) C(−1, 5) and D(6, 1)
(c) E(7, −2) and F(−3, −2)

6. The vertices of a rectangle are A(1, 5), B(−5, −1), C(−2, −4), and D(4, 2).

(a) Verify that the opposite sides of the rectangle are equal.
(b) Verify that the opposite sides of the rectangle are parallel.
(c) Verify that the angles of the rectangle are 90°.

YEAR-END REVIEW

1. Write the following numbers in words.

(a) 34 500　(b) 0.756
(c) 3 450 000　(d) 18.9
(e) 5611　(f) 67.76

2. Estimate, then calculate.

(a) $23\,450 + 9566 + 7777$
(b) $3001 - 2134$
(c) 670×46
(d) $713 \div 31$
(e) $456 + 98 + 307$
(f) $23\,099 - 13\,002$
(g) 43×19
(h) $1107 \div 9$

3. Carl earns $25/h as a consultant. Last year he worked seven hours per day and six days per week for forty-seven weeks. How much did he earn?

4. Estimate, then calculate.

(a) $32 \times (23 - 8) + 56$
(b) $84 \div 14 + 11 \times 12$
(c) $(45 - 32)(56 - 25)$
(d) $4^2 + 8^2 - 5^2$
(e) $\frac{2}{3}$ of $66 + 17$
(f) $\dfrac{52 + 18}{14 \div 2} + 7^2 - 3$

5. Estimate, then calculate.

(a) $12.4 + 45.67 + 78.9$
(b) $23.68 - 9.78$
(c) 9.3×7.5
(d) $16.2 \div 4.5$
(e) $2.009 - 0.786$
(f) $11.408 \div 12.4$
(g) $45.6 + 123.9 + 99.7$

6. Estimate the cost of each, then calculate.

(a) eleven theatre tickets at $35.50 each
(b) twenty-one cans of apple juice at $1.78 each
(c) forty-four hours of work at $13.75/h
(d) sixteen books at $31.80 each
(e) thirty-seven T-shirts at $11.95 each

7. Multiply.

(a) $3\frac{1}{2} \times \frac{1}{3}$　(b) $5 \times 2\frac{1}{4}$
(c) $1\frac{1}{8} \times 6$　(d) $3\frac{3}{5} \times 1\frac{1}{2}$

8. Divide.

(a) $\frac{3}{4} \div \frac{3}{5}$　(b) $1\frac{3}{4} \div \frac{1}{2}$
(c) $2\frac{1}{4} \div 2$　(d) $3 \div 3\frac{1}{3}$

9. Add.

(a) $\frac{5}{6} + \frac{2}{3}$　(b) $1\frac{3}{8} + 2\frac{3}{4}$
(c) $4\frac{7}{10} + 2\frac{1}{5}$　(d) $3\frac{1}{4} + 4\frac{1}{3}$

10. Subtract.

(a) $\frac{7}{8} - \frac{3}{4}$　(b) $4 - 3\frac{3}{8}$
(c) $3\frac{1}{2} - 2\frac{2}{5}$　(d) $5\frac{5}{6} - 2\frac{1}{4}$

11. Simplify.

(a) $-3 + 7 + 5$　(b) $-8 - (-4) + 6$
(c) $-15 \div 5 + 4$　(d) $(-6)(-3) + 7$
(e) $9 \times (-2) - 1$　(f) $-2 + (-4)(-2)$

12. Express each as a decimal.

(a) $\frac{3}{8}$　(b) $\frac{5}{6}$

13. Express each as a fraction.

(a) 0.55　(b) $0.\overline{5}$

14. Simplify.

(a) $1\frac{1}{4} - 2\frac{3}{8}$　(b) $-\frac{1}{2} - \frac{1}{4} - 2$
(c) $-2\frac{1}{3} \div \frac{1}{3}$　(d) $\frac{3}{4} \times \frac{-1}{2} \div \frac{2}{-3}$

15. If $x = 3$, find the value of each expression.

(a) x^2　(b) $2x^2$
(c) $-x^2$　(d) $-4x^3$
(e) $(2x)^2$　(f) $(-2x)^2$
(g) $-(2x)^3$　(h) $(-2x)^3$
(i) x^0　(j) $3x^2$
(k) x^{-1}　(l) $2x^{-2}$
(m) $(3x)^{-1}$　(n) $-x^{-2}$

16. If $x = 3$ and $y = -2$, find the value of each expression.

(a) $x^2 + y^2$
(b) $x^2 - y^3$
(c) $(x + y)^2$
(d) $(x - y)^2$
(e) $2(x^2 - y^3)^2$
(f) $-2x(x + 3y)^2$

17. Simplify.

(a) $(3x^2)(2x^3)$
(b) $(-2x^2)(x^3)$
(c) $(-x^3)(-3x^3)$
(d) $(3x^2)(-2x^3)$
(e) $-15x^4 \div 3x$
(f) $-20x^5 \div (-2x^3)$
(g) $(5a^3b^2)(-4ab) \div (-2a^2b^3)$

18. Write in scientific notation.

(a) 253.4
(b) 36 200
(c) 45.6
(d) 1 356 000
(e) 0.356
(f) 0.0256
(g) 0.005 75
(h) 0.000 789

19. Calculate.

(a) $(2.65 \times 10^4) \times (1.08 \times 10^3)$
(b) $(4.25 \times 10^3) \times (5.85 \times 10^{-2})$
(c) $(3.65 \times 10^{-3}) \times (7.05 \times 10^5)$
(d) $(1.25 \times 10^4) \times (3.65 \times 10^{-2})$

20. Evaluate.

(a) $\sqrt{121}$
(b) $\sqrt{169}$
(c) $\sqrt{324}$
(d) $\sqrt{400}$
(e) $\sqrt{625}$
(f) $\sqrt{1024}$

21. Approximate to the nearest tenth.

(a) $\sqrt{500}$
(b) $\sqrt{350}$
(c) $\sqrt{150}$
(d) $\sqrt{75}$
(e) $\sqrt{5.7}$
(f) $\sqrt{3.8}$

22. The formula for the speed of a car, immediately before braking, is

$$S = \sqrt{250\,d\,f}$$

where S is the speed in kilometres per hour, d is the length of the skid in metres, and f is the road condition factor.
What is the speed of a car with length of skid 30 m and road condition factor 0.75?

23. A ride in a horse-drawn carriage costs \$10 for the first kilometre and \$6.50 for each additional kilometre.

(a) Write a formula for the cost of a ride according to the distance travelled.
(b) Use the formula to determine the cost of a seven kilometre ride.

24. Simplify.

(a) $4a + 3b - 7a - 6b$
(b) $3x - 2y - 7 - 4x - y + 11$
(c) $4xy - 3x - 2y + 6xy - x - y$

25. If $a = 2$, $b = -2$, and $c = -1$, evaluate the following.

(a) $3ab - 2bc - 4ac$
(b) $a^2 + b^2 + c^2$
(c) $3a^2 - 2b^2 - 3ac$

26. Simplify.

(a) $2(x - 4) + 5(x + 7) - (x - 4)$
(b) $2a(a + 2) - 3a(a + 1) - 7a$
(c) $9 - 3x(x - 2) - x(x - 1)$
(d) $3(x^2 - 2x - 1) - 2(x^2 + 3x + 2)$

27. Solve and check.

(a) $3(m + 1) = 7 + m$
(b) $4(x - 1) - 2 = 2(x + 1)$
(c) $4(t - 3) - (t + 1) = 2$
(d) $\frac{x + 1}{2} = 5$
(e) $\frac{m + 2}{3} = \frac{m - 2}{6}$

28. The sum of three consecutive whole numbers is 186.
Find the numbers.

29. Expand.

(a) $(x + 4)(x + 7)$
(b) $(m - 4)^2$
(c) $(b - 3)(b + 3)$
(d) $(2x + 1)(3x + 4)$

30. Expand and simplify.

(a) $2(x - 2)(x + 3) - 3(x + 5)$
(b) $(t - 4)(t + 5) - (t + 6)(t - 9)$
(c) $2(m + 4)(m - 1) + (3m + 2)(m + 1)$
(d) $4(s + 2)^2 - 3(s + 5)(s - 6)$

31. Divide.

(a) $\frac{8x^2 - 4x}{4x}$
(b) $\frac{3xy - 6x^2y + 9xy^2}{3xy}$

32. Factor.

(a) $x^2 - 16$
(b) $x^2 - 7x + 12$
(c) $m^2 + 8m + 15$
(d) $y^2 + y - 6$
(e) $s^2 - 100$
(f) $a^2 + a - 30$

33. In the ordered pair (7, 4) the second element is 3 less than the first.
List four other ordered pairs with this characteristic and draw the graph.

34. The table gives the distance travelled by a search-and-rescue plane travelling at 300 km/h.

Time (h)	Distance (km)
1	300
2	600
3	900

(a) Display this information on a graph.
(b) Interpolate to find the distance travelled in 1.25 h.
(c) Extrapolate to find the distance travelled in 3.75 h.

35. The cost for a guide who takes you on a tour of the Everglades varies directly as the number of hours of the tour.
If the cost is $185 to hire a guide for five hours, how much will the cost be to hire a guide for eight hours?

36. A computer technician charges $70 for a house call and $45 for each hour of work.
(a) Write an equation relating the cost and the number of hours.
(b) What would a 6.5 h house call cost?

37. Graph each of the following.
(a) $y = 3x - 4$ (b) $x + y = 6$
(c) $2x - 3y = 6$ (d) $4 - y = 2x$
(e) $y = x^2 - 4$ (f) $x^2 + y^2 = 9$

38. Find the missing number in each ordered pair so that the ordered pair satisfies the given equation.
(a) $y = 2x - 7$
(4, ■) (■, −5) (−2, ■)
(b) $2x - 3y = 6$
(1, ■) (■, 4) (−3, ■)

39. Solve each pair of equations graphically.
(a) $y = 2x - 3$, $y = x + 3$ (b) $x + y = 5$, $y = x - 3$
(c) $x - y = 2$, $2x + 3y = 9$ (d) $3x + 2y = 7$, $2x + 3y = 8$

40. Solve each pair of equations.
(a) $4x - 3y = 1$, $2x + 3y = 5$ (b) $3x + 2y = 1$, $x - 2y = 3$
(c) $5x + 2y = 4$, $2x - 3y = 13$ (d) $-2x - 3y = 0$, $x + y = 5$

41. The sum of two numbers is 7. One number is 1 more than the other.
Find the numbers.

42. Twice one number plus another number is 1. The sum of the numbers is −2.
What are the numbers?

43. A picture frame is 7 cm longer than it is wide. The perimeter of the frame is 34 cm.
What are the dimensions of the frame?

44. One number is twice another. Four times the smaller number added to twice the larger is 16.
Find the numbers.

45. A soccer pitch is 27 m longer than it is wide. The perimeter of the pitch is 346 m.
Find the dimensions of the soccer pitch.

46. How would you collect data to determine the amount of money that tourists from the United States spend in your province each year?

47. The Pick company makes ballpoint pens. They make 50 000 every day, six days a week. A sample of 400 pens were tested in one week and 7 were found to be defective.
Predict how many pens in the week's production will be defective.

48. Determine the mean, median, and mode of the following test marks.
74, 70, 76, 82, 79, 78, 74, 78, 74

49. Determine the mean, median, and mode of each set of data.
(a) 10, 20, 19, 17, 10, 14, 15, 21
(b) 3, 4, 5, 4, 3, 5, 3, 4, 5, 4
(c) 23, 24, 25, 26, 25, 24, 23, 25, 23

50. The following table gives Sarah's bowling scores for four weeks.

Week	Game 1	Game 2	Game 3
1	200	203	200
2	198	185	196
3	174	193	188
4	204	202	185

Determine her average at the end of each week.

(a) Week 1 (b) Week 2
(c) Week 3 (d) Week 4

51. The table gives the number of students in Central High School by grade.
Display this information in a circle graph.

Grade	9	10	11	12
Number	256	245	220	201

52. The table gives the population of Canada, rounded to the nearest tenth of a million since 1871 when the first census was taken.

Year	Population (millions)
1871	3.7
1881	4.3
1891	4.8
1901	5.4
1911	7.2
1921	8.8
1931	10.4
1941	11.5
1951	14.0
1961	18.2
1971	21.6
1981	24.3

(a) Display this information in a line graph.
(b) In what ten year span was the population growth the largest?
(c) In what ten year span was the population growth the smallest?

53. Complete the following simple interest table. Interest is calculated at a per annum rate.

	I ($)	P ($)	r (%)	t
(a)	675.00	7500.00	■	4 months
(b)	98.00	■	9	6 months
(c)	957.00	97 000	12	■
(d)	120.00	32 000	■	60 d
(e)	104.00	■	11	30 d
(f)	■	175 000	8.5	1 a
(g)	■	225 000	10.7	6 months

54. Mort deposits $20 000 into an account that pays 9.5% interest compounded semi-annually.
What is the amount after 2.5 a?

55. Joseph puts $25 000 into a term deposit that pays 12% compounded quarterly.
What is the amount after 15 months?

56. Meredith buys $14 000 of Compound Interest Canada Savings Bonds which pay 9.5% interest.

(a) What amount will Meredith receive if the bonds are cashed in 3 a?
(b) How much interest is earned?

57. Use the table to find an estimate of the monthly payment for each automobile loan.

Monthly Payment Factor 15%	
Number of Payments	Factor
18	0.062 384 8
24	0.048 486 6
30	0.040 178 5
36	0.034 665 3

(a) $32 463.00 at 15% for 24 months
(b) $28 465.00 at 15% for 30 months
(c) $18 650.00 at 15% for 18 months
(d) $33 000.00 at 15% for 30 months
(e) $19 250.00 at 15% for 36 months
(f) $10 428.00 at 15% for 24 months

58. Compare the monthly payments on the following car loans.
* $25 000 for 24 months at 15%
* $18 750 for 30 months at 15%
* $16 666 for 36 months at 15%

For which loan would you pay the least amount of interest?

59. Find the missing term in each.

(a) $x : 8 = 3 : 4$ (b) $5 : 6 = y : 12$

(c) $12 : x : y = 4 : 3 : 2$

(d) $x : 10 : 20 = 10 : y : 40$

60. Denzil and Leslie invest in an antique car. The cost of the car is $25 000. Denzil and Leslie share the cost in the ratio 3 : 2. How much does each of them pay?

61. The Alpine Arena sells three types of seats: gold, red, and blue. The numbers of seats are in the ratio 2 : 3 : 7. If there are 4800 seats in the arena, how many of each type are there?

62. Calculate the unit rate for each of the following.

(a) $36.75 for seven litres of syrup

(b) $88.20 for nine hours of work

(c) 498 km driven in 6 h

63. The building is drawn to a scale of 1 : 1000. How tall is the building?

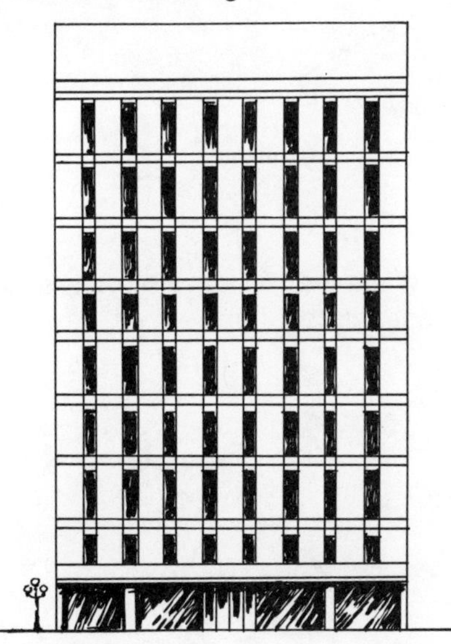

64. Find the measure of the unknown angles.

(a) (b)

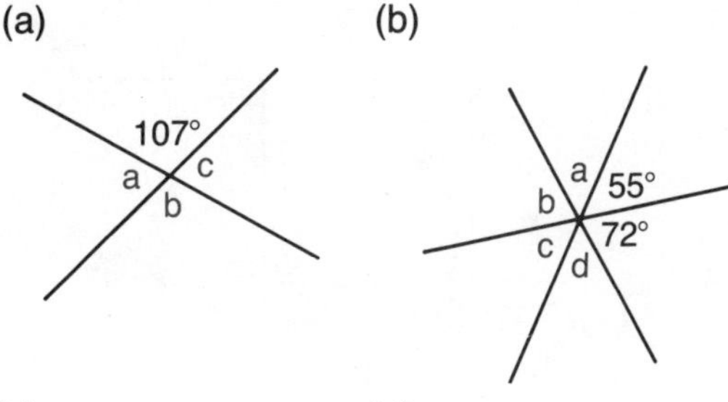

(c) (d)

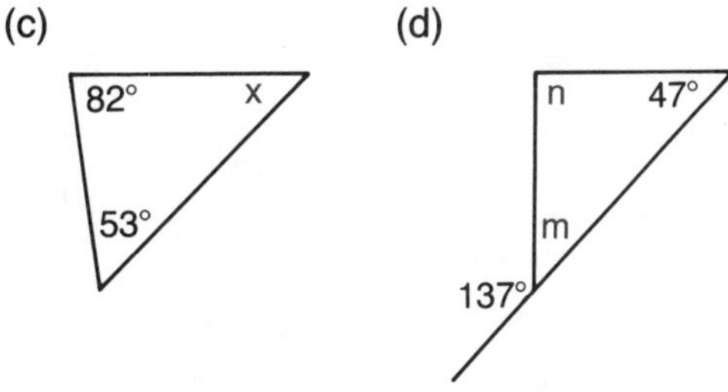

65. Find the measure of the unknown angles.

(a) (b)

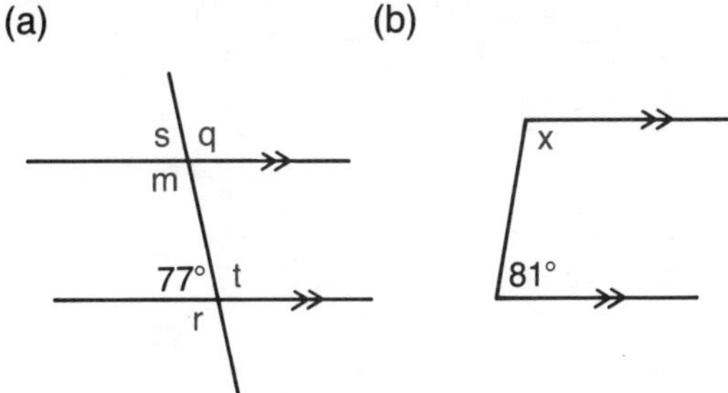

(c) (d)

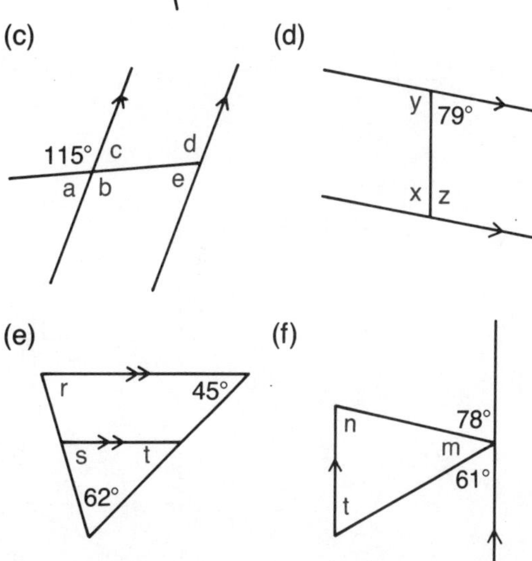

(e) (f)

66. Find the measure of the missing angles.

(a)

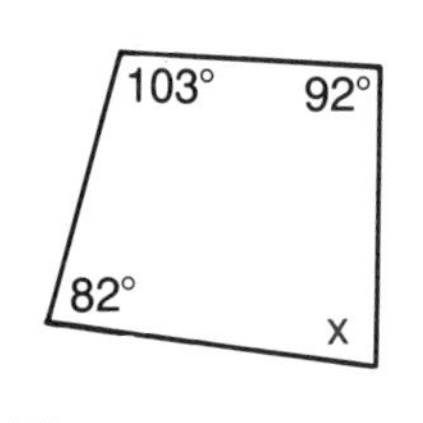

(b)

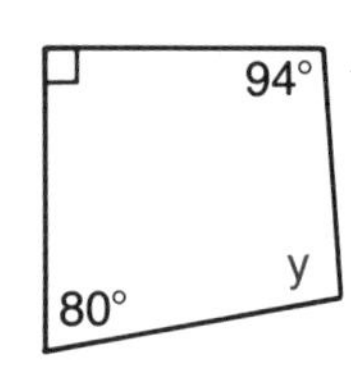

(c)

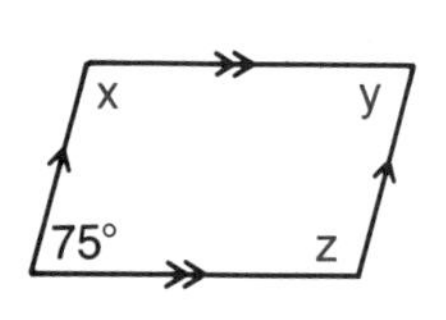

(d)

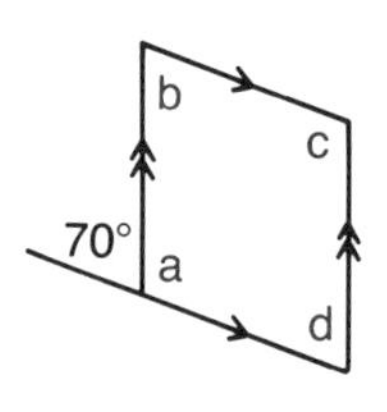

67. Find the unknown lengths and angles.

(a)

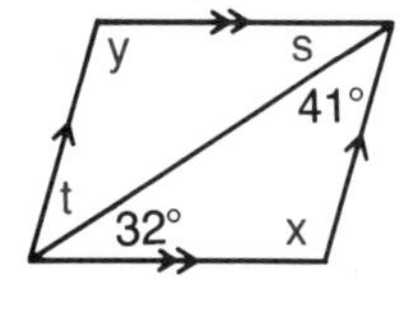

(b)

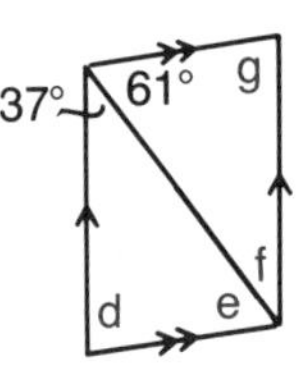

(c)

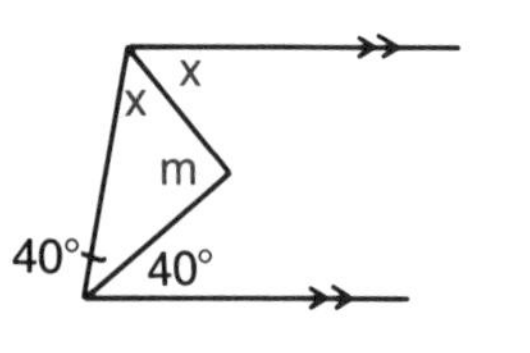

(d)

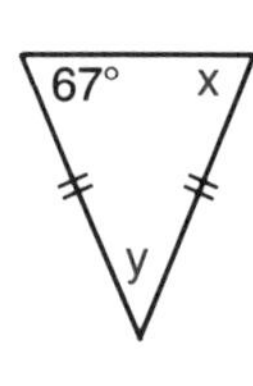

(e)

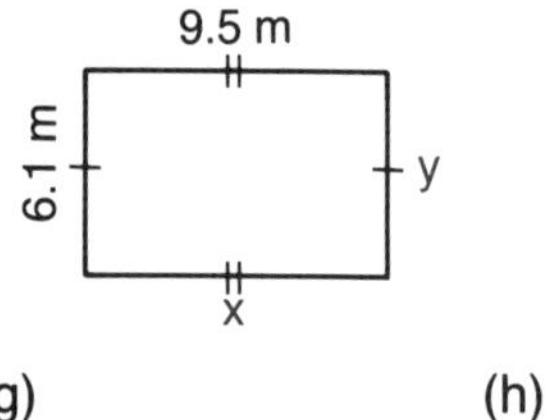

(f)

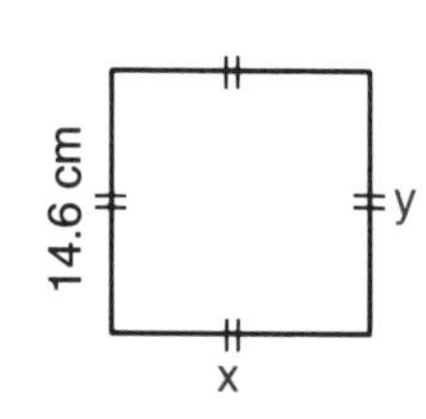

(g)

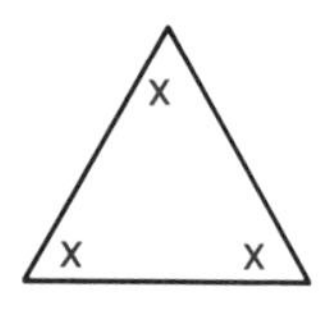

(h)

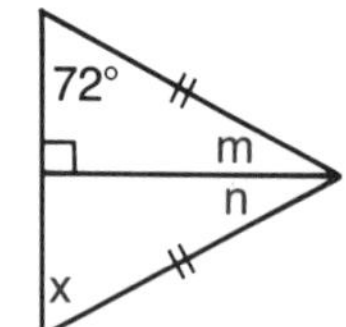

68. Draw angles similar to the ones shown and bisect each angle.

(a) (b)

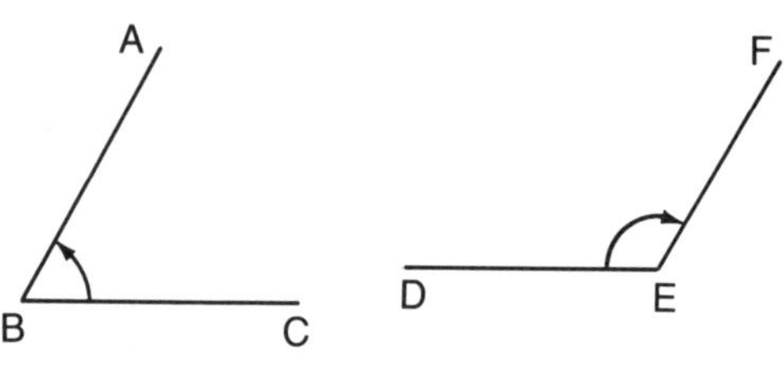

69. Draw line segments similar to the ones shown and draw the right bisector of each.

(a) (b)

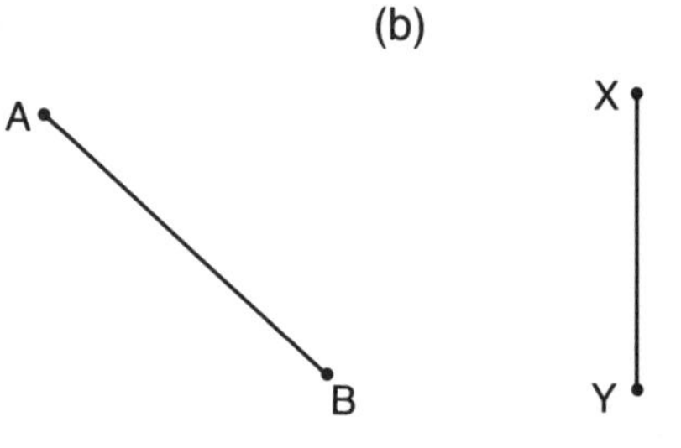

70. Construct a perpendicular from A to the line segment BC.

A •

71. Construct a perpendicular to the line segment ST at R.

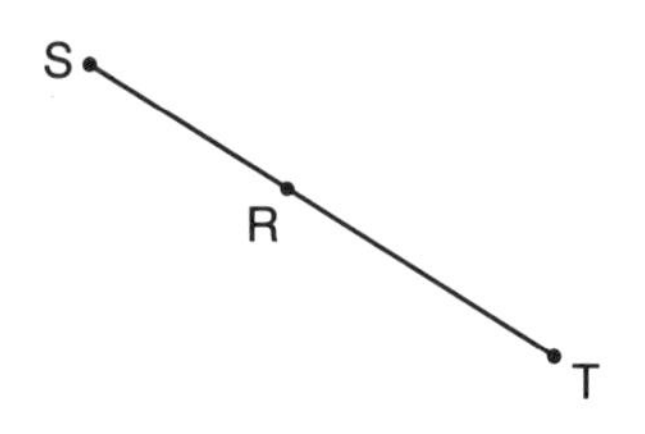

72. Draw angles similar to the ones shown and construct equal angles.

(a) (b)

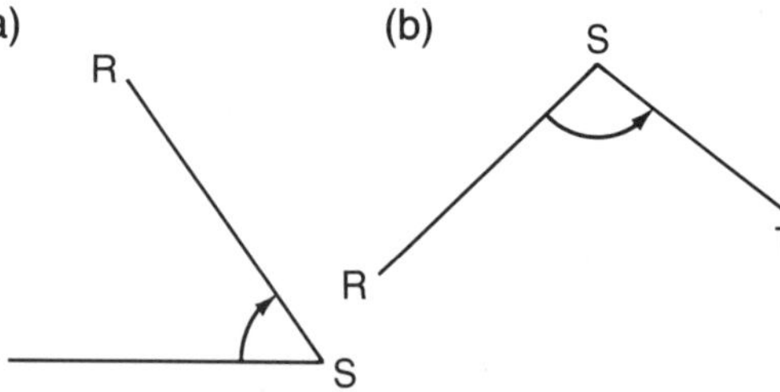

73. Calculate the perimeter of each figure.
(a) (b) (c) (d)

74. Calculate the area of each figure.
(a) (b) (c) (d)

75. Calculate the volume of each solid.
(a) (b) (c) (d)

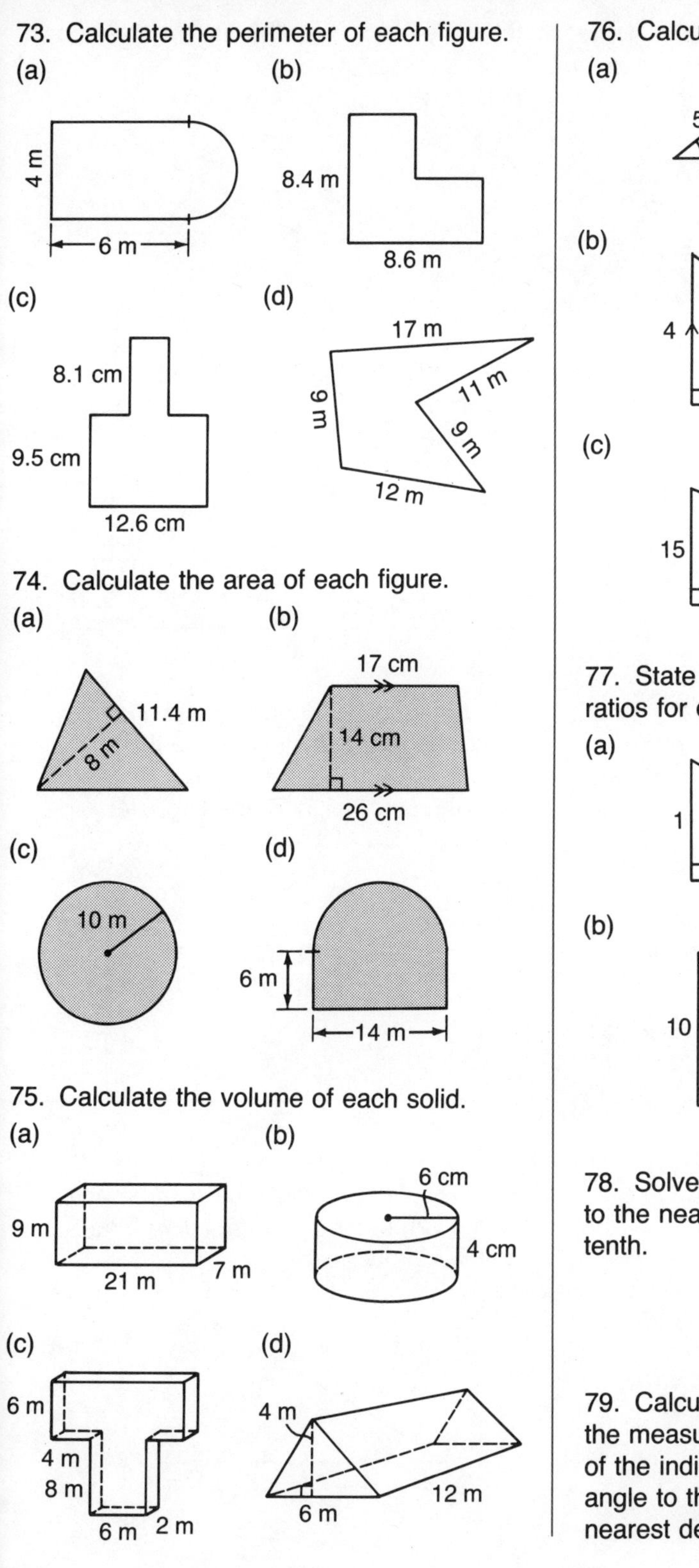

76. Calculate the lengths of each side.

(a)

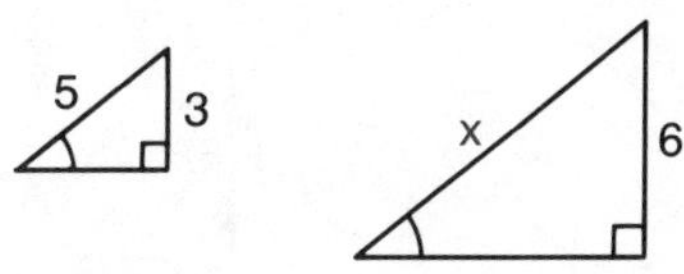

(b)

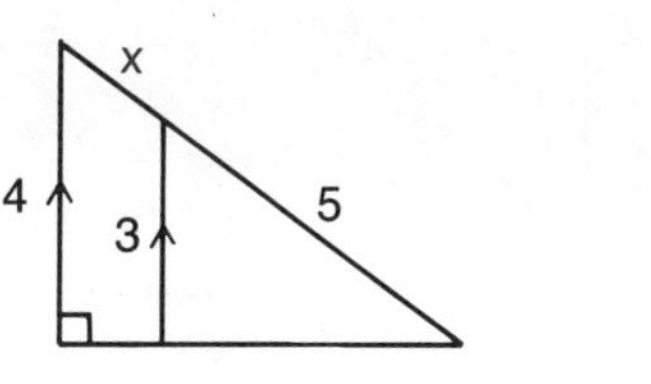

(c)

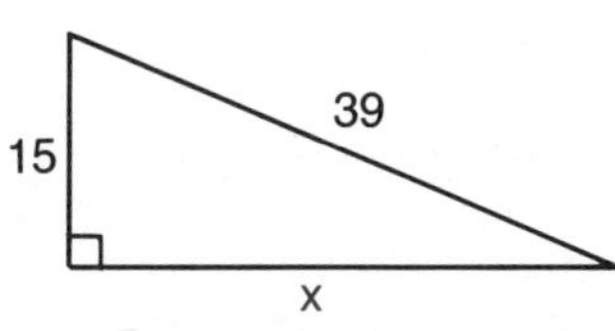

77. State the sine, cosine, and tangent ratios for each indicated angle.

(a)

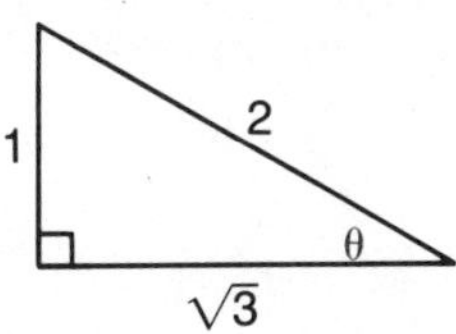

(b)

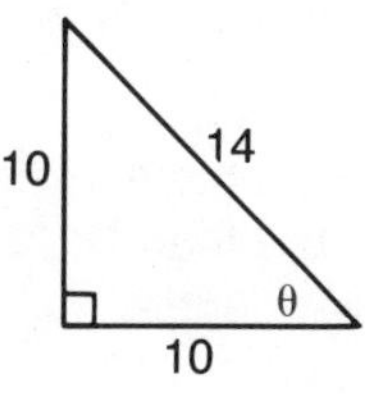

78. Solve for a to the nearest tenth.

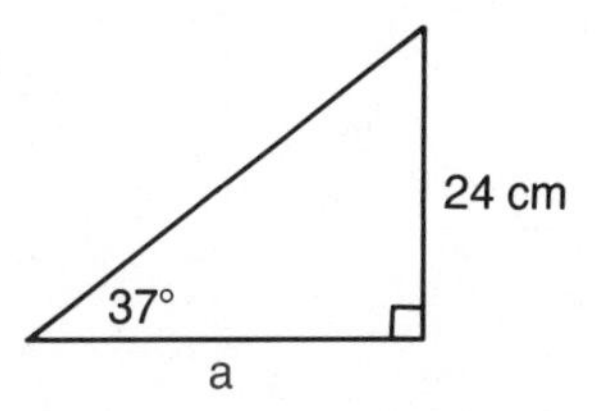

79. Calculate the measure of the indicated angle to the nearest degree.

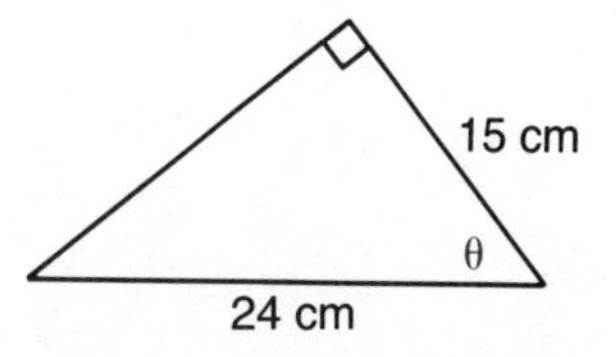

80. Solve for the variable in each triangle.

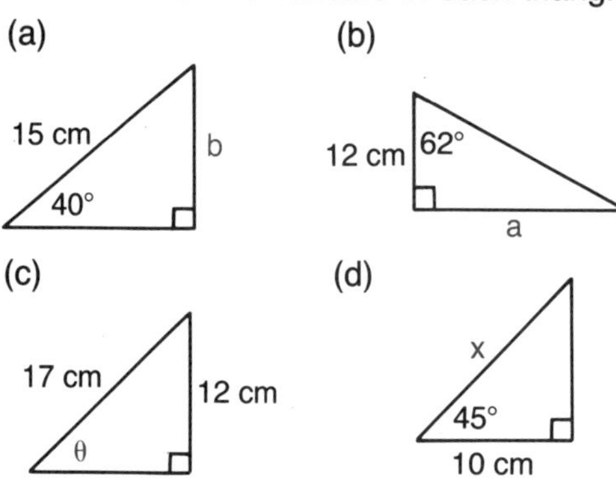

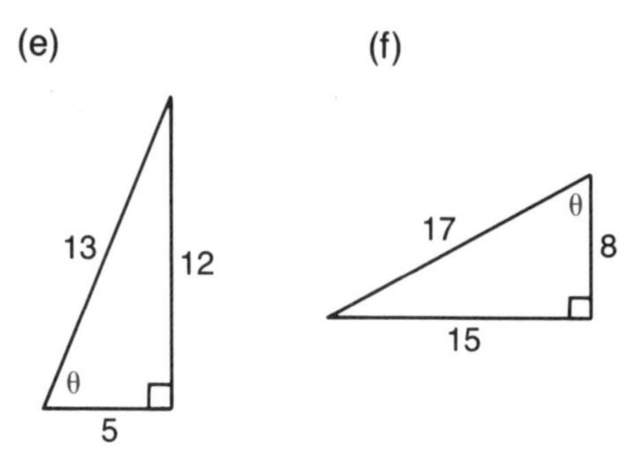

81. From the top of a building 102 m high the angle of depression of the riverbank is 32°.
How far is the riverbank from the foot of the building?

82. A utility pole is 14.7 m high and casts a shadow 21.2 m long.
What is the angle of elevation to the sun?

83. From a point 124 m from the foot of a tree the angle of elevation to the top of the tree is 52°.
What is the height of the tree?

84. A corner lot is 30 m by 52 m.
How much shorter is it to walk across the lot diagonally, rather than to walk around the corner?

85. Calculate the value of tan θ in each of the following triangles and then find the value of θ to the nearest degree.

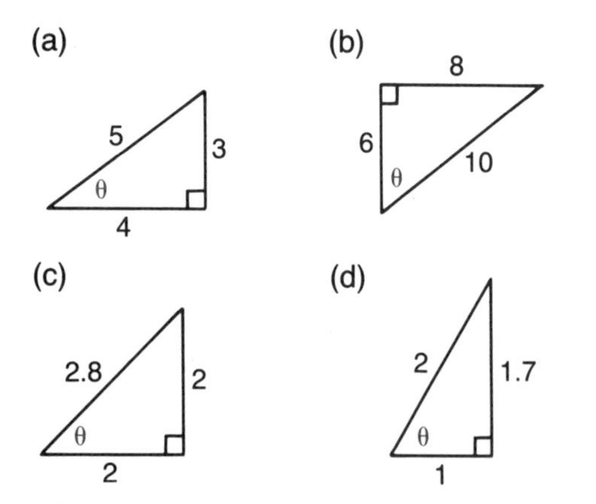

86. A building 11.0 m high casts a 27.0 m shadow.
Calculate θ.

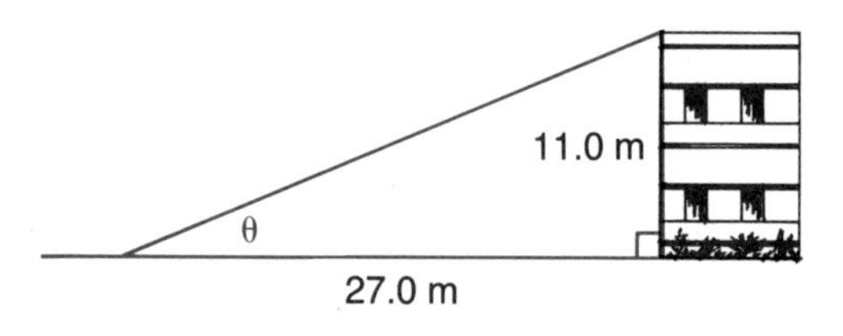

87. A flagpole casts a shadow of 22.0 m when the angle of elevation to the sun is 40°.
Calculate the height of the flagpole.

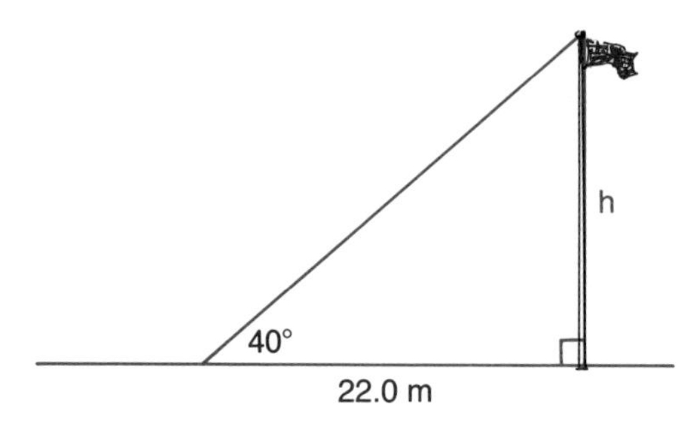

88. Calculate the distance AB from the data given in the diagram. (*Hint*: Find AD and BD separately.)

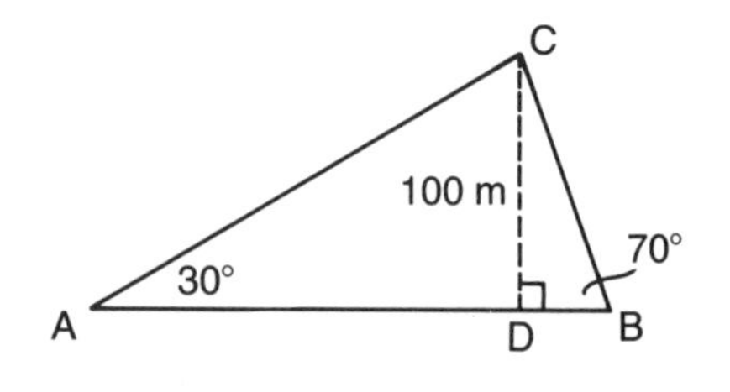

89. Find the slope of each line segment.

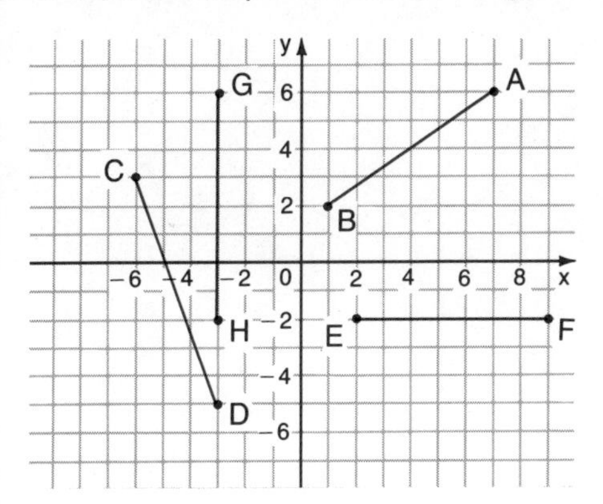

90. Calculate the slope of the line passing through each pair of points.

(a) (3, 4) and (1, 2)
(b) (0, −5) and (−5, 6)
(c) (5, −7) and (5, 3)
(d) (−6, −2) and (4, −2)
(e) (−3, −1) and (−8, −2)

91. Determine if the line segments are perpendicular, parallel, or neither.

(a) A(4, 5), B(7, 6) and C(−2, 1), D(1, 2)
(b) M(2, 2), N(4, 5) and R(1, 6), T(−2, 8)
(c) P(−3, 0), Q(5, −1) and G(2, 1), H(3, 9)
(d) E(4, −2), F(6, 1) and V(3, 4), W(1, 7)
(e) S(0, 5), T(4, 6) and A(−1, 4), B(0, 8)
(f) D(−1, 3), E(4, 3) and F(2, 1), G(5, 1)

92. Calculate the length of each line segment.

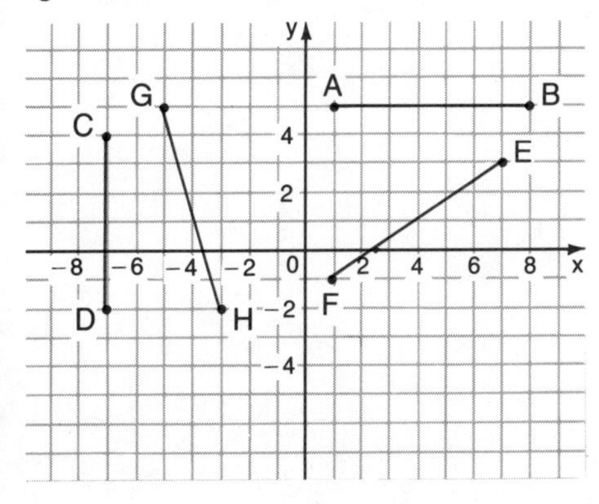

93. Calculate the length of each line segment to the nearest tenth.

(a) A(4, 5) to B(7, 9)
(b) D(−3, 5) to E(0, 0)
(c) F(−2, −1) to G(−4, −6)
(d) R(−7, 0) to S(0, 2)
(e) M(5, 5) to N(−2, −4)
(f) P(4, 2) to Q(−3, 2)
(g) S(−1, 7) to T(−1, −3)

94. The vertices of a triangle are P(−2, 1), Q(3, 4), and R(1, 6).
Verify that the triangle is isoceles.

95. The vertices of a quadrilateral are A(2, 4), B(−5, −1), C(−7, −4), and D(0, 1).
Verify that the quadrilateral is a parallelogram.

96. (a) Plot the points A(3, 2), B(−1, 5), C(−7, −3), and D(−3, −6).
(b) Join the points in the order
A–B–C–D–A.
(c) Verify that the opposite sides of the figure are parallel.
(d) Verify that all the angles of the figure ABCD are right angles.
(e) Verify that the opposite sides of the figure are equal in length.

97. Copy and complete each table of values, and then graph the equation.

(a)

x	$y = 2x + 3$
−1	
0	
1	
2	
3	

(b)

x	$y = -2x + 7$
−2	
−1	
0	
1	
2	

APPENDIX

TABLE I SQUARE ROOTS

n	$\sqrt{n}$	n	$\sqrt{n}$	n	$\sqrt{n}$	n	$\sqrt{n}$
1	1.000	51	7.141	101	10.050	151	12.288
2	1.414	52	7.211	102	10.010	152	12.329
3	1.732	53	7.280	103	10.149	153	12.369
4	2.000	54	7.349	104	10.198	154	12.410
5	2.236	55	7.416	105	10.247	155	12.450
6	2.450	56	7.483	106	10.296	156	12.490
7	2.646	57	7.550	107	10.344	157	12.530
8	2.828	58	7.616	108	10.392	158	12.570
9	3.000	59	7.681	109	10.440	159	12.610
10	3.162	60	7.746	110	10.488	160	12.649
11	3.317	61	7.810	111	10.536	161	12.689
12	3.464	62	7.874	112	10.583	162	12.728
13	3.606	63	7.937	113	10.630	163	12.767
14	3.742	64	8.000	114	10.677	164	12.806
15	3.873	65	8.062	115	10.724	165	12.845
16	4.000	66	8.124	116	10.770	166	12.884
17	4.123	67	8.185	117	10.817	167	12.923
18	4.243	68	8.246	118	10.863	168	12.961
19	4.359	69	8.307	119	10.909	169	13.000
20	4.472	70	8.367	120	10.954	170	13.038
21	4.583	71	8.426	121	11.000	171	13.077
22	4.690	72	8.485	122	11.045	172	13.115
23	4.796	73	8.544	123	11.091	173	13.153
24	4.899	74	8.602	124	11.136	174	13.191
25	5.000	75	8.660	125	11.180	175	13.229
26	5.099	76	8.718	126	11.225	176	13.266
27	5.196	77	8.775	127	11.269	177	13.304
28	5.292	78	8.832	128	11.314	178	13.342
29	5.385	79	8.888	129	11.358	179	13.379
30	5.477	80	8.944	130	11.402	180	13.416
31	5.568	81	9.000	131	11.446	181	13.454
32	5.657	82	9.055	132	11.489	182	13.491
33	5.745	83	9.110	133	11.533	183	13.528
34	5.831	84	9.165	134	11.576	184	13.565
35	5.916	85	9.220	135	11.619	185	13.601
36	6.000	86	9.274	136	11.662	186	13.638
37	6.083	87	9.327	137	11.705	187	13.675
38	6.164	88	9.381	138	11.747	188	13.711
39	6.245	89	9.434	139	11.790	189	13.748
40	6.325	90	9.487	140	11.832	190	13.784
41	6.403	91	9.539	141	11.874	191	13.820
42	6.481	92	9.592	142	11.916	192	13.856
43	6.557	93	9.644	143	11.958	193	13.892
44	6.633	94	9.695	144	12.000	194	13.928
45	6.708	95	9.747	145	12.042	195	13.964
46	6.782	96	9.798	146	12.083	196	14.000
47	6.856	97	9.849	147	12.124	197	14.036
48	6.928	98	9.900	148	12.166	198	14.071
49	7.000	99	9.950	149	12.207	199	14.107
50	7.071	100	10.000	150	12.247	200	14.142

TABLE II ONTARIO TAX DEDUCTIONS

ONTARIO

WEEKLY TAX DEDUCTIONS

Basis — 52 Pay Periods per Year

WEEKLY PAY Use appropriate bracket — PAIE PAR SEMAINE *Utilisez le palier approprié*	IF THE EMPLOYEE'S "NET CLAIM CODE" ON FORM TD1 IS *SI LE CODE DE DEMANDE NETTE DE L'EMPLOYÉ SELON LA FORMULE TD1 EST DE*										
	0	1	2	3	4	5	6	7	8	9	10
From - *De* Less than *Moins que*	DEDUCT FROM EACH PAY — *RETENEZ SUR CHAQUE PAIE*										
228.- 232.	60.60	27.70	23.90	16.40	6.15	.90					
232.- 236.	61.65	28.70	24.95	17.40	7.85	1.55					
236.- 240.	62.70	29.75	25.95	18.45	9.55	2.25					
240.- 244.	63.75	30.75	26.95	19.45	11.25	2.90					
244.- 248.	64.80	31.75	27.95	20.45	12.90	3.55					
248.- 252.	65.90	32.75	29.00	21.45	13.90	4.25					
252.- 256.	66.95	33.75	30.00	22.50	14.95	4.90					
256.- 260.	68.00	34.80	31.00	23.50	15.95	5.60	.60				
260.- 264.	69.05	35.80	32.00	24.50	16.95	7.10	1.25				
264.- 268.	70.10	36.80	33.00	25.50	17.95	8.75	1.95				
268.- 272.	71.15	37.80	34.05	26.50	19.00	10.45	2.60				
272.- 276.	72.20	38.85	35.05	27.55	20.00	12.15	3.30				
276.- 280.	73.25	39.85	36.05	28.55	21.00	13.45	3.95				
280.- 284.	74.30	40.85	37.05	29.55	22.00	14.45	4.65				
284.- 288.	75.35	41.85	38.10	30.55	23.05	15.50	5.30	.30			
288.- 292.	76.40	42.90	39.10	31.60	24.05	16.50	6.35	.95			
292.- 296.	77.45	43.90	40.10	32.60	25.05	17.50	8.05	1.65			
296.- 300.	78.50	44.90	41.10	33.60	26.05	18.50	9.75	2.30			
300.- 304.	79.60	45.90	42.10	34.60	27.05	19.55	11.45	3.00			
304.- 308.	80.65	46.90	43.15	35.60	28.10	20.55	13.05	3.65			
308.- 312.	81.70	47.95	44.15	36.65	29.10	21.55	14.05	4.30			
312.- 316.	82.75	48.95	45.15	37.65	30.10	22.55	15.05	5.00			
316.- 320.	83.80	49.95	46.15	38.65	31.10	23.60	16.05	5.65	.65		
320.- 324.	84.85	50.95	47.20	39.65	32.15	24.60	17.05	7.30	1.35		
324.- 328.	85.90	52.00	48.20	40.70	33.15	25.60	18.10	8.95	2.00		
328.- 332.	86.95	53.00	49.20	41.70	34.15	26.60	19.10	10.65	2.70		
332.- 336.	88.00	54.00	50.20	42.70	35.15	27.60	20.10	12.35	3.35		
336.- 340.	89.05	55.00	51.25	43.70	36.15	28.65	21.10	13.60	4.05		
340.- 344.	90.10	56.00	52.25	44.75	37.20	29.65	22.15	14.60	4.70		
344.- 348.	91.15	57.05	53.25	45.75	38.20	30.65	23.15	15.60	5.40	.35	
348.- 352.	92.25	58.05	54.25	46.75	39.20	31.65	24.15	16.60	6.55	1.05	
352.- 356.	93.30	59.05	55.25	47.75	40.20	32.70	25.15	17.60	8.25	1.70	
356.- 360.	94.35	60.05	56.30	48.75	41.25	33.70	26.20	18.65	9.95	2.40	
360.- 364.	95.40	61.10	57.30	49.80	42.25	34.70	27.20	19.65	11.60	3.05	
364.- 368.	96.45	62.10	58.30	50.80	43.25	35.70	28.20	20.65	13.15	3.70	
368.- 372.	97.50	63.10	59.30	51.80	44.25	36.70	29.20	21.65	14.15	4.40	
372.- 376.	98.55	64.10	60.35	52.80	45.25	37.75	30.20	22.70	15.15	5.05	.05
376.- 380.	99.60	65.10	61.35	53.85	46.30	38.75	31.25	23.70	16.20	5.80	.75
380.- 384.	100.65	66.15	62.35	54.85	47.30	39.75	32.25	24.70	17.20	7.45	1.40
384.- 388.	101.70	67.15	63.35	55.85	48.30	40.75	33.25	25.70	18.20	9.15	2.05
388.- 392.	102.75	68.15	64.35	56.85	49.30	41.80	34.25	26.75	19.20	10.85	2.75
392.- 396.	103.80	69.15	65.40	57.85	50.35	42.80	35.30	27.75	20.25	12.55	3.40
396.- 400.	104.85	70.20	66.40	58.90	51.35	43.80	36.30	28.75	21.25	13.70	4.10
400.- 404.	105.95	71.20	67.40	59.90	52.35	44.80	37.30	29.75	22.25	14.70	4.75
404.- 408.	107.00	72.20	68.40	60.90	53.35	45.80	38.30	30.75	23.25	15.70	5.45
408.- 412.	108.05	73.20	69.45	61.90	54.40	46.85	39.30	31.80	24.25	16.75	6.70
412.- 416.	109.10	74.25	70.45	62.95	55.40	47.85	40.35	32.80	25.30	17.75	8.40
416.- 420.	110.15	75.25	71.45	63.95	56.40	48.85	41.35	33.80	26.30	18.75	10.10
420.- 424.	111.20	76.25	72.45	64.95	57.40	49.85	42.35	34.80	27.30	19.75	11.75
424.- 428.	112.25	77.25	73.45	65.95	58.40	50.90	43.35	35.85	28.30	20.75	13.25
428.- 432.	113.30	78.25	74.50	66.95	59.45	51.90	44.40	36.85	29.35	21.80	14.25
432.- 436.	114.35	79.30	75.50	68.00	60.45	52.90	45.40	37.85	30.35	22.80	15.25
436.- 440.	115.40	80.30	76.50	69.00	61.45	53.90	46.40	38.85	31.35	23.80	16.25
440.- 444.	116.45	81.30	77.50	70.00	62.45	54.95	47.40	39.85	32.35	24.80	17.30
444.- 448.	117.50	82.30	78.55	71.00	63.50	55.95	48.45	40.90	33.35	25.85	18.30

TABLE III CANADA PENSION PLAN CONTRIBUTIONS

WEEKLY PAY PERIOD

188.83 — 325.96

Remuneration *Rémunération*		C.P.P. *R.P.C.*	Remuneration *Rémunération*		C.P.P. *R.P.C.*	Remuneration *Rémunération*		C.P.P. *R.P.C.*	Remuneration *Rémunération*		C.P.P. *R.P.C.*
From-*de*	To-*à*		From-*de*	To-*à*		From-*de*	To-*à*		From-*de*	To-*à*	
188.83 -	189.30	2.88	223.12 -	223.58	3.60	257.40 -	257.87	4.32	291.69 -	292.15	5.04
189.31 -	189.77	2.89	223.59 -	224.06	3.61	257.88 -	258.34	4.33	292.16 -	292.63	5.05
189.78 -	190.25	2.90	224.07 -	224.53	3.62	258.35 -	258.82	4.34	292.64 -	293.11	5.06
190.26 -	190.72	2.91	224.54 -	225.01	3.63	258.83 -	259.30	4.35	293.12 -	293.58	5.07
190.73 -	191.20	2.92	225.02 -	225.49	3.64	259.31 -	259.77	4.36	293.59 -	294.06	5.08
191.21 -	191.68	2.93	225.50 -	225.96	3.65	259.78 -	260.25	4.37	294.07 -	294.53	5.09
191.69 -	192.15	2.94	225.97 -	226.44	3.66	260.26 -	260.72	4.38	294.54 -	295.01	5.10
192.16 -	192.63	2.95	226.45 -	226.91	3.67	260.73 -	261.20	4.39	295.02 -	295.49	5.11
192.64 -	193.11	2.96	226.92 -	227.39	3.68	261.21 -	261.68	4.40	295.50 -	295.96	5.12
193.12 -	193.58	2.97	227.40 -	227.87	3.69	261.69 -	262.15	4.41	295.97 -	296.44	5.13
193.59 -	194.06	2.98	227.88 -	228.34	3.70	262.16 -	262.63	4.42	296.45 -	296.91	5.14
194.07 -	194.53	2.99	228.35 -	228.82	3.71	262.64 -	263.11	4.43	296.92 -	297.39	5.15
194.54 -	195.01	3.00	228.83 -	229.30	3.72	263.12 -	263.58	4.44	297.40 -	297.87	5.16
195.02 -	195.49	3.01	229.31 -	229.77	3.73	263.59 -	264.06	4.45	297.88 -	298.34	5.17
195.50 -	195.96	3.02	229.78 -	230.25	3.74	264.07 -	264.53	4.46	298.35 -	298.82	5.18
195.97 -	196.44	3.03	230.26 -	230.72	3.75	264.54 -	265.01	4.47	298.83 -	299.30	5.19
196.45 -	196.91	3.04	230.73 -	231.20	3.76	265.02 -	265.49	4.48	299.31 -	299.77	5.20
196.92 -	197.39	3.05	231.21 -	231.68	3.77	265.50 -	265.96	4.49	299.78 -	300.25	5.21
197.40 -	197.87	3.06	231.69 -	232.15	3.78	265.97 -	266.44	4.50	300.26 -	300.72	5.22
197.88 -	198.34	3.07	232.16 -	232.63	3.79	266.45 -	266.91	4.51	300.73 -	301.20	5.23
198.35 -	198.82	3.08	232.64 -	233.11	3.80	266.92 -	267.39	4.52	301.21 -	301.68	5.24
198.83 -	199.30	3.09	233.12 -	233.58	3.81	267.40 -	267.87	4.53	301.69 -	302.15	5.25
199.31 -	199.77	3.10	233.59 -	234.06	3.82	267.88 -	268.34	4.54	302.16 -	302.63	5.26
199.78 -	200.25	3.11	234.07 -	234.53	3.83	268.35 -	268.82	4.55	302.64 -	303.11	5.27
200.26 -	200.72	3.12	234.54 -	235.01	3.84	268.83 -	269.30	4.56	303.12 -	303.58	5.28
200.73 -	201.20	3.13	235.02 -	235.49	3.85	269.31 -	269.77	4.57	303.59 -	304.06	5.29
201.21 -	201.68	3.14	235.50 -	235.96	3.86	269.78 -	270.25	4.58	304.07 -	304.53	5.30
201.69 -	202.15	3.15	235.97 -	236.44	3.87	270.26 -	270.72	4.59	304.54 -	305.01	5.31
202.16 -	202.63	3.16	236.45 -	236.91	3.88	270.73 -	271.20	4.60	305.02 -	305.49	5.32
202.64 -	203.11	3.17	236.92 -	237.39	3.89	271.21 -	271.68	4.61	305.50 -	305.96	5.33
203.12 -	203.58	3.18	237.40 -	237.87	3.90	271.69 -	272.15	4.62	305.97 -	306.44	5.34
203.59 -	204.06	3.19	237.88 -	238.34	3.91	272.16 -	272.63	4.63	306.45 -	306.91	5.35
204.07 -	204.53	3.20	238.35 -	238.82	3.92	272.64 -	273.11	4.64	306.92 -	307.39	5.36
204.54 -	205.01	3.21	238.83 -	239.30	3.93	273.12 -	273.58	4.65	307.40 -	307.87	5.37
205.02 -	205.49	3.22	239.31 -	239.77	3.94	273.59 -	274.06	4.66	307.88 -	308.34	5.38
205.50 -	205.96	3.23	239.78 -	240.25	3.95	274.07 -	274.53	4.67	308.35 -	308.82	5.39
205.97 -	206.44	3.24	240.26 -	240.72	3.96	274.54 -	275.01	4.68	308.83 -	309.30	5.40
206.45 -	206.91	3.25	240.73 -	241.20	3.97	275.02 -	275.49	4.69	309.31 -	309.77	5.41
206.92 -	207.39	3.26	241.21 -	241.68	3.98	275.50 -	275.96	4.70	309.78 -	310.25	5.42
207.40 -	207.87	3.27	241.69 -	242.15	3.99	275.97 -	276.44	4.71	310.26 -	310.72	5.43
207.88 -	208.34	3.28	242.16 -	242.63	4.00	276.45 -	276.91	4.72	310.73 -	311.20	5.44
208.35 -	208.82	3.29	242.64 -	243.11	4.01	276.92 -	277.39	4.73	311.21 -	311.68	5.45
208.83 -	209.30	3.30	243.12 -	243.58	4.02	277.40 -	277.87	4.74	311.69 -	312.15	5.46
209.31 -	209.77	3.31	243.59 -	244.06	4.03	277.88 -	278.34	4.75	312.16 -	312.63	5.47
209.78 -	210.25	3.32	244.07 -	244.53	4.04	278.35 -	278.82	4.76	312.64 -	313.11	5.48
210.26 -	210.72	3.33	244.54 -	245.01	4.05	278.83 -	279.30	4.77	313.12 -	313.58	5.49
210.73 -	211.20	3.34	245.02 -	245.49	4.06	279.31 -	279.77	4.78	313.59 -	314.06	5.50
211.21 -	211.68	3.35	245.50 -	245.96	4.07	279.78 -	280.25	4.79	314.07 -	314.53	5.51
211.69 -	212.15	3.36	245.97 -	246.44	4.08	280.26 -	280.72	4.80	314.54 -	315.01	5.52
212.16 -	212.63	3.37	246.45 -	246.91	4.09	280.73 -	281.20	4.81	315.02 -	315.49	5.53
212.64 -	213.11	3.38	246.92 -	247.39	4.10	281.21 -	281.68	4.82	315.50 -	315.96	5.54
213.12 -	213.58	3.39	247.40 -	247.87	4.11	281.69 -	282.15	4.83	315.97 -	316.44	5.55
213.59 -	214.06	3.40	247.88 -	248.34	4.12	282.16 -	282.63	4.84	316.45 -	316.91	5.56
214.07 -	214.53	3.41	248.35 -	248.82	4.13	282.64 -	283.11	4.85	316.92 -	317.39	5.57
214.54 -	215.01	3.42	248.83 -	249.30	4.14	283.12 -	283.58	4.86	317.40 -	317.87	5.58
215.02 -	215.49	3.43	249.31 -	249.77	4.15	283.59 -	284.06	4.87	317.88 -	318.34	5.59
215.50 -	215.96	3.44	249.78 -	250.25	4.16	284.07 -	284.53	4.88	318.35 -	318.82	5.60
215.97 -	216.44	3.45	250.26 -	250.72	4.17	284.54 -	285.01	4.89	318.83 -	319.30	5.61
216.45 -	216.91	3.46	250.73 -	251.20	4.18	285.02 -	285.49	4.90	319.31 -	319.77	5.62
216.92 -	217.39	3.47	251.21 -	251.68	4.19	285.50 -	285.96	4.91	319.78 -	320.25	5.63
217.40 -	217.87	3.48	251.69 -	252.15	4.20	285.97 -	286.44	4.92	320.26 -	320.72	5.64
217.88 -	218.34	3.49	252.16 -	252.63	4.21	286.45 -	286.91	4.93	320.73 -	321.20	5.65
218.35 -	218.82	3.50	252.64 -	253.11	4.22	286.92 -	287.39	4.94	321.21 -	321.68	5.66
218.83 -	219.30	3.51	253.12 -	253.58	4.23	287.40 -	287.87	4.95	321.69 -	322.15	5.67
219.31 -	219.77	3.52	253.59 -	254.06	4.24	287.88 -	288.34	4.96	322.16 -	322.63	5.68
219.78 -	220.25	3.53	254.07 -	254.53	4.25	288.35 -	288.82	4.97	322.64 -	323.11	5.69
220.26 -	220.72	3.54	254.54 -	255.01	4.26	288.83 -	289.30	4.98	323.12 -	323.58	5.70
220.73 -	221.20	3.55	255.02 -	255.49	4.27	289.31 -	289.77	4.99	323.59 -	324.06	5.71
221.21 -	221.68	3.56	255.50 -	255.96	4.28	289.78 -	290.25	5.00	324.07 -	324.53	5.72
221.69 -	222.15	3.57	255.97 -	256.44	4.29	290.26 -	290.72	5.01	324.54 -	325.01	5.73
222.16 -	222.63	3.58	256.45 -	256.91	4.30	290.73 -	291.20	5.02	325.02 -	325.49	5.74
222.64 -	223.11	3.59	256.92 -	257.39	4.31	291.21 -	291.68	5.03	325.50 -	325.96	5.75

TABLE III (*Continued*)

WEEKLY PAY PERIOD

325.97 — 463.11

Remuneration *Rémunération* From-*de*	To-*à*	C.P.P. *R.P.C.*	Remuneration *Rémunération* From-*de*	To-*à*	C.P.P. *R.P.C.*	Remuneration *Rémunération* From-*de*	To-*à*	C.P.P. *R.P.C.*	Remuneration *Rémunération* From-*de*	To-*à*	C.P.P. *R.P.C.*
325.97 -	326.44	5.76	360.26 -	360.72	6.48	394.54 -	395.01	7.20	428.83 -	429.30	7.92
326.45 -	326.91	5.77	360.73 -	361.20	6.49	395.02 -	395.49	7.21	429.31 -	429.77	7.93
326.92 -	327.39	5.78	361.21 -	361.68	6.50	395.50 -	395.96	7.22	429.78 -	430.25	7.94
327.40 -	327.87	5.79	361.69 -	362.15	6.51	395.97 -	396.44	7.23	430.26 -	430.72	7.95
327.88 -	328.34	5.80	362.16 -	362.63	6.52	396.45 -	396.91	7.24	430.73 -	431.20	7.96
328.35 -	328.82	5.81	362.64 -	363.11	6.53	396.92 -	397.39	7.25	431.21 -	431.68	7.97
328.83 -	329.30	5.82	363.12 -	363.58	6.54	397.40 -	397.87	7.26	431.69 -	432.15	7.98
329.31 -	329.77	5.83	363.59 -	364.06	6.55	397.88 -	398.34	7.27	432.16 -	432.63	7.99
329.78 -	330.25	5.84	364.07 -	364.53	6.56	398.35 -	398.82	7.28	432.64 -	433.11	8.00
330.26 -	330.72	5.85	364.54 -	365.01	6.57	398.83 -	399.30	7.29	433.12 -	433.58	8.01
330.73 -	331.20	5.86	365.02 -	365.49	6.58	399.31 -	399.77	7.30	433.59 -	434.06	8.02
331.21 -	331.68	5.87	365.50 -	365.96	6.59	399.78 -	400.25	7.31	434.07 -	434.53	8.03
331.69 -	332.15	5.88	365.97 -	366.44	6.60	400.26 -	400.72	7.32	434.54 -	435.01	8.04
332.16 -	332.63	5.89	366.45 -	366.91	6.61	400.73 -	401.20	7.33	435.02 -	435.49	8.05
332.64 -	333.11	5.90	366.92 -	367.39	6.62	401.21 -	401.68	7.34	435.50 -	435.96	8.06
333.12 -	333.58	5.91	367.40 -	367.87	6.63	401.69 -	402.15	7.35	435.97 -	436.44	8.07
333.59 -	334.06	5.92	367.88 -	368.34	6.64	402.16 -	402.63	7.36	436.45 -	436.91	8.08
334.07 -	334.53	5.93	368.35 -	368.82	6.65	402.64 -	403.11	7.37	436.92 -	437.39	8.09
334.54 -	335.01	5.94	368.83 -	369.30	6.66	403.12 -	403.58	7.38	437.40 -	437.87	8.10
335.02 -	335.49	5.95	369.31 -	369.77	6.67	403.59 -	404.06	7.39	437.88 -	438.34	8.11
335.50 -	335.96	5.96	369.78 -	370.25	6.68	404.07 -	404.53	7.40	438.35 -	438.82	8.12
335.97 -	336.44	5.97	370.26 -	370.72	6.69	404.54 -	405.01	7.41	438.83 -	439.30	8.13
336.45 -	336.91	5.98	370.73 -	371.20	6.70	405.02 -	405.49	7.42	439.31 -	439.77	8.14
336.92 -	337.39	5.99	371.21 -	371.68	6.71	405.50 -	405.96	7.43	439.78 -	440.25	8.15
337.40 -	337.87	6.00	371.69 -	372.15	6.72	405.97 -	406.44	7.44	440.26 -	440.72	8.16
337.88 -	338.34	6.01	372.16 -	372.63	6.73	406.45 -	406.91	7.45	440.73 -	441.20	8.17
338.35 -	338.82	6.02	372.64 -	373.11	6.74	406.92 -	407.39	7.46	441.21 -	441.68	8.18
338.83 -	339.30	6.03	373.12 -	373.58	6.75	407.40 -	407.87	7.47	441.69 -	442.15	8.19
339.31 -	339.77	6.04	373.59 -	374.06	6.76	407.88 -	408.34	7.48	442.16 -	442.63	8.20
339.78 -	340.25	6.05	374.07 -	374.53	6.77	408.35 -	408.82	7.49	442.64 -	443.11	8.21
340.26 -	340.72	6.06	374.54 -	375.01	6.78	408.83 -	409.30	7.50	443.12 -	443.58	8.22
340.73 -	341.20	6.07	375.02 -	375.49	6.79	409.31 -	409.77	7.51	443.59 -	444.06	8.23
341.21 -	341.68	6.08	375.50 -	375.96	6.80	409.78 -	410.25	7.52	444.07 -	444.53	8.24
341.69 -	342.15	6.09	375.97 -	376.44	6.81	410.26 -	410.72	7.53	444.54 -	445.01	8.25
342.16 -	342.63	6.10	376.45 -	376.91	6.82	410.73 -	411.20	7.54	445.02 -	445.49	8.26
342.64 -	343.11	6.11	376.92 -	377.39	6.83	411.21 -	411.68	7.55	445.50 -	445.96	8.27
343.12 -	343.58	6.12	377.40 -	377.87	6.84	411.69 -	412.15	7.56	445.97 -	446.44	8.28
343.59 -	344.06	6.13	377.88 -	378.34	6.85	412.16 -	412.63	7.57	446.45 -	446.91	8.29
344.07 -	344.53	6.14	378.35 -	378.82	6.86	412.64 -	413.11	7.58	446.92 -	447.39	8.30
344.54 -	345.01	6.15	378.83 -	379.30	6.87	413.12 -	413.58	7.59	447.40 -	447.87	8.31
345.02 -	345.49	6.16	379.31 -	379.77	6.88	413.59 -	414.06	7.60	447.88 -	448.34	8.32
345.50 -	345.96	6.17	379.78 -	380.25	6.89	414.07 -	414.53	7.61	448.35 -	448.82	8.33
345.97 -	346.44	6.18	380.26 -	380.72	6.90	414.54 -	415.01	7.62	448.83 -	449.30	8.34
346.45 -	346.91	6.19	380.73 -	381.20	6.91	415.02 -	415.49	7.63	449.31 -	449.77	8.35
346.92 -	347.39	6.20	381.21 -	381.68	6.92	415.50 -	415.96	7.64	449.78 -	450.25	8.36
347.40 -	347.87	6.21	381.69 -	382.15	6.93	415.97 -	416.44	7.65	450.26 -	450.72	8.37
347.88 -	348.34	6.22	382.16 -	382.63	6.94	416.45 -	416.91	7.66	450.73 -	451.20	8.38
348.35 -	348.82	6.23	382.64 -	383.11	6.95	416.92 -	417.39	7.67	451.21 -	451.68	8.39
348.83 -	349.30	6.24	383.12 -	383.58	6.96	417.40 -	417.87	7.68	451.69 -	452.15	8.40
349.31 -	349.77	6.25	383.59 -	384.06	6.97	417.88 -	418.34	7.69	452.16 -	452.63	8.41
349.78 -	350.25	6.26	384.07 -	384.53	6.98	418.35 -	418.82	7.70	452.64 -	453.11	8.42
350.26 -	350.72	6.27	384.54 -	385.01	6.99	418.83 -	419.30	7.71	453.12 -	453.58	8.43
350.73 -	351.20	6.28	385.02 -	385.49	7.00	419.31 -	419.77	7.72	453.59 -	454.06	8.44
351.21 -	351.68	6.29	385.50 -	385.96	7.01	419.78 -	420.25	7.73	454.07 -	454.53	8.45
351.69 -	352.15	6.30	385.97 -	386.44	7.02	420.26 -	420.72	7.74	454.54 -	455.01	8.46
352.16 -	352.63	6.31	386.45 -	386.91	7.03	420.73 -	421.20	7.75	455.02 -	455.49	8.47
352.64 -	353.11	6.32	386.92 -	387.39	7.04	421.21 -	421.68	7.76	455.50 -	455.96	8.48
353.12 -	353.58	6.33	387.40 -	387.87	7.05	421.69 -	422.15	7.77	455.97 -	456.44	8.49
353.59 -	354.06	6.34	387.88 -	388.34	7.06	422.16 -	422.63	7.78	456.45 -	456.91	8.50
354.07 -	354.53	6.35	388.35 -	388.82	7.07	422.64 -	423.11	7.79	456.92 -	457.39	8.51
354.54 -	355.01	6.36	388.83 -	389.30	7.08	423.12 -	423.58	7.80	457.40 -	457.87	8.52
355.02 -	355.49	6.37	389.31 -	389.77	7.09	423.59 -	424.06	7.81	457.88 -	458.34	8.53
355.50 -	355.96	6.38	389.78 -	390.25	7.10	424.07 -	424.53	7.82	458.35 -	458.82	8.54
355.97 -	356.44	6.39	390.26 -	390.72	7.11	424.54 -	425.01	7.83	458.83 -	459.30	8.55
356.45 -	356.91	6.40	390.73 -	391.20	7.12	425.02 -	425.49	7.84	459.31 -	459.77	8.56
356.92 -	357.39	6.41	391.21 -	391.68	7.13	425.50 -	425.96	7.85	459.78 -	460.25	8.57
357.40 -	357.87	6.42	391.69 -	392.15	7.14	425.97 -	426.44	7.86	460.26 -	460.72	8.58
357.88 -	358.34	6.43	392.16 -	392.63	7.15	426.45 -	426.91	7.87	460.73 -	461.20	8.59
358.35 -	358.82	6.44	392.64 -	393.11	7.16	426.92 -	427.39	7.88	461.21 -	461.68	8.60
358.83 -	359.30	6.45	393.12 -	393.58	7.17	427.40 -	427.87	7.89	461.69 -	462.15	8.61
359.31 -	359.77	6.46	393.59 -	394.06	7.18	427.88 -	428.34	7.90	462.16 -	462.63	8.62
359.78 -	360.25	6.47	394.07 -	394.53	7.19	428.35 -	428.82	7.91	462.64 -	463.11	8.63

TABLE IV UNEMPLOYMENT INSURANCE PREMIUMS

For minimum and maximum insurable earnings amounts for various pay periods see Schedule II. For the maximum premium deduction for various pay periods see bottom of this page.

Les montants minimum et maximum des gains assurables pour diverses périodes de paie figurent en annexe II. La déduction maximale de primes pour diverses périodes de paie figure au bas de la présente page.

Remuneration *Rémunération* From-*de*	To-*à*	U.I. Premium *Prime d'a.-c.*	Remuneration *Rémunération* From-*de*	To-*à*	U.I. Premium *Prime d'a.-c.*	Remuneration *Rémunération* From-*de*	To-*à*	U.I. Premium *Prime d'a.-c.*	Remuneration *Rémunération* From-*de*	To-*à*	U.I. Premium *Prime d'a.-c.*
147.95	148.46	2.89	184.88	185.38	3.61	221.80	222.30	4.33	258.72	259.23	5.05
148.47	148.97	2.90	185.39	185.89	3.62	222.31	222.82	4.34	259.24	259.74	5.06
148.98	149.48	2.91	185.90	186.41	3.63	222.83	223.33	4.35	259.75	260.25	5.07
149.49	149.99	2.92	186.42	186.92	3.64	223.34	223.84	4.36	260.26	260.76	5.08
150.00	150.51	2.93	186.93	187.43	3.65	223.85	224.35	4.37	260.77	261.28	5.09
150.52	151.02	2.94	187.44	187.94	3.66	224.36	224.87	4.38	261.29	261.79	5.10
151.03	151.53	2.95	187.95	188.46	3.67	224.88	225.38	4.39	261.80	262.30	5.11
151.54	152.05	2.96	188.47	188.97	3.68	225.39	225.89	4.40	262.31	262.82	5.12
152.06	152.56	2.97	188.98	189.48	3.69	225.90	226.41	4.41	262.83	263.33	5.13
152.57	153.07	2.98	189.49	189.99	3.70	226.42	226.92	4.42	263.34	263.84	5.14
153.08	153.58	2.99	190.00	190.51	3.71	226.93	227.43	4.43	263.85	264.35	5.15
153.59	154.10	3.00	190.52	191.02	3.72	227.44	227.94	4.44	264.36	264.87	5.16
154.11	154.61	3.01	191.03	191.53	3.73	227.95	228.46	4.45	264.88	265.38	5.17
154.62	155.12	3.02	191.54	192.05	3.74	228.47	228.97	4.46	265.39	265.89	5.18
155.13	155.64	3.03	192.06	192.56	3.75	228.98	229.48	4.47	265.90	266.41	5.19
155.65	156.15	3.04	192.57	193.07	3.76	229.49	229.99	4.48	266.42	266.92	5.20
156.16	156.66	3.05	193.08	193.58	3.77	230.00	230.51	4.49	266.93	267.43	5.21
156.67	157.17	3.06	193.59	194.10	3.78	230.52	231.02	4.50	267.44	267.94	5.22
157.18	157.69	3.07	194.11	194.61	3.79	231.03	231.53	4.51	267.95	268.46	5.23
157.70	158.20	3.08	194.62	195.12	3.80	231.54	232.05	4.52	268.47	268.97	5.24
158.21	158.71	3.09	195.13	195.64	3.81	232.06	232.56	4.53	268.98	269.48	5.25
158.72	159.23	3.10	195.65	196.15	3.82	232.57	233.07	4.54	269.49	269.99	5.26
159.24	159.74	3.11	196.16	196.66	3.83	233.08	233.58	4.55	270.00	270.51	5.27
159.75	160.25	3.12	196.67	197.17	3.84	233.59	234.10	4.56	270.52	271.02	5.28
160.26	160.76	3.13	197.18	197.69	3.85	234.11	234.61	4.57	271.03	271.53	5.29
160.77	161.28	3.14	197.70	198.20	3.86	234.62	235.12	4.58	271.54	272.05	5.30
161.29	161.79	3.15	198.21	198.71	3.87	235.13	235.64	4.59	272.06	272.56	5.31
161.80	162.30	3.16	198.72	199.23	3.88	235.65	236.15	4.60	272.57	273.07	5.32
162.31	162.82	3.17	199.24	199.74	3.89	236.16	236.66	4.61	273.08	273.58	5.33
162.83	163.33	3.18	199.75	200.25	3.90	236.67	237.17	4.62	273.59	274.10	5.34
163.34	163.84	3.19	200.26	200.76	3.91	237.18	237.69	4.63	274.11	274.61	5.35
163.85	164.35	3.20	200.77	201.28	3.92	237.70	238.20	4.64	274.62	275.12	5.36
164.36	164.87	3.21	201.29	201.79	3.93	238.21	238.71	4.65	275.13	275.64	5.37
164.88	165.38	3.22	201.80	202.30	3.94	238.72	239.23	4.66	275.65	276.15	5.38
165.39	165.89	3.23	202.31	202.82	3.95	239.24	239.74	4.67	276.16	276.66	5.39
165.90	166.41	3.24	202.83	203.33	3.96	239.75	240.25	4.68	276.67	277.17	5.40
166.42	166.92	3.25	203.34	203.84	3.97	240.26	240.76	4.69	277.18	277.69	5.41
166.93	167.43	3.26	203.85	204.35	3.98	240.77	241.28	4.70	277.70	278.20	5.42
167.44	167.94	3.27	204.36	204.87	3.99	241.29	241.79	4.71	278.21	278.71	5.43
167.95	168.46	3.28	204.88	205.38	4.00	241.80	242.30	4.72	278.72	279.23	5.44
168.47	168.97	3.29	205.39	205.89	4.01	242.31	242.82	4.73	279.24	279.74	5.45
168.98	169.48	3.30	205.90	206.41	4.02	242.83	243.33	4.74	279.75	280.25	5.46
169.49	169.99	3.31	206.42	206.92	4.03	243.34	243.84	4.75	280.26	280.76	5.47
170.00	170.51	3.32	206.93	207.43	4.04	243.85	244.35	4.76	280.77	281.28	5.48
170.52	171.02	3.33	207.44	207.94	4.05	244.36	244.87	4.77	281.29	281.79	5.49
171.03	171.53	3.34	207.95	208.46	4.06	244.88	245.38	4.78	281.80	282.30	5.50
171.54	172.05	3.35	208.47	208.97	4.07	245.39	245.89	4.79	282.31	282.82	5.51
172.06	172.56	3.36	208.98	209.48	4.08	245.90	246.41	4.80	282.83	283.33	5.52
172.57	173.07	3.37	209.49	209.99	4.09	246.42	246.92	4.81	283.34	283.84	5.53
173.08	173.58	3.38	210.00	210.51	4.10	246.93	247.43	4.82	283.85	284.35	5.54
173.59	174.10	3.39	210.52	211.02	4.11	247.44	247.94	4.83	284.36	284.87	5.55
174.11	174.61	3.40	211.03	211.53	4.12	247.95	248.46	4.84	284.88	285.38	5.56
174.62	175.12	3.41	211.54	212.05	4.13	248.47	248.97	4.85	285.39	285.89	5.57
175.13	175.64	3.42	212.06	212.56	4.14	248.98	249.48	4.86	285.90	286.41	5.58
175.65	176.15	3.43	212.57	213.07	4.15	249.49	249.99	4.87	286.42	286.92	5.59
176.16	176.66	3.44	213.08	213.58	4.16	250.00	250.51	4.88	286.93	287.43	5.60
176.67	177.17	3.45	213.59	214.10	4.17	250.52	251.02	4.89	287.44	287.94	5.61
177.18	177.69	3.46	214.11	214.61	4.18	251.03	251.53	4.90	287.95	288.46	5.62
177.70	178.20	3.47	214.62	215.12	4.19	251.54	252.05	4.91	288.47	288.97	5.63
178.21	178.71	3.48	215.13	215.64	4.20	252.06	252.56	4.92	288.98	289.48	5.64
178.72	179.23	3.49	215.65	216.15	4.21	252.57	253.07	4.93	289.49	289.99	5.65
179.24	179.74	3.50	216.16	216.66	4.22	253.08	253.58	4.94	290.00	290.51	5.66
179.75	180.25	3.51	216.67	217.17	4.23	253.59	254.10	4.95	290.52	291.02	5.67
180.26	180.76	3.52	217.18	217.69	4.24	254.11	254.61	4.96	291.03	291.53	5.68
180.77	181.28	3.53	217.70	218.20	4.25	254.62	255.12	4.97	291.54	292.05	5.69
181.29	181.79	3.54	218.21	218.71	4.26	255.13	255.64	4.98	292.06	292.56	5.70
181.80	182.30	3.55	218.72	219.23	4.27	255.65	256.15	4.99	292.57	293.07	5.71
182.31	182.82	3.56	219.24	219.74	4.28	256.16	256.66	5.00	293.08	293.58	5.72
182.83	183.33	3.57	219.75	220.25	4.29	256.67	257.17	5.01	293.59	294.10	5.73
183.34	183.84	3.58	220.26	220.76	4.30	257.18	257.69	5.02	294.11	294.61	5.74
183.85	184.35	3.59	220.77	221.28	4.31	257.70	258.20	5.03	294.62	295.12	5.75
184.36	184.87	3.60	221.29	221.79	4.32	258.21	258.71	5.04	295.13	295.64	5.76

Maximum Premium Deduction for a Pay Period of the stated frequency. Déduction maximale de prime pour une période de paie d'une durée donnée.				
Weekly - Hebdomadaire	11.80		10 pp per year - 10 pp par année	61.35
Bi-Weekly - Deux semaines	23.60		13 pp per year - 13 pp par année	47.19
Semi-Monthly - Bi-mensuel	25.56		22 pp per year - 22 pp par année	27.89
Monthly - Mensuellement	51.12			

TABLE IV (*Continued*)

For minimum and maximum insurable earnings amounts for various pay periods see Schedule II. For the maximum premium deduction for various pay periods see bottom of this page.

Les montants minimum et maximum des gains assurables pour diverses périodes de paie figurent en annexe II. La déduction maximale de primes pour diverses périodes de paie figure au bas de la présente page.

Remuneration *Rémunération* From-*de*	To-*à*	U.I. Premium *Prime d'a.-c.*	Remuneration *Rémunération* From-*de*	To-*à*	U.I. Premium *Prime d'a.-c.*	Remuneration *Rémunération* From-*de*	To-*à*	U.I. Premium *Prime d'a.-c.*	Remuneration *Rémunération* From-*de*	To-*à*	U.I. Premium *Prime d'a.-c.*
295.65 -	296.15	5.77	332.57 -	333.07	6.49	369.49 -	369.99	7.21	406.42 -	406.92	7.93
296.16 -	296.66	5.78	333.08 -	333.58	6.50	370.00 -	370.51	7.22	406.93 -	407.43	7.94
296.67 -	297.17	5.79	333.59 -	334.10	6.51	370.52 -	371.02	7.23	407.44 -	407.94	7.95
297.18 -	297.69	5.80	334.11 -	334.61	6.52	371.03 -	371.53	7.24	407.95 -	408.46	7.96
297.70 -	298.20	5.81	334.62 -	335.12	6.53	371.54 -	372.05	7.25	408.47 -	408.97	7.97
298.21 -	298.71	5.82	335.13 -	335.64	6.54	372.06 -	372.56	7.26	408.98 -	409.48	7.98
298.72 -	299.23	5.83	335.65 -	336.15	6.55	372.57 -	373.07	7.27	409.49 -	409.99	7.99
299.24 -	299.74	5.84	336.16 -	336.66	6.56	373.08 -	373.58	7.28	410.00 -	410.51	8.00
299.75 -	300.25	5.85	336.67 -	337.17	6.57	373.59 -	374.10	7.29	410.52 -	411.02	8.01
300.26 -	300.76	5.86	337.18 -	337.69	6.58	374.11 -	374.61	7.30	411.03 -	411.53	8.02
300.77 -	301.28	5.87	337.70 -	338.20	6.59	374.62 -	375.12	7.31	411.54 -	412.05	8.03
301.29 -	301.79	5.88	338.21 -	338.71	6.60	375.13 -	375.64	7.32	412.06 -	412.56	8.04
301.80 -	302.30	5.89	338.72 -	339.23	6.61	375.65 -	376.15	7.33	412.57 -	413.07	8.05
302.31 -	302.82	5.90	339.24 -	339.74	6.62	376.16 -	376.66	7.34	413.08 -	413.58	8.06
302.83 -	303.33	5.91	339.75 -	340.25	6.63	376.67 -	377.17	7.35	413.59 -	414.10	8.07
303.34 -	303.84	5.92	340.26 -	340.76	6.64	377.18 -	377.69	7.36	414.11 -	414.61	8.08
303.85 -	304.35	5.93	340.77 -	341.28	6.65	377.70 -	378.20	7.37	414.62 -	415.12	8.09
304.36 -	304.87	5.94	341.29 -	341.79	6.66	378.21 -	378.71	7.38	415.13 -	415.64	8.10
304.88 -	305.38	5.95	341.80 -	342.30	6.67	378.72 -	379.23	7.39	415.65 -	416.15	8.11
305.39 -	305.89	5.96	342.31 -	342.82	6.68	379.24 -	379.74	7.40	416.16 -	416.66	8.12
305.90 -	306.41	5.97	342.83 -	343.33	6.69	379.75 -	380.25	7.41	416.67 -	417.17	8.13
306.42 -	306.92	5.98	343.34 -	343.84	6.70	380.26 -	380.76	7.42	417.18 -	417.69	8.14
306.93 -	307.43	5.99	343.85 -	344.35	6.71	380.77 -	381.28	7.43	417.70 -	418.20	8.15
307.44 -	307.94	6.00	344.36 -	344.87	6.72	381.29 -	381.79	7.44	418.21 -	418.71	8.16
307.95 -	308.46	6.01	344.88 -	345.38	6.73	381.80 -	382.30	7.45	418.72 -	419.23	8.17
308.47 -	308.97	6.02	345.39 -	345.89	6.74	382.31 -	382.82	7.46	419.24 -	419.74	8.18
308.98 -	309.48	6.03	345.90 -	346.41	6.75	382.83 -	383.33	7.47	419.75 -	420.25	8.19
309.49 -	309.99	6.04	346.42 -	346.92	6.76	383.34 -	383.84	7.48	420.26 -	420.76	8.20
310.00 -	310.51	6.05	346.93 -	347.43	6.77	383.85 -	384.35	7.49	420.77 -	421.28	8.21
310.52 -	311.02	6.06	347.44 -	347.94	6.78	384.36 -	384.87	7.50	421.29 -	421.79	8.22
311.03 -	311.53	6.07	347.95 -	348.46	6.79	384.88 -	385.38	7.51	421.80 -	422.30	8.23
311.54 -	312.05	6.08	348.47 -	348.97	6.80	385.39 -	385.89	7.52	422.31 -	422.82	8.24
312.06 -	312.56	6.09	348.98 -	349.48	6.81	385.90 -	386.41	7.53	422.83 -	423.33	8.25
312.57 -	313.07	6.10	349.49 -	349.99	6.82	386.42 -	386.92	7.54	423.34 -	423.84	8.26
313.08 -	313.58	6.11	350.00 -	350.51	6.83	386.93 -	387.43	7.55	423.85 -	424.35	8.27
313.59 -	314.10	6.12	350.52 -	351.02	6.84	387.44 -	387.94	7.56	424.36 -	424.87	8.28
314.11 -	314.61	6.13	351.03 -	351.53	6.85	387.95 -	388.46	7.57	424.88 -	425.38	8.29
314.62 -	315.12	6.14	351.54 -	352.05	6.86	388.47 -	388.97	7.58	425.39 -	425.89	8.30
315.13 -	315.64	6.15	352.06 -	352.56	6.87	388.98 -	389.48	7.59	425.90 -	426.41	8.31
315.65 -	316.15	6.16	352.57 -	353.07	6.88	389.49 -	389.99	7.60	426.42 -	426.92	8.32
316.16 -	316.66	6.17	353.08 -	353.58	6.89	390.00 -	390.51	7.61	426.93 -	427.43	8.33
316.67 -	317.17	6.18	353.59 -	354.10	6.90	390.52 -	391.02	7.62	427.44 -	427.94	8.34
317.18 -	317.69	6.19	354.11 -	354.61	6.91	391.03 -	391.53	7.63	427.95 -	428.46	8.35
317.70 -	318.20	6.20	354.62 -	355.12	6.92	391.54 -	392.05	7.64	428.47 -	428.97	8.36
318.21 -	318.71	6.21	355.13 -	355.64	6.93	392.06 -	392.56	7.65	428.98 -	429.48	8.37
318.72 -	319.23	6.22	355.65 -	356.15	6.94	392.57 -	393.07	7.66	429.49 -	429.99	8.38
319.24 -	319.74	6.23	356.16 -	356.66	6.95	393.08 -	393.58	7.67	430.00 -	430.51	8.39
319.75 -	320.25	6.24	356.67 -	357.17	6.96	393.59 -	394.10	7.68	430.52 -	431.02	8.40
320.26 -	320.76	6.25	357.18 -	357.69	6.97	394.11 -	394.61	7.69	431.03 -	431.53	8.41
320.77 -	321.28	6.26	357.70 -	358.20	6.98	394.62 -	395.12	7.70	431.54 -	432.05	8.42
321.29 -	321.79	6.27	358.21 -	358.71	6.99	395.13 -	395.64	7.71	432.06 -	432.56	8.43
321.80 -	322.30	6.28	358.72 -	359.23	7.00	395.65 -	396.15	7.72	432.57 -	433.07	8.44
322.31 -	322.82	6.29	359.24 -	359.74	7.01	396.16 -	396.66	7.73	433.08 -	433.58	8.45
322.83 -	323.33	6.30	359.75 -	360.25	7.02	396.67 -	397.17	7.74	433.59 -	434.10	8.46
323.34 -	323.84	6.31	360.26 -	360.76	7.03	397.18 -	397.69	7.75	434.11 -	434.61	8.47
323.85 -	324.35	6.32	360.77 -	361.28	7.04	397.70 -	398.20	7.76	434.62 -	435.12	8.48
324.36 -	324.87	6.33	361.29 -	361.79	7.05	398.21 -	398.71	7.77	435.13 -	435.64	8.49
324.88 -	325.38	6.34	361.80 -	362.30	7.06	398.72 -	399.23	7.78	435.65 -	436.15	8.50
325.39 -	325.89	6.35	362.31 -	362.82	7.07	399.24 -	399.74	7.79	436.16 -	436.66	8.51
325.90 -	326.41	6.36	362.83 -	363.33	7.08	399.75 -	400.25	7.80	436.67 -	437.17	8.52
326.42 -	326.92	6.37	363.34 -	363.84	7.09	400.26 -	400.76	7.81	437.18 -	437.69	8.53
326.93 -	327.43	6.38	363.85 -	364.35	7.10	400.77 -	401.28	7.82	437.70 -	438.20	8.54
327.44 -	327.94	6.39	364.36 -	364.87	7.11	401.29 -	401.79	7.83	438.21 -	438.71	8.55
327.95 -	328.46	6.40	364.88 -	365.38	7.12	401.80 -	402.30	7.84	438.72 -	439.23	8.56
328.47 -	328.97	6.41	365.39 -	365.89	7.13	402.31 -	402.82	7.85	439.24 -	439.74	8.57
328.98 -	329.48	6.42	365.90 -	366.41	7.14	402.83 -	403.33	7.86	439.75 -	440.25	8.58
329.49 -	329.99	6.43	366.42 -	366.92	7.15	403.34 -	403.84	7.87	440.26 -	440.76	8.59
330.00 -	330.51	6.44	366.93 -	367.43	7.16	403.85 -	404.35	7.88	440.77 -	441.28	8.60
330.52 -	331.02	6.45	367.44 -	367.94	7.17	404.36 -	404.87	7.89	441.29 -	441.79	8.61
331.03 -	331.53	6.46	367.95 -	368.46	7.18	404.88 -	405.38	7.90	441.80 -	442.30	8.62
331.54 -	332.05	6.47	368.47 -	368.97	7.19	405.39 -	405.89	7.91	442.31 -	442.82	8.63
332.06 -	332.56	6.48	368.98 -	369.48	7.20	405.90 -	406.41	7.92	442.83 -	443.33	8.64

Maximum Premium Deduction for a Pay Period of the stated frequency. Déduction maximale de prime pour une période de paie d'une durée donnée.				
	Weekly - Hebdomadaire	11.80	10 pp per year - 10 pp par année	61.35
	Bi-Weekly - Deux semaines	23.60	13 pp per year - 13 pp par année	47.19
	Semi-Monthly - Bi-mensuel	25.56	22 pp per year - 22 pp par année	27.89
	Monthly - Mensuellement	51.12		

TABLE V THE NUMBER OF EACH DAY OF THE YEAR

Day of Month	Jan.	Feb.	Mar.	Apr.	May	Jun.	Jul.	Aug.	Sept.	Oct.	Nov.	Dec.	Day of Month
1	1	32	60	91	121	152	182	213	244	274	305	335	1
2	2	33	61	92	122	153	183	214	245	275	306	336	2
3	3	34	62	93	123	154	184	215	246	276	307	337	3
4	4	35	63	94	124	155	185	216	247	277	308	338	4
5	5	36	64	95	125	156	186	217	248	278	309	339	5
6	6	37	65	96	126	157	187	218	249	279	310	340	6
7	7	38	66	97	127	158	188	219	250	280	311	341	7
8	8	39	67	98	128	159	189	220	251	281	312	342	8
9	9	40	68	99	129	160	190	221	252	282	313	343	9
10	10	41	69	100	130	161	191	222	253	283	314	344	10
11	11	42	70	101	131	162	192	223	254	284	315	345	11
12	12	43	71	102	132	163	193	224	255	285	316	346	12
13	13	44	72	103	133	164	194	225	256	286	317	347	13
14	14	45	73	104	134	165	195	226	257	287	318	348	14
15	15	46	74	105	135	166	196	227	258	288	319	349	15
16	16	47	75	106	136	167	197	228	259	289	320	350	16
17	17	48	76	107	137	168	198	229	260	290	321	351	17
18	18	49	77	108	138	169	199	230	261	291	322	352	18
19	19	50	78	109	139	170	200	231	262	292	323	353	19
20	20	51	79	110	140	171	201	232	263	293	324	354	20
21	21	52	80	111	141	172	202	233	264	294	325	355	21
22	22	53	81	112	142	173	203	234	265	295	326	356	22
23	23	54	82	113	143	174	204	235	266	296	327	357	23
24	24	55	83	114	144	175	205	236	267	297	328	358	24
25	25	56	84	115	145	176	206	237	268	298	329	359	25
26	26	57	85	116	146	177	207	238	269	299	330	360	26
27	27	58	86	117	147	178	208	239	270	300	331	361	27
28	28	59	87	118	148	179	209	240	271	301	332	362	28
29	29		88	119	149	180	210	241	272	302	333	363	29
30	30		89	120	150	181	211	242	273	303	334	364	30
31	31		90		151		212	243		304		365	31

TABLE VI TRIGONOMETRIC RATIOS

0°	sin θ	cos θ	tan θ	0°	sin θ	cos θ	tan θ
0	0.0000	1.0000	0.0000	**46**	0.7193	0.6947	1.0355
1	0.0175	0.9999	0.0175	**47**	0.7314	0.6820	1.0724
2	0.0349	0.9994	0.0349	**48**	0.7431	0.6691	1.1106
3	0.0523	0.9986	0.0524	**49**	0.7547	0.6561	1.1504
4	0.0698	0.9976	0.0699	**50**	0.7660	0.6428	1.1917
5	0.0872	0.9962	0.0875	**51**	0.7772	0.6293	1.2349
6	0.1045	0.9945	0.1051	**52**	0.7880	0.6157	1.2799
7	0.1219	0.9926	0.1228	**53**	0.7986	0.6018	1.3270
8	0.1392	0.9903	0.1405	**54**	0.8090	0.5878	1.3764
9	0.1564	0.9877	0.1584	**55**	0.8192	0.5736	1.4281
10	0.1737	0.9848	0.1763	**56**	0.8290	0.5592	1.4826
11	0.1908	0.9816	0.1944	**57**	0.8387	0.5446	1.5399
12	0.2079	0.9782	0.2126	**58**	0.8481	0.5299	1.6003
13	0.2250	0.9744	0.2309	**59**	0.8572	0.5150	1.6643
14	0.2419	0.9703	0.2493	**60**	0.8660	0.5000	1.7320
15	0.2588	0.9659	0.2680	**61**	0.8746	0.4848	1.8040
16	0.2756	0.9613	0.2867	**62**	0.8830	0.4695	1.8807
17	0.2924	0.9563	0.3057	**63**	0.8910	0.4540	1.9626
18	0.3090	0.9511	0.3249	**64**	0.8988	0.4384	2.0503
19	0.3256	0.9455	0.3443	**65**	0.9063	0.4226	2.1445
20	0.3420	0.9397	0.3640	**66**	0.9136	0.4067	2.2460
21	0.3584	0.9336	0.3839	**67**	0.9205	0.3907	2.3558
22	0.3746	0.9272	0.4040	**68**	0.9272	0.3746	2.4751
23	0.3907	0.9025	0.4245	**69**	0.9336	0.3584	2.6051
24	0.4067	0.9136	0.4452	**70**	0.9397	0.3420	2.7475
25	0.4226	0.9063	0.4663	**71**	0.9455	0.3256	2.9042
26	0.4384	0.8988	0.4877	**72**	0.9511	0.3090	3.0777
27	0.4540	0.8910	0.5095	**73**	0.9563	0.2924	3.2708
28	0.4695	0.8830	0.5317	**74**	0.9613	0.2756	3.4874
29	0.4848	0.8746	0.5543	**75**	0.9659	0.2588	3.7320
30	0.5000	0.8660	0.5774	**76**	0.9703	0.2419	4.0108
31	0.5150	0.8572	0.6009	**77**	0.9744	0.2250	4.3315
32	0.5299	0.8481	0.6249	**78**	0.9782	0.2079	4.7046
33	0.5446	0.8387	0.6494	**79**	0.9816	0.1908	5.1445
34	0.5592	0.8290	0.6745	**80**	0.9848	0.1737	5.6713
35	0.5736	0.8192	0.7002	**81**	0.9877	0.1564	6.3137
36	0.5878	0.8090	0.7265	**82**	0.9903	0.1392	7.1154
37	0.6018	0.7986	0.7536	**83**	0.9926	0.1219	8.1443
38	0.6157	0.7880	0.7813	**84**	0.9945	0.1045	9.5144
39	0.6293	0.7772	0.8098	**85**	0.9962	0.0872	11.430
40	0.6428	0.7660	0.8391	**86**	0.9976	0.0698	14.301
41	0.6561	0.7547	0.8693	**87**	0.9986	9.0523	19.081
42	0.6691	0.7431	0.9004	**88**	0.9994	0.0349	28.636
43	0.6820	0.7314	0.9325	**89**	0.9999	0.0175	57.290
44	0.6947	0.7193	0.9657	**90**	1.0000	0.0000	—
45	0.7071	0.7071	1.0000				

ANSWERS

GETTING TO KNOW YOUR TEXT

1. 12 2. 455 3. 47 4. 9 5. 314
6. 3 7. 9 8. $45 9. 12.3 10. 190
11. 262 12. 328 13. 5 14. 10 15. 5920

REVIEW AND PREVIEW TO CHAPTER 1

POWERS OF TEN

1. (a) 230 (b) 230 (c) 2300 (d) 2300 (e) 23 000 (f) 23 000
   (g) 2.3 (h) 2.3 (i) 0.23 (j) 0.23 (k) 0.023 (l) 0.023
2. (a) 230 (b) 560 (c) 17.9 (d) 6.7 (e) 45.6 (f) 0.752
   (g) 12 (h) 78 (i) 0.0267 (j) 340 000 (k) 60 (l) 3.49
3. (a) 4.56 (b) 200 (c) 16.2 (d) 4500 (e) 20 000 (f) 34 000
   (g) 0.89 (h) 0.0045 (i) 8 (j) 1340

EXPONENTS

1. (a) 9 (b) 32 (c) 1000 (d) 625 (e) 343 (f) 256
   (g) 1 (h) 216 (i) 4096 (j) 6.25 (k) 0.008 (l) 35.937
2. 1024

FRACTIONS AND DECIMALS

1. (a) $\frac{1}{4}$ (b) $\frac{7}{10}$ (c) $\frac{3}{4}$ (d) $\frac{1}{2}$ (e) $\frac{14}{25}$ (f) $\frac{13}{20}$
   (g) $\frac{1}{8}$ (h) $\frac{7}{8}$ (i) $\frac{9}{25}$
2. (a) 0.625 (b) 0.375 (c) 0.4375 (d) 0.84 (e) 0.86 (f) 0.7125
3. (a) $\frac{21}{4}$ (b) $\frac{18}{5}$ (c) $\frac{11}{6}$ (d) $\frac{19}{8}$ (e) $\frac{19}{2}$ (f) $\frac{67}{10}$

EXERCISE 1.1

1. (a) two hundred thirty-six (b) forty-five thousand six hundred seventy
   (c) four hundred fifty-six thousand (d) twenty-three and six tenths
   (e) four and twelve hundredths (f) two million three hundred thousand
   (g) two hundred thirty-four thousandths (h) three thousandths
   (i) three thousand one (j) three and eighty-nine hundredths
2. (a) 10, 70 (b) 10 000, 70 000 (c) 0.1, 0.7
   (d) 0.001, 0.007 (e) 1000, 7000 (f) 100 000, 700 000
3. (a) 3, 100, 300 (b) 1, 10 000, 10 000 (c) 3, 1, 3
   (d) 8, 10, 80 (e) 7, 0.01, 0.07 (f) 1, 0.1, 0.1
   (g) 3, 1 000 000, 3 000 000 (h) 8, 0.01, 0.08 (i) 9, 0.001, 0.009
   (j) 2, 100 000, 200 000
4. (a) three thousand four hundred seventy-eight
   (b) fifty-six thousand seven hundred eight
   (c) nine and fifty-eight hundredths
   (d) nine hundred sixty-six thousandths
   (e) eight million nine hundred seventy thousand
   (f) one thousandth (g) seven and nine tenths
   (h) ten thousand one hundred one
5. (a) 12 563 (b) 36 361 (c) 13 342 (d) 1543 (e) 3249 (f) 4334
   (g) 23 683
6. (a) 5491 (b) 20 500 (c) 1279 (d) 41 825 (e) 3577
7. (a) 17 680 (b) 281 274 (c) 42 024 (d) 46 800 (e) 504 000 (f) 3045
8. (a) 237 (b) 342 (c) 612 (d) 84
9. 25 920 000 h 10. $68 940 300 11. 45
12. (a) $27 + 54 + 432 = 513$ (b) $96 + 1536 = 1632$ (c) $488 + 976 + 1952 = 3416$

EXERCISE 1.2

1. (a) 18 (b) 11 (c) 21 (d) 8 (e) 99 (f) 5
   (g) 3 (h) 5 (i) 6 (j) 1 (k) 9 (l) 9
2. (a) 228 (b) 65 (c) 152 (d) 63 (e) 613 (f) 21
   (g) 397
3. (a) 389 (b) 147 (c) 3 (d) 5 (e) 20 (f) 24
   (g) 0 (h) 4
4. (a) 52 (b) 48 (c) 76 (d) 2 (e) 45 (f) 56
   (g) 12 (h) 1 (i) 17 (j) 3
5. (a) 8 (b) 5 (c) 9

EXERCISE 1.3

1. (a) exact (b) estimate (c) estimate (d) estimate (e) exact
2. (a) 300 (b) 700 (c) 600 (d) 800
3. (a) 4000 (b) 6000 (c) 6000 (d) 5000
4. (a) 0.7 (b) 0.4 (c) 0.1 (d) 0.4
5. (a) 0.06 (b) 0.08 (c) 0.02 (d) 0.05
6. (a) 20 000 (b) 0.5 (c) 40 (d) 8000 (e) 0.6 (f) 0.05
   (g) 0.004 (h) 0.008 (i) 700 000 (j) 8 (k) 70 000 (l) 1
7. Answers vary.
   (a) 1100 (b) 4000 (c) 5 (d) 20 (e) 2.2 (f) 40
   (g) 60 (h) 20 (i) 15 000
8. Answers vary.
   (a) $200 (b) 60 000 (c) Answers vary. (d) 7 h
   (e) 700 km (f) $14 (g) $290 (h) $40

EXERCISE 1.4

1. Estimates vary.
   (a) 45.562 (b) 112.868 (c) 8.2 (d) 75.188 (e) 281.34 (f) 5.65
2. Estimates vary.
   (a) 495.34 (b) 141.243 (c) 64.24 (d) 323.625 (e) 36.3
3. Estimates vary.
   (a) 22.11 (b) 76.81 (c) 4.158 (d) 635.23 (e) 73.04
4. Estimates vary.
   (a) 140.76 (b) 4.165 (c) 27.44 (d) 8.052 (e) 63.758
5. Estimates vary.
   (a) 3.2 (b) 0.3 (c) 10.7 (d) 1610 (e) 200
6. (a) 12.9 (b) 13.76 (c) 0.014 (d) 2.7 (e) 769.23
7. $508.88
8. (a) 162.87 km/h (b) 2449.3 km/h (c) 1836.98 km/h
   (d) 123 812.1 km/h (e) Mach 1.89

EXERCISE 1.5

1. 16, 31, 496
2. 5, 14, 53
3. 10, 8, 6
4. 5.9, 9.5, 302
5. 8.8, 1.6, 16.04
6. 5, 18, 22
7. (a) 21 (b) 1 (c) 17 (d) 60 (e) 21 (f) 9
   (g) 2 (h) 57
8. 177, 186, 213
9. (a) 239 (b) 40 (c) 110 (d) 34 (e) 2100 (f) 0.34
10. (a) 78 m (b) 16.0 cm (c) 8.5 mm (d) 764 m
11. (a) $775 (b) $1030 (c) $1600 (d) $300
12. (a) $478 (b) $822.60 (c) $1665.80 (d) $1364.60

EXERCISE 1.6

1. (a) $\frac{4}{5}$ (b) $\frac{5}{6}$ (c) $\frac{2}{3}$ (d) $\frac{1}{3}$ (e) $\frac{1}{5}$ (f) $\frac{1}{6}$
2. (a) $\frac{2}{1}$ or 2 (b) $\frac{4}{3}$ (c) $\frac{8}{7}$ (d) $\frac{16}{9}$ (e) $\frac{1}{2}$ (f) $\frac{3}{8}$
   (g) $\frac{8}{1}$ or 8 (h) $\frac{2}{3}$ (i) $\frac{4}{7}$
3. (a) $1\frac{3}{4}$ (b) $5\frac{1}{2}$ (c) $2\frac{2}{3}$ (d) $1\frac{4}{5}$ (e) $2\frac{5}{6}$ (f) $2\frac{1}{10}$
   (g) $3\frac{3}{4}$ (h) $5\frac{1}{5}$ (i) $3\frac{3}{10}$
4. (a) $\frac{1}{3}$ (b) $\frac{1}{10}$ (c) $\frac{2}{3}$ (d) $2\frac{1}{4}$ (e) $\frac{1}{10}$ (f) $\frac{1}{4}$
5. (a) $1\frac{1}{2}$ (b) $\frac{2}{3}$ (c) $\frac{3}{8}$ (d) $7\frac{1}{2}$ (e) $\frac{9}{10}$ (f) $\frac{3}{4}$
6. (a) $4\frac{1}{8}$ (b) $15\frac{2}{5}$ (c) $12\frac{3}{4}$ (d) 34 (e) $\frac{4}{5}$ (f) $2\frac{9}{20}$
   (g) $5\frac{1}{24}$ (h) 5
7. (a) $2\frac{1}{4}$ (b) $2\frac{1}{6}$ (c) $3\frac{3}{5}$ (d) $\frac{7}{8}$ (e) $\frac{52}{95}$ (f) $2\frac{1}{4}$
   (g) $2\frac{3}{4}$ (h) $\frac{5}{8}$
8. (a) $\frac{3}{8}$ (b) $\frac{3}{8}$ (c) $\frac{1}{10}$ (d) 8 (e) $1\frac{7}{8}$ (f) 2
9. (a) $\frac{1}{3}$ (b) $1\frac{1}{2}$ (c) $\frac{2}{3}$ (d) $\frac{1}{12}$
10. $45\frac{1}{2}$ 11. \$1312.50 12. $31\frac{1}{8}$ 13. 25 000

EXERCISE 1.7

1. (a) 6 (b) 4 (c) 15 (d) 20 (e) 24 (f) 18
2. (a) $\frac{13}{4}$ (b) $\frac{8}{3}$ (c) $\frac{8}{5}$ (d) $\frac{15}{2}$ (e) $\frac{11}{7}$ (f) $\frac{23}{10}$
3. (a) $\frac{1}{2}$ (b) $\frac{3}{4}$ (c) $1\frac{1}{6}$ (d) $1\frac{3}{10}$ (e) $7\frac{1}{2}$ (f) $7\frac{3}{8}$
4. (a) $\frac{2}{3}$ (b) $\frac{1}{4}$ (c) $\frac{3}{10}$ (d) $\frac{1}{4}$ (e) $4\frac{1}{2}$ (f) $2\frac{1}{8}$
5. (a) $7\frac{3}{4}$ (b) $6\frac{1}{4}$ (c) $6\frac{9}{10}$ (d) $2\frac{11}{12}$
   (e) $2\frac{13}{18}$ (f) $10\frac{19}{20}$ (g) $8\frac{16}{21}$ (h) $12\frac{1}{9}$
6. (a) $4\frac{1}{4}$ (b) $5\frac{3}{8}$ (c) $1\frac{1}{2}$ (d) $6\frac{1}{3}$
   (e) $6\frac{9}{20}$ (f) $2\frac{2}{3}$ (g) $6\frac{7}{24}$ (h) $1\frac{4}{5}$
7. (a) $3\frac{1}{8}$ (b) $5\frac{17}{20}$ (c) $9\frac{8}{9}$ (d) $2\frac{7}{12}$
   (e) $4\frac{5}{12}$ (f) $3\frac{11}{24}$ (g) $2\frac{17}{20}$ (h) $3\frac{1}{20}$
8. (a) $\frac{17}{24}$ (b) $1\frac{23}{24}$ (c) $6\frac{5}{8}$ (d) $15\frac{3}{20}$
   (e) $2\frac{8}{15}$ (f) 1 (g) $3\frac{7}{12}$ (h) $12\frac{3}{20}$
9. (a) $1\frac{1}{12}$ (b) $\frac{5}{12}$ (c) 2 (d) $\frac{1}{24}$ (e) $\frac{5}{12}$ (f) $3\frac{1}{3}$
10. \$637.50 11. $2\frac{1}{2}$ h 12. $26\frac{1}{4}$ 13. $\frac{5}{12}$

EXERCISE 1.8

1. (a) 0.45 (b) 0.67 (c) 0.92 (d) 0.06 (e) 0.01 (f) 1
   (g) 2 (h) 1.5 (i) 0.336 (j) 0.459 (k) 0.073 (l) 0.092
   (m) 0.004 (n) 0.001 (o) 0.005

2. (a) 23% (b) 56% (c) 79% (d) 90% (e) 70% (f) 5%
(g) 1% (h) 13.7% (i) 23.5% (j) 50.9% (k) 7.5% (l) 1.6%
(m) 0.6% (n) 120% (o) 250%

3. (a) $\frac{3}{5}$ (b) $\frac{3}{4}$ (c) $\frac{1}{4}$ (d) $\frac{1}{2}$ (e) $\frac{1}{10}$ (f) $\frac{1}{20}$
(g) $\frac{4}{5}$ (h) 1 (i) 2 (j) $1\frac{1}{2}$ (k) $\frac{1}{100}$ (l) $\frac{1}{3}$
(m) $\frac{2}{3}$ (n) $\frac{13}{200}$ (o) 10

4. (a) 50.0% (b) 75.0% (c) 37.5% (d) 44.4% (e) 80.0% (f) 42.9%
(g) 125.0% (h) 33.3% (i) 266.7% (j) 83.3% (k) 430.0% (l) 46.0%

5. (a) 50 (b) 6500 (c) $32 (d) $0.40

6. (a) 60% (b) 74% (c) 25% (d) 10%

7. $205.20 8. $207 9. $255.06 10. 42 cm 11. 90% 12. $31.50

13. 38.5% 14. $67.50 15. $46\frac{2}{3}$ g 16. $600 loss

EXERCISE 1.9

1. (a) −7 (b) +5 (c) −6 (d) +7 (e) −10

2. (a) +10 (b) +4 (c) +3 (d) −7 (e) −7

3. (a) 13 (b) 2 (c) −3 (d) −8 (e) 2 (f) −14
(g) −13 (h) 2

4. (a) 2 (b) 5 (c) 5 (d) −6 (e) 33 (f) −3
(g) −30 (h) 20

5. (a) 2 (b) 7 (c) −9 (d) −1 (e) 12 (f) −3
(g) −33 (h) 8

6. (a) 13 (b) −9 (c) 2 (d) 7 (e) −9 (f) −4
(g) 2 (h) 9

7. (a) 13 (b) −3 (c) −10 (d) −13 (e) 23 (f) −2
(g) −44

8. (a) 13 (b) −7 (c) 11 (d) −3 (e) −13 (f) 16
(g) −7 (h) −100

9. (a) $+8 - 6 = +2$ (b) $+3 - 5 = -2$ (c) $+40 - 15 - 13 = +12$
(d) $+7 - 8 = -1$ (e) $+100 - 200 + 400 = +300$ (f) $+2 - 1 + 0 = +1$

EXERCISE 1.10

1. (a) −12 (b) −48 (c) 10 (d) −24 (e) 24 (f) −21
(g) 12 (h) 0 (i) −6 (j) 12 (k) 99 (l) −3000

2. (a) −5 (b) 5 (c) −3 (d) 3 (e) −5 (f) 9
(g) −3 (h) −6 (i) −12 (j) 10

3. (a) 4 (b) −27 (c) −64 (d) 10 000 (e) −25 (f) −1

4. (a) −24 (b) 45 (c) −13 (d) 134 (e) −5 (f) 104
(g) 11 (h) 13 (i) 10 (j) −72 (k) 0 (l) 33
(m) 1

5. (a) 5 (b) −8 (c) 14 (d) 51 (e) 0

EXERCISE 1.11

1. Answers vary.

2. (a) $\frac{2}{5}$ (b) $\frac{1}{4}$ (c) $\frac{1}{10}$ (d) $\frac{14}{25}$ (e) $\frac{2}{25}$ (f) $\frac{1}{8}$

3. (a) 0.75 (b) 0.375 (c) −0.5 (d) −0.7 (e) −0.625 (f) −0.6875

4. (a) $-\frac{3}{10}$ (b) $-\frac{9}{4}$ (c) $-\frac{11}{3}$ (d) $\frac{3}{4}$ (e) $\frac{7}{5}$ (f) $-\frac{8}{5}$

5. (a) $0.\overline{1}$ (b) $0.\overline{27}$ (c) $0.8\overline{3}$ (d) $0.\overline{8}$ (e) $0.\overline{63}$ (f) $0.\overline{285714}$

6. (a) $\frac{1}{3}$ (b) $\frac{2}{9}$ (c) $\frac{2}{3}$ (d) $\frac{2}{11}$ (e) $\frac{1}{11}$ (f) $\frac{8}{15}$

7. (a) $\frac{7}{8}$ (b) $-\frac{5}{8}$ (c) $\frac{6}{35}$ (d) $\frac{8}{15}$ (e) -9 (f) 7

8. (a) $\frac{3}{2}$ (b) $\frac{4}{-3}$ (c) $\frac{-6}{5}$ (d) $\frac{1}{2}$ (e) $-\frac{1}{3}$ (f) -3

9. (a) $\frac{2}{3}$ (b) $-\frac{3}{4}$ (c) $-3\frac{1}{3}$ (d) $-\frac{3}{7}$ (e) $16\frac{2}{3}$ (f) $\frac{5}{12}$

10. (a) $1\frac{5}{12}$ (b) $1\frac{1}{3}$ (c) $-1\frac{1}{12}$ (d) $-1\frac{5}{8}$ (e) $2\frac{1}{6}$ (f) $2\frac{1}{6}$

11. (a) $\frac{2}{3}$ (b) -1 (c) $1\frac{5}{8}$ (d) $\frac{7}{12}$ (e) $-3\frac{1}{2}$ (f) $-4\frac{1}{3}$

12. (a) $\frac{7}{12}$ (b) $\frac{1}{12}$ (c) $\frac{13}{24}$ (d) -1 (e) $\frac{1}{12}$ (f) $-8\frac{1}{4}$
(g) $1\frac{2}{3}$ (h) $-\frac{1}{6}$ (i) $-1\frac{5}{6}$ (j) $-5\frac{19}{20}$ (k) $-1\frac{1}{2}$ (l) $-4\frac{1}{5}$

EXERCISE 1.12

1. 4
2. 2
3. Answers vary. Beginning at the top and then moving from left to right, place the numbers in the following order: 2, 6, 8, 5, 4, 1, 3, 7.
4. 67, 68, 69, 70
5. Answers vary.
6. \$3520.20
7. (a) $5 + 19 = 24$ (b) $7 + 31 = 38$ (c) $23 + 31 = 54$
8. 76
9. 82
10. 3322.4 km
11. Aisha
12. 8
13. Answers vary.
14. B, C, D
15. beginning at the left and moving clockwise: 21, 26, 25
16. \$7098.35
17. (a) 23:30, Jan 22 (b) 03:30, Jan 23 (c) 21:30, Jan 22 (d) 15:30, Jan 22
(e) 08:30, Jan 23 (f) 10:30, Jan 23 (g) 09:30, Jan 23 (h) 04:00, Jan 23
18. $13 \times 13 = 169$
19. (a) 21 (b) 61
20. 121, 123
21. (a) \$3.50 (b) \$7.00
22. 7
23. 2922 d
24. 10
25. $9567 + 1085 = 10\,652$
26. 6
27. \$10.88
28. 21
29. 04:15
30. Answers vary.
31. (a) 16, 20, 22 (b) 64, 128, 256 (c) 32, 44, 58

EXERCISE 1.13

NETWORKS

Networks a, b, c, and e are traversable.

Euler's rules for traversability:

Note first that odd vertices always occur in pairs and so there will always be an even number of odd vertices.

(i) If a network has more than 2 odd vertices, it is not traversable.
(ii) If a network has exactly 2 odd vertices, it is traversable, but you must start at an odd vertex and end at an odd vertex.
(iii) If all vertices are even, then the network is traversable.

1.14 REVIEW EXERCISE

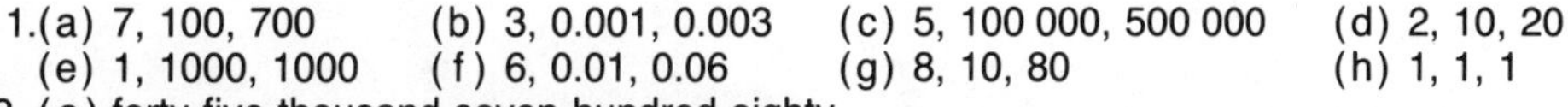

1. (a) 7, 100, 700 (b) 3, 0.001, 0.003 (c) 5, 100 000, 500 000 (d) 2, 10, 20
(e) 1, 1000, 1000 (f) 6, 0.01, 0.06 (g) 8, 10, 80 (h) 1, 1, 1
2. (a) forty-five thousand seven hundred eighty
(b) five and seventy-two hundredths
(c) eight million nine hundred twenty-six thousand
(d) eight hundred sixty-seven thousandths
(e) five thousandths
(f) fifty-six and eight hundred one thousandths
(g) three thousand four hundred sixty-nine
(h) two hundred four and two tenths

3. (a) 17 202 (b) 549 (c) 71 463 (d) 42 (e) 54 585 (f) 3544
(g) 603 180 (h) 156
4. 2000 h
5. (a) 459 (b) 150 (c) 123 (d) 364 (e) 12 (f) 19
(g) 96 (h) 29 (i) 73
6. (a) 6600 (b) 2.3 (c) 58 000 (d) 9.35
7. (a) 600 (b) 0.5 (c) 200 000 (d) 0.006 (e) 5000 (f) 0.002
8. Answers vary.
(a) 800 (b) 27 (c) 20 (d) 300 (e) 6100
9. Estimates vary.
(a) 65.45 (b) 2.546 (c) 123.39 (d) 2.1 (e) 93.88 (f) 0.02
(g) 21.711 (h) 144.7
10. Estimates vary.
(a) \$320.25 (b) \$123.03 (c) \$319.80 (d) \$691.60
11. 9, 14.4, 3.9
12. (a) \$147 500 (b) \$218 750 (c) \$248 750
13. (a) $\frac{3}{8}$ (b) $\frac{5}{6}$ (c) $4\frac{1}{4}$ (d) $11\frac{1}{2}$
14. (a) $\frac{2}{3}$ (b) 8 (c) $1\frac{1}{2}$ (d) $2\frac{1}{4}$
15. (a) $1\frac{1}{6}$ (b) $5\frac{13}{15}$ (c) $6\frac{17}{18}$ (d) $10\frac{3}{4}$
16. (a) $\frac{1}{14}$ (b) $1\frac{7}{8}$ (c) $2\frac{2}{5}$ (d) $1\frac{5}{12}$
17. (a) $\frac{5}{12}$ (b) $\frac{7}{12}$ (c) $2\frac{1}{2}$ (d) $\frac{19}{80}$
(e) $3\frac{5}{24}$ (f) $5\frac{5}{12}$
18. (a) 288 (b) 351 (c) 360 000
19. 101
20. 35%
21. (a) -1 (b) -14 (c) -25 (d) -5 (e) 0 (f) 0
22. (a) 7 (b) 21 (c) -17 (d) -61
23. (a) 8 (b) 9 (c) -1
24. (a) 0.875 (b) $0.\overline{27}$
25. (a) $\frac{33}{50}$ (b) $\frac{7}{9}$
26. (a) -5 (b) -15 (c) $-\frac{5}{8}$ (d) $-1\frac{3}{4}$ (e) $-\frac{13}{18}$ (f) $\frac{5}{8}$
(g) -1 (h) $-5\frac{3}{10}$
27. (a) 1 (b) 4 (c) 5 (d) 14 (e) 204
28. (a) 39, 44, 50 (b) 8, 4, 2 (c) 15, 9, 2

1.15 CHAPTER 1 TEST

1.(a) 42 741 (b) 1019 (c) 14 574 (d) 85
2.(a) 92 (b) 30 (c) 90 (d) 6 (e) 20
3.(a) 6000 (b) 0.9 (c) 10 000 (d) 800 (e) 0.07 (f) 8
4.(a) 279.17 (b) 10.28 (c) 3.06 (d) 5.4
5. 30.5 m (nearest tenth)
6.(a) $6\frac{1}{6}$ (b) $4\frac{1}{10}$ (c) $15\frac{3}{10}$ (d) $\frac{9}{22}$
7. 435
8.(a) 17 (b) -19 (c) -2 (d) 16
9.(a) $1\frac{5}{6}$ (b) $1\frac{2}{5}$
10. 13:00

REVIEW AND PREVIEW TO CHAPTER 2

FRACTIONS AND DECIMALS

ADDITION AND SUBTRACTION

1. (a) 2 (b) 5 (c) 8 (d) 5 (e) 2 (f) 9 (g) 9 (h) 38 (i) 97
2. (a) $\frac{3}{4}$ (b) $\frac{3}{5}$ (c) $\frac{4}{5}$ (d) 1 (e) $2\frac{5}{12}$ (f) $1\frac{1}{6}$ (g) $5\frac{17}{20}$ (h) $\frac{1}{24}$ (i) $2\frac{3}{8}$ (j) $1\frac{13}{18}$ (k) $2\frac{5}{28}$ (l) $\frac{23}{56}$
3. (a) $\frac{3}{4}$ (b) $2\frac{11}{15}$ (c) 1 (d) $3\frac{23}{24}$ (e) $\frac{4}{15}$ (f) $4\frac{7}{15}$ (g) 1 (h) $3\frac{3}{4}$
4. (a) $1\frac{1}{2}$ (b) $4\frac{7}{8}$ (c) $2\frac{5}{8}$ (d) $-\frac{1}{8}$ (e) $6\frac{1}{8}$ (f) $5\frac{1}{4}$ (g) 1 (h) $4\frac{1}{3}$

MULTIPLICATION AND DIVISION

1. (a) $\frac{10}{21}$ (b) $\frac{1}{10}$ (c) $\frac{4}{25}$ (d) $\frac{5}{6}$ (e) 3 (f) $\frac{2}{3}$ (g) $3\frac{1}{8}$ (h) 1 (i) $4\frac{1}{8}$ (j) $\frac{1}{3}$
2. (a) $1\frac{7}{8}$ (b) $\frac{15}{16}$ (c) $2\frac{1}{12}$ (d) $\frac{5}{9}$ (e) $1\frac{7}{25}$ (f) 4 (g) $\frac{7}{16}$ (h) $2\frac{1}{8}$ (i) $\frac{4}{9}$
3. (a) $3\frac{15}{16}$ (b) $\frac{3}{4}$ (c) $\frac{11}{20}$ (d) $1\frac{7}{48}$ (e) $1\frac{7}{8}$ (f) $\frac{13}{24}$ (g) $1\frac{11}{30}$ (h) $\frac{5}{16}$ (i) $\frac{169}{225}$
4. (a) $\frac{27}{32}$ (b) $\frac{27}{32}$ (c) $15\frac{15}{32}$ (d) $25\frac{7}{8}$ (e) $8\frac{3}{4}$ (f) $\frac{5}{8}$

OPERATIONS WITH DECIMALS

1. (a) 13.31 (b) 192.9 (c) 10.29 (d) 27.2 (e) 2.756 (f) 6.33 (g) 15.98
2. (a) 54.48 (b) 19.5 (c) 9.1 (d) 367.1 (e) 27.185 (f) 588.55 (g) 5.675
3. (a) 44.935 (b) 37.05 (c) 463.044 (d) 0.399 (e) 9.063 (f) 14.274 (g) 1.825
4. (a) 5.63 (b) 2.5 (c) 59.8 (d) 0.715 (e) 6.8 (f) 5.61 (g) 5.61
5. 9.77 km

ESTIMATION

1. (a) 300 000 (b) 8000 (c) 2 000 000 (d) 50 000 (e) 0.6 (f) 0.09 (g) 0.1 (h) 3
2. (a) 5 863 000 (b) 131.9 (c) 2108 (d) 0.34
3. (a) 17.86 (b) 104.3 (c) 45 467 (d) 0.607
4. (a) 120 000 (b) 100 (c) 24 000 000 (d) 15 (e) 28 (f) 12 000 (g) 1.8 (h) 35
5. (a) 12 (b) 12 (c) 5 (d) 10 (e) 10 000 (f) 12 (g) 1000

EXERCISE 2.1

1. (a) 3^4 (b) 4^3 (c) 5^4 (d) $(-3)^5$ (e) 7^5 (f) 9^3 (g) 5^1
2. (a) 27 (b) 25 (c) 49 (d) 343 (e) 125 (f) 64 (g) 16 (h) 16 (i) 64 (j) 10 000 (k) 1 (l) 0

3. (a) $2^2, 3$ (b) $2^2, 3^2$ (c) $2^5, 5^2$ (d) $2^4, 5^2$ (e) $2, 3^2, 5$ (f) $2^4, 3^2$
(g) $3^2, 7$ (h) $3^2, 5^2$ (i) $2^2, 3^3$ (j) $2^2, 5^4$ (k) $2^5, 3$ (l) $2^4, 3^2, 5$

4. (a) 25 (b) 91 (c) 169 (d) 1853 (e) 625 (f) 14 167
(g) 289 (h) 3887

5. (a) $5 \times x \times x \times x \times x$ (b) $3 \times y \times y$ (c) $2 \times 3 \times x \times x \times x$
(d) $x \times x \times y \times y$ (e) $2 \times 5 \times 5 \times x \times x$ (f) $2 \times 7 \times x \times x \times x$
(g) $x \times y \times y$ (h) $2 \times x \times x \times y \times y \times y$ (i) $-2 \times x \times x \times y \times y \times y$

6. (a) x^4 (b) y^6 (c) $5x^2y^3$ (d) $(5x)^3$ (e) $-3x^2y^2$ (f) $(-3xy)^3$
(g) $(x + y)^3$

7. (a) 25 (b) 250 (c) 127 (d) 343 (e) 13 (f) 1000

8. (a) 1 (b) 5 (c) 5 (d) -5 (e) -125 (f) 25

9. (a) 625 (b) 216 (c) 256 (d) 45

10. (a) -1 (b) 9 (c) 242 (d) -26 (e) -117 (f) 133

11. (a) $x^2 + y^2$ (b) $(x - y)^2$ (c) equal

EXERCISE 2.2

1. (a) 8 (b) 9 (c) 16 (d) 16 (e) 25 (f) 32

2. (a) 12 (b) 7 (c) 10 (d) 14 (e) 7 (f) 8
(g) 2 (h) 5

3. (a) 3 (b) 3 (c) 2 (d) 5
(e) 3 (f) 0 (g) 5 (h) 1

4. (a) 5 (b) 3 (c) 3 (d) 6 (e) 2 (f) 3
(g) 3 (h) 6

5. (a) x^8 (b) a^{12} (c) b^3 (d) m^6 (e) a^2b^5 (f) a^7b^4

6. (a) a^4 (b) b (c) n^9 (d) x^3 (e) a^4 (f) x

7. (a) $15a^4$ (b) $-8x^5$ (c) $6a^3b^5$ (d) $10x^4$ (e) $15m^2n$ (f) $12x^5$
(g) $30y^4$ (h) $-6a^8$ (i) $-14x^8$ (j) $-6x^3$

8. (a) $4a^2$ (b) $3xy^4$ (c) a (d) $8m$ (e) -5 (f) $9b^2$
(g) $7q$ (h) -3 (i) $-3x^9y$ (j) $4x^2$

9. (a) 6^9 (b) 2^8 (c) 7^{10} (d) 8^6 (e) 5^9 (f) 2^6
(g) 3^5 (h) 2^6 (i) 4^6 (j) 5^3

10. (a) x^9 (b) y^6 (c) m^5 (d) t^4 (e) x^7 (f) y
(g) t^9 (h) x^7

11. (a) 100 (b) 54 (c) 24 (d) 36 (e) 7 (f) 8
(g) 7 (h) 5

EXERCISE 2.3

1. (a) 6 (b) 12 (c) 15 (d) 16 (e) 10 (f) 20
(g) 6 (h) 6

2. (a) 6 (b) 12 (c) 12 (d) 3 (e) 3 (f) 2
(g) 1 (h) 4

3. (a) 2 (b) 15 (c) 9 (d) 3 (e) 15 (f) 1
(g) 3 (h) 3, 3

4. (a) x^8 (b) a^9 (c) a^6b^3 (d) x^5y^{15} (e) $a^5b^5c^5$ (f) b^{24}
(g) $8x^9$ (h) $a^{15}b^6$ (i) $9a^{10}$ (j) $27x^3y^6$ (k) $125a^{24}$ (l) $64x^6y^3z^3$

5. (a) $\frac{x^5}{y^5}$ (b) $\frac{a^4}{9}$ (c) $\frac{x^6}{y^3}$ (d) $\frac{a^4}{b^{20}}$ (e) $\frac{9x^2}{y^2}$ (f) $\frac{8x^6}{w^3}$
(g) $\frac{25a^4}{4b^6}$ (h) $\frac{9a^2}{b^6}$

6. (a) $-8a^9$ (b) $9x^6$ (c) $9x^4$ (d) $6a^7b$ (e) $\frac{a^{15}b^3}{c^{12}}$ (f) $4a^8$
(g) $-64x^6y^{18}$ (h) $-64a^{15}$ (i) $\frac{-27a^6}{b^3}$

7. (a) $16x^{11}$ (b) $2b^3$ (c) $3x^5y^{13}$ (d) $-4n^6$ (e) $-8xy$ (f) $-4m^4n^2$

EXERCISE 2.4

1. (a) x^{-2} (b) a^7 (c) y^4 (d) b^{-4} (e) a^8 (f) x (g) 1 (h) 1
2. (a) x^5 (b) b^{-3} (c) a^{-3} (d) x^{-5} (e) m^{-5} (f) n^7 (g) 0 (h) m^{-4}
3. (a) 1 (b) $\frac{1}{2}$ (c) 1 (d) $\frac{1}{10}$ (e) $\frac{1}{9}$ (f) 1 (g) $\frac{1}{16}$ (h) $\frac{1}{8}$ (i) $\frac{1}{16}$ (j) $\frac{1}{32}$
4. (a) $15x^3$ (b) a^5b^{-3} (c) 6 (d) x^2 (e) $9m^{-2}$ (f) $15a^{-6}$ (g) m^7 (h) $8a^{15}b^{-9}$ (i) x^5y^2 (j) $35x^4y^{-2}$ (k) $7m^{-2}n^{-2}$ (l) 15
5. (a) a^2 (b) $3x^{-5}$ (c) $4b^{-10}$ (d) $m^{-3}n$ (e) 1 (f) $3a^2b^{-2}$ (g) b^4 (h) y^{-4} (i) $16x^{-4}$ (j) x^7 (k) $16b^8$ (l) $\frac{a^{-14}}{2}$
6. (a) $1\frac{1}{5}$ (b) $\frac{7}{12}$ (c) $\frac{1}{25}$ (d) $\frac{1}{16}$ (e) 2 (f) 4 (g) 1 (h) $\frac{9}{20}$ (i) 10 (j) $\frac{1}{100}$ (k) 9 (l) $\frac{1}{100}$

EXERCISE 2.5

1. (a) 1000 (b) 100 000 000 (c) 100 000 (d) 100 (e) 1 (f) 10 000 000 (g) 10 (h) 100 000 000 000 (i) 1 000 000
2. (a) 10^4 (b) 10^5 (c) 10^2 (d) 10^6 (e) 10^1 (f) 10^0 (g) 10^8 (h) 10^3
3. (a) 0.01 (b) 0.1 (c) 0.000 001 (d) 0.000 000 01 (e) 0.0001 (f) 1 (g) 0.000 01 (h) 0.001 (i) 0.000 001
4. (a) 10^{-3} (b) 10^{-4} (c) 10^{-1} (d) 10^{-6} (e) 10^{-2} (f) 10^{-5}
5. (a) 10 000 000 (b) 100 000 (c) 1 000 000 000 000 (d) 10 000 (e) 100 000 000 (f) 100 000 (g) 1000 (h) 100
6. (a) 0.000 001 (b) 1 (c) 0.000 01 (d) 0.000 000 01 (e) 0.1 (f) 100 (g) 0.000 001 (h) 1 000 000 (i) 1000 (j) 1
7. (a) 10^{-11} (b) 10^{-6} (c) 10^1 (d) 10^0 or 1 (e) 10^1 (f) 10^3 (g) 10^3 (h) 10^{-3}
8. (a) 10^6 (b) 10^2 (c) 10^{-1} (d) 10^1 (e) 10^0 or 1 (f) 10^4
9. (a) 10^4 g (b) 10 g (c) 10^{-1} g (d) 10^6 g (e) 1 g (f) 10 g (g) 1 g (h) 100 g
10. (a) 1 L (b) 10^{-1} L (c) 10^4 L (d) 10^6 L (e) 10^{-3} L (f) 10^3 L (g) 10^2 L (h) 10^9 L
11. (a) 10^1 (b) 10^1 (c) 10^2 (d) 10^0 (e) 10^9 (f) 10^5 (g) 10^0 (h) 10^0 (i) 10 (j) 10^3 (k) 10^{-2}

EXERCISE 2.6

1. (a) 425 000 (b) 5640 (c) 156 750 (d) 25 (e) 750 (f) 1.2 (g) 3 650 000 (h) 45 750 (i) 0.3
2. (a) 67.41 (b) 0.562 (c) 3.8975 (d) 3.65 (e) 0.000 025 (f) 0.0525 (g) 392.75 (h) 1.253 75 (i) 2750
3. (a) 0.6425 (b) 1.573 (c) 3.458 36 (d) 250 (e) 0.000 035 (f) 0.000 005 (g) 0.000 04 (h) 0.000 002 4 (i) 236.45
4. (a) 456 (b) 2568 (c) 6 572 900 (d) 27.58 (e) 350.07 (f) 27.045 (g) 0.25 (h) 0.004 (i) 1.243

5. (a) 275 000 (b) 1 520 000 (c) 570 000 (d) 3 250 000
(e) 3600 (f) 65 000 (g) 5.39 (h) 0.0051
(i) 0.0625 (j) 1.25 (k) 45 000 (l) 0.000 027
6. (a) 275 (b) 5650 (c) 563 000 (d) 520 000
(e) 250 (f) 3.6 (g) 0.0685 (h) 0.0538
(i) 6.452 (j) 0.0256 (k) 0.000 000 052 85 (l) 52.7
7. (a) 0.0525 (b) 0.053 27 (c) 6.582 (d) 0.000 49
(e) 0.0639 (f) 0.000 36 (g) 457 (h) 26 500
(i) 250 (j) 57 000 (k) 0.0575 (l) 375
(m) 2 750 000
8. (a) 5.26 (b) 52.6 (c) 526 (d) 0.000 005 26
(e) 0.000 000 526 (f) 0.000 526
9. (a) 0.016 75 (b) 0.000 167 5 (c) 0.000 001 675 (d) 16.75
(e) 1675 (f) 1 675 000
10. (a) 3 568 000 (b) 3568 (c) 35 680 000 (d) 3.568
(e) 35.68 (f) 0.003 568
11. (a) 0.002 358 (b) 0.000 023 58 (c) 0.2358 (d) 2.358
(e) 23 580 (f) 23.58 (g) 235.8 (h) 0.000 235 8
(i) 0.023 58
12. (a) 24.5 m (b) 250 m (c) 0.125 m (d) 12.5 m
(e) 355.5 m (f) 1500 m (g) 0.0375 m (h) 250 m
13. (a) 2.5 L (b) 6500 L (c) 0.375 L (d) 2.5 L
(e) 0.000 875 L (f) 475 000 L
14. (a) 3.5 kg (b) 0.0035 kg (c) 2.5 kg (d) 45 kg
(e) 0.0005 kg (f) 500 kg
15. Answers vary.

EXERCISE 2.7

1. (a) 5.76×10^{2} (b) 2.435×10^{1} (c) 2.48×10^{-2} (d) 4.265×10^{9}
(e) 2.5×10^{-4} (f) 7.89×10^{-1} (g) 8.1×10^{-2} (h) 5.28×10^{5}
(i) 3.60×10^{-2} (j) 5.075×10^{2} (k) 2.5×10^{3} (l) 2.75×10^{-7}
(m) 5.75×10^{0} (n) 4.03×10^{5} (o) 2.05×10^{-4} (p) 5.205×10^{4}
2. (a) 5300 (b) 0.51 (c) 0.000 95 (d) 526 500
(e) 0.041 (f) 7.5 (g) 0.75 (h) 0.000 000 010 8
(i) 10 050 (j) 0.000 208 (k) 0.030 25 (l) 0.005 125
(m) 52 500 (n) 0.000 414 (o) 20 050 000 (p) 0.000 500 8
3. (a) 2.50×10^{6} (b) 9.53×10^{1} (c) 1250 (d) 2.5×10^{-3}
(e) 0.5625 (f) 5.1×10^{0} (g) 2.570×10^{-2} (h) 0.000 058
4. (a) 4.8×10^{14} (b) 1.9×10^{19} km

EXERCISE 2.8

1. (a) 9 (b) 5 (c) 3 (d) 11 (e) 10 (f) 2
(g) 8 (h) 7 (i) 1 (j) 6 (k) 4 (l) 12
2. c, d, e, g, and h

EXERCISE 2.9

1. (a) 15.7 (b) 3.1 (c) 0.62 (d) 8.6 (e) 76.1 (f) 0.28
(g) 19.3 (h) 1.9
2. (a) 5.7 (b) 7.7 (c) 4.6 (d) 6.2 (e) 24.4 (f) 27.4
(g) 38.5 (h) 76.9 (i) 0.65
3. (a) 9.3 (b) 3.6 (c) 7.6 (d) 2.5 (e) 24.0 (f) 37.2

EXERCISE 2.10

1. (a) 62 km (b) 73 km (c) 76 km (d) 85 km
2. (a) 4.4 s (b) 4.9 s (c) 5.5 s (d) 7.8 s (e) 9.2 s (f) 5.3 s
3. (a) 2.0 s (b) 1.0 s
4. 1.3 m

EXERCISE 2.11

1. (a) 5.00 cm (b) 7.01 m (c) 3.50 cm (d) 5.39 m (e) 4.18 cm (f) 5.89 km
   (g) 6.46 m (h) 14.13 cm (i) 4.34 cm (j) 3.54 m
2. (a) 134.6 cm (b) 1.8 m (c) 111.8 cm (d) 14.1 cm
3. 8.8 m
4. (a) 4.9 m (b) 7.6 m (c) 2.9 m (d) 11.8 m (e) 9.5 m
5. 157 m
6. 6.3 m
7. (a) $d_1 = 14.4$ m, $d_2 = 11.3$ m, $d_3 = 15.9$ m, $d_4 = 13.2$ m

EXERCISE 2.12

1. 11 min gained
2. 5.8 h
3. 3750
4. 36, 37, 38
5. 12, 15
6. 11
7. Answers for parts a, b, and c vary.
8. 60 km/h
9. Ingrid
10. 12 km/h
11. 5
12. 10
13. approximately 4.8×10^7 mL
14. Jean should first travel to Ford, then, in order, to Etain, Cartier (via Drouin and Bisonville), Almont, and finally Bisonville (via Cartier).
15. Answers vary.
16. Answers vary.
17. Answers vary.
18. approximately 3.7 times
19. Answers vary.
20. approximately 5 times
21. $118\frac{8}{9}$
22. Answers vary.

EXERCISE 2.13

2. (a) yes (b) no
3. They are straight lines terminating in the lower right corner.
4. They terminate in the lower left corner (after 1 rebound).
5. (a) The lengths of corresponding sides are proportional.
   (b) yes
6. (a) The length is 1 unit longer than the width.
   (b) They cross each square.

2.14 REVIEW EXERCISE

1. (a) 27 (b) 8 (c) 64 (d) 25 (e) 16 (f) 16
   (g) 36 (h) 32 (i) 9
2. (a) $\frac{1}{2}$ (b) $\frac{1}{8}$ (c) $\frac{1}{32}$ (d) $\frac{1}{3}$ (e) $\frac{1}{9}$ (f) $\frac{1}{27}$
   (g) $\frac{1}{16}$ (h) $\frac{1}{16}$ (i) $\frac{1}{25}$ (j) 1 (k) -1 (l) 1
   (m) $\frac{5}{6}$ (n) $\frac{17}{72}$
3. (a) 100 000 (b) 1000 (c) 100 (d) 100 (e) 10 000 000 (f) 10
4. (a) 3265 (b) 0.046 24 (c) 54 600 (d) 36 800 (e) 0.6825 (f) 45 300
   (g) 456 200 (h) 3.25
5. (a) 3.456×10^2 (b) 3.478×10^1 (c) 2.56×10^5 (d) 4.2×10^7
   (e) 1.2×10^4 (f) 6.3×10^1 (g) 2.56×10^{-3} (h) 2.35×10^{-7}
   (i) 2.46×10^{-2} (j) 5.75×10^{-1} (k) 4.56×10^{-4} (l) 2.51×10^{-5}
6. (a) 22 500 (b) 375 000 (c) 9 300 000 (d) 52 500
   (e) 0.005 25 (f) 0.000 125 (g) 0.000 062 5 (h) 0.775
   (i) 1.25 (j) 0.041 25

7. (a) 8.1 (b) 8.5 (c) 9.2 (d) 11.2 (e) 12.2 (f) 7.1
(g) 2.6 (h) 2.7 (i) 4.4 (j) 0.5 (k) 0.3 (l) 0.9

8. (a) 81 (b) 243 (c) 18 (d) 27 (e) 9 (f) 3
(g) $\frac{1}{9}$ (h) $\frac{1}{3}$ (i) $\frac{1}{27}$ (j) 1 (k) 3 (l) 1

9. (a) -18 (b) 81 (c) 18 (d) 81 (e) $\frac{1}{81}$ (f) $\frac{2}{27}$
(g) 18 (h) -27

10. (a) 1 (b) 125 250 (c) $1\frac{3}{25}$ (d) 1

11. (a) 4 (b) 1 (c) 4 (d) 8 (e) 5 (f) 4
(g) 2 (h) 5 (i) 0 (j) -1

12. (a) x^6 (b) y^7 (c) x^4 (d) x^3 (e) $15x^6$ (f) $2y^4$
(g) $4x$ (h) $6y^5$

13. (a) $6x^7$ (b) $-4x^5$ (c) $6x^8$ (d) $-15x^4y^4$ (e) $10x^7$ (f) $6x^7$
(g) $16x^8$ (h) $125x^6$ (i) x^9y^3 (j) $2x^8$

14. (a) $6x^9$ (b) $12x^6y^4$ (c) $-2x^2$ (d) $6x^5$ (e) $\frac{-20x^2}{3}$ (f) $18x^3y^5$
(g) $10x^4y^2$ (h) $4x^3y^5$

15. (a) $\frac{x^3}{y^3}$ (b) $\frac{8x^3}{27y^3}$ (c) $\frac{x^6}{y^4}$ (d) $\frac{8x^{12}}{27y^6}$ (e) $\frac{25x^2}{9y^2}$ (f) $\frac{-27x^6}{64y^6}$
(g) $\frac{-4x^6}{25y^2}$ (h) $\frac{-9x^6}{4y^4}$ (i) $\frac{-4x^2}{9y^2}$ (j) $\frac{-8x^6}{125y^3}$

16. (a) $18x^2y^{-4}$ (b) $6x^{-5}$ (c) $-6x^{-1}y^{-1}$ (d) 45 (e) $4xy^2$ (f) $3x^{-5}y$
(g) $\frac{1}{2}$ (h) $5x^{-4}$ (i) $3x^{-1}y^{-6}$ (j) $3xy^2$

17. (a) 28.3 cm (b) 7.1 cm

18. (a) 25.0 cm (b) 14.1 cm (c) 6.7 cm (d) 11.2 cm

19. 7.14 m

20. 10.6 m

21. (a) 1540 m (nearest ten) (b) 20 m

2.15 CHAPTER 2 TEST

1. (a) 125 (b) $\frac{1}{8}$ (c) 1 (d) 5 (e) 9 (f) 11

2. (a) 8.7 (b) 11.2 (c) 17.3

3. (a) 3.25×10^5 (b) 5.675×10^1 (c) 2.45×10^{-3}

4. (a) 51 500 (b) 0.006 39

5. (a) $-15x^5$ (b) $25x^4y^6$ (c) 1 (d) $\frac{-1}{3x^2y}$ (e) $8y^5$ (f) $-9xy$

6. 2 m

7. (a) 15.6 cm (b) 17.3 cm

8. (a) 1.4 m (b) 1.7 m

REVIEW AND PREVIEW TO CHAPTER 3

INTEGERS

1. (a) 6 (b) -11 (c) -8 (d) 7 (e) 4 (f) 12
(g) 30 (h) -10 (i) 6 (j) 30 (k) -9 (l) -3
(m) 2 (n) 15 (o) -2 (p) -24

2. (a) -3 (b) 1 (c) -1 (d) 4 (e) -8 (f) -9
(g) -4 (h) 9

3. (a) -4 (b) 10 (c) 21 (d) -54 (e) -1 (f) 13
(g) -9

THE DISTRIBUTIVE PROPERTY

1. (a) $2x + 12$ (b) $3m + 21$ (c) $4t - 4$ (d) $5 - 5m$
(e) $6x - 10$ (f) $-2s - 8$ (g) $-24t + 42$ (h) $-3m + 2$
(i) $24w + 18t$ (j) $-10m + 15n$ (k) $-5a + 5b + 5c$

EXPONENTS

1. (a) x^9 (b) a (c) b^6 (d) x^{-5} (e) z^{-3} (f) a^{-17} (g) b^7 (h) w^{-2}
2. (a) 12 (b) 13 (c) 3 (d) −18 (e) 36 (f) 5
3. (a) $15a^7$ (b) $28y^3$ (c) $-12x^8$ (d) $35a^8$ (e) $7a^8b$ (f) $-15a^3b^5$ (g) $-20x^2y^3$ (h) $15a^{-9}$
4. (a) $3b^2$ (b) $5x^6$ (c) $3y^2$ (d) $4x^7$ (e) $9b^2$ (f) $6x^5$ (g) $17y^4$ (h) $-8w^5$
5. (a) $5x^4$ (b) x^6y^{10} (c) $4x^3$ (d) $25a^2$ (e) $15a^4$
6. (a) $\frac{1}{9}$ (b) 4 (c) $\frac{5}{6}$ (d) 4 (e) $1\frac{1}{3}$ (f) $1\frac{1}{2}$

SUBSTITUTION

1. (a) 7 (b) 9 (c) 10 (d) 12 (e) 0 (f) 5
2. (a) 3000 (b) 500 (c) 48 (d) 1200
3. (a) 104 (b) 378 (c) 796

EQUATIONS

1. (a) 9 (b) 3 (c) 4 (d) 14 (e) 7 (f) 19
2. (a) 5 (b) 5 (c) 4 (d) −4 (e) 7 (f) 5
3. (a) 8 (b) 15 (c) 24 (d) 6 (e) 30 (f) 14

MENTAL MATH

1. (a) 23 000 (b) 4300 (c) 120 (d) 450 (e) 0.6 (f) 765 (g) 34.3 (h) 120 000
2. (a) 4.5 (b) 45 (c) 12 (d) 200 (e) 0.07 (f) 0.3 (g) 0.004 (h) 0.56 (i) 70 (j) 0.066
3. (a) 0.78 (b) 9.8 (c) 0.048 (d) 0.012 (e) 0.04 (f) 0.009 (g) 12 (h) 0.2
4. (a) 430 (b) 2000 (c) 7000 (d) 65 000 (e) 3 000 000 (f) 4 (g) 8900 (h) 1 (i) 2 000 000 (j) 7
5. (a) 600 (b) 6600 (c) 6000 (d) 1400 (e) 2000 (f) 12 000 (g) 4600 (h) 9000
6. (a) 60 (b) 30 (c) 200 (d) 111 (e) 222 (f) 4000 (g) 13 (h) 121 (i) 323 (j) 2002
7. (a) 45 (b) 20 (c) 44 (d) 200 (e) 30 (f) 330 (g) 1 (h) 750
8. (a) 50% (b) 25% (c) 75% (d) 10%

EXERCISE 3.1

1. (a) $c = 0.75n$ (b) $d = 80t$ (c) $a = \frac{h}{n}$ (d) $C = 400 + 35m$
2. (a) $C = 60 + 0.22(d - 150)$ (b) $104
3. (a) $d = \frac{m}{70} \times D$ (b) $d = \frac{a}{a + 12} \times D$ (c) 1.4 grains for both (d) 25.1 mg (Clark's), 23.5 mg (Young's)
4. (a) $s = 9.5h$ (b) $199.50
5. (a) $P = 5x + 3y + 16$ (b) $P = 10w + 12$
6. (a) $A = 20m$ (b) $A = 15st - x^2 - m^2$

EXERCISE 3.2

1. (a) 7 (b) 8 (c) −4 (d) −5 (e) 1 (f) −1
2. (a) a (b) x (c) t (d) m (e) none (f) n
3. (a) like (b) like (c) unlike (d) unlike (e) like (f) unlike
4. (a) $3x$ and $8x$, $4y$ and $-3y$ (b) $4x^2$ and $6x^2$, $-3x$ and $5x$ (c) $7a$ and $4a$, $5b$ and $3b$, $-2ab$ and $6ab$ (d) $-2m^2$ and $7m^2$, $3n^2$ and $-n^2$, $3mn$ and $-mn$ (e) 12 and −4, x and $5x$, $-3xy$ and xy

5. (a) $11x$ (b) $3y$ (c) $5t$ (d) $-4b$ (e) $9x$ (f) $-8d$
(g) $-4x$ (h) $3a$
6. (a) $14x$ (b) y (c) $-8a$ (d) $-9y$ (e) $15t$ (f) $-2r$
7. (a) $7x + 11y$ (b) $10a + 6b$ (c) $3s + 5t$ (d) $2m - 5n$ (e) $4x - 4y$ (f) $-7a - b$
(g) $4a + 7b - 3$
8. (a) $3x - 2y$ (b) $7r + 3t - 9s$ (c) $6a + b + 4c$ (d) $9x + 4y - 16$
(e) $a + b + 4c$
9. (a) $xy - 4x + 5y$ (b) $-8x^2 - 3x$ (c) $t^3 - 3t$ (d) $-4x^2 + 2x + 7$
(e) $-2a^2 + 11a - 11$
10. (b) $a + 3b + 7$ 11. (a) $14a - 2$ (b) $20x + 10y$ (c) $46y$

EXERCISE 3.3

1. (a) 13 (b) 6 (c) 6 (d) 1
2. (a) 6 (b) 36 (c) 10 (d) 35
3. (a) 0 (b) −13 (c) 5 (d) 0 (e) 16 (f) 11
4. (a) 6 (b) 21 (c) −12 (d) −13
5. (a) −12 (b) $\frac{5}{6}$ (c) $-4\frac{3}{4}$ (d) −6
6. (a) $560 (b) $720 (c) $720

EXERCISE 3.4

1. (a) $7x + 9$ (b) $6s + 11$ (c) $5w + 2$ (d) $9m + n$ (e) $11x - 9y$ (f) $9y - 8w$
(g) $-8x - 7$ (h) $-7t - 4$ (i) $-7s$
2. (a) $-2x - 5$ (b) $-7 + 3m$ (c) $-6t + 8$ (d) $-x^2 - 3x - 4$
(e) $-2t^2 + 4t + 5$ (f) $5 + 3m + m^2$
3. (a) $7x + 9$ (b) $9t - 4$ (c) $13s - 2t$ (d) $9m - 5m$
4. (a) $4x + 9$ (b) $3m + 5$ (c) $-2x + 10y$ (d) $2m + 2n$
5. (a) $x - 1$ (b) $m - 1$ (c) $s - t$ (d) $2w + 8$
(e) $4a + 1$ (f) $-x - 2y$ (g) $-m + n$
6. (a) $7x^2 + 7x + 13$ (b) $7m^2 - 8m - 3$ (c) $t^2 + 3t + 8$ (d) $-2y^2 + 2y - 2$
(e) $m^2 + 10m - 18$ (f) $3x^2 - 2x$ (g) $3t^2 + 5t - 5$ (h) $-4m^2 - 2m + 8$
7. (a) $10x + 10$ (b) $9t - 10$ (c) $-9a + 12$ (d) $2m + 8n$
(e) $3y$ (f) $5s - 16t$ (g) $2x + 2y$
8. (a) 27 (b) −4 (c) 7 (d) 1
9. (a) 48 (b) −6 (c) −16 (d) 10
10. (a) $P = 2n - 700$ (b) $1700 (c) $500 loss (d) yes, no
11. (a) $P = 17n - 5000$ (b) $1800 (c) $97 000 (d) no ($1600 loss)
(e) 294 — loss $2, 295 — profit $15

EXERCISE 3.5

1. (a) $6xy$ (b) $20a^7$ (c) $8x^5$ (d) $-21a^6$ (e) $5x^4$ (f) $-21mn$
(g) $24st$ (h) $-15x^5$ (i) $4x^2y$ (j) $28mn$ (k) $-30a^6$ (l) $9x^2$
2. (a) $15xy$ (b) $24a^3b^3$ (c) $-6x^2y^2$ (d) $-28x^3y^2$ (e) $-15a^4b^5$ (f) $-16x^4y^5$
(g) $-12s^6t^3$ (h) $-14a^3b^3c^4$ (i) $20p^3q^2$ (j) $35a^5b^5$ (k) $-24a^2b^2c^2$ (l) $6x^3y^3z^3$
3. (a) $2x + 6$ (b) $6a + 3$ (c) $4 + 8b$ (d) $-15x^2 + 10$
(e) $-2x + 10$ (f) $-6 + 18a$ (g) $-x^2 - 6x - 7$ (h) $-3x + 3y$
4. (a) $3x^2 + x$ (b) $2a + 6a^2$ (c) $-2b^3 - 6b^2 + 2b$ (d) $-3m + 6m^2$
(e) $-2a^2 + 4a$ (f) $ab - abc$ (g) $2x^3 - 4x^2 + 2x$
5. (a) $2x^2 - 2x - 6$ (b) $-3a^2 + 6a + 21$ (c) $-6x^3 + 6x^2 + 3x$ (d) $4b^3 + 4b^2 + 8b$
(e) $-3m^2 + m + 4$ (f) $2x - 2x^2 - 2x^3$
6. (a) $5x + 21$ (b) $23a - 9$ (c) $9m - 5$ (d) $-3d - 2$
(e) $11 - 18x + 8x^2$ (f) $a - 5b$ (g) $10a + 7b - 14c$
7. (a) $x - 29$ (b) $9x - 16y$ (c) $22a - 14$ (d) $-12x^2 + x + 17$
(e) $11m + 9$ (f) $13a - 12b + 6c$

8. (a) $-4x^2 - 17x$ (b) $5a^2 - 4a$ (c) $2x^3 - x^2 - 2x + 6$ (d) $b^2 - 28b$
(e) $a^2 + a + 8$ (f) $8ab - 4a^2$
9. (a) $14x + 22$ (b) $24m - 12$ (c) $23a - 6$ (d) $12x + 12$
10. (a) $20x^2 - 6x$ (b) $12x^2 + 17xy$

EXERCISE 3.6

1. (a) 4 (b) 3 (c) 9 (d) 17 (e) 8 (f) −7
(g) −7 (h) 2
2. (a) 3 (b) 9 (c) −8 (d) −3 (e) 6 (f) 0
(g) 7 (h) −3
3. (a) 2 (b) −11 (c) 2 (d) −5 (e) $-\frac{1}{2}$ (f) 3
(g) 3 (h) −3 (i) −4
4. (a) 4 (b) −4 (c) 2 (d) $4\frac{4}{5}$ (e) 6
5. (a) 8 (b) −9 (c) −4 (d) 1
6. (a) 22 (b) 5 (c) −2 (d) 1
7. (a) $5\frac{1}{2}$ (b) $\frac{1}{10}$ (c) $-2\frac{1}{2}$ (d) $\frac{-5}{14}$
8. $m - 46.73 = 158.60$; \$205.33
9. $m + 95.50 = 238.65$; \$143.15

EXERCISE 3.7

1. (a) 12 (b) 6 (c) 20 (d) 30 (e) 14 (f) 20
(g) 24 (h) 15 (i) 20
2. (a) 15 (b) 4 (c) 12 (d) 35 (e) 6 (f) 15
3. (a) 12 (b) 15 (c) 4 (d) 15 (e) 8 (f) 14
4. (a) $3\frac{1}{3}$ (b) $2\frac{1}{4}$ (c) −3 (d) −6 (e) 7 (f) 17
5. (a) 8 (b) 29 (c) −1 (d) −8 (e) 5 (f) $\frac{3}{4}$
6. (a) 5 (b) 7 (c) $-\frac{5}{11}$ (d) $\frac{14}{15}$ (e) 7
7. 14 L/100 km
8. 13.5 L/100 km

EXERCISE 3.8

1. $n + 8 = 20$
2. $n - 4 = 30$
3. $12n = 72$
4. $\frac{n}{4} = 11$
5. $t + (t + 2) = 32$ (t is Tom's age.)
6. $n + (n + 1) = 65$
7. $8 + 3n = 41$
8. $2z + z = 78$ (z is the distance Zee ran.)
9. 41
10. 59
11. 18
12. 147
13. 36 cows, 21 horses
14. \$150, \$200, \$235
15. 81, 82, 83
16. \$269.39
17. 58 km
18. 8
19. 88
20. \$33
21. $\ell = 26$ m, $w = 21$ m
22. 17
23. 0.9 m, 3.6 m

EXERCISE 3.9

1. (a) $x^2 + 4x + 3$ (b) $a^2 + 12a + 35$ (c) $m^2 + 9m + 18$ (d) $t^2 + 14t + 33$
(e) $d^2 + 13d + 42$ (f) $8 + 6m + m^2$ (g) $r^2 + 16r + 60$ (h) $s^2 + 13s + 12$
2. (a) $m^2 - m - 6$ (b) $x^2 - 11x + 30$ (c) $a^2 + a - 30$ (d) $b^2 + 3b - 18$
(e) $y^2 - 6y + 9$ (f) $m^2 - 4m - 60$ (g) $d^2 - 9$ (h) $s^2 - 15s + 56$
3. (a) $x^2 + 6x + 9$ (b) $y^2 - 4y + 4$ (c) $t^2 + 8t + 16$ (d) $a^2 - 10a + 25$
(e) $s^2 + 10s + 25$ (f) $x^2 + 14x + 49$ (g) $m^2 - 2m + 1$ (h) $n^2 + 2n + 1$
4. (a) $2x^2 + 13x + 6$ (b) $6a^2 - 11a + 4$ (c) $2m^2 - 3m - 14$ (d) $9x^2 - 1$
(e) $4b^2 - 12b + 9$ (f) $2d^2 + 7d - 30$ (g) $5 - 13r - 6r^2$ (h) $4r^2 - 7r - 15$
(i) $6t^2 - t - 15$ (j) $-15x^2 + 26x - 7$

5. (a) $7x^2 - 19x + 10$ (b) $3m^2 + m - 14$ (c) $1 - 25b^2$ (d) $9x^2 - y^2$
(e) $6a^2 + 5ab + b^2$ (f) $a^2 + 2ab + b^2$ (g) $3x^2 - 5xy + 2y^2$ (h) $6b^2 - bc - 2c^2$
6. (a) $9m^2 + 12m + 4$ (b) $25a^2 + 10a + 1$ (c) $9a^2 - 12a + 4$ (d) $16b^2 + 16b + 4$
(e) $4x^2 - 28x + 49$ (f) $36x^2 - 12x + 1$ (g) $4y^2 + 12y + 9$ (h) $25t^2 - 30t + 9$
(i) $36m^2 + 12m + 1$ (j) $1 - 4r + 4r^2$ (k) $1 + 12a + 36a^2$ (l) $4s^2 - 4s + 1$
7. (a) $2x^2 - 3x + 1$ (b) $6a^2 - 5a - 6$ (c) $4m^2 + 12m + 9$ (d) $2x^2 + 7xy + 6y^2$
8. (b) $x^3 + 5x^2 + 9x + 6$ (c) $2m^3 - 5m^2 + m + 2$
(d) $2s^3 - 3s^2 - 12s + 9$ (e) $2x^3 - 7x^2 - 5x + 4$
(f) $4b^3 + 3b + 18$ (g) $6a^3 + a^2 - 9a + 10$
(h) $x^4 + x^3 - 6x^2 - 5x - 1$ (i) $2m^4 - m^3 - 8m^2 + 10m - 3$

EXERCISE 3.10

1. (a) $2x + 4$ (b) $6x - 3$ (c) $a - 7$ (d) $2 - 6c$
(e) $-8x + 4$ (f) $-x + 7$ (g) $-6x - 12y$ (h) $2x^2 - 8x$
(i) $4a - 20$ (j) $-2y^2 - 3y$
2. (a) $7x + 26$ (b) $-a - 47$ (c) $6t - 60$ (d) $4x - 20$
(e) $12m + 2$
3. (a) $x^2 + 7x + 12$ (b) $x^2 - 11x + 30$ (c) $t^2 + 2t - 48$ (d) $a^2 + 5a - 14$
(e) $m^2 - 8m + 15$ (f) $y^2 - 7y + 6$ (g) $w^2 + 4w + 4$ (h) $a^2 - 10a + 25$
4. (a) $3x^2 + 5x + 2$ (b) $4y^2 - 25y + 25$ (c) $8y^2 + 18y + 7$ (d) $5m^2 + 23m + 12$
(e) $9a^2 - 1$ (f) $12s^2 + 28s + 15$ (g) $12x^2 - 20x + 7$ (h) $2m^2 + 19m + 24$
5. (a) $3x^2 + 27x + 60$ (b) $2a^2 - 6a + 4$ (c) $12b^2 + 20b - 8$ (d) $15m^2 + 35m + 10$
(e) $-2x^2 + 16x - 24$ (f) $-3x^2 - 18x - 15$
6. (a) $2a^2 + 6a - 16$ (b) $2x^2 - 5x + 22$ (c) $m^2 + 7m + 5$ (d) $2d^2 - 5d - 42$
(e) $-14b + 7$ (f) $-9m^2 - 4m - 7$
7. (a) $2x^2 + 4x - 15$ (b) $6a^2 - 21a - 30$ (c) $-4t^2 - 13t + 35$ (d) $9b^2 + 9b - 7$
(e) $-x^2 + 42x - 165$
8. (a) $4x^2 + 4x + 13$ (b) $10m^2 - 13m + 1$ (c) $8x - 10x^2$ (d) $8x - 6$
(e) $7m^2 + m + 2$ (f) $-23b + 24$
9. (a) $2x^2 + 2x + 13$ (b) $2a^2 + 4a + 34$ (c) $13x^2 - 2x + 2$ (d) $5m^2 + 2m + 58$
(e) $13b^2 + 14b + 26$ (f) $3x^2 + 12x + 14$
10. $125.6\ cm^2$

EXERCISE 3.11

1. (a) $2ab$ (b) xyt (c) $10b$ (d) $2bc$ (e) nt (f) $-2dem$
2. (a) xy (b) $3b$ (c) $2x$ (d) $-4c$ (e) mn (f) x^4
(g) $-6mn$ (h) $4x^4y$ (i) $-3a^2b$
3. (a) $4x + 3$ (b) $a - 2$ (c) $m + 1$ (d) $ab - 3b$ (e) $2a - b$ (f) $-3x - 2$
4. (a) $3x - 2$ (b) $2n + 3t$ (c) $2m^2 - 1$ (d) $3y - 2$ (e) $2r^3 + 4r$ (f) $3t + 5$
5. (a) $2x^2 - 3x + 4$ (b) $4 + a - 3a^2$ (c) $1 - 2b + 3a$ (d) $-1 + 2t - 3t^2$
6. (a) $x^3 + 4x^2$ (b) $2a^2 + 3a + 1$ (c) $p^3 + 2q^2$ (d) $2a^2 - 3ab - b^2$

EXERCISE 3.12

1. (a) 2 (b) 6 (c) $2x$ (d) $6a^2$
(e) $5b$ (f) $12a^2b$ (g) $10x^2y^2$ (h) $7s^2$
2. (a) 3 (b) 4 (c) 2 (d) $3a$ (e) $-2x$
3. (a) $4x + 7$ (b) $2a - b$ (c) $5t - 2$ (d) $3m + 4$
(e) $4x^2 + 5x - 6$ (f) $4y - 2z + y$ (g) $x^2 - 2x$ (h) $6x^2 - 2x - 4$
4. (a) $2(x + 3)$ (b) $3(a + 3)$ (c) $5(y - 2)$ (d) $4(s - 3)$
(e) $6(2m + n)$ (f) $7(y - 5w)$ (g) $2(4a + 3b + c)$ (h) $3(x - 4y + 5z)$
(i) $10(2d - 3e + 6f)$ (j) $a(b + c)$
5. (a) $2x(2x + 3)$ (b) $a(5b - 6c)$ (c) $3x(2y + t + 3a)$ (d) $2a(a - 2)$
(e) $6a(2a - b)$ (f) $5x(2 - x)$ (g) $2b(4b^2 - 2b + 3)$ (h) $3m(1 - 2m - 3m^2)$
6. (a) $xy(3 + 7t)$ (b) $mn(7 + 6m)$ (c) $ab(12a - 6b + 1)$ (d) $2rt(t + 3t^2 - 2r)$

7. (a) $9 \times (6 + 4) = 90$ (b) $5 \times (16 + 4) = 100$ (c) $16 \times (3 + 7) = 160$
(d) $24 \times (14 - 4) = 240$ (e) $63 \times (35 - 25) = 630$

EXERCISE 3.13

1. (a) 4 (b) 6 (c) t (d) 3m (e) 5x (f) 7k
(g) $2m^3$ (h) $8x^4$ (i) $10y^2$
2. (a) $x^2 - 4$ (b) $t^2 - 25$ (c) $a^2 - 16$ (d) $m^2 - 36$ (e) $r^2 - 49$ (f) $y^2 - 64$
3. (a) $x - 2$ (b) $a + 3$ (c) $m - 6$ (d) $m + 5$ (e) $2x - 1$ (f) $4 + n$
4. (a) $x - 7, x + 7$ (b) $m - 4, m + 4$ (c) $a - 12, a + 12$ (d) $b - 11, b + 11$
(e) $c - 10, c + 10$ (f) $d - 8, d + 8$ (g) $m - 1, m + 1$ (h) $x - 2, x + 2$
(i) $t - 7, t + 7$ (j) $r - 3, r + 3$ (k) $6 - x, 6 + x$ (l) $10 - x, 10 + x$
(m) $11 - t, 11 + t$ (n) $1 - b, 1 + b$ (o) $3 - a, 3 + a$
5. (a) $(2x - 4)(2x + 4)$ (b) $(1 - y)(1 + y)$ (c) $(4m - 1)(4m + 1)$ (d) $(3a - 5)(3a + 5)$
(e) $(10 - a)(10 + a)$ (f) $(3t - 6)(3t + 6)$ (g) $(2m - 7)(2m + 7)$ (h) $(1 - 6x)(1 + 6x)$
(i) $(7 - 3a)(7 + 3a)$
6. (a) $(2a - 3b)(2a + 3b)$ (b) $(4x - 5y)(4x + 5y)$ (c) $(a - 4b)(a + 4b)$
(d) $(12m - 4b)(12m + 4b)$ (e) $(5b - 7x)(5b + 7x)$ (f) $(1 - 20t)(1 + 20t)$
(g) $(5a - 4b)(5a + 4b)$ (h) $(12b - 6t)(12b + 6t)$
7. (a) $(4a^2 - 3)(4a^2 + 3)$ (b) $(5x^3 - 1)(5x^3 + 1)$ (c) $(6y^4 - 7x)(6y^4 + 7x)$
(d) $(3xy - 5)(3xy + 5)$ (e) $(2x - 3x^3)(2x + 3x^3)$ (f) $(10t^4 - 1)(10t^4 + 1)$
8. (a) $(9 - 4)(9 + 4)$, 65 (b) $(6 - 3)(6 + 3)$, 27 (c) $(2 - 1)(2 + 1)$, 3
(d) $(7 - 2)(7 + 2)$, 45 (e) $(10 - 8)(10 + 8)$, 36 (f) $(12 - 2)(12 + 2)$, 140
(g) $(99 - 98)(99 + 98)$, 197 (h) $(75 - 73)(75 + 73)$, 296 (i) $(27 - 20)(27 + 20)$, 329
9. (a) $(30 + 2)(30 - 2) = 900 - 4$, 896
(b) $(40 + 4)(40 - 4) = 1600 - 16$, 1584
(c) $(100 + 5)(100 - 5) = 10\,000 - 25$, 9975
(d) $(300 + 10)(300 - 10) = 90\,000 - 100 = 89\,900$

EXERCISE 3.14

1. (a) 2, 5 (b) 1, 3 (c) 2, 3 (d) −2, 4 (e) 2, −4 (f) 3, −4
(g) −3, 4
2. part c 3. part b 4. part d 5. part a
6. (a) 4 (b) 5 (c) 3 (d) 4 (e) 5 (f) m
(g) b (h) 5 (i) t
7. (a) x, 2 (b) 7, x (c) 5, a (d) m, 4 (e) 3, b (f) 2, t
(g) d, 3 (h) 11, x (i) 10, n
8. (a) $x + 5$ (b) $a + 3$ (c) $b + 1$ (d) $m - 5$ (e) $x + 5$ (f) $r + 2$
(g) $t + 4$ (h) $b + 3$
9. (a) $x + 3, x + 2$ (b) $a - 4, a + 3$
(c) $m - 3, m - 3$ (d) $-6, 3; r - 6, r + 3$
(e) $25; -5, -5; n - 5, n - 5$ (f) $2; -15; -3, 5; b - 3, b + 5$
(g) $5; 4; 1, 4; t + 1, t + 4$ (h) $9; 14; 2, 7; x + 2, x + 7$
(i) $-4; -21; -7, 3; c - 7, c + 3$ (j) $-8; 16; -4, -4; y - 4, y - 4$
10. (a) $(x + 1)(x + 2)$ (b) $(a + 3)(a + 5)$ (c) $(x + 5)^2$ (d) $(m + 1)(m + 6)$
(e) $(r - 3)(r + 6)$ (f) $(d - 3)(d - 4)$ (g) $(a - 5)(a + 3)$ (h) $(r - 2)^2$
(i) $(t + 2)(t + 4)$ (j) $(n - 1)(n - 5)$
11. (a) $(x - 4)(x - 5)$ (b) $(a + 2)(a + 12)$ (c) $(t + 4)(t + 9)$ (d) $(d + 10)^2$
(e) $(x - 1)(x + 3)$ (f) $(a - 11)(a + 8)$ (g) $(t - 5)(t + 4)$ (h) $(h - 5)(h - 6)$
(i) $(m - 11)(m + 2)$ (j) $(m - 4)(m + 5)$
12. (a) $(x - 10)(x + 7)$ (b) $(a - 7)^2$ (c) $(m - 3)(m - 8)$ (d) $(b + 4)^2$
(e) $(d - 5)(d + 7)$ (f) $(m - 11)(m + 3)$ (g) $(h - 5)(h - 10)$ (h) $(r - 7)(r + 8)$
(i) $(r + 12)^2$ (j) $(x - 4)(x + 10)$
13. (a) $2(x + 2)(x + 3)$ (b) $5(a - 4)(a + 2)$ (c) $10(x - 3)(x + 5)$ (d) $7(x - 3)(x + 1)$
(e) $4(r - 2)^2$ (f) $3(m + 3)(m + 4)$ (g) $2(t - 3)^2$ (h) $3(b + 5)^2$

EXERCISE 3.15

1. 12
2. 6
3. 78 cm
4. Beginning in the top left corner and proceeding clockwise, place the numbers in the following order: 10, 3, 9, 6.
5. 43
6. $12 288
7. 320 cm
8. 216
9. -45°C
10. 1
11. 24 km/h
12. 52, 53, 54
13. 72 m^2
14. 7
15. Since $13 \times 11 \times 7 = 1001$, any 3 digit number multiplied by 13, 11, and then by 7 will produce a 6 digit number with the original 3 digits repeated in order twice.
16. The most efficient method to find the counterfeit coin would require at most 5 weighings. It is possible to find the coin after the first weighing.
17. 800 cm^2

3.17 REVIEW EXERCISE

1. (a) $P = 18 + 14y$ (b) $P = 8x + 16t$
2. (a) $A = 35x$ (b) $A = 12xy - m^2$
3. (a) $c = 0.80 + 0.56(t - 1), t \geqslant 1$ min (b) $5.28
4. (a) $a + 11b$ (b) $-x - 3y$ (c) $-2x^2 - 2x$ (d) $-6a + b$ (e) $-3xy + 3y - x$ (f) $-m - 5n + 1$
5. (a) $22a + 8$ (b) $42x + 16y$
6. (a) 24 (b) 4 (c) 0 (d) 60 (e) 1 (f) -9 (g) 4 (h) -7
7. (a) $7x + 9$ (b) $12a - 14$ (c) $8x^2 + x - 3$ (d) $m - 5$ (e) $5x + 8$ (f) $7a + 4b$ (g) $5x + 8y$ (h) $-2t^2 - 7t - 5$
8. (a) $7x - 5$ (b) $-8a^2 + 8a + 3$ (c) $-a - b + 3c$ (d) $5m^2 - 3m + 6$ (e) $3b^2 - 5b - 7$ (f) $-7x^2 + 11x$ (g) $-10c + 17$
9. (a) $26t + 9$ (b) $16m + 46$
10. (a) 5 (b) -8 (c) 4 (d) -4 (e) -15
11. (a) 15 (b) 11 (c) 7 (d) 9
12. 37
13. 54, 55, 56
14. (a) $x^2 + 5x + 6$ (b) $m^2 + 6m + 9$ (c) $b^2 - 6b + 8$ (d) $t^2 - 9$ (e) $r^2 - 4r + 4$ (f) $1 + 2m + m^2$ (g) $b^2 - b - 12$ (h) $d^2 - 9d + 20$ (i) $x^2 - 2x + 1$ (j) $a^2 - 25$
15. (a) $2a^2 + 9a - 18$ (b) $6x^2 + 15x - 9$ (c) $2m^2 + 28m + 98$ (d) $-2 - 4x + 30x^2$ (e) $9b^2 - 18b + 37$ (f) $x^2 + 7x - 19$ (g) $-a^2 - 29a - 26$ (h) $9x^2 - 13x - 7$
16. (a) $2x^2 + 25x + 12$ (b) $4x^2 + 28x + 49$ (c) $3x^2 + 12x + 12$
17. (a) $2x + 1$ (b) $3b - c + 2d$ (c) $3x^3 - 2x^2 + x$ (d) $5ab^2 - 2b + 3b^2$
18. (a) $4(x - 3)$ (b) $3a(a + 3)$ (c) $10(2m^2 - a)$ (d) $xy(8 - 7z)$ (e) $2b^2(b^2 - 3b + 1)$ (f) $3abc(4a - 2b + 1)$ (g) $3xy(1 - 2y + 4x)$
19. (a) $(m - 11)(m + 11)$ (b) $(x + 3)(x + 6)$ (c) $(b - 6)^2$ (d) $(2r - 7)(2r + 7)$ (e) $(c - 9)(c + 3)$ (f) $(a - 2)(a + 6)$ (g) $(1 - 7ab)(1 + 7ab)$ (h) $(1 - 2d)(1 - 6d)$ (i) $(b - 4)(b + 14)$ (j) $(y + 5)(y + 16)$ (k) $(6a - 11b)(6a + 11b)$ (l) $(4x^2 - y)(4x^2 + y)$ (m) $(x - 10)(x + 9)$ (n) $(a - 4)(a + 23)$
20. (a) $x + y$ by $x + y$; $(x + y)^2$ (b) $x + 2y$ by $x + 2y$; $(x + 2y)^2$

3.18 CHAPTER 3 TEST

1. (a) $c = 500 + 40n$ (b) $14 100
2. (a) $x + y$ (b) $11t - 7s$ (c) $11a - 8b - 5c$
3. (a) 9 (b) -23
4. (a) $x + 10$ (b) $2b$ (c) $14y$
5. (a) $17x + 22$ (b) $-x^2 - 18x - 23$ (c) $5t^2 - 7t$
6. (a) $t^2 + t - 20$ (b) $2x^2 - 7x - 15$ (c) $12a^2 - ab - 6b^2$ (d) $9x^2 - 4$
7. (a) $2x^2 - x - 25$ (b) $7t^2 - 3t - 22$
8. (a) $t^2 - 2t$ (b) $2x^3 - 3x^2 + 4$
9. (a) $3y(x - 2z)$ (b) $4x(2x^2 - x + 4)$
10. (a) $(x - 3)(x + 3)$ (b) $(x + 3)(x + 4)$ (c) $(t - 4)(t + 5)$ (d) $(s - 6)^2$

REVIEW AND PREVIEW TO CHAPTER 4

ESTIMATING

Answers vary.

1. (a) $12 (b) $13 (c) $10 (d) $7 (e) $10 (f) $8
2. (a) $100 (b) $32 (c) $5 (d) $65 (e) $7
3. (a) $15 (b) $11 (c) $5

TABLES OF VALUES

1. $25, $37.50, $68.75, $118.75, $162.50
2. 90 km, 180 km, 382.5 km, 630 km, 967.5 km
3. $14.40, $21.60, $44.16, $60.96, $163.20
4. $544.50, $486.75, $676.50, $763.12

INVESTIGATING TWIN PRIMES

1. (3, 5), (5, 7), (11, 13), (17, 19), (29, 31), (41, 43)
2. (59, 61), (71, 73)
3. (a) 17, 18, 19; 29, 30, 31; 41, 42, 43; etc.
   (b) They are separated by a number which is a multiple of 6.

INVESTIGATING DIAGONALS

1. 5
2. 9
3. 5, 9, 14, 20, 27, 35
4. $d = 2 + 3 + ... + (n - 2)$, where n is the number of sides

EXERCISE 4.1

1. b
2. a
3. b
4. The starting temperature of the water is close to room temperature and increases to the temperature of the water in the hot water tank. As the tank empties, the temperature of the water decreases to the temperature of the water in the underground feed pipes.

EXERCISE 4.2

1. A(4, 2), B(2, 4), C(0, 3), D(−2, 3), E(−3, 1), F(−6, 3), G(−6, 0), H(−5, −2), I(−4, −4), J(−2, −2), K(−2, −5), L(0, −3), M(1, −2), N(3, −5), P(5, −3), Q(6, −2), R(4, 0)
2. (a) M (b) W (c) V (d) T (e) P (f) R
   (g) S (h) K (i) U (j) N (k) Z (l) Q
3. (a) IV (b) III (c) y-axis (d) x-axis (e) I (f) II
4. (a) parallelogram (b) triangle (c) square (d) rectangle (e) trapezoid
5. (b) D(2, −3)
6. (b) S(3, −3)
7. They may be connected by a straight line.
8. (a) 50°N, 97°W (b) 44°N, 80°W (c) 26°N, 80°W (d) 56°N, 37°E
   (e) 21°N, 158°W (f) 22°N, 114°E (g) 34°S, 151°E

EXERCISE 4.3

1. (1, 1), (2, 1), (3, 2), (5, 3), (6, 4), (7, 5), (8, 6), (9, 7)

EXERCISE 4.4

1. (b) 450 km (c) 900 km
2. (b) approximately 53 km/h (c) approximately 77 km/h
3. (b) approximately $11 (c) approximately $18.75 (d) approximately 5.5 kg
4. (b) approximately 30.6 m (c) approximately 60.0 m

EXERCISE 4.5

1. (a) yes, 70 (b) no (c) no
2. (a) $p = 8h$ (b) $s = 30b$ (c) $d = 80h$

3.(a) (i) x : 5, y : 6, 16 (ii) 2 (iii) $y = 2x$
(b) (i) x : 7, y : 12, 27 (ii) 3 (iii) $y = 3x$
(c) (i) x : 6, y : 5, 7 (ii) $\frac{1}{2}$ (iii) $y = \frac{1}{2}x$

4. (a) $k = 3$, $x = 3y$ (b) 57
5. $10 000 000
6. 525
7. $18
8. $17.50
9. 198 cm^3

EXERCISE 4.6

1. (b) $500, air fare (c) $C = 200d + 500$ (d) $3300
2. (b) $C = 20h + 80$ (c) $210
3. (a) $C = 125(n - 3) + 600, n \geqslant 3$ (b) $1975
4. (a) $C = 225n + 600$ (b) $8925
5. (a) $C = 41h + 56$ (b) $240.50 (c) $12 800, $160
6. (a) $C = 150n + 800$ (b) $2300

EXERCISE 4.7

1. (a) x : 5, y : 3, 0 (b) x : 5, y : 1, 6
2. (a) 8, 14, 17, 26 (b) 1, 5, 11, 13 (c) 9, −3, 5, 17, 25 (d) 7, 12, 22, 32, 47
3. part a, c, d, and f
4. part a, c, d, and f

EXERCISE 4.8

1. (a) $m = 2, b = 3$ (b) $m = -3, b = -2$ (c) $m = 4, b = 6$
(d) $m = -\frac{1}{2}, b = 0$ (e) $m = -3, b = 8$ (f) $m = 2, b = -7$
2. (a) $y = -3x + 7$ (b) $y = 4x + 6$ (c) $y = -3x + 9$
(d) $y = 4x + 6$ (e) $y = 5x - 3$ (f) $y = 3x + 2$
3. (a) $y = 2x + 3$ (b) $y = -5x + 4$ (c) $y = -\frac{3}{2}x + 2$
(d) $y = -\frac{7}{2}x + 6$ (e) $y = 2x + 3$ (f) $y = -3x + 4$
4. (a) $y = \frac{3}{2}x - 4$ (b) $y = -\frac{4}{5}x + \frac{7}{5}$ (c) $y = -\frac{8}{3}x + 2$
(d) $y = \frac{3}{5}x + \frac{6}{5}$ (e) $y = -\frac{4}{5}x - \frac{6}{5}$ (f) $y = -\frac{3}{2}x + \frac{17}{2}$
5. (a) $y = -3x - 2$ (b) $y = 4x + 7$ (c) $y = -2x - 6$
(d) $y = -8x + 3$ (e) $y = 3x - 12$ (f) $y = -5x + 7$
6. (a) $y = -2x + 4$ (b) $y = \frac{3}{2}x - 3$ (c) $y = \frac{7}{3}x - 2$
(d) $y = -\frac{3}{4}x + \frac{3}{2}$ (e) $y = \frac{8}{3}x + \frac{2}{3}$ (f) $y = \frac{3}{4}x - 3$
7. Ordered pairs vary.
(a) $y = -3x + 6$ (b) $y = -\frac{1}{2}x + 3$ (c) $y = \frac{1}{2}x + \frac{3}{2}$ (d) $y = 2x - 4$
(e) $y = \frac{1}{2}x - \frac{3}{2}$ (f) $y = -\frac{3}{2}x + \frac{1}{2}$ (g) $y = \frac{1}{3}x - \frac{2}{3}$ (h) $y = -\frac{1}{3}x$
(i) $y = \frac{1}{4}x - \frac{3}{4}$ (j) $y = x + 2$ (k) $y = -x - 3$ (l) $y = x - 1$
(m) $y = -x$ (n) $y = x$ (o) $y = -2x + 3$ (p) $y = x$
10. b is the y-coordinate of the point where the line crosses the y-axis.
13. (b) (i) positive values (ii) negative values
(c) The larger the magnitude of m, the steeper the line.
14. (a) T (b) M (c) S (d) N (e) P
15. (a) R (b) G (c) H (d) Q (e) P

EXERCISE 4.9

1. (a) 4, 3 (b) 3, 2 (c) 2, 5 (d) 5, 3 (e) 4, 2 (f) 2, 6
(g) 3, −6 (h) 2, −3 (i) 3, −4 (j) 9, −3

EXERCISE 4.10

1. (a) (0, 2), (1, 3), (−1, 3), (2, 6), (−2, 6), (3, 11), (−3, 11)
   (b) (0, −3), (1, −2), (−1, −2), (2, 1), (−2, 1), (3, 6), (−3, 6)
2. (a) (0, −5), (0, 5), (3, −4), (3, 4), (4, −3), (4, 3), (5, 0), (−5, 0)
   (b) (0, −6), (0, 6), (2, −5.7), (2, 5.7), (−2, −5.7), (−2, 5.7), (4, −4.5), (4, 4.5), (6, 0), (−6, 0)

EXERCISE 4.11

1. (a) 37% (b) 63% or 640 2. 406 3. 640 4. 315
6. Answers to Test: 1(b), 2(a), 3(c), 4(e), 5(c), 6(a)

EXERCISE 4.12

1. 192 2. Answers vary. 3. $20 \times 21 = 420$ 4. 961 5. $140 000
6. Beginning at the top and proceeding clockwise, place the numbers as follows: 14, 11, 19, 16, 17, 12, 15, 13, 18.
7. $660
8. The first two numbers of the last two products are given by $7 \times 6 = 42$ and $8 \times 7 = 56$, respectively.
9. 19
10. Turn over each of the first two coins on the first move and then each of the second two coins on the second move.
11. Answers vary. 12. 6 13. 12 14. 4
15. 9 cm 16. D only 17. Answers vary.
18. Beginning at the left and proceeding clockwise, place the numbers as follows: 18, 17, 20.
19. (a) 72 L (b) Answers vary.

4.13 REVIEW EXERCISE

4. A(4, 3), B(−3, 2), C(−7, −3), D(2, −5), E(6, 0), F(0, 1), G(−5, 0), H(0, −3), I(2, 4), J(−8, 4), K(−4, −2), L(5, −3)
5. (a) rectangle (b) triangle (c) square
6. A(3, 4), B(1, −1), C(6, −3), P(−7, 3), Q(−3, 3), R(−2, −3), S(−6, −3)
7. (−7, 1), (−5, 2), (−3, −2), (2, 4), (4, −3), (6, −1)
8. (a) 150 km (b) 212.5 km (c) 5.5 h
9. (b) 60 m 10. (b) 28 km (c) 52 km
11. (a) $P = 13h$ (b) $d = 18t$ 12. 14 13. $10
14. (a) $E = 0.05n + 125$ (b) $225 (c) $300
15. (b) $c = 0.085d + 1200$ (c) $2177.50

4.14 CHAPTER 4 TEST

2. parallelogram 3. (c) 15 km 4. 156 5. $143
6. (a) $C = 50h + 500$ (b) $900

4.15 CUMULATIVE REVIEW CHAPTERS 1 TO 4

1. (a) 100, 7, 700 (b) 100 000, 5, 500 000 (c) 10, 9, 90
   (d) 0.1, 5, 0.5 (e) 0.01, 8, 0.08 (f) 1, 6, 6
2. (a) 6859 (b) 547 (c) 16 064 (d) 45 (e) 19 684 (f) 52 491
3. (a) 98 (b) 14 (c) 32 (d) 30
4. (a) 700 (b) 800 (c) 50 000 (d) 40 000 (e) 0.05 (f) 0.07
5. (a) 130 (b) 450 (c) 200 (d) 300
6. (a) $\frac{1}{6}$ (b) $\frac{5}{12}$ (c) $\frac{1}{3}$ (d) $\frac{3}{8}$ (e) $8\frac{1}{6}$ (f) 8
   (g) $6\frac{3}{4}$ (h) $3\frac{2}{5}$

7. (a) 1058 (b) 469 (c) 65 (d) 1.36
8. (a) 1 (b) 146 (c) 18 (d) −27
9. (a) 0.375 (b) $0.\overline{142857}$
10. (a) 16 (b) $\frac{1}{2}$ (c) $\frac{1}{16}$ (d) 1 (e) 64 (f) $\frac{1}{8}$
(g) 216 (h) $\frac{1}{5}$ (i) 1
11. (a) 3.4×10^4 (b) 9.8×10^{-5} (c) 1.2×10^5 (d) 7.6×10^{-4}
12. (a) 7.7 (b) 12.0 (c) 2.9
13. (a) y^5 (b) t^2 (c) $12m^9$ (d) $6r^4$ (e) x^6y^4 (f) $8m^6n^9$
14. (a) $40x^9$ (b) $6t^{11}$ (c) $-10m^7$ (d) $-20x^7y^5$ (e) $-2m$ (f) $2s^4t^3$
(g) $5m^3$ (h) $-9x^2y$ (i) $\frac{9m^4}{4n^2}$ (j) $\frac{-27x^6}{125y^3}$
15. (a) $-18m^2$ (b) $8x^2$ (c) $2x^3y^{-5}$ (d) $4m^2n^2$
16. (a) 7.8m (b) 8.1m
17. (a) −48 (b) −24 (c) 48 (d) −3
18. (a) $-5x - 22$ (b) $-a^2 + 5a$ (c) $16t^2 - 17t + 1$
19. (a) $9x + 19$ (b) $10x + 12y$
20. (a) 6 (b) 7 (c) 9
21. (a) $x^2 + 9x + 6$ (b) $t^2 - 3t - 7$ (c) $-m^2 - 11m$
22. (a) $2t^2 + 9t + 10$ (b) $9t^2 + 24t + 16$
23. (a) $x + 2x^2$ (b) $1 - 3a + 4b^2$
24. (a) $(m + 3)(m - 1)$ (b) $(t - 4)^2$ (c) $(y - 4)(y + 4)$ (d) $(x - 3)(x - 5)$
27. (b) approximately $18.50 (c) approximately $25.50 (d) approximately 3.5 kg
28. $162
29. (b) $C = 20n + 100$ (c) $1560

REVIEW AND PREVIEW TO CHAPTER 5

ALGEBRA: ADDITION AND SUBTRACTION

1. (a) $7x + 9y$ (b) $8a + 10b + 7c$ (c) $13x - y$ (d) $13p - 7q - 3r$
(e) $x - 2y$ (f) $7a - 7b - 6c$ (g) $5s + 3t - 3$ (h) $14a - 4b - 3c$
2. (a) $6x + 7y + 6z$ (b) $a - 4b + 13c$ (c) $9x - 3y + 11z$ (d) $7a + 3b + 4c$
3. (a) $3x + 2y$ (b) $3a + b + 3c$ (c) $-x + 7y$ (d) $p + 8q + r$
(e) $8a - 2b$ (f) $-y + z$ (g) $-2x + 2y$ (h) $16a - 13b + 3c$
4. $-x - 12y$
5. $-a - b$

SOLVING EQUATIONS

1. (a) 7 (b) 7 (c) 10 (d) −4 (e) 0 (f) −1
(g) −4 (h) 1 (i) −9 (j) 16 (k) 27 (l) −20
(m) −7 (n) 0 (o) 2
2. (a) 5 (b) 4 (c) −3 (d) 4 (e) 15 (f) 2
(g) 2 (h) 4
3. (a) 19 (b) 2 (c) 36 (d) 4 (e) 11 (f) 4
(g) 26 (h) −5

MENTAL MATH

1. 3h 2. $26 3 $104 4. $37 5. shoe store 6. 06:00

EXERCISE 5.1

1. (a) (3, 5), (6, 2), (7, 1) (b) (5, 2), (7, 4) (c) (2, 3), (5, 9)
2. (a) 3, 0, 5, 8, 9 (b) 1, 5, 8, 6, −1
3. (a) (3, 1) (b) (1, 5) (c) (2, 5) (d) (4, 2)
4. (a) (4, 3) (b) (5, 1) (c) (4, 1) (d) (3, 0)

EXERCISE 5.2

1. (a) (1, 4) (b) (1, 3) (c) (2, 1) (d) (−2, −1) (e) (−2, 6) (f) (1, 2)
2. (a) (2, 1) (b) (3, 3) (c) (6, 0) (d) (4, 5) (e) (−3, −2) (f) (−2, 2)

3. (a) $(\frac{1}{2}, 4)$ (b) $(\frac{1}{2}, \frac{1}{2})$ (c) $(-\frac{1}{2}, 2)$ (d) $(2\frac{1}{2}, -3)$

4. (a) no solutions, no points of intersection
   (b) no solutions, no points of intersection
   (c) no solutions, no points of intersection
   (d) The lines coincide: infinite number of solutions and an infinite number of points of intersection.

EXERCISE 5.3

1. (a) 2y (b) s (c) 2a (d) 2p (e) −2d (f) 4x
   (g) 8m (h) −x
2. (a) (1, 1) (b) (2, 3) (c) (3, 4) (d) (4, 2) (e) (6, 2) (f) (4, 3)
   (g) (−1, −2) (h) (−2, 3) (c) (2, 1) (j) (1, 2)
3. (a) (9, 1) (b) $(-\frac{1}{2}, 3)$ (c) (−1, −6) (d) (4, 5)

EXERCISE 5.4

1. (a) 7x (b) 16s (c) 2b (d) −12n (e) −3x (f) 2c
2. (a) (3, 2) (b) (2, 1) (c) (5, 4) (d) (2, −2) (e) (0, −4) (f) (−3, −4)
   (g) (−4, −1) (h) (−2, −1) (i) (−1, 2) (j) (1, 1) (k) (−3, −4)

EXERCISE 5.5

1. (a) (3, 1) (b) (6, 4) (c) (2, 2) (d) (3, 3) (e) (2, 0) (f) (1, 1)
   (g) (1, 1) (h) (1, 3)
2. (a) (4, 2) (b) (3, 2) (c) (3, 3) (d) (4, 2) (e) (7, 2) (f) (2, 3)
   (g) (2, 2) (h) (2, 1)
3. (a) (3, 2) (b) (1, 2) (c) (2, 3) (d) (1, 2) (e) (2, 2) (f) (2, 1)
   (g) (5, 1) (h) (4, 2)
4. (a) (−1, −2) (b) (−2, 1) (c) (−2, 5) (d) (−1, 5) (e) (−2, −3) (f) (−1, −2)
   (g) (6, 2) (h) (−3, −3) (i) (1, 4) (j) (6, −2)
5. (a) (2, 3) (b) (2, 5) (c) (3, −3) (d) (−4, −2) (e) (3, 2) (f) (4, 3)
6. (a) (5, 2) (b) (1, −1)

EXERCISE 5.6

1. (a) (−1, −3) (b) (1, 9) (c) (−12, 11) (d) (8, 6) (e) (3, 1) (f) $(2\frac{1}{2}, -1\frac{3}{4})$
   (g) (4, −3) (h) (−1, −4) (i) (1, −2) (j) (7, 11)
2. (a) (1, 2) (b) (10, −2)

EXERCISE 5.7

1. (a) 3x (b) $\frac{1}{2}x$ (c) x + 7 (d) y + 12 (e) b + 12 (f) 2m
   (g) m − 4 (h) x − 8 (i) 4x + 6 (j) 2x − 7 (k) b + 7 (l) 3x − 6
2. (a) x + y = 12 (b) x + y = 26 (c) x − y = 6 (d) x + y = 28
   (e) x + y = 156 (f) x − y = 4 (g) 2x + 6y = 88 (h) 5x − 3y = 18
   (i) x − y = 56 (j) x + y = 77 (k) x + y = 96

3. 24, 17
4. 29, 23
5. 63, 49
6. 124 pies, 48 strudels
7. 10, 14
8. 3, 9
9. 27
10. 78, 53
11. 15
12. 16
13. 2, 3
14. 8, 32
15. $0.17
16. 15 m, 7 m
17. 168 ha, 144 ha
18. 32
19. 100 m by 73 m
20. $30, $17/h
21. 65
22. 89
23. 320
24. 38°, 52°
25. 18 m by 12 m
26. (15, 85)
27. 350
28. 200
29. 15
30. (a) C = 40 000 + 2.5d (b) A = 5d (c) 16 000

EXERCISE 5.9

1. 21
2. 9; 40, 96; 9, 12; 45, 70, 45
3. 60 km/h
4. 12.5%
5. (a) 2 (b) 3
6. 32
7. S
8. 65 000 L
9. (a) $66.69 (b) $13.31
10. $10^2 - 2^2 = 96$, $11^2 - 5^2 = 96$, $14^2 - 10^2 = 96$
11. 13:55
12. 3 quarters, 1 nickel, 5 pennies; 2 quarters, 7 nickels; 1 quarter, 4 dimes, 4 nickels; 8 dimes, 1 nickel

5.10 REVIEW EXERCISE

1. (a) 5, 1, 9 (b) 2, 4, 2 (c) 13, −1, 5 (d) 2, −19, −25
2. (a) (2, 5) (b) (5, 11) (c) (2, 0) (d) (−2, 0) (e) (4, 1) (f) (8, 4) (g) (−9, 24) (h) (11, 3)
3. (a) (2, 3) (b) (7, 1) (c) (2, 1) (d) (2, 3) (e) (3, 4) (f) (5, 2)
4. (a) (9, 1) (b) (3, 0) (c) (−1, 3) (d) (3, −1) (e) (0, 4)
5. (a) (3, −2) (b) (1, 0) (c) (3, 2) (d) (2, 2) (e) (0, 2) (f) (3, 4) (g) (3, 3) (h) (−1, 3)
6. (a) (1, 2) (b) (1, 3) (c) (1, 2) (d) (−3, 0) (e) (2, −4) (f) ($2\frac{1}{2}$, 5)
7. 14, 25
8. 22, 30
9. 9, 14
10. 15, 18
11. 14
12. 60 m, 50 m
13. 26 m by 14 m
14. 2000
15. 300

5.11 CHAPTER 5 TEST

1. (a) (i) 4 (ii) 1 (iii) 8 (b) (i) 3 (ii) 8 (iii) −6
2. (3, 5)
3. (a) (4, 5) (b) (−1, −3)
4. (3, 2)
5. 8, 13
6. 18 m by 12 m
7. 1250

REVIEW AND PREVIEW TO CHAPTER 6

PERCENT

1. (a) 0.36 (b) 0.25 (c) 0.1 (d) 0.08 (e) 1.23 (f) 3.45 (g) 1 (h) 2 (i) 0.005 (j) 0.001
2. (a) 25% (b) 75% (c) 50% (d) 10% (e) 56% (f) 78% (g) 7% (h) 130% (i) 0.3% (j) 5.1%
3. (a) 50% (b) 25% (c) 30% (d) 15% (e) 16% (f) 68% (g) 78% (h) 37.5%
4. (a) $\frac{1}{4}$ (b) $\frac{1}{2}$ (c) $\frac{3}{4}$ (d) 1 (e) $\frac{1}{10}$ (f) $\frac{23}{100}$ (g) $\frac{9}{100}$ (h) $\frac{3}{100}$
5. (a) 60 (b) 45.6 (c) 135 (d) 12.8 (e) 7 (f) 70 (g) 67 (h) 176 (i) 2.5 (j) 1.3 (k) 0.9 (l) 14.1
6. (a) 50% (b) 25% (c) 10%
7. (a) $83.46 (b) $678.40 (c) $16.96 (d) $497.04
8. 80%
9. 67.8%
10. 3864
11. $65 to $78
12. (a) 35% (b) 53% (c) 12%
13. (a) 2100
14. (a) (i) 126 000 (ii) 3000 (iii) 81 000
15. (a) 49 (b) 25 or 26

EXERCISE 6.1

1. Answers vary.
2. Answers vary.
3. (a) Testing destroys the product.
4. 24 boys, 26 girls
5. Answers vary.

EXERCISE 6.2

1. (a) R: 20%, S: 40%, L: 40% (b) R: 400, S: 800, L: 800
2. Mon: 800, Tues: 1200, Wed: 300, Thurs: 400, Fri: 600
3. (a) 1500 (b) 55.5% (c) 44.5% or 667

EXERCISE 6.3

1. (a) 15 decibels (b) 4 (c) whisper, park, blender, factory, jackhammer, jet
2. (a) 90 (b) 90 (c) 10th and 11th
   (d) between the evening of the 12th and the morning of the 13th
   (e) during the day on the 14th
3. (a) $9 000 000 (b) $3 000 000 (c) $2 000 000
   (d) $2 400 000 (e) $1 200 000 (f) $2 400 000
4. Mid-size: 1750, Full-size: 1450, Compact: 1050, Pick-up: 400, Sub-compact: 200, Van: 150

EXERCISE 6.4

1. Answers vary. 2. Answers vary.

EXERCISE 6.5

1. (a) 9 (b) 9, 10 (c) no mode (d) 21
2. (a) 11, 10, 9 (b) 41, 41.5, no mode (c) 34, 34, 31
   (d) 40.5, 39, 9, and 68 (e) 88.4, 90, 91
3. (a) 9 (b) The numbers do not refer to measurements or observations.
   (c) Gordie Howe
4. 192.8 5. 6.8, 7, 7 6. 1, 2 7. 2.30

EXERCISE 6.6

1. (a) Dennis (b) Dennis: 4, 4, 5, Tyson: 4.1, 3, 2
   (c) Dennis (d) Tyson (e) Tyson (f) median
2. (a) 10.3, 7.5, 6 (b) median (c) mean (d) Answers vary.
3. (a) 3, 2, 0 (b) mean (c) mode (d) 180
4. (a) 168.5, 167, 162 (b) median 5. none 6. 94

EXERCISE 6.7

2. (b) 130 and 145 4. (b) 21 and 51

EXERCISE 6.8

1. (a) Sep 16 : 21, Sep 23 : 20, Sep 30 : 21, Oct 7 : 30 (b) high increase in absences
2. Answers vary.
3. (a) Sales increased by a factor of 8.
   (b) The lengths of the edges of the suitcases have doubled.

EXERCISE 6.9

1. one 2. 11
3. under the quarters (1¢ in first row, 10¢ in second row, 5¢ in third row)
4. 06:25 5. Jeffreys 6. 15
7. (a) 5 days + 48 hours = 1 week (b) 1 day − 60 minutes = 23 hours
8. 15, 21 9. equal 10. 4 11. Answers vary.
12. 4 13. $4\frac{1}{2}$ units 14. 32

EXERCISE 6.10

1. \$15 238.08   2. 36

6.11 REVIEW EXERCISE

1. Grade nine: boys — 16, girls — 15
   Grade ten: boys — 13, girls — 13
   Grade eleven: boys — 11, girls — 12
   Grade twelve: boys — 10, girls — 10
2. 414   3. 1500   4. (a) 150   (b) 235   (c) 60
5. (a) June   (b) January and November, March and April, July and August
   (c) 4   (d) 68
10. (a) 17, 16, 15   (b) 128, 131, no mode   (c) 45, 45, 50
11. (a) 202   (b) 206   (c) 204   (d) 204   (e) 205
12. (a) no (for example, 42, 44)   (b) no (same example as part a)
    (c) yes (The mode is the value that occurs most often.)
13. (c) 71.5

6.12 CHAPTER 6 TEST

1. CFCF: 120 000, CEED: 80 000, CJAM: 200 000   3. 45, 45, 47
5. (a) 18°C   (b) 13:00 - 14:00   (c) 15°C

REVIEW AND PREVIEW TO CHAPTER 7

PERCENT

1. (a) 0.07   (b) 0.05   (c) 0.125   (d) 1   (e) 0.085   (f) 0.0325
   (g) 0.0075   (h) 0.025   (i) 0.015   (j) 0.0475
2. (a) 25%   (b) 75%   (c) $7\frac{1}{2}$%   (d) 14%   (e) $4\frac{3}{4}$%   (f) $6\frac{1}{10}$%
   (g) $1\frac{1}{4}$%   (h) $3\frac{7}{10}$%
3. (a) \$28.20   (b) \$45.00   (c) \$54.00   (d) \$33.75   (e) \$8.94   (f) \$0.70
   (g) \$1.31   (h) \$10.65   (i) \$266.25   (j) \$669.50

CIRCLE GRAPHS

1. (a) $\frac{1}{3}$ (b) $\frac{1}{4}$ (c) $\frac{1}{24}$
   (d) Sleeping: $33\frac{1}{3}$%, Reading: $6\frac{1}{4}$%, Exercise: $4\frac{1}{6}$%, Recreation: $12\frac{1}{2}$%, Other: $6\frac{1}{4}$%,
   Commuting: $6\frac{1}{4}$%, Working: 25%, Eating: $6\frac{1}{4}$%
3. (a) luxury cars: 10%, standard cars: 40%, compact cars: 30%, import cars: 20%

SUBSTITUTING INTO FORMULAS

1. (a) 216 cm², 60 cm   (b) 375 cm², 80 cm   (c) 750 cm², 110 cm
2. 9.0 cm   3. 15
4. (a) 37.7 cm, 113.0 cm²   (b) 62.8 cm, 314 cm²
   (c) 314 cm, 7850 cm²   (d) 157 cm, 1962.5 cm²
5. (a) 630 km   (b) 1 080 000 km   (c) 96 m
6. (a) 10   (b) 15   (c) 14

MENTAL MATH

1. 4	2. 22	3. 4	4. 16	5. 4	6. 3
7. 7	8. 4	9. 17	10. −5	11. 6	12. 6
13. −1	14. 5	15. 6	16. 1	17. 4	18. 4
19. 9	20. 4	21. 6	22. $3\frac{1}{2}$	23. 4	24. 55
25. $6\frac{2}{3}$	26. 12	27. 12	28. 20	29. 10	30. 5

EXERCISE 7.1

1. (a) $66.50 (b) $250 (c) $97.50 (d) $150.00
2. (a) $9.00/h (b) $15.00/h (c) $21.00/h (d) $18.00/h
   (e) $9.75/h (f) $13.50/h (g) $14.25/h (h) $23.25/h
3. (a) $364.00 (b) $665.60 (c) $1073.00 (d) $218.40
4. $2643.74
5. (a) $157.55 (b) $742.50 (c) $528.13 (d) $377.50
   (e) $97.50 (f) $950.25 (g) $106.80 (h) $261.25
   (i) $174.20 (j) $934.50 (k) $6660.45 (l) $14 170.50
6. $3880.00
7. $484.62 versus $469.80
8. $1240.00
9. $553.05
10. (a) $822.00 (b) $91.33
11. Downtown Motors: $37 464.00, Country Leasing: $37 336.00
12. (a) $22.14 (b) $39.75
13. $22 301.25, $245.31

EXERCISE 7.2

1. (a) $4.16 (b) $4.72 (c) $6.15 (d) $5.75
2. (a) $5.36 (b) $6.02 (c) $5.44 (d) $6.53
3. (a) $33.75 (b) $52.00 (c) $44.90 (d) $57.05
4. (a) $5.52 (b) $6.14 (c) $48.95 (d) $74.11 (e) $240.89
5. (a) $5.13 (b) $5.77 (c) $44.90 (d) $240.20
6. (a) C.P.P.: $5.46, U.I.: $6.08, income tax: $48.95
   (b) $87.84 (c) $224.16
7. (a) $320.30 (b) $5.64 (c) $6.25 (d) $50.95 (e) $90.84 (f) $229.46
8. (a) $7.40/h (b) 36 (c) 6 (d) $333.00 (e) $85.69 (f) $247.31

EXERCISE 7.3

1. (a) Estimates vary.
   (b) (i) $33.68 (ii) $44.90 (iii) $44.90
   (iv) $56.13 (v) $22.45 (vi) $22.45
2. (a) Rent/utilities: 41.4% ($9.17), Food: 13.5% ($11.40), Clothes: 9.4% (−$12.80), Personal: 5.1% (−$9.20), Acting lessons: 16.8% ($0.00), Furniture loan: 12.7% ($0.00), Contributions: 0.8% (−$3.50), Savings: 0.3% ($5.00)
   (c) $0.07 (underspend)
3. (a) Estimates vary.
   (b) (i) $22 250 (ii) $5340 (iii) $5340
   (iv) $19 580 (v) $6230 (vi) $26 700
4. (a) Week 1: $60, Week 2: $60, Week 3: $60, Week 4: $60; Bus: $40.00, Lunches: $21.15, Clothes: $77.40, Entertainment: $59.40, Miscellaneous: $18.05, Savings: $24.00
   (b) Bus: $10.00, Lunches: $5.29, Clothes: $19.35, Entertainment: $14.85, Miscellaneous: $4.51, Savings: $6.00
   (c) Bus: 17%, Lunches: 9%, Clothes: 32%, Entertainment: 25%, Miscellaneous: 8%, Savings: 10%
5. (a) Housing: $7680, Food: $8960, Clothing: $2560, Transportation: $3520, Insurance: $2560, Loans: $3840, Entertainment: $1600, Miscellaneous: $640, Savings: $640
   (b) Housing: $7200, Food: $8280, Clothing: $4320, Transportation: $3600, Insurance: $2520, Loans: $5040, Entertainment: $2160, Miscellaneous: $1440, Savings: $1440
6. (a) Total income: $1690.90; 31.6%, 4.3%, 17.8%, 2.5%, 2.0%, 7.2%, 14.1%, 7.1%, 3.5%; Total: $1521.87, 90.1%; Balance: + $169.03, 9.9%
   (b) $169.03 (under) (c) $289.03
7. (a) 26% (b) 17%
8. (a) Lunches: $2.25, Entertainment: $7.00, Transportation: $0.00, Magazines: −$0.75, Clothes: −$1.50
   (b) $19.50

EXERCISE 7.4

1. (a) \$593.50 (b) \$493.50
2. (a) \$133.50 (b) \$300.00 (c) \$366.50
3. (a) two hundred forty-five and $\frac{50}{100}$
   (b) one thousand six hundred thirty-five and $\frac{75}{100}$
   (c) twenty-five thousand and $\frac{xx}{100}$
   (d) sixteen thousand five hundred thirty-five and $\frac{xx}{100}$
4. (a) \$57.50 (b) \$111.25
5. \$265.74, \$390.74, \$314.94, \$299.94, \$347.44, \$371.44, \$306.44, \$250.44, \$265.44, \$155.44, \$179.44, \$124.44, \$99.44 (balance)
6. \$350.35
7. \$1292.06
8. \$545.73, \$495.73, \$620.73, \$562.93, \$537.93, \$522.18, \$482.18, \$607.18, \$367.18, \$332.18, \$407.18, \$364.68, \$339.68, \$294.68, \$419.68, Interest: \$0.74, Service charge: \$1.50, August 1 balance: \$418.92
9. (a) Interest: \$1.39, Service charge: \$1.20, April 1 balance: \$1103.41
   (b) Interest: \$0.37, Service charge: \$2.40, June 1 balance: \$462.30
10. \$1348.51, \$1148.51, \$1113.51, \$1055.01, \$1042.26, \$842.26, \$799.99, \$583.65, \$1746.33, \$1621.33, \$1596.33, \$1560.68, \$1541.88, \$851.88, \$765.98, \$530.98, \$280.35, \$180.35, \$61.53, \$1299.21, Interest: \$0.18, Service charge: \$4.95, August 1 balance: \$1294.44

EXERCISE 7.5

1. (a) Net deposit: \$724.70 (b) Net deposit: \$239.74
   (c) Cash: \$364.00, Net deposit: \$1438.27
2. (a) Net deposit: \$911.50 (b) Net deposit: \$1588.78
3. \$1506.83, \$1406.83, \$1531.83, \$1511.33, \$1561.33, \$1686.33, \$1601.33, \$1556.33, \$1731.33
4. Corrected balances from August 6 to August 31: \$3232.48, \$3482.48, \$3357.48, \$3157.48, \$3382.48, \$3437.48, \$3362.48, \$3312.48, \$3062.48, \$3187.48, \$3186.49, \$3386.49

EXERCISE 7.6

1. (a) \$541.43, \$491.43, \$916.43, \$659.15, \$1059.15, \$1051.65
   (b) \$277.56, \$168.56, \$1140.96, \$898.26, \$682.26
2. (b) \$754.83 + \$250.00 − \$575.46 = \$429.37
3. (a) \$227.55, \$1790.75, \$3114.75, \$2682.75, \$1432.75, \$1400.75, \$1378.60, \$1350.26, \$1342.76
   (b) \$1393.25 + \$0.00 − (\$22.15 + \$28.34) = \$1342.76
4. (a) The balances do not agree.
   (b) Karl's personal chequing book contains errors due to incorrect calculations on the dates July 12 and July 31. The final balance should read: \$985.95
   Summary of reconciliation: \$925.95 + \$300.00 − \$240.00 = \$985.95

EXERCISE 7.7

3. (a) \$604.63 (b) \$1116.96 (c) \$516.29 (d) \$668.01
4. (a) \$357.50 (b) \$934.05 (c) \$392.84 (d) \$1203.15 (e) \$213.27 (f) \$5801.66

EXERCISE 7.8

1. (a) 4 (b) \$56.00 (c) July 4 (d) \$1965.48 (e) \$358.63 (f) \$49.00
   (g) August 25
2. (a) \$18.00 (b) \$10.00 (c) \$22.00 (d) \$10.00 (e) \$25.00 (f) \$16.00
   (g) \$10.00 (h) \$11.00 (i) \$20.00 (j) \$20.00 (k) \$25.00 (l) \$10.00

3. (a) $49.00 (b) $65.00 (c) $34.00 (d) $37.00 (e) $55.00 (f) $93.00
   (g) $77.00 (h) $59.00
4. (a) $7.51 (b) $12.52 (c) $13.19 (d) $0.97 (e) $3.30 (f) $6.50
5. (a) $248.70 (b) $13.00
6. (a) $310.41 (b) $16.00
7. (a) $12.18 (b) $690.75
8. (a) $0.00 (b) $168.35 (c) $10.00
9. (a) $0.00
   (b) Total debits: $751.38, Total credits: $628.50
   (c) $751.38 (d) $38.00
10. (i) (a) $12.57 (b) Total debits: $1272.60, Total credits: $935.00 (c) $1213.42 (d) $61.00
    (ii) (a) $0.00 (b) Total debits: $1289.74, Total credits: $798.32 (c) $1082.66 (d) $52.00

EXERCISE 7.9

1. one 50¢, seven 5¢; one 50¢, one 25¢, one 5¢, five 1¢; two 25¢, one 10¢, five 5¢; one 25¢, five 10¢, two 5¢
2. X = 7, Y = 8
3. no
4. 12 + 13 + 14 + 15 = 54
5. Beginning at the top and proceeding clockwise, place the numbers as follows: 7, 1, 6, 9, 2, 4, 8, 3, 5.
6. 1010
7. no
8. 10
9. 22, 27, 31, 36, 40, 45, 54, 63, 72, 79, 81, 88, 90, 97
10. 42 857
11. 80 km/h
12. $66.67
13. $23.03
14. 490
15. 40
16. 4.5
17. 4
18. Paula: $17.50, Monique: $32.50
19. (a) 13 900
    (b) Expos v. Cubs, Mets v. Giants, Braves v. Reds, Astros v. Pirates

EXERCISE 7.10

F5 and J2

7.12 REVIEW EXERCISE

1. (a) $288.65 (b) $630.75 (c) $188.48 (d) $1008
2. (i) (a) $4.96 (b) $5.62 (c) $42.90; $234.52
   (ii) (a) $4.70 (b) $5.37 (c) $38.85; $226.58
   (iii) (a) $5.30 (b) $5.93 (c) $46.90; $246.12
   (iv) (a) $6.43 (b) $6.98 (c) $60.05; $284.56
   (v) (a) $8.21 (b) $8.63 (c) $81.30; $344.54
   (vi) (a) $7.08 (b) $7.59 (c) $68.15; $306.38
   (vii) (a) $7.91 (b) $8.36 (c) $78.25; $334.13
   (viii) (a) $6.67 (b) $7.21 (c) $63.10; $292.77
3. (a) $425.00 (b) $7.83 (c) $8.29 (d) $77.25 (e) $124.47 (f) $300.53
4. Rent: $280, Food: $280, Transportation: $40, Miscellaneous: $200
5. (a) Estimates vary.
   (b) Rent: $428.75, Food & Clothes: $297.50, Supplies & Books: $35, Transportation: $52.50, Entertainment: $61.25
6. (a) Lunches: $5.58, Clothes: $13.14, Supplies: $10.73, Leisure: $10.29, Bus: $4.88, Savings: $5.40
   (b) Lunches: 11%, Clothes: 26%, Supplies: 21%, Leisure: 21%, Bus: 10%, Savings: 11%
7. Interest: $3.08, Service charge: nil, March 1 balance: $1408.75
8. (a) $563.74, $398.50, $710.50, $638.50, $612.04, $937.04, $677.04, $659.04, $651.54
   (b) (i) $929.54 (ii) $0.00 (iii) $929.54 (iv) $278.00
   (v) $651.54 (vi) $651.54 (vii) $0.00

9. (a) $4.15 (b) Total debits: $1344.81; Total credits: $1600.00
(c) $556.22 (d) $28.00

7.13 CHAPTER 7 TEST

1. (a) $423.85 (b) $7.81 (c) $8.27 (d) $76.25 (e) $123.93 (f) $299.92
2. Rent: 50%, Food & Clothes: 31%, Transportation: 6%, School supplies: 5%, Entertainment: 8%
3. (a) $113.72 (b) $600.00 (c) $713.72 (d) $0.00 (e) $713.72 (f) $713.35 (g) $0.37
4. $185.26 + $0.00 + $204.65 − $185.26 = $204.65, $11.00

REVIEW AND PREVIEW TO CHAPTER 8

PERCENT

1. (a) 0.45 (b) 0.065 (c) 0.0775 (d) 0.75 (e) 0.091 25 (f) $0.05\overline{3}$
2. (a) $62\frac{1}{2}\%$ (b) 6% (c) $5\frac{1}{2}\%$ (d) 55% (e) $45\frac{1}{2}\%$ (f) $\frac{3}{8}\%$
3. (a) $15.00 (b) $107.50 (c) $34.00 (d) $770.00 (e) $1127.50
4. (a) $35.71 (b) $338.18 (c) $236.67
5. (a) $74.60, $5.97 (b) $1118.95, $89.52 (c) $12 620.70, $1009.66
6. (a) $279.85, $27.99 (b) $291.05, $29.11 (c) $3118.35, $311.84 (d) $46.23, $4.62

PROFIT AND LOSS

1. (a) $10.63 (profit) (b) $82.88 (profit) (c) $263.48 (profit) (d) $34.50 (profit) (e) $8.46 (profit)
2. (a) $8.45 (b) 53%
3. (a) $118 (b) $177
4. (a) $23 175 (b) $675
5. $30.50

SUBSTITUTION

1. (a) 1 (b) 0 (c) −18
2. (a) 32 (b) 60 (c) −3 (d) −4 (e) 0 (f) 12
3. (a) 5 (b) −3 (c) 3 (d) 3 (e) 1 (f) 4
4. (a) 29 (b) 25 (c) 25 (d) 25
5. (a) 1,2 (b) 0,1 (c) −1,2 (d) −2,1
6. (a) 25 (b) 4 (c) 0 (d) 0 (e) 9
7. (a) 1.953 125 (b) 1.3225 (c) 9.378 906 25 (d) 1.762 341 683 (e) 3.138 428 377 (f) 8.916 100 448

MENTAL MATH

1. (a) 5 (b) 15 (c) 9 (d) 4 (e) 9 (f) 15 (g) 0 (h) 7 (i) 11 (j) 24 (k) −9 (l) 5 (m) 2.0 (n) 6.2 (o) 5.0 (p) 0.7 (q) 0.3 (r) 0.4 (s) 5.4 (t) 10.5
2. (a) (i) $2.50 (ii) $12.50 (iii) $525
(b) (i) $5 (ii) $25 (iii) $1100
(c) (i) $30 (ii) $7.50 (iii) $250
(d) (i) $110 (ii) $550 (iii) $880
3. (a) $15 (b) $60 (c) $150 (d) $180 (e) $360 (f) $450
4. (a) 50 (b) 120 (c) 2410 (d) 35 (e) 47 (f) 68 (g) 49.5 (h) 622.5 (i) $109.5
5. (a) $5 (b) $0.75 (c) 20% of 49.95 (d) $20 (e) $25

EXERCISE 8.1

1. (a) $450.60 (b) $574.60 (c) $3114.60
2. (a) $2400 (b) $1200 (c) $3600 (d) $20 000 (e) $1800
3. (a) $2200 (b) $1450 (c) $3250 (d) $2300 (e) $1700

4. (a) $500 (b) $800 (c) $220 (d) $725 (e) $1250
5. (a) $1614.60 (b) $1114.60 (c) $1243.20 (d) $1743.20 (e) $128.60
6. (a) $16 060 (b) $19 432.80 (c) $24 432.80 (d) $3372.80
7. $43.48 8. $21.72 9. $42.85
10. (a) $175 000 (b) $466 500 (c) $666 500 (d) $291 500

EXERCISE 8.2

1. (a) $9.03 (b) $17.65 (c) $3.52 (d) $6.10 (e) $4.12 (f) $5.19
2. (b) $100 at 24% for 12 months
3. (a) $88.85 (b) $44.65 (c) $49.92 (d) $93.11 (e) $480.10 (f) $366.60 (g) $3467
4. 24 months 5. 12 months
6. (a) $0.18 (b) $0.09 (c) $0.06 (d) $0.05 (e) $0.04 (f) $0.04
7. $110.12 8. $64.21 9. (a) $2000 (b) $80.38
10. (a) $500 (b) $44.90 (c) $681.40 (d) $38.80
11. (a) $2987.28 (b) $2000 (c) $121.96 (d) $3182.56 (e) $195.28

EXERCISE 8.3

1. (a) $1400 (b) $2200 (c) $2900
2. (a) $267.48 (b) $77.48 (c) $664.80 (d) $67.36
3. 18.5% 4. $1860
5. (a) 9.8% (b) 18.8% (c) 17.2% (d) 14.6%
6. 20.4% 7. (a) $449.56 (b) $17.61 (c) 6.7%
8. (a) $321.52 (b) $1645.60 (c) 15.9%
9. 5.2% (bank) v. 9.6% (dealership)

EXERCISE 8.4

1. (a) $55 (b) $225 (c) $2700 (d) $450 (e) $1120 (f) $337.50 (g) $945 (h) $1200 (i) $127.19
2. $7.43 3. 10 a
4. 0.75a or 9 months
5. (a) $7950 (b) $12 950 6. $892.86

EXERCISE 8.5

1. (a) $12 (b) $23.33 (c) $18.98
2. (a) $513.33 (b) $854.29 (c) $1338.23
3. $1331 4. $2207.63
5. (a) $2332.80 (b) $2339.72 (c) $2343.32
6. (a) $1166.40 (b) $2604.52 (c) $4432.37
7. (b) 7.5 a 8. (a) $46.92 (b) $39.81

EXERCISE 8.6

7. (a) 162 (b) 116 (c) 282 (d) 147 (e) 213 (f) 186 (g) 92
8. (a) $1392 v. $1210 (b) $110.70 v. $122.40 (c) $1500 v. $1645 (d) $2500 v. $2640
9. (b) $180 10. (b) $39.06 (c) $45.57
11. (a) 04 15 to 04 30, $20.32, $3052.95; 04 30 to 05 15, $17.56, $2870.51; 05 15 to 05 31, $17.62, $2388.13; 05 31 to 06 15, $13.74, $2201.87; 06 15 to 06 30, $12.67, $1714.54; 06 30 to 07 15, $9.86, $824.40; 07 15 to 07 31, $5.06, $829.46
(b) $824.40 (c) $829.46 (d) $129.46 (e) $4329.46

12. (a) 03 12 to 03 31, $78.08, $9078.08; 03 31 to 04 30, $111.92, $8190.00;
04 30 to 05 31, $104.34, $7294.34; 05 31 to 06 30, $89.93, $6384.27;
06 30 to 07 31, $81.33, $5465.60; 07 31 to 08 31, $69.63, $4535.23;
08 31 to 09 30, $55.91, $3591.14; 09 30 to 10 31, $45.75, $2636.89;
10 31 to 11 30, $32.51, $1669.40; 11 30 to 12 31, $21.27, $690.67;
12 31 to 01 31, $8.80, $699.47
(b) January 31 (c) $699.47 (d) $699.47 (e) $23 474.47

EXERCISE 8.7

5. (a) $142.45 (b) $39.99 (c) $505.97 (d) $100.67 (e) $112.50
6. Account B, by $21.22
8. (a) $80 (b) $680 (c) $19.73 (d) $73.97 (e) $1881 (f) $493.15
9. (a) $23.01 (b) $36.51 (c) $116.44 (d) $87.67 (e) $199.03 (f) $209.59
(g) $2188.30 (h) $232.88 (i) $360.00 (j) $54.79
10. (a) by $15.50
11. (a) $31.68 (b) $2831.68 (c) $2891.03
12. $2350
13. $666.94
14. (a) $95 (b) $1095 (c) $109.50 (d) $204.50
15. Bonds: $11 816.39, Savings account: $11 655.26
16. $11 744.80
17. $5872.40
18. (a) $902.13 (b) $941.94 (c) $976.52 (d) $975.35 (e) $950.25 (f) $949.39
19. (a) 11.0% (b) 10.5% (c) 10.5% (d) 9.5%
20. (a) (i) $51 699.68 (ii) $55 945.13 (iii) $52 851.60
(b) term deposit

EXERCISE 8.8

1. (a) $1492.40 (b) $3423.60 (c) $2000 (d) $2920
(e) $3968 (f) $4320
2. (a) $2289.02 (b) $774.31 (c) $1436.55 (d) $2081.96
(e) $1772.31 (f) $1665.57
3. (a) $748.32 (b) $1374.60 (c) $774.77 (d) $1031.58
(e) $1386.44 (f) $1466.52
4. (a) $1211.16 (b) $865.85 (c) $894.36 (d) $1925.81
(e) $640.61 (f) $1248.10
5. (a) $29 317.50 (b) $43 650 (c) $23 395.50
6. (a) $597.14 (b) $752.27 (c) $730.13
7. $965.99
8. $1007.68
9. (a) $1030.34 (b) $53 092.24 (c) $8592.24
10. (a) $1343.13 (b) $63 352.68 (c) $11 200.68
11. (a) $779.48 (b) $45 461.28 (c) $5575.28
12. (a) $13 050 (b) $23 075 (c) $34 996.50 (d) $11 850
13. (a) $21 200 (b) $15 125 (c) $16 100 (d) $10 498.25
14. $26 501.60
15. $182.30
16. $969.42
17. $490.05
18. (a) $13 030 (b) $931.60 (c) $33 537.60 (d) $47 610
19. (a) $24 200 (b) $19 250 (c) $543.81 (d) $23 851.44
20. (a) $38 150 (b) $802.50 (c) $14 000 (d) $434.09
(e) $38 150, $43 890, $45 887.24 (including tax on the lease end purchase) (f) $1997.24
(g) Answers vary.
21. (a) $35 370 (b) $1220.22 (c) $793.86
22. (a) $3433.40 (b) $4166.30 (c) $732.90 variance
23. (a) $2719.75 (b) $3412.45 (c) $4120.39

EXERCISE 8.9

1. no
2. noon, July 29
3. 3580
4. 5 min
5. 125, 130, 135
6. B, D, F, G
7. the 28th
8. May 2
9. (a) approximately 2 m (b) 48 828 000 (c) 0.000 000 02 cm
10. 57
12. 3524 × 16
13. (a) at the corner of 6th Avenue and 5th Street
    (b) at any of the corners represented by the following pairs, (Avenue, Street): (3, 5), (4, 5), (5, 5), (6, 5), (3, 6), (4, 6), (5, 6), (6, 6), (3, 7), (4, 7), (5, 7), (6, 7)
14. English: 78, Science: 80

EXERCISE 8.10

1. 6
2. 3rd (lettuce)
3. sled
4. 2
5. K
6. (a) 10, 5 (b) 4, 6 (c) G, J
7. B
8. 9
9. MRENVTO (or VERMONT)
10.
11. ○○ ⊗●
12. 5
13. 6, 7, 9
14. 3 or 66
15. T
16. LIAMESS (or MELISSA)
17. 4
18.
19. 10 : 40
20. 3, 8
21. 7

8.11 REVIEW EXERCISE

5. (a) \$3121.20 (b) \$2121.20 (c) \$2486.40 (d) \$3486.40 (e) \$365.20
6. (a) \$100.48 (b) \$328.32 (c) \$847.30 (d) \$180.75
7. \$61.47
8. 10%
9. 33.7%
10. (a)

Payment (\$)	Interest Owing (\$)	Balance Owing (\$)
		2500.00
85	18.75	2433.75
85	18.25	2367.00
85	17.75	2299.75
85	17.25	2232.00
85	16.74	2163.74

(b)

Interest (\$)	Payment (\$)	Balance Owing (\$)
		2500
18.75	103.75	2415
18.11	103.11	2330
17.48	102.48	2245
16.84	101.84	2160
16.20	101.20	2075

(c) \$2163.74 v. \$2075

11. (a) \$2.47 (b) \$25 (c) \$912.50 (d) \$632.67 (e) 1.5% (f) 10% (g) 3 d (h) 16 d (i) 5 months
12. \$12 155.06
13. (a) \$9291.21 (b) \$791.21
14. \$4815.89
15. (a) \$1454.60 (b) \$1145.08 (c) \$649.97 (d) \$995.18 (e) \$1733.27
16. (a) \$29 560 (b) \$1024.71 (c) \$56 889.56 (d) \$7329.56
17. (a) \$530.36 (b) \$2710.80
18. (a) \$17 875 (b) \$20 099.40
19. (a) \$14 700 (b) \$9000 (c) \$346.64 (d) \$29 999.04
20. (a) \$3181.08 (b) \$3990.90 (c) \$4818.96

8.12 CHAPTER 8 TEST

1. \$5002.00 2. \$124.41 3. 12% 4. \$52.50
5. \$905.74 6. \$7440 7. \$1115.38

8.13 CUMULATIVE REVIEW FOR CHAPTERS 5 TO 8

1. (a) (3, 7) (b) (3, 1)
2. (a) (2, −1) (b) (0, −5) (c) (0, 3) (d) (5, −1) (e) ($\frac{1}{2}$, −2)
3. (a) (7, 3) (b) (2, 1) 4. 14, 26 5. 5000 6. 200
9. (a) 12.1, 12, 11 (b) 22.8, 22.5, 25
11. (a) \$335.32 (b) \$66.49 (c) \$268.83
12. Clothes: \$123.50, Rent: \$446.50, Food: \$228, Transportation: \$57, School supplies: \$38, Entertainment: \$57
13. Summary of reconciliation:
(a) \$433.89 (b) \$350 (c) \$783.89 (d) \$375.24 (e) \$408.65
(f) \$408.65 (g) \$0.00
14. (a) 20%/a (b) 32 d (c) 6% (d) \$4950 (e) \$1250
15. \$2604.52 16. \$5970.26 17. (a) \$5251.73 (b) \$1251.73
18. (a) \$990.68 (b) \$897.00 (c) \$580.64 (d) \$1357.62 (e) \$753.97
(f) \$1590.36 (g) \$3024.55 (h) \$1968.75
19. (a) (i) \$969.73 (ii) \$602.68 (iii) \$462.19 (b) iii (c) i

REVIEW AND PREVIEW TO CHAPTER 9

THE RULE OF THREE

1. 760 km 2. \$105 3. 196 4. \$89.60 5. 42.5 km 6. \$130
7. 20.5 8. 4950 9. \$14.25 10. 90 11. 868 000

MENTAL MATH

1. 8 2. 12 3. 15 4. 34 5. 9 6. 8
7. 10 8. 12 9. 6.6 10. 7.2 11. 5 12. 75
13. 17 14. 163 15. 20 16. 4 17. 20 18. 50
19. 7.6 20. 35 21. 37 22. $\frac{1}{2}$ 23. 8 24. 15
25. 21 26. $\frac{1}{4}$ 27. 3.5 28. 101

EXERCISE 9.1

1. (a) 5 : 6 (b) 6 : 5 (c) 10 : 6 (d) 10 : 5 (e) 6 : 10
(f) 6 : 5 : 10 (g) 5 : 10 : 6 (h) 5 : 6 : 10
2. (a) 2 : 1 (b) 6 : 1 (c) 1 : 7 (d) 4 : 3 (e) 5 : 2
(f) 3 : 2 (g) 3 : 1 (h) 5 : 2
3. (a) 4 : 2 : 1 (b) 2 : 1 : 1 (c) 1 : 2 : 4 (d) 6 : 2 : 3
(e) 3 : 1 : 2 (f) 2 : 5 : 8 (g) 2 : 2 : 5 (h) 4 : 2 : 1
(i) $\frac{2}{3}$ (j) $\frac{5}{6}$ (k) $\frac{5}{4}$ (l) $\frac{3}{5}$
(m) $\frac{2}{1}$ (n) $\frac{7}{2}$ (o) $\frac{1}{6}$ (p) $\frac{1}{5}$
4. Answers vary.
5. (a) 8 (b) 30 (c) 10 (d) 15
(e) 8 (f) 10, 15 (g) 5, 28
6. (a) 5 : 6 (b) 3 : 4 (c) 4 : 5 (d) 3 : 4
(e) $\frac{3}{10}$ (f) $\frac{3}{8}$ (g) 6 : 8 (h) 3 : 7
7. (a) 2 : 5 : 3 (b) $\frac{3}{10}$ (c) $\frac{1}{2}$ (d) $\frac{1}{5}$

8. (a) 15 : 250 (cm) (b) 3000 : 400 (mL) (c) 500 : 23 (cm) (d) 3000 : 120 (g)
(e) 48 : 30 (cm^2) (f) 9 : 4 (ha)

EXERCISE 9.2

1. Ivan: $3440, Ronald: $2580
2. 36 km, 9 km
3. Olivia: $480 000, Carmella: $400 000
4. $51 200, $76 800, $128 000
5. (a) 0.356 (b) 26% (c) 0.406
6. $32 760 000, $35 880 000, $40 560 000, $46 800 000
7. Ho: $214 285.71, Marvin: $285 714.29, Al: $357 142.86, Margaret: $142 857.14

EXERCISE 9.3

1. a, c, d, f, g, i, and j
2. (a) 15 (b) 12 (c) 10 (d) 12 (e) 18 (f) 12
(g) 2 (h) 20 (i) 22 (j) 12.5
3. 4.5 L
4. 112
5. 5
6. 588
7. 300
8. (a) 100 (b) 28.6%
9. 7 or 8
10. $7.20
11. (a) 40 L (b) 27.5 L (c) 35 L (d) 64 L

EXERCISE 9.4

1. (a) factor of 5 (b) factor of $\frac{1}{2}$ (c) factor of 3 (d) factor of $\frac{1}{10}$
(e) factor of $\frac{3}{2}$
2. a, c, and e
3. (a) $x = 4, y = 8$ (b) $x = 7, y = 8$ (c) $m = 16, n = 13.5$ (d) $r = 20, s = 16$
(e) $a = 9, b = 15$ (f) $s = 60, t = 16$ (g) $x = 20, y = 12$
4. 70 shovels gravel, 21 shovels cement
5. (a) 175 (b) $14 910
6. (a) 12 L orange juice, 24 L ginger ale
(b) 8 L grapefruit juice, 16 L orange juice

EXERCISE 9.5

1. (a) 63 wpm (b) $8.45/h (c) $4.75/kg (d) 50 km/h
(e) $35.50/ticket (f) 33 points/game (g) $5.95/cassette
2. (a) 70 km/h (b) $4.50/L (c) $9.60/h
3. (a) 90 km/h (b) 5.5 h
4. (a) $8.70/h (b) $348
5. Vince
6. 15 kg for $130.50
7. 12 L for $11.28
8. (a) $68.75 (b) 20 kg
9. $2.33
10. Clearwater by $1/d
11. 750 mL for $6.45
12. 7 L/100 km

EXERCISE 9.6

1. (a) 6.9 m by 4.0 m (b) 2.3 m by 1.8 m (c) 25 m^2
2. (a) 435 km (b) 340 km (c) 295 km (d) 380 km
3. 30 m
4. (a) 380 km (b) 205 km (c) 145 km (d) 240 km (e) 480 km

EXERCISE 9.7

1. (a) $180 (b) $340 (c) $100
2. (a) $4650 (b) $2054 (c) $6520 (d) $114
3. (a) 225 km (b) 30 s (c) 155 km (d) 220 km (e) 170 km (f) 4 s, 26 s
4. (a) 18 h (b) 34 h (c) 35 (d) same cost

EXERCISE 9.8

1. (a) 41.9 km/h (b) Alaska (c) Argentina (d) approx 30 h (e) 13
2. 6
3. 18:05
4. Use the 3 L container three times to fill the 8 L container, leaving 1 L of water in the 3 L container. Empty the 8 L container and pour the remaining 1 L of water from the 3 L container into the 8 L container. Then use the 3 L container to pour an additional 3 L of water into the 8 L container.
5. (a) 18 (b) 4 (c) 18
6. 187
7. 24
8. 23, 24, 25
9. 6
10. 100
11. 400
12. Answers vary.
13. Answers vary.
14. (a) Doc (b) Dopey (c) Sleepy (d) Bashful (e) Grumpy (f) Happy (g) Sneezy

EXERCISE 9.9

1. (a) 3 628 800 000 m^3 (b) 254 016 000 m^3 (c) 3 374 784 000 m^3 (d) 1 min 51.7 s
2. (a) 19 520 m^2 (b) 36 342 m^2
3. (a) 1.4 m
4. (a) 116.8 m^2 (b) 38.3 m
5. (a) 311 040 (b) $1 088 640
6. strait
7. goldminer from Kirkland Lake

9.10 REVIEW EXERCISE

1. (a) 5 : 8 (b) 6 : 5 (c) 8 : 6 (d) 6 : 8 (e) 5 : 6 : 8 (f) 8 : 5 : 6
2. (a) 2 : 1 (b) 1 : 3 (c) 3 : 4 (d) 3 : 2 (e) 2 : 3 : 1 (f) 3 : 2 : 5
3. Answers vary.
4. 7 : 1 and 14 : 2, 7 : 1 and 56 : 8, 14 : 2 and 56 : 8, 9 : 6 and 27 : 18
5. (a) 9 (b) 18 (c) 35 (d) 5 (e) 1 (f) $1\frac{2}{3}$ (g) $x = 4, y = 2$ (h) $a = 6, b = 16$
6. Esther: $15 000, Helga: $9000
7. Adam: $75 000, Aaron: $50 000, Aden: $125 000
8. (a) 24 (b) 14 (c) 6 (d) 11.25 (e) 4.9 (f) 3.2
9. 273
10. (a) 36 kg (b) 125 kg (c) 100 kg (d) 52.5 kg
11. (a) $x = 10, y = 15$ (b) $x = 20, y = 35$ (c) $x = 12, y = 1.5$
12. 1500 student, 300 children
13. 390 elm, 195 birch
14. (a) $11.75/h (b) $0.89/L (c) 73 km/h (d) $0.73/loaf (e) 80 km/h
15. 7 kg for $8.75
17. 1.5 m
18. (a) $1325 (b) $2700 (c) $1650 (d) 35
19. (a) $75 (b) $42.71 (c) 74 km/h

9.11 CHAPTER 9 TEST

1. (a) 3 : 1 (b) 1 : 6 (c) $\frac{3}{2}$
2. Answers vary.
3. (a) 4 (b) 96 (c) 28 (d) 12 (e) 3
4. Robyn: $15 000, Emil: $6000
5. 30 L
6. (a) $9.75 (b) $2.60/L
7. 4 kg for $6.08
8. 12 m
9. (a) $75 (b) $115

REVIEW AND PREVIEW TO CHAPTER 10

BASIC TERMS IN GEOMETRY

1. Answers vary.
2. Answers vary.

ANGLES

1. (a) 28° (b) 79° (c) 89° (d) 50° (e) 24° (f) 101° (g) 151° (h) 130° (i) 101° (j) 156°

2. (a) 42° (b) 100° (c) 66° (d) 115° (e) 12° (f) 65°
(g) 80° (h) 168° (i) 138° (j) 180°
4. (a) 90° (b) 180° (c) 60° (d) 30°
5. (a) 360° (b) 180° (c) 90° (d) 135° (e) 60° (f) 30°
(g) 210° (h) 330°

EXERCISE 10.1

1. (a) obtuse (b) right (c) reflex (d) straight
2. (a) isosceles (b) right (c) obtuse (d) equilateral
(e) acute (f) right isosceles
3. (a) complementary (b) 52°
4. (a) supplementary (b) 123°
5. (a) complementary (b) 42°
6. (a) supplementary (b) 41°
7. (a) vertically opposite (b) 43°
8. (a) acute (b) 40°
9. (a) right (b) 34°
10. (a) isosceles (b) 40°
11. (a) equilateral (b) 60° each
12. (a) $x = 75°$, $y = 105°$, $z = 75°$ (b) 30°
13. (a) 59° (b) 38° (c) $x = 57°$, $y = 123°$ (d) $x = 40°$, $y = 50°$
14. (a) $m = 50°$, $n = 45°$, $s = 50°$, $t = 85°$ (b) 36°
(c) $x = 37°$, $y = 48°$, $t = 80°$ (d) $x = 37°$, $y = 72°$, $t = 72°$, $s = 24°$
15. (a) $x = 30°$ (b) $x = 40°$, $m = 60°$
(c) $x = 81°$, $y = 56°$ (d) $x = 30°$
16. (a) $x = 70°$, $y = 30°$, $z = 40°$ (b) $m = 80°$, $x = 130°$, $y = 50°$
(c) $x = 45°$ (d) $x = 20°$, $y = 80°$, $m = 55°$
(e) $x = 45°$, $m = 135°$, $n = 45°$ (f) $x = 50°$, $y = 40°$, $z = 50°$

EXERCISE 10.2

1. (i) (a) ∠3 and ∠6, ∠4 and ∠5
(b) ∠1 and ∠6, ∠2 and ∠5, ∠3 and ∠8, ∠4 and ∠7
(c) ∠3 and ∠5, ∠4 and ∠6
(ii) (a) ∠1 and ∠5, ∠4 and ∠8
(b) ∠1 and ∠7, ∠2 and ∠8, ∠3 and ∠5, ∠4 and ∠6
(c) ∠1 and ∠8, ∠4 and ∠5
(iii) (a) ∠3 and ∠5, ∠4 and ∠6
(b) ∠1 and ∠5, ∠2 and ∠6
(c) ∠3 and ∠6, ∠4 and ∠5
(iv) (a) ∠1 and ∠3, ∠2 and ∠4
(b) none
(c) ∠1 and ∠2, ∠3 and ∠4
(v) (a) ∠4 and ∠5, ∠6 and ∠8
(b) ∠2 and ∠5, ∠6 and ∠10
(c) ∠3 and ∠5, ∠6 and ∠7
2. (a) ∠a = 105°, ∠b = 75°, ∠d = 75°, ∠f = 75°
(b) ∠r = 72°, ∠s = 108°, ∠p = 72°, ∠u = 72°, ∠v = 108°
(c) ∠d = 123°, ∠e = 57°, ∠f = 123°, ∠m = 57°, ∠n = 123°
(d) ∠c = 91°, ∠e = 91°, ∠f = 89°, ∠g = 91°, ∠h = 91°
3. (a) ∠a = 63°, ∠b = 63°, ∠c = 117° (b) ∠r = 73°, ∠s = 107°, ∠t = 107°
4. (a) ∠a = 72°, ∠b = 72°, ∠c = 108° (b) ∠d = 98°, ∠e = 82°, ∠f = 98°
(c) ∠a = 60°, ∠b = 60° (d) ∠a = 95°, ∠b = 85°, ∠c = 95°
5. (a) ∠x = 82°, ∠y = 52°, ∠z = 46° (b) ∠m = 51°, ∠n = 39°, ∠t = 90°
(c) ∠a = 46°, ∠b = 50°, ∠c = 50° (d) ∠a = 55°, ∠b = 62°, ∠x = 63°, ∠y = 63°

EXERCISE 10.3

1. (a) rectangle (b) parallelogram (c) trapezoid (d) square
   (e) rhombus (f) quadrilateral (g) rectangle (h) parallelogram
   (i) trapezoid
2. (a) 64° (b) 100° (c) $\angle x = 122°$, $\angle y = 148°$
   (d) $\angle x = 59°$, $\angle y = 87°$
3. (a) (i) $\angle a = 105°$, $\angle b = 75°$, $\angle c = 105°$
   (ii) $\angle x = 65°$, $\angle y = 115°$, $\angle z = 65°$
   (iii) $\angle m = 98°$, $\angle n = 82°$, $\angle t = 98°$
   (iv) $\angle r = 80°$, $\angle s = 100°$, $\angle p = 80°$, $\angle q = 100°$
   (b) equal
4. (a) x = 6.3 cm, y = 3.5 cm
   (b) $\angle m = 108°$, $\angle n = 115°$
   (c) m = 5 cm, n = 5 cm, t = 5 cm, $\angle x = 90°$
   (d) x = 7 cm, y = 7 cm, $\angle s = 65°$, $\angle t = 115°$
5. (a) $\angle x = 32°$, $\angle y = 41°$, $\angle z = 107°$, $\angle t = 107°$
   (b) x = 25°
   (c) $\angle x = 99°$, $\angle y = 99°$, $\angle m = 81°$, $\angle r = 81°$
   (d) $\angle x = 59°$, $\angle y = 51°$
6. (a) $\angle x = 87°$ (b) x = 60°
   (c) x = 15° (d) x = 10°, $\angle y = 160°$
7. hexagon: 720°, septagon: 900°, octagon: 1080°, nonagon: 1260°, decagon: 1440°

EXERCISE 10.5

6. side, side, side (SSS); two sides and a contained angle (SAS); two angles and a contained side (ASA); two corresponding sides of a right triangle
7. $\triangle ABC \cong \triangle RPQ$ (SAS), $\triangle JKL \cong \triangle NOM$ (SSS), $\triangle HGI \cong \triangle XYW$ (ASA)

EXERCISE 10.6

1. (a) 183 cm (b) 18.6 m
2. (a) 36.6 m (b) 37.6 m
3. (a) 44.0 m (b) 113.0 cm
4. (a) 38.6 m (b) 27.7 m
5. (a) 60 m (b) 113 cm
6. 92.6 m
7. 94.4 cm
8. 81.6 m
9. 429.84 m
10. (a) 426 m (b) $7135.50
11. 175.8 cm
12. 826.8 m
13. (a) 65 km (b) 6.5 h

EXERCISE 10.7

1. (a) 25 m^2 (b) 18 cm^2 (c) 24 cm^2 (d) 54 m^2 (e) 28 m^2 (f) 24 m^2
2. (a) 169 m^2 (b) 1092 cm^2 (c) 1176 m^2 (d) 252 cm^2 (e) 77 m^2 (f) 314 cm^2
3. (a) 81.8 m^2 (b) 64.5 m^2 (c) 55.9 m^2 (d) 213 m^2
4. 10 410.3 m^2
5. 1764 cm^2
6. 225.7 m^2
7. (a) 2275 (b) 45 625
8. $14.36
9. (a) 78.5 cm^2 (b) 113.0 cm^2

EXERCISE 10.8

1. (a) 24 (b) 27 (c) 24 (d) 70
2. (a) 5434 cm^3 (b) 276 m^3 (c) 16.4 m^3 (d) 4710 m^3 (e) 480 cm^3 (f) 352.8 m^3
3. (a) 9000 m^3 (b) 600 m^3
4. (a) 1575 m^3 (b) 4200 m^3
5. (a) 78 m^2 (b) 117 m^3

EXERCISE 10.9

1. (a) yes (b) yes (c) yes (d) no (e) no (f) yes
   (g) yes (h) yes

2. (a) 5, $\sqrt{32}$, 13, 10 (b) 24, 15, $\sqrt{3}$, 1
3. (a) 4, 5 (b) 7, 8 (c) 9, 10 (d) 12, 13 (e) 24, 25 (f) 13, 14
4. (a) 45, 6.71 (b) 52, 7.21 (c) 34, 5.83 (d) 14, 3.74 (e) 89, 9.43 (f) 149, 12.21
5. 21.2 cm 6. 45 m 7. 8.9 m 8. 3.9 m 9. (a) 5.66 m (b) 0.46 m
10. 583 km 11. (b) 13 cm 12. 65.6 cm
13. AJ $\doteq$ 6.08 cm, DK $\doteq$ 5.39 cm, EL = 5 cm, FM = 5 cm, GN = 5.39 cm, HB $\doteq$ 6.08 cm
14. 13 cm 15. 2.24 m 16. 22.9 m
17. (a) $x \doteq 1.73$, (8.66) (b) $y = 1.41$, (7.07)

EXERCISE 10.11

1. (a) 25 (b) 53 2. $95.98 3. 3 4. 2029 5. 12 h
6. 3 bags of 10, 6 bags of 20 7. 96 cm, 576 cm^2
8. 17
9. (a) British Columbia, Manitoba (b) Saskatchewan (c) New Brunswick
(d) Nova Scotia (e) British Columbia, Prince Edward Island
(f) Ontario (g) Alberta, Quebec (h) Prince Edward Island
10. 18: 21 11. 18 + 19 + 20 = 57 or 7 + 8 + 9 + 10 + 11 + 12 = 57
12. one $10 bill, one $2 bill, one 50¢ piece, two dimes and one penny
14. $25 000 15. 100 m by 50 m
16. The cylinder in (a) has twice the volume.

10.12 REVIEW EXERCISE

1. (a) isosceles (b) right (c) obtuse
(d) equilateral (e) right isosceles (f) obtuse isosceles
2. (a) $\angle x = 61°$, $\angle y = 119°$, $\angle z = 61°$
(b) $\angle m = 45°$, $\angle n = 42°$, $\angle s = 93°$, $\angle t = 45°$
(c) $\angle x = 30°$ (d) $\angle y = 79°$ (e) $\angle m = 41°$, $\angle n = 79°$
(f) $\angle r = 46°$, $\angle s = 53°$, $\angle t = 81°$ (g) $\angle t = 40°$, $\angle x = 84°$, $\angle y = 59°$
(h) $\angle r = 60°$, $\angle s = 60°$, $\angle t = 60°$, $\angle u = 120°$, $\angle v = 120°$
3. (a) $\angle m = 106°$, $\angle t = 74°$, $\angle w = 74°$, $\angle x = 74°$
(b) $\angle r = 101°$, $\angle x = 79°$, $\angle y = 79°$
(c) $\angle a = 119°$, $\angle b = 119°$, $\angle c = 61°$, $\angle d = 119°$, $\angle e = 61°$, $\angle f = 119°$, $\angle g = 61°$
(d) $\angle x = 100°$, $\angle y = 80°$, $\angle z = 100°$
4. (a) 73° (b) $\angle x = 82°$, $\angle y = 92°$
(c) $\angle x = 108°$, $\angle y = 72°$, $\angle z = 108°$ (d) x = 8 cm, y = 8 cm
(e) $\angle x = 99°$, $\angle y = 61°$ (f) x = 8.6 m, y = 5.1 m
5. (a) 7.8 m (b) 9.7 m
12. (a) 21.2 m (b) 39 cm (c) 188.4 m (d) 46.2 m
13. (a) 936 m^2 (b) 294.9 cm^2 (c) 336 m^2 (d) 5024 m^2
14. (a) 1287 m^3 (b) 3454 m^3

10.13 CHAPTER 10 TEST

1. (a) $\angle x = 72°$, $\angle y = 72°$, $\angle z = 108°$ (b) 74 °
(c) $\angle r = 27°$, $\angle s = 69°$, $\angle t = 84°$
3. (a) 251.2 m, 5024 m^2 (b) 114 m, 498 m^2 4. (a) 235 500 m^3 (b) 260 m^3
5. (a) 9.2 m (b) 7.5 m

REVIEW AND PREVIEW TO CHAPTER 11

RATIO AND PROPORTION
1. (a) 0.4 (b) 0.3 (c) 0.3 (d) 6.2 (e) 0.7 (f) 9.5
(g) 31.3 (h) 14.4 (i) 34.0 (j) 2.4
2. (a) 3.7 (b) 7.6 (c) 5.2 (d) 186.7 (e) 34.9 (f) 2.2
(g) 14.3 (h) 20.5 (i) 233.6 (j) 2.3

PYTHAGOREAN THEOREM

1. (a) 50.0 (b) 4.0 (c) 1.4 (d) 13.0
2. (a) 15.0 (b) 5.6 (c) 2.0 (d) 2.2

ANGLE CALCULATIONS

1. (a) 63° (b) x = 30° (c) x = 45° (d) 26° (e) 73° (f) 26°
2. (a) a = 70°, b = 70°, c = 70°
   (b) a = 68°, b = 112°, c = 68°, d = 112°

MENTAL MATH

1. 4	2. 15	3. 5	4. 16	5. 3	6. 3
7. 5	8. 5	9. 8	10. −3	11. 6	12. 5
13. −2	14. 3	15. 3	16. 2	17. 2	18. 4
19. 8	20. 6	21. 5	22. 6	23. 2	24. 28
25. 5	26. 8	27. 12	28. 15	29. 15	30. 20
31. 100	32. 7	33. 18	34. 21	35. 0.3	36. 100
37. 16	38. 2				

EXERCISE 11.1

1. (a) (i) △ABC, △DEF (ii) △EDC, △EAB (iii) △ABX, △DCX
2. (a) △HGI (Corresponding angles are equal.)
   (b) △DFE (Corresponding angles are equal.)
   (c) △DBE (Corresponding angles are equal.)
3. (a) $\frac{5}{10} = \frac{3.2}{6.4} = \frac{3.8}{7.6}$ (b) $\frac{a}{p} = \frac{b}{q} = \frac{c}{r}$ (c) $\frac{5}{9} = \frac{10}{18}$ (d) $\frac{4}{6} = \frac{5}{7.5}$
   (e) $\frac{4}{5.6} = \frac{5}{7}$ (f) $\frac{a}{b} = \frac{c}{d}$
4. (a) 12 cm (b) 2.5 cm (c) 10 cm (d) 4 m (e) 5 m
5. DE = 3.2 cm, GH = 9.9 cm
6. (a) Corresponding angles are equal. (b) a = 5.0 m, b = 6.7 m
7. 10.4 m 8. 21 m 9. 15 m 10. 13.5 m

EXERCISE 11.2

1.

	Hypotenuse	Opposite	Adjacent
(a)	x	3	5
(b)	6	y	4
(c)	12	x	8
(d)	c	b	a
(e)	z	y	x
(f)	10	8	6

2. (a) $\frac{3}{6}$ (b) $\frac{5}{6}$ (c) $\frac{12}{5}$ (d) $\frac{15}{8}$ (e) $\frac{3}{4}$ (f) $\frac{y}{x}$
   (g) $\frac{10}{10}$ (h) $\frac{5}{h}$
3. (a) 0.3249 (b) 0.7265 (c) 1.2799 (d) 0.4877 (e) 1.0000 (f) 0.4663
   (g) 0.8391 (h) 5.6713 (i) 0.1763 (j) 0.3640 (k) 0.5317 (l) 1.4826
   (m) 0.5774 (n) 1.7321
4. (a) 12° (b) 29° (c) 36° (d) 45° (e) 30° (f) 75°
   (g) 60° (h) 40° (i) 52° (j) 29° (k) 68° (l) 84°
   (m) 68° (h) 45°
5. (a) 37° (b) 74° (c) 62° (d) 67°
6. (a) 90.0 m (b) 14.9 m (c) 210.1 m (d) 86.6 m (e) 35.8 cm (f) 23.4 m
7. (a) 238.4 m (b) 8.1 m (c) 5 cm (d) 72.8 m
8. 14 m 9. 22.6 m 10. 5.0 m 11. 46.1 m 12. 700 m 13. 58°

EXERCISE 11.3

1. (a) $\frac{8}{17}$ (b) $\frac{8}{10}$ (c) $\frac{1.0}{2.0}$ (d) $\frac{x}{10}$ (e) $\frac{y}{r}$ (f) $\frac{b}{c}$
2. (a) $\sin 42° = \frac{h}{12}$ (b) $\sin 48° = \frac{a}{25}$ (c) $\sin 63° = \frac{b}{15}$ (d) $\sin 60° = \frac{h}{60}$
   (e) $\sin 65° = \frac{a}{10}$ (f) $\sin 45° = \frac{h}{14}$
3. (a) 13.1 m (b) 29.0 cm (c) 48.1 m (d) 9.2 cm (e) 40.5 cm (f) 12.5 cm
4. (a) 43° (b) 26° (c) 56° (d) 25° (e) 42° (f) 49°
5. 4.5 m 6. 79° 7. 175 m 8. 4°

EXERCISE 11.4

1. (a) $\frac{12}{13}$ (b) $\frac{8}{10}$ (c) $\frac{8}{17}$ (d) $\frac{a}{c}$ (e) $\frac{x}{r}$ (f) $\frac{q}{r}$
2. (a) $\cos 40° = \frac{a}{25}$ (b) $\cos 38° = \frac{b}{18}$ (c) $\cos 35° = \frac{x}{32}$ (d) $\cos 70° = \frac{r}{16}$
   (e) $\cos 63° = \frac{h}{10}$ (f) $\cos 50° = \frac{h}{14}$
3. (a) 13.8 cm (b) 13.0 cm (c) 33.1 m (d) 22.1 m (e) 17.7 m (f) 26.0 cm
4. (a) 41° (b) 32° (c) 31° (d) 46° (e) 37° (f) 39°
5. 76° 6. 2.1 km 7. 94 m

EXERCISE 11.6

1. (a) 0.7314 (b) 2.7475 (c) 0.8192 (d) 0.9004 (e) 0.4226 (f) 5.6713
   (g) 0.9659 (h) 0.9397 (i) 0.6428 (j) 0.9063 (k) 0.6745 (l) 0.7660
   (m) 0.9962 (n) 0.9659 (o) 1.3764 (p) 0.3249 (q) 0.5736 (r) 0.8192
2. (a) 15° (b) 20° (c) 75° (d) 49° (e) 30° (f) 30°
   (g) 40° (h) 63° (i) 19° (j) 3° (k) 33° (l) 27°
   (m) 49° (n) 76° (o) 32° (p) 46° (q) 90° (r) 45°

EXERCISE 11.7

1. (a) $\sin 55° = \frac{x}{15}$ (b) $\cos 50° = \frac{x}{20}$ (c) $\tan 37° = \frac{x}{32}$ (d) $\cos 29° = \frac{x}{42}$
2. (a) $\sin \theta = \frac{5}{8}$ (b) $\cos \theta = \frac{5}{7}$ (c) $\tan \theta = \frac{8}{13}$ (d) $\sin \theta = \frac{20}{23}$
   (e) $\cos \theta = \frac{10}{15}$ (f) $\cos \theta = \frac{12}{18}$
3. (a) 50° (b) 40° (c) 9.6 cm
4. (a) 52° (b) 15.4 cm (c) 19.7 cm
5. 15.3 m 6. 134 m 7. (a) 67° (b) 3.4 m
8. (a) 39° east of north (b) 661 km

EXERCISE 11.8

1. 8° south of west at 212 km/h 2. 12° into the current
3. 453 km/h 4. 6° into the current
5. 6° 6. (a) 256 km/h (b) 12° east of north

EXERCISE 11.9

1. 22° 2. 18.0 m 3. 25.6 m 4. 17° 5. 437 m 6. 53°
7. 116 m 8. 960 m 9. 2° 10. 14.8 m 11. 71° 12. 97 m

EXERCISE 11.10

1. (a) i, v, and vi
3. 36
4. $\sqrt{128} \doteq 11.3$ cm
6. 15
7. 15
8. 30 chickens, 20 rabbits
9. 37
10. 14 min
11. 45.5 km
12. 1 5 10 10 5 1, 1 6 15 20 15 6 1, 1 7 21 35 35 21 7 1, 1 8 28 56 70 56 28 8 1, 1 9 36 84 126 126 84 36 9 1
13. Manager: Arnie, Cook: Kenneth, Waiter: Sal

EXERCISE 11.11

1. (a) 17 885 (b) 160 965 (disregarding leap years)
2. (a) 492 (b) 891 (c) 70.3
3. Hart
4. 24
5. (a) 59 (b) 138 (c) 47
8. (a) 1.0 (b) 0.6 (c) Greyhounds (d) Racers

11.12 REVIEW EXERCISE

1. (a) 19.2 cm (b) 26.25 cm (c) 7.3 cm
2. (a) $\sin\theta = \frac{100}{200}$, $\cos\theta = \frac{173}{200}$, $\tan\theta = \frac{100}{173}$
   (b) $\sin\theta = \frac{85}{110}$, $\cos\theta = \frac{70}{110}$, $\tan\theta = \frac{85}{70}$
   (c) $\sin\theta = \frac{68}{108}$, $\cos\theta = \frac{84}{108}$, $\tan\theta = \frac{68}{84}$
3. $b \doteq 8.4$, $c \doteq 19.9$
4. (a) 14.1 cm (b) 14.7 cm (c) 49° (d) 39° (e) 21.4 cm (f) 16.4 cm
5. 667 m
6. 29.3 m
7. 56°
8. 5°
9. 4.17 cm
10. 1.3 m
11. 9.6 cm
12. (a) 280 km (b) 27° east of north
13. 31.4 m

11.12 CHAPTER 11 TEST

1. 2.5 cm
2. (a) $\sin\theta = \frac{15}{17}$, $\cos\theta = \frac{8}{17}$, $\tan\theta = \frac{15}{8}$
   (b) $\sin\theta = \frac{12}{13}$, $\cos\theta = \frac{5}{13}$, $\tan\theta = \frac{12}{5}$
3. (a) 7.5 cm (b) 41° (c) 2.9 cm
4. 14.1 m
5. 80.5 m
6. 54°

REVIEW AND PREVIEW TO CHAPTER 12

GRAPHING

1. A(2, 3), B(−1, 4), C(2, 0), D(5, 4), E(−6, 0), F(5, −2), G(0, −2), H(−2, −1), J(−1, 1), K(−1, −4), L(3, −4), M(0, 5), N(−5, 4), O(0, 0), P(−5, −4)
3. (a) C (b) Q, R
4. (a) $v = \frac{d}{t}$ (b) $r = \frac{C}{2\pi}$ (c) $b = P - a - c$ (d) $P = \frac{I}{rt}$

THE PYTHAGOREAN THEOREM

1. (a) 13.0 m (b) 3.0 m (c) 10.0 m (d) 12.0 m (e) 5.7 m (f) 3.6 m
   (g) 5.7 m (h) x = 5.0 m, y = 13.0 m
   (i) x = 2.2 m, y = 2.4 m, z = 2.6 m
   (j) x = 3.9 m, y = 5.7 m (k) x = 6.0 m, y = 6.7 m
   (l) x = 12.0 m, y = 16.5 m

EXERCISE 12.1

1. (a) $\frac{2}{5}$ (b) 1 (c) $\frac{1}{3}$ (d) $\frac{7}{4}$ (e) 3 (f) $\frac{7}{13}$
2. (a) $\frac{4}{5}$ (b) $\frac{3}{2}$
3. (a) $\frac{3}{7}$ (b) $\frac{3}{2}$
4. (a) $\frac{3}{5}$ (b) $\frac{4}{3}$ (c) $\frac{2}{7}$ (d) $\frac{5}{2}$
5. (a) 1 (b) $\frac{1}{2}$ (c) $\frac{1}{7}$ (d) 2

EXERCISE 12.2

1. (a) AB, KL (b) EF, GH, NM (c) CD (d) IJ
2. (a) $\frac{1}{2}$ (b) $\frac{2}{3}$ (c) -1 (d) $-\frac{4}{7}$
3. (a) $\frac{6}{5}$ (b) $-\frac{6}{5}$
4. AB : $-\frac{3}{2}$, CD : $\frac{3}{2}$, EF : 0, GH : no slope, MN : 5
5. (a) 2 (b) 2 (c) $\frac{3}{4}$ (d) 0 (e) 5 (f) $\frac{7}{6}$
   (g) $\frac{1}{3}$ (h) $\frac{8}{3}$ (i) $-\frac{1}{2}$ (j) $-\frac{3}{4}$ (k) $\frac{2}{3}$ (l) $-\frac{5}{2}$
   (m) $\frac{7}{10}$ (n) -1 (o) $-\frac{4}{15}$ (p) -1
10. (a) 45 m (b) 2.7 km
11. the first

EXERCISE 12.3

1. (a) $\frac{3}{2}$ (b) $\frac{3}{2}$ (c) yes
2. (a) -2 (b) $-\frac{4}{3}$ (c) no
3. a, d, and f
4. (a) Both slopes are zero. (b) parallel (c) 0
5. (a) no slope (b) parallel (c) undefined
6. AB and DC both have a slope of $-\frac{3}{2}$ and so are parallel. AD and BC both have a slope of $\frac{1}{3}$ and thus are also parallel. ABCD is a parallelogram.
7. (a) 6 (b) 3 (c) 4 (d) 9 (e) -1 (f) 15

EXERCISE 12.4

1. (a) -2 (b) $-\frac{1}{3}$ (c) $-\frac{5}{2}$ (d) $\frac{4}{3}$ (e) $\frac{1}{4}$ (f) $\frac{5}{6}$
   (g) undefined (h) 1
2. (a) $\frac{1}{2}$ (b) -1 (c) $\frac{5}{2}$ (d) $\frac{7}{3}$ (e) $-\frac{4}{3}$ (f) $\frac{2}{5}$
3. (a) $-\frac{4}{3}$ (b) $\frac{3}{4}$ (c) yes
4. (a) $-\frac{4}{5}$ (b) $\frac{5}{4}$ (c) yes
5. (a) $\angle$PRQ (b) $\angle$BAC (c) $\angle$MNO

EXERCISE 12.5

1. AB : 5, CD : 5, EF : 3, GH : 4, JK : 4
2. (a) 5.0 (b) 7.2 (c) 2.0 (d) 2.2 (e) 6.4 (f) 8.1
3. AB and DC : 4.2, AD and BC : 6.3 (nearest tenth)
4. PM : 5.4, QM : 5.4 (nearest tenth)

EXERCISE 12.6

1. (a) All indicated distances are $\sqrt{18} \doteq 4.2$
   (b) slope of AB and CD : 1, slope of AD and BC : -1
   (c) The product of the slopes of any two indicated intersecting lines is -1, and hence perpendicular.
2. (a) Each diagonal has length $\sqrt{50} \doteq 7.1$
   (b) The slope of QS is -1, the slope of PR is 1, and the product is -1.
3. (a) EF = DG = $\sqrt{18} \doteq 4.2$, ED = FG = $\sqrt{72} \doteq 8.5$
   (b) FE = GD = 1, ED = FG = -1 (slopes)
   (c) The product of the slopes of intersecting lines is -1.
   (d) EG = FD = $\sqrt{90} \doteq 9.5$
4. (a) BA = CD = 7, BC = AD = $\sqrt{34} \doteq 5.8$
   (b) BA = CD = 0, CB = DA = $\frac{5}{3}$ (slopes)
5. (a) BA = $\frac{4}{5}$, AC = $-\frac{5}{4}$ (slopes), $\angle$BAC = 90°
   (b) AB = AC = $\sqrt{41} \doteq 6.4$
   (c) BC = $\sqrt{82} \doteq 9.1$, $(AB)^2 + (AC)^2 = (BC)^2$
6. (a) EM = FM = $\sqrt{29} \doteq 5.4$ (b) DN = FN = $\sqrt{40} \doteq 6.3$
   (c) MN = 5, ED = 10 (d) MN = ED = $-\frac{3}{4}$ (slopes)
7. AB: (2, 3), CD: (7, 3), BD: (4.5, 5), AC: (4.5, 1)
8. (a) E (-1, 3), F(-5, -2), G(2, -4), H(6, 1)
   (b) AE = BE = $\sqrt{26} \doteq 5.1$, BF = CF = $\sqrt{17} \doteq 4.1$, CG = DG = $\sqrt{40} \doteq 6.3$, AH = DH = $\sqrt{13} \doteq 3.6$
   (c) EH = FG = $\sqrt{53} \doteq 7.3$
   (d) EF = HG = $\sqrt{41} \doteq 6.4$
   (e) EF = HG = $\frac{5}{4}$ (slopes)
   (f) EH = FG = $-\frac{2}{7}$ (slopes)
9. (a) AC = BC = $\sqrt{65} \doteq 8.1$ (b) no

EXERCISE 12.7

1. (a) 3274 km (b) 3799 km (c) 4001 km (d) 3452 km
4. (a) \$2.39 (b) \$7.67

EXERCISE 12.8

1. (a) 60 cm² (b) 24 cm 2. C 3. 26
4. (a) 132 m (b) \$2191.20 (c) 44 (d) \$561 (e) \$3352.20
5. 4 6. 74 × 653 = 48 322 7. 9 km 8. 750
10. (a) 3350 km (b) 469 11. 5, 12, 6 12. 11 13. 34
14. 5
15. Make a straight line cut; then each of the 3 subsequent cuts must cross all previous cuts.
16. 14

12.9 REVIEW EXERCISE

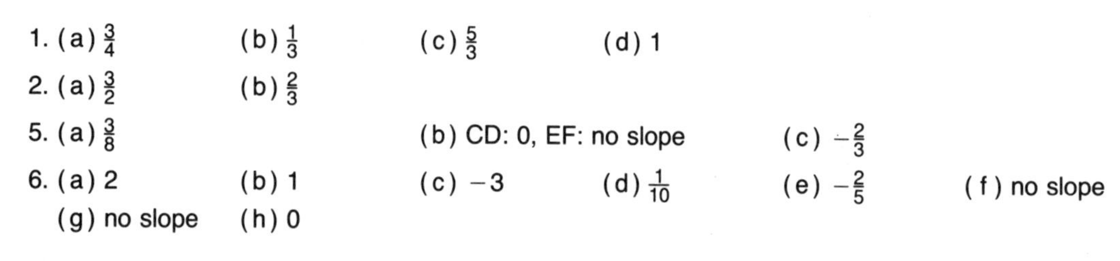

1. (a) $\frac{3}{4}$ (b) $\frac{1}{3}$ (c) $\frac{5}{3}$ (d) 1
2. (a) $\frac{3}{2}$ (b) $\frac{2}{3}$
5. (a) $\frac{3}{8}$ (b) CD: 0, EF: no slope (c) $-\frac{2}{3}$
6. (a) 2 (b) 1 (c) -3 (d) $\frac{1}{10}$ (e) $-\frac{2}{5}$ (f) no slope
   (g) no slope (h) 0

9. (a) $\frac{3}{4}$ (b) $-\frac{4}{3}$ (c) perpendicular
10. (a) 3 (b) 3 (c) parallel
11. (a) parallel (b) perpendicular (c) perpendicular
(d) parallel (e) parallel
12. (a) 12 m (b) 8 m
13. AB : 5, CD : 4, EF : 10, GH : 6
14. AB : 5, CD : $\sqrt{17} \doteq 4.1$, EF : 5, GH : $\sqrt{37} \doteq 6.1$
15. (a) 5 (b) 13 (c) 10 (d) 10
(e) $\sqrt{2} \doteq 1.4$ (f) $\sqrt{58} \doteq 7.6$ (g) $\sqrt{45} \doteq 6.7$
16. PR and PQ both have length $\sqrt{68} \doteq 8.2$
17. (a) BA = CD = 0, CB = DA = 3 (slopes)
(b) BA = CD = 8, BC = AD = $\sqrt{40} \doteq 6.3$
(c) AM = CM = $\sqrt{34} \doteq 5.8$, BM = DM = $\sqrt{18} \doteq 4.2$

12.10 CHAPTER 12 TEST

1. (a) $\frac{2}{3}$ (b) $\frac{1}{3}$
2. (a) $\frac{4}{9}$ (b) $\frac{7}{4}$
3. (a) $\frac{4}{7}$ (b) $-\frac{7}{5}$ (c) 0
4. (a) (i) $\frac{4}{3}$ (ii) -2 (iii) $\frac{4}{3}$ (iv) $-\frac{3}{4}$
(b) (i) and (iii) (c) (ii) and (iv)
5. (a) 9 (b) $\sqrt{65} \doteq 8.1$ (c) 10

YEAR-END REVIEW

1. (a) thirty-four thousand five hundred
(b) seven hundred fifty-six thousandths
(c) three million four hundred fifty thousand
(d) eighteen and nine-tenths
(e) five thousand six hundred eleven
(f) sixty-seven and seventy-six hundredths
2. Estimates vary.
(a) 40 793 (b) 867 (c) 30 820 (d) 23 (e) 861 (f) 10 097
(g) 817 (h) 123
3. $49 350
4. Estimates vary.
(a) 536 (b) 138 (c) 403 (d) 55 (e) 61 (f) 56
5. Estimates vary.
(a) 136.97 (b) 13.9 (c) 69.75 (d) 3.6 (e) 1.223 (f) 0.92
(g) 269.2
6. Estimates vary.
(a) $390.50 (b) $37.38 (c) $605.00 (d) $508.80 (e) $442.15
7. (a) $1\frac{1}{6}$ (b) $11\frac{1}{4}$ (c) $6\frac{3}{4}$ (d) $5\frac{2}{5}$
8. (a) $1\frac{1}{4}$ (b) $3\frac{1}{2}$ (c) $1\frac{1}{8}$ (d) $\frac{9}{10}$
9. (a) $1\frac{1}{2}$ (b) $4\frac{1}{8}$ (c) $6\frac{9}{10}$ (d) $7\frac{7}{12}$
10. (a) $\frac{1}{8}$ (b) $\frac{5}{8}$ (c) $1\frac{1}{10}$ (d) $3\frac{7}{12}$
11. (a) 9 (b) 2 (c) 1 (d) 25 (e) -19 (f) 6
12. (a) 0.375 (b) $0.8\overline{3}$
13. (a) $\frac{11}{20}$ (b) $\frac{5}{9}$
14. (a) $-1\frac{1}{8}$ (b) $-2\frac{3}{4}$ (c) -7 (d) $\frac{9}{16}$

15. (a) 9 (b) 18 (c) −9 (d) −108 (e) 36 (f) 36
(g) −216 (h) −216 (i) 1 (j) 27 (k) $\frac{1}{3}$ (l) $\frac{2}{9}$
(m) $\frac{1}{9}$ (n) $-\frac{1}{9}$
16. (a) 13 (b) 17 (c) 1 (d) 25 (e) 578 (f) −54
17. (a) $6x^5$ (b) $-2x^5$ (c) $3x^6$ (d) $-6x^5$ (e) $-5x^3$ (f) $10x^2$
(g) $-10a^2$
18. (a) 2.534×10^2 (b) 3.62×10^4 (c) 4.56×10^1 (d) 1.356×10^6
(e) 3.56×10^{-1} (f) 2.56×10^{-2} (g) 5.75×10^{-3} (h) 7.89×10^{-4}
19. (a) 2.862×10^7 (b) $2.486\,25 \times 10^2$ (c) $2.573\,25 \times 10^3$ (d) 4.5625×10^2
20. (a) 11 (b) 13 (c) 18 (d) 20 (e) 25 (f) 32
21. (a) 22.4 (b) 18.7 (c) 12.2 (d) 8.7 (e) 2.4 (f) 1.9
22. 75 km/h
23. (a) $C = 3.50 + 6.50d, d \geqslant 1$ (b) $49
24. (a) $-3a - 3b$ (b) $-x - 3y + 4$ (c) $10xy - 4x - 3y$
25. (a) −8 (b) 9 (c) 10
26. (a) $6x + 31$ (b) $-a^2 - 6a$ (c) $-4x^2 + 7x + 9$ (d) $x^2 - 12x - 7$
27. (a) 2 (b) 4 (c) 5 (d) 9 (e) −6
28. 61, 62, 63
29. (a) $x^2 + 11x + 28$ (b) $m^2 - 8m + 16$ (c) $b^2 - 9$ (d) $6x^2 + 11x + 4$
30. (a) $2x^2 - x - 27$ (b) $4t + 34$ (c) $5m^2 + 11m - 6$ (d) $s^2 + 19s + 106$
31. (a) $2x - 1$ (b) $1 - 2x + 3y$
32. (a) $(x - 4)(x + 4)$ (b) $(x - 3)(x - 4)$ (c) $(m + 3)(m + 5)$ (d) $(y - 2)(y + 3)$
(e) $(s - 10)(s + 10)$ (f) $(a - 5)(a + 6)$
33. Answers vary.
34. (b) 375 km (c) 1125 km
35. $296
36. (a) $C = 70 + 45h$ (b) $362.50
38. (a) 1, 1, −11 (b) $-\frac{4}{3}$, 9, −4
39. (a) (6, 9) (b) (4, 1) (c) (3, 1) (d) (1, 2)
40. (a) (1, 1) (b) (1, −1) (c) (2, −3) (d) (15, −10)
41. 3, 4
42. −5, 3
43. 12 cm by 5 cm
44. 2, 4
45. 100 m by 73 m
46. Answers vary.
47. 5250
48. 76.1, 76, 74
49. (a) 15.8, 16, 10 (b) 4, 4, 4 (c) 24.2, 24, 23 and 25
50. (a) 201 (b) 197 (c) 193 (d) 194
52. (b) 1951-1961 (c) 1881-1891
53. (a) 27% (b) $2177.78 (c) 30d (d) 2.3%
(e) $11 503.03 (f) $14 875 (g) $12 037.50
54. $25 223.20
55. $28 981.85
56. (a) $18 381.05 (b) $4381.05
57. (a) $1574.02 (b) $1143.68 (c) $1163.48 (d) $1325.89 (e) $667.31 (f) $505.62
58. $1212.17/month, $753.35/month, $577.73/month; $18 750 for 30 months at 15%
59. (a) 6 (b) 10 (c) x = 9, y = 6 (d) x = 5, y = 20
60. Denzil: $15 000, Leslie: $10 000
61. 800 gold, 1200 red, 2800 blue
62. (a) $5.25/L (b) $9.80/h (c) 83 km/h
63. 70 m
64. (a) ∠a = 73°, ∠b = 107°, ∠c = 73°
(b) ∠a = 53°, ∠b = 72°, ∠c = 55°, ∠d = 53°
(c) 45° (d) ∠m = 43°, ∠n = 90°
65. (a) ∠m = 103°, ∠q = 103°, ∠r = 103°, ∠s = 77°, ∠t = 103°
(b) 99°
(c) ∠a = 65°, ∠b = 115°, ∠c = 65°, ∠d = 115°, ∠e = 65°
(d) ∠x = 79°, ∠y = 101°, ∠z = 101°
(e) ∠r = 73°, ∠s = 73°, ∠t = 45°
(f) ∠m = 41°, ∠n = 78°, ∠t = 61°
66. (a) 83° (b) 96°
(c) ∠x = 105°, ∠y = 75°, ∠z = 105°
(d) ∠a = 110°, ∠b = 70°, ∠c = 110°, ∠d = 70°

67. (a) $\angle x = 107°$, $\angle y = 107°$, $\angle s = 32°$, $\angle t = 41°$
(b) $\angle d = 82°$, $\angle e = 61°$, $\angle f = 37°$, $\angle g = 82°$
(c) $\angle x = 50°$, $\angle m = 90°$
(d) $\angle x = 67°$, $\angle y = 46°$
(e) $x = 9.5$ m, $y = 6.1$ m
(f) $x = 14.6$ cm, $y = 14.6$ cm
(g) $\angle x = 60°$
(h) $\angle x = 72°$, $\angle m = \angle n = 18°$

73. (a) 22.3 m (b) 34 m (c) 60.4 cm (d) 58 m
74. (a) 45.6 m^2 (b) 301 cm^2 (c) 314 m^2 (d) 160.9 m^2
75. (a) 1323 m^3 (b) 452.2 cm^3 (c) 264 m^3 (d) 144 m^3
76. (a) 10 (b) $x = 1\frac{2}{3}$ (c) 36
77. (a) $\sin\theta = \frac{1}{2}$, $\cos\theta = \frac{\sqrt{3}}{2}$, $\tan\theta = \frac{1}{\sqrt{3}}$
(b) $\sin\theta = \frac{5}{7}$, $\cos\theta = \frac{5}{7}$, $\tan\theta = 1$
78. 31.8 cm 79. 51°
80. (a) 9.6 cm (b) 10.6 cm (c) 45° (d) 14.1 cm
81. 163 m 82. 35° 83. 159 m 84. 22 m
85. (a) $\tan\theta = \frac{3}{4}$, 37° (b) $\tan\theta = \frac{8}{6}$, 53° (c) $\tan\theta = 1$, 45° (d) $\tan\theta = 1.7$, 60°
(e) $\tan\theta = \frac{12}{5}$, 67° (f) $\tan\theta = \frac{15}{8}$, 62°
86. 22° 87. 18.5 m 88. 209.6 m
89. AB : $\frac{2}{3}$, CD : $-\frac{8}{3}$, EF : 0, GH : no slope
90. (a) 1 (b) $-\frac{11}{5}$ (c) no slope (d) 0 (e) $\frac{1}{5}$
91. (a) parallel (b) perpendicular (c) perpendicular (d) neither
(e) neither (f) parallel
92. AB : 7, CD : 6, EF : $\sqrt{52} \doteq 7.2$, GH : $\sqrt{53} \doteq 7.3$
93. (a) 5.0 (b) 5.8 (c) 5.4 (d) 7.3 (e) 11.4 (f) 7.0
(g) 10.0
94. $PQ = PR = \sqrt{34} \doteq 5.8$
95. $AB = CD = \frac{5}{7}$, $AD = BC = \frac{3}{2}$ (slopes)
96. (c) $BA = CD = -\frac{3}{4}$, $DA = CB = \frac{4}{3}$ (slopes)
(d) $AB \perp AD$, $AD \perp CD$, $CD \perp CB$, $CB \perp AB$
(e) $AB = CD = 5$, $BC = AD = 10$
97. (a) 11, 9, 7, 5, 3 (b) 1, 3, 5, 7, 9

GLOSSARY

account statement Lists cheques and deposits for a specific period of time.
acute angle Measures between 0° and 90°.
acute triangle Has only acute angles.
adjacent angles Share a common vertex and a common side.
alternate angles Two angles on opposite sides of a transversal that intersects two lines. Each line is a side of each angle.

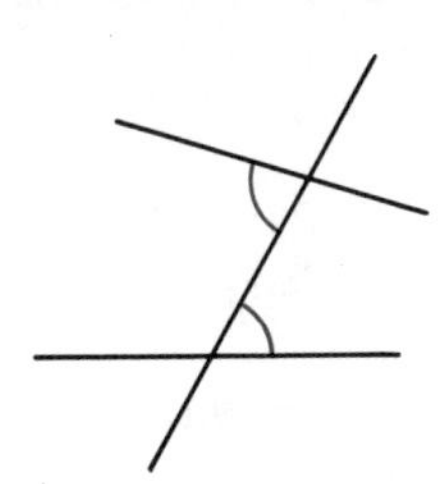

altitude of a triangle A perpendicular line from the vertex to the opposite side.

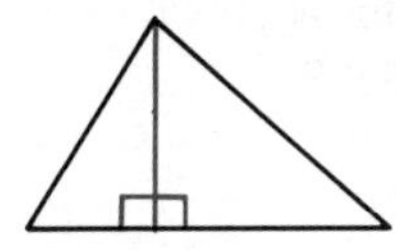

amortization schedule Shows the repayment of a loan as a series of equal payments, breaking the payments down into principal and interest.
angle A figure formed by two rays with a common endpoint, called the vertex.
angle bisector A ray dividing an angle into two equal angles.
angle of depression (elevation) Is measured from an observer, that is, the horizontal plane of the observer, to a point below (above) the observer.

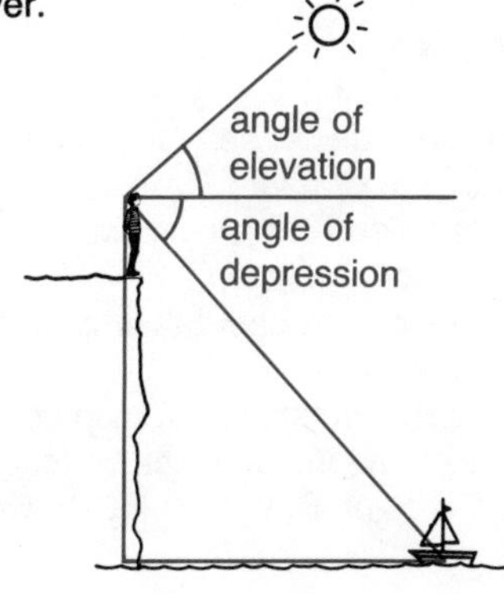

annual A period of 12 months or 1 a.
area The number of unit squares contained in a region.
assessed value The value at which a property is taxed.
average The sum of n numbers divided by n.
axis A number line used to locate points on a coordinate plane.

bar graph Uses bars to represent and compare data.
bases of a trapezoid The parallel sides in a trapezoid.

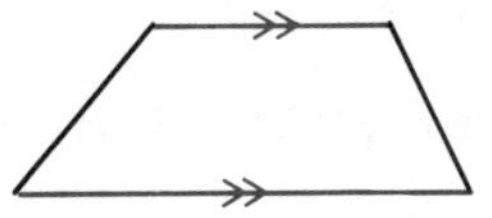

BASIC (Beginners All-purpose Symbolic Instruction Code) is a symbolic computer language.
binomial A polynomial consisting of two terms.
bond A debt certificate issued by a government or a corporation, promising to pay the holder a fixed sum of money on a given date, or maturity, with interest.
broken-line graph Uses line segments to represent data and show change in the data.
broker An agent who buys and sells investments for others.
broker's fee The commission charged by a broker for buying and selling investments.
budget A financial plan for managing personal income.

carrying charge Or finance charge, is the cost to borrow money.
cash discount A reduction on an invoice price if the invoice is paid within a specified time.
central angle of a circle Has its vertex at the centre of a circle and its sides are radii of the circle.

centroid The point at which the three medians, the lines that join the vertex to the midpoints of the sides, of a triangle intersect.
chartered bank A financial institution that provides savings, loan, and chequing services, as well as other financial services.
chord of a circle A line segment that has its endpoints on the circumference of a circle.

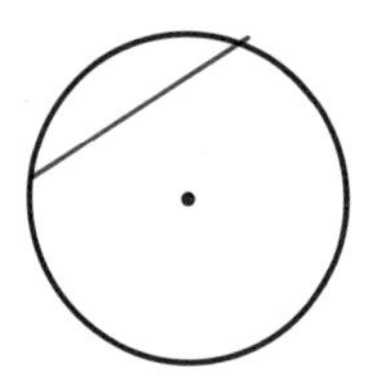

circle The set of points in the plane that are equidistant from a fixed point or centre.
circle graph Uses sectors of a circle to represent data and to show how one thing is divided.
circumcentre The centre of the circle which passes through the three vertices of a triangle.
circumference The perimeter of a circle.
circumscribed circle A circle is circumscribed about a polygon if all the vertices of the polygon lie on the circle.
collinear points Lie on the same straight line.
commission A percentage of sales that a salesperson receives as pay or an incentive.
complementary angles Sum to 90°.
compound interest Is calculated on the principal plus previous interest.
concentric circles Share the same centre.

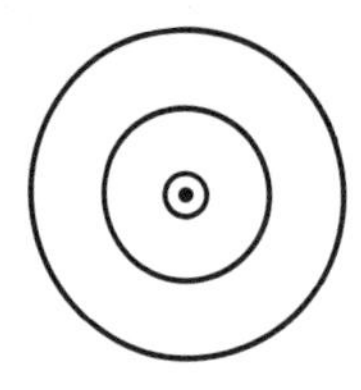

congruent angles Have the same measure.
congruent figures Are equal in all respects.

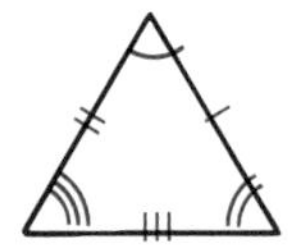

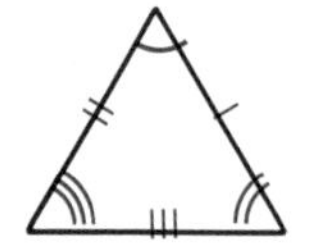

consecutive even (odd) numbers Are obtained by counting by twos from a given even (odd) number.
consecutive numbers Are obtained by counting by ones from a given number.
consistent equations Equations in a system that have at least one solution.
construction The process of drawing a geometric figure using only a ruler and compass.
coordinate A real number paired with a point on a number line.
coordinate plane Also called the Cartesian coordinate plane, is a one-to-one pairing of all ordered pairs of real numbers with the points of a plane.
cosine ratio The ratio of the length of the adjacent side to the length of the hypotenuse in a right triangle.
coterminal angles Have the same initial and terminal rays or arms.
credit cards Are issued by financial institutions, large stores, and companies, and enable the holders to make purchases and then delay payment for about 30 days, or to pay by instalments with interest.
cubic polynomial Has the form $ax^3 + bx^2 + cx + d$, and $a \neq 0$.

deductions Are subtracted by an employer from an employee's pay cheque; deductions subtracted from gross income equals taxable income.
degree A unit of angular measure equal to $\frac{1}{360}$ of a rotation.
degree of a monomial The sum of the exponents of the variables.
degree of a polynomial The degree of its greatest term, providing the polynomial is simplified.
dependent equations Have an infinite number of solutions in a system.
depreciation The decrease in the value of a property due to age or use.
diagonal A line segment joining two non-adjacent vertices of a polygon.
diameter of a circle A chord that contains the centre of the circle.
dilatation A transformation that maps each point of a figure to an image point, so that for a centre, C, and a point, P, $CP' = k(CP)$, where k is the scale factor.

direct variation A function in the form $y = kx$, where k is a constant factor.
discount A deduction from the original price of an article.
distance from a point to a line The perpendicular distance from a point to a line.
dividend The return or yield on an investment.
domain of a variable The set of values for a variable.
down payment The amount of money which is paid at the time that a purchase is made.

END statement The last statement in a computer program.
equation An open sentence formed by two expressions separated by an equal sign.
equidistant Means at the same distance.
equilateral triangle Has all sides equal.
equivalent equations Have the same solution over a given domain.
estimate Means to arrive at an approximate calculation by rounding all numbers to the highest place value and then calculating mentally.
event Any possible outcome for an experiment in probability.
exponent The number of times the base is multiplied in a power.
expressing a fraction in simplest form Means to divide the numerator and denominator of a fraction by the greatest common factor.
exterior angle of a polygon Is formed by extending one side of a polygon.

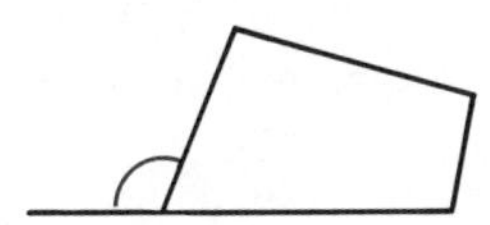

factors Numbers that are multiplied to give a product.
fixed costs The regular expenses, such as rent, which do not change despite changes in production levels.
formula An equation that states the relationship among quantities that are represented by variables.
FOR–NEXT statement Used to loop through the same set of statements several times in a computer program.
frequency of an event The number of times an event takes place.
function A rule that assigns to each element in the domain a single element in the range.

glide reflection The combination of a translation and a line reflection.
greatest common factor The greatest integer that is a factor of two or more integers.
greatest monomial factor The factor of two or more monomials that has the greatest coefficient and the greatest degree.
gross profit The difference between net sales and the cost of goods sold.

hypotenuse The side opposite the right angle in a right triangle.

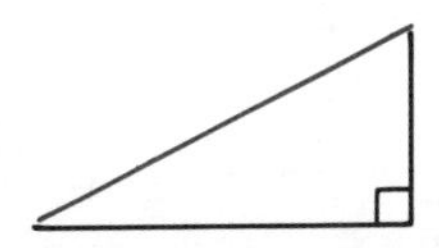

identity elements The identity element for addition is 0 since $a + 0 = a$. The identity element for multiplication is 1 since $a \times 1 = a$.
included angle Contains the sides of a triangle.
inconsistent equations Have no solutions in a system.
inequality A statement that one expression is greater, or less, than another expression.
inscribed angle An angle subtended by an arc of a circle with its vertex on the circumference.

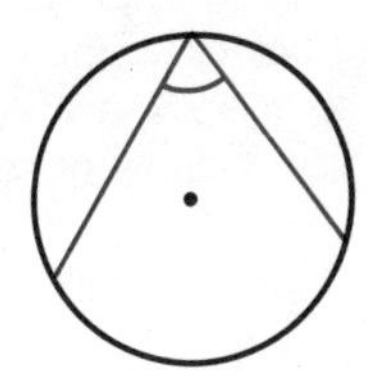

inscribed polygon Is inside a circle with its vertices intersecting the circle.

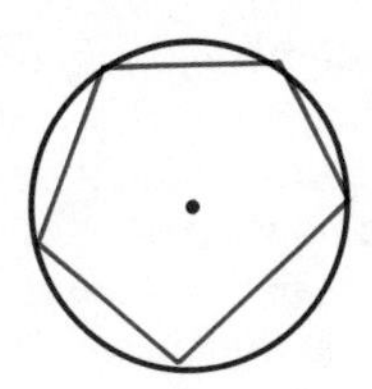

integer A member of the set $\{..., -3, -2, -1, 0, 1, 2, 3, ...\}$.
interest The amount of money paid.
intersection The elements that two sets have in common.
inventory A listing of items that are on hand, and their value.
inverse variation A function in the form $xy = k$, where $k \neq 0$ and is a constant factor.
invest To put money into a bank account, stocks, bonds, mutual funds, or real estate to obtain a gain or interest.
irrational number A real number that cannot be expressed in the form $\frac{a}{b}$, where $a, b \in I$, and $b \neq 0$.
isosceles triangle Has only two sides equal.

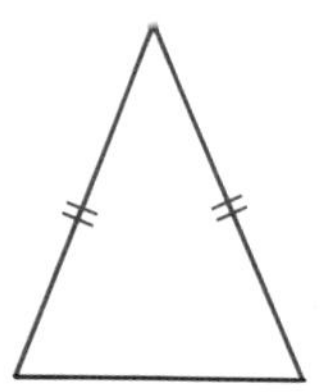

lateral area The sum of the areas of the faces of a polyhedron other than the base.
LET statement Assigns a value or an expression to a variable in a computer program.
linear equation Is an equation in which each term is either a constant or has degree 1.
line segment Two points on a line and the points between them.
line symmetry A figure has line symmetry if there is a line such that the figure coincides with its reflection image over the line.
list price The price that the manufacturer advises the retailer to put on the tag.
lowest common multiple The monomial with the smallest positive coefficient; also the smallest degree that is a multiple of several monomials.

market price The current price on the stock exchange of a current stock, bond, or mutual fund.
mass The amount of matter in an object; the base unit for measuring mass is the kilogram.
maturity value The total amount due on the date of payment of a note.
mean The sum of a set of values divided by the number of values in the set.
median The middle number in a set of numbers that are arranged in order from smallest to largest, or largest to smallest.
midpoint The point which divides a line segment into two equal parts.
mixed number Part whole number and part fraction, such as $4\frac{1}{3}$.
mode The number that occurs most often in a set of numbers.
monomial A number, a variable, or a product of numbers and variables.
mortgage A contract pledging specific property as a security for a loan.
mutually exclusive events Cannot occur at the same time.

natural numbers The set of numbers $\{1, 2, 3, 4, 5, 6, ...\}$.
net A pattern for constructing a polyhedron.
net loss Occurs when the cost of goods sold and the operating expenses are greater than the gross income.
net profit (net income) Occurs when the gross income is greater than the cost of the goods sold and the operating expenses.
NEW statement The first statement in a computer program.
nonagon A polygon with 9 sides.
number line A pictorial representation of a set of numbers.

obtuse angle Is greater than 90° but less than 180°.
obtuse triangle Has one obtuse angle.
octagon A polygon with 8 sides.
octahedron A polyhedron with 8 faces.
ordered pair A pair of numbers which names a point on a graph or coordinate grid.
order of operations The rules to be followed when simplifying expressions. These rules are also referred to as BODMAS or BEDMAS.
origin The intersection of the horizontal and vertical axes on a graph. It is described by the ordered pair (0, 0).
orthocentre The point where the altitudes of a triangle intersect.

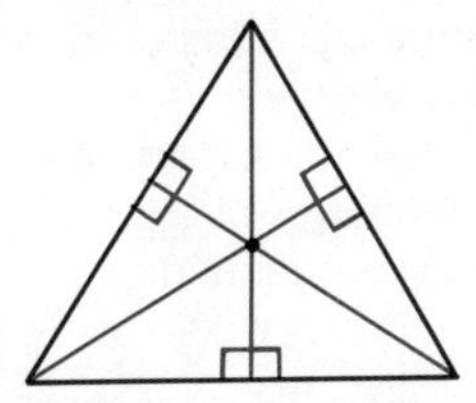

outcome The result of an experiment or a trial.

palindrome A number, such as 232, that is read the same forward or backward.
parabola A set of points in a plane that are equidistant from a fixed point and a fixed line.
parallel lines Lines in the same plane that never meet.
parallelogram A quadrilateral with opposite sides parallel.
parameter An arbitrary constant.
partial variation A relation describing a fixed amount plus a variable amount, such as $C = nd + 15$.
pentagon A polygon with 5 sides.
percent A fraction or ratio in which the denominator is 100.
perimeter The distance around a polygon.
periodic To occur at regular time intervals.
perpendicular bisector A line that cuts a line segment into 2 equal parts at right angles.
perpendicular lines Intersect at right angles.
pi (π) The result when the circumference of a circle is divided by the diameter. We use $\pi = 3.14$
pictograph Uses pictures to represent data.
polygon A closed figure formed by line segments.
polyhedron A three-dimensional solid having polygons as faces.
polynomial A monomial or the sum of monomials.
population The set of all possible outcomes of an experiment, for example, the number of cattle in a herd, or the length of time required to complete a homework assignment.
portfolio Investments, such as bonds, stocks, or certificates, held by an investor.
power A product represented by a base and an exponent. The base is a factor, and the exponent indicates the number of times the factor is multiplied.

E.g., $2^3 = 2 \times 2 \times 2$ (power: 2^3; base: 2; exponent: 3)

prime number A number with exactly 2 factors — itself and 1.
principal The amount of money borrowed.
prism A polyhedron with two parallel and congruent bases.
probability of an event occurring The ratio of the number of favourable outcomes to the number of all possible outcomes.
proportion An equation showing that 2 ratios are equal.
pyramid A polyhedron with 3 or more triangular faces and the base is a polygon.
Pythagorean Theorem The area of a square drawn on the hypotenuse of a right-angled triangle is equal to the sum of the areas of the squares drawn on the other 2 sides of this triangle.

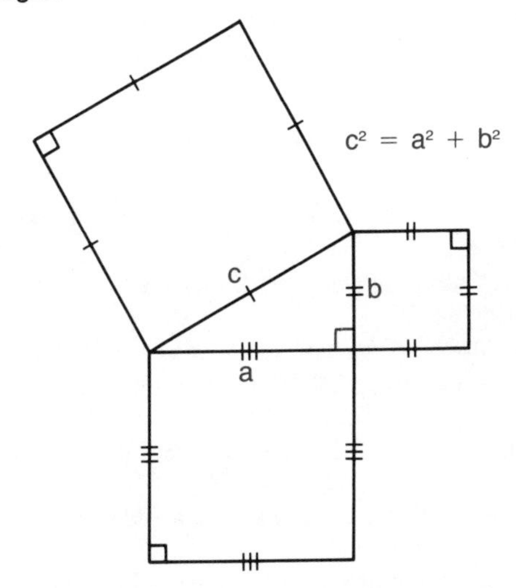

quadrant One of the four regions formed by the intersection of the x-axis and y-axis in a coordinate grid.
quadrilateral A polygon with 4 sides.
quotient The result of division.

radian Is equal to a central angle in a circle such that the length of the arc subtended by the angle is equal to the radius.

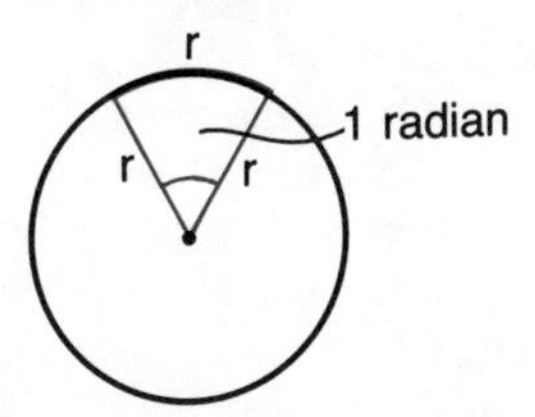

radical sign The symbol $\sqrt{}$.
radius The length of the line segment that joins the centre of a circle to a point on the circumference.
random sample A sample in which each member of the population has the same chance of being selected.
range The set of all values of a function; the set of all second coordinates of the ordered pairs of a relation.
rate A ratio of two measurements that have different units.
rate of return An indicator, expressed as a percent, of the potential profit of an investment.
ratio A comparison of two numbers. E.g., 2 : 4
rational number Expressed as the ratio of two integers.
ray Part of a line extending in one direction without end.

real numbers The set of rational and irrational numbers.
reciprocals Two numbers that have a product of 1.
rectangle A parallelogram with 4 right angles.
reflection A transformation that maps an object into a reflected image.
reflex angle Is greater than 180° and less than 360°.
regular polygon Has all equal sides and angles.
relation A set of ordered pairs.
repeating decimal Has one, or more, digits that repeat without end.
retail price Price at which goods are sold to consumers.
rhombus A parallelogram having all sides equal.
right angle Measures 90°.
right cone A cone in which the axis is perpendicular to the base.
right cylinder A cylinder in which the sides are perpendicular to the bases.
right prism A prism in which the lateral edges are perpendicular to the bases.
right triangle Has one right angle.
root of an equation A solution of the equation.
rotation A transformation that maps an object onto its image by a rotation about a point.
rotational symmetry Describes a figure if it maps onto itself after one turn.
rounding To replace a number by an approximate number.

scale drawing Has dimensions which are proportional to actual dimensions.
scale factor The multiplication factor used in dilatations (enlargements and reductions).
scalene triangle Has all unequal sides.
scientific notation Indicates that a number is written as a product of a number x, such that $1 \leqslant x \leqslant 10$, and a power of ten.
E.g., $2700 = 2.7 \times 10^3$
sector angle An angle in a circle with its vertex at the centre and which is subtended by an arc.

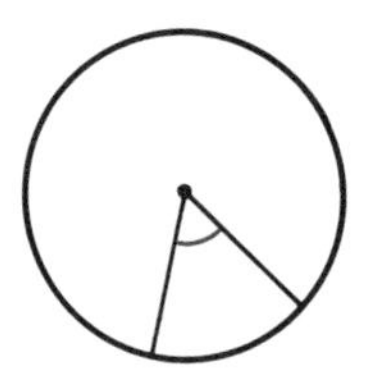

sector of a circle A region bounded by two radii and an arc.

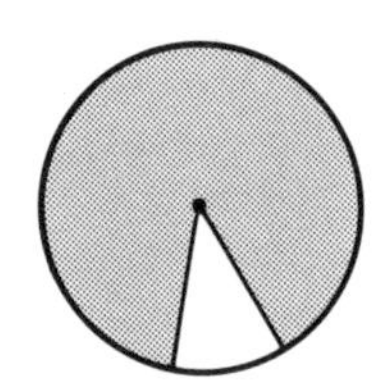

securities The written statements of ownership of an investment in stocks or bonds.
segment of a circle A region bounded by a chord and an arc.

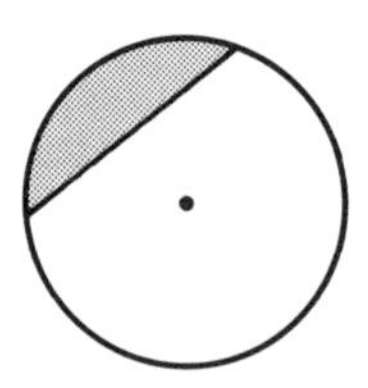

semi-annual Twice per year or every 6 months.
sequence An ordered list of numbers.
set A collection of objects.

shell A three-dimensional object whose interior is empty.
share (of stock) Represents ownership in a corporation.
similar figures Have equal corresponding angles and proportional corresponding sides.
sine ratio The ratio of the length of the opposite side to the length of the hypotenuse in a right triangle.
skeleton A representation of the edges of a polyhedron.
slope of a line The value m

$$m = \frac{y_2 - y_1}{x_2 - x_1}$$

where (x_1, y_1) and (x_2, y_2) are two distinct points on a non-vertical line.
solid A three-dimensional object whose interior is completely filled.
solution set The set of all values that make a statement, equation, or inequality true.
sphere The set of all points in a plane that are a given distance from a specific point.
spreadsheet The electronic spreadsheet is a computer application that allows the storing of information in cells and the performing of a variety of computations using formulas.
square A quadrilateral with 4 congruent sides and 4 right angles.
square root A number multiplied by itself.
standard form of a linear equation An equation in the form $ax + by + c = 0$.
statistics The science of collecting and analysing numerical information.
stem-and-leaf plot A graph using digits of numbers to display data.
stock exchange A place where securities, such as shares and bonds, are bought and sold.
straight angle Measures 180°.
supplementary angles Two angles whose sum is 180°.
surface area The sum of the total surface area of a polyhedron.

tangent ratio The ratio of the length of the opposite side to the length of the adjacent side in a right triangle.
tangent to a circle A line in the plane of the circle that intersects the circle in exactly one point.
tax-refund The amount of overpayment that has been made and is returned to the tax payer.
term of a polynomial The product of one or more numerical factors and variable factors.
terminating decimal Expressed as a finite number. E.g., 2.134
tessellation A pattern of geometric figures that completely covers a surface.
tetrahedron A polyhedron with four triangular faces.
theorem A mathematical statement that can be proved.
trade discount The reduction in the list price given by a manufacturer to retailers.
transformation A mapping which maps the points of a plane onto other points of the same plane, according to a specific pattern.
translation A transformation that maps a figure onto its image so that each point in the figure is moved the same distance and direction.
transversal A line which intersects two lines in the same plane at two distinct points.
trapezoid A quadrilateral with one pair of parallel sides.
tree diagram Illustrates the possible outcomes of consecutive events.
triangle A polygon with 3 sides.
trinomial A polynomial with 3 terms.

union of sets The set of all elements that belong to the sets, without repetition of elements.
unit price The price for a single unit of the item.

variable A letter or symbol used to represent a number.
variable costs Such as labour and materials, are flexible and change according to production levels and other factors.
vector A directed line segment.
vertex of an angle The common endpoint of 2 rays.
vertex of a polygon The point where two adjacent sides meet in a polygon.
volume The number of cubic units contained in a solid.

whole numbers Numbers in the set {0, 1, 2, 3, 4, 5, ...}.

x-axis The horizontal line used as a scale for the independent variable in the Cartesian coordinate system.

x-intercept The x-coordinate of the point where a curve crosses the x-axis.

y-axis The vertical line used as a scale for the dependent variable in the Cartesian coordinate system.

y-intercept The y-coordinate of the point where a curve crosses the y-axis.

zero-product property If $ab = 0$, then $a = 0$ or $b = 0$.

INDEX

PHOTOGRAPH CREDITS AND NOTES

p. 3: Courtesy of the Toronto Blue Jays Baseball Club. p. 4: Compliment of Molson Indy. p. 8: Ontario Ministry of Transportation. p. 10: Media Centre, Ontario Veterinary College, University of Guelph. p. 11: British Airways. p. 12: Canadian Forces photo. p. 14: (top) F.K.L. Kartworld Group, Ajax, Ontario; (bottom) Canadian Forces photo. p. 16 (also pp. 20, 42, 46, 57, 59, 62, 80, 85, 100, 108, 115, 118, 119, 142, 144, 154, 155, 156 (left), 160, 162, 168, 169, 173, 179, 198, 232, 236, 242 (top), 258, 270, 273, 282, 301, 310 (top and bottom), 357 (left and right), 366 (left), 382, 383): Photo by Don Ford. p. 18: Photo of the hit musical GOOD NEWS by Schwab, de Sylva, Henderson and Brown, which appeared on the Royal George stage at the Shaw Festival, 1989. Photo by David Cooper. p. 19: Courtesy of Young Drivers of Canada. p. 22: Courtesy of AMCU Credit Union Inc. p. 25: Photo courtesy of the Royal Canadian Mint. p. 30: Nova Scotia Department of Tourism & Culture. p. 31: H. Armstrong Roberts/Miller Comstock Inc. p. 56: H. Armstrong Roberts/Miller Comstock Inc. p. 61: (left) Novosti Press Agency; (right) Photo Courtesy of Georgian College, Barrie. p. 70: (top) H. Armstrong Roberts/Miller Comstock Inc.; (bottom) Courtesy Metropolitan Toronto Police Force. p. 72: Courtesy of NASA. p. 75: Photo courtesy of Art McIntyre. p. 84: Photo courtesy of Marineland, Niagara Falls, Canada. p. 92: H. Armstrong Roberts/Miller Comstock Inc. p. 99: Nissan Canada. p. 107: H. Armstrong Roberts/Miller Comstock Inc. p. 110: Transport Canada Photo. p. 113: H. Armstrong Roberts/Miller Comstock Inc. p. 130: (left) The Public Archives of Canada (C-11351); (right) The Public Archives of Canada (C-3207). p. 131: National Archives of Canada C 4501. p. 132: STRATFORD FESTIVAL THEATRE, THRUST STAGE Photographer: Jane Edmonds. p. 133: K. Feres-Patry RDH — Photo courtesy of CDHA. p. 156: (right) Transport Canada Photo. p. 161: (left) Transport Canada Photo; (right) Courtesy of ATMOSPHERIC ENVIRONMENT SERVICE. p. 163: Reprinted with the permission of Fitness and Amateur Sport Canada. p. 174: Courtesy of the Montreal Baseball Club. p. 176: Lakehead University Staff photo. p. 185: Picture of Newsroom 11 Team, courtesy of CHCH-TV.

p. 186: (left and right) Photo: by Marko Shark. p. 194: (top) Photo supplied courtesy of Canada Post Corporation; (bottom) Levi Strauss & Co. (CANADA) Inc. p. 197: RED LOBSTER CANADA. p. 203: Toronto Dominion Bank. p. 209: Toronto Dominion Bank. p. 216: Central Guaranty Trust Company. p. 221: Courtesy, Bank of Montreal. p. 223: Winston Lee, Intermediate Clearing Administrator, and Gillian Forsyth, Customer Service Representative, Toronto Main Branch, Lloyds Bank Canada. p. 242: (bottom) Courtesy of the Royal Bank of Canada; photograph by Patrick McCoy. p. 244: Courtesy General Motors of Canada. p. 256: Imperial Oil Limited. p. 257: Courtesy Ford of Canada. p. 259: H. Armstrong Roberts/Miller Comstock Inc. p. 263: Courtesy Ford of Canada. p. 272: TORONTO BLUE JAYS. p. 274: TRAVEL ALBERTA. p. 278: Photograph courtesy of Mitsubishi Electric Sales Canada Inc. p. 283: Photo by Chuck Lewis. p. 286: Photo courtesy of Ontario Hydro. p. 294: (left) Novosti Press/TASS; (right) Transport Canada Photo. p. 298: Novosti Press/TASS. p. 323: H. Armstrong Roberts/Miller Comstock Inc. p. 330: (top) H. Armstrong Roberts/Miller Comstock Inc.; (bottom) OCAM, The Ontario Centre for Advanced Manufacturing, Cambridge, Ontario, N1R 7H7, Contact: Tony Spoore (519) 622-3100. p. 344: Photograph courtesy of College Pro Painters Limited. p. 356: Photo courtesy of Canadian National Railways. p. 358: Edmonton Oilers. p. 366: (right) Photo courtesy of Nova Scotia Tourism & Culture. p. 369: Ontario Ministry of Transportation, Communications Services Branch. p. 372: Courtesy of David Silbert. p. 381: (top) Place d'Armes: Notre-Dame Basilica facing the statue of Montréal's founder (1642) Paul de Chomedey, Sieur de Maisonneuve (Cidem-Ville de Montréal); (bottom) Courtesy Greater Hamilton Visitor & Convention Services.

p. 130/31: The Gallup history test (*The Toronto Star*, August 13, 1988) is reproduced with the permission of Gallup Canada Inc.

p. 197: Portions of the tables illustrating income tax deduction that appear on this page and also in the Appendix are reproduced with the permission of the Minister of Supply and Services.

pp. 280/81: Portions of the maps are taken from THE MACMILLAN SCHOOL ATLAS, by Ronald C. Daly, B.A., M.Ed. Maps and illustrations by John R. Waller. Copyright © 1982 Gage Publishing Limited. Reproduced by permission of Gage Educational Publishing Company.

p. 298: The B.C. cartoon "Did you know that parallel lines never meet?" is reproduced by permission of Johnny Hart and Creators Syndicate, Inc.

p. 230: The quotation introducing section 10.10 is from "Tessellating the Plane with Convex Polygon Tiles" by Martin Gardiner (*Scientific American*, July 1975, pp. 112–17).